# RECENT ADVANCES IN TRYPTOPHAN RESEARCH

## Tryptophan and Serotonin Pathways

# ADVANCES IN EXPERIMENTAL MEDICINE AND BIOLOGY



---

Recent Volumes in this Series

Volume 390
ANTIMICROBIAL RESISTANCE: A Crisis in Health Care
Edited by Donald L. Jungkind, Joel E. Mortensen, Henry S. Fraimow, and Gary B. Calandra

Volume 391
NATURAL TOXINS 2: Structure, Mechanism of Action, and Detection
Edited by Bal Ram Singh and Anthony T. Tu

Volume 392
FUMONISINS IN FOOD
Edited by Lauren S. Jackson, Jonathan W. DeVries, and Lloyd B. Bullerman

Volume 393
MODELING AND CONTROL OF VENTILATION
Edited by Stephen J. G. Semple, Lewis Adams, and Brian J. Whipp

Volume 394
ANTIVIRAL CHEMOTHERAPY 4: New Directions for Clinical Application and Research
Edited by John Mills, Paul A. Volberding, and Lawrence Corey

Volume 395
OXYTOCIN: Cellular and Molecular Approaches in Medicine and Research
Edited by Richard Ivell and John A. Russell

Volume 396
RECENT ADVANCES IN CELLULAR AND MOLECULAR ASPECTS OF ANGIOTENSIN RECEPTORS
Edited by Mohan K. Raizada, M. Ian Phillips, and Colin Sumners

Volume 397
NOVEL STRATEGIES IN THE DESIGN AND PRODUCTION OF VACCINES
Edited by Sara Cohen and Avigdor Shafferman

Volume 398
RECENT ADVANCES IN TRYPTOPHAN RESEARCH: Tryptophan and Serotonin Pathways
Edited by Graziella Allegri Filippini, Carlo V. L. Costa, and Antonella Bertazzo

---

# RECENT ADVANCES IN TRYPTOPHAN RESEARCH

## Tryptophan and Serotonin Pathways

Edited by

Graziella Allegri Filippini
Carlo V. L. Costa
Antonella Bertazzo

University of Padova
Padova, Italy

PLENUM PRESS • NEW YORK AND LONDON

Library of Congress Cataloging-in-Publication Data

Recent advances in tryptophan research : tryptophan and serotonin pathways / edited by Graziella Allegri Filippini, Carlo V.L. Costa, Antonella Bertazzo.
p. cm. -- (Advances in experimental medicine and biology ; v. 398)
"Proceedings of the Eighth International Meeting on Tryptophan Research, held June 25-29, 1995, in Padova, Italy"--T.p. verso.
Includes bibliographical references and index.
ISBN-13: 978-1-4613-8026-9
1. Tryptophan--Congresses. 2. Serotonin--Congresses. I. Filippini, Graziella Allegri. II. Costa, Carlo V. L. III. Bertazzo, Antonella. IV. International Meeting on Tryptophan Research (8th : 1995 : Padua, Italy) V. Series.
[DNLM: 1. Tryptophan--congresses. 2. Serotonin--congresses. W1 AD559 v.398 1996 / QU 60 R294 1996]
QP562.T7R43 1996
612'.01575--dc20
DNLM/DLC
for Library of Congress

96-20315
CIP

Proceedings of the Eighth International Meeting on Tryptophan Research, held June 25 – 29, 1995, in Padova, Italy

ISBN-13: 978-1-4613-8026-9 e-ISBN-13: 978-1-4613-0381-7
DOI: 10.1007/978-1-4613-0381-7



Softcover reprint of the hardcover 1st edition 1996

A Division of Plenum Publishing Corporation
233 Spring Street, New York, N. Y. 10013

10 9 8 7 6 5 4 3 2 1

# PREFACE

Wide and fascinating is the field of research on tryptophan, a most versatile amino acid, transformed, as it is, in our organism into many biologically active substances.

This volume contains the proceedings of the Eighth International Meeting on Tryptophan Research, held at the University of Padova, Padova, Italy, from June 25 to 29 1995, under the auspices of the University of Padova, National Research Council, Italian Chemical Society-Division of Pharmaceutical Sciences, Region of the Veneto and City of Padova.

The meeting was held in Padova to commemorate Prof. Luigi Musajo twenty years after his death and the editors dedicate this book to him in recognition of his pioneering work in tryptophan metabolism. Prof. Osamu Hayaishi delivered the Musajo Memorial Award Lecture: *Tryptophan oxygenase, and sleep.* Figure 1 shows the ISTRY President Prof. Simon N. Young presenting the Musajo Memorial Medal to Prof. Hayaishi during the Opening Ceremony.

Two hundred scientists from twenty two countries participated in the meeting. These proceedings contain 121 papers encompassing a variety of topics and disciplines. The plenary lectures dealt with links between tryptophan, serotonin and brain, and clinical implications of tryptophan biology. The other papers have been divided into nine specific chapters: 1. Tryptophan and serotonin: neurobiology and neurochemistry, 2. Immunosystem and tryptophan catabolism, 3. Kynurenines and derivatives, 4. Indole derivatives as antioxidants, 5. Metabolic and clinical aspects, 6. Enzymes in tryptophan degradation, 7. Tryptophan and eating behaviour and nutritional aspects, 8. Tryptophan in peptides and proteins, 9. Chemical and analytical aspects of tryptophan and related compounds. For few contributors who were unable to provide a manuscript, we reproduced the early abstracts.

At the Conference Dinner the President of ISTRY gave Emeritus Professors R.R. Brown, G. Curzon, R. Kido a Medal Award to mark their outstanding contributions.

The meeting was sponsored by: Università degli Studi di Padova, Ministero Università e Ricerca Scientifica e Tecnologica-Roma, Consiglio Nazionale delle Ricerche-Roma, Regione Veneto, Comune di Padova, Fondazione Cassa di Risparmio Padova e Rovigo, Banca Popolare Veneta, Pharmacia-Milano, and Polifarma-Roma. We gratefully acknowledge their support. We would also thank the University of Padova for providing excellent facilities for the Opening Ceremony, our colleagues of the programme committee and all who helped in the preparation of and during the meeting.

Last but not least, we would like to express our gratitude to all participants, invited speakers and chairpersons, for the outstanding quality of their contributions to the success of the meeting.

**Figure 1.** Prof. Simon N. Young, the President of ISTRY gives the Musajo Memorial Award Medal to Prof. Osamu Hayaishi.

Finally, we wish to thank the Plenum Press, especially Mr. Robert Wheeler, Bertha O'Keefe, and Greg Safford, for their helpful cooperation.

The next ISTRY Meeting will be organized by Dr. F. Artigas and will be held in Barcelona, Spain, in 1998.

Graziella Allegri Filippini
Carlo V.L. Costa
Antonella Bertazzo
Padova, Italy, 1995

# CONTENTS

## IMMUNOSYSTEM AND TRYPTOPHAN CATABOLISM

## KYNURENINES AND DERIVATIVES

## INDOLE DERIVATIVES AS ANTIOXIDANTS

## METABOLIC AND CLINICAL ASPECTS

## ENZYME IN TRYPTOPHAN DEGRADATION

## TRYPTOPHAN AND EATING BEHAVIOUR AND NUTRITIONAL ASPECTS

## TRYPTOPHAN IN PEPTIDES AND PROTEINS



## CHEMICAL AND ANALYTICAL ASPECTS OF TRYPTOPHAN AND RELATED COMPOUNDS

# OPENING ADDRESS BY THE PRESIDENT OF ISTRY

Simon N. Young

McGill University
Department of Psychiatry
1033 Pine Avenue, West Montreal H3A 1A1, Quebec, Canada

The meeting in Padova, Italy from June 25 to 28, 1995 was the eighth triennial meeting of the International Study Group for Tryptophan Research (ISTRY), and marked the 21st year of ISTRY's existence. One of the founder's of ISTRY, Dr. R.R. Brown, has written about its early years: "Prior the establishment of ISTRY there had been sporadic conferences on tryptophan-related topics over the years, organized by individuals or groups as seemed appropriate and timely. Such meetings occurred at irregular intervals, were not usually well publicized, and sometimes conflicted with related meetings. As a result of discussions during a tryptophan conference at Schloss Hohenried near Munich in 1971, it was suggested by Professors W. Kochen and H. Schievelbein that an international organization be formed with the purpose of furthering tryptophan research by improving communication and collaboration among such workers and through the organization of regular international meetings. Through the initiative of Drs. Kochen and Schievelbein, an informal election was held and the first President and Executive Committee of ISTRY was formed. This consisted of Professor Kochen as President and Professors O. Hayaishi, L. Musajo, R. Brown, H. Schievelbein, and C. Streffer as Executive Committee. The Committee felt it was most appropriate that the first ISTRY meeting be held in Padova, Italy to honor the pioneering work of Professor Luigi Musajo and his colleagues. He graciously agreed to host and organize the inaugural meeting of ISTRY, with the assistance of a local organizing committee consisting of Drs. G. Allegri, C. Costa, and A. De Antoni." The meeting in 1974 in Padova was followed by meetings in 1977 in Madison, Wisconsin, U.S.A., in 1980 in Kyoto, Japan, in 1983 in Martinsried near Munich, Germany, in 1986 in Cardiff, Wales, in 1989 in Baltimore, Maryland, U.S.A., and in 1991 in Nagoya, Japan. In some cultural traditions a child becomes an adult at 21 years of age, and it is appropriate that ISTRY returned to its birthplace in Padova to celebrate its 21st year in 1995.

The passage into adulthood is an appropriate time for introspection, and I would like to take some time to speculate on the reasons for the success of ISTRY. Most would agree that an organization based solely on a triennial meeting is unlikely to survive. I was present at the first meeting of ISTRY as a postdoctoral fellow, and at the opening reception I listened, with some dismay, to a conversation between two senior researchers who were certain that no further tryptophan meetings would be held because most of the interesting discoveries in the field had already been made. My intuition that they must be wrong was confirmed as soon as the first scientific session began. In the two decades since then tryptophan and its metabolites have continued to fascinate all ISTRY members because of the diversity and complexity of their actions. I am sure that we would all agree that this field merits many more decades of research. Certainly one reason for the success of ISTRY is the seemingly endless series of intriguing research problems associated with tryptophan.

When I mention to my colleagues that I attend meetings devoted solely to tryptophan and its metabolites, they often express surprise that a society could exist devoted to such a topic. On hearing about the diversity of topics at ISTRY meetings, they often express even greater surprise that such a meeting could be a success. This reflects the current trend for ever greater specialization. New journals proliferate dealing with ever more specialized topics. The same is true for scientific meetings. Given the explosion of knowledge, this is necessary. However, for some the need to look at research from a broader perspective accompanies the move to greater specialization. One of the great attractions of ISTRY meetings is the diversity of topics, ranging from the control of aggression to the pigments of butterfly wings, from diurnal rhythms to analytical chemistry, and from enzymology to food chemistry. At ISTRY we can approach new and unexplored topics, that, at the same time, have a ring of familiarity, because of the connection to tryptophan. Certainly, this type of meeting is not for everyone, but ISTRY meetings are a delight for those with the intellectual curiosity to venture beyond their own research topic. The insights obtained from those tackling varied problems, but with a unified theme, must be a help to all ISTRY meeting participants. Thus, the great strength of ISTRY is the intellectual curiosity of ISTRY members and the diversity of topics in a meeting unified by a single theme–tryptophan. It is this that has caused the success of ISTRY in the past and will continue to do so in the future.

# *MUSAJO MEMORIAL AWARD LECTURE*

# 1

# TRYPTOPHAN, OXYGENASE, AND SLEEP

O. Hayaishi

Osaka Bioscience Institute
6-2-4 Furuedai, Suita, Osaka 565, Japan

In May 1974, or about 21 years ago, when the first International Meeting of ISTRY was held in this beautiful city of Padova, I had the great pleasure of making personal acquaintance with Prof. Luigi Musajo, whom I had heard about so much while I was in Japan. At this meeting, I presented two lectures, one on oxygenases and the other on indoleamine dioxygenase. I still remember his warm and encouraging remarks and was greatly impressed by his science as well as his personality. It is indeed a great honor and pleasure for me to visit Padova again and to present the Luigi Musajo Award lecture to this distinguished audience. In memory of Prof. Luigi Musajo, I respectfully dedicate this lecture, and describe briefly our past work on tryptophan and oxygenases and the on-going work concerning the mystery of sleep.

## I. ENCOUNTER WITH TRYPTOPHAN

I graduated from Osaka University School of Medicine in 1942 during the Second World War, and served as a medical officer in the Japanese Navy for 3 years. When the war ended in 1945, I returned to Osaka to find the entire city almost totally demolished. The home of my family had completely disappeared. Food was scarce, and supplies of most commodities, gas, electricity, and even water, were extremely limited. Osaka University was one of the centers of biochemical research in Japan, and Yashiro Kotake and his associates were well-known for their research on tryptophan metabolism. You may recall that Kotake discovered kynurenine, the key intermediate of tryptophan metabolism. When I was a medical student at the Osaka University School of Medicine, I had the good fortune of attending the biochemistry course given by Professor Kotake. In one of his many fluent and exhaustive lectures, Kotake explained how kynurenic acid was discovered in the urine of dogs by Liebig and how another carboxylic quinolinic acid, named xanthurenic acid because of its yellow color, was discovered by Professor Musajo in the urine of rats fed a high-protein diet.

When I started my research career in 1945, Kotake kindly gave me several grams of tryptophan, kynurenine, and other related chemicals, and encouraged me to use these valuable compounds for our experiments, but research funds were extremely limited, facilities were poor, chemicals were scarce, and there were no laboratory animals. So I went out to the back yard of the university, took some soil samples from the ruins, mixed them with tryptophan and water in a test tube, and waited. It was indeed a very simple and inexpensive experiment, but after a few days, I observed some cloudiness in the test tube, indicating the growth of bacteria. Thus, by the so-called enrichment culture technique, I was able to isolate several strains of soil bacteria that could grow on tryptophan as their sole source of carbon and nitrogen.

It was already known, then, that in mammals tryptophan is degraded through kynurenine to kynurenic acid, anthranilic acid, and xanthurenic acid, all of which are excreted in the urine as the end metabolites. However, in my soil bacteria, I was able to demonstrate that anthranilate was further metabolized to catechol, then to muconic acid, and ultimately oxidized to $CO_2$, ammonia and $H_2O$. Encouraged by this unexpected discovery, I proceeded to isolate and purify the enzyme that catalyzed the oxidative cleavage of catechol to produce muconic acid as the reaction product.

## II. DISCOVERY OF OXYGENASE

This enzyme was named pyrocatechase. In 1955, I was able to demonstrate by the use of a heavy oxygen isotope, that the molecular oxygen is directly incorporated into the product of the reaction, oxygen-18 (Hayaishi et al., 1955). At that time, because of the famous dehydrogenation theory of Professor Heinrich Wieland, the 1927 Novel Laureate in chemistry, it was generally believed that the enzymatic incorporation of molecular oxygen into substrates can never happen in living systems. According to Wieland's theory, molecular oxygen accepts hydrogen atoms and is reduced to water or hydrogen peroxide, but, it is never incorporated into the substrate. However, our findings along with subsequent work by Mason, Bloch and others established that "oxygen fixation" did indeed occur in biological systems, and I proposed to name this new group of enzymes "oxygenases".

Fig.1 shows a metabolic map of tryptophan, in which the thick oxygen atoms are incorporated into these substrates, and black arrows indicate oxygenase reactions. The ubiquitous presence of oxygenase reactions is clearly indicated. For example, the two most important metabolic pathways for tryptophan in mammals, one leading to the formation of pyridine nucleotide coenzymes and poly (ADP-ribose), and the other, to the formation of various indoleamines, are initiated by two well-known oxygenases, tryptophan 2, 3-dioxygenase and tryptophan 5-monooxygenase, respectively.

When I presented this slide at the first ISTRY meeting in 1974, Prof.Musajo liked it so much that upon his request, I presented a copy of this slide which, he told me, he would like to use in a lecture to his students.

## III. PROSTAGLANDINS AND SLEEP

Prostaglandin (PG) was discovered in semen by Kurzrok and Lieb in 1930, and was shown to be produced by the action of a unique oxygenase, cyclooxygenase, an enzyme that incorporates two molecules of oxygen into its substrate, arachidonic acid. Prostaglandins are ubiquitously distributed in virtually all mammalian tissues and organs and exhibit numerous and diverse biological effects on a wide variety of physiological and pathological activities, such as contraction and relaxation of smooth muscles, aggregation and disaggregation of blood platelets, inflammation, pain, and so forth. However, little had been known about PGs in the mammalian brain until recently. In the late 1970's, we showed that $PGD_2$ and $E_2$ are the major prostanoids in the brain of rats and other mammals including humans. Until that time, $PGD_2$ had long been considered as a minor and biologically inactive prostanoid or

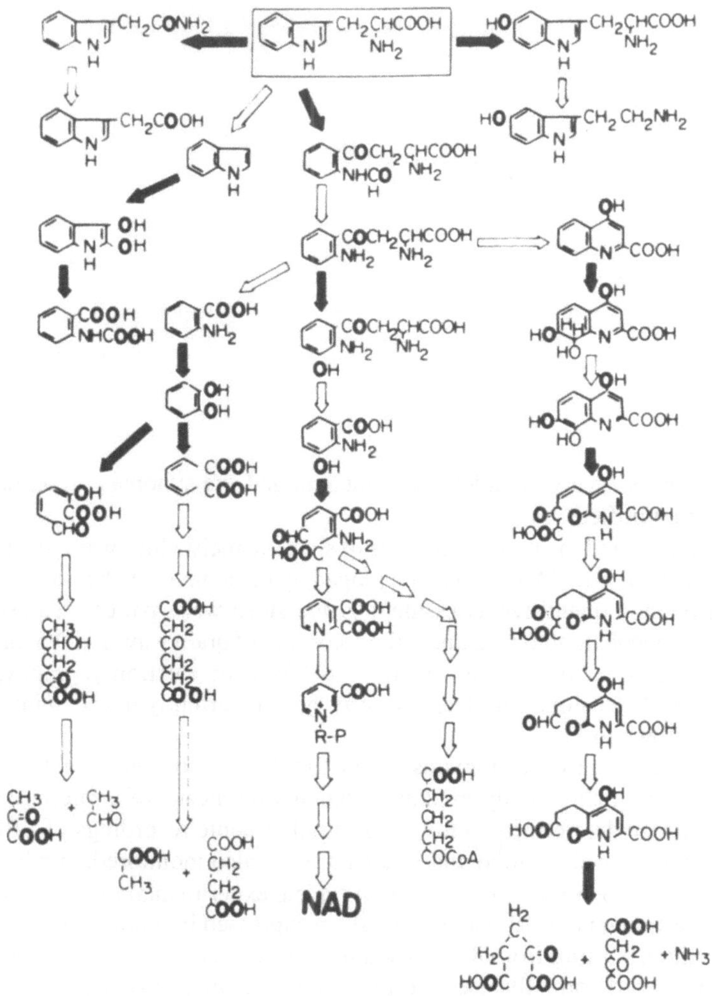

Figure 1

perhaps even as a non-enzymatic decomposition product of $PGH_2$. However, $PGD_2$ appeared to be unique among the PGs in being present in relatively high concentrations in the mammalian brain, and I wondered what $PGD_2$ was doing in the brain.

We repeat sleep-wake cycles every day and night, and spend almost one-third of our life time sleeping in bed or somewhere else. However, "the molecular mechanisms of sleep-wake regulation" have thus far been hardly understood. Sleep is one of the most important and yet least understood biological phenomena of the brain. In spite of the intensive effort by numerous investigators since the beginning of this century, we still can not answer simple questions such as: what is sleep?, why do we need to sleep?, how is sleep regulated?, and so forth. In the remainder of my lecture, I shall talk about three short and seemingly simple stories; first, prostaglandin $D_2$ makes you sleep. Second, prostaglandin D synthase is the key enzyme in sleep regulation. And third, prostaglandin $E_2$ makes you wake up.

Before I proceed further, I would first like to say a few words about sleep and how sleep is determined for the benefit of the nonspecialists in the audience. As all of you are

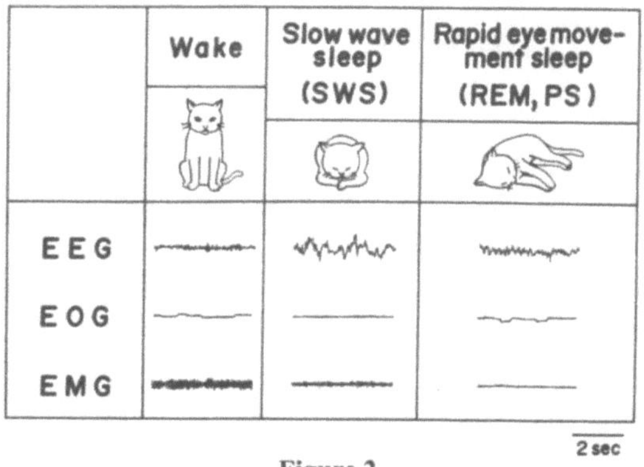


Figure 2

well aware, sleep is a very complex phenomenon, and the synopsis presented in Fig.2 is obviously oversimplified.

Wakefulness and the two major types of sleep, namely, slow-wave sleep or SWS and rapid eye movement or REM sleep, can be judged by the animal's behavior. However, more precise qualitive and quantitive assessment of the sleep-wake pattern requires the use of EEG, or electroencephalography, namely the recording of brain waves, EOG, or electro-oculography, the recording of eye movement, and EMG, or electromyography, that is, the recording of muscle tension. The sleep bioassay system currently in use in my laboratory is schematically shown in Fig. 3.

Samples are infused continuously and slowly into the third ventricle of a freely moving rat through a chronically implanted cannula by means of a microinjection pump. The sleep stages are determined on the basis of polygraphic recordings of EEG and EMG. Brain temperature as well as food and water intake is also monitored, and behavior of the rat is recorded by a video-recording system. A typical experimental result is shown in Fig.4.

Here the amounts of SWS and REM sleep expressed in minutes per hour are plotted against clock time, starting from 8 o'clock a.m. until 8 o'clock the next morning. Rats are nocturnal animals and normally sleep during the day, but are awake most of the time during the night as shown by the dotted lines. When $PGD_2$ was infused at a rate of 0.6 pmol per

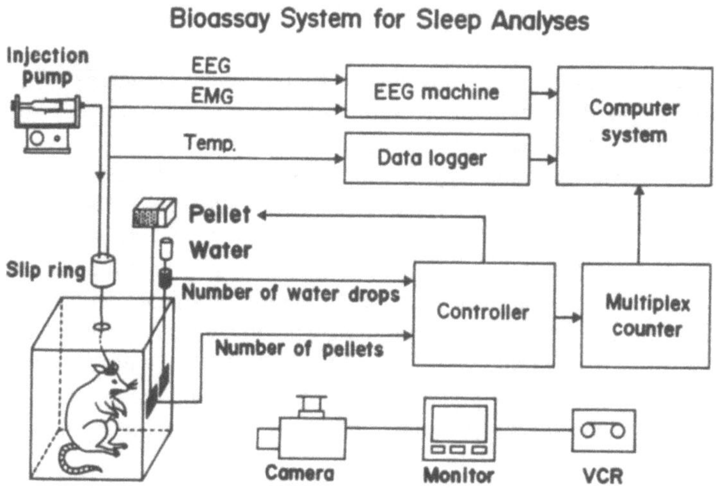


**Figure 3.** Bioassay system for sleep analyses.

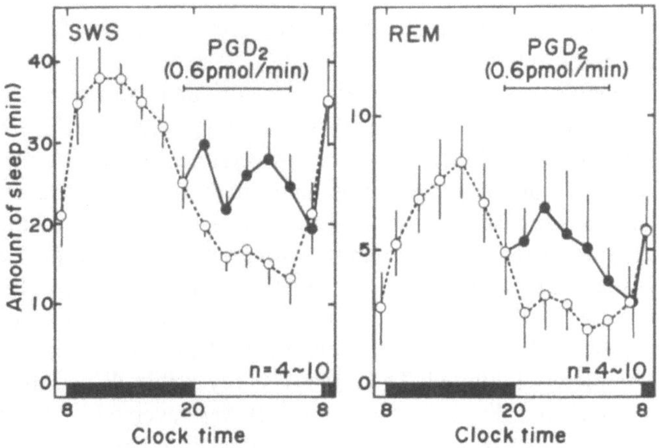


**Figure 4.** Effect of PGD2 on sleep patterns.

minute during the night, both SWS and REM sleep increased significantly as indicated by the solid lines. The effect was specific for $PGD_2$ and dose-dependent. As little as one femtomole ($10^{-15}$ mol) of $PGD_2$ per second was effective in inducing excess sleep. Most importantly, sleep induced by $PGD_2$ was indistinguishable from physiological sleep as judged by EEG, EMG, locomotor activities, brain temperature, heart rate, and general behavior of the rat. In order to further prove that the effect of $PGD_2$ was not pharmacological, but physiological, we determined the specific activity of the enzyme that produces $PGD_2$ in the rat brain, namely PGD synthase, during the sleep-wake cycle. The PGD synthase activity in the rat brain exhibited a circadian fluctuation in parallel with the sleep/wake cycle. In contrast, the specific activities of other related enzymes, for example, $PGD_2$ dehydrogenase, remained almost the same during the day and night. More recently, We showed that the concentration of $PGD_2$ itself in the rat CSF also exhibits a similar circadian fluctuation in parallel with the sleep/wake cycle. If PGD synthase is indeed involved in the regulation of physiological sleep, then, when the synthase activity during the day is inhibited by a specific enzyme inhibitor, sleep should also be inhibited and the amount of diurnal sleep should decrease. In order to find a specific inhibitor of PGD synthase, we then purified this enzyme and studied its properties. PGD synthase produces $PGD_2$ from its substrate $PGH_2$. This enzyme is a monomeric glycoprotein with a molecular weight of approximately 27, 000. We purified the enzyme from human and rat brains, isolated cDNAs encoding PGD synthase, and determined the nucleotide and amino acid sequences. These experiments produced several lines of important and unexpected information, which, as I shall tell you later, gave us a new insight into our understanding of the mechanism of the somnogenic activity of $PGD_2$.

## IV. PROSTAGLANDIN D SYNTHASE

The tertiary structure of the brain PGD synthase was then computergraphically deduced. The free SH group due to cysteine residue 65 is found only in the human and rat PGD synthase; and it is not present in the other lipocalins. This sulfhydryl group is in the hydrophobic pocket and is considered to be in the active site of the enzyme. In fact, replacement of cysteine 65 by serine or alanine by site-directed mutagenesis resulted in complete loss of the enzyme activity. Islam and co-workers in my laboratory discovered that inorganic quadrivalent ($4^+$) selenium compounds are potent, specific, non-competitive, and reversible inhibitors of brain prostaglandin D synthase, presumably by interacting with this free sulfhydryl group in the active center.

In vitro kinetic studies revealed that various quadrivalent selenium compounds are inhibitory, with Ki values of about 10 mM. However, hexavalent and organic selenium compounds are hardly effective. This inhibition is reversible and can be reversed by the addition of excess amounts of sulfhydryl compounds such as glutathione or DTT. Other enzymes in the arachidonate cascade system are not inhibited by any of these selenium compounds. We then administered these selenium compounds into the third ventricle of a freely moving rat to see if they would have any inhibitory effect on sleep. When selenium chloride was infused into the third ventricle of a rat during day, sleep was inhibited time- and dose-dependently. After about 2 hours from the start of the infusion, both SWS and REM sleep were almost completely inhibited. The effect was reversible. When the infusion was interrupted, sleep was restored. Furthermore, the effect was reversed by the simultaneous infusion of sulfhydryl compounds such as DTT and reduced glutathione, as in the case of the in vitro enzyme activity. These results therefore clearly show that PGD synthase is the key enzyme in sleep regulation and that $PGD_2$ is the endogenous sleep-regulating substance in the rat under physiological conditions (Matsumura et al., 1991).

If this be the case, other questions arise, like "in what region is the PGD synthase located in the mammalian brain?" and "in what type of cells?" In an attempt to answer these questions, we have employed three independent approaches, namely, in situ hybridization, immunohistochemical staining and the enzyme assay. The results were quite remarkable and unexpected.

Typical results of in situ hybridization of a rat brain and spinal cord clearly indicate that mRNA for PGD synthase is expressed faintly and diffusely in the white matter of the brain. However, a much stronger signal is seen in the leptomeninges, namely the pia mater and arachnoid membrane, and in the choroid plexus, CP, in the ventricles (Urade et al., 1993). As revealed by in situ hybridization and by the immunohistochemical technique, the leptomeninges and choroid plexus are loaded with PGD synthase mRNA and protein ; the oligodendrocytes are also positively stained but the other cells including neurons show little activity.

Based on these morphological observations, we then determined the enzyme activity in these tissues and also in the cerebrospinal fluid of rats and humans. The results were consistent with those obtained with in situ hybridization and immunohistochemistry. The specific activities in the leptomeninges and choroid plexus were much higher than that activity in the whole brain. Furthermore, both rat and human CSF were shown to contain a remarkably large amount of PGD synthase activity. Recently, Dr. Murphy of the University of Tennessee, Knoxville, isolated a protein from human urine, the partial sequence of which was identical to that of PGD synthase. This urinary protein was also shown to have a significant enzyme activity. In addition to this protein, β-trace, which is the major protein in CSF, showed a very high specific activity, almost comparable to that of the most purified enzyme preparation from rat brain.

β-trace is the second most abundant protein in human CSF after albumin. It was discovered in 1961 as a protein specific to the CSF, but its function has not yet been elucidated. Recently several groups of investigators in Germany reported that the N-terminal partial sequence of human β-trace is highly homologous to that of human brain PGD synthase except that β- trace has no signal peptide. We then purified β-trace from human CSF and found that β-trace and PGD synthase were structurally, enzymatically and immunologically identical (Watanabe et al., 1994), which means that $PGD_2$ and PGD synthase circulate in the CSF and extracellular space. How then does $PGD_2$ work ? Where is its site of action ? In an attempt to determine the exact site of action of the sleep-promoting effect of $PGD_2$, we

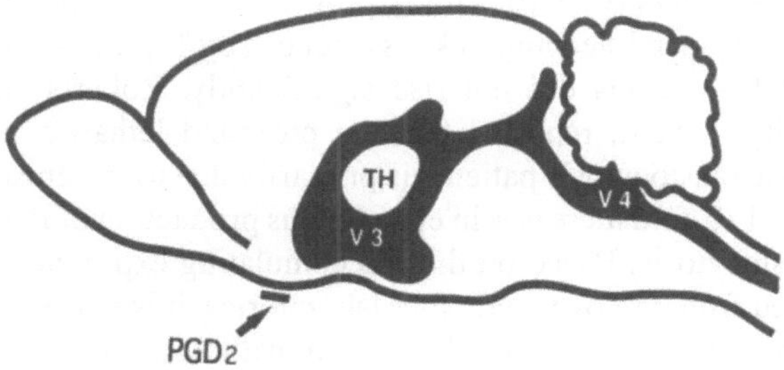


Figure 5

recently infused a small amount of $PGD_2$ in nearly 300 different sites in the rat brain. To our big surprise, bilateral infusion of $PGD_2$ into the subarachnoid space in a surface area of the medial ventral region of the rostral basal forebrain was found to be the most effective. SWS was increased more than 120 minutes above the base line value of 100 minutes during the 6 hours of $PGD_2$ infusion. It has been known for some time that the chemosensitive regions involved in the regulation of blood pressure and respiration are located in the ventrolateral surface region of the brain stem, and these regions have been termed Rostral, Intermediate, and Caudal areas by Loeschke and co-workers. These chemosensors are reportedly activated by various ligands, such as hydrogen ion, $CO_2$, acetylcholine, serotonin, endothelin, and other compounds. It would not be surprising, therefore, to find a chemosensitive locus for $PGD_2$ in these surface areas. All these experimental observations taken together led us to arrive at the tentative postulate schematically shown in Fig.5.

This is a highly oversimplified structure of a rat brain. PGD synthase, the enzyme that produces $PGD_2$ in the rat brain, is present mainly, if not exclusively, in the membrane system surrounding the brain rather than in the brain parenchyma. As I mentioned earlier, PGD synthase is a secretory protein. It is therefore reasonable to assume that it is secreted into the CSF to become β-trace protein, a major protein component of the CSF. PGD synthase, as well as the $PGD_2$ thus produced, circulates in the ventricular system, subarachnoidal space, and extracellular space in the brain system, as indicated here.

$PGD_2$ interacts with chemosensors or receptors on the ventro-medial surface of the rostral basal forebrain, thus initiating the signal to make the brain sleep. Recently, two principal modes of electrochemical communication in the brain, namely volume transmission through the brain extracellular fluid and CSF, as opposed to the conventional "wiring transmission through the nerve cells and synapses", were advocated by Fuxe and Agnati of the Karolinska Institute. The volume transmission was suggested to have advantages for global regulations such as consciousness, sleep, pain, emotion and possibly other behaviors. Our findings are consistent with their concept in the sense that $PGD_2$ is not a typical neurotransmitter but rather a "neurohormone" or an "informational substance" that circulates through the CSF in the ventricular system and the extracellular space and transmits certain chemical messages.

We further extended these studies to the Macaca mulatta, that is, the Rhesus monkey. Essentially similar results were obtained, thus indicating that prostaglandin $D_2$ induces physiological sleep in primates also (Onoe et al., 1988).

So far I have emphasized that the sleep induced by $PGD_2$ is indistinguishable from natural sleep and that $PGD_2$ controls physiological sleep in rats and monkeys. How then is pathological sleep controlled ? Are human sleep disorders, such as sleeping sickness, narcolepsy, and so forth, caused by aberrant metabolism of prostaglandins?

Pentreath and coworkers (1990) at Sulford University, U.K., reported that the $PGD_2$ concentration was markedly ($10^2 \sim 10^3$ - fold), progressively and selectively elevated in the CSF of patients with advanced sleeping sickness induced by Trypanosoma brucei gambiense, whereas $PGE_2$ and IL-1 levels did not rise significantly. Roberts and Oates (1980) at Vanderbilt University, U.S.A., reported that the profound lethargy, namely, deep sleep, exhibited by systemic mastocytosis patients is primarily due to the episodic release of mast cell mediators. Up to 150-fold increases in endogenous production of $PGD_2$ have been found in patients with mastocytosis. Based on these accumulating experimental data and clinical observations reported from my own and other laboratories, it is reasonable to conclude that $PGD_2$ induces sleep in humans also and that certain pathological types of sleep are caused by excessive endogenous production of $PGD_2$. All of these results taken together clearly show that $PGD_2$ is the endogenous sleep-regulating substance under physiological as well as pathological conditions and that PGD synthase is the key enzyme in sleep regulation.

## V. PROSTAGLANDIN $E_2$

So far, I have talked about the sleep-promoting effect of $PGD_2$. Lastly, I would like to briefly talk about the waking effect of $PGE_2$. $PGE_2$ and $PGD_2$ are positional isomers and have been shown to sometimes exhibit opposite biological activities. For example, $PGD_2$ lowers the body temperature, suppresses secretion of luteinizing hormone-releasing hormone, and decreases ion transport across rat colon mucosal membrane, whereas $PGD_2$ shows the opposite effects.

In examining the neural functions of $PGE_2$, we found that intraventricular infusion of prostaglandin $E_2$ reduced the amount of sleep or increased the time of wakefulness. When prostaglandin $E_2$ was infused at a rate of 10 pmoles /min into the third ventricle of a rat during the day, the amounts of both SWS and REM sleep were reduced significantly. After the termination of infusion, there was a rebound, and nocturnal sleep was increased to a significant extent; but the sleep pattern returned to normal during the next day, indicating that the effect of $PGE_2$ is not due to its toxicity.

## VI. SUMMARY

In summary, then, $PGD_2$ and $E_2$ are the major endogenous sleep-regulating substances, the former inducing physiological sleep and the latter, wakefulness, in rats, monkeys and probably in humans as well. Second, PGD synthase, the key enzyme in the regulation of sleep, is present mainly in the leptomeninges and choroid plexus, from which it is secreted into the CSF to become β-trace protein. Third, $PGD_2$ exerts its somnogenic activity at the ventro-rostral surface of the basal forebrain.

## REFERENCES

Hayaishi O., Katagiri M., and Rothberg S., 1955, Mechanism of the pyrocatechase reaction, *J.Amer. Chem. Soc.*, 77: 5450.

Matsumura H., Takahata R., and Hayaishi O., 1991, Inhibition of sleep in rats by inorganic selenium compounds, inhibitors of prostaglandin D synthase, *Proc. Natl. Acad.Sci. USA*, 88: 9046.

Onoe H., Ueno R., Fujita I., Nishino H., Oomura Y., and Hayaishi O., 1988, Prostaglandin $D_2$, a cerebral sleep-inducing substance in monkeys, *Proc. Natl.Acad. Sci. USA*, 85: 4082.

Pentreath V.W., Rees K., Owalabi O.A., Philip K.A., and Doua F., 1990, The somnogenic T lymphocyte suppressor prostaglandin $D_2$ is selectively elevated in cerebrospinal fluid of advenced sleeping sickness patients, *Trans. Royal. Soc., Trop. Med. Hyg.*, 84: 795.
Roberts J.L., II., Sweetman B.J., Lewis R.A., Austen K.F, and Oates J.A., 1980, Increased production of prostaglandin $D_2$ in patients with systemic mastocytosis, *New Engl. J. Med.*, 303: 1400.
UradeY., Kitahama K., Ohishi H., Kaneko T., Mizuno N., and Hayaishi O., 1993, Dominant expression of mRNA for prostaglandin D synthase in leptomeninges, choroid plexus, and oligodendrocytes of the adult rat brain, *Proc. Natl. Acad. Sci. USA.*, 90: 9070.
Watanabe K., Urade Y., Mäder M., Murphy C., and Hayaishi O., 1994, Identification of β- trace as prostaglandin D synthase, *Biochem. Biophys. Res. Commun.*, 203:1110.

# *PLENARY LECTURES*

# 2

# METABOLISM AND BIOLOGY OF TRYPTOPHAN

## Some Clinical Implications

R. R. Brown

Professor Emeritus, Department of Human Oncology
University of Wisconsin Medical School
Madison, Wisconsin 53792

The age of 21 years is usually considered the age at which a person becomes an adult. So it is that the International Study Group for Tryptophan Research (ISTRY), which is now 21 years old becomes a mature organization. The first meeting was held 21 years ago here in beautiful and historic Padua organized by Professor L. Musajo and his able assistants Professors G. Allegri C. Costa and A. De Antoni. Intervening triennial meetings were held in Madison, Kyoto, Martinsried, Cardiff, Baltimore, and Nagoya. It is appropriate that this meeting again be in Padua to celebrate the maturity of ISTRY, and we are grateful to Professors Allegri-Filippini and Costa for hosting this conference.

It is important to remember the many pioneers in biochemistry who laid the solid foundations upon which our present knowledge is based. An incomplete list of such pioneers would include Hopkins, Musajo, the Kotakes, Berg, Butenandt, Hayaishi, Udenfriend, Quagliariello, Knox, Henderson, Hankes, Price, and many others too numerous to list here. It is also important to remember some of the old facts about tryptophan (Trp) and essential amino acids which may be forgotten in the rush to present day molecular biology.

1. Trp provides a major part of niacinamide cofactors (NAD, NADP) for most animal species.
2. Trp is essential for protein synthesis. Inadequate dietary intake of Trp (or any other essential amino acid) leads quickly to negative nitrogen balance, weight loss (mostly muscle mass), and possibly death.
3. If any one essential amino acid is absent at the time it is needed for protein synthesis on the ribosome, the whole process of peptide synthesis is stopped, the incomplete peptide is rejected from the ribosome, and the ribosome complex is disrupted. Provision of the deficient amino acid only a few minutes after it is needed is too late.

*Recent Advances in Tryptophan Research*, edited by Graziella Allegri Filippini et al.
Plenum Press, New York, 1996

4. Significant reutilization of Trp occurs. Estimates of total protein synthesis per day suggest that 5 to 10 times more Trp is incorporated into protein than is provided by the daily requirement for Trp. Therefore a large percentage of Trp released from breakdown of cells and proteins is actually reutilized.

The discovery of the kynurenine-niacin pathway and the description of liver Trp dioxygenase and its induction by glucocorticoids and stabilization by elevated Trp levels were helpful in trying to interpret early findings of elevated metabolite levels in a variety of apparently unrelated clinical conditions. However, the findings in Professor Hayaishi's laboratory of a second enzyme which could also initiate the kynurenine pathway, namely indoleamine dioxygenase (IDO) (Shimizu, et al., 1978; Yamazaki et al., 1985) and many elegant studies showing that IDO was highly inducible by interferon-gamma (IFN-G) (Yoshida et al., 1979; Yoshida, et al., 1981; Ozaki, et al., 1987; Takikawa, et al., 1990; Werner, et al., 1989) suddenly permitted new interpretations of much of the clinical data: infection, inflammatory processes, immune stimulation of any type which could induce production of IFN-G resulted in IDO induction and enhanced catabolism of Trp along the kynurenine pathway (Brown, 1991; Carlin et al., 1989; Taylor and Feng, 1989; Feng and Taylor, 1991). Because of its high Km, liver Trp dioxygenase is probably active only when levels of Trp are high such as after a meal or a loading dose, but this enzyme is probably not very effective at normal blood levels of Trp. However, IDO has a low Km and once induced is capable of causing very low Trp levels in blood or tissues or cultures. For example, administration of IFN-G or interleukin-2 (which induces IFN-G) as experimental therapy to cancer patients resulted within 24 hours, in blood Trp levels as low as 20-50 per cent of pre-treatment values while other blood amino acids were virtually unchanged. (Brown, et al., 1989; Byrne, et al., 1986; Carlin et al., 1989). Corresponding values of blood or urine kynurenine were remarkably elevated.

This enhanced activity along the kynurenine pathway must have a negative effect on the serotonin (5-hydroxytryptamine, 5-HT) pathway. Indeed many patients receiving IFN-G may develop evidence of depression, dementia as well as fatigue and weakness, conditions not uncommon in other conditions having enhanced immune system activity such AIDS, infection, autoimmune diseases and related conditions.

## TRYPTOPHAN METABOLISM IN AIDS

Several laboratories have reported that patients infected with HIV-1 have low blood Trp and 5-HT and correspondingly increased levels of kynurenine pathway metabolites including quinolinic acid in blood or cerebrospinal fluid (Gisslen, et al., 1994; Launay, et al., 1988; Hortin, et al., 1994; Heyes, 1993). Since quinolinic acid is neuroexcitatory and neurotoxic for certain neurones, its elevation in AIDS patients is likely related to the dementia seen in this disease. In pediatric HIV-1 patients, we find a strong association between serum Trp levels and failure-to-thrive (E. Chadwick and R.R. Brown, unpublished data). Pediatric AIDS patients who continue to thrive have more normal Trp levels, Figure 1.

It seems likely that the wasting or failure to thrive in AIDS patients is due, at least in part, to a chronic depletion of Trp, and that the dementia aspects may be due in part to excess brain quinolinic acid (Heyes, et al., 1992)) and low levels of serotonin. Although blood Trp levels are low in AIDS patients, other amino acids are not reduced to the same extent (Hortin, et al., 1994), a pattern generally similar to that found in pellagra, due to poor Trp intake (Truswell, et al., 1968). Nutrition and anthropometric studies in AIDS patients indicate that protein is lost but fat is little changed. Administration of excess dietary Trp may relieve the Trp deficiency, but would increase quinolinic acid levels and likely worsen the dementia of

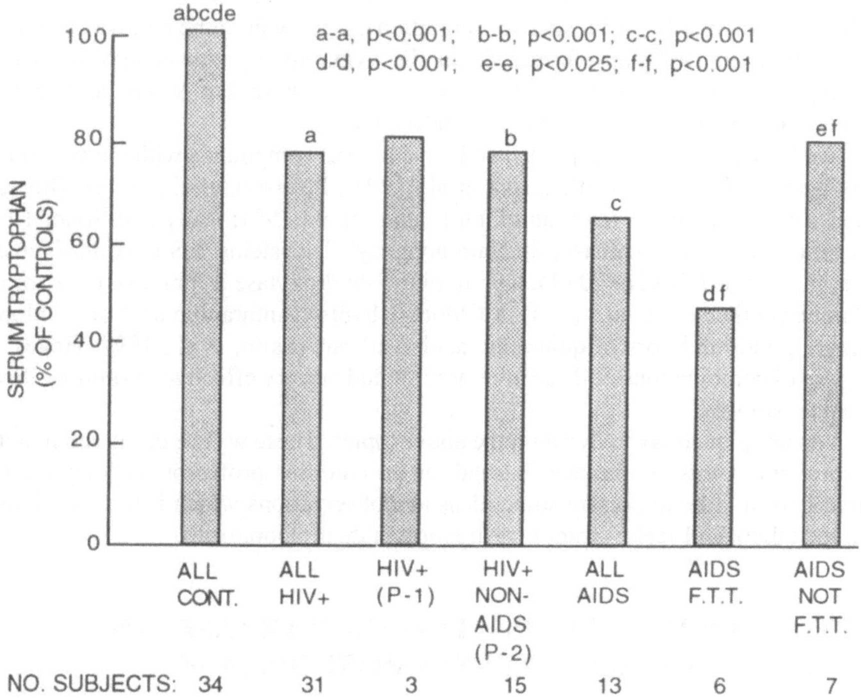


**Figure 1.** Serum Trp levels in control and HIV-1 children expressed as a percent of the control group. Abbreviations used include: ALL CONT., healthy HIV- age-matched children mostly born of HIV+ mothers; ALL HIV+, 31 HIV+ children with all classifications of disease; HIV+ (P-1), 3 patients, stage P-1; HIV+ NON-AIDS (P-2), 15 patients stage P-2; ALL AIDS, 13 patients classified as having AIDS; AIDS F.T.T., 6 AIDS patients with failure-to-thrive; AIDS NOT F.T.T., 7 AIDS patients who were thriving essentially normally. Columns bearing the same letter superscripts were significantly different with the p values indicated.

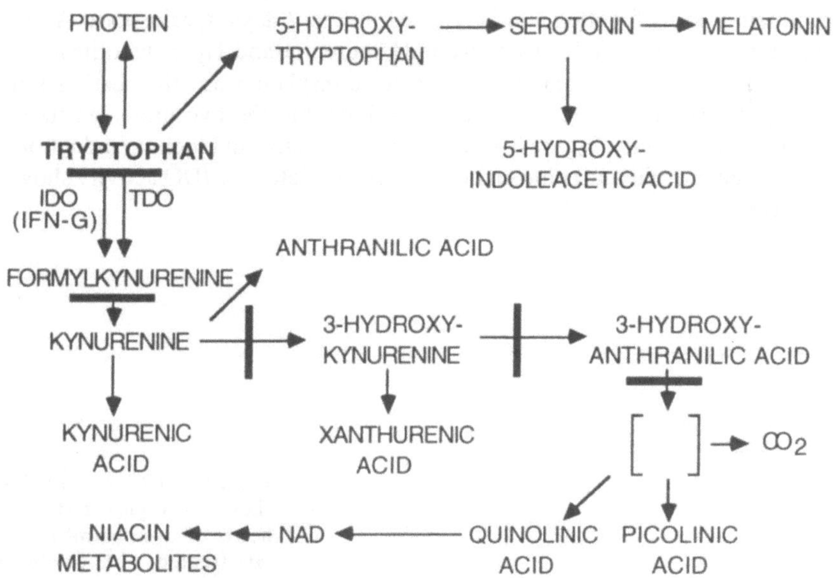


**Figure 2.** Pathway of Trp metabolism. Heavy bars indicate potentially useful sites for inhibition along the kynurenine pathway.

AIDS. A more effective approach to restore Trp levels would be to develop inhibitors of IDO. Several potentially useful sites for inhibition of the kynurenine pathway are shown in Figure 2. Inhibition of steps further down the kynurenine pathway may be useful for decreasing levels of quinolinic acid, but would not restore Trp levels, and may cause accumulation of metabolites, desirable or undesirable.

We have pursued this approach and have several compounds with modest inhibitory activity against IDO in vitro (Peterson, et al., 1993;, Peterson, et al., 1994). Others have reported similar studies of IDO inhibition (Salter, et al., 1991; Cady and Sono, 1991). A potent inhibitor of kynureninase, S-(2-aminophenyl)-L-cysteine S,S-dioxide, is available (Dua, et al., 1993) with Ki of 70 nM. Kynurenine-3-hydroxylase is inhibited by *m* -nitrobenzoylalanine (Pellicciari, et al., 1994). 4-Chloro-3-hydroxyanthranilic acid or 6-chlorotryptophan are good inhibitors of quinolinic acid synthesis (Saito, et al., 1994; Heyes, et al., 1993). Use of combinations of these inhibitors should be very effective in shutting down the kynurenine pathway.

I do not plan today to dwell on the above topics. These will be discussed in detail by other speakers at this conference. Instead, as an emeritus professor no longer active in research, I would like to present some ideas and observations which I think are important and I hope others will feel warrant investigation in their laboratories.

## TRYPTOPHAN METABOLISM IN ORGAN TRANSPLANT REJECTION AND GRAFT VERSUS HOST DISEASE

The immunological consequences of organ graft rejection and of graft versus host disease (GVHD) undoubtedly result in induction of IDO in or near the tissues involved. The importance of this enzyme induction and the resulting Trp breakdown may be important in the rejection process. One report indicates that serum kynurenine rises early in the rejection process (Holmes, 1994). We have done preliminary studies with Dr. S. Knechtle and Dr. Y. Ozaki, University of Wisconsin, using a transplanted rat heart model which leaves the recipients original heart in place; thus, the rat has two functioning hearts. The transplanted heart reproducibly loses function and is rejected in 5 to 7 days. Cyclosporin A effectively blocks rejection. We examined IDO activity and kynurenine and Trp metabolite levels in this system at 5 days after transplantation when the transplant was still beating but losing function. Original hearts had very low levels of IDO, but the two untreated transplanted hearts had very high activity, Figure 3. CsA was very effective in blocking induction of IDO activity. Measurement of kynurenine in these hearts paralleled the IDO activity showing that IDO was active *in vivo*, Figure 4.

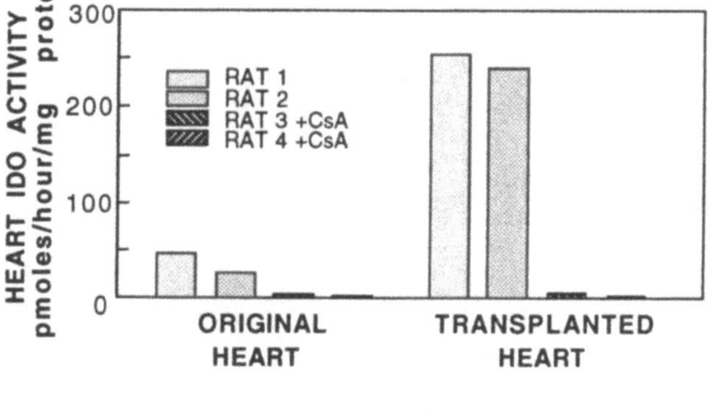


**Figure 3.** IDO activity levels are shown in original and transplanted hearts of two control rats and two rats treated with Cyclosporin A (CsA). Values are at 5 days after transplantation. Note that Cyclosporin very effectively blocks IDO activity.

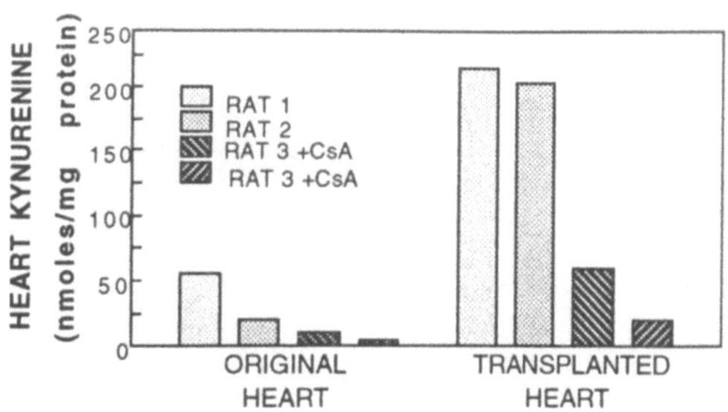


**Figure 4.** Kynurenine values in original and transplanted hearts 5 days after transplant. Values are appreciably lower in rats treated with Cyclosporin (CsA), consistent with the low IDO levels shown in Figure 3.

Surprisingly, Trp levels in all four transplanted hearts were higher than in the original hearts despite evidence that IDO was elevated and functioning in the untreated transplants, Figure 5. Further, CsA had no effect on Trp levels even though IDO activity was virtually undetectable in the CsA treated rats. These data suggest that factors not influenced by CsA may play a role in enhancing Trp transport or uptake into the transplanted organs.

We have also done some preliminary studies in children who developed graft versus host disease (GVHD) after receiving bone marrow transplants. Such subjects show low serum Trp and elevated kynurenine levels which correlate with episodes of acute or chronic GVHD (R.R. Brown and P.M. Sondell, unpublished data).

## IS IDO INVOLVED IN PRODUCTION OF A CHEMOTACTIC FACTOR?

At an earlier ISTRY meeting in 1983, it was reported that *Blepharisma japonicum* under suitable culture conditions produced a metabolite of 5-hydroxytryptophan, gamone II, which was chemotactic towards type I Blepharisma, and required an IDO-like reaction for its synthesis, Figure 6 (Myake, 1984; Jaenicke, 1984). This metabolite, a derivative of 5-hydroxy-formylkynurenine, was active at 1 x 10 E-9 M and became essentially inactive when deformylated.

Since IFN-gamma-stimulated human cells probably have all the enzymes needed to synthesize Gamone II, including IDO, we should ask the question whether human immune-stimulated cells make Gamone II and whether it may be involved in attracting inflammatory

**Figure 5.** Trp levels in heart tissues 5 days after transplant. Even though IDO and kynurenine levels were very high in untreated transplants, indicating Trp catabolism, Trp levels were actually higher than in the original hearts, and were equally high in hearts treated with cyclosporin (CsA).

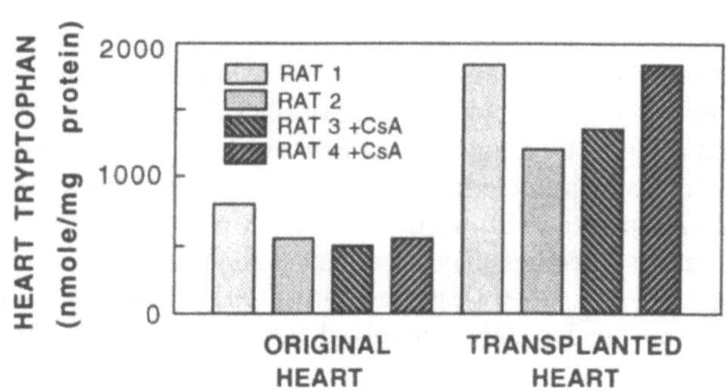

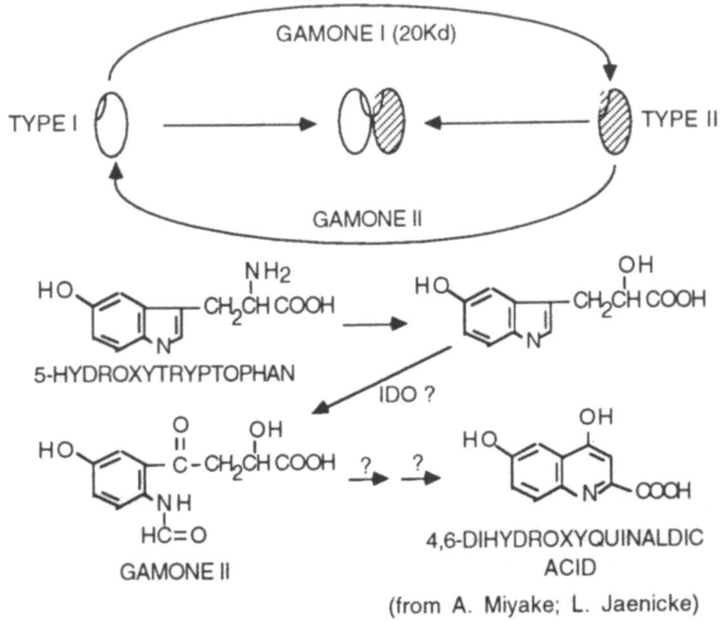


**Figure 6.** This figure shows the relationship between type I and type II *Blepharisma japonicum* and their production of gamone I and II. When cultured under suitable conditions, the organism differentiates into two distinct types (I and II). Type I produces a 20 Kd protein (gamone I) which type II recognizes and responds by producing gamone II. Under these conditions, the two types couple in a mating-like action, suggesting a chemotactic activity for Gamone II. Gamone II is derived from 5-hydroxytryptophan by the reactions shown in the lower part of this slide. 4,6-Dihydroxyquinaldic acid is a predicted metabolite of gamone II resulting from oxidative cyclization of deformylated gamone II.

cells to the site of its production. The wide distribution of kynurenine formamidase in tissues would provide a ready mechanism for inactivating such a factor once it had accomplished its function. It seems to me that a search for Gamone II or its predicted metabolites in immune-stimulated systems would be worthwhile. The literature is replete with reports of unidentified chemotactic factors (Esposito, et al., 1989.).

## SOME REPORTED FUNCTIONS OF TRP IN PROTEINS AND ENZYMES

There are many recent reports that Trp is not just a structural amino acid in proteins, but that specific Trp locations play an essential role in the function of many proteins or enzymes. Such roles include receptor binding of protein hormones such as insulin-like growth factor-1 (Blakesley, et al. 1995), human growth hormone (Clackson and Wells, 1995), binding of the 2-acyl fatty acid of phospholipids to Annexin V (Meers and Mealy, 1994), and receptor binding of 25-hydroxy-vitamin D3 (Swamy et al., 1995). Trp-50 in diphtheria toxin is a major determinant of NAD affinity to this toxin, and replacement of this Trp with another amino acid inactivates the toxin (Wilson, et al., 1994). Such non-toxic forms of toxins as vaccines may be of great importance. The importance of Trp-62 in egg white lysozyme for protein-carbohydrate interaction has been reported (Maenaka, et al., 1994). A Trp residue, Trp-221, is involved with GTP/GDP binding in elongation factor eEF-2 and is important for its function on the ribosome (Guillot, et al., 1993, Guillot, et al., 1993a). Similarly, Trp-207 in transducin-t*a* is involved in binding of the G-protein activation switch (Faurobert, et al., 1993). Binding of the immunosuppressant cyclosporin A to its receptor cyclophylins depends upon a specific Trp residue in the receptor (Liu, et al., 1991). Mouse P-glycoprotein, involved with multi-drug resistance of cancer cells, contains 5 tryptophan residues, one of which is involved in the function of this protein (Baubichon-Cortay, et al., 1994).

These data indicate that specific Trp residues in various proteins may be involved in binding to (a) guanine or adenine nucleotide cofactors, (b) phospholipids (c) certain glycosides, (d) cyclosporin A, and possibly other xenobiotics, (e) growth hormone, (f) insulin-like growth factor , and (g), vitamin D.

There are numerous reports of binding of the indole ring of Trp with adenine or guanine nucleotides and with nucleic acids (Raszka and Mandel, 1971; Toulme, et al., 1974; Sigel and Naumann, 1976; Helene and Dimicoli, 1972; Dimicoli and Helene, 1974) by sharing of pi electrons (stacking, charge-transfer complexing). The energy of such associations may be as high as 7 Kcal/mole, i.e. stronger than hydrogen bonds. Thus it seems likely that exposed Trps in certain proteins may function to orient and bind various purine-containing cofactors (e.g. NAD, NADP, FAD, ATP, GTP, etc.) and to bind to DNA or RNA. It is also of interest to note that certain proteins have their Trp residues concentrated at one end of the protein chain, and that the Trp residues are often in a regular repeating configuration.

Table 1 indicates such sequences in several proteins (Intracel, 1995; Stahl and Prusiner, 1991). In view of the relatively strong binding between Trp sidechains and purine bases, it is tempting to speculate that the Trp residues in such proteins may be important in orienting and attaching such proteins to nucleic acids of virus nucleic acid cores. Similarly, binding of Trp residues to phospholipids (Meers and Mealy, 1994) may be important in orienting certain receptor proteins in phosphpolipid membranes of cells. The repeated Trp-containing sequences found in prion proteins suggests that these proteins may bind

**Table 1.** High tryptophan content of some protein segments. The Trp content of several proteins is shown. The high Trp sequences are at the amino-terminal end of the protein with very little Trp elsewhere in the molecule. Standard single letter abbreviations for amino acids are used. Trp (**W**) is shown in bold type

| | | | | | |
|---|---|---|---|---|---|
| HIV-1 | MRVKEKYQHL | **W**R**W**G**W**K**W**GTM | LLGILMICSA | TEKL**W**VTVYY | GVPV**W**KEATT |
| (gp120iii) | TLFCASDAKA | YDTEVHNV**W**A | THACVPTDPN | PQEVVLVNVT | ENFNM**W**KNDM |
| | VEQMHEDIIS | L**W**DQSLKPCV | KLTPLCVSLK | CTDLGNATNT | NSSN..... |
| HIV-1 | HMLQLTV**W**GI | KQLQARVLAV | ERYLKDQQLL | GP**W**GCSGKLI | CTTTVP**W**NAS |
| (gp41mn) | **W**SNKSLDDI**W** | NNMT**W**MQ**W**ER | EIDNYTSSRG | DRPEGIEEEG | GERDRDTSGR |
| | LVHGFLAII**W** | VDLRSLFLFS | YHH....... | | |
| HIV-1 | .......... | HTNDVKQLTE | AVQKITTESI | VI**W**GKTPKFK | LPIQKET**W**ET |
| (RT-iiib) | **WW**TEY**W**QAT**W** | IPE**W**EFVNTP | PLVKL**W**YQLE | KEPIVGAETF | YVDGAASRET |
| | KLGKAGYVTN | RGRQKVVTLT | DTTNQKTELQ | AIHLALQDSG | ....... |
| HIV-1 | MENR**W**QVMIV | **W**QVDRMRIRT | **W**KSLVKHHMY | VSGKARG**W**FY | RHHYESPHPR |
| (vif-iiib) | ISSEVHIPLG | DARLVITTY**W** | GLHTGERD**W**H | LGQGVSIE**W**R | KKRYSTQVDP |
| | ELADQLIGLY | YFDCFSDSAI | RKALLGHIVS | PRCEYQAGHN | KVG...... |
| PRION | man..LAY**W**I | LALFVAM**W**TD | VGLCKKRPKP | G-G**W**NTGGSR | YPGQGSPGNR |
| (SHaPrP) | YPPQGGGT**W**G | QPHGGG**W**GQP | HGGG**W**GQPHG | GG**W**GQPHGGG | -**W**GQGGGTNN |
| | Q**W**NKPSKPKT | NMKHMAGAAA | AGAVVGGLGG | YMLGSAMSRP | MHFQN..... |

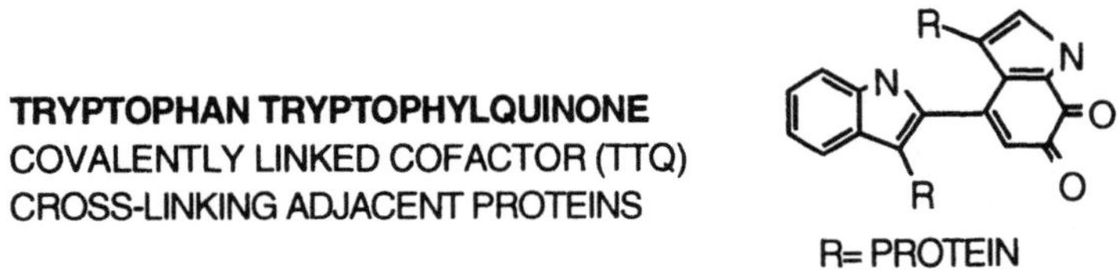


**Figure 7.** Structure of tryptophan tryptophylquinone (TTQ) showing crosslinking of protein chains.

strongly to nucleic acids and that such binding may be important in the as yet unclear mechanism of prion pathogenesis (Stahl and Prusiner, 1991).

## TRYPTOPHAN-CONTAINING QUINOPROTEINS

The family of recently described quinoproteins includes the tryptophan-tryptophylquinone (TTQ) enzymes methylamine dehydrogenase and aromatic amine dehydrogenase (Page and Jones, 1993; Govindaraj, et al., 1994; McIntire, 1994). The structure of TTQ is shown in Figure 7 Trp residues on separate protein chains are oxidatively crosslinked, possibly by a c-type cytochrome reaction (Page and Jones, 1993). Thus separate protein chains are covalently crosslinked. To date, this family of enzymes has been found only in microorganisms, but a somewhat analogous crosslinking of proteins through tyrosine residues has been observed in human tissues, catalyzed by tissue peroxidases, i.e. myeloperoxidase. We find associations between levels of protein dityrosine and inflammatory lung disease in bronchoalveolar lavage samples from humans (K.C. Meyer and R.R. Brown, unpublished data).

## MISCELLANEOUS TRYPTOPHAN-RELATED OBSERVATIONS

Several laboratories have shown that the antiproliferative effect of IFN-G on some, but not all, human tumor cells in culture was due to depletion of Trp in the media (Ozaki, et al., 1988; Aune and Pogue, 1989; Takikawa, et al., 1988; Taylor and Feng, 1991). In other tumor cell lines, inhibition by IFN-G was not due to Trp depletion, but due to NAD and ATP depletion resulting from induction of high levels of poly-ADP ribosyl transferase, an enzyme which uses large quantities of NAD. Inhibitors of this enzyme restored NAD and ATP levels and protected from IFN-G inhibition (Aune and Pogue, 1989). A recent study using human tumor cells growing in nude mice showed that human IFN-G suppressed tumor growth, but that Trp depletion was probably not a major part of the inhibitory mechanism (Burke, et al. 1995). Since poly-ADP ribosyl transferase is thought to play an important role in DNA repair and in programmed cell death (apoptosis ) (Carson et al., 1986), the role of IFN-G-induced Trp metabolism and effects on NAD biosynthesis from Trp need to be studied in detail.

Trp levels in human fibroblasts also seem to be important in regulating levels of collagenase synthesis but not collagen production in cultures treated with IL-1 ( Varga, this proceedings). Such an imbalance between collagen synthesis and breakdown may be important in various inflammatory and fibrotic disease processes.

Superantigens are microbial enterotoxins which are capable of using alternative mechanisms to activate large numbers of human lymphocytes and are suggested as playing important roles in induction of human autoimmune diseases (Taub, D.D. and Blank, K.J., 1993; Silverman, 1994). Toxic shock syndrome toxin-1 (TSST-1) is such a superantigen produced predominantly by Trp-requiring strains (auxotrophs) of of*Staphylococcus aureus* but not by those strains which do not require Trp (prototrophs) (Chu, et al., 1985). Since the Trp-associated disease eosinophilia myalgia syndrome (EMS) is almost certainly an autoimmune disease, I suggest the possibility that prolonged use of L-Trp may select for toxin-producing strains of *S. aureus* in the Trp user. The simultaneous presence of immunogenic impurities in the L-Trp, together with chronic superantigen production, may be the mechanism for development of EMS.

IFN-G is reported to inhibit oxidation of low density lipoprotein (LDL) by human mononuclear cells by inducing IDO and the formation of 3-hydroxyanthranilic acid, an aminophenol with antioxidant properties (Christen, et al., 1994). 3-Hydroxykynurenine, although not tested, would probably function similarly. Since IDO utilizes superoxide anion in metabolizing Trp, this enzyme may also serve as a scavenger of active oxygen radicals.

## REFERENCES

Aune, T.M., and Pogue, S.L. 1989. Inhibition of tumor cell growth by interferon-gamma is mediated by two distinct mechanisms dependent upon oxygen tension: induction of tryptophan degradation and depletion of intracellular nicotinamide adenine dinucleotide. *J. Clin. Invest.* 84:863.

Baubichon-Cortay, H., Baggetto, L.G., Dayan, G. et al. 1994. 1994. Overexpression and purification of the carboxyl-terminal nucleotide-binding domain of mouse P-glycoprotein: Strategic location of a tryptophan residue. *J. Biol. Chem.* 269:22983

Blakesley, V.A., Kato, H., Roberts, C.T. Jr., et al., 1995. Mutation of a conserved amino acid residue (tryptophan 1173) in the tyrosine domain of the IGF-I receptor abolishes autophosphorylation but does not eliminate biologic function. *J. Biol. Chem.* 270:2764.

Brown, R.R., Lee, C.M., Kohler, P.C., et al. 1989. Altered tryptophan metabolism in cancer patients treated with recombinant interleukin 2. *Cancer Res.* 49:4911.

Brown, R.R., Ozaki, Y., Datta, S.P., et al. 1991. Implications of interferon-induced tryptophan catabolism in cancer, autoimmune diseases and AIDS. *Adv. Expt. Med. Biol.* 294:425.

Burke, F., Knowles, R.G., East, N., et al.1995. The role of indoleamine 2,3-dioxygenase in the anti-tumor activity of human interferon-gamma *in vivo* . *Int. J. Cancer* 60:115.

Byrne, G.I., Lehmann, L.K., Kirschbaum, J.G., et al. 1986. Induction of tryptophan degradation in vitro and in vivo, a gamma-interferon stimulated activity. *J. Interferon Res.* 6:389.

Carlin, J.M., Ozaki, Y., Byrne, G.I., et al. 1989. Interferons and indoleamine 2,3-dioxygenase: role in antimicrobial and antitumor effects. *Experientia* 45:535.

Carson, D.A., Seto, S., Wasson D.B., et al., 1986. DNA strand breaks, NAD metabolism and programmed cell death. *Exp. Cell Res.* 164:273.

Christen, S., Thomas, S.R., Garner, B., et al. 1994. Inhibition by interferon-gamma of human mononuclear cell-mediated low density lipoprotein oxidation. Participation of tryptophan metabolism along the kynurenine pathway. *J. Clin. Invest.* 93:2149.

Chu, M.C., Melish, M.E., and James, J.F. 1985. Tryptophan auxotypy associated with *Staphylococcus aureus* that produce toxic-shock-syndrome toxin. *J. Infectious Dis.* 151:1157

Clackson, T., and Wells, J.A. 1995. A hot spot of binding energy in a hormone-receptor interface. *Science* 267:383

Cady, S.G., and Sono, M. 1991. 1-Methyl-DL-tryptophan, B-(3-benzofuranyl)-DL-alanine (the oxygen analog of tryptophan), and B-[3-benzo(b)thienyl]-DL-alanine are competitive inhibitors for indoleamine 2,3-dioxygenase. *Arch. Biochem. Biophys.* 291:326.

Dimicoli, J-L., and Helene, C. 1974. Interactions of aromatic residues of proteins with nucleic acids. I. Proton magnetic resonance studies of the binding of tryptophan-containing peptides of poly(adenylic acid) and deoxynucleic acid. *Biochemistry*, 13:714.

Dua, R.K., Taylor, E.W., and Phillips, R.S. 1993. S-aryl-L-cysteine S,S-dioxides: Design, synthesis and evaluation of a new class of inhibitors of kynureninase. *J. Am. Chem. Soc.* 115:1264

Esposito, A., L., Poirier, W.J., Clarck, C.A., et al., 1989. The release of neutrophil chemoattractant activity by bronchoalveolar macrophages from adult and senescent mice. *J. Gerontology: Biological Sciences*, 44:B93

Faurobert, E., Otto-Bruc, A., Chardin, P., et al. 1993. Tryptophan W207 in transducin t*a* is the fluorescence sensor of the G protein activation switch and is involved in the effector binding. *EMBO J.* 12:4191.

Feng, G., and Taylor, M.W. 1989. Interferon-gamma-resistant mutants are defective in the induction of indoleamine 2,3-dioxygenase. *Proc. Nat. Acad. Sci.*, U.S.A. 86:7144.

Gisslen, M., Larsson, M., Norkrans, G., et al. 1994. Tryptophan concentrations increase in cerebrospinal fluid and blood after Zidovudine treatment in patients with HIV type 1 infection. *AIDS Res. Human Retroviruses*. 10:947.

Govindaraj, S., Eisenstein, E., Jones, L.H., et al. 1994. Aromatic amine dehydrogenase, a second tryptophan tryptophylquinone enzyme. *J. Bacteriol.* 176:2922.

Guillot, D., Lavergne, J-P., and Reboud, J-P. 1993. Trp-221 is involved in the protective effect of elongation factor eEF-2 on the ricin/*a* -sarcin site of the ribosome. *J. Biol. Chem.* 268:26082.

Guillot, D., Penin, F., Di Pietro, A., et al. 1993a. GTP binding to elongation factor eEF-2 unmasks a tryptophan residue required for biological activity. *J. Biol. Chem.* 268:20911.

Helene, C., and Dimicoli, J-L. 1972. Interactions of oligopeptides containing aromatic amino acids with nucleic acids. Fluorescence and proton magnetic resonance studies. *FEBS Letters*. 26:6.

Heyes, M.P., Brew, B.J., Saito, S., et al., 1992. Inter-relationships between quinolinic acid, neuroactive kynurenines, neopterin and B2-microglobulin in cerebrospinal fluid and serum of HIV-1 infected patients. *J. Neuroimmunol.* 40:71

Heyes, M.P., Saito, K., Major, E.O., et al. 1993. A mechanism of quinolinic acid formation by brain in inflammatory neurological disease. Attenuation of synthesis from L-tryptophan by 6-chlorotryptophan and 4-chloro-3-hydroxyanthranilate. *Brain* 116:1425.

Holmes, E.W., Russell, P.M., Kinzler, G.J., et al. 1994. Inflammation-associated changes in the cellular availability of tryptophan and kynurenine in renal transplant recipients. *Clin. Chim. Acta* 227:1.

Hortin, G.L., Landt, M., and Powderly, W.G. 1994. Changes in plasma amino acid concentrations in response to HIV-1 infection. *Clin. Chem.* 40:785.

Intracel. 1995. Intracel retroviral products catalog. Intracel Corp., 359 Allston St., Cambridge MA 02139, pp.13-23.

Jaenike, L. 1994. Biological activity of blepharismone derivatives. In: Progress in Tryptophan and Serotonin Research, Eds. Schlossberger, H.G., Kochen, W., Linzen, B., and Steinhart, H., W. de Gruyter, Publ. Berlin, 1994, pp.815-826.

Launay, J-M., Copel, L., Callebert, J., et al. 1988. Decreased whole blood 5-hydroxytryptamine(serotonin) in AIDS patients. *J. Acquired Immune Def. Syndromes.* 1:324.

Liu, J., Chen, C-M., and Walsh C.T. 1991. Human and *Escherichia coli* cyclophylins: Sensitivity to inhibition by the immunosuppressant cyclosporin A correlates with a specific tryptophan residue. *Biochemistry* 30:2306.

Maenaka, K., Kawai, G., Watanabe, K., et al., 1994. Functional and structural role of a tryptophan generally observed in protein-carbohydrate interaction: Trp-62 of hen egg white lysozyme. *J. Biol. Chem.* 269:7070.

McIntire, W.S. 1994. Quinoproteins. *FASEB J.* 8:513.

Meers, P., and Mealy, T. 1994. Phospholipid determinants for annexin V binding sites and the role of tryptophan 187. *Biochemistry* 33:5829.

Myake, A. 1994. Blepharismone: A conjugation-inducing tryptophan derivative in the ciliate *Blepharisma.* In: Progress in Tryptophan and Serotonin Research, Eds. Schlossberger, H.G., Kochen, W., Linzen, B., and Steinhart, H., W. de Gruyter, Publ. Berlin, 1994, pp807-813.

Ozaki, Y., Edelstein, M., and Duch, D. 1987. The actions of interferon and antiinflammatory agents on induction of indoleamine 2,3-dioxygenase in human peripheral blood monocytes. *Biochem. Biophys. Res. Commun.* 144:1147

Ozaki, Y., Edelstein, M., and Duch, D. 1988. Induction of indoleamine dioxygenase: a mechanism of the antitumor activity of interferon-gamma. *Proc. Nat. Acad. Sci. U.S. A.* 85:1242.

Page, M.D., and Jones, S.J., 1993. Mutants of *Methylobacterium extorquens* and *Paracoccus denitrificans* deficient in *c* -type cytochrome biogenesis synthesize the methylamine-dehydrogenase polypeptides but cannot assemble the tryptophan-tryptophylquinone group. *Eur. J. Biochem.* 218:711.

Pellicciari, R., Natalini, B., Constantino, G., et al., 1994. Modulation of the kynurenine pathway in search for new neuroprotective agents. Synthesis and preliminary evaluation of (*m* -nitrobenzoyl)alanine, a potent inhibitor of kynurenine-3-hydroxylase. *J. Med. Chem.* 37:647.

Peterson A.C., La Loggia, A.J., Hamaker, L.K., 1993. Evaluation of substituted B-carbolines as noncompetitive indoleamine 2,3-dioxygenase inhibitors. *Med. Chem. Res.* 3:473.

Peterson, A.C., Migawa, M.T., La Loggia, A.J., et al. 1994. Evaluation of functionalized tryptophan derivatives and related compounds as competitive inhibitors of indoleamine 2,3-dioxygenase. *Med. Chem. Res.* 3:531.

Raszka, M., and Mandel, M. 1971. Interaction of aromatic amino acids with neutral polyadenylic acid. *Proc. Nat. Acad. Sci. U.S.A.* 68:1190.

Saito, K., Markey, S.P. and Heyes M.P. 1994. 6-Chloro-D,L-tryptophan, 4-chloro-3-hydroxyanthranilate and dexamethasone attenuate quinolinic acid accumulation in brain and blood following systemic immune activation. *Neurosci. Letters.* 178:211.

Salter, M., Beams, R.M., Critchley, M.A.E., et al., 1991. Effects of tryptophan 2,3-dioxygenase inhibitors in the rat. *Adv. Expt. Med. Biol.* 294:281.

Shimizu, T., Nomiyama, S. Hirata, F., et al. Indoleamine 2,3-dioxygenase. Purification and some properties. *J. Biol. Chem.* 253:4700.

Sigel, H., and Naumann, C.F. 1976. Ternary complexes in solution XXIV. Metal ion bridging of stacked purine-indole adducts. The mixed-ligand complexes of adenosine 5'-triphosphate, tryptophan, and manganese(II), copper (II), or zinc (II). *J. Am. Chem. Soc.* 98:730.

Silverman, G., 1994. Superantigens and the spectrum of unconventional B-cell antigens. *The Immunologist* 2:51.

Stahl, N., and Prusiner, S.B. 1991. Prions and prion proteins. *FASEB J.* 5:2799.

Swamy, N., Brisson, M., and Ray, R. 1995. Trp-145 is essential for the binding of 25-hydroxyvitamin $D_3$ to human serum vitamin D-binding protein. *J. Biol. Chem.* 270:2636.

Takikawa, O., Kuroiwa, T., Yamazaki, F., et al. 1988. Mechanism of interferon-gamma action. *J. Biol. Chem.* 263:2041.

Takikawa, O., Habara-Ohkubo, A., and Yoshida, R. 1990. IFN-gamma is the inducer of indoleamine 2,3-dioxygenase in allografted tumor cells undergoing rejection. *J. Immunol.* 145:1246.

Taub, D.D., and Blank, K.J. 1993. Superantigens and microbial pathogenesis. *Annals of Internal Med.* 119:89.

Taylor, M.W. and Feng G. 1991. Relationship between interferon-gamma, indoleamine 2,3-dioxygenase, and tryptophan catabolism. *FASEB J.* 5:2516

Toulme, J-J., Charlier, M., and Helene, C. 1974. Specific recognition of single-stranded regions in ultraviolet-irradiated and heat-denatured DNA by tryptophan-containing peptides. *Proc. Nat. Acad. Sci.*, U.S.A. 71:3185.

Truswell, A.S., Hansen J.D.L., and Wannenburg, P. 1968. Plasma tryptophan and other amino acids in pellagra. *Am. J. Clin. Nutr.* 21:1314.

Werner, E.R., Werner-Felmayer, G., Fuchs, D., et al. 1989. Parallel induction of tetrahydrobiopterin biosynthesis and indoleamine 2,3-dioxygenase activity in human cells and cell lines by interferon-gamma. *Biochem. J.* 262:861.

Wilson, B.A., Blanke, S.R., Reich, K.A., et al., 1994. Active-site mutations of diphtheria toxin: Tryptophan 50 is a major determinant of NAD affinity. *J. Biol. Chem.* 269:23296.

Yamazaki, F., Kuroiwa, T., Takikawa, O., et al. 1985. Human indolylamine 2,3-dioxygenase. its tissue distribution and characterization of the placental enzyme. *Biochem J.* 230:635.

Yoshida, R., Imanishi, J., Oku, K.T, et al. 1981. Induction of pulmonary indoleamine 2,3-dioxygenase by interferon. *Proc. Nat. Acad. Sci., U.S.A.* 78:129.

Yoshida, R., Urade, Y., Tokuda, M., et al., 1979. Induction of indoleamine 2,3-dioxygenase in mouse lung during virus infection. *Proc. Nat. Acad. Sci. U.S.A.* 76:4084

3

# BRAIN TRYPTOPHAN

## Normal and Disturbed Control

G. Curzon

Institute of Neurology
Queen Square
London, WC 1N 3BG, United Kingdom

## 1. INTRODUCTION

I am both pleased and honoured to have been invited to lecture here where the modern age might reasonably be said to have begun. It was in Padua that Galileo wrote a book, Siderius Nuncius - in English, "The Messengers of the Stars" - which set the seal on a revolution in how we visualise the Universe and our place in it. I am also pleased to be at a place that has had a great influence on British scolarship - where William Harvey was led towards discovering the circulation of the blood and where the English ambassador to Venice, Sir Henry Wotton was so involved with its intellectual life that he sent Galileo's Messenger of the Stars to King James the First on its actual day of publication (13/3/1610). The covering letter (quoted by Smith, 1907) is quite amusing; Wotton is impressed but cautious and dismisses science in favour of more important matters.

"I send — his Majesty the strangest piece of news — that he hath ever yet received— [a book published] this very day of the Mathematical Professor at Padua, who by the help of an optical instrument— hath overthrown all former astronomy —[and] must become either exceeding famous or exceeding ridiculous. Now to descend from these novelties to those which do more concern the wise men of this place." Wotton then goes on to discuss political topics.

It is also a pleasure to be invited to give a plenary lecture at ISTRY as I have attended all the ISTRY meetings since the first one held here 21 years ago and I have been involved in some way with tryptophan (TRP) since 1948 when I began working for a PhD at Leeds in the Biochemistry Department then chaired by Professor Frank Happold. His main interest at that time was the bacterial enzyme tryptophanase which hydrolyses TRP into indole and serine. My PhD was not on this topic but I must have breathed in an awareness of TRP. Rather similarly, when years later, Simon Young the present President of the executive committee of ISTRY, did his PhD at my laboratory, in London at the Institute of Neurology, though he did not work on TRP, he also seems to have suffered an air-borne infection with it.

The pattern of my own work was largely determined back in 1964 by John Cumings when I was a junior member of his department at the Institute of Neurology. Cumings had

*Recent Advances in Tryptophan Research,* edited by Graziella Allegri Filippini et al.
Plenum Press, New York, 1996 

planned a book on biochemical aspects of neurology and psychiatry to be based on a series of lectures. When he told me that I was to do the one on the biochemistry of depression I said that it was a subject I knew nothing about. He replied: "well you will have to go off and find out," which was how I first heard about the liver enzyme TRP 2,3-dioxygenase, (then called TRP pyrrolase) and had what I then thought was an original idea on how depression might be precipitated, i.e that an "alteration of TRP metabolism occurs [in depression] of which 5-hydroxytryptamine changes are a part. [As] the main pathway of TRP metabolism by TRP pyrrolase to kynurenine, etc. is increased by stress — it may be that the result of long periods of stress is the diversion of TRP from the 5-hydroxytryptamine— pathways to the kynurenine pathway" (Curzon, 1965).

Similar ideas were in fact proposed by a number of people around that time (see Curzon, 1969 for references). Examples were subsequently reported of reciprocal associations between activity of the liver enzyme and brain TRP concentration (e.g. Badawy et al., 1989; Litman and Correia, 1985) and recently developed inhibitors of the enzyme have been shown to strikingly increase central TRP concentration (Salter et al., 1995). However, though there is now substantial evidence of moderate TRP deficiency in depression, how it occurs and how important it is in the illness remain incompletely resolved questions (see below). Much of the work of my own laboratory on the *normal* control of brain TRP and the effect of TRP on 5-HT synthesis was first stimulated by the interest in these questions.

## 2. NORMAL CONTROL OF BRAIN TRP

When we began our work in 1965 there was what now seems surprisingly little interest in the influence of TRP availability on brain 5-HT synthesis. Eccleston et al. (1965) had reported in the same year that TRP increased rat brain 5-HT and its metabolite 5-HIAA but they had used very large doses of amino acid so that brain TRP rose far beyond the physiologically relevant range. It was only later that work on the effects of dietary intake on brain TRP concentration was stimulated by the important studies of Fernstrom and Wurtman (1971, a,b) on the effect on brain 5-HT synthesis of relatively small changes of TRP availability. At about the same time, we were depriving rats of food for 24 hr, expecting that brain TRP and hence 5-HT synthesis would decrease. Instead they both increased (Curzon et al., 1972) as fasting elevated plasma unesterified fatty acid which freed plasma TRP from albumin so that it became more available to the brain (Knott and Curzon, 1972). This findings, together with the increase of brain TRP after a single large TRP-free high carbohydrate meal (Fernstrom and Wurtman, 1972) which also might have been expected to decrease it, imply that the brain is effectively protected in these rather extreme circumstances against low TRP intake. Similarily, intrauterinally malnourished newborn rats have raised brain TRP values and corresponding humans have raised plasma free TRP concentrations and raised plasma ratios of free (but not of total) TRP to the sum of the large neutral amino acids (LNAAs) which compete with it for transport to the brain (Hernandez et al., 1989).

In an investigation on the dietary control of human plasma TRP, Eriksson et al. (1989) found that TRP/LNAA ratios in subjects on standard hospital meals only varied slightly over a 24 hr period. In related animal work, exposure of rats to diets od 0-40% protein for 2 hr after 12 hr food deprivation had only moderate effects on brain TRP concentration and though there was a significant correlation between plasma total TRP/LNAA ratios and brain TRP this only accounted for 20% of the variance of the brain values (Voog and Eriksson, 1992). Predictability of brain values from plasma ratios was increased in studies in which calculations were based on plasma free TRP (Smith et al., 1987., Takada et al., 1993). Similarly, plasma free TRP appeared to be the form of the amino acid that was most readily available to the human brain (Gillman et al., 1981., Delgado et al., 1990; Cangiano et al., 1990). Other workers reported that quite wide human dietary variation had little effect on total plasma TRP/LNAA ratios (Fernstrom, 1987) and human CSF TRP and 5-HIAA concentrations were unaffected by protein or carbohydrate breakfasts (Teff et al.1989). These findings are consistent with the stability of brain amino acid pools under many conditions (Harper and Tews, 1988).

However, more extreme influences can increase brain TRP e.g. prolonged exercise (Chaouloff, 1989; Wilson and Marsden, 1993) via increased plasma free TRP; immobilization by enhancing the kinetics of TRP transport to the brain (Kennett et al., 1986); a large high carbohydrate meal by decreasing the plasma concentrations of other LNAAs (Fernstrom and Wurtman, 1971; Sarna et al., 1985); ageing by a combination of these influences (Sarna et al., 1982). It is remarkable that brain TRP increases in so many circumstances but resists decraese except on prolonged malnourishment (Dickerson and Pao, 1975; Fernstrom and Hirsch, 1977) or pharmacological treatments.

The kinetics of transport of TRP across the blood-brain barrier vary with species (Pardridge and Fierer, 1990). Enhanced transport during immobilization stress (Kennett et al., 1986) is propably due to action of secreted catecholamine at β-adrenergic receptors (Dunn and Welch, 1991), the brain appearing to strip TRP from its binding to albumin under these conditions.

## 3. DISTURBED CONTROL OF BRAIN TRP

### 3.1 Depression

Evidence on the availability of TRP to the brain in depressive illness is mostly based on plasma determinations. Low plasma free TRP in depression was first reported by Coppen et al. (1973) and confirmed by DeMyer et al. (1981) and Schmid-Burgk et al. (1981). Unlike the other two groups, DeMyer et al. also found low plasma total TRP as did Joseph et al. (1984) Lucca et al. (1992) and Karege et al. (1994). These three groups did not report free TRP data. Coppen, DeMyer, Karege and their co-workers all found that values returned to normal on recovery. Moller et al. (1979) obtained normal plasma TRP concentrations but Moller et al. (1990) later reported that lower TRP or TRP/LNAA values predicted % improvement on treatment with 5-HT reuptake inhibitors. Mathis et al. (1988) also found normal TRP but reported low TRP/LNAA values. Taken together, these results strongly suggest that low availability of plasma TRP to the brain is associated with depressive symptoms. Whether it has some responsability for them is another question though the reversal of the antidepressant effect of 5-HT reuptake inhibitors when plasma TRP is rapidly decreased by an oral load of LNAAs (Delgado et al., 1990) is suggestive.

There are also reports on subjects with postpartum depression of low plasma free (but not total) TRP (Stein et al., 1976), low plasma total TRP/LNAA ratios (Maes et al.,1992) and absence of the normal postpartum rise in plasma TRP (Handley et al., 1980). The exacerbation of premenstrual depression and irritability when plasma TRP is rapidly decreased by oral LNAA loading (Menkes et al., 1994) is also of interest. However, low plasma TRP may be related to retarded affect rather than to depressive illness as such as both total and free values of a group of patients with high retardation scores were significantly low independently of their final diagnoses (Curzon et al., 1979). It is also worth noting that subjects without significant anxiety or depression but with presumed stress as they were full-time carers of patients with Alzheimer's disease had low plasma free (but not total) TRP (Hockney et al., 1992).

The following observations point to different (but not necessarily mutually exclusive) ways low plasma TRP could occur in depression.

a. *Enhanced metabolism, enhanced uptake by tissues or defective absorption.* Mechanisms of this kind are indicated by the relatively small rises of plasma TRP after oral loading (Russ et al., 1990). Enhanced metabolism is suggested by the high urinary excretion of metabolites on the tryptophan-2,3-dioxygenase or indoleamine-2,3-dioxigenase (IDO) pathways (Curzon and Bridges, 1970; Maes et al., 1987). Increased metabolism by indoleamine-2,3-dioxygenase is conceivably consistent with various indices of altered immune function that are reported to correlate with the low TRP values (Maes et al., 1993).
b. *Effects of loss of appetite.* This is suggested by reports that low plasma TRP in depression was associated with both female sex and low body weight (Anderson et al., 1990) and that plasma TRP decreased in normal women (but not men) after 3 weeks of moderate dieting (Walsh et al., 1995).

Low plasma TRP in depression is paralleled to some extent by CNS findings. Inefficient uptake of plasma TRP by the brain is perhaps suggested by PET scan studies showing abnormally low uptake of the related amino acid. [$^{1}$C]-5-HTP when given intravenously (Agren et al., 1991). Also, TRP concentrations in lumbar CSF were found to be low by Coppen et al. (1972) though this was not confirmed by Ashcroft et al. (1993) and Banki et al. (1981) was unable to show significant correlations between lumbar values and intensity of symptoms.

That low brain TRP concentrations would lead not only to low concentrations of 5-HT in brain tissue but also to low concentrations at receptors is indicated by in vivo dialysis studies on rats with TRP availability to the brain impaired by administration of LNAAs (Heslop et al., 1991; Gartside et al., 1992).

## 3.2 Appetite Disorders

Starvation can cause anorexia (Hunsicker et al., 1992), and appetite is impaired by activation of hypothalamic postsynaptic 5-HT receptors (rev. Curzon et al., 1993). Therefore, the rise of brain TRP concentration and 5-HT metabolism on 24 hr food deprivation (Curzon et al., 1972); Fuenmayor and Garcia, 1984) is of interest with respect to anorexia nervosa especially as in vivo dialysis data suggests that the increase of 5-HT at receptors in the lateral hypothalamus on giving TRP is greater after food deprivation (Schwartz et al. 1990). In normal subjects, any appetite suppressant effect of this increase is presumably effectively opposed by appetite stimulation by other substances released on deprivation e.g. neuropeptide Y (Lambert et al., 1994) and noradrenaline (Stanley et al., 1989). However, it is conceivable that in some subjects the 5-HT changes are dominant and lead to anorexia nervosa. Such changes might well be provokable by slimming regimes and exacerbated by stress (Knott et al., 1977). It is relevant that female but not male volunteers on low calorie intake have an increased 5-HTergic response to tryptophan as indicated by the rise of plasma prolactin on TRP loading (Goodwin et al., 1987) though whether this reflects enhanced hypothalamic 5-HT function or altered availability of the TRP load to the brain is not entirely clear (Anderson et al., 1990). Cancer anorexia could also involve increased 5-HT function as plasma free TRP and CSF TRP both increase (Cangiano et al., 1990). However, the hypothesis for precipitation of anorexia nervosa through excessive release of 5-HT to receptors is not strengthened by data on subjects with established anorexic illness as evidence pointing to increased brain 5-HT is lacking (Ploog and Pirke, 1987). On the other hand, metabolic abnormalities resulting from the illness make interpretation difficult.

Increased urge to eat appears to be diametrically opposite to anorexia, though subjects can fluctuate between anorexia nervosa and bulimia nervosa (i.e. weight control by self-induced vomiting after binge eating). However, there is some evidence of contrasting neurochemical abnormalities in the two conditions as bingeing terminates when plasma TRP/LNAA ratios rise (Kaye et al., 1988) so that brain 5-HT synthesis presumably increases. This agrees with the causal role for defective hypothalamic 5-HT function in the illness which is indicated by reduced prolactin responses to 5-HTergic drugs (Brewerton et al., 1992; Jimerson et al., 1994).

## 3.3 Other Disturbances

Brain TRP changes have been described in a number of the other disorders. A large and proliferating literature now exists on the enhanced induction of IDO in a wide range of inflammatory neurological disorders with excessive production of potentially neurotoxic quinolinic acid (e.g. Heyes et al., 1992; Holzman, 1993). Justice cannot be done to this topic within the scope of the present lecture.

There is also a substantial literature on the increase of brain TRP and 5-HT metabolism in human and experimental (shunt) liver failure (e.g. Record et al., 1976; Bloxam and Curzon, 1978; Bengtsson et al., 1991). The TRP changes are explicable partly by a rise of plasma free TRP and partly by increased uptake across the blood-brain barrier (Bloxam and Curzon, 1978) which is, in turn, at least in part, dependent on hyperammonaemia (Jessy et al., 1990). Brain 5-HT metabolism increases largely as a result of the rise of brain TRP though increases of tryptophan hydroxylase activity are also reported in some brain regions (Bengtsson et al., 1991). Resultant increases of the availability of 5-HT to receptors could conceivably be implicated in the development of symptoms. However, recent in vivo dialysis experiments indicate that 5-HT release did not increase on chronic liver failure and decreased on acute shunt (Bergqvist et al., 1995). This finding suggests that the symptoms of hepatic encephalopathy may not involve increased 5-HT release. Whether these symptoms and those associated with hyperammonaemia depend on increased production of quinolinic acid is controversial (Robinson et al., 1992).

Increased brain TRP and 5-HT metabolism are also indicated in human renal disease with encephalopathy (Sullivan et al., 1980) as CSF TRP and 5-HIAA both rose. The central changes were explicable by increased plasma free TRP and were reversed by kidney dialysis (Sullivan et al., 1980).

## REFERENCES


Agren, H., Reibring, L., Hartvig P., Tedroff, J., Bjurling, P.,Hornfeldt, K., Andersson, Y., Lundqvist, H. and Langstrom, B., 1991, Low brain uptake of L-[$^{11}$C]5-hydroxytryptophan in major depression: a positron emission study on patients and healthy volunteers., *Acta Psychiat. Scand.*, 83: 449.

Anderson, I.M., Parry-Billings, M., Newsholme, E.A., Poortmans, J.R., and Cowen, P. J., 1990, Decreased plasma tryptophan concentration in major depression : relationship to melanchonia and weight loss, *J.Affect.Dis.*, 20:185.

Anderson, I.M., Parry-Billings, M., Newsholme, E.A., Fairburn, C.G., and Cowen P.J., 1990, Dieting reduces plasma tryptophan and alters brain 5-HT function in women, *Psychol. Med.*, 20: 785.

Ashcroft, G.W., Blackburn, I.M., Eccleston, D., Glen, A.I.M., Hartley, W., Kinloch, N.E., Lonergan, M., Murray, L.G., and Pullen, I.A., 1973, Changes on recovery in the concentartion of tryptophan and the biogenic amine metabolites in the cerebrospinal fluid of patients with affective illness, *Psychol. Med.*, 3: 319.

Badawy, A.A-B., Morgan, C.J., Lane, J., Dhaliwal, K., and Bradley, D.M., 1989, Liver tryptophan pyrrolase. A major determinant of the lower brain 5-hydroxytryptamine concentration in alcohol-preferring C57BL mice, *Biochem. J.* 264: 597.

Banki, C.M., Vojnik, M., and Molnar, G., 1981, Cerebrospinal fluid amine metabolites, tryptophan and clinical parameters in depression, *J.Affect., Dis.*, 3:81.

Bengtsson, F., Bugge, M., Johansen, K.H., and Butterworth, R.F., 1991, Brain tryptophan hydroxylase in the portocaval shunted rat: a hypothesis for the regulation of serotonin, *J.Neurochem.*, 56: 1069.

Bergqvist, P.B.F., Vogels, B.A.P.M., Bosman, D.K., Maas, M.A.W., Hjorth, S., Chamuleau, R.A.F.M., and Bengtsson, F., 1995, Neocortical dialysate monoamines of rats after acute, subacute and chronic liver shunt, *J.Neurochem.* 64: 1238.

Bloxam, D.L. and Curzon, G., 1978, A study of proposed determinants of brain tryptophan concentration in rats after portocaval anastomosis or sham operation, *J.Neurochem.*, 31:1255.

Brewerton, T.D., Mueller, E.A., Lesem, M.D., Brandt, H.A., Querry, B., George, D.T., Murphy, D.L., and Jimerson, D.C., 1992, Neuroendocrine response to m-chlorophenylpiperazine and L-tryptophan in bulimia, *Arch. Gen. Psychiat.*, 49: 852.

Cangiano, C., Cascino, A., Ceci, F., Laviano, A., Mullerl, M., Muscaritoll, M., and Rossi-Fanelli, F., 1990, Plasma and CSF tryptophan in cancer anorexia, *J.Neural.Transm.* [Gen Sect.], 81 : 225.

Chaouloff, F., Kennett, G.A., Serrurrier, B., Merino, D., and Curzon, G., 1986, Amino acid analysis demonstrates that increased plasma free tryptophan causes the increase of brain tryptophan during exercise in the rat, *J.Neurochem.*, 46: 1647.

Coppen, A., Brooksbank, B.W.L., and Peet, M., 1972, Tryptophan concentration in the cerebrospinal fluid of depressive patients, *Lancet*, 1:1393.

Coppen, A., Eccleston, E.G., and Peet, M., 1973, Total and free tryptophan concentration in the plasma of depressive patients, *Lancet*, ii: 60.

Curzon, G., 1965, The biochemistry of depression in :"Biochemical Aspects of Neurological Disorders, 2nd series" J.N.Cumings and M.Kremer, eds., Blackwell, Oxford, pp257.

Curzon, G., 1969, Tryptophan pyrrolase- a biochemical factor in depressive illness?, *Br.J.Psychiat.*, 115: 1367.

Curzon, G., and Bridges, P.K., 1970, Tryptophan metabolism in depression, *J.Neurol. Neurosurg. Psychiat.*, 33: 698.

Curzon, G., Joseph, M.H. and Knott, P.J., 1972, Effcets of immobilization and food deprivation on rat brain tryptophan metabolism, *J.Neurochem.*, 19: 1967.

Curzon, G., Kantamaneni, B.D., Lader, M.H., and Greenwood, M.H., 1979, Tryptophan disposition in psychiatric patients before and after stress, *Psychol.Med.*, 9:457.

Curzon, G., Kantamaneni, B.D., Van Boxel, P., Gillman, P.K., Bartlett, J.R. and Bridges, P.K., 1980, Substances related to 5-hydroxytryptamine in plasma and in lumbar and ventricular fluids of psychiatric patients, *Acta Psychiat., Scand.* 61 Suppl. 280:3.

Curzon, G., Gibson, E.L., Kennedy, A.J., and Oluyomi, A.D., 1993, Serotoninergic pharmacology of appetite, in "Serotonin. From cell biology to pharmacology and therapeutics", P.M. Vanhoutte, P.R. Saxena, R.Paoletti, and N.Brunello, eds., Kluwer, Dordrecht, pp 393.

Delgado, P.L., Charney, D.S., Price, L.H., Aghajanian, G.K., Landis, H., and Heninger, G.R., 1990, Serotonin function and the mechanism of antidepressant action. Reversal of antidepressant-induced remission by rapid depletion of plasma tryptophan, *Arch.Gen.Psychiat.*, 47:411.

De Myer, M.K., Shea, P.A., Hendrie, H.L., and Yoshimura, N.N. 1981, Plasma tryptophan and five other amino acids in depressed and normal subjects, *Arch.Gen.Psychiat.*, 38:642.

Dickerson, J.W.T., and Pao, S-K., 1975, The effect of a low protein diet and exogenous insulin on brain tryptophan and its metabolites in the weanling rat, *J. Neurochem.*, 25:555.

Dunn, A.J., and Welch, J., 1991, Stress-and endotoxin-induced increases in brain tryptophan and serotonin metabolism depend on sympathetic nervous system activity, *J.Neurochem.*, 57: 1615.

Eccleston, D., Ashcroft, G.W., and Crawford, T.B.B., 1965, 5-Hydroxyindole metabolism in rat brain. A study of intermediate metabolism using the technique of tryptophan loading-II, *J.Neurochem.*, 12:493.

Eriksson, T., Voog, L., Walinder, J., and Eriksson, T.E. 1989, Diurnal rhythm in absolute and relative concentrations of large neutral amino acids in human plasma, *J. Psychiat. Res.*, 23:241.

Fernstrom, J.D., 1987, Food induced changes in brain serotonin synthesis: is there a relationship to appetite for specific macronutrients?, *Appetite*, 8:163.

Fernstrom, J.D., and Hirsch, M.J., 1977, Brain serotonin synthesis: reduction in corn-malnourished rats, J. *Neurochem.*, 28:877.

Fernstrom, J.D., and Wurtman, R.J. 1971a, Brain serotonin content: physiological dependence on plasma tryptophan levels, *Science*, 173: 249.

Fernstrom, J.D., and Wurtman, R.J., 1971b, Brain serotonin content : increase following ingestion of carbohydrate diet, *Science*, 174: 1023.

Fernstrom, J..D., and Wurtman, R.J., 1972, Brain serotonin content: physiological regulation by plasma neutral amino acids, *Science*, 178: 414.

Fuenmayor, L.D., and Garcia, S., 1984, The effect of fasting on 5-hydroxytryptamine metabolism in brain regions of the albino rat, *Br.J.Pharmacol.*, 83:357.

Gartside, S.E., Cowen, P.J., and Sharp, T.,1992, Evidence that the large neutral amino acid L-valine decreases electrically-evoked release of 5-HT in the hippocampus, *Psychopharmacology*, 109:251.

Gillman, P.K., Bartlett, J.R., Bridges, P.K., Hunt, A., Patel, A.J., Kantamaneni, B.D., and Curzon, G., 1981, Indolic substances in plasma, cerebrospinal fluid, and frontal cortex of human subjects infused with saline or tryptophan, *J.Neurochem.*, 37: 410.

Goodwin, G.M., Fairburn, C.G., and Cowen, P.J., 1987, The effect of dieting and weight loss on neuroendocrine responses to tryptophan, clonidine and apomorphine in volunteers, *Arch. Gen. Psychiat.*, 44:952.

Handley, S.L., Dunn, T.L., Waldron, G., and Baker, J.M., 1980, Tryptophan, cortisol and puerperal mood, *Br.J.Psychiat.*, 136:498.

Harper, A.E., and Tews, J.K., 1988, Nutritional and metabolic control of brain amino acid concentrations, In:"Amino acid availability and brain function in health and disease". G.Huether, ed., Springer, Berlin, pp3.

Hernandez, J., Manjarrez, G.G., and Chagoya, G., 1989, Newborn humans and rats manourished in utero: free plasma L-tryptophan, neutral amino acids and brain serotonin synthesis, *Brain Res.*, 488:1.

Heslop, K., Portas, C.M., and Curzon, G., 1991, Effect of altered tryptophan availability on tissue and extracellular serotonin in the rat cortex, Proc. 5th Int. Conf. In Vivo Methods, 259.

Heyes, M.P., Quearry, B.J., and Markey, S.P., 1989, Systemic endotoxin increases L-tryptophan, 5-hydroxy-indolacetic acid, 3-hydroxykynurenine and quinolinic acid content of mouse cerebral cortex, *Brain Res.*, 491: 173.

Heyes, M.P., et al., 1992, Quinolinic acid and kynurenine pathway metabolism in inflammatory and non.in-flammatory neurological disease, *Brain* 115: 1249.

Hockney, R.A., Campling, G.M., Laver, D.A., Castell, L.M., Da Roza Davis, J.M., and Cowen, P.J., 1992, Biochemical stress in carers, J.Psychopharmacol., 5: Br.Assoc. Psychopharmacol/. Eur. Biol. Psy-chiat. Soc. Meeting, Abstracts, A16.

Holzman, D., 1993, Inflamed debate over neurotoxin, *Science*, 259: 25.

Hunsicker, K.D., Mullen, B.J., and Martin, R.J., 1992, *Physiol. Behav.*, 51: 325.

Jessy, J., Mans, A.M., DeJoseph, M.R., and Hawkins, R.A., 1990, Hyperammonaemin causes many of the changes found after portocaval shunting, *Biochem. J.*, 272: 311.

Jimerson, D.C., Wolfe, B.E., Metzer, E.D., Levine, J.M., and Cooper, T.B., 1994, Blunted neuroendocrine responses to serotonergic challenge in bulimia nervosa, *Neuropsychopharmacology*, 10:79S.

Joseph, M.S., Brewerton, T.D., Reuss, V.I., and Stebbins G.T., 1984, Plasma L-tryptophan/neutral amino acid ratio and dexamethasone suppression in depression, *Psychiatry Res.*, 11:185.

Karege, F., Widmer, J., Bovier, P., and Gaillard, J-M., 1994, Platelet serotonin and plasma tryptophan in depressed patients: effects of drug treatment and clinical outcome, *Neuropharmacology*, 10: 207.

Kaye, W.H., Gwirtsman, H.E., Brewerton, T.D., George, D.T., and Wurtman, R.T., 1988, Bingeing behaviour and plasma amino acids: a possible involvement of brain serotonin in bulimia nervosa, *Psychiat.Res*, 23: 31.

Kennett, G.A., Curzon, G., Hunt, A., and Patel, A.J., 1986, Immobilization decreases amino acid concentrations in plasma but maintains or increases them in brain, *J.Neurochem.*, 46: 208.

Knott, P.J., and Curzon, G., 1972, Free tryptophan in plasma and brain tryptophan metabolism, *Nature*, 239: 452.

Knott, P.J., Hutson, P.H., and Curzon, G., 1977, Fatty acid and tryptophan changes on disturbing groups of rats and caging them singly, *Pharmacol. Biochem. Behav.*, 7: 245.

Lambert, P.D., Wilding, J.P.H., Turton, M.D., Ghatei, M.A., and Bloom, S.R., 1994, Effect of food deprivation and streptozotocin- induced diabetes on hypothalamic neuropeptide Y release as measured by a radioimmunoassay-linked microdialysis procedure, *Brain Res.*, 656, 135.

Litman, D.A., and Correia, M.A., 1985, Elevated brain tryptophan and enhanced 5-hydroxytryptamine turnover in acute hepatic herme deficiency: clinical implications, *J.Pharmacol. Exp. Ther.*, 232: 337.

Lucca, A., Lucini, V., Piatti, E., Ronchi, P., and Smeraldi, E., 1992, Plasma tryptophan levels and plasma tryptophan/neutral amino acid ratio in patients with mood disorder, patients with obsessive-compul-sive disorder, and normal subjects, *Psychiatry Res.*, 44: 85,

Maes, M., DeRuyter, M., and Suy, E., 1987, The renal excretion of xanthurenic acid following L-tryptophan loading in depressed patients, *Human Psychopharmacology*, 2: 231.

Maes, M., Claes, M., Schotte, C., Delbeke, L., Jacquemyn, Y., Verkerk, R., De Meester, I., and Scharpe, S., 1992, Disturbances in dexamethasone suppression test and lower availability of L-tryptophan and tyrosine in early puerperium and women under contraceptive *Therapy. J.Psychosomat. Res.*, 191-197.

Maes, M., Meltzer, H.Y., Scharpe, S., Bosmans, E., Suy, E., De Meester, I., Calabrese, J., and Cosyns, P., 1993, Relationships between lower plasma L-tryptophan levels and immune-inflammatory variables in depression, *Psychiatry Res.*, 49: 151.

Mathis, P., Schmitt, L., Benatia, M., Granier, F., Ghisolfi, J., and Moron, P., 1988, Perturbations des acids amines plasmatiques et depression, *L'Encephale*, 14: 77.

Menkes, D.B., Coates, D.L., and Fawcett, J.P., 1994, Acute tryptophan depletion aggravates premenstrual syndrome, *J.Affect. Dis.*, 32: 37.

Moller, S.E., Kirk, L., and Honore, P., 1979, Free and total plasma tryptophan in endogenous depression, *J.Affect. Dis.*, 1:69.

Moller, S.E., Bech, P., Bjerrum, H., Bojholm, S., Butler, B., Folker, H., Gram, L.F., Larsen, J.K., Lauritzen, L., Loldrup, D., Munk-Andersen, E., Odum, K., and Rafaelsen, O.J. 1990, Plasma ratio tryptophan/neutral amino acids in relation to clinical response to paroxetine and clomipramine in patients with major depression, *J.Affect. Dis.*, 18: 59.

Partridge, W.M., and Fierer, G., 1990, Transport of tryptophan into brain from the circulating albumin-bound pool in rats and rabbits, *J. Neurochem.*, 54: 971.

Ploog, D.W., and Pirke, K.M., 1987, Psychobiology of anorexia nervosa, *Psychol. Med.*, 17: 843.

Record, C.O., Buxton, B., Chase, R.A., Curzon, G., Murray-Lyon, I.M., and Williams, R., 1976, Plasma and brain amino acids in fulminant hepatic failure and their relationship to hepatic encephalopathy, *Eur. J. Clin.Invest.*, 6: 387.

Robinson, M.B., Heyes, M.P., Anegawa, N.J., Gorry, E., Sina Djali, E., Mellits, D., and Batshaw, M.L., 1992, Quinolinate in brain and cerebrospinal fluid in rat models of congenital hyperammonemia, *Pediatr. Res.*, 32: 483.

Russ, M.J., Ackerman, S.H., Banay-Schwartz, M., Shindledecker, R.D., and Smith, G.P., 1990, Plasma tryptophan to large neutral amino acid ratios in depressed and normal subjects, *J. Affect. Dis.*, 19: 9.

Salter, M., Hazelwood, R., Pogson, C.I., Iyer, R., Madge, D.J., Jones, H.T., Cooper, B.R., Cox, R.F., Wang, C.M., and Wiard, R.P., 1995, The effects of inhibitors of tryptophan 2,3-dioxygenase and combined inhibitors of tryptophan 2,3-dioxygenase and 5-HT reuptake in the rat, *Neuropharmacology*, 34: 217.

Sarna, G.S., Tricklebank, M.D., Kantamaneni, B.D., Hunt, A., Patel, A.J., and Curzon, G., 1982, Effect of age on variables influencing the supply of tryptophan to the brain, *J.Neurochem.*, 39: 1283.

Sarna, G.S., Kantamaneni, B.D., and Curzon G., 1985, Variables influencing the effect of a meal on brain tryptophan, *J.Neurochem.*, 44: 1575.

Schmid-Burgk, W., Kim, J.S., Lischewski, R., and Rassmann, W., 1981, Levels of total and free tryptophan in the plasma of endogenous and neurotic depressives, *Arch. Psychiat. Nervenkr.*, 231: 35.

Schwartz, D.H., Hernandez, L., and Hoebel, B.G., 1990, Tryptophan increases extracellular serotonin in the lateral hypothalamus of food-deprived rats, *Brain Res. Bull.*, 25: 803.

Smith L.P., 1907, "Life and Letters of Sir Henry Wotton", Oxford University Press, Oxford, 1: pp 485-487.

Smith, Q.R., Momma, S., Aoyagi, M., and Rapoport, S.I., 1987, Kinetics of neutral amino acid transport across the blood-brain barrier, *J.Neurochem.*, 49: 1651.

Stanley, B.G., Schwartz, D.H., Hernandez, L., Hoebel, B.G., and Leibowitz, S.F., 1989, Patterns of extracellular norepinephrine in the paraventricular hypothalamus: relationship to circadian rhythm and deprivation-induced eating behaviour, *Life Sci.*, 45: 275.

Stein, G., Milton, F., Bebbington, P., Wood, K., and Coppen, A., 1976, Relationship between mood disturbances and free and total plasma tryptophan in postmortem women, *Br. Med.J.*, 2: 457.

Sullivan, P.A., Murnaghan, D., Callaghan, N., Kantamaneni, B.D., and Curzon, G., 1978, Cerebral transmitter precursors and metabolites in advanced renal disease, *J. Neurol. Neurosurg. Psychiat.*, 41: 581.

Sullivan, P.A., Murnaghan, D., Callaghan, N., Kantamaneni, B.D., and Curzon, G., 1980, Effect of dialysis on plasma and CSF tryptophan and CSF 5-hydroxyindoleacetic acid in advanced renal disease, *J.Neurol. Neurosurg. Psychiat.*, 43: 739.

Takada, A., Grdisa, M., Diksic, M., Gjedde, A., and Yamamoto, Y.L., 1993, Rapid steady-state analysis of blood-brain transfer of L-Trp in rat, with special reference to the plasma protein binding, *Neurochem. Int.*, 23: 351.

Teff, K.L., Young, S.N., Marchand, L., and Botez, M.I., 1989, Acute effect of protein or carbohydrate breakfasts on human cerebrospinal fluid monoamine precursor and metabolite levels, *J.Neurochem.*, 52: 235.

Voog, L., and Eriksson, T., 1992, Relationship between plasma and brain large neutral amino acids (LNAAs) in rats fed diets with different compositions at different times of the day, *J.Neurochem.*, 59: 1868.

Walsh, A.E.S., Oldman, A.D., Franklin, M., Fairburn, C.G., and Cowen, P.J., 1995, Dieting decreases plasma tryptophan and increased the prolactin response to d-fenfluramine in women but not men, *J.Affect.Dis.*, 33: 89.

Wilson, W., and Marsden, C.A., 1994, The effect of treadmill running on brain serotonin (5-HT), Proc. 6th Inst.Conf. In Vivo Methods, 223-224.

# 4

# BRAIN SEROTONIN, CARBOHYDRATE-CRAVING, OBESITY AND DEPRESSION*

R. J. Wurtman† and J. J. Wurtman

Department of Brain and Cognitive Sciences and
Clinical Research Center Massachusetts Institute of Technology
Cambridge, Massachusetts

## 1. ABSTRACT

Serotonin in the brain, and perhaps elsewhere, is involved in mechanisms that affect macronutrient selection, generate feelings of satiety, and, if malfunctioning, predispose to obesity. Drugs which increase the quantities of serotonin present within synapses can cause weight reduction. Such drugs presently include those that release the neurotransmitter by a direct action on nerve terminals (e.g., dexfenfluramine's metabolite dexnorfenfluramine) or by activating serotoninergic neurons (e.g., nicotine); by activating post-synaptic serotonin receptors (e.g., dexnorfenfluramine); or by prolonging serotonin's existence within synapses by blocking its reuptake (e.g., dexfenfluramine; fluoxetine, sertraline; paroxetine). Additional ways are known by which intrasynaptic serotonin levels can be augmented (e.g. increasing its synthesis with tryptophan; inhibiting its destruction by monoamine oxidase), and it can be anticipated that drugs acting at these loci will also become candidates for treating obesity.

Serotoninergic drugs act in at least three ways to facilitate weight loss: They accelerate the onset of satiety (Blundell, 1986), and enhance basal metabolic rate by about 100 calories per day (Fernstrom, 1989). They also inhibit the "carbohydrate craving" manifested by many people who are overweight or are becoming so (c.f., Wurtman and Wurtman, 1989). There is reason to believe that this inappropriate eating behavior actually constitutes a "serotonin hunger" by the brain, perhaps related to the number or activity of raphe neurons, in which case giving the serotoninergic drug might constitute a specific therapy for the etiologic process causing the obesity.

---

*Modified from: Wurtman, R.J., & Wurtman, J.J., Brain serotonin, carbohydrate craving, obesity, & depression. *Obesity Research* (in press).

† Dr. Richard J. Wurtman, M.I.T., E25-604, 77 Massachusetts Avenue, Cambridge, MA 02139.

*Recent Advances in Tryptophan Research*, edited by Graziella Allegri Filippini et al.
Plenum Press, New York, 1996

Two composite case reports are presented which typify syndromes that patients can present.

*Laura*, a 33 year old administrator, has suffered from severe premenstrual swings in appetite and mood for a decade. During the first 21 days of her menstrual cycle, her mood is normal and her food intake well balanced; 6 or 7 days past mid-cycle, however, she finds herself becoming irritable, anxious, angry, confused and clumsy. Although she pushes herself to continue daily aerobic exercise (indoors on an exercycle or outdoors), her fatigue level rises and she becomes more sluggish when she exercises. Her appetite also changes so that the foods she consumes are largely those rich in carbohydrates. She stops eating fruits and vegetables, limits her protein intake to dairy products such as sweetened yogurt and cheese, and eats only muffins, doughnuts, bagels, or frozen yogurt during the day. As she approaches the end of her cycle, she especially craves chocolate and often eats a pint of chocolate ice cream or bag of chocolate chip cookies for supper. She also desires salty snacks such as potato chips but tries to avoid them because she thinks they make her retain water. When her food intake was examined, Laura was found to consume about 2200 calories per day during the early part of her menstrual cycle, but almost 3000 calories during the latter half: these consisted primarily of starchy carbohydrates at meals, and cookies and chocolate between meals. During the follicular phase of the menstrual cycle, her mood (as assessed by the Hamilton Rating Scale of Depressed Mood) was normal; during the luteal phase, however, this score rose from 3 to 26 (an abnormal value) and measurements of menstrual symptomatology using the Menstrual Distress questionnaire also showed significant increases in self-reported anger, confusion, irritability, carbohydrate craving, fatigue and anxiety.

Laura claimed that she was unable to lose weight because her eating was out of control for much of each month and she could not keep from overeating carbohydrates, no matter how hard she tried. She had been able to stay on a low carbohydrate diet only once, and her mood and subjective fatigue became worse as a result.

As a subject in a placebo-controlled study on the ability of dexfenfluramine to suppress premenstrual carbohydrate craving, weight gain, and mood disturbances, Laura happened to receive the drug (15 mg p.o., b.i.d.) for the 15 days of each month's luteal phase, over a period of three months. It completely suppressed her carbohydrate craving and premenstrual weight gain, as well as the abnormal test scores on the Hamilton Rating Scale and the Menstrual Distress questionnaire. During three additional months of the study, when she received placebo instead of drug, there was no beneficial effect.

*Phyllis*, a 55 year old housewife who is 45 lbs. over ideal body weight, has been unable to lose weight on conventional diets; although they work for a few weeks, she becomes agitated and anxious. She then reverts to her pre-diet intake and stops losing weight. She is sedentary except during spring and summer when she occasionally takes walks.

Phyllis usually starts each day with a low-calorie breakfast and a small lunch. By mid-afternoon, she feels her mood worsening, experiencing boredom, loneliness, frustration and often severe anxiety over unimportant problems. She starts to nibble on crackers, bread, cookies, cereal and ice cream. Dinner is a small meal, consisting of chicken or fish and vegetables. Later in the evening, however, she eats bread and cheese, ice cream, or a handful of cookies. This need to eat in the evening is again accompanied by feelings of loneliness or anxiety.

When asked about her eating pattern, Phyllis claims that she eats to feel better and that even though she feels guilty at the quantity of food she consumes and at its lack of nutrients, she is unable to stop herself. When studied in our research facility, she was found to consume 2100 calories per day at meals and 800 more as snacks, with these providing more than half of her total carbohydrate intake.

Like Laura, Phyllis received dexfenfluramine (15 mg p.o., b.i.d.) as part of a placebo-controlled, double-blind study on the use of this drug or of fluoxetine in treating obesity associated with carbohydrate craving. During the 12-week treatment period she lost 11.5 lbs.

## 2. DISCUSSION

An association between a mood disturbance (often an atypical depression characterized by sadness, lethargy, "muddle headedness", and social withdrawal), the inability to lose or to stop gaining weight, and a craving for carbohydrates, is observed in a number of syndromes which lead patients to solicit medical assistance (Wurtman and Wurtman, 1989). These include obesity per se (Wurtman et al., 1987); the premenstrual syndrome (or Late Luteal Phase Dysphoric Disorder) (Brzezinski et al., 1990); Seasonal Affective Disorder (Rosenthal et al., 1984; Rosenthal et al., 1989); and the nicotine-abstinence syndrome, in long-term smokers who try to abstain (Wack and Rodin, 1982; Spring et al., 1991). Data from animal studies suggest that this association of affective and appetitive symptoms arises from the neurochemical effects of eating carbohydrate-rich, protein-poor foods: These foods act via the hormone insulin to increase sequentially, the "plasma tryptophan/LNAA ratio", the uptake of circulating tryptophan in the brain; brain tryptophan levels and the production and release of the neurotransmitter serotonin (Fernstrom and Wurtman, 1971b). The resulting increase in serotonin levels within brain synapses (like the increase produced by administering most antidepressant drugs) tends to ameliorate mood disturbances, albeit transiently.

Patients overeat high-carbohydrate foods, particularly in the form of snacks, not because they are hungry or even because they like the taste, but for their tryptophan- and serotonin-mediated psychopharmacological effects. Unfortunately, many carbohydrate-rich foods also contain fat (e.g., ice cream, cookies, potato chips, bread with butter) and this leads to weight gain. Drugs that promote serotonin-mediated neurotransmission might be expected to benefit both the affective and the appetitive disturbances, and the clinical data currently available on two such drugs, dexfenfluramine and fluoxetine, indicate that this can, in fact, be the case.

A dietary carbohydrate's ability to modify the plasma amino acid pattern so as to enhance the uptake of circulating tryptophan into the brain depends on its glycemic index, i.e., its ability to promote the secretion of insulin; it is independent of whether the food happens to taste sweet (Lieberman et al., 1986). Insulin has only a small effect on plasma tryptophan levels [free or albumin-bound, both of which are accessible to the brain (Pardridge and Flere, 1990], however it markedly lowers the plasma levels of large neutral amino acids (LNAA), like leucine, isoleucine, and valine, which compete with tryptophan for passage across the blood-brain barrier. This decrease allows more tryptophan to enter the brain and resolves the paradox of why dietary carbohydrates, which *lack* tryptophan, should increase brain levels of this amino acid while protein-rich foods, which *do* contain some tryptophan (about 1%) fail to do so: Dietary proteins *do* raise plasma tryptophan levels, however this rise is small relative to the increases in the other, more abundant LNAA, since tryptophan tends to be the least abundant of the 22 amino acids in proteins.

The increase in brain serotonin synthesis - and levels - that accompanies the rise in tryptophan reflects an unusual property of the enzyme tryptophan hydroxylase, which catalyzes the initial and rate-limiting step in this process (i.e., its *low* affinity for its substrate, tryptophan) (Fernstrom and Wurtman, 1971a). At normal levels of brain tryptophan, the enzyme is quite unsaturated with substrate, so that an increase in tryptophan levels (as might follow a carbohydrate-rich meal or snack) speedily increases the enzyme's saturation and accelerates serotonin synthesis. It is interesting to note that the components of the mechanism coupling food consumption to serotonin synthesis are all "open-loop", and lack tight homeostatic regulation. Plasma amino acid levels rise and fall, some over a six-fold range

(Fernstrom et al., 1979), depending on what is being digested; tryptophan's transport into the brain varies with food-induced changes in the plasma amino acid pattern, and a change in brain tryptophan levels modulates the rate of serotonin synthesis by changing the saturation of tryptophan hydroxylase. (Unlike carbohydrates and proteins, dietary fats are not known to have a major effect on the production of any brain neurotransmitter.)

An increase in brain tryptophan levels on the order of that produced by eating a carbohydrate-rich, protein-poor meal causes parallel increases in the amounts of serotonin released into synapses, basally and when the neurons fire (Schaechter and Wurtman, 1990). This change in amplitude of a neurochemical signal provides the brain with on-line information about the individual's nutritional state and, particularly, about the macronutrient composition of the most-recently-eaten meal or snack. The brain can then use this information in deciding what should be eaten in the future. Although people tend to be unaware of the extent to which protein and carbohydrate intake is regulated, there is abundant evidence that this phenomenon occurs. Thus, in virtually all cultures where macronutrient intake has been measured, protein intake is found to constitute about 13 ± 2% of daily calorie consumption, and the *ratio* of carbohydrates to protein in the diet to vary narrowly, around 4-5:1 (Kim et al., 1987). This regulation is not manifest at every meal or snack, nor even every day; however, it *is* demonstrable in patterns of food consumption over longer periods.

Of course, many factors affect what we choose to eat: culture, taste and smell, appearance, and, especially in the contemporary United States where fewer and fewer meals are taken at home, what happens to be on the menu. However, it is clear that one such factor is the composition of the food that has been eaten previously. If a laboratory rat, which normally eats both high-protein and high carbohydrate foods, is allowed to choose from two test foods, one of which would tend to raise brain serotonin (because it is rich in carbohydrates and poor in protein) and one which would not, it will consume daily approximately equal amounts of the two foods every day. However, if the animal is given a drug that selectively releases, acts like, or prolongs the intrasynaptic actions of serotonin, consumption of the carbohydrate-rich food is selectively suppressed (Moses and Wurtman, 1984; Leibowitz et al., 1988). Why did the evolutionary process allow the production and release of an important neurotransmitter, serotonin, to rise or fall depending on what a person happens to choose for breakfast? Possibly so that he or she can make nutritionally-wiser choices for lunch. Unfortunately, brain serotonin also has additional functions, including the maintenance of a subjectively acceptable mood state. Thus, in certain syndromes, its role in macronutrient maintenance can become hostage to its effect on mood.

The hypothesis that there is a subset of obese patients whose weight problem is associated with uncontrolled carbohydrate intake, and with symptoms of atypical depression, is supported by the occurrence of this triad of symptoms in several other diseases. Two become symptomatic cyclically, in late fall and winter for patients with the Seasonal Affective Disorder, and premenstrually in women (like Laura) with the Late Luteal Phase Dysphoric Disorder. Other patients inflict these symptoms on themselves (and we encourage them to do so!) by making the medically-essential decision to give up smoking. The hypothesis that these syndromes are all related to inadequate serotonin release in the brain is supported by numerous findings: (1) the unique ability of serotonin-producing neurons to make and release more of their neurotransmitter after carbohydrate-rich foods are eaten (Fernstrom and Wurtman, 1971b); (2) animal studies demonstrating that serotonin-like drugs given systemically (Moses and Wurtman, 1984) [or serotonin itself, placed within the hypothalamus (Leibowitz et al., 1988)] can selectively suppress carbohydrate intake; (3) clinical studies showing that serotonin antagonists can cause weight gain (Kalucy, 1980); (4) the well-established generalization that drugs (such as the monoamine oxidase inhibitors and monoamine uptake blockers) which increase intrasynaptic serotonin levels tend to have antidepressant properties; (5) animal studies showing that systemic administration of nicotine increases serotonin release (Ribeiro et al., 1993); and (6) studies on long-term smokers attempting to free themselves from nicotine addiction. Such people typically exhibit mood disturbances, weight gain, and carbohydrate craving, often to the point of causing them to resume smoking (Wack and Rodin, 1982).

That the mechanism by which nicotine withdrawal produces this behavioral triad is associated with impaired serotonin release was initially suggested by the finding that serotonin neurons contain excitatory nicotinic cholinergic receptors, the apparent number of which changes after prolonged exposure to nicotine (Benwell et al., 1988). We could postulate that exposure to nicotine initially fosters positive subjective responses, partly by enhancing serotonin release (Ribeiro et al., 1993); however, habituation and tolerance might then occur, so that the nicotinic receptors on serotonin neurons could no longer be stimulated by the endogenous transmitter, acetylcholine, which normally activates them. Thereafter, the patient must smoke to maintain serotonin release even at pre-drug levels. Recovery from this adaptive response could take a long time; we observed that ex-smokers continued to snack excessively on carbohydrate-rich foods after 29 days of abstinence (Spring et al., 1991).

These hypotheses have been tested successfully by administering the serotoninergic drugs dexfenfluramine and fluoxetine to carbohydrate-craving obese patients, and those with the other three syndromes. (Tryptophan administration also helps perhaps half of all patients with carbohydrate-craving obesity, but its use is still proscribed in the United States, even in academic experimental studies). Dexfenfluramine is a serotonin uptake blocker; its chief metabolite, nordexfenfluramine, also releases serotonin directly into synapses and activates post-synaptic 5HT-2 receptors. The actions of dexfenfluramine do not require that the serotoninergic neurons continue to fire; hence, unlike uptake blockers, which ultimately slow neuronal firing and suppress serotonin release, the drug apparently retains its clinical activity for at least a year (Guy-Grand et al., 1989).

In rats, megadoses of either dexfenfluramine or fluoxetine, well beyond the clinical range, can cause prolonged decreases in brain serotonin levels (Fuller et al., 1988; Trouvin et al., 1993) and in the quantities of serotonin released into synapses upon depolarization (Laferrere and Wurtman, 1989; Gardier and Wurtman, 1991). Some investigators have proposed that these changes reflect "neurotoxicity" (Schuster et al., 1986), however, neither drug has been shown to produce any of the classic hallmarks of neurotoxicity (e.g., loss of cell bodies; axonal uptake of silver; "sprouting" of surviving neurons; or gliosis, identifiable histologically or by measuring characteristic glial proteins), nor any signs whatsoever of behavioral or neurological toxicity. Moreover the doses required generate brain fenfluramine levels which are far beyond those produced in humans by therapeutic doses of the drug (B. Sandage, Personal Communication, 1995). In all probability, the megadose-induced reductions in serotonin levels (and their immunohistochemical equivalent of reduced numbers of neurons visibly staining for serotonin) (Molliver and Molliver, 1990; Kalia, 1991) reflect exaggerations of the drugs' mechanisms of action, which cause serotonin to be lost from the neurons by releasing it and blocking its reuptake. Moreover a prolonged massive increase in intrasynaptic serotonin levels, acting via presynaptic serotonin receptors, probably modifies the mechanisms that regulate serotonin levels, without affecting its release or the neuronal functions it mediates. Indeed, the prolonged serotonin depletion that megadoses of either drug induce can be blocked by serotonin receptor antagonists acting external to the neuron (Gardier et al., 1992).

Many other new drugs thought to selectively enhance serotonin-mediated neurotransmission have been developed in the past decade, and several of these compounds - originally conceptualized as antidepressants - are being evaluated for the management of obesity. Since serotoninergic mechanisms reduce body weight by accelerating the onset of satiety (Blundell, 1986) and increasing the metabolic rate (Fernstrom, 1989) besides by suppressing excessive snacking of carbohydrate-rich foods, these drugs are more broadly useful in treating obesity or uncontrolled weight gain than their specific efficacy in carbohydrate-cravers might imply (Toornvliet et al., 1994). Epidemiologic evidence accumulated during the past few years indicates that obesity can significantly enhance morbidity and mortality even among patients who weigh only 20 percent above normal, and is thus important to treat (McCue, 1981; Pi-Sunyer, 1991). Obviously, patients whose obesity can be effectively treated using just such "life-style" manipulations as increased exercise and reduced calorie and fat intake should follow that course. However, treatment failures following such regimens abound. Hence, the availability of effective drugs which may be administered over and over again, when uncontrolled eating leads to weight gain, might have the same salutary effect on the natural history of obesity and its sequelae that the development of antihypertensive agents had on blood pressure and stroke three decades ago. And if, in the process, the drugs also ease the affective symptoms which afflict many of these patients, so much the better.

## ACKNOWLEDGMENT

These studies were supported by a grant from the Center for Brain Sciences and Metabolism Charitable Trust

## 3. REFERENCES

Benwell M.E., Balfour D.J. and Anderson J.M., 1988, Evidence that tobacco smoking increases the density of (-)-[$^3$H] nicotine binding sites in human brain, *J. Neurochem.* 50:1243.

Blundell J.E., 1986, Serotonin manipulations and the structure of feeding behavior, *Appetite.* (suppl):39-56.

Brzezinski A.A, Wurtman J.J., Wurtman R.J., Gleason R., Nader T., and Laferrere B., 1990, d-Fenfluramine suppresses the increased calorie and carbohydrate intakes and improves the mood of women with premenstrual syndrome, *Obstet. Gynecol.* 76:296.

Fernstrom J.D. and Wurtman R.J.,. 1971 a, Brain serotonin content: Physiological dependence on plasma tryptophan levels, *Science.* 173:149.

Fernstrom J.D. and Wurtman R.J., 1971 b, Brain serotonin content: Increase following ingestion of carbohydrate diet, *Science* 174:1023.

Fernstrom J.D., Wurtman R.J., Hammarstrom-Wiklund B., Rand W.M., Munro H.N. and Davidson C.S., 1979, Diurnal variations in plasma concentrations of tryptophan, tyrosine, and other neutral amino acids: Effect of dietary protein intake, *Am. J. Clin. Nutr.* 32:1912.

Fernstrom M.H., 1989, Depression, antidepressants, and body weight, *Ann. N.Y. Acad. Sci.* 575:31.

Fuller R.W., Snoddy H.D. and Robertson D.W., 1988, Mechanisms of effects of d-fenfluramine on brain serotonin metabolism in rats: Uptake inhibition versus release, *Pharmacol. Biochem. Behav.* 30:715.

Gardier A. and Wurtman R.J., 1991, Persistent blockade of potassium-evoked serotonin release from rat frontocortical terminals after fluoxetine administration, *Brain Res.* 540:325.

Gardier A., Kaakkola S., Erfurth A. and Wurtman R.J., 1992, Effects of methiothepin on changes in brain serotonin release induced by repeated administration of high doses of anorectic drugs, *Brain Res.*588(1):67.

Guy-Grand G., Apfelbaum M., Crepaldi G., leFebvre P., Greis A. and Turner P., 1989, International trial of long-term dexfenfluramine in obesity. *Lancet* 2:1142.

Kalia M., 1991, Reversible, short-lasting, and dose-dependent effect of (+)-fenfluramine on neocortical serotonergic axons, *Brain Res.* 548:111.

Kalucy R.J., 1980, Drug-induced weight gain, *Drugs*19:268.

Kim W., Kelay J.,Judd J., Marshall M.W., Metz W. and Prather E.S., 1987, Evaluation of long-term dietary intakes of adults consuming self-selected diets, *Am. J. Clin. Nutr.* 40:1327.

Laferrere B. and Wurtman R.J., 1989, Effect of d-fenfluramine on serotonin release in brains of anesthetized rats, *Brain Res.* 504:258.

Leibowitz S.F., Weiss G.F. and Shor-Posner G., 1988, Hypothalamic serotonin: Pharmacological, biochemical and behavioral analyses of its feeding-suppressive action, *Clin. Neuropharmacol.* 11:S51.

Lieberman H.R., Caballero B. and Finer N., 1986, The composition of lunch determines afternoon plasma tryptophan ratios in humans, *J. Neural Transm.* 65:211.

Moses P.L. and Wurtman R.J., 1984, The ability of certain anorexic drugs to suppress food consumption depends on the nutrient composition of the test diet, *Life Sci.* 35:1297.

McCue H., Jr., 1981, The 1979 build and blood pressure study, *in:* " Medical Aspects of Mortality Statistics", H. Bostrom, N. Ljungstedt, eds., Almquist and Wiksell International, Stockholm.

Molliver D.C. and Molliver M.E., 1990, Anatomic evidence for a neurotoxic effect of (+)- fenfluramine upon serotoninergic projections in the rat, *Brain Res.* 551:165.

Pardridge W. and Flere G., 1990, Transport of tryptophan into brain from the circulating, albumin-bound pool in rats and in rabbits, *J. Neurochem.* 54:971.

Pi-Sunyer F., 1991, Health implications of obesity, *Am. J. Clin. Nutr.*53:1595S.

Ribeiro E., Bettiker R., Bogdanov M. and Wurtman R.J., 1993, Effects of systemic nicotine on serotonin release in rat brain, *Brain Res.* 621:311.

Rosenthal N.E., Sack D.A.,Gillin J.C, Lewey A.J., Goodwin F.K., Davenport Y., Mueller P.S., Newsome D.A., and Wehr T.A., 1984, Seasonal Affective Disorder: A description of the syndrome and preliminary findings with light therapy, *Arch. Gen. Psychiatry.*41:72.

Rosenthal N., Genhart M., Caballero B., Jacobsen F.M., Skwerer R.G., Coursey R.D., Rogers S., and Spring B.J., 1989, Psychobiological effects of carbohydrate- and protein-rich meals in patients with Seasonal Affective Disorder and normal controls, *Biol. Psychiatry.* 25:1029.

Schaechter J.D. and Wurtman R.J., 1990, Serotonin release varies with brain tryptophan levels, *Brain Res.*532:203.

Schuster C.R., Lewis M. and Seiden L.S., 1986, Fenfluramine neurotoxicity, *Psychopharmacol. Bull.* 22:148.

Spring B., Wurtman J., Gleason R. and Kessler K., 1991, Weight gain and withdrawal symptoms after smoking cessation: A preventive intervention using d-fenfluramine, *Health Psychol.* 10:216.

Toornvliet A.C., Pijl H., Hopman E., Pietero M. and Meinders A.E., 1994, The identification of obese patients sensitive to a serotoninergic drug, Abstract Presented at the 7th International Congress on Obesity, Toronto, Canada, August 20-25.

Trouvin J.H,. Gardier A.M, Chanut E., Pages N. and Jacquot C., 1993, Time course of brain serotonin metabolism after cessation of long-term fluoxetine treatment in the rat, *Life Sci.* 52:187.

Wack J. and Rodin J., 1982, Smoking and its effect on body weight and the systems of caloric regulation, *Am. J. Clin. Nutr.* 35:366.

Wurtman J.J., Wurtman R.J., Reynolds S., Tsay R., and Chew B., 1987, d-Fenfluramine suppresses snack intake among carbohydrate cravers but not among non carbohydrate cravers, *Int. J. of Eating Disord.* 6:687.

Wurtman R.J. and Wurtman J.J., 1989, Carbohydrates and depression, *Sci. Am.* January:68-75.

***TRYPTOPHAN AND SEROTONIN: NEUROBIOLOGY AND NEUROCHEMISTRY***

5

# THE EFFECT OF LOW BRAIN SEROTONIN ON MOOD AND AGGRESSION IN HUMANS

## Influence of Baseline Mood and Genetic Factors

S. N. Young,[1] R. O. Pihl,[2] C. Benkelfat,[1] R. Palmour,[1] M. Ellenbogen,[1] and D. Lemarquand[2]

[1] Department of Psychiatry
McGill University
1033 Pine Avenue West, Montréal, Québec, Canada, H3A 1A1
[2] Department of Psychology
McGill University
1205 Docteur Penfield Avenue, Montréal, Québec, Canada, H3A 1B1

## 1. INTRODUCTION

Most of the evidence relating low brain serotonin levels to psychopathology is circumstantial. For example, the relationship between low levels of 5-hydroxyindoleacetic acid in the cerebrospinal fluid of depressed or aggressive patients (van Praag, 1994) is an association that does not necessarily imply cause, while the ability of antidepressants to potentiate serotonin function (Blier and de Montigny, 1994) may be more relevant to the action of antidepressant drugs than the etiology of psychopathology. The acute tryptophan depletion (ATD) technique is designed to test directly the effects of low brain serotonin in humans. Subjects ingest an amino acid mixture containing all the essential amino acids except for tryptophan. This induces protein synthesis (Moja et al., 1991), and as tryptophan is incorporated into protein its level in tissues and plasma falls, by as much as 80 to 90%. Testing is usually done five hours after ingestion of the amino acid mixture, when plasma tryptophan reaches its lowest level. The control treatment is a similar amino acid mixture containing the correct amount of tryptophan for a good protein source.

The main advantage of the ATD technique is that cause and effect are clear. The disadvantages are that (i) the evidence that brain serotonin is lowered in humans is circumstantial, although the technique lowers the level of 5-hydroxyindoleacetic acid in the cerebrospinal fluid of vervet monkeys (Chamberlain et al., 1987), (ii) lowering tryptophan availability may lower serotonin, but it may alter other aspects of metabolism in the brain including protein synthesis, and the synthesis of other metabolites such as tryptamine, melatonin and quinolinic acid, and (iii) the putative lowering of brain serotonin occurs for only a brief period of time, and its effect may thus differ quantitatively or qualitatively from

*Recent Advances in Tryptophan Research*, edited by Graziella Allegri Filippini et al.
Plenum Press, New York, 1996

those that occur after longer term lowered brain serotonin in patients. Other problems associated with the technique are ethical and practical.

## 2. ETHICAL AND PRACTICAL ASPECTS OF TRYPTOPHAN DEPLETION

The ethical problems are due to the fact that the technique attempts to cause symptomatology that is similar in some ways to psychopathology. This is dealt with by paying close attention to the biochemistry and psychology of the subjects. Because our studies are done on non-hospitalized subjects, we terminate the depletion after five hours by giving the subjects a tryptophan supplement. Thus, their biochemistry is normalized before they are sent home. The subjects are also debriefed at the end of the study, and they are usually contacted the same evening and the next day by phone to ensure that there are no lingering adverse effects. In contrast, studies carried out at Yale University (e.g. Delgado et al., 1990) are on inpatients, who can be supervised more closely. In those studies the depletion lasts seven hours and the subjects' biochemistry is normalized more slowly by allowing them to start eating normal foods at the end of the seven hours.

The main practical difficulty in ATD studies is the palatability of the amino acid mixtures and their gastrointestinal side effects. The problems with palatability can be overcome in part by putting three of the amino acids with the worst taste (arginine, cysteine and methionine) in capsules, and including the fourth with an objectionable taste (lysine) as the monohydrochloride. Some subjects who ingest the amino acid mixtures happily still experience gastrointestinal side effects up to two hours after ingestion. Experience suggests that the main identifiable influence on the degree of discomfort of the subjects who ingest amino acid mixtures is the positive attitude of the personnel running the experiment.

Starting in 1989 there was a minor epidemic of a new disorder, Eosinophilia Myalgia Syndrome (EMS), which was associated with contaminants in various batches of tryptophan produced by a single manufacturer. This has resulted in tryptophan being removed from the market in many countries. In Canada, where tryptophan is sold as a prescription drug for use in affective disorders, there were no cases of EMS associated with the tryptophan sold as a drug (although there were a few cases in Canadians who bought it over the counter in the U.S.A.) (Wilkins, 1990). Tryptophan in its purified form is necessary both as a constituent of the nutritionally balanced control amino acid mixture, and as a means of terminating the effects of ATD at the end of the experiment, in our studies. The lack of availability of tryptophan in some countries is a problem that is not entirely insurmountable. One approach is to use a control mixture that contains one quarter the amount of the tryptophan deficient amino acid mixture. The mixture will cause a certain decline in tryptophan availability, but one that is not as large as with the full 100 g tryptophan deficient amino acid mixture (Young et al., 1989). The normal control nutritionally balanced amino acid mixture will also tend to decrease tryptophan levels in the brain. It has an amino acid composition similar to that of a balanced protein. Proteins contain higher levels of the other large neutral amino acids, which compete with tryptophan for entry into brain, than of tryptophan. As a result, the control mixture, like a normal protein, should tend to lower brain tryptophan and serotonin synthesis (Wurtman et al., 1981). However, there is one important difference between a control treatment that is merely 25% of the active mixture, and the proper control mixture that differs from the experimental mixture only in its tryptophan content. The real control mixture will contain the same caloric content as the experimental mixture, instead of only 25% of the calories. Studies on normal human subjects have shown differences in mental function after meals associated with different caloric loads (Craig, 1986). Nonetheless, the control mixture containing only 25% of the amino acids in the active treatment is an acceptable, if not ideal, control treatment where tryptophan is not available for human use.

## 3. DOES ACUTE TRYPTOPHAN DEPLETION LOWER SEROTONIN FUNCTION?

The ATD technique is meant to duplicate, in part, the low serotonin levels that have been hypothesized in some of the more simplistic hypotheses of psychopathology. One important question is whether lowering the rate of serotonin synthesis for a few hours will result in a decline in the level of serotonin in relevant areas of brain that is sufficient to cause a decline in serotonin release from neurons. Function of serotonin neurons is the result of release of serotonin from those neurons, and any manipulation that alters complex aspects of brain function through alterations in serotonin function must alter serotonin release.

Release of serotonin from neurons will depend in part on the firing rate of those neurons. Studies on experimental animals have shown that the rate of firing of serotonin neurons is dependent in part on the level of behavioral arousal (Trulson and Jacobs, 1979). If a neuron is firing at a very slow rate, then alterations in the level of the neurotransmitter stored in that neuron are unlikely to influence release. On the other hand, high rates of neuronal firing may cause alterations in the neurotransmitter level to alter the amount released. Thus, the level of arousal of the subject may influence the effect of ATD.

Evidence for lowered serotonin release in humans after ATD must remain circumstantial. Evidence from animals suggests that lowered serotonin should increase aggression and pain sensitivity, among other behaviors. After ATD in humans there was no increase in baseline pain sensitivity, but the manipulation completely abolished the analgesic effect of morphine (Abbott et al., 1992). This, and animal evidence that morphine may act in part through enhanced release of serotonin in the spinal cord (Matos et al., 1992), is circumstantial evidence for an effect of ATD on serotonin release in human CNS. Additional evidence comes from studies on aggression. Numerous animal studies suggest that lowered serotonin function increases aggression (Eichelman, 1993), and two human ATD studies suggest that decreased tryptophan availability can increase aggressive responding in laboratory tests of aggression (Pihl et al., 1995; Cleare and Bond, 1995).

## 4. ACUTE TRYPTOPHAN DEPLETION AND MOOD

In our first study, ATD lowered mood significantly in unscreened male subjects, and also influenced cognition of the subjects (Young et al., 1985); they paid selective attention to dysphoric stimuli after tryptophan depletion. The baseline depression scores of the subjects were in the high normal range, and the lowering of mood was not into the clinical range. However, in subsequent studies on carefully screened subjects, who had low normal depression scale scores, ATD had no effect on mood, suggesting that the subjects' susceptibility to lowered mood influences their response (Abbott et al., 1992; Benkelfat et al., 1994). On the other hand, Delgado et al. (1990) studied depressed patients who had responded to treatment, but were still on antidepressant drugs. In about three quarters of the patients ATD resulted in the temporary reappearance of the full syndrome of clinical depression. However, this finding may say more about the mechanism of antidepressant action than the etiology of depression. ATD tends to reverse the action of serotonin reuptake inhibitors, while inhibition of catecholamine synthesis with alpha-methyl-para-tyrosine reverses the action of noradrenaline reuptake inhibitors (Delgado et al., 1993). In depressed patients who were still depressed and had not yet started to receive antidepressant drugs, ATD did not result in any changes in mood on the same day. However, on the day following the manipulation, some patients showed a worsening of mood, while a somewhat higher percentage of the patients exhibited a significant improvement in mood (Delgado et al., 1994). The lack of effect on the day of the depletion may be due to a ceiling effect, but the changes in mood on the following day are not easy to understand.

Overall, the studies of ATD on mood do not give a consistent picture, but one pattern that emerges is that lowering of mood seems to occur only in subjects who have some susceptibility to lowered mood, as indicated by an episode of major depression, or by high normal depression scale scores. We have approached this hypothesis by using high risk subjects. Male subjects were recruited who were themselves normal, and had never had an episode of depression, but had a heavy genetic weighting for depression (family history positive or FH+ subjects). Control subjects were mentally normal and had no family history of depression (FH-). On ATD, the FH+ subjects, but not the FH- subjects, showed a significant lowering of mood (Benkelfat et al., 1994). Recently we have studied FH- women. Because women have a greater incidence of depression than men, FH- women might be considered more susceptible to depression than FH- men. We found that the response of FH- women was very similar to that of FH+ men; there was a significant lowering of mood (M.A. Ellenbogen, R.M. Palmour, S.N. Young, P. Dean, C. Benkelfat, unpublished data).

## 5. ACUTE TRYPTOPHAN DEPLETION AND AGGRESSION

An early ATD study on normal male subjects (Smith et al., 1986), and a more recent study on aggressive patients (Salomon et al., 1994), failed to show any increased aggression as a result of low serotonin levels. However, we studied the influence of ATD on aggression further because (i) two studies suggested that tryptophan supplementation has a beneficial effect when given for several weeks to aggressive patients (Morand et al., 1983; Volavka et al., 1990), and (ii) ATD in vervet monkeys did increase aggression (Chamberlain et al., 1987). In the study on aggression in vervet monkeys, ATD had only moderate effects on aggression during undisturbed social interactions in the group. However, at the end of the experiment, the monkeys, which had not received any food since the previous day, where shown their normal food bucket. This increased general levels of activity and aggression, but also increased the effects of altered tryptophan availability. These data are consistent with the idea, discussed above, that increased levels of arousal are associated with an increased effect of altered serotonin levels.

In a recent study we looked again at the effect of ATD on aggression in normal male subjects. The laboratory test of aggression was modified to increase the arousal level of the subjects as much as possible. The subjects engaged in a reaction time competition with a (non-existent) partner. When the subject won a trial they administered an electric shock to their "partner". When the "partner" (a computer program) won, they received an electric shock below their previously determined painful threshold. The level of shock chosen by the subject was the measure of aggression. Using this experimental design, (i) ATD increased aggression, (ii) alcohol increased aggression, and (iii) the effects of ATD and alcohol were additive (Pihl et al., 1995). This supports the idea that some subjects become violent when they drink alcohol because of low brain serotonin levels. Another recent laboratory test of aggression after ATD has demonstrated an increase with this manipulation (Cleare and Bond, 1995).

The susceptibility of subjects for aggression may influence their response to ATD, in the same way that their susceptibility to lowered mood influences their affective response. To test this we reanalyzed an old data set in which male subjects with only cursory screening underwent ATD. The experimental manipulation increased the irritability of the subjects, but this was seen only in those who had high scores (in the clinically significant range) on the Psychopathic deviate scale of the Minnesota Multiphasic Personality Inventory (Finn et al., 1995).

If low brain serotonin contributes to aggression, this may be due either to increased in aggressive intent, or to an increase in impulsivity which enhances the expression of aggression. Therefore, we looked at the effect of ATD on a test of impulsivity in which there was no aggressive component (D. LeMarquand, C. Benkelfat, S.N. Young, R.M. Palmour, R.O. Pihl, unpublished data). The measure used was a Go/No-go task in which the subjects had to respond to one set of numbers, and not respond to another set of numbers, displayed on a computer screen. The measure of impulsivity was the errors of commission, or passive avoidance errors i.e., errors in which the subject was not able to prevent a response when the number shown was one for which there should be no response. Thus, ATD increased impulsivity in a situation which minimized the emotional component of the response present in aggressive responding.

## 6. DISCUSSION AND CONCLUSIONS

ATD is capable of lowering mood, and increasing aggression, irritability and impulsivity in humans. These results are consistent with other data suggesting that low serotonin may be involved in the etiology of depression and aggression. Effects on mood were seen in the absence of any appreciable change in anxiety, suggesting that low brain serotonin has some specificity and does not enhance all negative emotions. Low brain serotonin seems to contribute directly to lowered mood. However, the lack of any response in subjects who had no susceptibility to lowered mood, suggests that serotonin is only one of several neurotransmitter systems regulating affect. Presumably in some subjects the homeostatic effect of other neurotransmitter systems prevents lowered serotonin from altering mood. In other subjects, in whom the other neurotransmitter system are in a state that produces a susceptibility to affective disturbances, ATD will lower mood. Thus, low brain serotonin can not, by itself, produce depression or aggression, but probably contributes to these disorders in some patients.

## ACKNOWLEDGMENT

Studies in the authors' laboratories were funded by the Medical Research Council of Canada.

## REFERENCES

Abbott, F.V., Etienne, P., Franklin, K.B.J., Morgan, M.J., Sewitch, M.J. and Young, S.N., 1992, Acute tryptophan depletion blocks morphine analgesia in the cold-pressor test in humans, *Psychopharmacology*. 108:60.

Benkelfat, C., Ellenbogen, M.A., Dean, P., Palmour, R.M. and Young, S.N., 1994, Mood-lowering effect of tryptophan depletion: Enhanced susceptibility in young men at genetic risk for major affective disorders, *Arch Gen Psychiatry*. 51:687.

Blier, P. and de Montigny, C., 1994, Current advances and trends in the treatment of depression, *Trends Pharmacol Sci*. 15:220.

Chamberlain, B., Ervin, F.R., Pihl, R.O. and Young, S.N., 1987, The effect of raising or lowering tryptophan levels on aggression in vervet monkeys, *Pharmacol Biochem Behav*. 28:503.

Cleare, A.J. and Bond, A.J., 1995, The effect of tryptophan depletion and enhancement on subjective and behavioural aggression in normal male subjects, *Psychopharmacology*. 118:72.

Craig, A., 1986, Acute effects of meals on perceptual and cognitive efficiency, *Nutr Rev*. 44 (supplement):163.

Delgado, P.L., Charney, D.S., Price, L.H., Aghajanian, G.K., Landis, H. and Heninger, G.R., 1990, Serotonin function and the mechanism of antidepressant action: reversal of antidepressant-induced remission by rapid depletion of plasma tryptophan, *Arch Gen Psychiatry*. 47:411.

Delgado, P.L., Miller, H.L., Salomon, R.M., Licinio, J., Heninger, G.R., Gelenberg, A.J. and Charney, D.S., 1993, Monoamines and the mechanism of antidepressant action: effects of catecholamine depletion on mood of patients treated with antidepressants, *Psychopharmacol Bull*. 29:389.

Delgado, P.L., Price, L.H., Miller, H.L., Salomon, R.M., Aghajanian, G.K., Heninger, G.R. and Charney, D.S., 1994, Serotonin and the neurobiology of depression: Effects of tryptophan depletion in drug-free depressed patients, *Arch Gen Psychiatry*. 51:865.

Eichelman, B., 1993, Bridges from the animal laboratory to the study of violent or criminal individuals, in: "Mental Disorders and Crime," S. Hodgins, ed., Sage, New York.pp. 194.

Finn, P.R., Young, S.N., Pihl, R.O. and Ervin, F.R., 1995, The influence of antisocial and low-anxious traits on altered mood, anxiety and hostility induced by acute tryptophan depletion, *Psychopharmacology*. in press:

Matos, F.F., Rollema, H., Brown, J.L. and Basbaum, A.I., 1992, Do opioids evoke the release of serotonin in the spinal cord? An in vivo microdialysis study of the regulation of extracellular serotonin in the rat, *Pain*. 48:439.

Moja, E.A., Restani, P., Corsini, E., Stacchezzini, M.C., Assereto, R. and Galli, C.L., 1991, Cycloheximide blocks the fall of plasma and tissue tryptophan levels after tryptophan-free amino acid mixtures, *Life Sci*. 49:1121.

Morand, C., Young, S.N. and Ervin, F.R., 1983, Clinical response of aggressive schizophrenics to oral tryptophan, *Biol Psychiatry*. 18:575.

Pihl, R.O., Young, S.N., Harden, P., Plotnick, S., Chamberlain, B. and Ervin, F.R., 1995, Acute effect of altered tryptophan levels and alcohol on aggression in normal human males, *Psychopharmacology*. in press:

Salomon, R.M., Mazure, C.M., Delgado, P.L., Mendia, P. and Charney, D.S., 1994, Serotonin function in aggression: the effect of acute plasma tryptophan depletion in aggressive patients, *Biol Psychiatry*. 35:570.

Smith, S.E., Pihl, R.O., Young, S.N. and Ervin, F.R., 1986, Elevation and reduction of plasma tryptophan and their effects on aggression and perceptual sensitivity in normal males, *Aggress Behav*. 12:393.

Trulson, M.E. and Jacobs, B.L., 1979, Raphe unit activity in freely moving cats: correlation with level of behavioral arousal, *Brain Res*. 163:135.

van Praag, H.M., 1994, Anxiety and aggression as pacemakers of depression: A dimensional variation of the serotonin hypothesis of depression, in: "Current Therapeutic Approaches to Panic and Other Anxiety Disorders," J. Mendlewicz, G. Racagni and N. Brunello, eds., Karger, Basel.pp. 144.

Volavka, J., Crowner, M., Brizer, D., Convit, A., van Praag, H.M. and Suckow, R.F., 1990, Tryptophan treatment of aggressive psychiatric inpatients, *Biol Psychiatry*. 28:728.

Wilkins, K., 1990, Eosinophilia-myalgia syndrome, *Can Med Assoc J*. 142:1265.

Wurtman, R.J., Hefti, F. and Melamed, E., 1981, Precursor control of neurotransmitter synthesis, *Pharmacol Rev*. 32:315.

Young, S.N., Smith, S.E., Pihl, R.O. and Ervin, F.R., 1985, Tryptophan depletion causes a rapid lowering of mood in normal males, *Psychopharmacology*. 87:173.

Young, S.N., Ervin, F.R., Pihl, R.O. and Finn, P., 1989, Biochemical aspects of tryptophan depletion in primates, *Psychopharmacology*. 98:508.

# ADAPTATIVE CHANGES OF THE SEROTONERGIC SYSTEM AFTER ANTIDEPRESSANT TREATMENTS

F. Artigas, N. Bel, J. Mª Casanovas, and L. Romero

Department of Neurochemistry, C.I.D.
Consejo Superior de Investigaciones Científicas
Jordi Girona 18-26, E-08034 Barcelona, Spain

## 1. INTRODUCTION

The search for an adequate treatment of major depression is one of the main challenges of Neuropharmacology. To date, no single drug is effective in all patients treated, probably due to the heterogeneity of the disease and to individual differences in the response to the agents used. Despite provoking pharmacological effects when given at once, all treatments need to be administered for several weeks before amelioration signs begin to emerge. This has led to the notion that —rather than the inhibition of uptake or monoamine oxidase (MAO) activity *per se*— adaptative changes of aminergic receptors mediate the clinical effects.

The involvement of the brain serotonergic (5-hydroxytryptamine, 5-HT) system in the mechanism of action of antidepresant drugs is supported by many preclinical and clinical evidences. The chronic treatment with antidepressants alters the density and/or sensitivity of serotonergic receptors in experimental animals, increasing the effectiveness of serotonergic transmission as a general rule (Blier and de Montigny, 1994). From a clinical point of view, the efficacy of antidepressant drugs is reduced or cancelled by treatments that impair 5-HT synthesis, like PCPA (p-chlorophenylalanine, irreversible inhibitor of tryptophan hydroxylase) or the administration of tryptophan-free aminoacid mixtures (Shopsin et al., 1975, 1976; Delgado et al., 1990),. Also, that the inhibition of 5-HT uptake plays a critical role in the relief of depressive symptoms is supported by the antidepressant activity of the selective serotonin reuptake inhibitors (SSRIs), that lack affinity for other amine receptors or transporters. Yet, as other antidepresant drugs, these have a delayed onset of action.

During the last five years, new aspects of the mechanism of action of the antidepressant drugs have emerged as a consequence of the application of *in vivo* microdialysis to study brain function. The present work will review these data, focusing on the critical role played by somatodendritic $5\text{-HT}_{1A}$ autoreceptors.

*Recent Advances in Tryptophan Research*, edited by Graziella Allegri Filippini et al.
Plenum Press, New York, 1996 

# 2. METHODS

## 2.1. Microdialysis Procedures

Concentric dialysis probes were made as previously described (Adell and Artigas, 1991). The length of membrane exposed to the brain tissue was 1.5 to 4 mm long (O.D. 0.25 mm). Probes were sterotaxically implanted in various brain regions of anesthetized rats according to the brain atlas of Paxinos and Watson (1982). A detailed description of the placement of the probes can be obtained in previous works from this laboratory (Adell and Artigas, 1991; Adell et al., 1991; Ferré et al., 1994). Animals were allowed to recover from surgery for approximately 20 h and then probes were perfused with artificial cerebrospinal fluid (CSF) (125 mM NaCl, 2.5 mM KCl, 1.26 mM $CaCl_2$ and 1.18 mM $MgCl_2$; pH = 6.2) at 0.25 - 0.5 µl/min. Sample collection started 60 min after the beginning of perfusion and 4-5 fractions were collected to obtain basal values before drug administration. Successive 20 - 30 min dialysate samples were collected. In some experiments, animals were implanted with a dialysis probe in dorsal striatum and a metal cannula in the dorsal raphe nucleus (DRN) for the local application of drugs (Romero et al., 1994).

## 2.2. Chromatographic Analyses

5-HT and 5-HIAA were analyzed by HPLC, using 3 µm ODS 2 columns (7.5 cm x 0.46 cm; Beckman, CA, U.S.A.) and were detected coulometrically with an ESA detector or amperometrically with a Hewlett Packard 1049 detector. The detection limit for 5-HT using the latter detector was typically 0.5-1 fmol.

## 2.3. Data Treatment

Microdialysis results are given as raw data (fmol per fraction for 5-HT, pmol per fractions for 5-HIAA, or as percentages of basal values, calculated as the individual means of four fractions before drug administration). Results are given as means±SEM of 4-9 rats/group. Statistical analysis of microdialysis data was performed using analysis of variance (ANOVA) for repeated measures of raw data and *t*-tests where appropriate.

# 3. RESULTS AND DISCUSSION

## 3.1. Single Administration of Antidepressant Drugs

*3.1.1. Local Applications.* Antidepressant drugs increase the concentration of 5-HT in dialysate samples from the midbrain raphe nuclei and the frontal cortex of conscious rats when locally applied by reverse dialysis. Drugs examined include the tricyclic uptake blockers imipramine and clomipramine and several MAO inhibitors. Figure 1 shows the effects of the local application of 10 µM of clomipramine, a potent 5-HT uptake blocker, on dialysate 5-HT in frontal cortex and the midbrain raphe nuclei. The net elevations produced are more marked in the latter region, where the cell bodies and dendrites of serotonergic neurones are located.

*3.1.2. Systemic Administration.* The administration of single doses of antidepressant drugs that inhibit 5-HT uptake or MAO activity leads to marked elevations of dialysate 5-HT, greater in the raphe nuclei than in frontal cortex (Adell and Artigas, 1991; Bel and Artigas,

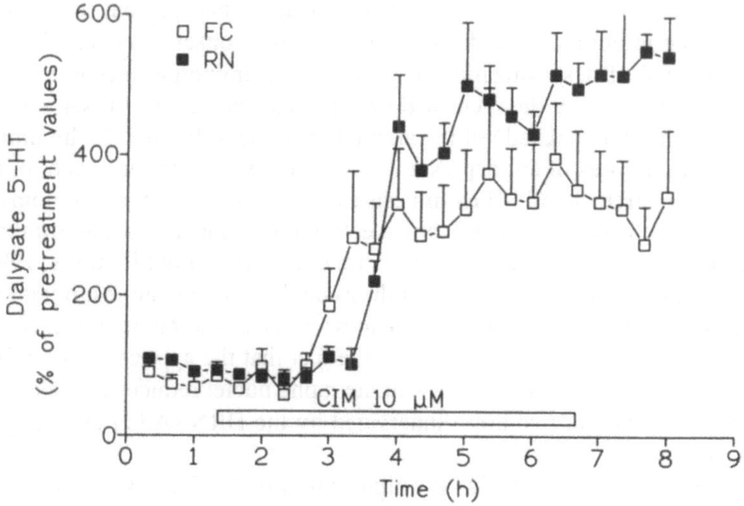


**Figure 1.** Increase of dialysate 5-HT in frontal cortex (FC) and the midbrain raphe nuclei (RN) caused by the local application of 10 μM of the uptake blocker clomipramine (CIM). Redrawn from data in refernce 5. Points are means+SEM of dialysate 5-HT concentrations expressed as percentages of basal values. Significant effects of the application of clomipramine were noted in both regions. Bar represents the period of drug application.

1992; Invernizzi et al., 1992; Celada and Artigas, 1993; Celada et al., 1994; Wozniak et al. 1992). With the exception of relatively high doses of imipramine, this is a general rule for all compounds and doses studied. This includes tricyclic drugs like clomipramine and imipramine (only at a low dose), selective serotonin reuptake inhibitors (SSRIs) and MAO inhibitors, either selective and reversible—like brofaromine— or irreversible, as clorgyline (MAO-A) and tranylcypromine (non-selective).

Figure 2 shows the effects of 1 mg/kg of the SSRI fluvoxamine and 4 mg/kg of imipramine on dialysate 5-HT in frontal cortex and raphe nuclei. At these doses, close to the clinical range used in depressed patients, the effect in the raphe nuclei remains but that in frontal cortex is marginal and transient.

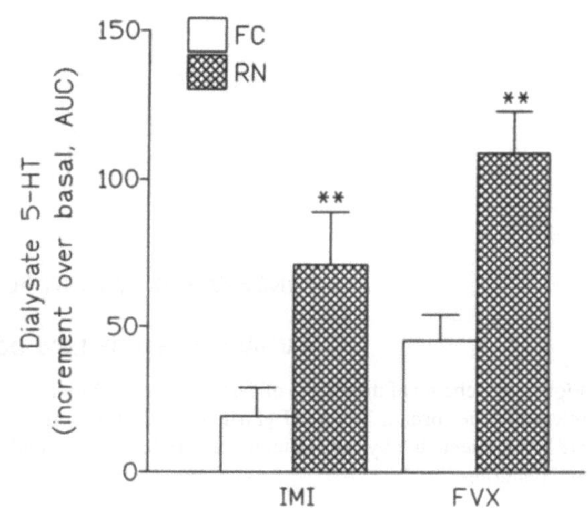


**Figure 2.** Regional selectivity of the systemic administration of antidepressants. Bars are means+SEM of the increases over basal of the AUC (area under the curve) after the i.p. administration of fluvoxamine (FVX, 1 mg/kg; AUC until 4 h post-administration) and imipramine (IMI, 4 mg/kg; AUC until 2 h post-administration). Data taken from reference 10 and unpublished results. **P<0.01 vs. frontal cortex.

*3.1.3. Implications for Antidepressant Treatments.* This regional selectivity of antidepressants may be important to understand their delayed onset of action. This has been attributed to the need of desensitization of postsynaptic aminergic receptors, notably the ß-adrenoceptors, produced by the increased synaptic concentrations of NA and 5-HT that would take place after the blockade of the amine transporters. In view of the limited effect of cllinically relevant doses of antidepressants, alternative explanations need to be considered. Indeed, the impairement of 5-HT synthesis exacerbates depressive symptoms (Shopsin et al., 1975; 1976; Delgado et al., 1990) and electrophysiological data suggest that chronic antidepressant treatments increase serotonergic function (Blier and Montigny, 1994). Thus, the delayed onset of action of antidepressant drugs could be due to an insuficient release of 5-HT by forebrain terminals, caused by the excess 5-HT in the raphe nuclei provoked by antidepresant treatments. This is supported by the fact that the application of 5-HT uptake inhibitors —clomipramine or citalopram— in the raphe nuclei reduces the release of 5-HT in frontal cortex or striatum, two areas innervated by the DRN (Adell and Artigas, 1991; Romero et al., 1994).

This is due to the existence of a very high density of 5-HT uptake sites and inhibitory 5-$HT_{1A}$ autoreceptors (Hrdina et al., 1990; Pazos and Palacios, 1985; Sotelo et al., 1990). The latter reduce cell firing, 5-HT synthesis and terminal 5-HT release when activated (Sprouse and Aghajanian, 1987; Hudson et al., 1989; Invernizzi et al., 1991) (Fig. 3). This mechanism may be also operant in depressed patients, as cell bodies of human serotonergic neurones also contain a high density of 5-$HT_{1A}$ autoreceptors and 5-HT uptake sites in the DRN (Palacios et al., 1987; Cortes et al., 1988). Overall, this denotes the midbrain raphe nuclei as the most sensitive targets for antidepressant drug action in the brain.

The negative feed-back mechanism operant at somatodendritic level in serotonergic neurones may not be equally effective to reduce 5-HT release in all brain regions. The dorsal

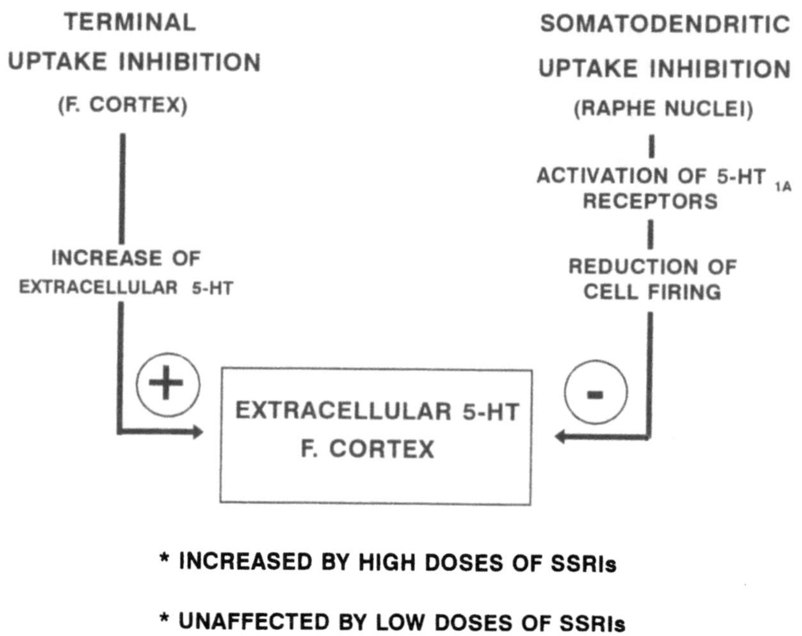


**Figure 3.** Scheme of the effects of 5-HT uptake inhibition in frontal cortex and midbrain raphe nuclei on 5-HT release in the former. The 5-HT-enhancing effect of uptake inhibitors in terminal regions innervated by the DRN are attenuated by the reduction of 5-HT release secondary to the inhibition of the uptake in cell bodies and dendrites.

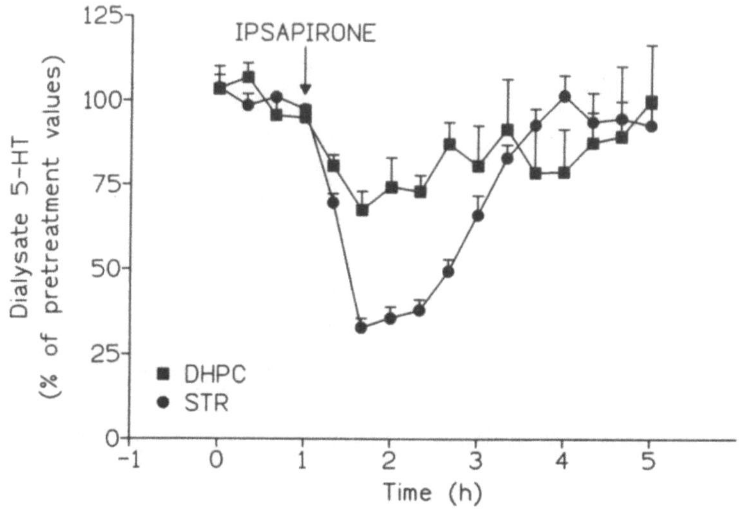


**Figure 4.** Effects of ipsapirone (3 mg/kg, s.c.) on 5-HT release in striatum (STR) and dorsal hippocampus (DHPC). Data are means±SEM, expressed as percentages of pretreatment values.

and median raphe nuclei (DRN and MRN) give rise to separate projections to forebrain (Azmitia and Segal, 1978). Thus, the dorsal striatum and the frontal cortex receive serotonergic fibres exclusively (striatum) or preferentially (frontal cortex) from the DRN, whereas the hippocampus is preferentially innervated by the MRN (Azmitia and Segal, 1978; Imai et al., 1986). As serotonergic neurones of the DRN are more sensitive to the inhibitory actions of 5-$HT_{1A}$ agonists systemically administered (Sinton and Fallon, 1988; Blier et al., 1990), 5-HT release may be selectively reduced in areas innervated by this nucleus. Despite initial results indicating a comparable sensitivity of DRN- and MRN-innervated regions to the systemic effects of 5-$HT_{1A}$ agonists (Hjorth and Sharp, 1991), we have found marked differences among them. Ipsapirone (3 mg/kg, s.c.), a selective 5-$HT_{1A}$ agonist, reduced 5-HT release in dorsal striatum (innervated by the DRN) to a larger extent than in dorsal hippocampus (Fig. 4).

The sensitivity of DRN-innervated areas to the inhibitory actions of 5-HT is also shown by the decreased striatal 5-HT release induced by the local application of citalopram (SSRI) in the DRN (Romero et al., 1994). This procedure induced a ca. 50 % decrease of 5-HT release in striatum which was prevented by the previous application of pertussis toxin in the DRN (uncoupling the 5-$HT_{1A}$ receptor protein from the effector $K^+$ channel) or by the treatment with (-)pindolol and (-)tertatolol, two ß-adrenoceptor/5-$HT_{1A}$ antagonists (Romero et al., 1994; Bel et al., 1994).

## 3.3. Chronic Antidepressant Treatments. Effects of Tryptophan Depletion

The administration of low doses of imipramine (4 mg/kg day), fluvoxamine (1 mg/kg day) or tranylcypromine (0.5 mg/kg day) given by osmotic minipumps for two weeks significantly increased extracellular 5-HT in frontal cortex (Bel and Artigas, 1993; Ferrer and Artigas, 1994) (Fig. 5). These doses do not modify extracellular 5-HT in cortex in a significant manner when given as a bolus (except for 1 mg/kg fluvoxamine, that induces a transient increase) (Bel and Artigas, 1992; 1993). This is a clear indication that the chronic — but not acute — antidepressant treatment with clinically relevant doses might result in an

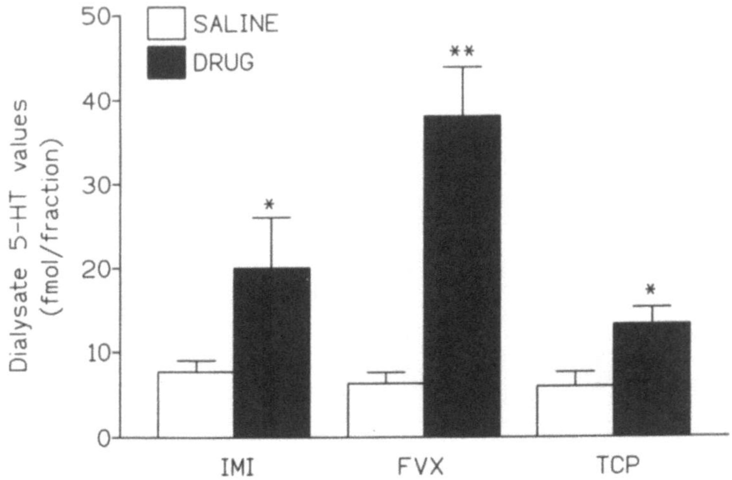


**Figure 5.** Effects of the administration of imipramine (IMI, 4 mg/kg day), fluvoxamine (FVX, 1 mg/kg day) and tranylcypromine (TCP, 0.5 mg/kg day) for two weeks (with minipumps) on basal dialyate 5-HT values. Bars are means±SEM. Data from references 29 and 30 and unpublished results. *P<0.05, **P<0.01 vs. saline.

increased serotonergic function in cortex, paralleling the clinical course of antidepressant treatments.

The hypothesis of an association between antidepressant effects and increased serotonergic function was assessed by the depletion of brain tryptophan. The administration of a tryptophan-free aminoacid mixture (Delgado et al., 1990) caused a significant and marked decline of extracellular 5-HT and 5-HIAA concentrations in frontal cortex (Fig. 6) that lasted for a minimum of 6 h.

Thus, a pharmacological treatment that severely impairs the mood of treated depressive patients dramatically reduced serotonergic function, as judged by the decreased output of 5-HT and 5-HIAA. This appears to support the view that the mood worsening effects of tryptophan depletion are due to an impaired 5-HT release in forebrain.

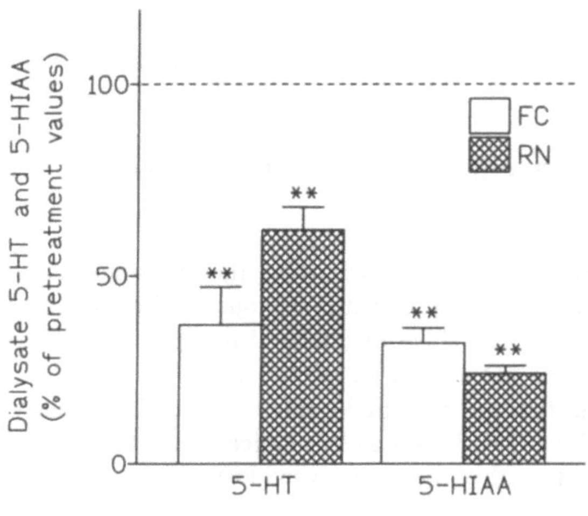


**Figure 6.** Maximal reductions of dialysate 5-HT and 5-HIAA concentrations following the administration to fluvoxamine-treated rats of a tryptophan-free aminoacid mixture with identical composition and at the same dose (weight/weight) than that used by Delgado and coworkers (1990). Bars are means±SEM expressed as percentages of pretreatment values. **P<0.01 vs. basal values.

## 3.4. Conclusions and Future Trends

In addition to chronic adaptative mechanisms involving the desensitization of terminal and somatodendritic autoreceptors, the use of the microdialysis technique has enabled to document the existence of rapid adaptative changes of serotonergic transmission after single treatment with antidepressants. These trigger a negative feed-back mechanism that offsets their effects in frontal cortex (and perhaps other brain areas) by inducing marked elevations of extracellular 5-HT near the cell bodies and dendrites. After chronic treatment, this self-inhibitory mechanism are probably less effective, enabling 5-HT terminals to normalize their release and increase 5-HT above normal values due to the inhibition of terminal uptake.

Thus, the prevention of the attenuation of 5-HT release by antidepressants using 5-$HT_{1A}$ antagonists can be a novel approach to treat major depression. In apparent agreement with this hypothesis (Artigas, 1993), the concurrent treatment of major depressive patients with paroxetine (SSRI) and pindolol (a ß-adrenoceptor/5-$HT_{1A}$ antagonist) led to rapid (< 1 week) full remission of depressive symptoms in about two thirds of the patients (Artigas et al., 1994). Rapid improvements were also noted in depressive patients resistant to uptake blockers and MAO inhibitors after the addition of pindolol to their current medication (Artigas et al., 1994). These results, already replicated by an independent group with comparable results (Blier and Bergeron, in press), are suggestive that the blockade of somatodendritic 5-$HT_{1A}$ receptors may accelerate and potentiate the action of serotonergic antidepressants. The use of an antagonist of somatodendritic 5-$HT_{1A}$ autoreceptors would render serotonergic neurones more active, permitting a net increase of the "synaptic" concentration of 5-HT in forebrain after the blockade of the transporter by SSRIs. As hippocampal (postsynaptic) 5-$HT_{1A}$ receptors are known to contribute to antidepressant effects, the use of a non-selective 5-$HT_{1A}$ antagonist might attenuate or cancel the benefits obtained by enhancing the presynaptic serotonergic function. However, the net effect of the combination of pindolol with an SSRI seems to be an increased activation of serotonergic receptors, as pindolol blocks the self-inhibitory effects of the SSRI in the DRN but not those of 5-HT in the area $CA_3$ of the dorsal hippocampus (Bel et al., 1994). Indeed, further confirmation in double-blind studies is needed to verify the usefulness of 5-$HT_{1A}$ antagonists in the treatment of major depression.

## ACKNOWLEDGMENTS

This work has been supported by a grant from the Fondo de Investigación Sanitaria (FIS 95/266). Financial support from Bayer-Tropon and CIBA-GEIGY S.A. is also acknowledged. Thank are given to Lundbeck A/S and Servier for their generous supply of drugs. The skillful technical assistance of Josefa Torres and Letizia Campa is acknowledged.

## REFERENCES

Adell A. and Artigas F., 1991, Differential effects of clomipramine given locally or systemically on extracellular 5-hydroxytryptamine in raphe nuclei and frontal cortex. An in vivo microdialysis study *Naunyn-Schmied. Arch. Pharmacol.* 343:237.

Adell A., Carceller A., and Artigas F., 1991, Regional distribution of extracellular 5-hydroxytryptamine and 5-hydroxyindoleacetic acid in the brain of freely moving rats, *J. Neurochem.* 56:709.

Artigas F., 1993, 5-HT and antidepressants:new views from microdialysis studies, *Trends Pharmacol. Sci.* 14:262.

Artigas F., Perez V., and Alvarez E., 1994, Pindolol induces a rapid improvement of depressed patients treated with serotonin reuptake inhibitors, *Arch. Gen. Psychiatry* 51:248.

Azmitia E.C. and Segal M., 1978, An autoradiographic analysis of the differential ascending projections of the dorsal and median raphe nuclei in the rat, *J. Comp. Neurol.* 179:641.

Bel N. and Artigas F., 1992, Fluvoxamine preferentially increases extracellular 5-hydroxytryptamine in the raphe nuclei:an in vivo microdialysis study, *Eur. J. Pharmacol.* 229:101.

Bel N. and Artigas F., 1993, Chronic treatment with fluvoxamine increases extracellular serotonin in frontal cortex but not in raphe nuclei, *Synapse* 15:243.

Bel N., Romero L., Celada P., De Montigny C., Blier P., and Artigas F., 1994, Neurobiological basis for the potentiation of the antidepressant effect of 5-HT reuptake inhibitors by the 5-$HT_{1A}$ antagonist pindolol, *in:* "Monitoring Molecules in Neuroscience" Proceedings of the 6th International Conference of *in vivo* Methods, A. Louilot, T. Durkin, U. Spampinato, M. Cador, eds., Université de Bordeaux, Bordeaux, 209.

Blier P., Serrano A., and Scatton B., 1990, Differential responsiveness of the rat dorsal and median raphe 5-HT systems to 5-$HT_1$ receptor agonists and p-chloroamphetamine, *Synapse* 5:120.

Blier P. and de Montigny C., 1994, Current advances and trends in the treatment of depression, *Trends Pharmacol. Sci.* 15:220.

Blier P., Seletti B., Bouchard C., Artigas F., and De Montigny C., 1994, Functional evidence for the differential responsiveness of pre-and postsynaptic 5-$HT_{1A}$ receptors in the rat brain, *Soc. Neurosci. Abs.* 20:1540.

Blier P. and Bergeron R., Effectiveness of pindolol with selected antidepressant drugs in the treatment of major depression, *J. Clin. Psychopharmacol.* 15: (in press)

Celada P. and Artigas F., 1993, Monoamine oxidase inhibitors increase preferentially extracellular 5-hydroxytryptamine in the midbrain raphe nuclei. A brain microdialysis study in the awake rat, *Naunyn-Schmied. Arch. Pharmacol.* 347:583.

Celada P., Bel N., and Artigas F., 1994, The effects of brofaromine, a reversible MAO-A inhibitor, on extracellular serotonin in the raphe nuclei and frontal cortex of freely moving rat, *J. Neural Transm.* (Suppl)41:357.

Cortes R., Soriano E., Pazos A., Probst A., and Palacios J.M., 1988, Autoradiography of antidepressant binding sites in the human brain:localization using [$^3$H]imipramine and [$^3$H]paroxetine, *Neuroscience* 27:473.

Delgado P.L., Charney D.S., Price L.H., Aghajanian G.K., Landis H., and Heninger G.R., 1990, Serotonin function and the mechanism of antidepressant action, *Arch. Gen. Psychiatry* 47:411.

Ferré S., Cortés R., and Artigas F., 1994, Dopaminergic regulation of the serotonergic raphe-striatalpathway:microdialysis studies in freely moving rats, *J. Neurosci.* 14:4839.

Ferrer A. and Artigas F., 1994, Effects of single and chronic treatment with tranylcypromine on extracellular serotonin in rat brain, *Eur. J. Pharmacol.* 263:227.

Hjorth S. and Sharp T., 1991, Effect of the 5-$HT_{1A}$ receptor agonist 8-OH-DPAT on the release of 5-HT in dorsal and median raphe-innervated rat brain regions as measured by in vivo microdialysis, *Life Sci.* 48:1779.

Hrdina P.D., Foy B., Hepner A., and Summers R.J., 1990, Antidepressant binding sites in brain: autoradiographic comparison of [$^3$H]paroxetine and [$^3$H]imipramine localization and relationship to serotonin transporter, *J. Pharmacol. Exp. Ther.* 252:410.

Hutson P.H., Sarna G.S., O'Connell M.T., and Curzon G., 1989, Hippocampal 5-HT synthesis and release in vivo is decreased by infusion of 8-OH-DPAT into the nucleus raphe dorsalis, *Neurosci. Lett.* 100:276.

Imai H., Steindler D.A., and Kitai S.T., 1986, The organization of divergent axonal projections from the midbrain raphe nuclei in the rat, *J. Comp. Neurol.* 243:363.

Invernizzi R., Carli M., Di Clemente A., and Samanin R., 1991, Administration of 8-Hydroxy-2-(Di-n-Propylamino)tetralin in raphe nuclei dorsalis and medianus reduces serotonin synthesis in the rat brain:differences in potency and regional sensitivity, *J. Neurochem.* 56:243.

Invernizzi R., Belli S., and Samanin R., 1992, Citalopram's ability to increase the extracellular concentrations of serotonin in the dorsal raphe prevents the drug's effect in the frontal cortex, *Brain Res.* 584:322.

Palacios J.M., Pazos A., and Hoyer D., 1987, Characterization and mapping of 5-$HT_{1A}$ sites in the brain of animals and man, *in*: "Brain 5-$HT_{1A}$ Receptors" C.T. Dourish, S. Ahlenius, P.H. hutson, eds., VCH-Ellis Horwood, Weinheim-Chichester, 67.

Paxinos G. and Watson C., 1982, "The Rat Brain in Stereotaxic Coordinates", Academic Press, Sydney.

Pazos A. and Palacios J.M., 1985, Quantitative autoradiographic mapping of serotonin receptors in the rat brain. I. Serotonin-1 receptors, *Brain Res.* 346:205.

Romero L., Celada P., and Artigas F., 1994, Reduction of in vivo striatal 5-hydroxytryptamine release by 8-OH-DPAT after inactivation of Gi/Go proteins in dorsal raphe nucleus, *Eur. J. Pharmacol.* 265:103.

Shopsin B., Gershon S., Goldstein M., Friedman E., and Wilk S., 1975, Use of synthesis inhibitors in defining a role for biogenic amines during imipramine treatment in depressed patients *Psychopharmacol. Commun.* 1:239.

Shopsin B., Friedman E., and Gershon S., 1976, Parachlorophenylalanine reversal of tranylcypromine effects in depressed patients, *Arch. Gen. Psychiatry* 33:811.

Sinton C.M. and Fallon S.L., 1988, Electrophysiological evidence for a functional differentiation between subtypes of the 5-$HT_1$ receptor, *Eur. J. Pharmacol.* 157:173

Sotelo C., Cholley B., Mestikawy S.E., Gozlan H., and Hamon M., 1990, Direct immunohistochemical evidence of the existence of 5-HT autoreceptors on serotonergic neurons in the midbrain raphe nuclei *Eur J Neurosci* 2:1144.

Sprouse J.S. and Aghajanian G.K., 1987, Electrophysiological responses of serotoninergic dorsal raphe neurons to 5-$HT_{1A}$ and 5-$HT_{1B}$ agonists, *Synapse* 1:3.

Wozniak K., Pert A., and Linnoila M., 1992, Alterations in raphe and frontal cortex serotonin overflow following focal and systemic administration of serotonin, *Soc. Neurosci. Abs.* 18:1525.

# THE EFFECTS OF A SELECTIVE INHIBITOR OF TRYPTOPHAN 2,3-DIOXYGENASE AND A COMBINED INHIBITOR OF TRYPTOPHAN 2,3-DIOXYGENASE AND 5-HT REUPTAKE ON SEROTONERGIC FUNCTION IN THE RAT

M. Salter, R. Hazelwood, C. I. Pogson, R. Iyer, D. J. Madge, H. T. Jones, B. R. Cooper, R. F. Cox, C. M. Wang, and R. P. Wiard

Wellcome Research
Beckenham, United Kingdom and Research Triangle Park, North Carolina

## 1. INTRODUCTION

Over the last few decades it has become clear that depression can be effectively treated by elevating serotonergic and/or noradrenergic function in the brain. The close interplay between the ascending serotonergic pathways from the raphe and the ascending noradrenergic pathways from the locus coeruleus at both somatic and presynaptic sites makes it likely that both of these monoaminergic pathways will be altered by drugs acting selectively at enzymes or receptors of either one of these pathways. The role of 5-HT in the amelioration of unipolar depression has gained recent support with the introduction of the clinically effective selective serotonin reuptake inhibitors (SSRIs). However, a problem still unsolved by the introduction of the SSRIs is the slow onset of therapeutic effectiveness of all antidepressants of several weeks. Several lines of evidence suggest that this slow onset of action, particularly for the SSRIs, may be due to an inhibitory autoregulation of the serotonergic pathway after an elevation of synaptic 5-HT caused by activation of the somatodendritic 5-HT1A autoreceptor (Artigas et al.,1994; see Blier and Montigny, 1994). Larger increases in 5-HT than those achieved by the SSRIs may overcome some of the consequences of this inhibitory autoregulation and may therefore hasten the onset of therapeutic effect.

The amino acid precursor of 5-HT, tryptophan, has been used as an antidepressant on its own or in combination with other antidepressants with varying degrees of success (see Young, 1991). The limited efficacy of tryptophan as an antidepressant may be due to its rapid metabolism by tryptophan 2,3-dioxygenase (TDO) in the liver (Moller et al., 1982; Salter et al., 1986) and/or the intraneuronal metabolism of the newly synthesised 5-HT (Sharp et al., 1992). In this study we examine the role of TDO as a regulator of tryptophan and serotonergic

*Recent Advances in Tryptophan Research*, edited by Graziella Allegri Filippini et al.
Plenum Press, New York, 1996

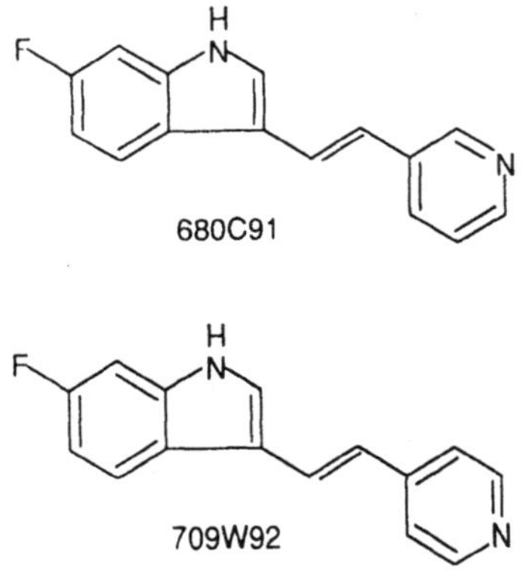


**IN VITRO ACTIVITY**

| | TDO (Ki) | 5-HT REUPTAKE ($IC_{50}$) |
|---|---|---|
| 680C91 | 42 nM | 51,000 nM |
| 709W92 | 40 nM | 700 nM |

**Figure 1.** *In vitro* activity of 680C91 and 709W92.

function by studying the effects of a novel and selective inhibitor of TDO and a novel inhibitor of both TDO and 5-HT reuptake in the rat. The results suggest that these classes of compounds may offer advantages over the present generation of antidepressants.

## 2. METHODS

All methods are described in Salter et al. (1995a, b).

## 3. RESULTS AND DISCUSSION

680C91 and 709W92 were potent inhibitors of TDO (fig. 1). 709W92 was also a potent inhibitor of 5-HT uptake into cerebral cortical slices (fig. 1) with approximately 50- and 100-fold selectivity against noradrenaline and dopamine uptake respectively.

680C91 and 709W92 were selective with respect to a number of other enzymes and receptors of the tryptophan and serotonergic pathways (MAO A and B, IDO, tryptophan hydroxylase, 5-HT1A, 1B, 1D, 2A and 2C). 680C91 and 709W92 were also effective inhibitors of TDO *in vivo* as demonstrated by their ability to totally inhibit the TDO-catalysed metabolism of L-[ring-2-$^{14}$C]tryptophan to $^{14}$C- formate *in vivo* (data not shown). To examine whether this inhibition of TDO would inhibit the catabolism of a load dose of tryptophan *in vivo*, 709W92 was administered 30min prior to a load dose of tryptophan and the temporal plasma profile of tryptophan was followed (fig. 2). 709W92 clearly sustained the increase in plasma tryptophan produced by the tryptophan load dose, consistent with TDO representing the major route of tryptophan disposal in the rat.

Inhibitors of 5-HT reuptake are known to potentiate head-twitch in the rat induced by the 5-HT precursor 5-hydroxytryptophan (5-HTP) and to inhibit head-twitch and the depletion in whole brain 5-HT elicited by ρ-chloroamphetamine. 709W92, but not 680C91, potentiated 5-HTP induced head-twitch and inhibited head-twitch and the depletion in whole brain 5-HT produced by ρ-chloroamphetamine (data not shown). In addition to this, inhibitors of 5-HT reuptake are known to produce a rapid decrease in raphe serotonergic neuronal firing by elevating 5-HT in the vicinity of the somatodendritic 5-HT1A autorecep-

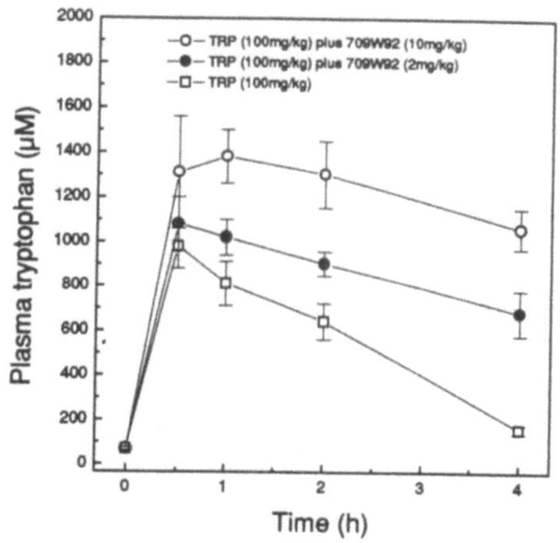


**Figure 2.** The effect of 709W92 on the plasma concentration of tryptophan after a load dose of tryptophan. Tryptophan was administered at zero time.

tor. 709W92 rapidly decreased dorsal raphe firing with an ID50 of 18μg/kg after intravenous administration (fig. 3). Taken together with the above data these results demonstrate that 709W92 is a potent inhibitor of 5-HT reuptake as well as TDO *in vivo*.

680C91 also inhibited dorsal raphe firing but only at doses 200-fold greater that those seen with 709W92 (fig. 3), consistent with the difference in their *in vitro* affinities for 5-HT uptake. Although tryptophan has been shown to inhibit raphe firing *in vivo* (presumably by elevating synaptic 5-HT), the decrease in dorsal raphe firing seen with 709W92 was too fast to be due to an elevation of tryptophan and therefore reflects its effect on 5-HT reuptake alone. From the data obtained so far it is clear that 680C91 and 709W92 are respective inhibitors of TDO and TDO plus 5-HT reuptake *in vivo*. The effects of these two molecules were then examined on a global measure of brain 5-HT that reflects synaptic rather than total brain levels of 5-HT, cerebrospinal fluid (CSF) 5-HT. Before this was carried out the effects of a number of serotonergically active drugs were examined on CSF 5-HT (fig. 4).

Maximally effective doses of tryptophan, pargyline (a monoamine oxidase inhibitor) and fluoxetine (a SSRI) and their combinations produced varying increases in CSF 5-HT

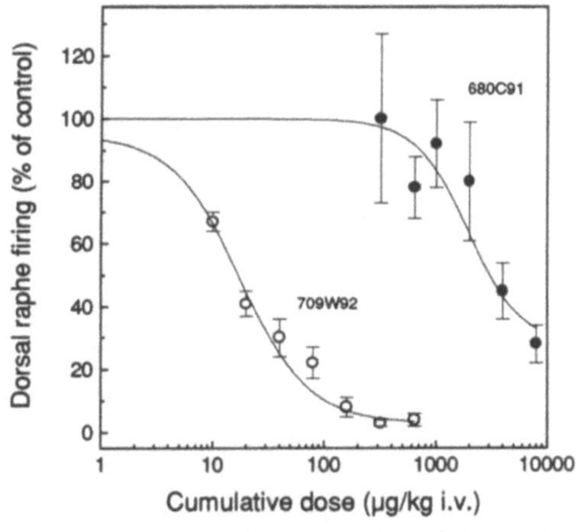


**Figure 3.** The effect of 709W92 and 680C91 on dorsal raphe firing.

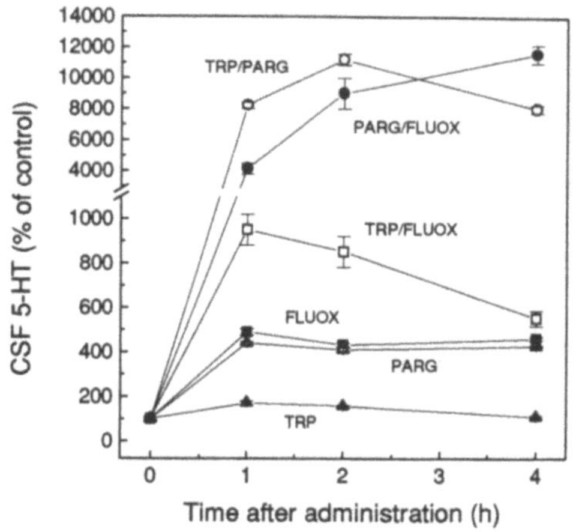


**Figure 4.** The effect of tryptophan (100mg/kg i.p.), pargyline (70mg/kg i.p.), and fluoxetine (10mg/kg i.p.) on CSF 5-HT.

consistent with their different sites of action within the serotonergic pathway. The tryptophan/pargyline and pargyline/fluoxetine combinations elicited the serotonin syndrome when the elevation of CSF 5-HT was greater than 3000% of control.

680C91 and 709W92 elevated CSF levels of tryptophan up to 250% of control (fig. 5). 680C91 increased CSF 5-HT to levels seen maximally after tryptophan administration but for a greater duration (fig. 5). 709W92 increased CSF levels of 5-HT to those seen maximally after the fluoxetine/tryptophan combination (compare figs. 4 and 5), but again with a superior duration of action. The effects of 680C91 and 709W92 on CSF 5-HT are consistent with their inhibitory activities against either TDO alone or TDO and 5-HT reuptake respectively.

This study describes the effect of two new types of inhibitor: selective inhibitors of TDO (680C91) and combined inhibitors of both TDO and 5-HT reuptake (709W92). It can be concluded from the data that TDO plays a key role in the homeostasis of tryptophan and 5-HT function in the body. These observations may be of significance considering the

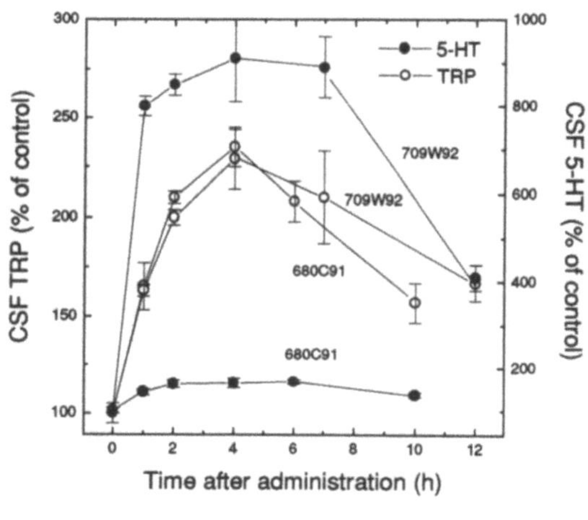


**Figure 5.** The effect of 680C91 and 709W92 (both 7.5 mg/kg free base equivalent) on CSF tryptophan and 5-HT.

elevation of cortisol in a large number of depressed patients, its ability to markedly induce the activity of TDO (Salter et al., 1986) and that depletion of tryptophan (and thus 5-HT) can precipitate depression (Heninger et al., 1992).

680C91 achieves similar maximal increases in CSF 5-HT to those seen with tryptophan but sustains those increases for a greater duration. If the antidepressant efficacy of tryptophan is limited by its metabolism then superior efficacy of treatment should be achieved by this mechanistic approach. However, comparison of the clinically effective doses of fluoxetine with our dose-response data on its effects on CSF 5-HT, suggest that fluoxetine is most efficacious towards the top of its dose-response curve. It is unlikely, therefore, that a TDO inhibitor alone (or indeed tryptophan itself) could produce the increases in 5-HT needed to achieve the therapeutic benefits seen with the SSRIs or the monoamine oxidase inhibitors (see figs. 4 and 5). However, superior antidepressant effects have been observed with combinations of antidepressant treatments such as tryptophan plus a monoamine oxidase inhibitor and tryptophan plus a reuptake inhibitor (see Young, 1991). In these studies we have observed that these combinations produce larger increases in the functional levels of 5-HT than monotherapy. Use of the former combination is problematic due to the very high levels of 5-HT achieved and the consequent precipitation of the serotonin syndrome (see Young, 1991). The latter combination does not increase 5-HT to levels associated with the serotonin syndrome but, as shown in fig. 4, is limited in its ability to increase 5-HT maximally for a sustained duration. Administration of a TDO inhibitor (such as 680C91) along with a SSRI or indeed administration of a combined TDO and 5-HT reuptake inhibitor (such as 709W92) can overcome this problem and produce sustained increases in 5-HT greater than those achievable with current antidepressant monotherapy. These larger increases in 5-HT may overcome to some extent the effects of the inhibitory autoregulation of the serotonergic system after acute antidepressant monotherapy and may therefore lead to a more rapid onset of antidepressant action.

## 4. REFERENCES

Artigas F., Perez V., and Alvarez E., 1994, Pindolol induces a rapid improvement of depressed patients treated with serotonin reuptake inhibitors, *Arch. Gen. Psychiat.* 51:248.

Blier P. and de Montigny C., 1994, Current advances and trends in the treatment of depression, *TIPS* 15:220.

Heninger G.R., Delgado P.L., Charney D.S., Price L.H., and Aghajanian G.K., 1992, Tryptophan-deficient diet and amino acid drink deplete plasma tryptophan and induce a relapse of depression in susceptible patients, *J. Chem. Neuroanat.* 5:347.

Moller S.E., Kirk L., and Honore P., 1982, Tryptophan tolerance and metabolism in endogenous depression, *Psychopharmacology* 76:79.

Salter M., Knowles R.G., and Pogson C.I., 1986, Quantitation of the importance of individual steps in the control of aromatic amino acid metabolism, *Biochem. J.* 234:635.

Salter M., Hazelwood R., Pogson C.I., Iyer R., and Madge D.J., 1995, The effects of a novel and selective inhibitor of TDO on tryptophan and serotonin metabolism in the rat, *Biochem. Pharmacol.* 49:1435.

Salter M., Hazelwood R., Pogson C.I., Iyer R., Madge D.J., Jones H.T., Cooper B.R., Cox R.F., Wang C.M., and Wiard R.P., 1995, The effects of an inhibitor of TDO and a combined inhibitor of TDO and 5-HT reuptake in the rat, *Neuropharmacology* 34:217

Sharp T., Bramwell S.R., and Grahame-Smith D.G., 1992, Effect of acute administration of L-tryptophan on the release of 5-HT in rat hippocampus in relation to serotonergic neuronal activity: an *in vivo* microdialysis study, *Life Sci.* 50:1215.

Young S., 1991, Use of tryptophan in combination with other antidepressant treatments: a review, *J. Psychiatr. Neurosci.* 16:241.

8

# EFFECT OF VALINE ON 5-HT NEUROTRANSMISSION AND MOOD

P. J Cowen, D. J. Williamson, and S. F. B. McTavish

Department of Psychiatry
Littlemore Hospital
Oxford OX4 4XN, United Kingdom

## 1. INTRODUCTION

Changes in brain serotonin (5-HT) function have been implicated in both the aetiology of mood disorders and the actions of antidepressant drugs (for reviews see Delgado et al, 1992; Cowen, 1993). Recently, Delgado et al (1990) have reported that a 100g amino acid mixture, free of the 5-HT precursor, L-tryptophan (TRP), causes an acute symptomatic relapse in remitted depressed patients maintained on antidepressant drugs. This relapse has been attributed to an abrupt decline in the availability of TRP for brain 5-HT synthesis which temporarily reverses the effects of the antidepressant drug treatment.

## 2. VALINE IN ANIMAL EXPERIMENTAL STUDIES

In animal studies, administration of a single neutral amino acid, such as valine, lowers the brain entry of TRP by competing for a common transport mechanism at the blood-brain barrier (see Lehnert and Wurtman, 1993). Functional investigations suggest that valine pre-treatment lowers stress-induced cortisol release in rodents, a neuroendocrine response mediated by brain 5-HT pathways. Direct evidence that valine can alter 5-HT release comes from studies using in vivo cerebral microdialysis where Gartside et al. found that valine (40mg/kg) decreased the 5-HT release evoked by both the 5-HT releasing agent, d-fenfluramine, and electrical stimulation of dorsal raphe pathways (Gartside et al, 1992a; 1992b).

We therefore decided to asssess whether valine might alter brain 5-HT function in humans, and in particular whether it might lower mood in recovered depressed patients maintained on 5-HT potentiating treatments.

*Recent Advances in Tryptophan Research,* edited by Graziella Allegri Filippini et al.
Plenum Press, New York, 1996

# 3. VALINE STUDIES IN HUMANS

All subjects were given a standard dose of valine (30gm) which was dissolved in 500mls tap water flavoured with pineapple. The dose of valine was based on that found to be effective at lowering electrically-evoked 5-HT release in rodents (40 mg/kg). All subjects were tested in the morning after an overnight fast.

## 3.1 Healthy Subjects

*3.1.1 Methods.* We tested a total of 22 healthy male subjects in two separate double-blind, placebo controlled, cross-over investigations. In the first study, 8 subjects received, on one occasion, valine (30g), and on the other occasion, the same volume of flavoured tap water (placebo). In the second study, 14 subjects received valine and tap water as above. In addition, one hour after valine/placebo administration they received 30mg (orally) of the 5-HT releasing agent, d-fenfluramine. During the test, subjects were asked to complete 10 centimetre visual analogue scales at 60-minute intervals for the following items "sad", "anxious", "difficulty concentrating", "nauseated", "hungry".

*3.1.2 Results.* In the eight subjects who received valine and placebo alone, there was a sevenfold elevation in plasma branch chain amino acid (BCAA) levels one hour following valine administration compared to placebo (3038 ± 284 μg/ml vs 411± 41 μg/ml) and BCAA concentrations were still elevated when sampling was discontinued at 6 hours (3414 ± 299 μg/ml. Valine did not alter plasma TRP.

In the eight subjects who received valine and placebo alone there was no significant effect of valine on plasma prolactin levels. Administration of d-fenfluramine was followed by an increase in plasma prolactin. Analysis of variance showed a significant valine by time interaction (F=2.66; df=7,91; p=0.015) where prolactin levels were lower following valine adminstration (Fig 1). Levels of d-fenfluramine and its active metabolite d-norfenfluramine were slightly but non-significantly lower following valine administration. We carried out an analysis of covariance whereby the change in prolactin levels following valine and placebo pre-treatment was re-analysed using d-fenfluramine and d-norfenfluramine levels as a varying covariate. The interaction of valine and time remained statistically significant (drug by time interaction F=3.35; df=3,35; p=0.03).

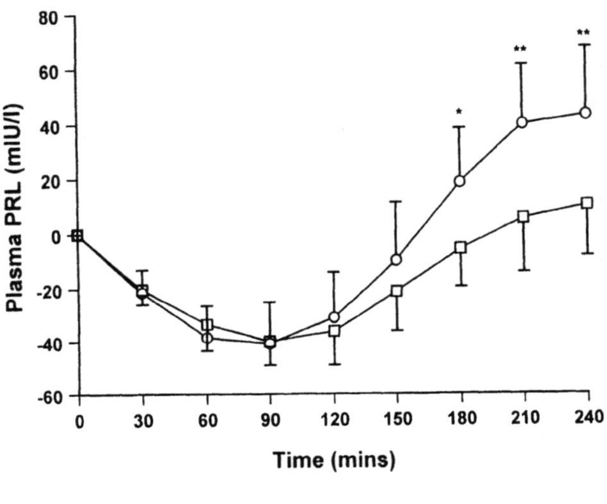


**Figure 1.** Change from baseline (time 0) of plasma PRL (mean ± SEM) in 14 healthy male subjects who were tested on two occasions receiving: a) valine (30g) at time -60 and d-fenfluramine (30mg) at time 0 (*squares*); b) placebo at time -60 and d-fenfluramine (30mg) at time 0 (*circles*). (Significantly different from valine * p <0.05, ** p<0.01, Fisher's test of least significant difference).

In the eight subjects who received valine and placebo alone, valine produced no significant effects on any of the visual analogue ratings of "sad" , "anxious", "difficulty concentrating", "hungry" or "nauseated" (all p values > 0.2).

## 3.2 Studies in Patients

*3.2.1 Methods.* We studied 12 patients (7 male, 5 female), mean age 39 (range 24-66). All patients had previously met DSM-III-R criteria for major depression but at the time of testing were clinically remitted and essentially asymptomatic. The mean length of recovery from depressive episode was 10.5 weeks (range 2-20). Ten of the patients were receiving treatment with selective serotonin re-uptake inhibitors (2 on paroxetine 20-40mg; 8 on sertraline 50-100mg) and two with a combination of lithium (600-800mg) and amitriptyline (100-150mg).

The study was a double-blind, placebo controlled, cross-over design with 1 week between the two tests. Subjects came to the Research Unit after an overnight fast. A blood sample was taken for plasma TRP and BCAA levels and then the valine or placebo was administered. Subjects then rested quietly in a room for the following 6 hours. They completed hourly visual analogue ratings as described above for the healthy volunteers and were also interviewed 3 hours and 5 hours following the drink by a clinician who had previous knowledge of their condition.

*3.2.2 Results.* As in the healthy subjects, plasma BCAA levels were markedly increased five hours following valine administration compared to placebo (4455 μg/ml vs 378 μg/ml), but there was no significant change in plasma TRP levels.

When interviewed following valine administration by a clinician with prior knowledge of their condition, 4 of the 12 patients, (all taking SSRIs) had developed recognisable symptoms characteristic of their depression. For example, one patient experienced depressed mood accompanied by intrusive thoughts that harm might befall a member of her family, thoughts which she had not experienced since her response to treatment. Another noted irritability and poor concentration, which had been prominent features of her illness. In three patients these changes though detectable were mild in degree but in a fourth subject, lowered mood was accompanied by psychomotor retardation and impaired insight, such that he was unable to attribute his symptoms to having had the valine drink. All symptoms had resolved by the end of the test.

In the patient group, visual ratings of "sad" were significantly greater following valine adminstration (0 vs 345; p=0.02) while other ratings on the valine day ("anxious", "difficulty concentrating", "hungry", "nauseated") did not significantly differ from placebo (all p > 0.2).

# 4. DISCUSSION

## 4.1 Valine and 5-HT-Mediated Prolactin Release

Fenfluramine increases plasma prolactin levels by facilitating presynaptic 5-HT release and thus indirectly activating post-synaptic 5-$HT_2$ receptors (see Cowen, 1993). The size of the prolactin response to d-fenfluramine therefore provides an index of the functional activity of brain 5-HT pathways). In animal studies, the increase in 5-HT release produced by d-fenfluramine and electrical stimulation of the dorsal raphe, is attenuated by a 60 minute pre-treatment with valine, suggesting that valine administration produces a rapid decrease of TRP availability for brain 5-HT synthesis, presumably by competition at the blood brain barrier (Gartside et al, 1992a; 1992b).

In normal subjects we found that valine lowered the prolactin response to d-fenfluramine which provides indirect evidence that valine may decrease brain 5-HT release in humans. Given by itself valine did not alter plasma prolactin levels suggesting that it is unlikely to produce a non-specific suppression of prolactin secretion. It did, however, tend to lower plasma concentrations of d-fenfluramine and d-norfenfluramine, perhaps by slowing their absorption. Nevertheless this effect did not seem to account for the ability of valine to lower d-fenfluramine-induced prolactin release because the significant effect of valine remained after the difference in drug levels had been partialled out with analysis of covariance.

## 4.2 Valine and Mood

In healthy subjects there was no effect of valine on ratings of sadness or on any of the other subjective measures. In the remitted depressed patients, however, there was evidence that valine produced a mild but significant lowering of mood on the various ratings used.

In four patients, the assessing clinicians could detect changes in mental state which bore some resemblance to the symptomatology of the original depressed state. However, the changes were modest and, with the possible exception of one case, clearly fell short of the full clinical relapse detected by Delgado et al (1990) using the TRP-free amino acid mixture. Our patients were probably less affected than those of Delgado et al (1990), because the technique of TRP depletion markedly lowers plasma TRP levels as well as attenuating the entry of TRP to the brain. Thus valine may simply produce a less potent attenuation of brain 5-HT function than TRP depletion. It is worth noting, however, that our subjects may have been less vulnerable to mood change than the patients of Delgado et al, (1990) many of whom had been treatment resistant before their eventual remission.

Our study with d-fenfluramine provides indirect evidence that valine may lower mood in remitted depressed patients by decreasing brain 5-HT neurotransmission. However, clearly other mechanisms such as an effect of valine to lower noradrenaline through impairment of brain tyrosine brain entry cannot be excluded (see Lehnert and Wurtman, 1993).

## 5. ACKNOWLEDGMENT

The study was supported by the Medical Research Council.

## 6. REFERENCES

Cowen, P.J., 1993, Serotonin receptor subtypes in depression: evidence from studies in neuroendocrine regulation, *Clin Neuropharmacol.* 16 (Suppl 3), S6-S18.

Delgado, P.L., Charney, D.S., Price, L.H., *et al,* 1990, Serotonin function and the mechanism of antidepressant action: reversal of antidepressant-induced remission by rapid depletion of plasma tryptophan, *Arch Gen Psychiat*, 47, 411.

Delgado, P.L., Price, L.H., Heninger, G.R., *et al,* 1992, Neurochemistry, *in* "Handbook of Affective Disorders," E.S. Paykel, Churchill Livingstone, Edinburgh.

Gartside, S.E., Cowen, P.J. & Sharp, T., 1992a, Evidence that the large neutral amino acid L-valine decreases electrically-evoked release of 5-HT in rat hippocampus in vivo. *Psychopharmacol*, 109, 251.

Gartside, S.E., Cowen, P.J. & Sharp, T., 1992b, Effect of amino acid loads on hippocampal 5-HT release *in vivo* evoked by electrical stimulation of the dorsal raphe nucleus and d-fenfluramine administration, *Brit J Pharmacol*, 107, 448P.

Joseph, G.A. and Kennett, G.A., 1983, Corticosteroid response to stress depends upon increased tryptophan availability, *Psychopharmacol*, 79, 79.

Lehnert, H. & Wurtman, R.J., 1993, Amino acid control of neurotransmitter synthesis and release: physiological and clinical implications, *Psychotherapeut Psychosomat*, 60, 18.

9

# TRYPTOPHAN AND 5-HT IN PSYCHIATRIC ILLNESS

P. L. Delgado, A. J. Gelenberg, F. Moreno, C. Laukes, and L. Strayer

Department of Psychiatry
University of Arizona College of Medicine
1501 N. Campbell Ave,Tucson, Arizona 85724

Brain serotonin (5-HT) content is dependent on plasma levels of tryptophan (TRP). Rapid, transient depletion of TRP in humans can be accomplished by oral administration of a TRP-free amino acid mixture. This procedure leads to a 30 to 50% decrease in brain 5-HT levels in laboratory animals and decreases human plasma TRP levels by 70-80% in 5 hours. TRP depletion alters human behaviour in different ways depending on the gender and clinical status of the subject. This presentation will review current experience with TRP depletion in healthy subjects, depressed patients, panic disorder patients, obsessive compulsive disorder (OCD) patients, and medication-free, psychiatrically-well subjects with a prior history of major depression. TRP depletion causes a transient return of anxiety and depressive symptoms in antidepressant-treated psychiatric patients, regardless of diagnosis. In contrast, TRP depletion causes minimal change in symptoms in healthy subjects or in actively ill, medication-free patients with OCD, panic disorder or major depression. These results suggest that antidepressant drugs treat depressive and anxiety symptoms by enhancing 5-HT neurotransmission but dysfunction of the 5-HT system may not be the immediate cause of OCD, panic disorder, or major depression.

*Recent Advances in Tryptophan Research*, edited by Graziella Allegri Filippini et al.
Plenum Press, New York, 1996

10

# TRYPTOPHAN METABOLISM AND DISPOSITION IN RELATION TO ALCOHOL AND ALCOHOLISM

A. A.-B. Badawy

Cardiff Community Healthcare NHS Trust
Biomedical Research Laboratory, Whitchurch Hospital
Cardiff CF4 7XB, Wales, United Kingdom

## SUMMARY

Tryptophan (Trp) metabolism and disposition in relation to alcohol and alcoholism are briefly reviewed. The changes observed could generally be classified into those: (1) exerted by acute or chronic alcohol administration and/or subsequent withdrawal; (2) already present in the absence of alcohol consumption, such as in naive alcohol-preferring animals or in abstinent alcoholics. In normal rats, acute ethanol administration activates liver Trp pyrrolase and exerts a biphasic effect on brain 5-HT (5-hydroxytryptamine or serotonin) synthesis, whereas chronic ethanol administration and subsequent withdrawal exert opposite effects on 5-HT synthesis mediated by corresponding changes in liver Trp pyrrolase activity. A cerebral 5-HT deficiency has been demonstrated in the alcohol-preferring C57BL mouse strain and in a number of alcohol-preferring rat lines, the mechanism of which is understood only in two models; the C57Bl mouse strain has a higher liver Trp pyrrolase activity and the P rat line from Indiana has a lower density of serotonergic fibres in cerebral cortex. In man, acute ethanol intake lowers circulating [Trp] and its availability to the brain, almost certainly by activating liver Trp pyrrolase. Some evidence exists for possible inhibition of pyrrolase activity in non-abstinent chronic alcoholics. Evidence in recently abstinent alcoholics suggests that Trp availability to the brain may be impaired and that this may be particularly so in patients with positive family history. Exploration of this latter possibility may be important in understanding the biological basis of predisposition to alcoholism.

## INTRODUCTION

Tryptophan (Trp) is the most extensively studied amino acid in relation to alcohol and alcoholism, because it is the precursor of the cerebral indolylamine 5-hydroxytryptamine (5-HT or serotonin). The study of serotonin metabolism itself in relation to alcohol and alcoholism has been motivated by the role of this indolylamine in the regulation of mood and certain other behaviours, disturbances of which are frequent clinical features of acute and chronic alcoholism and the alcohol-withdrawal states. In recent years, however, more effort in serotonin research on alcoholism has been directed towards identifying the role that this neuromodulator itself exerts in the regulation of alcohol intake. In the present short review, aspects of Trp metabolism and disposition related to this role of serotonin will be briefly reviewed. Because alcohol itself is capable of causing widespread disturbances in Trp and 5-HT metabolism, these effects will also be summarized. Emphasis in this review will be placed on Trp disposition aspects, rather than those directly involved in serotonin metabolism. Also, no attempt will be made here to describe the effects of ethanol on serotonin receptors and/or related function, nor clinical studies on the effects of modulators of this function on alcohol intake in man or experimental animals. Detailed accounts of all those aspects not covered in this short article can be seen in the excellent reviews by Sellers et al. (1992) and LeMarquand et al. (1994a,b).

## EFFECTS OF ALCOHOL ON TRYPTOPHAN METABOLISM AND DISPOSITION IN ANIMALS

These effects have been reviewed briefly (Badawy, 1988) and in greater detail more recently (LeMarquand et al., 1994b). They vary depending on many factors, the most important of which is whether ethanol is administered acutely or chronically and, in the latter case, whether studies are performed during chronic treatment or subsequent withdrawal. Other important factors include dose of ethanol, route of its administration, time interval of testing, animal species and/or strain used, nutritional status and method of assessment of parameters measured, e.g. 5-HT turnover rate. In general, however, in the most widely studied animal species, the rat, acute ethanol administration elicits a biphasic effect on cerebral 5-HT synthesis and turnover; an initial enhancement, followed by an inhibition (Badawy and Evans, 1976). The initial enhancement is caused by an increase in circulating free Trp availability to the brain secondarily to a non-esterified fatty acid-mediated catecholamine-dependent release of albumin-bound Trp, whereas the subsequent inhibition is due to a decrease in circulating Trp availability to the brain secondarily to a Trp-mediated activation of liver Trp pyrrolase. Changes in liver Trp pyrrolase activity also appear to mediate both the chronic and withdrawal effects of ethanol on brain 5-HT synthesis. Thus, inhibition of pyrrolase activity during chronic ethanol treatment and its enhancement during subsequent withdrawal lead to corresponding changes in circulating Trp availability to the brain and consequent enhancement and inhibition of 5-HT synthesis respectively (Badawy et al., 1979, 1980). From this account, it appears that liver Trp pyrrolase activity plays a major role in the effects of alcohol on cerebral 5-HT metabolism in the rat. As will be described below, there is also evidence that Trp pyrrolase may also be a target of ethanol's actions on Trp metabolism and disposition in man.

In our previous studies in rats, ethanol was chronically administered in their drinking water. Branchey and Lieber (1982) reported that chronic ethanol administration to rats by the liquid diet procedure actually enhances liver Trp pyrrolase activity. An explanation of the inability of these authors to demonstrate inhibition has been obtained in our recent studies using the liquid diet method of chronic ethanol administration. Previously, when we chronically administered ethanol in drinking water, we found that the enhancement of liver Trp pyrrolase activity during withdrawal, which follows the earlier inhibition, was of the hormonal type (almost certainly corticosterone-mediated) and reached a maximum (2.5-fold the basal enzyme activity) some 8 days after withdrawal. By contrast, a dramatic acceleration of this time-course of enhancement was observed upon withdrawal of ethanol-containing liquid diets. Thus we (Bano et al., 1996) found that pyrrolase activity rose significantly within 2h after withdrawal and reached a maximum at 7-8h. This enhancement was also accompanied by increased expression of Trp pyrrolase mRNA, as determined by the oligonucleotide solution hybridisation technique. The enhancement of pyrrolase activity in the study by Branchey and Lieber (1982) may therefore be explained by the rats having been withdrawn from ethanol, if their last intake of the liquid diet was only two or three hours before death. The increased conversion of a Trp load to kynurenine in these authors' study can be explained by the measurements having been made at 7h after withdrawal of the ethanol-containing liquid diets, i.e. at the peak of the pyrrolase activation. Another important difference between these two procedures of chronic ethanol administration is that, with the liquid diet, the pyrrolase enhancement during ethanol withdrawal is of the cofactor-type, whereas, as stated above, hormonal induction characterises that of the drinking water procedure.

In our studies using the liquid diet procedure, we were able to block the effects of alcohol withdrawal on liver Trp pyrrolase activity and other effects on Trp disposition by administration of the translational inhibitor cycloheximide during the withdrawal phase (Oretti et al., 1995). Concurrently, cycloheximide was also capable of blocking the alcohol-withdrawal-induced audiogenic seizures. This is the first demonstration of the ability of a translational inhibitor to block the alcohol-withdrawal syndrome in rats and strongly suggests that a rapidly-turning-over protein(s) mediates this syndrome. An interesting possibility worthy of investigation in this respect is that of activation of liver Trp pyrrolase during withdrawal increasing the availability to the brain of the excitotoxic Trp metabolite quinolinic acid, which exerts its effects in the brain by activating the N-methyl-D-aspartate (NMDA) receptors, which have also been implicated in the alcohol withdrawal syndrome. Further studies on the role of Trp metabolic disturbances in the alcohol-withdrawal syndrome in experimental animals may therefore throw light on the mechanism(s) of alcohol dependence at the behavioural, biochemical and molecular levels.

## TRYPTOPHAN METABOLISM AND DISPOSITION IN ALCOHOL-PREFERRING ANIMALS

A number of rat lines are known to prefer alcohol to water in the free choice drinking paradigm. In mice, the C57BL strain also prefers alcohol. This strain exhibits a central serotonin deficiency (Yoshimoto et al., 1985), as do the following alcohol-preferring rat lines: N/NIH (Murphy et al., 1986), P (Murphy et al., 1987) and Fawn-hooded rats (Rezvani et al., 1990 and references cited therein). The alcohol-preferring AA rat line from Finland has been suggested not to exhibit a serotonin deficiency (Korpi et al., 1988). Whether a central serotonin deficiency also exists in other alcohol-preferring rat lines (such as the Chilian UCha, the Sardinian SP and a few other lines) remains to be established.

The mechanism of the serotonin deficiency has been established to a reasonable degree in only two of the above animal models. In the case of the P rat line, the serotonin deficiency has been attributed to a lower density of serotonergic fibres in cerebral cortex (Zhou et al., 1991), whereas that of the C57BL mouse strain involves a decrease serotonin synthesis caused by a decrease in circulating Trp availability to the brain, secondarily to a higher liver Trp pyrrolase activity associated with a higher circulating corticosterone concentration (Badawy et al., 1989). In further unpublished experiments, we found that this mouse strain is more sensitive than the non-preferring CBA/Ca strain as regards inducibility of Trp pyrrolase by glucocorticoids. Work is currently in progress to establish whether expression of the Trp pyrrolase gene is different in these two mouse strains and, if not, what other factors determine the differences in Trp pyrrolase activity. We are also currently studying Trp metabolism and disposition in a number of alcohol-preferring rat lines. Another interesting question in relation to the serotonin deficiency in alcohol-preferring animals is that which line or strain could be considered the most suitable experimental model for human alcoholism? The answer to this question requires more work in humans first to establish a similarity with an animal model. Available evidence, however preliminary, suggests that the C57BL mouse strain may be such a model, because the serotonin deficiency of human alcoholism may also involve a decrease in circulating Trp availability to the brain (see the later sections).

## EFFECTS OF ALCOHOL ON TRYPTOPHAN METABOLISM AND DISPOSITION IN MAN

Less work has been done on the effects of ethanol on Trp metabolism and disposition in man. For ethical reasons, the effects of acute ethanol administration could only be studied in normal subjects or in non-abstinent alcoholics. In normal healthy volunteers, only a few studies have been performed. In two of these (Siegal et al., 1964; Pietraszek et al., 1991), no changes in plasma [Trp] were observed, whereas in the third (Eriksson et al., 1983), a decrease occurred in [Trp] and some, but not other, amino acids. In a recent study, we (Badawy et al., 1995) found that acute ethanol consumption decreased circulating free and total Trp concentrations by about 25%. Because of these proportionate decreases in free and total [Trp], Trp binding to albumin was not significantly altered. This suggests that the depletion of serum Trp by acute ethanol intake may be mediated by an enhancement of liver Trp pyrrolase activity. The mechanism of such a possible enhancement is unlikely to be cortisol- or Trp-mediated, because of the absence of an increase in circulating cortisol or Trp availability to the liver; it may therefore involve cofactor activation by haem, although further work is required to ascertain this point. Availability of circulating Trp to the brain, as determined by the Trp to its competitors ([Trp]/[CAA]) ratio, was also decreased after acute ethanol intake by normal subjects, because of the above decrease in [Trp], rather than a rise in the [CAA]. This effect is almost certain to lead to a decrease in cerebral 5-HT synthesis. These findings are the first demonstration of the ability of acute ethanol intake to lower circulating Trp availability to the brain in normal subjects, and may have important clinical implications in relation to alcohol-induced aggressive behaviour and also alcohol-induced depression. For example, it is possible that a greater depletion of circulating Trp may occur in susceptible individuals, which may trigger an aggressive episode or precipitate a depressive state. Further studies are therefore being planned to examine the response of depressed and/or aggressive subjects to an acute ethanol load. The response of non-abstinent chronic alcoholics to an acute dose of ethanol is also currently under investigation, both in relation to the above clinical aspects and also the effects of chronic ethanol intake itself on Trp metabolism and disposition.

Very little evidence also exists concerning the chronic and withdrawal effects of ethanol on Trp metabolism and disposition in man. Evidence for a possible defect in the hepatic kynurenine pathway of Trp degradation during chronic alcohol consumption was obtained by Walsh et al. (1966) when they demonstrated that urinary excretion of Trp metabolites of the above pathway following an oral Trp load was decreased in chronic alcoholics within 24h of cessation of alcohol consumption. Kynurenine production after a Trp load has more recently been demonstrated (Friedman et al., 1988) to be higher in chronic alcoholics one month after detoxification, in comparison with that at the time of admission, implying a possible higher liver Trp pyrrolase activity (or at least a higher rate of Trp oxidation in general) after detoxification. Because control data were not presented, it is not clear if these results also suggest that Trp oxidation is inhibited in chronic alcoholics before detoxification. These points require further investigation. The effects of acute alcohol withdrawal on Trp metabolism and disposition have also not received sufficient experimental attention. Buydens-Branchey et al. (1989) found that the [Trp]/[CAA] ratio was lowest 1 day after cessation of alcohol intake than subsequently. No detailed time-course examination of Trp metabolism and disposition during the acute alcohol withdrawal phase has been performed. Such a study is currently in progress in this Laboratory in relation to the alcohol-withdrawal syndrome. From this account, it is clear that much work is needed to ascertain the effects of acute and chronic ethanol consumption and subsequent withdrawal on various aspects of Trp metabolism and disposition in relation to the psychological and other behavioural aspects of alcohol dependence.

## THE TRYPTOPHAN AND SEROTONIN METABOLIC STATUS IN CHRONIC ALCOHOLISM

To establish the Trp and serotonin metabolic status in chronic alcoholism, studies must be performed in abstinent alcoholics. However, a major problem in interpretation of findings is that of the distinction between metabolic changes induced by previous long-term alcohol consumption and/or recent withdrawal and those which could be inherently present in chronic alcoholics, perhaps representing trait markers. Studies on Long-term abstinent alcoholics (i.e. abstinent for at least 6 months, but preferably longer than 1 or even 2 years) are therefore needed to address these issues. No such studies exist, as far as we could ascertain. Only data on Trp metabolism and disposition in short-term abstinent alcoholic patients are available. The findings, however, are encouraging and justify further detailed studies in longer-term abstinent populations.

As stated above, Friedman et al. (1988) found that serum kynurenine concentration following oral Trp loading was higher in 1-month abstinent alcoholics, than at admission, thus implying a higher rate of Trp oxidation, almost certainly via hepatic Trp pyrrolase, because there is no evidence at present to implicate increased indoleamine dioxygenase (IDO) or immune activation in abstinence. A more definite answer to these questions requires assessment of parameters of immune activation and IDO induction and also the use of controls to establish the Trp pyrrolase status before abstinence. That Trp pyrrolase activity may be inhibited by chronic ethanol consumption by chronic alcoholics (as is the case in the rat) is suggested from the urinary Trp metabolite data of Walsh et al. (1966) described above and also from additional data presented by Friedman et al. (1988), which showed that kynurenine production from an exogenous Trp load was again decreased in alcoholics who relapsed into drinking 3 months after abstinence, in comparison with those who continued to remain abstinent. Some additional supporting evidence is also provided from our current studies involving acute ethanol loading experiments in non-abstinent chronic alcoholics, in whom ethanol failed to decrease circulating [Trp] to the same extent as in normal subjects (see above).

Trp availability to the brain, as determined by the [Trp]/[CAA] ratio, has been shown (Branchey et al., 1984) to be significantly lower in recently abstinent (2-week) chronic alcoholics, than in control subjects. A further lowering of the ratio was observed if the alcoholics also exhibited depression and/or aggression. These results suggest that brain serotonin synthesis is likely to be decreased in recently abstinent alcoholics, in comparison with levels before detoxification. An overall serotonin deficiency in recently abstinent (5-week) chronic alcoholics is also suggested by the observed (Thomson and McMillen,

1987) decrease in the urinary ratios of concentrations of the major serotonin metabolite 5-HIAA (5-hydroxyindol-3-ylacetic acid) to metabolites of the tryptamine (indol-3-ylacetic acid) and kynurenine (anthranilic acid) pathways of Trp degradation. Extrapolation of urinary serotonin metabolite data to central serotonin metabolism is, however, unjustified. It is clear from the above accounts that disturbances in Trp metabolism and disposition leading to changes in central serotonin metabolism occur both during chronic alcohol intake and in short-term abstinence. However, more work is required to elucidate more fully the nature and mechanism(s) of these changes, and also to establish whether the changes observed during short-term abstinence are secondary to recent alcohol consumption and/or withdrawal, or represent long-term inherent (genetic) disturbances in alcoholism.

Few studies have addressed this latter question. In that by Buydens-Branchey et al. (1987) described above, it was also found that the decrease in Trp availability to the brain, as suggested by the lower [Trp]/[CAA] ratio, could be correlated with depressive and aggressive tendencies only in the group of alcoholics with early alcoholism onset [type 2 in the Cloninger et al. (1981) classification]. In a recent study just summarized (Morgan et al., 1995), details of which will be reported elsewhere, we found that chronic alcoholics with positive family history had a significantly lower serum total (free plus albumin-bound) [Trp], than those with negative family history. This difference was evident during the first week of detoxification and persisted during the 8-week detoxification study period. No significant differences between the two subgroups were observed with respect to free (ultrafiltrable) serum [Trp], Trp binding to albumin or serum [kynurenine]. The latter was, however, significantly higher at the 8-week detoxification time point, than during the first week, in the alcoholic group as a whole, implying an accelerated rate of Trp oxidation following detoxification, which lends support to the data obtained by Friedman et al. (1988) from their Trp loading experiments described above. In the study by Morgan et al. (1995), it was also found that the [Trp]/[CAA] ratio was significantly lower in alcoholics with positive family history, thus suggesting that Trp availability to the brain and hence serotonin synthesis are impaired in alcoholics with positive family history. This is therefore the first direct evidence in support of the serotonin deficiency hypothesis of predisposition to alcoholism. Further work is, however, required to confirm these findings in larger numbers of patients and across different populations, not only in short-term, but also in long-term, abstinence, in order to establish the Trp and serotonin metabolic status of chronic alcoholics in relation to family history.

## CONCLUSIONS AND COMMENTS

The present short review has shown clearly that the area of Trp metabolism and disposition continues to be a focus of interest in relation to alcohol and alcoholism research. Much work is required to address several important issues concerning the role of serotonin and related aspects of Trp metabolism and disposition in alcohol dependence. The recent preliminary evidence obtained from this Laboratory demonstrating a decrease in circulating tryptophan availability to the brain in alcoholics with positive family history is the first definite evidence in support of the serotonin deficiency hypothesis of predisposition to alcoholism, and merits further investigation.

## ACKNOWLEDGMENTS

Reviewed work from this Laboratory has been done in collaboration with many colleagues, to whom the author is indebted, and with the support of several organizations, notably the employing authority, the Welsh Office, the British Medical Research Council and the British Medical Council on Alcoholism.

## REFERENCES

Badawy, A. A.-B., 1988, Effects of alcohol on tryptophan metabolism, *Biochem. Soc. Trans.* 16: 254.

Badawy, A. A.-B., and Evans, M., 1976, The role of free serum tryptophan in the biphasic effect of acute ethanol administration on the cencentrations of rat brain tryptophan, 5-hydroxytryptamine and 5-hydroxyindol-3-ylacetic acid, *Biochem. J.* 160: 315.

Badawy, A. A.-B., Morgan, C. J., Lane, J., Dhaliwal, K., and Bradley, D. M., 1989, Liver tryptophan pyrrolase: a major determinant of the lower brain 5-hydroxytryptamine concentration in alcohol-preferring C57BL mice, *Biochem. J.* 264: 597.

Badawy, A. A.-B., Morgan, C. J., Lovett, J. W. T., Bradley, D. M., and Thomas, R., 1995, Decrease in circulating tryptophan availability to the brain after acute ethanol consumption by normal volunteers: implications for alcohol- induced aggressive behaviour and depression, *Pharmacopsychiat.* 28 (in press).

Badawy, A. A.-B., Punjani, N. F., and Evans, M., 1979, Enhancement of rat brain tryptophan metabolism by chronic ethanol administration and possible involvement of decreased liver tryptophan pyrrolase activity, *Biochem J.* 178:575.

Badawy, A. A.-B., Punjani, N. F., Evans, C. M., and Evans M., 1980, Inhibition of rat brain tryptophan metabolism by ethanol withdrawal and possible involvement of the enhanced liver tryptophan pyrrolase activity, *Biochem. J.* 192:449.

Bano S., Oretti R.G., Morgan C.J., Badawy A.A.-B., Bickland P., and McGuffin P., 1996, Effects of chronic ethanol administration by the liquid diet procedure on rat liver tryptophan pyrrolase and tryptophan metabolism, *Alcohol Alcohol.* 31: in press.

Branchey L., Branchey M., Shaw S., and Lieber C.S., 1984, Depression, suicide, and aggression in alcoholics and their relationship to plasma amino acids, *Psychiatry Res.* 12:219.

Branchey L., and Lieber C.S., 1982, Activity of tryptophan pyrrolase after chronic alcohol administration, *Substance Alcohol Actions/Misuse* 2:225.

Buydens-Branchey L., Branchey M.H., Noumair D., and Lieber C.S., 1989, Age of alcoholism onset: II. relationship to susceptibility to serotonin precursor availability, *Arch. Gen. Psychiat.* 46: 231.

Cloninger, C. R., Bohman, M., and Sigvardsson, S., 1981, Inheritance of alcohol abuse: cross-fostering analysis of adopted men, *Arch. Gen. Psychiat.* 38: 861.

Eriksson, T., Magnusson, T., Carlsson, A., Hagman, M.,and Jagenburg, R., 1983, Decrease in plasma amino acids in man after an acute dose of ethanol, *J Stud. Alcohol* 44: 215.

Friedman, M. J., Krstulovic, A. M., Severighaus, J. M., and Brown, S. J., 1988, Altered conversion of tryptophan to kynurenine in newly abstinent alcoholics, *Biol. Psychiat.* 23: 89.

Korpi, E. R., Sinclair, J. D., Kaheinen, P., Viitamaa, T., Hellevuo, K., and Kiianmaa, K., 1988, Brain regional and adrenal monoamine concentrations and behavioral responses to stress in alcohol-preferring AA and alcohol-avoiding ANA rats, *Alcohol.* 5: 417.

LeMarquand, D., Pihl, R. O., and Benkelfat, C., 1994a, Serotonin and alcohol intake, abuse, and dependence: clinical evidence, *Biol. Psychiat.* 36: 326.

LeMarquand, D., Pihl, R. O., and Benkelfat, C., 1994b, Serotonin and alcohol intake, abuse, and dependence: findings of animal studies, *Biol. Psychiat.* 36: 395.

Morgan, C. J., Badawy, A. A.-B., Den Dulk, A., Verbanck, P., and Pelc, I., 1995, Lower serum tryptophan concentrations in alcoholics with positive family history, *Alcohol Alcohol.* 30: 554 (Abstract).

Murphy, J. M., McBride, W. J., Lumeng, L., and Li, T.-K., 1986, Alcohol preference and regional brain monoamine contents of N/NIH heterogeneous stock rats, *Alcohol Drug Res.* 7: 33.

Murphy, J. M., McBride, W. J., Lumeng, L., and Li, T.-K., 1987, Contents of monoamines in forebrain regions of alcohol-preferring (P) and -nonpreferring (NP) lines of rats, *Pharmacol. Biochem. Behav.* 26: 389.

Oretti, R. G., Bano, S., Morgan, C. J., Badawy, A. A.-B., Buckland, P., and McGuffin, P., 1995, Prevention of alcohol withdrawal seizures and associated tryptophan metabolic disturbances in rats by cycloheximide, *Alcohol Alcohol.* 30: 539.

Pietraszek, M. H., Urano, T., Sumioshi, K., Serizawa, K., Takahashi, S., Takada, Y., and Takada, A., 1991, Alcohol-induced depression: involvement of serotonin, *Alcohol Alcohol.* 26: 155.

Rezvani, A. H., Overstreet, D. H., and Janowsky, D. S., 1990, Genetic serotonin deficiency and alcohol preference in the Fawn Hooded rats, *Alcohol Alcohol.* 25: 573.

Sellers, E. M., Higgins, G. A., and Sobell, M. B., 1992, 5-HT and alcohol abuse, *Trends Pharmacol. Sci.* 13: 69.

Siegal, F. L., Roach, M. K., and Pomeroy, L. R., 1964, Plasma amino acid patterns in alcoholism: the effects of ethanol loading, *Proc. Nat. Acad. Sci. USA* 51: 605.

Thomson, S. M., Jr., and McMillen, B. A., 1987, Test for decreased serotonin/tryptophan metabolite ratios in abstinent alcoholics, *Alcohol* 4: 1.

Walsh, M. P., Howarth, P. J. N., and Marks, V., 1966, Pyridoxine deficiency and tryptophan metabolism in chronic alcoholics, *Am. J. Clin. Nutr.* 19: 379.

Yoshimoto, K., Komura, S., Kano, A., and Mizohata, K., 1985, Alcohol preference and brain monoamines in five inbred strains of mice. *IRCS Med. Sci.* 13: 1192.

Zhou, F. C., Bledsoe, S., Lumeng, L., and Li, T.-K., 1991, Immunostained serotonergic fibers are decreased in selected brain regions of alcohol- preferring rats, *Alcohol* 8: 425.

# 11

# ENDOGENOUS KYNURENINE DERIVATIVES AND PAIN

J. Weil-Fugazza

Unité de Recherches de Physiopharmacologie du Système Nerveux
INSERM U 161
2 Rue d' Alésia 75014, Paris, France

## INTRODUCTION

Early behavioural and neurochemical studies suggested the involvement of tryptophan (TRP) in the mechanisms of analgesia. These studies were followed by an impressive number of experiments on the possible involvement of serotonin (5-HT) in this phenomenom. However the physiological role of the serotonin system(s) in pain and analgesia remained to be clarified. During the last decade the role of the excitatory amino acids in the transmisson of nociceptive messages at spinal level was clearly established. During the same period, parallel studies suggested that the second route of trytophan metabolism e.g the kynurenine pathway, is involved in the control of neuronal activity. Considered simultaneously these different studies lead to the idea that metabolites of kynurenine e.g kynurenic acid (KYN) and quinolinic acid (QUIN), endogenous ligands of excitatory amino acids receptors, might play a role in analgesia and nociception. This paper discusses some recent data related to the possible respective roles of 5-HT and kynurenine derivatives in pain mechanisms at the spinal level.

## SEROTONIN, PAIN AND ANALGESIA

It was shown in the seventies and eighties that morphine induces an increase in both TRP levels and 5-HT synthesis in several areas of the brain and spinal cord (Samamin et al., 1972; Messing et al., 1978; Weil-Fugazza et al., 1979). This effect is antagonized by the mu opioid receptor antagonist naloxone. It was thus suggested that morphine analgesia is partly related to an activation of the serotonergic system(s). An increase in TRP incorporation, leading to an increase in 5-HT synthesis, was also related to stress-induced analgesia. Inversely, it was proposed that hyperalgesia observed when TRP availibility is decreased in CNS was related to a decrease in 5-HT synthesis. Subsequently electrophysiological, pharmacological, anatomical and behavioral studies (see references in Le Bars, 1987) have suggested a key role for the bulbospinal serotonergic system which innervate the spinal cord.

*Recent Advances in Tryptophan Research*, edited by Graziella Allegri Filippini et al.
Plenum Press, New York, 1996

It was proposed that serotonergic fibers arising from the nucleus raphé magnus modulate the activity of the nociceptive neurons located in the dorsal horn of the spinal cord and that morphine activates this descending system. However recent studies based on sensitive biochemical methods allowing measurements of the monoamines and their metabolites in the spinal areas specifically involved in nociceptive transmission (laminae I-II and V of the dorsal horn) failed to demonstrate a relationship between an increase in 5-HT release in these areas and morphine analgesia. For instance, a study based on the microdiayalisis technique in freely moving rats, indicates that analgesic doses of morphine either induced an increase in the release of 5-HT or did not modify this release. In the same way Chiang and Xiang (1992), using the differential pulse voltametric method, did not observe in the rat a significant increase in the release of 5-HT in the dorsal horn of the cord after morphine administration. This does not preclude the possibilty that morphine increases 5-HT synthesis in other areas of the CNS. For example it has been observed, with HPLC-electrochemical detection, that an analgesic dose of morphine which does not modify 5-HT metabolism in laminae I-II and III-V of the dorsal horn, increases the 5-HIIAA/5-HT ratio in the motoneurone area of the ventral spinal cord (Weil-Fugazza, 1992). This indicates not only that some of the side effects of morphine might be under serotonergic control, but also that the role of 5-HT in analgesia must be reconsidered. In addition it is now established, with specific 5-HT receptor ligands, that 5-HT may exert excitatory effects on the spinal nociceptive neurons. Finally the nociceptive effect of 5-HT via 5-HT3 receptors at the periphery is also demonstrated (See references in Milan, 1994; Cesselin et al., 1994). All these observations indicate that the involvement of TRP in analgesia cannot be clearly related to modifications of the serotonergic metabolic pathway.

## KYNURENINE DERIVATIVES, PAIN AND ANALGESIA

The role of excitatory amino acids in nociception is now well documented. Together, with neuropeptides such as substance P, glutamate and aspartate are transmitters of the small primary afferents (Aδ and C) that relay nociceptive information from the periphery to nociceptive neurons of the dorsal horn in the spinal cord (see references in Wilcox, 1991; Ault, 1993; Tohyama et al., 1994). The potential utility of excitatory amino acid inhibitors as analgesics is suggested by pharmacological studies (see references in Coderre, 1994). Interestingly numerous studies strongly support the hypothesis of the involvement of the kynurenine pathway in the modulation of the activity of the excitatory amino acids transmitters in the CNS (see references in Swartz, K.J. et al., 1990; Moroni et al.,1991; Stone and Connick, 1991). QUIN and KYN are excitatory amino acid receptor ligands in the brain, respectively agonist and antagonist at the N-methyl-D-aspartate (NMDA) receptor. The putative role of modifications of the kynurenine metabolic pathway in neurological dysfunctions is supported by different studies (see references in Heyes et al., 1992). The following paragraph summarizes data suggesting an involvement of the kynurenic metabolic pathway in analgesia and nociception, at least at the spinal level.

### Possible Modulatory Role of Endogenous Kynurenic Acid on the Activity of the Spinal Dorsal Horn Nociceptive Neurons

Gannong et al. (1983) reported that KYN inhibits synaptic and amino acid induced responses of the rat hippocampus and spinal cord neurons, in addition Elmslie and Yoshimaki (1985) have shown that KYN blocks the synaptically mediated ventral root potential produced by the stimulation of the dorsal root. In the same way, Jahr and Jessel (1984) shown

**Table 1.** Tryptophan metabolite levels (in pmoles/mg protein) in the superficial layers of the dorsal horn of the rat lumbar spinal cord

| | |
|---|---|
| Tryptophan | 253.00 ± 25 |
| 5-Hydroxytryptophan | 1.02 ± 0.04 |
| Serotonin | 45.23 ± 2.3 |
| 5-HIAA | 17.6 ± 1.1 |
| Kynurenine | 4.12 ± 0.4 |
| 3-OH-kynurenine | traces |

that KYN antagonizes excitatory postsynaptic potentials, and glutamate excitation of rat spinal dorsal horn neurons in culture. More recently Ault and Hidebrand (1993), using the rat isolated spinal cord-tail preparation, have shown that KYN inhibits the nociceptive reflexes elicited by chemical excitation of unmyelinated peripheral sensory fibers. Antinociceptive effects of KYN was also reported by several groups (see references in Coderre and Van Empel, 1994). We have identified (Godefroy et al., 1991) kynurenine and 3-hydroxykynurenine in the superficial layers of the dorsal horn in the spinal cord (Table1) where C fibers terminate.

All these observations suggest that *endogenous kynurenic acid*, taken up and putatively synthetized in the spinal cord might play a modulatory role on the activity of spinal nociceptives neurons. From a pharmacological point of view a role for endogenous KYN in the effects of some analgesics was proposed. For instance it was suggested that the central analgesic properties of non-steroidal antii-inflammtory drugs (NSAIDs) might be, in part, related to their ability to modify tryptophan 2,3-dioxygenase activity and to increase KYN formation (see references in McKormack, 1994). This hypothesis must also be considered for the central analgesic activity of aspirin, originally related to its effect on 5-HT synthesis (Groppeti et al., 1998). It has long been known that salicylate increases free TRP level in blood (Tagliamonte et al., 1973)). Thus an increase in kynurenine metabolic pathway leading to an increase in KYN formation might be involved in the central effect of aspirin. Since morphine can also increase the levels of free TRP in the plasma (Weil-Fugazza et al., 1981), this opiate might also induce increases in kynurenic acid levels at both periphery and CNS.

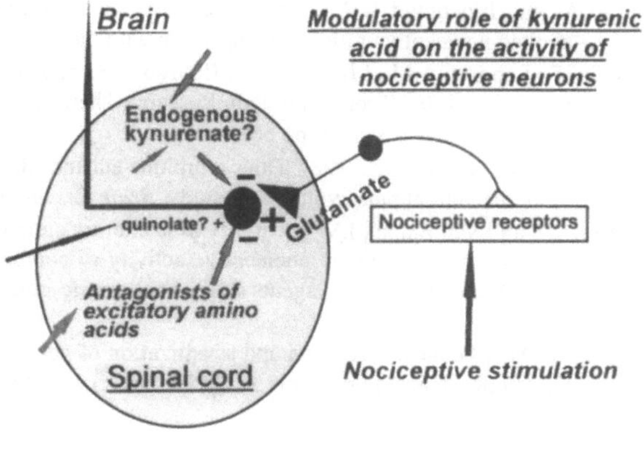


**Figure 1.** Role of excitatory amino acids in the transmission of nociceptive mesages from periphery to the spinal cord. Inhibitory effects of excitatory amino acid antagonists at the NMDA receptor including the putative role of endogenous kynurenic acid.

## Possible Role of Endogenous Quinolinic Acid in Inflammatory Pain

Animal studies have shown that stimulation of the immune system increases QUIN formation. In human, during diseases associated with endogenous or exogenous stimulation of the immune system, QUIN and KYN are taken up or synthetized in no-negligible amounts in the CNS and increased QUIN and KYN levels are observed in CSF (Heyes, 1992). For instance at the early stages of the infection with the human immunodeficency virus moderate increases in QUIN levels and normal KYN levels were observed. Following more prolonged immune activation, QUIN levels were markedly increased. KYN levels were also significantly increased but not to the same extend as the QUIN levels. Thus the increased ratio QUIN/KYN might favor both an excitatory effect, or the neurotoxicity of QUIN, in inflammatory neurological diseases. To this date no study has investigated the possible modifications of spinal KYN and QUIN in animal models of peripheral inflammation. However one can postulate an increase in QUIN levels elicited during peripheral inflammation: e.g rhumatoid diseases, burn, tissue injury, etc. This excitatory compound might participate, together with other factors, such as cytokynines, bradykinin, nitric oxide, serotonin, etc (Dray, 1994), to the activation of nociceptors at the periphery. QUIN might also be accumulated in the spinal cord therefore activating spinal nociceptives neurons. Finally spinal neuronal dysfunction and sensitization observed during peripheral inflammation, or after peripheral nerve injuries, might also be related to the neurotoxicity of QUIN.

# IN CONCLUSION

Kynurenine derivatives may play an important role in the mechanisms of the transmission of nociceptive messages at the spinal level. If this assertion is correct, then pharmacological studies of drugs modulating the kynurenine metabolism initiated for the treatment of neurological dysfunctions, could also be useful for pain therapy.

# ACKNOWLEDGMENTS

This work was supported by INSERM France and a grant from Bristol-Myers Squibb Foundation.

# REFERENCES

Ault, B.A. and Hildebrand, L. M., 1993, Effects of excitatory amino acid receptor antagonists on a capsaicin-evoked nociceptive reflex: a comparison with morphine, clonidine and baclofen, *Pain,* 52: 341.

Cesselin, F., Laporte, A.M., Miquel, M.C., Bourgoin, S., and Hamon, M., 1994, Serotonergic mechanisms of pain control, *in*: "Progress in Pain Research Therapy", G.F. Gebhart, L., Hammond and T.S. Jensen, edts, IASP Press, Seattle. pp 669.

Chiang, C.Y and Xiang, X.K., 1987, Does morphine enhance the release of 5-HT in the rat spinal cord? An in vivo differential pulse voltametry study, *Brain Research.*, 411: 259.

Coderre, T.J. and Van Empel, I.V. The utility of excitatory amino acid (EAA) antagonists as analgesic agents. II. Assesment of the antinociceptive activity of combinations of competitive and non competitive NMDA antagonists with agents acting at allosteric-glycine and polyamine receptor sites, *Pain.* 1994, 59: 353.

Dray, A., 1994, Chemical activation and sensitization of nociceptors, *in*: Peripheral neurons in nociception: physio-pharmacological aspects, J.M Besson, G. Guilbaut, H. Ollat, edts, John Libbey Eurotext, Paris. pp 49-70.

Elmslie, K.S. and Yoshikami, D., 1985, Effect of kynurenate on root potentials evoked by synaptic activity and amino acids in the frog spinal cord, *Brain Research*. 330: 265.

Gannong, A.H., Lanthorn, T.H. and Cotman, C. W., 1983, Kynurenic acid inhibits the synaptic and amino acid-induced responses in the rat hippocampus and spinal cord, *Brain Research* , 273: 170.

Godefroy, F., Matson, W.R., Gamache, P.H. and Weil-Fugazza, J., 1990, Simultaneous measurements of tryptophan and its metabolites, kynurenine and serotonin, in the superficial layers of the spinal dorsal horn. A study in normal and arthritic rats, *Brain Research,* 526: 169.

Groppeti, A., Braga, P.C., Biella, G, Parenti, M, Rusconi, L and Mantegazza, P., 1988, Effect of aspirin on serotonin and met-enkephalin in brain: correlation with the anyinociceptive activity of the drug, *Neuropharmacology*, 27: 499.

Heyes, M.P., Saito, K., Crowley, J.S., Davis, L.E., Demitrac, M.A.,Der, M., Dilling, A., Elia, J., Kruesi, M.J.P. et al., 1992, Quinolinic acid and kynurenine pathway metabolism in inflammatory and non-inflammatory neurological disease, *Brain*, 115: 1249.

Jahr, C.E. and Jessel, T.M.,1985, Synaptic transmission between dorsal root ganglion and spinal dorsal horn in cultures: antagonist of monosynaptic excitatory post synaptic potentials and glutamate excitation by kynurenate, *J. Neurosci.* 5: 2281.

Le Bars, D., 1988, Serotonin and pain, *in* : "Neuronal serotonin" , N.N. Osborne and M. Hamon, edts, John Wiley & Sons Ltd. pp: 171-119.

McKormack., K.M., 1994, Non-steroidal anti-inflammatory drugs and spinal nociceptive processing., *Pain*, 59: 9.

Matos, F.F., Rollema, H., Brown, J.L. and Basbaum, A.I.,1992, Do opioids evoke the release of serotonin in the spinal cord? An in vivo microdyalisis study on the regulation of the extracellular serotonin in the rat, *Pain*, 48: 439;.

Messing, R.B., Flincchbaugh, C. and Waymire. J.C., 1978, Tryptophan and 5-hydroxyindoles in different CNS regions fpllowing acute morphine, *Eur. J. Pharmacol.*, 48: 137.

Milan, M.J.,1994, Serotonin and Pain: evidence that activation of 5-$HT_{1A}$ receptors does not elicit antinociception against noxious thermal, mechanical and chemical stimuli in mice, *Pain*, 58: 45.

Moroni, F., Russi, P., Gallo-Mezo, M.A. Moneti, G. and Pellicciari, R., 1991, Modulation of quinolinic and kynurenic acid content in the rat brain: effects of endotoxins and nicotinylalanine, *J. Neurochem.*, 57: 1630.

Saito, K., Nowak, T.S., Suyama, K, Quearry, B.J., Saito, M., Crowley, J.S., Markey, P.S., and M.P., Heyes., 1993, Kynurenine pathway enzymzs in brain: responses to ischemic brain injury versus systemic immune activation, *Journal of Neurochemistry*, 61: 2061.

Samamin, R., Miranda, F. and Mennini, T., 1972, Serotonergic mechanismsof naecotic actin, *in* "Factors affecting the actions of narcotics", M.L. Adler, L. Manara and R. Samamin, edts., Raven Press New York. pp. 523-539.

Stone, T.W. and Connick, J.H. Effects of quinolinic and kynurenic acids on central neurons. 1991 *in*: "Kynurenine and serotonin pathways", R. Schwartz, S.N. Youg and R. Brown, edts, Plenum Press, New York and london, pp. 329-336.

Swartz, K. J., During, M.T., Freese, A, and Beal, M. F., 1990, Cerbral synthesis and release of kynurenic acid: an endogenous antagonist of excitatory amino acid receptors, *J. of Neuroscience,* 10: 2965.

Tagliamonte, A., Biggio, G., Vargiu, L. and Gessa, G.L., 1973, Increase of brain tryptophan and stimulation of serotonin synthesis by salicylate, *J. Neurochem.*, 20: 909.

Tohyama, M., Kitama, H., Satoh, K, Araki, T, Maeno, H, Oyamada, H. and Kondo, E. 1994, Glutamergic neurotransmission, proto-oncogenes, and pain. in: "Progress in Pain Research Therapy", G.F. Gebhart, D.L., Hammond and T.S. Jensen, edts, IASP Press, Seattle. pp: 395-410.

Weil-Fugazza, J., Godefroy, F. and Besson, J.M., 1979, Changes in brain and spinal tryptophan and 5-hydroxyindoleacetic acid levels following acute morphine administration in normal and arthritic rats, *Brain Res.*,175: 291.

Weil-Fugazza, J., Godefroy, F., Coudert. D. and Besson, J.M., 1981, Total and free tryptophan levels in serum following acute morphine administration in arthritic rats, *Neurochem. Intern.*, 3: 323.

Weil-Fugazza, J. and Godefroy.F., 1993, Dorsal and ventral dopaminergic innervation of the spinal cord: functional implications, *Brain Res. Bull.*, 30: 319.

Wilcox, G.L., 1991, Excitatory neurotransmitters in pain, *in* "Pain research and clinical management" M.R. Bond and J.E. Charlton, edts., Elsevier, Amsterdam. pp. 97-117.

# 12

# SEXUAL BEHAVIOR IN MALE AND FEMALE RATS AFTER A TRYPTOPHAN-FREE AMINO ACID MIXTURE

E. A. Moja and F. Benedetti

Medical Psychology
Medical School University of Milano
via F. Sforza 23, 20122 Milano, Italy

## 1. INTRODUCTION

While the decrease in brain serotonin (5-HT) is well known to enhance sexual activity in the male rat (e.g., Tagliamonte et al., 1969), the administration of antiserotonergic agents to female rats caused contradictory results (e.g., Segal and Whalen, 1970; Zemlan et al., 1973; Emery and Moss, 1984). Previous results by our laboratory (Moja et al., 1989) showed that the acute administration of an amino acid mixture lacking tryptophan (TRY) causes a 40% depletion in whole brain 5-HT. This is because the TRY-free amino acid mixture will induce new protein synthesis, and endogenous blood and brain TRY is incorporated into proteins (Moja et al., 1991).

Dietary manipulations of TRY were shown to cause in naive rats and rabbits a marked increase of homosexual mounting behavior (Carruba et al., 1977; Fratta et al., 1977). In the present experiments, we studied the effects of the acute administration of a TRY-free amino acid mixture on heterosexual drive in sexually experienced male and female rats.

## 2. METHOD

### 2.1 Animals and Treatment

Since the aphrodisiac effect of 5-HT depletion in male rats has been shown to depend on the basal sexual activity of the animals (e.g., Tagliamonte et al., 1971; Ginton, 1975), in a first experiment we studied two groups of six sexually experienced male rats which showed a high or a low basal level of spontaneous sexual behavior. In a second experiment (Benedetti and Moja, 1993) we studied six sexually experienced female rats, which were ovariectomized and then weekly injected with 0.5 mg estradiol valerate. In agreement with literature data (e.g., Hlinak and Madlafousek, 1981), this hormonal treatment elicited complete proceptive and receptive sexual behavior and precluded its decline following chronic treatment.

*Recent Advances in Tryptophan Research*, edited by Graziella Allegri Filippini et al.
Plenum Press, New York, 1996

Two hours before the experimental sessions, rats were treated per gavage with either an amino acid mixture lacking TRY or a control mixture, given at a dose of 1.32 g/kg (see Moja et al., 1989, for details on composition).

## 2.2 Testing Procedures

Sexual motivation was evaluated with an operant method. A Plexiglas cage was divided into three compartments by two removable partitions: the experimental animal was put in the central compartment, while each of the other two contained a sexually vigorous rat of the opposite sex. During three training sessions, the experimental animals were trained to remove the partition by operating one of two levers which protruded in the cage. Animals remained together until copulation occurred, then the partition was restored.

Sexual contacts were recorded either as mounts, intromissions, or ejaculations, and mean contact response latencies (defined as the time between the sexual contact and the successive press on the lever) were recorded as an index of sexual motivation.

# 3. RESULTS

Contact-response latencies recorded in male and female rats are summarized in Table 1. Inspection of the results shows that both in male and female rats the latencies to press the lever after ejaculation were longer than those after intromission, which, in turn, exceeded those after mount.

TRY-free amino acid mixtures significantly (Wilcoxon matched pairs, TRY-free vs control) enhanced sexual motivation in male rats with low basal sexual activity levels, while failed to modify sexual behavior in male rats with high activity levels and in female rats.

# 4. DISCUSSION

Our results confirm the importance of the initial state of the male rat in determining the response to 5-HT depletion: only the animals with low basal activity levels showed an increase in sexual drive following the administration of the TRY-free mixture.

In females, our data confirm the hypothesis that the role of 5-HT in the control of sexual behavior cannot be reduced to simple tonic inhibition, suggesting a possible inter sexual difference in the control of sexual motivation.

Alternatively, males with high basal activity levels and females after high doses of estradiol treatment could have reached a maximal activation which precluded any further increase due to 5-HT depletion.

**Table 1**

| | Males with low basal activity | | Males with high basal activity | | Females | |
|---|---|---|---|---|---|---|
| Latencies after | Control | TRY-free | Control | TRY-free | Control | TRY-free |
| Mount | 34.1±29.6 | 30.3±13.5 | 15.8±7.0 | 14.5±15.0 | 12.3±3.2 | 9.7±3.8 |
| Intromission | 53.9±23.4 | 30.8±19.8 | 19.2±5.9 | 24.6±19.3 | 22.8±14 | 21.3±6.8 |
| Ejaculation | 239.3±110.2 | 167.5±76.7 | 133.2±42.3 | 169.8±115.2 | 44.8±23.2 | 38.1±4.9 |

All latencies are in seconds. Values are means ± standard deviations.

By the way, it should be noted that the similarity in operant responding for a sexual reinforcer between male and female rats suggests that our method could be a useful tool in comparing sexual motivation between the two sexes.

## REFERENCES

Benedetti, F., and Moja, E.A., 1993 Failure of a tryptophan-free amino acid mixture to modify sexual behavior in the female rat, *Physiol. Behav.* 54:1235.

Carruba M.O., Picotti, G.B., Genovese, E., and Mantegazza, P., 1977. Stimulatory effect of a maize diet on sexual behavior of male rats. *Life Sci.* 20:159.

Emery, D.E., and Moss, R.L., 1984. P-cholorophenylalanine alters pacing of opulation in female rats. *Pharmacol. Biochem. Behav.* 20:337.

Fratta, W., Biggio, G., and Gessa, G.L., 1977. Homosexual mounting behavior induced in male rats and rabbits by a tryptophan-free diet. *Life Sci.* 21:379.

Ginton, A., 1975. Copulation in noncopulators: effect of PCPA in male rats. *Pharmacol. Biochem. Behav.* 4:357.

Hlinak, Z., and Madlafousek, J., 1981. Estradiol treatment and precopulatory behavior in ovariectomized female rats. *Physiol. Behav.* 26:171.

Moja, E.A., Cipolla, P., Castoldi, D., and Tofanetti, O., 1989. Dose-response decrease in plasma tryptophan and in brain tryptophan and serotonin after tryptophan-free amino acid mixtures in rats. *Life Sci.* 44:971.

Moja, E.A., Restani, P., Corsini, E., Stacchezzini, M.C., Assereto, R., and Galli, C.L., 1991. Cycloheximide blocks the fall of plasma and tissue tryptophan levels after tryptophan-free amino acid mixtures. *Life Sci.* 49:1121.

Segal, D.S, and Whalen, R.E., 1970. Effect of chronic administration of p-chlorophenylalanine on sexual receptivity of the female rat. *Psychopharmacologia* 16:434.

Tagliamonte, A., Tagliamonte, E., Gessa, G.L., and Brodie, B.B., 1969. Compulsive sexual activity induced by p-chlorophenylalanine in normal and pinealectomized male rats. *Science* 166:1433

Tagliamonte, A., Tagliamonte, P., and Gessa, G.L., 1971. Reversal of pargyline-induced inhibition of sexual behavior in male rats, by parachlorophenylalanine. *Nature* 230:244.

Zemlan, F.P., Ward, I.L., Crowley, W.R., and Margules, D.L., 1973. Activation of lordotic responses in female rats by suppression of serotonergic activity. *Science* 179:1010.

# 13

# DELIRIUM AFTER CARDIAC SURGERY

## The Possible Role of Tryptophan in Relation to the Other Neutral Aminoacids

R. C. van der Mast,[1] W. W. van den Broek,[1] D. Fekkes,[1]
L. Pepplinkhuizen,[1] and F. H. J. Roest[2]

[1] Department of Psychiatry
University Hospital-Dijkzigt, The Netherlands
[2] Center for Clinical Decision Sciences
Department of Public Health
Erasmus University Rotterdam, The Netherlands

## 1. INTRODUCTION

Delirium is a transient and fluctuating organic mental syndrome of acute onset, characterized by a global impairment of cognitive functions, a reduced level of consciousness, attentional abnormalities, increased or decreased psychomotor activity and a disordered sleep-wake cycle (Lipowski, 1990). Cardiac surgery with cardiopulmonary bypass (CPB) has traditionally been associated with a particularly high rate of postoperative delirium (Smith and Dimsdale, 1989), supposed to be due to cytotoxic oedema induced by cerebral microemboli, hypoperfusion or haemodilution (Editorial, 1989). This psychiatric complication is associated with increased morbidity, high mortality, prolonged hospital stay and poor functional recovery (van der Mast, 1994). Little is known about possible predictive factors and mediating pathophysiological mechanisms. Following on earlier research suggesting a role of reduced plasma tryptophan in postcardiotomy delirium (van der Mast, 1991), we prospectively examined the contribution of disturbances in the plasma concentrations of aminoacids, in relation to a catabolic state, among patients undergoing elective cardiac surgery.

## 2. PATIENTS AND METHODS

### 2.1 Patients

The final study sample consisted of 296 patients admitted between may 1991 and november 1992 to the Thoraxcentre of the University Hospital Rotterdam (Table 1).

**Table 1.** Age and sex of the overall population (n=296)

| | |
|---|---|
| Mean age (SD) | 63 (11) |
| Age range (years) | 26–83 |
| Sex | |
| Male | 192 (65%) |
| Female | 104 (35%) |

## 2.2 Preoperative Assessment

Patients were assessed, one to four weeks before surgery, according to a standardized psychiatric and psychometric evaluation, including somatic and psychiatric history, demographic and medication data and the following (self-rating) scales: the Mini Mental State Examination (MMSE) as a cognitive screening test (van der Mast, 1994), the 30-item version (being the least dependent on somatic items) of the General Health Questionnaire (GHQ) to screen for possible psychiatric disturbances (van der Mast, 1994), the Münchner Alcohol Test to detect possible alcohol abuse (van der Mast, 1994), and the Daily Activity List to measure functional status (van der Mast, 1994).

Besides, on the day before surgery patients were asked to fill in the Profile of Mood States (van der Mast, 1994). Just before induction of anaesthesia blood samples were drawn for the following laboratory investigations: 1. Levels of cortisol, (pre)albumin, $T_3$ and r(everse)$T_3$ as measures for catabolism. A low $T_3$ syndrome, associated with catabolism, is characterised by a decreased $T_3$, an elevated $rT_3$ and an increased ratio $rT_3$:$T_3$. Since this may cause diminished transport of aminoacids and carbohydrates into the cell, a low $T_3$ syndrome possibly gives rise to substrate deficiency, thus inhibiting the synthesis of the directly substrate dependent neurotransmitters, especially acetylcholine and serotonin; 2. Levels of total plasma tryptophan (TRP), the other large neutral aminoacids-ie, valine(VAL), leucine(LEU), isoleucine(ILE), tyrosine(TYR) and phenylalanine(PHE)-and methionine(MET), serine(SER) and taurine(TAU). The plasma ratios of TRP, TYR and PHE were calculated by dividing the plasma concentration of the concerning aminoacid by the sum of the other five large neutral aminoacids. The ratio TAU:METxSER was determined as a measure for the amount of methyl donors; and 3. Folic acid and vitamins $B_1$, $B_6$ and $B_{12}$, because of their major importance in the metabolism of both TRP and most of the other large neutral aminoacids.

## 2.3. Postoperative Assessment

On the first postoperative day 8 a.m. (fasting) blood samples were drawn for plasma levels of cortisol, (pre)albumin, the ratio $rT_3$:$T_3$, TRP and the other aminoacids. On day 3,4 and 5 or earlier, if requested because of behavioural problems, mental status examination and chart review was performed for the presence of delirium according to DSM-III-R criteria (Diagnostic and Statistical Manual, 1987). For a diagnosis of delirium symptoms had to be present for at least 24 hours and not only on the first postoperative day. Patients were rated on the Symptom Rating Scale of Delirium (van der Mast, 1994).

## 3. RESULTS

Delirium occurred in 40 (13.5%) patients and was strongly related to age. Univariate analysis, performed on all possible predictive variables, followed by entering the resulting

**Table 2.** Predictors of delirium after cardiac surgery

| Predictor | Odds Ratio (95% CI) | p-value |
|---|---|---|
| Age ≥ 65 years | 5.2 (2.1-13.3) | <.001 |
| Albumin ≤ 40 g/l | 3.5 (1.6- 7.6) | .002 |
| Use of nifedipine | 2.4 (1.0- 5.8) | .047 |
| PHE-ratio ≥ 13.35 | 2.5 (0.9- 6.5) | .070 |
| GHQ score > 7 | 2.0 (0.9- 4.2) | .071 |

bivariate correlates (p .10) in a backward stepwise logistic regression analysis, resulted in the independent predictors of delirium (Table 2).

Postoperative delirium was not related to intraoperative or anaesthesia variables. Under the influence of cardiac surgery the plasma levels of most of the aminoacids decreased significantly in relation to the already low levels preoperatively, for both delirious and non-delirious patients (p<.001). Only the plasma levels of TRP and PHE increased significantly (p<.001) over time for all patients. Multivariate repeated measurement analysis of variance showed a significant difference in the plasma aminoacid concentrations between delirious and non-delirious patients, mainly caused by tryptophan (p=.004). Postoperatively, delirium was associated with a lower plasma level of tryptophan and, also, with a lower tryptophan ratio (p=.011), a higher PHE ratio (p=.005) and an increased ratio TAU:SERxMET (p=.022). Furthermore, the increase in TRP ratio over time of surgery proved to be significantly less for delirious patients (p=.017). Also postoperatively, delirious patients were more catabolic as was shown by a higher, though not significantly, ratio $rT_3:T_3$ (p=.055).

## 4. DISCUSSION

Contrary to our expectations, the mean plasma concentration of TRP increased and those of the other neutral aminoacids, except PHE, reduced over time, resulting in a rise of the ratios of TRP and PHE, postoperatively. This decrease of the plasma levels of almost all aminoacids, except of TRP and PHE, may be in favour of the brain levels of these aminoacids. The rise in the plasma concentrations of TRP and PHE may be the result of the decreased conversion of these aminoacids in their metabolites under hypothermia. The converting enzymes TRP hydroxylase in the liver and serotoninergic neurons and PHE hydroxylase in the liver, pancreas and kidney are in need of both oxygen and tetrahydrobiopterin, which is synthesized from GTP, for their activity. The existence of a low $T_3$ syndrome causes a reduced synthesis of ATP and thus of GTP. The resulting decrease in tetrahydrobiopterin synthesis may, together with the diminished rate of metabolic processes in the liver, lead to the decreased conversion and increased plasma levels of both TRP and PHE. Moreover, a deficient tetrahydrobiopterin supply leading to diminished conversion of cerebral TRP may directly affect the synthesis of serotonin in the brain. Nevertheless, delirium appeared to be especially associated with a lower plasma level of TRP and a lower TRP ratio, and the increase in TRP ratio over time proved to be significantly less for delirious patients, supporting the hypothesis that a relationship between a reduced cerebral TRP availability from plasma and the occurrence of postoperative delirium may exist. Furthermore, it is probably of great importance that for all patients the levels of the aminoacids TRP, TYR, MET and SER were, already preoperatively, far below normal values indicating decreased intake and/or a catabolic state. Delirious patients, being more catabolic as reflected in a lower mean concentration of albumin and a higher ratio $rT_3$:$T_3$, may have been less able to compensate for the inadequate preoperative TRP levels, since the main difference between delirious and non-delirious patients was a lower plasma level of TRP and a smaller increase of TRP ratio in response to cardiac surgery. This inadequate compensation capacity may be due to the coexistence with a higher ratio of PHE in delirious patients, already found preoperatively, giving rise to an even more reduced cerebral availability of TRP, since the particular affinity of the blood-brain barrier for PHE in preference of the other aminoacids may cause a blockade of the uptake of the other neutral amino acids such as TRP, into the brain. This may hamper adequate function of the serotoninergic system. The normal plasma levels of PHE, being a precursor of TYR, may take care of an adequate or, in the presence of an increased ratio of PHE, (relatively) elevated cerebral dopaminergic and noradrenergic function. The association of delirium with a higher ratio TAU:SERxMET, indicating lower plasma levels of the methyl donors SER and MET, may signify an endangered production of cerebral acetylcholine due to a lack of these methyl precursors. Moreover, this may also lead to increased levels of dopamin and (nor)adrenaline, since the enzyme cathechol-O-methyl-transferase (COMT) needs S-adenyl-methionine (SAM), thus a methyl donor, for degradation of dopamin and (nor)adrenaline. In conclusion, a decreased TRP availability from plasma may be a risk factor for delirium in the presence of a (relatively) increased PHE. An imbalance of the cerebral neurotransmitter metabolism amounting to a decreased serotoninergic and cholinergic function on the one hand, and a (relatively) increased activity of the dopaminergic (and noradrenergic) system on the other hand, may underlie the pathogenesis of delirium after cardiac surgery.

## REFERENCES

Diagnostic and Statistical Manual of Mental Disorders, Third edition, Revised, American Psychiatric Association, Washington, DC (1987).

Editorial, Brain damage and open-heart surgery, *The Lancet* 338: 364 (1989).

Lipowski Z.J., 1990, "Delirium: Acute Confusional States," Oxford University Press, New York.

Smith L.W. and Dimsdale J.E., 1989, Postcardiotomy delirium: Conclusions after 25 years? *Am J Psychiatry* 146:452..

van der Mast R.C., Fekkes D., Moleman P., and Pepplinkhuizen L., 1991, Is postoperative delirium related to reduced plasma tryptophan?, *The Lancet* 338:851.

van der Mast R.C., 1994, "Delirium after Cardiac Surgery: A Prospective Study," Dissertation Erasmus University, Rotterdam.

14

# THE POSSIBLE ROLE OF TRYPTOPHAN IN MIGRAINE

S. Salmon, C. Curradi, E. Del Bene, and F. Sicuteri

Institute of Internal Medicine and Therapy IV
University of Florence
Firenze, Italy

## INTRODUCTION

For many years, it has been accepted that the serotoninergic system plays a major role in the modulation of nociception.

As concerns 5-hydroxytryptamine receptors, a bewildering multiplicity has now been characterised, only $5HT_1$ - $5HT_3$ have been examined for their potential role in the modulation of nociception.

The hypothesised vulnerability of the serotoninergic system in migraine sufferers has been tested with several 5-hydroxytryptamine depleting drugs (p-chlorophenilalanine, reserpine, nitro-glycerine, fenfluramine). All these induce pain in migraineurs (Sicuteri, 1971, Sicuteri et al., 1973).

Fenfluramine, a putative 5HT agonist, potentiates 5HT release from nerve terminals and inhibits its reuptake; its effect probably results from an enhancement in serotoninergic brain neurotransmission (Garattini et al., 1979).

Previous studies have shown that fenfluramine is a pain inducing agent not only in migraine sufferers (Sicuteri, 1976) but also in hereditary predisposed subjects (Poggioni et al., 1992). Fenfluramine administration in humans can provoke both a painful reaction and several symptoms not related to pain.

A highly significant increase of cerebrospinal fluid and plasma free tryptophan during the migraine attack was previously detected (Salmon et al., 1978, 1980).

The aim of the present study was therefore the evaluation of plasma tryptophan and platelet 5-hydroxytryptamine in migraine sufferers during a short chronic fenfluramine or placebo treatment.

## PATIENTS AND METHODS

*Patients:* Two groups of twelve hospitalized migraine sufferers, drug free for two days, were studied. Early in the morning twelve migraineurs (7 males, 5 females, mean age

*Recent Advances in Tryptophan Research,* edited by Graziella Allegri Filippini et al.
Plenum Press, New York, 1996

**Table 1.** Mean tryptophan plasma levels in migraine sufferers following oral placebo (saline) or fenfluramine (60 mg p. os pro die) four days treatment

| Try (mg/ml) | Days | Placebo | Fenfluramine |
|---|---|---|---|
| Total | 0 | 12.83 ± 1.30 | 14.25 ± 1.852 |
| Free | | 1.88 ± 0.20 | 2.02 ± 0.09 |
| Total | 1 | 12.96 ± 1.03 | 15.05 ± 2.07 |
| Free | | 1.94 ± 0.22 | 2.16 ± 0.16 |
| Total | 2 | 12.91 ± 1.11 | 15.65 ± 1.92 |
| Free | | 1.93 ± 0.18 | 2.12 ± 0.20 |
| Total | 3 | 12.93 ± 1.19 | 14.5 ± 1.78 |
| Free | | 1.8 ± 0.19 | 2.07 ± 0.21 |

Paired t test: N.S.

39± 4) were given placebo (saline p. os/die); the other twelve (6 males, 6 females, mean age 37±5) received fenfluramine (60 mg p. os/die); both treatments lasted four days.

*Methods:* Blood samples were collected from the antecubital vein, for the determination of blood, platelet and plasma 5-hydroxytryptamine and plasma total and free tryptophan.

The levels were measured by means of an ultrafiltration technique, spectrophotometric methods and high performance liquid chromatography (HPLC).

Headache intensity was measured by means of a four grade scale, as evaluated by the patients themselves

## RESULTS

A positive logarithmic correlation bewtween plasma total tryptophan and blood 5HT following fenfluramine treatment was demonstrated ($y = -28.06 + 58.54 Log(x)$; $r = 0.725$, $p < 0.05$).

Plasma total and free tryptophan levels are reported in Table 1.

The parametric statistical analysis did not indicate any significant difference.

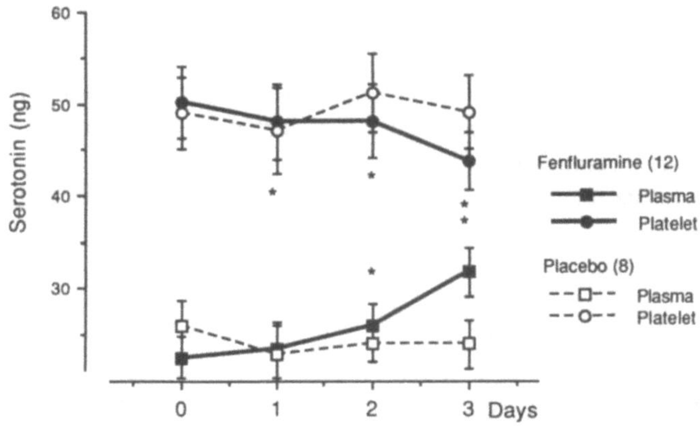


**Figure 1.** Plasma and platelet 5-HT following fenfluramine (60 mg pro die per os) or placebo. *Paired t test: p<0,05.

Blood, platelet (expressed as ng/$10^8$ platelets) and plasma 5-hydroxy-tryptamine levels are shown in Figure 1.

A significant decrease ($p < 0.05$) of platelet 5HT and a correspondent increase ($p < 0.05$) of plasma 5HT content appears after the second fenfluramine administration. No modifications were induced by the placebo.

No correlation between fenfluramine-induced pain and tryptophan modifications was found.

## CONCLUSIONS

A positive 5HT cerebral feed-back mechanism during spontaneous attacks of migraine and cluster headache was suggested by the highly significant increase in the availability of unbound tryptophan both in plasma and in cerebrospinal fluid (Salmon et al., 1982). An upregulation of pain transmission from deficient serotoninergic analgesia in migraine was postulated by Sicuteri et al. (1990).

The dynamics of plasma tryptophan following acute antimigraine drugs administration stressed a displacing activity on bound tryptophan, conferming the hypothesis of a central mechanism of action of these molecules (Salmon et al., 1984). Apomorphine, a dopamine agonist, in a two week increasing doses administration to migraine sufferers, induced a late increase of plasma free tryptophan (Caleri et al., 1984).

Fenfluramine, as well as nitro-glycerine (unpublished data), both pain inducing agents, did not increase plasma tryptophan in migraine sufferers, either with acute administration or with the sub chronic treatment reported above, as shown in animals (Raleigh, 1986).

The significant platelet 5HT release, induced by fenfluramine, confirms that platelet may be considered as a peripheral model of neuronal activity as suggested by Pletscher (1978).

Modulation of serotonin release in the central nervous system, mediated by presynaptic 5HT inhibitory autoreceptors, has been suggested by Goethert (1990).

Cerebrospinal fluid tryptophan levels in migraine and in endogenous depression emphasize the occurrence biochemical and receptorial differences between the two pathologies (Anselmi et al., 1980). This fact might justify the frequent discrepancies in clinical reports dealing with 5HT precursors (L-tryptophan and 5-hydroxytriptophan) both in the prophylaxis and the therapy of migraine, as previously reported in the literature (Bono et al., 1982).

Different endocrine responses to L-tryptophan and fenfluramine, the first mediated by post-synaptic $5HT_{1A}$ receptors, the second by $5HT_{2/1c}$ receptors are reported by Cowen (1993).

It is provocative and legitimate to ask whether the activation of an individual receptor type or uptake site (Clarke et al., 1992) would be sufficient to induce or relieve pain.

The physiological role and potential therapeutic implication of multiple 5HT receptors would seem of relevance in the development of a more specific monoaminergic strategies for pain relief while permitting a greater understanding of the role of endogenous tryptophan in migraine and in pain related deseases.

## ACKNOWLEDGMENTS

This research was partly supported by Grants from the Italian National Council of Research (CNR, Rome): Targeted Progects P.F. 72 CHIMICA FINE n. 93.02907 and P.F. 41 FATMA n. 94.00522

The authors wish to thank Mr. Vittorio Vivoli for his expert technical assistance.

## REFERENCES

Anselmi B., Baldi E., Casacci F., and Salmon S., 1980, Endogenous opioids in CSF and blood in idiopathic headache sufferers, *Headache*, 20: 294.

Bono G., Criscuoli M., Martignoni E., Salmon S., and Nappi G., 1982, Serotonin precursors in migraine patients, *in*: "Advances in Neurology", 33, Critchley M. et al. eds., Raven Press, N.Y.

Caleri D., Gatto G., Bonciani M., Curradi C., and Salmon S., 1984, Influence of apomorphine on plasma tryptophan and cAMP in migraine sufferers, *IRCS Med. Sci.*, 12: 142.

Clarke D.E., 1992, A synopsis of the pharmacology of clinically used drugs at 5-HT receptors and uptake sites, in: "5-Hydroxytryptamine Mechanisms in Primary Headaches", Olesen J., and Saxena P.R., eds., Raven Press, N.Y.

Cowen P.J., 1993, Serotonin receptor subtypes in depression, evidence from studies in neuroendocrine regulation, *Clin. Neuropharmacol. (U.S.),* 13 *suppl.* 3: 6.

Goethert M. and Schlicker E., 1990, Modulation of serotonin release in the central nervous system via presynaptic 5HT autoreceptors, *in*: "Serotonin: from cell biology to pharmacology and therapeutics", Paoletti R. et al. eds., Kluwer Academic Publishers, Dordrecht.

Garattini S., Caccia S., Mennini T., Samanin R., Consolo S. and Ladinsky H., 1979, Biochemical pharmacology of the anorectic drug fenfluramine: a revue. *Curr. Med. Res. Opin.* 6, *suppl.* 1: 15.

Pletscher A. 1978, Platelets as a model for monoaminergic neurons, *Neurochem. Neuropharmacol.,* 3: 49.

Poggioni M., Bonazzi A., Del Bene E., and Sicuteri F., 1992, Focus on latent nociceptive dysfunction in humans, *in*: "Adv. in Pain Res. and Therapy", 20, Sicuteri F. et al. eds., Raven Press, N.Y.

Raleygh M.J., Brammer G.L., Ritvo E.R., Geller E., McGuire M.T. and Yuwiler A., 1986, Effects of chronic fenfluramine on blood serotonin, cerebrospinal fluid metabolites, and behavior in monkeys, *Psychopharmacology*, 90: 503.

Salmon S., Fanciullacci M., Bonciani M., Sicuteri F., 1978, Plasma tryptophan in migraine, *Headache,* 17: 238.

Salmon S., Bonciani M., Fanciullacci M., Marianelli L., Michelacci S. and Sicuteri F., 1982, A putative 5-HT central feedback in migraine and cluster headache attacks. *in*: "Advances in Neurology", 33, Critchley M. et al. eds., Raven Press, N.Y.

Salmon S., Del Re P., Caleri D., Fanciullacci M., and Sicuteri F., 1984, Dynamics of plasma tryptophan following antimigraine drugs. *in*: "Progress in tryptophan and serotonin research", Schlossberger H.G. et al., eds., Walter de Gruyter, Berlin-NY

Sicuteri F., 1971, Pain syndrome in man following treatment with p-chlorophenilalanine, *Pharmacol Res. Commun.*, 3: 401.

Sicuteri F., Anselmi B., Del Bianco P.L., 1973, 5-hydroxytriptamine supersensitivity as a new theory of headache and central pain, *Psychopharmacology (Berl.)*, 29: 347.

Sicuteri F., Del Bene E., Anselmi B., Fenfluramine headache, 1976, *Headache* 16: 185.

Sicuteri F., Poggioni M., Panconesi A., Upregulation of pain transmission from deficient serotonergic analgesia in migraine. In: "Serotonin, from cell biology to pharmacology and therapeutics", Paoletti R. et al. eds., Kluwer Academic Publishers, Dordrecht.

# 15

# TRYPTOPHAN AND EPILEPSY

G. Lunardi,[1] P. Mainardi,[1] V. Rubino,[1] M. Fracassi,[1] F. Pioli,[2] S Cultrera,[1] and C. Albano[1]

[1] Dipartimento di Scienze Neurologiche
Università di Genova, Italy
[2] Centro di Macromolecole Sintetiche e Naturali del Centro Nazionale Ricerche (CNR) di Genova, Italy

## INTRODUCTION

An implication for Serotonin in the pathophysiology of epilepsy has been suggested in experimental and clinical papers.

Results of experimental epilepsy models had shown that:

- Pretreatment with parachlorophenylalanine, depleting Serotonin (5-HT) terminals, lowered the convulsive threshold induced by maximal electroshock (MES) (Scudder et al., 1966).
- Pretreatment of photosensitive baboons with 5-Hydroxy-tryptophan (5-HTP), blocked epileptogenic activity normally occurring after Imipramine or Chlorimipramine treatment (Trimble et al., 1977).

Fluoxetine has been recently shown to exert anticonvulsant effects in animal models of epilepsy (Prendeville and Gale, 1993.).

Results of clinical studies had shown that:

- In epileptic patients Serotonin (5-HT), Tryptophan (TRP) and 5-Hydroxyindole acetic acid (5-HIAA) CSF levels (Liao, 1991; Chadwick et al., 1975) and plasma TRP level (Liao, 1991) were found lowered.

5-Hydroxytryptophan had shown an anticonvulsant effect (De La Torre and Mullan, 1970).

The more recent SSRIs (Selective Serotonin Reuptake Inhibitors) have been shown both to potentiate the usual antiepileptic drugs and to have an antiepileptic effect by themselves (Leander, 1992; Favale et al., in press).

In this study we evaluate a Selective Serotonin Reuptake Inhibitor (SSRI) (Fluoxetine) administered as adjunctive drug in 17 refractory epileptic patients with daily seizures (Partial Complex or Partial Complex with secondary generalization).

Furthermore, the plasma levels of Tryptophan (TRP) and of the others amino acids competing for the same Blood-Brain Barrier (BBB) carrier were analysed in 14 out 17

*Recent Advances in Tryptophan Research*, edited by Graziella Allegri Filippini et al.
Plenum Press, New York, 1996

epileptic patients before fluoxetine administration and compared with 14 controls to study the TRP brain uptake.

## MATERIALS AND METHODS

Fluoxetine 20 mg/day was administered as adjunctive drug in refractory epileptic patients as open trial. Clinical evaluation and drug plasma level assays were performed monthly. Before Fluoxetine treatment plasma levels of Tryptophan and of the other amino acids competing for the same BBB carrier were determined.

Plasma antiepileptic drug levels were determined by Fluorescence Polarization Immuno Assay method (FPIA) supplied by Abbott (TDx-FLx). Plasma amino acids levels were determined by Pico-TAG method on HPLC.

The rates of blood-brain passage of Tryptophan and of the other amino acids competing for the same BBB carrier were evaluated as reported by Smith et al., (1987).

## CONCLUSION

The good clinical results reached with Fluoxetine, as adjunctive drug in refractory epileptic patients, without plasma level modification of the antiepileptic drugs, lead to attribute to fluoxetine itself a synergic effect with antiepileptic drugs. On the other hand, both TRP and the other AAs competing for the same BBB carrier were found lowered in 14 out 17 epileptic patients, compared with 14 controls.

Despite their lower plasma levels, the rates of BBB passage by Tyrosine, Valine, Isoleucine, Leucine and Phenylalanine were found to be similar in patients and in controls, while a 30% decrease was found in TRP rate of brain uptake in patients in the respect of controls.

Our results stress the role of serotonin in pathophysiology of epilepsy and suggest that the serotonin reuptake inhibition, exerted by Fluoxetine, might compensate the lower brain uptake of TRP, the serotonin precursor.

## REFERENCES


Chadwick, D., Janner, P. and Reynolds, E.H., 1975, Amines, anticonvulsants and epilepsy. *Lancet,* 1:473.

De La Torre, J.C. and Mullan S., 1970, A possible role for 5-Hydroxytryptamine in drug-induced seizures. *J.Pharm.Pharmacol.* 22:858.

Favale, E., Albano, C., Lunardi, G., Rubino, V., Mainardi P. Anticonvulsant effect of Fluoxetine in man. *Neurology*, in press.

Leander, J.D., 1992, Fluoxetine, a selective serotonin uptake inhibitor, enhances the anticonvulsant effects of Phenytoin, Carbamazepine and Ameltolide (LY201116). *Epilepsia*, 33 (3):573.

Liao, W.P., 1991, Altered monoamine metabolism in children with epilepsy. *Chung-Hwa-Shen-Ching-Ching-Shen-Ko-Tsa-Chia.* 24(1):0.

Prendeville S. and Gale K., 1993, Anticonvulsant effect of Fluoxetine on focally evoked limbic motor seizures. *Epilepsia*, 34 (2):381.

Scudder, C.L., Karczmar, A.G., Everett, G.M., Gibson, J.E. and Rifkim, M., 1966, Brain catecholamines and Serotonin levels in various strains and genera of mice and a possible interpretation for the correlations of amine levels with electroshock latency and behavior. *Int.J.Neuropharmacol.* 5:343.

Smith, Q.R., Momma S., Aoyagi M., and Rapoport S.I., 1987, Kinetics of neutral amino acid transport across the blood brain barrier. *J.Neurochem.*, 49:5.

Trimble, M., Anlezark, G. and Meldrum, B., 1977, Seizures activity in photosensitive boboons following antidepressant drugs and the role of serotoninergic mechanisms. *Psychopharmacology,* 51:159.

16

# PLASMA TRYPTOPHAN LEVELS IN CHILDREN WITH EPILEPSY BEFORE AND AFTER ANTIEPILEPTIC TREATMENT

D. D. Milovanovic,[1] H. L. Stevanovic,[2] and N. Jovic[1]

[1] Clinic for Neurology and Psychiatry for Children and Young People
11000 Belgrade, Yugoslavia
[2] Institute of Physiology
Faculty of Medicine
11000 Belgrade, Yugoslavia

## 1. INTRODUCTION

People with epilepsy have fits or seizures, caused by sudden disturbances in the normal functioning of the brain (Hazeldine, 1986). Epilepsy is a chronic disease of central nervous system which allowes subjects to have normal living activity between seizures. However a lower percentage of epileptics (e. g. with Lennox-Gastaut syndrome) are unresponsive to drugs therapy. A boy was described (Milovanovic, 1985) with epilepsy and mental retardation (intelligence quotient=37; social quotient=63). During medical treatment for 12.5 years he has been 32 times to his neurologist and he has takan 31637 different anticonvulsive tablets. He still has fits and he is still taking tablets.

There are data on increasing serotonin in epilepsy (Witkop, 1991). In the group of patients with epilepsy and anticonvulsive drugs somewhat higher concentrations of serotonin in cerebrospinal fluid were found (Milovanovic and Stevanovic, 1987). However, quinolinic acid, member of kynurenine pathways, has convulsive properties (Stone, 1987). It has been found that quinolinic acid lesioned hyppocampus bears a very close resemblance to most hyppocampi which are removed during the surgical treatment of temporal lobe epilepsy (Schwarcz and Du, 1991). It was described that quinolinic acid and L-kynurenine were convulsant, while kynurenic acid was an antagonist of exitatory amino acid receptors. Imbalance in the concentrations of these metabolites has been implicated in the etiology of human seizure disorders (Heyes et al., 1994).

In this work investigation of plasma tryptophan was limited to the patients with primary generalized seizures and partial seizures.

## 2. MATERIAL AND METHODS

For the examination of plasma tryptophan (total and free) 28 children with epilepsy (females 12, males 16, aged 7-16 years, mean age 11.82) were selected. Indications for taking

*Recent Advances in Tryptophan Research*, edited by Graziella Allegri Filippini et al.
Plenum Press, New York, 1996

samples and treatment were made after investigating each child, including EEG and examination of the fundus oculi. The somatic, neurologic and psychiatric status of the children studied were normal. Electroencephalographic records (EEG) showed epileptic discharges in 85.71%, while pathological EEG abnormalities were found in remaining 14.29%.

The patients and healthy subjects were allowed free diet and activity. Patients were divided in four groups: the group (a) was formed of 10 untreated patients (females 3, males 7) with primary generalized epilepsy (PGE). 11 children (females 5, males 6) with partial epilepsy with or without secondary generalization (PE) were included in the group (b). Two other groups consisted of 22 patients treated either with valproate or with carbamazepine. In patients with PGE (group c, females 4, males 7) daily doses of valproate were 14.65 ± 2.71 ($\bar{X} \pm SD$) mg/kg b. w. In patients with PE (the group d, females 5, males 6) daily doses of carbamazepine were 10.08 ± 3.07 mg/kg b. w.

Serum levels of valproate and carbamazepine were in optimal therapeutic range. A group of 14 healthy children (females 6, males 8, aged 7-14 years, mean age 12.36) were used as a control group. Plasma concentrations of tryptophan (TRP) were measured in 43 venous blood samples from patients before anticonvulsant medication and 37.00 ± 13.32 days after receiving anticonvulsant monotherapy.

In 7 patients with PGE and in 8 patients with PE the blood samples were taken twice, and according to it these patients were included in untreated groups (a,b) of patients, as well as in treated groups (c,d) of those after initiating therapy. The blood samples for analysis were taken from cubital vein on an empty stomach. Venous blood (10 ml) was collected in heparinized tubes. The plasma was separated by centrifugation (3000 g for 15 min) at room temperature and these samples in the plastic tubes were placed on ice and stored at -20°C. Total and free plasma TRP levels were measured fluorometrically (Bloxam and Warren, 1974). Statistical analysis of resusts were performed by using Student's t-test.

## 3. RESULTS

The values of plasma TRP are presented in Tables 1 and 2. As it can be seen from these findings untreated children with both type of seizures had somewhat higher albumin bound plasma TRP concentrations than controls. Plasma total TRP was somewhat higher in the group (b) in patients with PE. Plasma free TRP was reduced in both groups of untreated epileptics; higher decrease was in the group with PGE (Table 1).

Patients treated with valproate or carbamazepine had higher total and albumin bound plasma TRP, compared to the control group. The drug, carbamazepine, enhanced total and albumin bound plasma TRP levels (for bound TRP, P<0.05), more than valproate, compared

**Table 1.** Plasma tryptophan in children with epilepsy

| | N | Total | Bound | Free |
|---|---|---|---|---|
| a: Untreated pat with PGE | 10 | 64.57 ± 10.13 | 59.19 ± 10.84 | 5.78 ± 2.55 |
| T-test | | 0.02 | 0.45 | $1.47^{r}$ |
| b: Untreated pat with PE | 11 | 68.94 ± 8.82 | 62.33 ± 9.56 | 6.61 ± 2.15 |
| T-test | | 1.34 | 1.47 | $0.55^{r}$ |
| c: Treated pat with valproate, PGE | 11 | 68.93 ± 12.20 | 63.36 ± 12.63 | 5.58 ± 1.84 |
| T-test | | 1.08 | 1.46 | $1.92^{*,r}$ |
| d: Treated pat with carbamazepine, PE | 11 | 71.86 ± 12.17 | 66.73 ± 13.49 | $6.17 \pm 1.63^{n}$ |
| T-test | | 1.82 | $2.16^{**}$ | $1.18^{r}$ |
| Control group | 14 | 64.64 ± 6.55 | 57.56 ± 5.89 | 7.08 ± 1.88 |

Pat: Patients; n: Plasma free TRP in a blood sample not determined; **: P<0.05; *: P ≅0.05. Plasma TRP levels are expressed as μmol/l; r: Reduced plasma free TRP.

**Table 2.** Comparison plasma tryptophan in treated with untreated children with epilepsy

| | | N | Total | Bound | Free |
|---|---|---|---|---|---|
| c: | Valproate, PGE vs | 11 | | | |
| a: | Untreated pat with PGE | 10 | | | |
| | T-test | | 1.72 | 1.32 | 0.17[r] |
| d: | Carbamazepine, PE vs | 11 | | | |
| b: | Untreated pat with PE | 11 | | | |
| | T-test | | 0.75 | 0.64 | 0.02[n] |

Pat: Patients; n: Plasma free TRP in a blood sample not determined; r: Reduced plasma free TRP

to the controls. In contrast to these above findings valproate reduced more ($p \cong 0.05$) plasma free TRP levels than carbamazepine in comparison with the controls.

Using t-test of pairs (Table 2) it was found that both drugs increased total and albumin bound plasma TRP levels, which are no statistically significant compared to the control group a and b. In contrast, valproate prodused a slight decrease in plasma free TRP, while there was no carbamazepine effect on free TRP.

## 4. DISCUSSION

It has been found that carbamazepine and phenytoin had opposite effects on plasma tryptophan levels in epileptic patients by an unknown mechanism (Pratt et al., 1984); Carbamazepine elevating and phenytoin depressing them. These observations may be related to their different effects on mental function and mood. We could find a tendency that serum valproate levels (50-65 μg/ml) increased plasma total and albumin bound TRP, while higher valproate levels (70-85 μg/ml) decreased them in treated patients with primary generalized epilepsy.

Opposite, serum valproate levels (50-65 μg/ml) slightly decreased, while valproate levels from 70-85 μg/ml slightly increased plasma free TRP. Serum concentrations of carbamazepine (4-9 μg/ml) increased total and bound plasma TRP. We could not find any connection between serum carbamazepine levels and plasma free TRP in treated patients with partial epilepsy.

These data support our previous study that there was a certain regularity between serum phenobarbitone levels and urinary 5-HIAA examined in a large group of children suffering from epilepsy. One of the posible explanations was that higher values of phenobarbitone had a repressing effect on tryptophan metabolism, which, at lower concentrations of the drug in the blood, was not observed (Milovanovic et al., 1984).

Although the studied population was relatively small, these findings show that the presence of seizures (=anxiety, stress, nervousness, expectation etc.) in PGE and PE, the treatment with valproate and carbamazepine, and their serum concentrations in both types of epilepsy (c, d) are associated with changes plasma TRP (Table 1) in either direction. These data indirectly suggest an underlying disturbance of TRP metabolism in both types of seizures. It was recently showed that the reductions in serum quinolinic acid and L-kynurenine were correlated to blood antiepileptic medication (Heyes et al., 1994). Tryptophan levels also tended to be lower in both cerebrospinal fluid and serum of the seizure patients.

## 5. CONCLUSION

In this paper the levels of total, albumin bound and free plasma tryptophan were investigated in 28 children with epilepsy aged 7-16 years.

The present results have demonstrated that there were some changes in plasma tryptophan concentrations. These changes are connected with epileptic fits, and also with sort of drug and it's serum concentrations. Period between fits is accompanied with stress, anxiety, nervousness, expectation etc. Both valproate and carbamazepine increased total and tryptophan bound to albumin. Carbamazepine significantly increased levels of bound tryptophan in treated patients compared to the control group. Valproate and carbamazepine have opposite influence on free plasma tryptophan. Valproate slightly or more reduced plasma free tryptophan in different groups, while effect of carbamazepine on free plasma tryptophan was unchanged or produced a slight decrease.

## ACKNOWLEDGMENTS

This study was subsidized by Bureau for International Scientific, Educational, Cultural and Technical Cooperation of the Republic of Serbia and by Hipozal bank and its general manager, Mr Jovan Jovanoviæ. We also gratefully acknowledge Prof. Slavko Simeunoviæ, the Vice Dean of the Faculty of Medicine in Belgrade and Miss Sunèica Borozan.

## REFERENCES

Bloxam D.L. and Warren W.H., 1974, Error in the determination of tryptophan by the method of Denka and Dewey. A revised procedure, *Anal Biochem* 60:621.

Hazeldine P., 1986, "Epilepsy," Thorsons Publishing Group Wellingborough, New York.

Heyes M.P., Saito K., Devinsky O., and Nadi N. S., 1994, Kynurenine pathway metabolites in cerebrospinal fluid and serum in complex partial seizures, *Epilepsia* 35:251.

Milovanovic D., Stankovic B., Milovanovic Lj., Koricanac Z., and Trpkovic D., 1984, Urinary 5-hydroxyindoleacetic acid and serum phenobarbital of children suffering from epilepsy, in: "Progress in Tryptophan and Serotonin Research," H.G. Schlossberger, W. Kochen, B. Linzen, and H. Steinhart, ed, Walter de Gruyter, Berlin and New York pp. 473-478.

Milovanovic D.D., 1985, Epilepsy - disease and social aspect, *Socijalna politika*, 40:26.

Milovanovic D.D. and Stevanovic H.L., 1987, Serotonin and 5-hydroxyindoleacetic acid csf levels and plasma tryptophan in neurological diseases, in: "Progress in Tryptophan and Serotonin Research," D. A. Bender, M. H. Joseph, W. Kochen, and H. Steinhart, ed, Walter de Gruyter, Berlin and New York pp. 325-332.

Pratt J., Jenner P., Reynolds E.H., and Shorvon S.D., 1984, Anticonvulsant drugs produce different effects on plasma tryptophan concentrations in epileptic patients, in: "Progress in Tryptophan and Serotonin Research, " H. G. Schlossberger, W. Kochen, B. Linzen, and H. Steinhart, ed, Walter de Gruyter, Berlin and New York pp. 479-482.

Schwarcz R. and Du F., 1991, Quinolinic acid and kynurenic acid in the mammalian brain, in: "Kynurenine and Serotonin Pathways," R. Schwarcz, S. N. Young, and R. R. Brown, ed, Plenum Press, New York and London pp. 185-199.

Stone T.W., 1987, Neurophysiology of kynurenines, in: "Progress in Tryptophan and Serotonin Research, " D. A. Bender, M. H. Joseph, W. Kochen, and H. Steinhart, ed, Walter de Gruyter, Berlin and New York pp.111-115.

Witkop B., 1991, Retro-, intro- and perspectives of trypto-fun, in: "Kynurenine and Serotonin Pathways," R. Schwarcz, S. N. Young, and R. R. Brown, ed, Plenum Press, New York and London pp.3-13.

# 17

# PLASMA TRYPTOPHAN LEVELS IN THE PATIENTS WITH CERVICOBRACHIAL AND LUMBOSACRAL PAIN SYNDROMES

D. D. Milovanovic,[1] H. L. Stevanovic,[2] and J. Pavlovic[1]

[1] Health Center
26204 Opovo, Yugoslavia
[2] Institute of Physiology
Faculty of Medicine
11000 Belgrade, Yugoslavia

## INTRODUCTION

Pain is either very useful or very dangerous indicator for the health of human beings. Pain perception depends on the intensity and duration of noxious stimulation (Melzack and Wall, 1965), pain threshold as well as some personality factors. Therefore, pain perception = 1/pain threshold. If the pain threshold is higher then pain perception will be lower and pain stimuli ought to be raised and vice versa. A physiological function serotonin (5-hydroxytryptamine - 5-HT) is related to the increase of pain threshold (Witkop, 1991), and according to it persons with normal or increased levels of 5-HT in the central nervous system (CNS) have an optimal state decreasing pain perception (normal hypalgesia). A descending serotoninergic system, located mainly in the midline raphe of the brainstem participate in modulation pain impulses in the dorsal horns of the spinal cord (Brain, 1985; Angevine and Cotman, 1981). In the brain stem there is an endogenous analgesia system. Hyperpathia unpleasant character of pain could be found in thalamic lesions. Somatosensory and association cortex take a part in receiving pain impulses and its modulation.

It has been reported that there is an important role of tryptophan (TRP) and serotoninergic system in modulation of pain in patients with lumbar discopathy or prolapsed disc (Costa, 1984). Oral L-TRP administration decreases the perception of pain (elevated pain threshold), appearing to act synergistically with the enkephalins and endorphins (Haze, 1991). In contrast with above findings, moderate doses of TRP given by the mouth did not affect postoperative pain or consumption of analgesics (Ekblom, 1991).

In this work investigation of plasma TRP was limited to the patients suffering from acute or recurrent pain symptoms which are rather radicular.

# MATERIAL AND METHODS

For the determination of plasma tryptophan (total and free) the patients were selected by neurologist in acute or recurrent stage of pain symptoms. Nineteen patients of both sexes (aged 26-73 years, mean age 47.21) were included in this investigation. 10 patients with right or left lumboischialgia, 2 with cervicobrachial and lumbosacral pain syndrome, 2 with lumbago, 2 with ischiadic neuralgia, 2 with cervicobrachial pain or cervical neuralgia and one patient with cervicobrachial pain syndrom have been included in this study. In seven patients diagnoses were established by orthopedist or neurosurgeon. Disc hernia was operated at the L5/S1 level in one patient seven year ago. Pain symptoms in patients were estimated by two persons (Table 1).

Before taking venous blood for analysis of TRP, ten patients were drugs free three or more days, while the drugs free time were 12-24 hours for the other nine patients.

The control group consisted of 10 helthy persons (females 6, meles 4, aged 27-67 years, mean age 38.5). No dietary restriction was imposed on the subjects.

The blood samples for analysis was taken from the cubital vein on an empty stomach. Venous blood (10 ml) were collected in heparinized tubes. The plasma was separeted by centrifugation (3000 g for 15 min.) at room temperature and these samples in the plastic tubes were placed on ice and stored at -20°C. The plasma was ultrafiltrated (Amicon cell, 50 ml volume PM 10000 membrane) in controlled atmosphere (95% $N_2$, 5% $CO_2$) for determination of free plasma TRP.

Total and plasma free TRP levels were measured fluorometrically (Bloxam and Warren, 1974). Statistical analysis of results was performed by using Student's t-test

# RESULTS

The results of these examinations are shown in Tables 1. and 2. and Figure 1. slight to severe pain symptoms were found in all of patients. From Table 1. we could see that eleven patients had some of psychiatric findings. Neurologic signs were found in 18 of patients and it correlated with X-ray findings in nine patients.

Total and bound plasma TRP in patients were significantly increased ($P<0.01$), while plasma free TRP shown decreases tendency ($P>0.05$), compared to the control group (Table 2 and Figure 1). These results indicate that there is a dynamic balance between plasma free TRP and albumin bound plasma TRP (Figure 1).

**Table 1.** Clinical characteristics and signs in examined patients

| Patients/Sex | N=19, females 6, males 13 |
|---|---|
| Duration of disease | 4-45 days (n=5); 2-10 months (n=7); 3-7 years (n=6); 13 year (n=1) |
| Pain symptoms | Slight ± (n=2); moderate +(n=13); severe ++ (n=4) |
| Neurologic findings | Slight ±(n=11); moderate +(n=7); normal (n=1) |
| Psychiatric findings | Nervous (n=5); sleep disorders (n=3); nervous or anxiety an sleep disorders (n=3); normal (n=8) |
| X-ray findings | Spondylosis slight ±(n=4); spondylosis marked +(n=1);narrowing of the disc space at the L5/S1 level (n=2); arthritic changes at the C6/C7 level (n=1); spondylolisthesis (n=1); normal (n=3); not examined (n=8) |

**Table 2.** Plasma tryptophan in patients with acute or recurrent pain symptoms

| | N | $\overline{X}$ | SD | T-test | P | CV |
|---|---|---|---|---|---|---|
| Patients | | | | | | |
| Total TRP | 19 | 64.08 ± | 12.21 | 3.79 | <0.01 | 19.06 |
| Bound TRP | 19 | 57.23 ± | 12.50 | 3.69 | <0.01 | 21.84 |
| Free TRP | 19 | 6.85 ± | 1.92 | 0.15 | >0.05 | 27.96 |
| Controls | | | | | | |
| Total TRP | 10 | 46.46 ± | 9.91 | | | 21.33 |
| Bound TRP | 10 | 39.80 ± | 9.87 | | | 24.80 |
| Free TRP | 10 | 6.66 ± | 2.43 | | | 36.43 |

Plasma TRP levels are expressed as μmol/l

# DISCUSSION

Our data suggest that organic neural pain in the patients change equilibrium between albumin bound TRP levels, which are segnificantly increased, and plasma free TRP levels without significant decrease, compared with the control group (Table 2 and Figure 1). As it can be seen from these findings there are decrease plasma free TRP levels in the pain syndromes and according to it lower TRP cross through blood-brain barrier, lower neurotransmitter 5-HT synthesis, lower pain threshold, higher pain perception (hyperalgesia) and vice versa. It has been reported that animals having a reduced level of brain serotonin demonstrate a heightened sensitivity to pain stimuli; this pain sensitivity can be restored to normal values by a single injection of L-TRP, wich rapidly elevates brain serotonin (Fern-

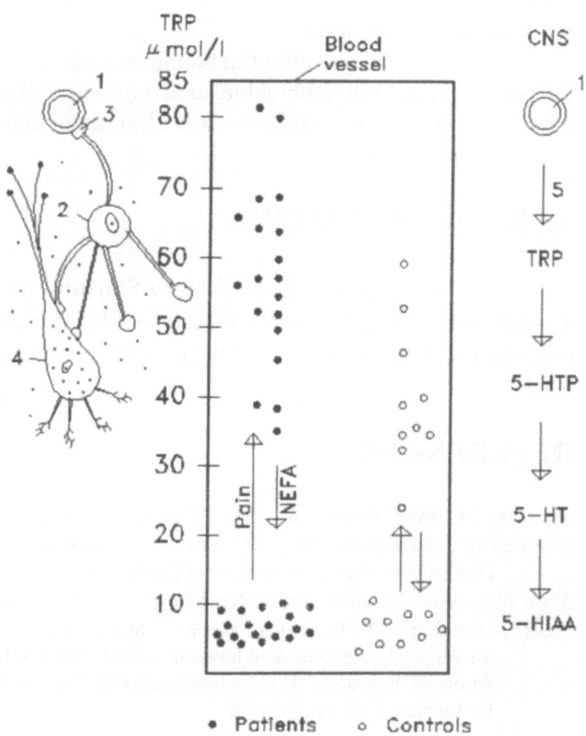


**Figure 1.** Relationship between plasma free TRP (below 15 μmol/l) and albumin bound TRP (above 20). BBB: Capillary with tight junctions of specialized endothelium (1) and astrocyte (2) with feet like processes in apposition with the capillary (3) and the wall of the neuron (4). NEFA: Unesterified fatty acids. 5: Minor pathway metabolism of the TRP via the 5-hydroxy-indole route (Udenfriend, 1955).

strom, 1977). The brain change may reflect that of plasma free TRP or its ratio to that of the competers but it cannot reflect plasma total TRP (Curzon and Sarna, 1984). The changes of plasma free TRP can be expected in the increase binding of TRP to plasma albumin which is observed in the acute and the recurrent phase of pain symptoms. In stress the binding of free TRP to plasma albumin is reduced (Joseph et al., 1984).

It is known that unesterified fatty acids are able to displace TRP binding to albumin. Both stress and elevated concentrations fatty acids increase plasma free TRP. It is also important to mention the role of food, certain drugs and other factors which are able to changes plasma free and albumin bound TRP (Fernstrom, 1977; Curzon and Sarna, 1984). Blood-brain barrier, (BBB), prevented many blood-borne compounds from entering the brain (Schade and Ford, 1973).

It has been postulated that TRP uptake into the brain, can be altered by changing either plasma TRP levels, or concentrations of one or more of its brain transport competitive large neutral amino acids, or both (Fernstrom and Wurtman, 1972). It was observed that there was a tendency of plasma free TRP to decrease in patients with demyelinating diseases and in patients with the chronic course of the diseases, as well as in patients with the blood-brain barrier damage (Milovanovic and Stevanovic, 1992). It has been found that increased catabolism of 5-HT in chronic pain patients, support an involvement of serotonin in antinociception and insomniac situation (Costa et al. 1984),. We agree with the authors that chronic pain represents an enduring challenge to man's capacity to function productively (Seltzer et al., 1982/83).

## CONCLUSION

The results of the present investigation indicate that organic neural pain reduced plasma free tryptophan concentrations in acute and recurrent phase of the pain symptoms. This finding supports the role of plasma free tryptophan and indirectly of serotonin in regulated pain symptoms.

Plasma total and albumin bound tryptophan were significantly increased ($P<0.01$) in 19 patients with cervicobrachial and lumbosacral pain syndromes. It may be seen that equilibrium between plasma free and albumin bound tryptophan is changed.


## ACKNOWLEDGMENTS

We would like to thank Biserka Stojanov, a laboratory technician, for taking blood samples and preparing them for analysis. We also thank Professor Dr. Zivorad Dodevic, a neurosurgeon, for help in the clinical diagnosis in a case.


## REFERENCES


Angevine J.B. and Cotman C.W., 1981, "Principles of Neuroanatomy," New York Oxford University Press.

Bloxam D.L., and Warren W.H., 1974, Error in the determination of tryptophan by the method of Denkla and Dewey. A revised procedure, *Anal Biochem* 60:621.

Brain W.R., 1985, "Brain's Clinical Neurology," 6th ed, Revised by R. Bannister, Oxford University Press.

Costa C., Ceccherelli F., Bettero A., Marin G., Mancusi L., and Allegri G., 1984, Tryptophan, serotonin and 5-hydroxyindoleacetic acid levels in human CSF in relation to pain, in: "Progress in Tryptophan and Serotonin Research," H.G. Schlossberger, W. Kochen,B. Linzen, H. Steinhart, ed, Walter de Gruyter, Berlin·New York pp. 413-416.

Curzon G. and Sarna G.S., 1984, Tryptophan transport to the brain: newer findings and older ones reconsidered, in: "Progress in Tryptophan and Serotonin Research," H.G. Schlossberger, W. Kochen, B. Linzen, H. Steinhart, ed, Walter de Gruyter, Berlin·New York pp. 145-160.

Ekblom A., Hansson P., and Thomsson M., 1991, L-Tryptophan supplementation does not affect postoperative pain intensity or consumption of analgesics, *Pain* 44:249.

Fernstrom J.D. and Wurtman R.J., 1972, Brain serotonin content: physiological regulation by plasma neutral amino acids, *Science* 178:414.

Fernstrom J.D., 1977, Effects of the diet on brain neurotransmitters, *Metabolism* 26:207.

Haze J.J., 1991, Toward an understanding of the rationale for the use of dietary supplementation for chronic pain management: the serotonin model, *Cranio* 9:339 1. Melzack R. and Wall P.D., 1965, Pain mechanism: a new theory, *Science* 150:971.

Joseph M.H., Johnson J.A., and Kennet G.A., 1984, Increased availability of tryptophan to the brain in stress is not mediated via changes in competing amino acids, in: "Progress in Tryptophan and Serotonin Research," H.G. Schlossberger, W. Kochen, B. Linzen, H. Steinhart, ed, Walter de Gruyter, Berlin·New York pp. 387-390.

Melzack R. and Wall P.D., 1965, Pain mechanism: a new theory, *Science* 150:971.

Milovanovic D.D., and Stevanovic H.L., 1992, Cerebrospinal fluid serotonin and 5-hydroxyindoleacetic acid and plasma tryptophan in demyelinating diseases and some neurological disorders, in: "Advances in Tryptophan Research," I. Ishiguro, R. Kido, T. Nagatsu, Y. Nagamura, Y. Ohta, ed, Fujita Health University Press, Toyoake, Japan pp. 101-106.

Schade J.P. and Ford D.H., 1973, "Basic Neurology," 2nd ed, Elsevier Scientific Publishing Company, Amsterdam, London, New York.

Seltzer S., Dewart D., Pollack R.L., and Jackson E., 1982/83, The effects of dietary tryptophan on chronic maxillofacial pain and experimental pain tolerance, *J Psychiat Res* 17:181.

Udenfriend S., Titus E., and Weissbach H., 1955, The identification of 5-hydroxy-3-indoleacetic acid in normal urine and a method for its assay, *J Biol Chem* 216:499.

Witkop B., 1991, Retro-, intro- and perspectives of trypto-fun, in: "Kynurenine and Serotonin Pathways," R. Schwarcz, S.N. Young, R.R. Brown, ed, Plenum Press, New York and London pp. 3-13.

# 18

# MELATONIN IN INSOMNIAC PATIENTS AND EFFECTS OF SLEEP MEDICATION

A. Rodenbeck, G. Hajak, and G. Huether

Department of Psychiatry
University of Göttingen
Von Siebold Str. 5, 7075 Göttingen, Germany

Sleep complaints, especially difficulties in initiating and/or maintaining sleep, are wide-spread in industrial countries and sleep complaints often lead to chronic insomnia (Holzrichter et al., 1994, in press). Although neurochemical processes are generally assumed to mediate the various external and internal factors causing sleep disorders, neuroregulatory mechanisms and neuroendocrine correlates of insomnia are still unknown. Recent research assumes a functional link between sleep and the pineal hormone melatonin (Dawson and Encel, 1994). The circadian secretion pattern of melatonin is widely accepted as a reliable tool to measure human circadian clock functions (Rosenthal, 1991). Considering that melatonin promotes internal and external synchronization in humans (Arendt et al., 1986), altered circadian sleep-wake functions in insomniac patients has be related of disturbances in those central control mechanisms which also influence the circadian secretion pattern of melatonin. The sleep promoting properties of melatonin are well-known. Sedation and reduced sleep latency are seen either after direct melatonin administration (Dawson and Encel, 1994) or after application of L-tryptophan which causes a rapid increase of the melatonin plasma concentration (Hajak et al., 1991). The possible alterations of the melatonin secretion pattern in insomniac patients with difficulties in maintaining sleep, however, has not yet been studied in some detail.

We therefore investigated sleep parameters and nocturnal plasma melatonin secretion in 10 patients (41.3 ± 9.5 years) suffering from primary insomnia with frequent nocturnal awakenings as the predominating complaint and in 5 healthy male controls (27.2 ± 0.7 years) (Hajak et al., in press). Primary insomnia was diagnosed following DSM-III-R (APA 1987) and polysomnographic recordings for two entire nights. After an adaptation night, the investigation started 4 hours before individual bed time. Blood was obtained every hour via indwelling catheter and through-the-wall tube system up to 9 hours following lights off. Melatonin plasma levels were measured by direct radioimmunoassay (Fraser et al., 1983), and group differences and interactions were calculated by analysis of variance and subsequent analysis with Mann-Whitney U-Test. Sleep was measured and analysed by polysomnographic standard criteria (Rechtschaffen and Kales, 1968).

Compared to young healthy male controls, patients showed a decreased sleep efficiency, increased wake percentages and more nocturnal awakenings, but did not differ

*Recent Advances in Tryptophan Research*, edited by Graziella Allegri Filippini et al.
Plenum Press, New York, 1996

in terms of sleep onset latencies. Nocturnal plasma melatonin levels at three to six hours after lights off were significantly lower in insomniac patients compared to controls, while evening melatonin levels were higher. Since the patients did not take any sleep medication up to two weeks prior the investigation, their reduced nocturnal melatonin levels could not be due to pharmacological mechanisms. Reduced nocturnal melatonin levels were only observed in those patients which had a long history of sleep-maintainance insomnia (15.3 ± 5.9 years). Patients with a less than five years duration of insomnia (3.8 ± 1.5 years) showed rather normal nocturnal melatonin plasma concentrations (fig. 1). The analysis of sleep parameters indicated that sleep was equally disturbed in both groups.

The pronounced reduction of nocturnal plasma melatonin concentration in only long-term insomniac patients indicates that low nocturnal melatonin secretion is not a characteristic feature disturbed sleep. It rather seems to represent the final result of very slowly developing neuroendocrine dysregulation caused by chronic insomnia. The results support the hypothesis that even in insomniac patients mechanisms of sleep regulation are independent from those regulating nocturnal melatonin secretion. Since melatonin promotes internal and external synchronization in humans (Arendt et al., 1986), the low nocturnal melatonin secretion in long-lasting insomnia may be an additional internal factor maintaining disturbed sleep continuity. Low nocturnal plasma melatonin levels might be used as a state marker for chronic insomnia. Since nocturnal melatonin was lowered only in chronic insomniac patients, treatment of insomnia should be started as soon as possible in order to prevent the progression of complex neuroendocrine dysregulations.

In the therapy of insomnia, benzodiazepine hypnotics (BZ) are currently accepted as the medication of first choice. The effects of BZ on sleep parameters are well studied and BZ have been shown to reduce nocturnal melatonin secretion (McIntyre et al., 1988, Monteleone et al., 1989). Tricyclic antidepressants (TCA) are known to improve sleep and to increase circulating melatonin levels in healthy controls and depressed patients (for references see Rodenbeck et al., 1994), but use of TCA in the treatment of primary insomnia is still discussed because of TCA side effects. A direct comparison of BZ and TCA effects

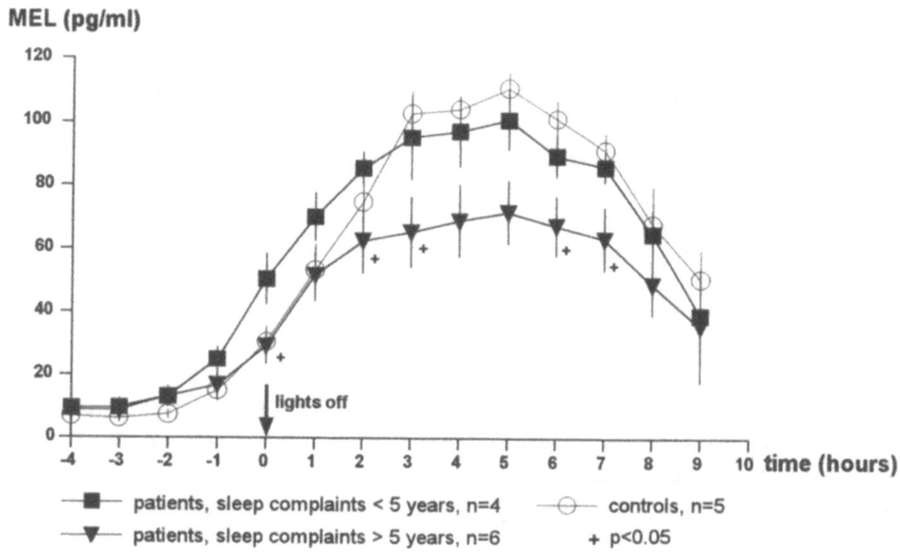


**Figure 1.** Plasma melatonin concentration in 4 patients with a history of primary insomnia of less than 5 years, 6 patients suffering from primary insomnia since more than 5 years (mean values and standard error of the mean). Plasma melatonin concentration of 5 healthy male controls are additionally shown.

on sleep and melatonin levels was made by us in patients with primary insomnia and healthy controls.

In a first trial we studied the effects of 25 mg doxepin administered intravenously (i.v.) and placebo in a doubleblind randomized order in the same patients and controls described above (Rodenbeck et al., 1994). The acute i.v. administration of doxepin significantly improved nonREM-sleep parameters like sleep effiency and slow wave sleep latency in patients and controls, but increased REM-sleep latency. However, nocturnal plasma melatonin levels remained unaffected in patients (fig. 2) and controls (data not shown). These findings show that the acute sleep improving effects are independent of nocturnal melatonin secretion. The increase of REM-sleep latency after acute doxepin administration is typical for TCA's and is due to their anticholinergic action.

In order to study the effects of long-term doxepin treatment all patients received 25 mg doxepin orally over a three week period. At the end of the treatment the improvement of nonREM-sleep parameters persisted and REM-sleep latency was normalized compared to placebo indicating a normalisation of the ultradian sleep structure. Following the chronic intake of doxepin, nocturnal plasma melatonin concentration was significantly increased compared to placebo and reached almost normative values (fig. 2).

The reduction of REM-sleep latency following chronic doxepin administration in patients indicates that long-term pharmacological effects are involved in reinstallation of physiological ultradian sleep patterns. The increased nocturnal melatonin secretion after chronic TCA intake seems to reflect the normalisation of complex physiological rhythms. An enhancement of melatonin secretion after chronic TCA administration has also been reported in depressed patients (Checkley and Palazidou, 1988). The normalisation of nocturnal melatonin plasma levels in patients with primary insomnia seems to reflect a general improvement of circadian clock functions by long-term administration of doxepin.

In a second trial we studied the simultanous effects of BZ on sleep and melatonin plasma concentration. Flunitrazepam (1 mg) and placebo were intravenously administered in a doubleblind and randomized order in five younger (25.6 ± 1.1 years) and five older (49.4 ± 5.4 years) healthy male controls (Rodenbeck et al., in press). While sleep latency was

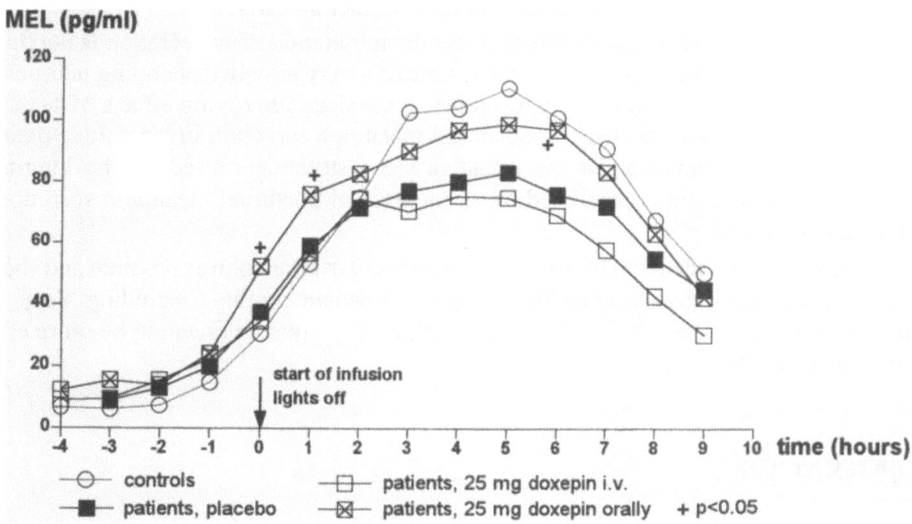


**Figure 2.** Plasma melatonin concentration in 10 patients with primary insomnia after placebo load and 25 mg doxepin administered acute-intravenously (i.v.) and orally over a three-week period, respectively. Normative values of 5 healthy controls are addionally shown.

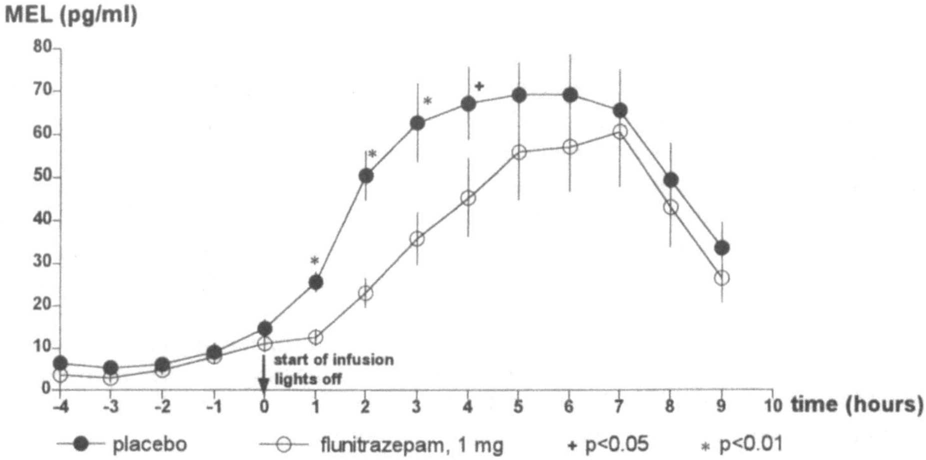


**Figure 3.** Plasma melatonin concentration in 10 controls after placebo load and intravenous application of 1 mg flunitrazepam (mean values and standard error of the mean).

decreased in both groups, no other sleep parameter was affected by flunitrazepam, excepted a massive decline of the percentage slow wave sleep in the older controls. In contrast to the acute effect of doxepin, plasma melatonin concentrations were significantly decreased in both groups compared to placebo (fig. 3). This effect was more pronounced in the older controls with significantly lower values between one and seven hours after the start of infusion. Younger controls showed only a signifcant decrease between two and four hours after start of infusion (data not shown in detail).

These results showed again that sleep promoting effects are independent from nocturnal melatonin secretion. Although the reduced melatonin plasma concentration may indicate a destabilisation of circadian function, negative feed-back mechanisms of BZ on melatonin receptors can not be excluded and need further investigation.

In summary, the results indicate that low nocturnal melatonin secretion is not due to disturbed sleep, but rather represent the final state of a very slowly developing neuroendocrine dysregulation caused by chronic insomnia, acute sleep improving effects of TCA and BZ are independend from changes in nocturnal melatonin secretion in insomniac patients and controls,the normalisation of the ultradian sleep structure caused by the long-term administration of doxepin is paralleled by an increase of nocturnal melatonin secretion in insomniac patients.

In conclusion, treatment of insomnia should be started as soon as possible and should aim at normalising and stabilizing the complex circadian rhythms including sleep and neuroendocrine parameters. In this respect tricyclic antidepressants seem to be more effective than benzodiazepines.

## REFERENCES

American Psychiatric Association (APA), 1987, Diagnostic and statistical manual of mental disorders. Third edition, revised. American Psychiatric Association (ed), Washington, DC

Arendt J., Aldhous M., Marks V., 1986, Alleviation of jet lag by melatonin: Preliminary results of controlled double blind trial, *Br Med J* 292:1170

Checkley S.A., Palazidou E., 1988, Melatonin and antidepressant drugs: Clinical pharmacology, in: "Melatonin. Clinical perspectives", Miles A., Philbrick D.S., Thompson C., eds. Oxford University Press, Oxford, pp 190-204

Dawson D., Encel N., 1993, Melatonin and sleep in humans, *J Pineal Res* 15:1

Fraser S., Cohen P., Franklin U., Franey C., Arendt J., 1983, A direct radioimmunoassay for melatonin, *Clin Chem* 29:799

Hajak G., Huether G., Blanke J., Bloemer M., Freyer C., Poeggeler B., Reimer A., Rodenbeck A., Schulz-Varszegi M., Rüther E., 1991, The influence of intravenous l-tryptophan on plasma melatonin and sleep in men, *Pharmacopsychiat* 24:17

Hajak G., Rodenbeck A., Staedt J., Bandelow B., Huether G., Rüther E., in press, Nocturnal plasma melatonin levels in patients suffering from primary insomnia, *J Pineal Res*

Holzrichter S., Hajak G., Rudolph G., Schlaf G., Westenhöfer J., Rodenbeck A., Pudel V., Rüther E., 1994, Wie schlafen die Deutschen - eine Repräsentativumfrage, *Wien Med Wschr Sonderheft:* 62-73

Holzrichter S., Hajak G., Schlaf G., Westenhöfer J., Rodenbeck A., Bandelow B., Pudel V., Rüther E., in press, Chronifizierung von Schlafbeschwerden - Ergebnisse einer Repräsentativumfrage, *Nervenarzt*

McIntyre I.M., Burrows G.D., Norman T.R., 1988, Suppression of plasma melatonin by a single dose of the benzodiazepine alprazolam in humans, *Biol Psychiatry* 24:105

Monteleone P., Forziati D., Orazzo C., Maj M., 1989, Preliminary observations on the suppression of nocturnal plasma melatonin levels by short-term administration of diazepam in humans, *J Pineal Res* 6:253

Rechtschaffen A., Kales A., 1968, A manual for standard terminology, techniques and scoring system for sleep stages of human sleep, Public Health Service, ed., Printing Office, Washington DC

Rodenbeck A., Hajak G., Herrendorf G., Staedt J., Adler L., Biutmann V., Müller-Ißberner R., Huether G., Rüther E., 1994, Schlafpolygraphische und neuroendokrine Aspekte zur Wirkung trizyklischer Antidepressiva in der Behandlung von Insomnie-Patienten, *Wien Med Wschr Sonderheft*:125-128

Rodenbeck A., Hajak G., Staedt J., Herrendorf G., Holzrichter S., Huether G., Rüther E., in press, Erniedrigte Melatonin-Plasma-Spiegel als Ausdruck einer chronischen Insomnie und dessen pharmakologische Beeinflussung, in: "Jahrbuch der Schlafmedizin 1994", Deutsche Gesellschaft für Schlafmedizin, ed., MMW

Rosenthal N.E., 1991, Plasma melatonin as a measure of the human clock, *J Clin Endocrinol Metab* 73(2):225

# TYROSINE AND TRYPTOPHAN DERIVATIVES IN PIG LUMBAR CEREBROSPINAL FLUID

## Effects of Subchronic Administration of Dopa Associated with Benzeraside

S. Banting,[1] G. Achilli,[2] A. Banting,[1] D. Le Bars,[3] and J. Weil-Fugazza[3]

[1] CEBIPHAR
BP 8, 37230 Fondettes, France
[2] ESA
Piazza Maggiolini 3, 20015 Parabiago, Milano, Italia
[3] INSERM, U. 161
2, rue d'Alésia, 75014 Paris, France

## INTRODUCTION

The role of the kynurenine metabolic pathway in neurodegenerative diseases was suggested on the basis of measurements of tryptophan derivatives in post mortem microdissected brain structures or in cerebrospinal fluid (CSF) from patients suffering from various neurological dysfunctions (Beal et al., 1990, Heyes et al., 1992, Ikarashi et al., 1992, Ogawa et al.,1992, Tohgi et al., 1993). Decreased levels of both kynurenine and kynurenine metabolites were observed in Parkinson's disease and the results of one report suggest that L-dopa therapy might further reduce these levels in some areas of the brain (Ogawa et al., 1992). To further investigate the possible effects of L-dopa administration on tryptophan metabolic pathways we determined the levels of tryptophan metabolites in pig lumbar cerebro-spinal fluid (CSF) before and following subchronic treatment with L-dopa associated with benzeraside.

## METHODS

Pigs (7 females and 3 males; 4-5 months; 25-35 kg) were fed with a commercial food (Athlon 100, Loire Elevage Aliments). Lumbar punctures (500 µl ) were performed at L2-L3 level. 4 females and 1 male were used as controls. In the other five animals (3 females, 2 males), punctures were made before and following per os administration of L-dopa (100 mg)

*Recent Advances in Tryptophan Research*, edited by Graziella Allegri Filippini et al.
Plenum Press, New York, 1996

and benzeraside (25 mg) twice a day for consecutive 4 days. The lumbar punctures after L-dopa therapy were performed two hours after the last administration. The chromatographic conditions were previously described in detail (Svendsen, 1993, Achilli et al., 1994). Statistical analyses were made with non parametric tests, namely Kryskall-Wallis and Wilcoxson test.

## RESULTS

The effect of per os administration of L-dopa (100 mg) and benzeraside (25 mg) twice a day for 4 consecutive days in pigs are summarized in Figure 1. The treatment resulted in both a significant decrease in tyrosine and normetanephrine levels and significant increases in 3-o-M-dopa levels HVA levels tended to increase but the difference was not significant. No modifications were observed regarding the other compounds, except for kynurenine levels which decreased drastically.

## DISCUSSION

Regarding the tyrosine metabolic pathway our results following subchronic treatment with L-dopa + benzerazide in pig are in agreement with previous reports in human. The marked increase in 3-*o*-M-dopa formation in periphery after L-dopa administration with an involvement in the side effects of this therapy is well established (Feuerstein et al., 1977, Pletscher, 1993). Increases in 3-*o*-M-dopa and HVA levels in human brain (Ogawa et al., 1992) and CSF (Ikarashi et al., 1992, Toghi et al., 1993) were reported in Parkinsonian patients treated with L-dopa. The present observation of a decrease in tyrosine and nor-

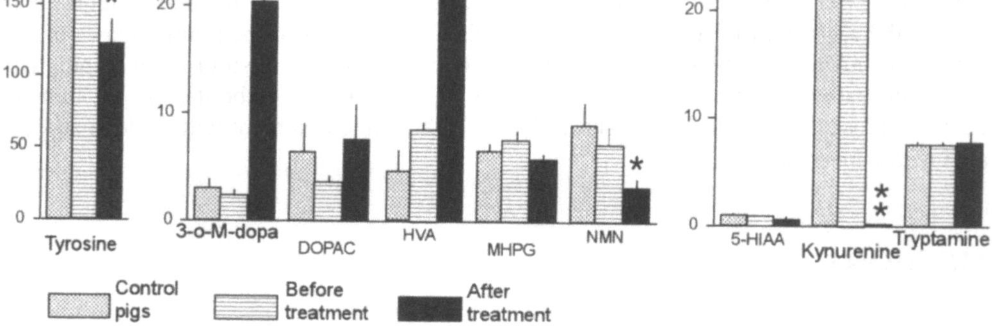


**Figure 1.** Effect of subchronic (twice a day for consecutive 4 days) per os administration of L-dopa (100 mg) and benzeraside (25 mg) on tryptophan and tyrosine metabolites levels in pig cerebrospinal fluid. Lumbar puncture (L2-L4) were performed in a group of control pigs and, in the group of treated pigs, before the treatment and 2 hours after the last administration. Each column represents the mean ± SEM of 5 determinations. * $p < 0.05$, ** $p < 0.01$ vs control or vs before treatment.

metanephrine levels is in keeping with the report of Maruyama et al. (1992) that 3-*o*-M-dopa produced from L-dopa might both inhibit the transport of tyrosine to neurons and decrease tyrosine hydroxylase activity. It appears also likely that dopa and probably 3-*o*-M-dopa are neuromodulators of norepinephrine release from cathecholaminergic terminals (Misu and Goshima, 1993). The report of Marauyama et al. (1992) indicated that dopa and 3-o-M-dopa may also alter tryptophan metabolism by both competing with the uptake of the amino acid in neurons and inhibiting tryptophan hydroxylase activity. Ogawa et al. (1992) reported decreased levels of 5-HT, kynurenine and kynurenic acid levels in cortex, putamen and substantia nigra of patients suffering from Parkinson's disease. They also described a greater decrease in patients treated with L-dopa. This suggests that L-dopa modifies not only the 5-hydroxytryptamine but also the kynurenic metabolic pathway. The present observation of a decrease of kynurenine levels in CSF of pigs treated with L-dopa also supports such an hypothesis. However, Tohgi et al. (1993) observed lower 5-HT levels in CSF from patients treated with L-dopa than in untreated patients, while the kynurenine levels were of the same magnitude wether the patients were treated or not. It is possible that the kynurenine metabolic pathway is more sensitive to L-dopa administration in pig than in human. It should however be noted that our study was conducted in normal animals submitted to L-dopa treatment for a short period, a situation which differs notably from treatment of patients for several months or years. Further studies in adult animals submitted to chronic treatment are therefore needed for a definite conclusion to be drawn on this point. In addition one cannot preclude the possibility that benzeraside may modify kynurenine metabolism. Thus the effect of this drug administered alone must also be investigated.

## REFERENCES


Achilli G., Grégoire S., Banting A., Lanteri-Minet M., Le Bars D. and Weil-Fugazza J., 1994, High-performance Liquid Chromatography with multielectrode array measurements of tyrosine and tryptophan derivatives in human and pig lumbar cerebrospinal fluid. in '*Monitoring molecules in Neuroscience*', A.Louilot et al. Edts., Copy Top, Draguignan, France. pp 35-36.

Beal F.M., Matson W.R., Swartz K.J., Gamache P.H. and Bird ED., 1990, Kynurenine pathway measurements in Huntington 's disease striatum: evidence for reduced formation of kynurenic acid., *J Neurochem.*, 55: 1327.

Feuerstein C., Serre F., Gavend M., Pellat J., Perret J and Tanche M.,1977, Plasma o-ethyldopa in levodopa-induced dyskinesia, *Acta Neurol. Scandinav.*, 56: 508.

Heyes M.P., Saito K., Crowley S., Davis E., Demitrac M.A. et al. ,1992, Quinolinic and kynurenine pathway metabolism in inflammatory and non-inflammatory neurological disease, *Brain*, 115: 1249

Ikarashi Y., Le Roy Blank C., Inoue H.K. and Maruyama Y.,1992, Measurements of neurochemical levels in ventricular cerebrospinal fluid of Parkinson's patients using multicolumn, multielectrode liquid chromatography with electrochemical detection, *Biogenic Amines,* 30: 175.

Maruyama W., Naoi M., Takahashi A., Watanabe H., Konagaya Y., Mokumo K., Hasegawa S. and Nakahara D.,1992, The mechanism of perturbation in monoamine metabolism by L-DOPA therapy: in vivo and in vitro studies, *J. Neural Transm.* , 90: 183.

Misu, Y. and Goshima, Y.,1993, Is L-dopa an endogenous transmitter?., *Trends in Pharmacological Sciences*, 14: 119.

Ogawa T., Matson W.R., Beal M.F., Myers R.H., Bird E.D., Milbury P. and Saso S. ,1992, Kynurenine pathway abnormalities in Parkinson's disease., *Neurology,* 42: 1702.

Pletscher A. and Da Prada M.,1993, Pharmacotherapy of Parkinson's disease: Research from 1960-1991, *Acta Neurol. Scand.,* 46 suppl. 87: 26

Svendsen C.N., 1993, Multi-electrode array detectors in High-Performance Liquid Chromatography: A new dimension in electrochemical analysis, *Analyst,* 118: 123

Tohgi H., Abe T., Takahshi S., Takahashi J. and Hamato H.,1993, Alteration in the concentration of serotonergic and dopaminergic substances in the cerebrospinal fluid of patients with Parkinson's disease, and their changes after L-dopa therapy., *Neuroscience Letters* ,159: 135.

# *IMMUNOSYSTEM AND TRYPTOPHAN CATABOLISM*

# 20

# THE KYNURENINE PATHWAY AND NEUROLOGIC DISEASE

## Therapeutic Strategies

M. P. Heyes*

Section on Analytical Biochemistry
Laboratory of Clinical Science
National Institute of Mental Health
Bethesda, Maryland 20892

## INTRODUCTION

Quinolinic acid (QUIN) is a neurotoxic metabolite of the tryptophan - kynurenine pathway (Bender, 1982) that selectively injures and kills vulnerable neurons via activation of the N-methyl-D-aspartate class of excitatory amino acid receptors (Stone, 1993). Accumulations of QUIN have been hypothesized to cause a broad spectrum of human neurologic disease.

## CLINICAL STUDIES

CNS QUIN levels are not elevated in many neurologic disorders, including Huntington's disease, Alzheimer's disease, olivopontocerebellar atrophy, progressive supra nuclear palsy, myoclonus epilepsy, temporal lobe epilepsy, schizophrenia, bulemia nervosa, anorexia nervosa, depression and obsessive-compulsive behavior (see Heyes, et al., 1992). In contrast, substantial elevations do occur in patients and animal models with inflammatory neurologic conditions, including physical trauma to the CNS, microbial infections, various forms of meningitis and autoimmune conditions. In the case of retrovirus infection, increases in CSF QUIN levels are quantitatively correlated to the severity of neurologic impairments and decrease following anti-retroviral therapy (Brouwers, et al., 1993; Heyes, et al., 1991).

* Address Correspondence to: Melvyn P. Heyes, Section on Analytical Biochemistry, Laboratory of Clinical Science, Building 10, Room 3D40, National Institute of Mental Health, Bethesda, MD 20892. Telephone #: (301)-496-3628. FAX #: (301)-402-0188.

## MECHANISMS OF FORMATION

Accumulations of QUIN in CNS-restricted inflammation are generally higher than the increases associated with systemic immune activation, and have been explained by increased QUIN production in macrophage infiltrates and perhaps activated microglia within the CNS. Increased activities of the enzymes indoleamine-2,3-dioxygenase and kynurenine-3-hydroxylase, with concomitant accelerated catabolism of L-tryptophan through the kynurenine pathway are critical biochemical events that increase CNS QUIN levels (Heyes, et al., 1992, 1993). By contrast, peripheral synthesized QUIN and QUIN precursors are important for the moderate elevations in CNS levels during systemic immune activation, such as septicemia (Saito, et al., 1993b). Many systemic tissues appear to be able to convert L-tryptophan to QUIN, and contribute to the elevations in blood and brain QUIN (Saito, et al., 1993b). Cytokines appear to be important links between the presence of immune activation and increased QUIN production (Heyes, et al., 1991; Saito, et al., 1993a).

While the neurotoxic effects of QUIN are well established, and circumstances where QUIN levels within the CNS are elevated have been identified, it is not known whether such accumulations are of neuropathogenic or physiologic significance. At least two tools are required to investigate this question. Firstly, agents are needed to specifically reduce QUIN synthesis, lower CNS levels, or attenuate the neurotoxic effects of QUIN. Secondly, an appropriate animal model is needed that has substantial accumulations of QUIN in CNS tissue with which to determine the drugs' effects on QUIN metabolism and evaluate whether modulation of QUIN synthesis has any effects on neurologic outcome.

Potential therapeutic targets include strategies to reduce macrophage activation, decrease the stimulatory effects of cytokines, reduce the availability of L-tryptophan for QUIN synthesis, as well as drugs which block NMDA receptors. However, these approaches have widespread effects that are independent of QUIN. Consequently, if these strategies were to be effective, it would not be possible to attribute their effects to QUIN rather than other mechanisms. A more specific approach is to use inhibitors of kynurenine pathway enzymes.

## SCREENING STRATEGIES FOR KYNURENINE PATHWAY INHIBITORS

We have used macrophages in culture to evaluate the ability of drugs to block QUIN production (Heyes, et al., 1993; Saito, et al., 1993a). The system has the advantage of simplicity, a high throughput and represents a major source of QUIN in the inflammatory conditions where QUIN levels are actually elevated. The system does not replicate issues of bioavailability, blood-brain barrier permeability and drug metabolism, and so additional models are used to evaluate potential candidates. Ischemic brain injury in gerbils (Heyes and Nowak, 1990; Saito, et al., 1993c), poliovirus infection of non-human primates (Heyes, et al., 1993) and contusion injury of the spinal cord of guinea pigs (Blight, et al., 1993, 1995) are used as models of CNS-restricted inflammation. Intraperitoneal injections of endotoxin, pokeweed mitogen or cytokines (Saito, et al., 1991, 1992, 1993b) are used as models of systemic immune activation.

We have found that 4-chloro-3-hydroxyanthranilate (Parli, et al., 1980) is a potent inhibitor of 3-hydroxyanthranilate-3,4-dioxygenase in macrophages (Heyes, et al., 1993; Saito, et al., 1993a). 3-Hydroxyanthranilate-3,4-dioxygenase is a non-regulatory enzyme that converts 3-hydroxyanthranilate to the unstable intermediate, 2-amino-3-carboxymuconate semialdehyde, which then spontaneously cyclyses to form QUIN. 4-Chloro-3-hydroxyanthranilate is also active in the CNS when administered systemically (Saito, et al., 1994) or applied directly to the CNS (Heyes, et al., 1988, 1993).

## EVALUATION OF 4-CHLORO-3-HYDROXYANTHRANILATE

Since the earliest organized studies, the functional deficits that result from damage to the spinal cord in contusion or compression trauma have been interpreted to be not entirely instantaneous and mechanical, but to evolve by delayed pathophysiologic mechanisms over the course of hours or days. Understanding such secondary mechanisms has important implications for development of acute therapy. Brief experimental compression injury of the spinal cord of guinea pigs results in a limited primary loss of spinal cord function that is followed over the course of two to three days by a distinct secondary loss (Blight, 1991, 1993). This model allows some aspects of post-traumatic pathophysiology to be monitored, so that the timing and extent of the secondary processes can be measured accurately. Large numbers of macrophages invade the injured spinal cord within days of injury, and their activity has been implicated in secondary demyelination and loss of axons following trauma (Blight, 1985, 1992, 1993). Circumstantial evidence for the role of macrophages in trauma has been augmented by the observations that intraperitoneal injection of silica, a macrophage "toxin", immediately after injury attenuated the secondary functional and histopathologic deficits (Blight, 1994).

We have also determined that tissue levels of QUIN are significantly elevated in the lesion site at 5 days post injury, in association with increased activity of indoleamine-2,3-dioxygenase and L-KYN accumulation, and are potentially sufficient to produce direct neural damage and contribute to functional deficits (Blight, et al., 1993). The guinea pig model offers several advantages in the context of testing modulators of QUIN metabolism relative to other neurologic conditions in which QUIN elevations have been observed. These include the ability to quantify the severity of functional and neurophysiologic deficits, the relatively short time course and the clearly secondary nature of the functional pathologic changes, the relatively small size of the animal, and the demonstrated involvement of macrophages. Spinal cord neurons have also been reported to be particularly sensitive to the neurotoxic effects of QUIN and other excitotoxins (Giulian, et al., 1990)

In our most recent study (Blight, et al., 1995), intraperitoneal injections of 4-chloro-3-hydroxyanthranilate (100 mg/kg every 12 hours) attenuated QUIN accumulations in spinal cord following injury, and reduced the severity of delayed functional deficits. Intraperitoneal injections of the macrophage toxin, silica also reduced QUIN levels and attenuated functional deficits. A direct subdural infusion of 4-chloro-3-hydroxyanthranilate into the injured spinal cord (50 μM, 1 μL/h) promptly exacerbated functional impairments, which suggests that 4-chloro-3-hydroxyanthranilate at this concentration had direct toxic effects. These results indicate that guinea pigs with spinal cord injury constitute a useful model to study the mechanisms that increase CNS QUIN levels in conditions of CNS inflammation, and to evaluate the neurochemical and neurological effects of agents designed to reduce the accumulations of QUIN and other potential pathogenic mediators within the CNS. The results are consistent with a contributory role for QUIN in the pathogenesis of secondary functional impairments following spinal cord injury, although the possibility that 4-chloro-3-hydroxyanthranilate had additional effects independent of QUIN cannot be excluded. Further studies are required to determine whether the beneficial effects of 4-chloro-3-hydroxyanthranilate are sustained. While it is unknown whether secondary inflammatory processes contribute significantly to neurologic deficits in humans spinal cord injury, strategies that reduce the accumulation of QUIN are worthy of consideration and evaluation as a therapeutic target.

## SUMMARY AND CONCLUSION

The neurotoxic effects of QUIN have been well established. Clinical conditions have been identified where substantial elevations in CNS QUIN levels occur. There is a relationship between the severity of neurologic impairments and macrophage activation, with the magnitude of the increases in QUIN. The magnitude of QUIN increases in experimental immune activation, and macrophages *in vitro,* are highest in non-human primates, intermediate in gerbils and guinea pigs, and lowest in mice and rats. Macrophages *in vitro* are a useful screening system to evaluate potential inhibitors of the kynurenine pathway. Several models of CNS inflammation are available, including brain injury in post-ischemic gerbils and spinal cord injury in guinea pigs. 4-Chloro-3-hydroxyanthranilate is a potent inhibitor of QUIN production by macrophages and reduces QUIN accumulations in spinal cord injury. Such reductions are associated with significant neurologic improvements in the early post-injury period. The results support further investigation of QUIN as a mediator of neurologic dysfunction and damage in neurologic diseases.

## REFERENCES

Bender, D.A., 1982, Biochemistry of tryptophan in health and disease. *Molec. Aspects Med.* 6: 101.

Blight, A.R., 1985, Delayed demyelination and macrophage invasion: a candidate for secondary cell damage in spinal cord injury. *Cent. Nerv. Syst. Trauma* 2: 299.

Blight, A.R., 1991, Morphometric analysis of a model of spinal cord injury in guinea pigs, with behavioral evidence of delayed secondary pathology. *J. Neurol. Sci.* 103: 156.

Blight, A.R., 1992, Macrophages and inflammatory damage in spinal cord injury. *J. Neurotrauma* 9:S83.

Blight, A.R., 1993, Remyelination, revascularization and recovery of function in experimental spinal cord injury, in Neural Injury and Regeneration, F. J. Seil, eds, Raven Press, New York, pp. 91-104.

Blight, A.R., 1994, Effects of silica on the outcome of experimental spinal cord injury: Implication of macrophages in secondary tissue damage. *Neuroscience* 60:263.

Blight, A.R., Cohen, T.I., Saito, K. and Heyes, M.P., 1995, Quinolinic acid accumulation and functional deficits following experimental spinal cord injury. *Brain* In Press, .

Blight, A.R., Saito, K. and Heyes, M.P., 1993, Increased levels of the excitotoxin quinolinic acid in spinal cord following contusion injury. *Brain Res.* 632:314.

Brouwers, P., Heyes, M.P., Moss, H., Wolters, P., El-Amin, D., Poplack, D., Markey, S.P. and Pizzo, P., 1993, Quinolinic acid in the cerebrospinal fluid of children with symptomatic human immunodeficiency virus type-1 disease: relationships to clinical status and therapeutic response. *J. Infect. Dis.* 168:1380.

Giulian, D., Vaca, K. and Noonan, C.A., 1990, Secretion of neurotoxins by mononuclear phagocytes infected with HIV-1. *Science* 250:1593.

Heyes, M.P., Brew, B.J., Martin, A., Price, R.W., Salazar, A.M., Sidtis, J.J., Yergey, J.A., Mouradian, M.M., Sadler, A.E., Keilp, J., Rubinow, D. and Markey, S.P., 1991, Quinolinic acid in cerebrospinal fluid and serum in HIV-1 infection: Relationship to clinical and neurologic status. *Ann. Neurol.* 29:202.

Heyes, M. P., Hutto, B. and Markey, S.P., 1988, 4-chloro-3-hydroxyanthranilic acid inhibits 3-hydroxyanthranilic acid oxidase in brain. *Neurochem. Int.* 13:405.

Heyes, M.P. and Nowak, T.S.J., 1990, Delayed increases in regional brain quinolinic acid follow transient ischemia and the gerbil. *J. Cerebral Blood Flow Metabol.* 10:660.

Heyes, M.P., Saito, K., Crowley, J., Davis, L.E., Demitrak, M.A., Der, M., Dilling, L., Kruesi, M.J.P., Lackner, A., Larsen, S.A., Lee, K., Leonard, H., Markey, S.P., Martin, A., Milstien, S., Mouradian, M.M., Pranzatelli, M.R., Quearry, B.J., Salazar, A., Smith, M., Straus, S.E., Sunderland, T., Swedo, S. and Tourtellotte, W.W., 1992, Quinolinic acid and kynurenine pathway metabolism in inflammatory and non-inflammatory neurologic disease. *Brain* 115:1249.

Heyes, M.P., Saito, K., Major, E.O., Milstein, S., Markey, S.P. and Vickers, J.H., 1993, A mechanism of quinolinic acid formation by brain in inflammatory neurologic disease: Attenuation of synthesis from L-tryptophan by 6-chloro-tryptophan and 4-chloro-3-hydroxyanthranilate. *Brain* 116:1425.

Parli, C.J., Krieter, P. and Schmidt, B., 1980, Metabolism of 6-chlorotryptophan to 4-chloro-3-hydroxyanthranilic acid: A potent inhibitor of 3-hydroxyanthranilic acid oxidase. *Arch. Biochem. Biophys.* 203:161.

Saito, K., Chen, C.Y., Masana, M., Crowley, J., Markey, S.P. and Heyes, M.P., 1993a, 4-Chloro-3-hydroxyanthranilic acid, 6-chlorotryptophan and norharmane attenuate quinolinic acid formation by interferon-γ stimulated monocytes (THP-1) cells. *Biochem. J.* 291:11.

Saito, K., Crowley, J.S., Markey, S.P. and Heyes, M.P., 1993b, A mechanism for increased quinolinic acid formation following acute systemic immune stimulation. *J. Biol. Chem.* 268:15496.

Saito, K., Markey, S.P. and Heyes, M.P., 1991, Chronic effects of gamma-interferon on quinolinic acid and indoleamine-2,3-dioxygenase in brain of C57BL6 mice. *Brain Research* 546:151.

Saito, K., Markey, S.P. and Heyes, M.P., 1992, Effects of immune activation on quinolinic acid and kynurenine pathway metabolism in the mouse. *Neuroscience* 51:25.

Saito, K., Markey, S.P. and Heyes, M.P., 1994, 6-Chloro-tryptophan, 4-chloro-3-hydroxyanthranilate and dexamethasone attenuate quinolinic acid accumulation in brain and blood following systemic immune activation. *Neurosci. Lett.* 178:211.

Saito, K., Nowak, T.S.,Jr., Markey, S.P. and Heyes, M.P., 1993c, Mechanism of delayed increases in kynurenine pathway metabolism in damaged brain regions following transient cerebral ischemia. *J. Neurochem.* 60:180.

Stone, T.W., 1993, Neuropharmacology of quinolinic and kynurenic acids. *Pharmacol. Rev.* 45:309.

# 21

# INCREASE OF TRYPTOPHAN IN SERUM AND IN CEREBROSPINAL FLUID OF PATIENTS WITH HIV INFECTION DURING ZIDOVUDINE THERAPY

D. Fuchs,[1] M. Gisslén,[1] M. Larsson,[2] G. Norkrans,[2] L. Hagberg,[2] and H. Wachter[1]

[1] Institute of Medical Chemistry and Biochemistry, University of Innsbruck and
Ludwig Boltzmann Institute of AIDS Research
Innsbruck, Austria
[2] Department of Infectious Diseases and Department of Clinical Neuroscience
University of Göteburg
Göteburg, Sweden

## ABSTRACT

A high percentage of patients with human immunodeficiency virus infection presents with decreased tryptophan concentrations in serum and cerebrospinal fluid. In parallel degradation products of tryptophan like kynurenine and quinolinic acid are increased. We investigated the behavior of tryptophan concentrations in 14 patients with HIV infection before and during treatment with zidovudine, and we found a significant increase of tryptophan in serum and cerebrospional fluid after 4-14 months of therapy. In parallel, neopterin concentrations decreased signficantly. Moreover, an association existed in cerebrospinal fluid between the degree of tryptophan increase and neopterin decrease. Thus, treatment with zidovudine contributes to a gradual normalization of tryptophan metabolism in patients with HIV-1 infection. The data imply that zidovudine therapy is associated not only with a reduction of virus replication but also immune activation is reduced.

## INTRODUCTION

Decreased serum and cerebrospinal fluid (CSF) tryptophan concentrations occur frequently in human immunodeficiency virus (HIV) infected individuals (Werner et al., 1988; Larsson et al., 1989; Fuchs et al., 1990). Usually a correlation exists between the

*Recent Advances in Tryptophan Research,* edited by Graziella Allegri Filippini et al.
Plenum Press, New York, 1996

disturbances of tryptophan metabolism and enhanced formation of neopterin (Fuchs et al., 1990a; Fuchs et al., 1990b). This observation supports the view that chronic immune activation being associated with degradation of tryptophan is likely to be the underlaying mechanism of tryptophan disturbances: Interferon-g, a cytokine produced during various types of infections, including that of HIV, has been shown to activate indoleamine-2,3-dioxygenase (IDO), an enzyme degrading tryptophan to form N-formyl-kynurenine which is further converted to nicotinamide adenine dinucleotide (Freese et al., 1990). Interferon-$\gamma$ also activates monocytes/ macrophages for neopterin production (Fuchs et al., 1988).

HIV infection is frequently associated with neurological complications, neuropsychiatric symptoms, and, in advanced cases of AIDS, with dementia. The pathogenic mechanisms behind the devolopment of dementia are still unknown, but the process is probably multifactorial. Because of its neurotoxicity, the tryptophan degradation products kynureninic acid and quinolinic acid have been suggested as possible mediators for development of AIDS dementia complex (Lipton, 1992).

## PATIENTS AND METHODS

We analyzed concentrations of tryptophan, kynurenine and neopterin in blood and CSF from 14 HIV-seropositive individuals before antiviral treatment with zidovudine and after 4-14 months of treatment (Gisslén et al., 1994). Nine patients were asymptomatic, Centers for Disease Control (CDC) stage II, and five had symptomatic HIV infection, CDC stage IV. Zidovudine (Retrovir; Wellcome, Research Triangle Park, NC) was given three or four times a day, with daily doses varying between 400 and 1200mg (mean, 668mg). Tryptophan concentrations were determined by high performance liquid chromatography with fluorescence detection (Larsson et al., 1989). Neopterin was measured in CSF and in

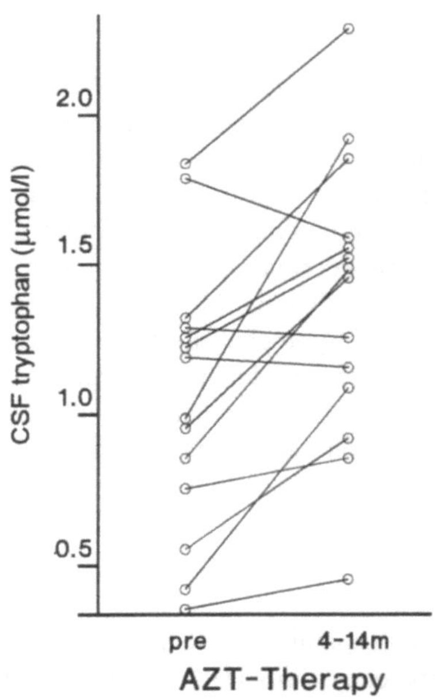


**Figure 1.** Individual tryptophan concentrations in cerebrospinal fluid of patients with HIV infection before and after 4-14m of zidovudine therapy.

serum employing a commercially available radioimmunassay (IMMUtest; Henning-Berlin, Germany) (Fuchs et al., 1992).

## RESULTS

The median pretreatment concentrations of tryptophan in CSF and blood were low (224 ng/ml and 6.0 mg/ml, respectively) but increased after 4-14 months of zidovudine treatment ($p<0.0$, Figs.1 and 2). In contrast the neopterin concentrations in CSF decreased after 4-14 months of zidovudine treatment (Table 1). Serum neopterin concentrations behaved similarly.

The pretreatment serum concentrations of neopterin and of tryptophan correlated inversely ($p=0.02$). After zidovudine treatment, the degree of the tryptophan increase and the neopterin decrease in CSF also showed an inverse correlation ($p<0.01$).

## DISCUSSION

Tryptophan is a nutritionally essential amino acid with important roles in serveral metabolic pathways, such as the synthesis of nicotinic acid and 5-hydroxy-tryptamine (serotonin). A possible explanation of the tryptophan deficiency in HIV-1-infected individuals is an increase in tryptophan degradation caused by immune stimulation. Interferon-γ, a cytokine produced during various types of infections, including that of HIV, has been shown to activate IDO which degrades tryptophan to N-formyl-kynurenine and kynurenine. Kynurenine is further degraded to quinolinic acid. Both substances have been reported as increased in CSF from patients with HIV-1 infection (Fuchs et al., 1990a). Because of its neurotoxicity quinolinic acid has been suggested as a possible mediator for development of

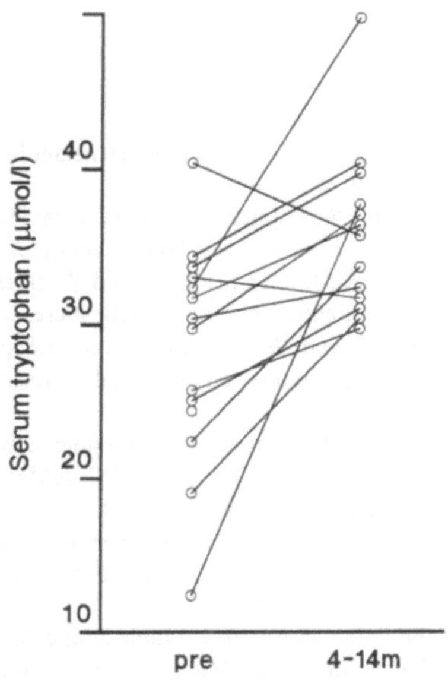


**Figure 2.** Individual tryptophan concentrations in serum of patients with HIV infection before and after 4-14m of zidovudine therapy.

**Table 1.** Neopterin concentrations in serum and cerebrospinal fluid of patients with HIV infection before and 4-14 months after start of zidovudine therapy (means ± S.D. are shown)

| | Before | After 4–14 m of therapy |
|---|---|---|
| Serum | 20.7 ± 8.7 | 13.6 ± 5.5* |
| Cerebrospinal fluid | 20.0 + 13.6 | 10.1 ± 4.5* |

*significantly lower than pretreatment values (serum: U=2.89, p=0.004; cerebrospinal fluid: U=2.76, p=0.006).

AIDS dementia complex (Lipton, 1992). The association between increased neopterin and decreased tryptophan concentrations found in HIV infected patients adds support to the theory of increased tryptophan degradation ascribable to immune stimulation. After initiation of zidovudine treatment in HIV seropositive individuals the concentrations of tryptophan increased and those of neopterin decreased in CSF and blood. The present study shows that tryptophan deficiency is reversible by zidovudine treatment. Again, the parallelism between the trends towards normalization of both analytes, the increase of tryptophan and the decrease of neopterin, points to the view that zidovudine therapy also achieves some sort of down-regulation of the stimulated immune system in patients with HIV infection. Further studies will demonstrate if the changes are persistent or whether the development of zidovudine-resistant HIV strains or clinical signs of HIV dementia may be linked to any eventual reduction in tryptophan concentration.

## ACKNOWLEDGMENT

This work was supported by the Austrian Ministry of Science and Research, "Sektion Forschung".

## REFERENCES

Freese A., Swartz K.J., During M.J. and Martin J.B., 1990, Kynurenine metabolites of tryptophan: Implications for neurologic diseases, *Neurology* 40: 691.

Fuchs D., Hausen A., Reibnegger G., Werner E.R., Dierich M.P. and Wachter H., 1988, Neopterin as a marker for activated cell-mediated immunity: Application in HIV infection, *Immunol Today* 9:150.

Fuchs D., Möller A.A., Reibnegger G., Stöckle E., Werner E.R., and Wachter H., 1990, Decreased serum tryptophan in patients with HIV-1 infection correlates with increased serum neopterin and with neurologic/ psychiatric symptoms, *J Acquir Immune Defic Syndr* 3:873.

Fuchs D., Forsman A., Hagberg L., et al., 1990, Immune activation and decreased tryptophan in patients with HIV-1 infection, *J Interferon Res* 10:599.

Fuchs D., Werner E.R. and Wachter H., 1992, Soluble products of immune activation: Neopterin. In: Manual of Clinical Laboratory Immunology, 4th edition, R.R. Rose, E.C. deMacario , J.L. Fahey , H. Friedman and G.M. Penn, eds., American Society for Microbiology 251.

Gisslén M., Larsson M., Norkrans G., Fuchs D., Wachter H. and Hagberg L., 1994, Tryptophan concentrations increase in cerebrospinal fluid and blood after zidovudine treatment in patients with HIV type 1 infection, AIDS *Res Human Retroviruses* 10:947.

Larsson M., Hagberg L., Norkrans G., and Forsman A., 1989, Indole amine deficiency in blood and cerebrospinal fluid from patients with human immunodeficiency virus infection, *J Neurosci Res* 23:441.

Lipton S.A., 1992, Models of neuronal injury in AIDS: another role for the NMDA receptor? *Trends Neurosci* 15:75.

Werner E.R., Fuchs D., Hausen A., et al., 1988, Tryptophan degradation in patients infected by human immunodeficiency virus, *Biol Chem Hoppe Seyler* 369:337.

22

# IMMUNOBIOLOGY OF PICOLINIC ACID

G. Melillo, M. C. Bosco, T. Musso, and L. Varesio

Macrophage Cell Biology Section, Lab. Exp. Immunol., BRMP, and
Biological Carcinogenesis and Development Program, PRI/DynCorp
National Cancer Institute-FCRDC, NIH
Frederick, Maryland 21702

## 1. PICOLINIC ACID AND MACROPHAGE ACTIVATION

Exposure of various cell types to IFN results in increased enzymatic activities leading to the breakdown of tryptophan and evidence exists to suggest that the catabolism of tryptophan may be connected to the biological effects of interferon (Carlin et al., 1989. The first and rate-limiting enzymatic step in the degradation of tryptophan is the conversion of tryptophan to formylkynurenine. Two different enzymes are deputed to this function. The tryptophan 2,3-dioxygenase (TDO) is found only in the liver and the indoleamine 2,3-dioxygenase (IDO) is found in all tissues examined so far and inducible by IFN and IFN-inducers such as LPS and viruses. Tryptophan catabolism can be demonstrated *in vivo* in experimental murine models as well as in human subjects (Yoshida et al, 1979). We have confirmed the expression of IDO activity in human monocytes activated by IFN-γ and in the monocytic leukemia cell line THP-1 activated by IFN-γ or by TPA (Musso et al., 1994).

We have extensively studied picolinic acid, an end product of tryptophan metabolism, and we have demonstrated that picolinic acid is biologically active on macrophages(Varesio et al., 1990). Experiments in which macrophages were incubated with IFN-γ plus picolinic acid demonstrated that picolinic acid is a potent costimulator of macrophage activation that functions as a second signal [for review see (Varesio et al., 1992; 1993; Sharon et al., 1990; Rebello et al., 1982). Picolinic acid alone does not activate cytotoxic macrophages, however, it acts synergistically with IFN-γ in activating macrophages from C57BL/6 mice. In addition, picolinic acid is a second signals in the activation of macrophages from C3H/HeJ and C3H/HeN mice, which do not become cytotoxic in response to IFN-γ alone (Varesio et al., 1990).

The possibility that picolinic acid is a physiological stimulus for macrophages is suggested by several considerations. The body can catabolize tryptophan to picolinic acid that may be released in the extracellular milieu or act intracellularly on the producing cell. Moreover, we have found that picolinic acid may be involved *in vivo* in the process of macrophage activation. Intraperitoneal injection of picolinic acid into mice induced antitumor activity in macrophages (Ruffmann et al., 1984; 1987) suggesting that this compound may enhance the response of macrophages to low levels of cytokines present in the peritoneal

*Recent Advances in Tryptophan Research,* edited by Graziella Allegri Filippini et al.
Plenum Press, New York, 1996

cavity. Macrophage activation could account for the observed antitumor effect of picolinic acid in tumor bearing mice. In conclusion, the demonstration that picolinic acid is a macrophage costimulator *in vitro*, taken together with the evidence that picolinic acid is produced and active *in vivo*, support the hypothesis that picolinic acid can function as a physiologic signal for the activation of macrophages to a cytotoxic stage.

Macrophages respond to picolinic acid with changes in several biochemical pathways. We have initially characterized the changes in rRNA metabolism. Picolinic acid affects macrophage rRNA metabolism by inhibiting the accumulation of *de novo* synthesized 28S but not 18S rRNA and inducing accumulation of rRNA precursors (Varesio et al., 1990). We have previously demonstrated the existence of a direct correlation, and most likely of a causal relationship, between macrophage activation and accumulation of rRNA precursors (Varesio, 1985; 1986; Varesio et al., 1992) that occurs only in macrophages expressing tumoricidal activity in response to *in vitro* or *in vivo* activation. The changes in rRNA metabolism induced by picolinic acid are remarkably similar to those occurring in activated murine macrophages by other agents and suggest that the costimulatory activity of picolinic acid may be due to its effects on rRNA metabolism which, in turn is necessary for macrophage activation. The molecular bases for the involvement or accumulation of rRNA precursors and macrophage functional activity has been discussed elsewhere.

## 2. PICOLINIC ACID AND THE CATABOLISM OF ARGININE

We have found that picolinic acid induces the expression of the nitric oxide synthase gene and this response is directly related to the activation of macrophages to a tumoricidal stage (Melillo et al., 1994). NO is a gas, produced during the oxidation of arginine through the nitric oxide synthase (NOS), that functions as intracellular messenger or as effector molecule depending upon the tissue and the concentration. Endothelium-dependent relaxation, neurotransmission and cell mediated immune effector functions are well documented effects of NO (Nathan and Hibbs, 1991; Nathan, 1992; Lowenstein and Snyder, 1992). NO diffuses into the target cells with a passive, receptor independent mechanism, and virtually every cell is a potential target of NO. Thus, the diversity of NO demand in the various tissues is met by the existence and differential expression of various isoforms of NOS uniquely regulated in terms of Ca and calmodulin requirement and control of gene expression (Xie et al., 1992; Lorsbach et al., 1993; Nathan and Xie, 1994). The acute requirement of NO as effector molecule as in the case of inflammation, bacterial infection, tumor rejection or wound is achieved by mobilizing macrophages that will accumulate in the lesion and, upon appropriate stimulation, will release high levels of NO. This mechanism ensures that the toxic consequences of high NO output be limited in space and time by the transient nature of the infiltrate and stage of macrophage activation. Murine macrophages express a $Ca^{2+}$ and calmodulin independent isoform of NOS (iNOS) whose enzymatic levels are controlled primarily at the level of transcription of the iNOS gene (Nathan and Xie, 1994). The mobility of macrophages and hence the diversity of the conditions/tissues in which macrophages are required to transcribe iNOS that range from brain to skin, from hypoxic necrotic areas to oxygenated lung tissues, suggest that iNOS gene expression is induced by stimuli that differ substantially in nature and perhaps mechanism of action.

We have demonstrated that picolinic acid is a costimulus, together with IFN-γ, for the induction of L-arginine-dependent NO production in murine macrophages (Melillo et al., 1993). This observation provided the first evidence of a possible connection between tryptophan and arginine metabolism. Activation of NOS gene expression by IFN-γ plus picolinic acid is due to increased rate of transcription (Melillo et al., 1994).

Elevated levels of tryptophan metabolites, such as kynurenine and quinolinic acid, have been detected *in vivo* in cerebrospinal fluid in pathologic conditions (Heyes et al., 1992; Saito et al., 1993) as well as in cancer patients treated with IL-2 (Brown et al., 1989). Although there is only one report of RNI production *in vitro* by human monocyte-derived macrophages (Denis, 1991), elevation of nitrate levels in urine and plasma of IL-2-treated patients also has been demonstrated (Hibbs et al., 1992). Therefore, in pathologic conditions or following therapy with cytokines, an increased production of IFN-γ may be responsible for the concomitant activation of tryptophan and arginine metabolic pathways. Under the same conditions, activation of IDO activity by IFN-γ might lead to accumulation of picolinic acid, which in localized compartments could feed back on macrophages amplifying the production of RNI.

IL-4 down-regulated, although did not completely abolish, IFN-γ plus picolinic acid-induced NOS mRNA expression (Melillo et al., 1994). In addition to IL-4 (Liew et al., 1991), other cytokines, such as TGF-β (Ding et al., 1990), have been described to exert a negative influence on $NO_2^-$ production by murine macrophages. Interestingly, also the induction of IDO by IFN-γ in human monocytes was susceptible to inhibition by IL-4 (Musso et al., 1994). It is intriguing that one of the enzymes catabolizing tryptophan and one of those catabolizing arginine are susceptible to the same kind of cytokine control. Both are induced by IFN-γ and inhibited by IL-4. Thus, one of the common events associated with the response of the organism to IFN-γ is the induction of tryptophan- and arginine-degrading enzymes and the release of many catabolites mediating a variety of biological effects on different tissues besides macrophages. The demonstration that picolinic acid, although it does not induce RNI production by itself, can exert synergistic activity in combination with IFN-γ suggests the existence of a network of amino acid catabolites that are important for the activation and the expression of effector functions by macrophages, and it provides a good indication that the spectrum of action of the biological effects of cytokines may be extended to amino acid metabolism.

## 3. PICOLINIC ACID STIMULATES THE TRANSCRIPTION OF THE NO SYNTHASE GENE THROUGH THE HYPOXIA RESPONSIVE ELEMENT

The diversity of the pathological conditions in which macrophage NO is produced may be based on the existence of several signals inducing iNOS and implies the need for multiple regulatory elements to control its transcription (Nathan and Xie, 1994). One pathway of iNOS induction in murine macrophages is initiated by IFN-γ plus bacterial lipopolysaccharide (LPS) (Xie et al., 1992); Lorsbach et al., 1993). A 1.6 kb fragment of the 5′-flanking region of the mouse iNOS gene is sufficient to confer inducibility by IFN-γ plus LPS (Xie et al., 1993; Lowenstein et al., 1994). Additionally, the consensus sequences for the binding of interferon regulatory factor-1 (IRF-1) (Kamijo et al., 1994; Martin et al., 1994) and NF-kB (Xie et al., 1994) of the iNOS promoter are required for activation of transcription by IFN-γ and LPS, respectively. Picolinic acid is a macrophage activating agent structurally and biologically different from LPS and we reasoned that it could induce iNOS transcription utilizing promoter elements different from those upon which LPS acts.

To test the hypothesis that picolinic acid acted on distinct regions of iNOS promoter, we performed functional studies using several deletion mutant of the murine iNOS promoter linked to the chloramphenicol acetyltransferase (CAT) reporter gene in the ANA-1 murine macrophage cell line . Cells transfected with the full-length 1.6 Kb 5′-flanking region of the iNOS gene expressed little or no CAT activity either constitutively or after treatment with

IFN-γ. Picolinic acid alone induced low but appreciable levels of CAT expression, whereas high expression of CAT reporter gene was achieved by stimulation with IFN-γ plus picolinic acid, similar to those elicited by IFN-γ plus LPS. Analysis of deletion mutants of the iNOs promoter, allowed us to map the element involved in the responsiveness of IFN-γ treated ANA-1 macrophages to picolinic acid to the position -328 to -201 of the iNOS promoter (5′-GTGAC**TACGTGCT**GCCTAG-3′). These data demonstrate the existence of a new region important for the control of iNOS expression, located in between IFN-γ and LPS responsive regions and define one element target of the transcriptional activation by picolinic acid.

GeneBank search indicated the presence of 100% homology between this region of the iNOS promoter with 13 bp of a hypoxia-responsive enhancer element, originally described in the 3′ untranslated region of the erythropoietin (Epo) gene at a site required for transcriptional activation of Epo gene (Semenza and Wang, 1994). Hypoxia-inducible enhancer activity of this element has been demonstrated in Epo producing as well as non-Epo producing cells from different species, suggesting its general involvement in the regulation of transcriptional activity (Maxwell et al., 1993; Wang and Semenza, 1994). Beyond Epo gene, a hypoxia enhancer sequence has been recently characterized in the transcriptional regulation of several glycolytic enzymes, whose expression is increased by low $O_2$ levels (Semenza et al., 1994; Firth et al., 1994). Molecular $O_2$ is a cosubstrate of NOS but there is no information on the role, if any, of $O_2$ tension on iNOS expression. The presence of a functional hypoxia responsive sequence in the context of iNOS promoter prompted us to test the possibility that iNOS itself might be a hypoxia inducible gene. We found that incubation of macrophages in low oxygen tension (1% oxygen) caused a tremendous increase of iNOS mRNA expression in IFN-γ treated cells, compared to cells cultured in regular $O_2$ concentration (20% $O_2$). Nuclear run-on experiments demonstrated that hypoxia, ineffective by itself, synergistically induces iNOS transcription in combination with IFN-γ. Our data establish that a novel region of iNOS promoter, containing a hypoxia-responsive enhancer, is critical for the regulation of iNOS transcription and target of picolinic acid. The signal transduction pathway of hypoxia is not fully understood. Macrophages are very likely to be exposed during several pathological conditions (i.e. wound repair, tumor associated mac, inflammation) to changes of oxygen tension, thereby supporting the role of tissue oxygen level as a critical modulator of macrophage activation and the outcome of inflammatory responses. It will be important to investigate the role of this alternative pathway of activation of iNOS expression in different tissues, in light of the potential implications for the understanding of the biology of NO.

## 4. CONCLUDING REMARKS

The finding that picolinic acid induces iNOS transcription through a novel regulatory sequence that is not needed for the response to LPS, provides evidence of the existence of a distinct pathway of iNOS transcriptional activation, that may be important during a microbial-independent activation of the immune response. The catabolism of L-tryptophan has been demonstrated under conditions of immune activation and it might play a role in the induction of iNOS expression. However the presence of picolinic acid in inflammatory conditions or during therapy with cytokines has not been fully investigated and remains to be established.

The ability of picolinic acid, produced via an oxidative metabolic pathway, to activate a hypoxia-responsive sequence is intriguing. Besides hypoxia, other non-physiologic stimuli, including cobalt chloride ($CoCl_2$) and desferrioxamine (DFX) , induce the activity of the HRE, as demonstrated by transient transfection experiments (Wang and Semenza, 1993; 1994; Pugh et al., 1991). Because picolinic acid is a natural product derived from L-trypto-

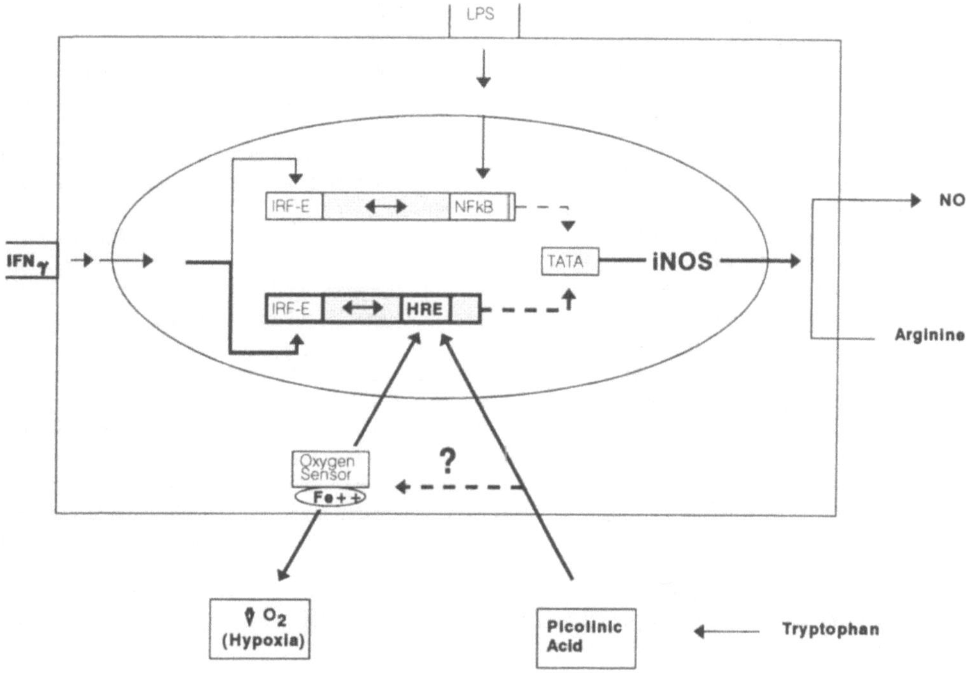


**Figure 1.** Activation of iNOS promoter in murine macrophages.

phan catabolism, it is possible to speculate that picolinic acid might be an alternative, biological activator of the HRE, thereby regulating the expression of a broader range of genes. In addition, the activation of the HRE by the iron chelator DFX has been attributed to its ability to bind to the iron moiety of a heme-protein that might function as intracellular oxygen sensor (Goldberg et al., 1988). Picolinic acid shares with DFX the property of being an iron chelator (Fernandez-Pol, 1977) and the ability of both agents to activate the HRE suggests the possibility of a common mechanism transducing the signal (see Figure 1).

Schematic representation of the signals inducing iNOS promoter activation and of the promoter elements target of the various stimuli. IFN-γ is a needed signal that is insufficient by itself to induce iNOS transcription and requires the cooperation of LPS or picolinic acid or hypoxia. IFN-γ acts on the IRF-E region; LPS acts on the NFkB site; picolinic acid and hypoxia act on the HRE. Transcriptional activation of the gene is achieved by cooperation of IRF-E with either NFkB or HRE. Picolinic acid and hypoxia may act on the same promoter element because they share the signal transduction mechanism. It is proposed, as a working hypothesis, that picolinic acid binds the iron of the oxygen sensor that transduces the hypoxia stimulus. Binding of picolinic acid to the hypoxia sensor may displace the oxygen and, by mimicking at a biochemical level the state of hypoxia, deliver the signal leading to the induction of the hypoxia responsive factor. Subsequent binding of the Hypoxia responsive factor to the HRE would than initiate iNOS transcription.

## REFERENCES

Brown R.R., Lee C.M., Kohler P.C., Hank J.A., Storer B.E., and Sondel P.M., 1989, Altered tryptophan and neopterin metabolism in cancer patients treated with recombinant interleukin 2. *Cancer Res.* 49:4941.

Carlin J.M., Ozaki Y., Byrne G.I., Brown R.R., and Borden E.C., 1989,, Interferons and indoleamine 2,3-dioxygenase: role in antimicrobial and antitumor effects. *Experientia* 45:535

Denis M., 1991, Tumor Necrosis factor and granulocyte macrophage colony-stimulating factor stimulate human macrophages to restrict growth of virulent Mycobacterium avium and to kill avirulent M. avium: killing effector mechanism depends on the generation of reactive nitrogen intermediates. *J. Leukoc. Biol.* 49:380.

Ding A., Nathan C.F., Graycar J., Derynck R., Stuehr D.J., and Srimal S., 1990, Macrophage deactivating factor and tranforming growth factors-beta 1, -beta 2, and -beta 3 inhibit induction of macrophage nitrogen oxide synthesis by IFN-gamma. *J. Immunol.* 145:940.

Fernandez-Pol J.A., 1977, Iron: possible cause of the G1 arrest induced by picolinic acid. *Biochem. Biophys. Res. Commun.* 78:136.

Firth J.D., Ebert B.L., Pugh C.W., and Ratcliffe P.J., 1994, Oxygen-regulated control elements in the phosphoglycerate kinase 1 and lactate dehydrogenase A genes: Similarities with the erythropoietin 3′ enhancer. *Proc. Natl. Acad. Sci. USA* 91:6496.

Goldberg M.A., Dunning S.P., and Franklin Bunn H., 1988, Regulation of the erythropoietin gene: evidence that the oxygen sensor is a heme protein. *Science* 242:1412.

Heyes M.P., Saito K., Jacobowitz D., Markey S.P., Takikawa O., and Vickers J.H., 1992, Poliovirus induces indoleamine-2,3-dioxygenase and quinolinic acid synthesis in macaque brain. *FASEB J.* 6:2977.

Hibbs J.B., Jr., Westenfelder C., Taintor R., Vavrin Z., Kablitz C., Baranowski R.L., Ward J.H., Menlove R.L., McMurry M.P., Kushner J.P., and et al, 1992, Evidence for cytokine-inducible nitric oxide synthesis from L-arginine in patients receiving interleukin-2 therapy [published erratum appears in J Clin Invest 1992 Jul;90(1):295]. *J. Clin. Invest.* 89:867.

Kamijo R., Harada H., Matsuyama T., Bosland M., Gerecitano J., Shapiro D., Le J., Koh S.I., Kimura T., Green S.J., Mak T.W., Taniguchi T., and Vilcek J., 1994, Requirement for transcription factor IRF-1 in NO synthase induction in macrophages. *Science* 263:1612.

Liew F.Y., Li Y., Severn A., Millott S., Schmidt J., Salter M., and Moncada S., 1991, A possible novel pathway of regulation by murine T helper type-2 (Th2) cells of a Th1 cell activity via the modulation of the induction of nitric oxide synthase on macrophages. *Eur. J. Immunol.* 21:2489.

Lorsbach R.B., Murphy W.J., Lowenstein C.J., Snyder S.H., and Russell S.W., 1993, Expression of the nitric oxide synthase gene in mouse macrophages activated for tumor cell killing. Molecular basis for the synergy between interferon-γ and lipopolysaccharide. *J. Biol. Chem.* 268:1908.

Lowenstein C.J. and Snyder S.H., 1992, Nitric oxide, a novel biologic messanger. *Cell* 70:705.

Lowenstein C.J., Alley E.W., Raval P., Snowman A.M., Snyder S.H., Russel S.W., and Murphy W.J., 1994, Macrophage nitric oxide synthase gene: two upstream regions mediate induction by interferon t and lipopolysaccharide. *Proc. Natl. Acad. Sci. USA* 90:9730.

Martin E., Nathan C., and Xie Q., 1994, Role of interferon regulatory factor 1 in induction of nitric oxide synthase. *J. Exp. Med.* 180:977.

Maxwell P.H., Pugh C.W., and Ratcliffe P.J., 1993, Inducible operation of the erythropoietin 3′ enhancer in multiple cell lines: Evidence for a widespread oxygen-sensing mechanism. *Proc. Natl. Acad. Sci. USA* 90:2423.

Melillo G., Cox G.W., Radzioch D., and Varesio L., 1993, Picolinic acid, a catabolite of L-tryptophan, is a costimulus for the induction of reactive nitrogen intermediate production in murine macrophages. *J. Immunol.* 150:4031.

Melillo G., Cox G.W., Biragyn A., Sheffler L., and Varesio L., 1994, Regulation of nitric oxide synthase mRNA expression by interferon-gamma and picolinic acid. *J. Biol. Chem.* 269:8128.

Musso T., Gusella G.L., Brooks A., Longo D.L., and Varesio L., 1994,, Interleukin-4 inhibits indoleamine 2,3,-dioxygenase expression in human monocytes. *Blood* 83:1408.

Nathan C.F. and Hibbs J.B., Jr., 1991, Role of nitric oxide synthesisin macrophage antimicrobial activity. *Curr. Opin. Immunol.* 3:65.

Nathan C.F., 1992, Nitric oxide as a secretory product of mammalian cells. *FASEB J.* 6:3051.

Nathan C.F. and Xie Q., 1994, Nitric oxide synthases: roles, tolls, and controls. *Cell* 78:915.

Nathan C.F. and Xie Q., 1994, Regulation of biosynthesis of nitric oxide. *J. Biol. Chem.* 269:13725.

Pugh C.W., Tan C.C., Jones R.W., and Ratcliffe P.J., 1991, Functional analysis of an oxygen-regulated transcriptional enhancer lying 3′ to the mouse erythropoietin gene. *Proc. Natl. Acad. Sci. U.S.A.* 88:10553.

Rebello T., Lonnerdal B., and Hurley L.S., 1982, Picolinic acid in milk, pancreatic juice and intestine: inadequate for role in zinc absorption. *Am. J. Clin. Nutr.* 35:1.

Ruffmann R., Welker R.D., Saito T., Chirigos M.A., and Varesio L., 1984, In vivo activation of macrophages but not natural killer cells by picolinic acid (PLA). *J. Immunopharmacol.* 6:291.

Ruffmann R., Schlick R., Chirigos M.A., Budzynsky W., and Varesio L., 1987, Antiproliferative activity of picolinic acid due to macrophage activation. *Drugs. Exp. Clin. Res.* 13:607.

Saito K., Crowley J.S., Markey S.P., and Heyes M.P., 1993, A mechanism for increased quinolinic acid formation following acute systemic immune stimulation. *J. Biol. Chem.* 268:15496.

Semenza G.L., Roth P.H., Fang H., and Wang G.L., 1994, Transcriptional regulation of genes encoding glycolytic enzymes by hypoxia-inducible factor 1. *J. Biol. Chem.* 269:23757.

Semenza G.L. and Wang G.L., 1994, A nuclear factor induced by hypoxia via de novo protein synthesis binds to the human erythropoietin gene enhancer at a site required for transcriptional activation. *Mol. Cell. Biol.* 12:5447.

Sharon M., Gnarra J.R., and Leonard W.J., 1990, A 100-kilodalton protein is associated with the murine interleukin 2 receptor: Biochemical evidence that p100 is distinct from the a and b chains. *Proc. Natl. Acad. Sci. USA*, 87:4869

Varesio L., 1985, Induction and expression of tumoricidal activity by macrophages. "Mononuclear Phagocytes: Physiology and Pathology," R.T. Dean and W. Jessup eds., Elsevier Science Publishers, Amsterdam, pp. 381-407.

Varesio L., 1986, Molecular bases for macrophage activation. *Ann. Inst. Pasteur. Immunol.* 137:235.

Varesio L., Clayton M., Blasi E., Ruffmann R., and Radzioch D., 1990, Picolinic acid, a catabolite of tryptophan, as the second signal in the activation of IFN-gamma primed macrophages. *J.Immunol.*, 145: 4265.

Varesio L., Cox G.W., Pulkki K., Brooks A., and Melillo G., 1992, Arginine and tryptophan catabolism: the picolinic acid connection. "Advances in tryptophan Research," I. Ishiguro, R. Kido, T. Nagatsu, Y. Nagamura and Y. Otha eds., Fujita Health University Press, Toyko, Japan pp. 309-313.

Varesio L., Radzioch D., Bottazzi B., and Gusella G.L., 1992, Ribosomal RNA metabolism in macrophages. *Curr. Top. Microbiol. Immunol.* 181: 209.

Varesio L., Cox G.W., Pulkki K., Musso T., Latham P.S., Gusella G.L., and Espinoza-Delgado I., 1993, Activation of macrophages and monocytes by interferon-gamma and interleukin-2. "Mononuclear Phagocytes in Cell Biology," J. Klostergaard and G. Lopez-Berestein eds., CRC Press, Inc., Boca Raton, Ann Arbor, London, Tokio pp. 119-146.

Wang G.L. and Semenza G.L., 1993, Desferrioxamine induces erythropoietin gene expression and hypoxia-inducible factor 1 DNA-binding activity: implications for models of hypoxia signal transduction. *Blood* 82:3610.

Wang G.L. and Semenza G.L., 1994, General involvement of hypoxia-inducible factor 1 in transcriptional response to hypoxia. *Proc. Natl. Acad. Sci. USA* 90:4304.

Xie Q., Cho H.J., Calaycay J., Mumford R.A., Swiderek K.M., Lee T.D., Ding A., Troso T., and Nathan C.F., 1992, Cloning and characterization of inducible nitric oxide synthase from mouse macrophages. *Science* 256:225.

Xie Q., Whisnant R., and Nathan C.F., 1993, Promoter of the mouse gene encoding calcium-independent nitric oxide synthase confers inducibility by interferon t and bacterial lipopolysaccharide. *J. Exp. Med.* 177:1779.

Xie Q., Kashiwabara Y., and Nathan C., 1994, Role of transcription factor NF-kB/Rel in induction of nitric oxide synthase. *J. Biol. Chem.* 269:4705.

Yoshida R., Urade Y., Tokuda M., and Hayaishi O., 1979,, Induction of indoleamine 2,3-dioxygenase in mouse lung during virus infection. *Proc. Natl. Acad. Sci. U.S.A.* 76:4084.

# 23

# CONTROL OF EXTRACELLULAR MATRIX DEGRADATION BY INTERFERON-γ

## The Tryptophan Connection

J. Varga,[1] T. Yufit,[1] E. Hitraya,[1] and R. R. Brown[2]

[1] Division of Rheumatology
Jefferson Medical College
Philadelphia, Pennsylvania 19107 and
[2] University of Wisconsin Cancer Center
Madison, Wisconsin 53706

## SUMMARY

Interleukin-1β (IL-1β) is a potent signal for the induction of the matrix-degrading enzymes collagenase and stromelysin. These metalloproteinases (MMP) play a critical role in physiologic and pathologic connective tissue remodeling, and are potential targets for therapeutic manipulation. Treatment of human dermal fibroblasts with interferon-γ inhibited Type I collagen gene expression, and abrogated the effect of IL-1β on MMP expression. Interferon-γ also caused a dramatic dose-dependent increase in indoleamine 2,3-dioxygenase mRNA, with consequent depletion of tryptophan and accumulation of kynurenine in the culture media. To examine the role of tryptophan metabolism in the effects of interferon-γ on matrix-degrading enzymes, exogenous tryptophan was added to tryptophan-depleted media, followed by stimulation of the cultures with IL-1β. Supplementation with tryptophan completely overcame the inhibitory effects of interferon-γ on MMP mRNA expression and metalloproteinase secretion into the media. In contrast, mRNA levels for Type I collagen remained profoundly depressed in interferon-γ-treated cultures in spite of addition of exogenous tryptophan. These results indicate that oxidative tryptophan metabolism mediates the effects of interferon-γ on MMP gene expression in human fibroblasts.

## INTRODUCTION

Interferon-γ (IFN-γ) is a pleiotropic immunomodulatory cytokine produced by activated $T_h1$ lymphocytes and natural killer cells with a major role in human antimicrobial host defense and in anti-tumor activity. In addition to its ability to modulate MHC Class I and Class II and FcgRI expression, induces the expression of genes for a several important

*Recent Advances in Tryptophan Research,* edited by Graziella Allegri Filippini et al.
Plenum Press, New York, 1996

enzymes in mammalian cells. These include nitric oxide synthetase, 2′-5′-oligoadenylate synthetase, and two enzymes involved in tryptophan metabolism, tryptophanyl-tRNA synthetase and indoleamine 2,3-dioxygenase (IDO) (Reviewed in Vilcek and Oliveira, 1994). These gene products mediate many of the biological activities associated with IFN-γ. Activation of IDO, which catalyzes the initial and rate-limiting step in tryptophan metabolism, accounts for the anti-proliferative effects of IFN-γ on tumor cells and the inhibition of intracellular replication of microbial pathogens (Taylor and Feng, 1991). These effects of IFN-γ are due to depletion of tryptophan, and are reversible by repletion of this essential amino acid.

The extracellular matrix is a complex aggregate of collagens and non-collagenous molecules, which maintains the structural integrity of connective tissue. A major function of IFN-γ, in addition to the immune/inflammatory response, is modulation of connective tissue metabolism. For instance, in fibroblasts IFN-γ inhibits constitutive production of Type I collagen (Varga et al., 1990). IFN-γ also inhibits the ability of fibroblasts to degrade the surrounding extracellular matrix in response to inflammatory stimuli (Brinckeroff, 1991). The mechanisms involved in these important effects of IFN-γ have not been characterized. In the present study, we examined whether modulation of IDO-mediated cellular tryptophan catabolism played a role in mediating these responses to IFN-γ in dermal fibroblasts. The results indicate that tryptophan depletion was responsible for the inhibitory effects of IFN-γ on the inducible expression of MMP genes, but not of the constitutively expressed Type I collagen gene.

## METHODS

Primary cultures of adult human dermal fibroblasts were maintained in modified Eagle's medium (MEM, approx. 50 μM tryptophan) until confluence, when IL-1β, IFN-γ (both Boehringer-Mannheim), tryptophan (Sigma) or other reagents were added as indicated. Total RNA was isolated by CsCl centrifugation and examined by Northern analysis using radiolabeled human cDNAs. Tryptophan and kynurenine in the culture media were assayed by high performance liquid chromatography. Secreted collagenase was quantitated by a competitive-binding ELISA which measures both active and latent forms of collagenase.

## RESULTS

### IFN-γ Abrogates IL-1β-Induced MMP mRNA Expression, and Enhances Tryptophan Metabolism

IFN-γ and IL-1β were added to confluent fibroblasts, and following a 24 h incubation mRNA levels were examined by Northern analysis. IL-1β caused a marked increase in MMP mRNAs (Figure 1A). Pretreatment with IFN-γ abolished the stimulatory effect of IL-1β. In contrast, IFN-γ added to the cultures 24 h after IL-1β had only minimal inhibitory effect.

Identical results were obtained when the tumor promoter phorbol myristate acetate (PMA) was used instead of IL-1β to induce MMP expression. IDO mRNA was induced by IFN-γ in a time- and dose-dependent fashion. IFN-γ (3200 U/ml for >24 h) caused a >90% decrease in tryptophan, and a >30-fold increase in kynurenine concentrations in the culture media (Figure 1B). These findings indicate that IFN-γ abrogated the stimulatory effect of IL-1β or PMA on MMP expression, and also caused a sustained increase of cellular IDO gene expression, resulting in enhanced catabolism of tryptophan and consequent depletion of this essential amino acid from the culture media.

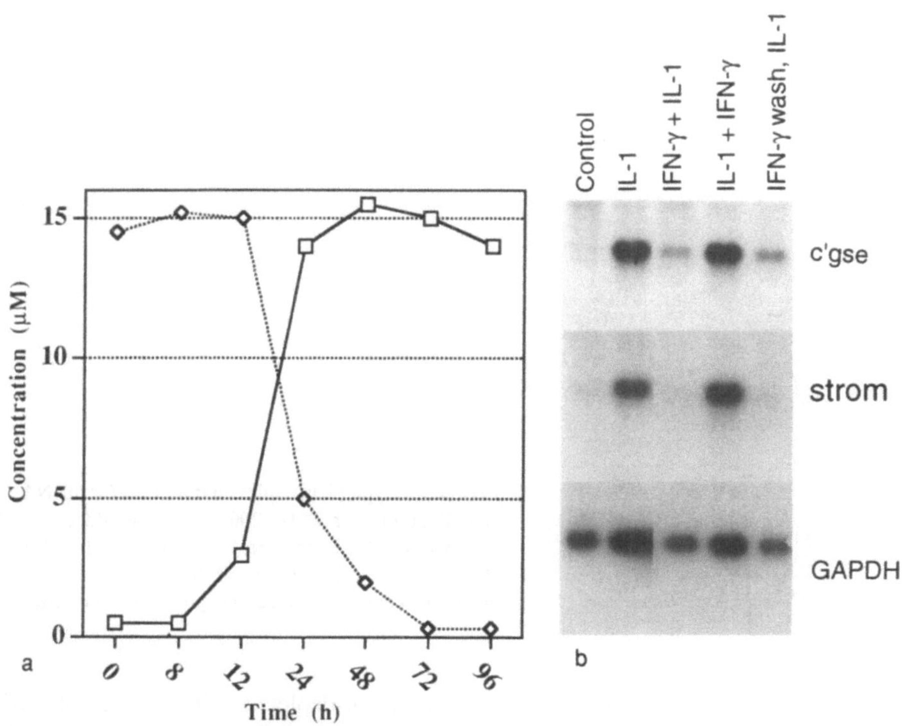


**Figure 1.** IFN-γ abrogates IL-1β-induced MMP expression, and enhances tryptophan catabolism. **A)** IFN-γ (300 U/ml) was added to confluent dermal fibroblasts 24 h prior to (lane 3) or 24 h following IL-1β (1 ng/ml) (lane 4). In lane 5, cultures were washed following a 24 h exposure to IFN-γ, and then incubated with fresh medium containing IL-1β but no IFN-γ for a further 24 h. mRNA levels were determined by Northern hybridization (10 μg/lane) to collagenase (**c'gse**), stromelysin (**strom**), and GAPDH cDNAs and quantitated by scanning densitometry. Values are shown after correction for variations in GAPDH mRNA levels (lower panel). **B)** Confluent cultures were incubated with IFN-γ (500 U/ml). At the indicated times, tryptophan (diamonds) and kynurenine (squares) in the culture media was quantitated.

## Tryptophan Depletion Inhibits IL-1β-Induced MMP Gene Expression

To examine the influence of tryptophan concentration on MMP gene expression, fibroblasts were switched from MEM to tryptophan-depleted media at confluence, and then treated with IL-1β. Stimulation by IL-1β of MMP mRNA expression and collagenase enzyme secretion was markedly reduced in the tryptophan-depleted cultures. In contrast, GAPDH mRNA levels, total protein synthesis and cellular viability were similar in tryptophan-containing and tryptophan-depleted cultures. These results demonstrate that depletion of tryptophan selectively interfered with the ability of IL-1β to induce MMP expression.

Activation of MMP transcription involves the interaction of the AP-1 DNA-binding protein with its cognate *cis*-acting recognition sites (TPA-response element) in the promoter. AP-1 is a heterodimeric complex of the c-*fos* and c-*jun* oncogenes. Because down-regulation of these oncogenes results in diminished AP-1 DNA binding and consequently reduced MMP transcription, we examined whether the negative effects of tryptophan depletion on the expression of the MMP genes involved modulation of the oncogene levels. In fibroblasts cultured in MEM, PMA caused a stimulation of c-*fos* mRNA expression, with maximal elevation at 60 min (Figure 2). In tryptophan-depleted media, c-*fos* mRNA levels increased for 40 min following exposure to PMA, but declined thereafter. The apparently shorter

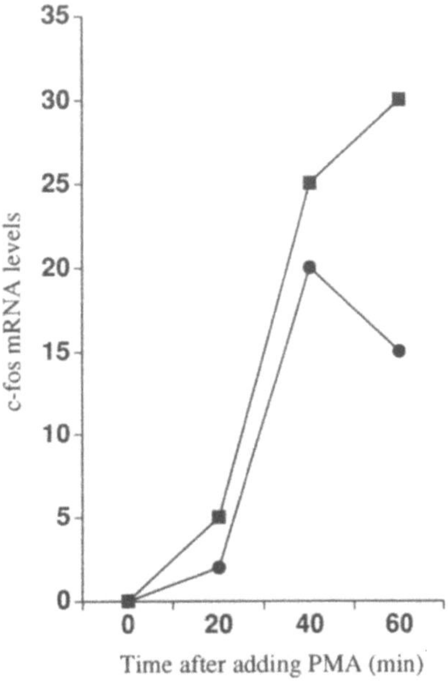


**Figure 2.** Tryptophan depletion inhibits c-*fos* mRNA expression. Phorbol ester (PMA, 200 ng/ml) was added to confluent fibroblasts in MEM (squares) or tryptophan-free MEM (circles). Cells were harvested and mRNA levels for c-*fos* were determined at the indicated time points by Northern analysis.

half-life of PMA-induced c-*fos* mRNA in tryptophan-depleted cultures suggests reduced mRNA stability. The effects of tryptophan depletion on c-*fos* mRNA expression in fibroblasts are strikingly similar to the previously described effects of IFN-γ on c-*fos* mRNA expression in macrophages (Radzioch and Varesio, 1991).

## Tryptophan Supplementation of IFN-γ-Treated Cultures Restores Response to IL-1β

IFN-γ induced the expression of IDO with consequent depletion of tryptophan from the cultures, and tryptophan-starved fibroblasts displayed reduced responsiveness to IL-1β. We therefore speculated that abrogation of IL-1β-induced MMP expression by IFN-γ was mediated through tryptophan depletion. To test this hypothesis directly, we examined whether tryptophan supplementation of the IFN-γ-treated (tryptophan-depleted) cultures could restore fibroblast responsiveness to IL-1β. Exogenous tryptophan was added to IFN-γ-treated fibroblasts, followed 24 h later by IL-1β. In the presence of tryptophan, addition of IL-1β to IFN-γ-treated cultures caused an increase in MMP mRNA and protein secretion, and 250 μg/ml of tryptophan completely overcame the IFN-γ-induced inhibition (Figure 3). In the absence of IL-1β, addition of tryptophan to untreated or IFN-γ-treated cultures had no effect on MMP gene expression.

## Tryptophan Supplementation of IFN-γ-Treated Cultures Fails to Restore Type I Collagen mRNA Levels

Since tryptophan supplementation overcame the inhibitory effects of IFN-γ on IL-1β-induced MMP stimulation, we investigated whether other negative responses associated with IFN-γ also involved tryptophan depletion. IFN-γ was added to confluent fibroblasts, followed by various concentrations of tryptophan and incubation was continued for 28 h. Addition of tryptophan, in quantities sufficient to attain a final concentration ≥50

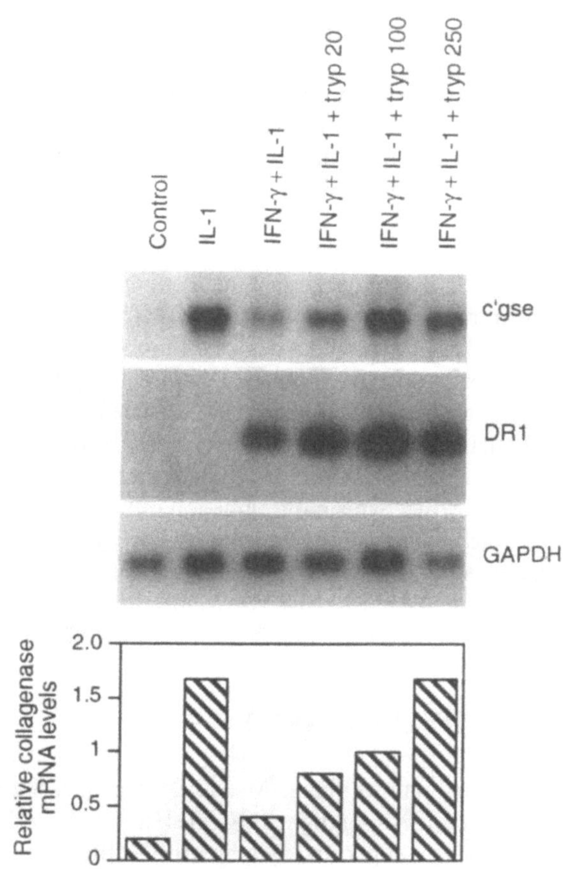


**Figure 3.** Tryptophan supplementation of IFN-γ-treated cultures restores IL-1β response. IFN-γ (300 U/ml) was added to confluent fibroblasts. Forty-eight h later, the indicated amounts of tryptophan were added, followed after 24 h by IL-1β. Control cultures were left untreated, or exposed to tryptophan only. At the end of a further 32 h of incubation, cultures were harvested, and total RNA was analyzed by Northern hybridization.

μM, to IFN-γ-treated cultures failed to overcome the inhibition of Type I collagen mRNA expression.

## DISCUSSION

By inducing the expression of matrix-degrading enzymes, inflammatory cytokines play a central role in tumor invasion, rheumatoid and osteoarthritis, and other pathologic processes (Brinckeroff, 1991). The present results demonstrate that IFN-γ blocked IL-1β-induced MMP gene expression. The inhibitory effect was maximal in cultures treated with IFN-γ prior to IL-1β, indicating that fibroblasts needed to be "primed" for IL-1β resistance. IFN-γ by itself, or in combination with IL-1β, caused a marked increase in IDO expression and tryptophan degradation. The effect is specific for tryptophan, as neither the intracellular transport nor the degradation of other amino acids is significantly effected.

Exogenous tryptophan reversed the inhibitory effects of IFN-γ on MMP, but not Type I collagen, gene expression. Taken together, these results strongly suggest that tryptophan is necessary for maximal stimulation of MMP gene expression. Tryptophan is a precursor for the synthesis of biologically important amines, and at high concentrations, directly stimulates MMP gene expression (Varga et al., 1994). Since tryptophan is important in several biological processes, it is not surprising that stimulation of IDO expression and consequent cellular tryptophan starvation have pleiotropic effects. Overexpression of IDO in transformed fibroblasts markedly inhibited their proliferation, and this effect was shown to be correlated with depletion of tryptophan from the media (Gupta et al., 1994). Conversely, in mutant ME180 cervical carcinoma lines deficient in IDO activity, and therefore unable to catabolize tryptophan, IFN-γ failed to inhibit cellular proliferation or the intracellular growth of *Chlamydia* or *Toxoplasma* (Feng and Taylor, 1989; Thomas et al., 1993). The present findings extend these observations by demonstrating that IFN-γ-induced IDO activation and consequent tryptophan depletion also accounts for inhibition of matrix-degrading enzyme production.

How does depletion of tryptophan selectively interfere with the inducible expression of MMP genes? Tryptophan may be required for the synthesis, or the DNA-binding activity of nuclear transcription factors such as AP-1. Consistent with this hypothesis, the present results indicate that in tryptophan-depleted fibroblasts, c-*fos* induction by PMA is markedly reduced. Varesio et al. showed previously that in macrophages, IFN-γ reduced inducible c-*fos* mRNA levels (Radzioch and Varesio, 1991). Since IFN-γ causes selective tryptophan depletion, it is likely that the findings of Varesio et al. can also be attributed to tryptophan starvation.

Regardless of the molecular mechanisms involved, the present findings shed new light on the mechanism of regulation of extracellular matrix degradation by IFN-γ Furthermore, these observations raise the possibility that pharmacological modulation of tryptophan concentration in tissues undergoing cytokine-driven connective tissue breakdown (such as rheumatoid cartilage and synovium) may reduce the local expression and activity of matrix-degrading enzymes, and therefore inhibit the destruction of these structures.

## ACKNOWLEDGMENT

The authors thank Ms. K. Fliszar (Washington University) for performing collagenase ELISA's, and Dr. B. Cronstein (New York University) for valuable suggestions. This study was supported by grants to J.V. from the National Institutes of Health (AR-42309 and AR-01817), and from Showa-Denko Tokyo.

## REFERENCES

Brinckerhoff, C., 1991, Joint destruction in arthritis: metalloproteinases in the spotlight. *Arthritis Rheum.* 34:1073.

Feng G.S., and Taylor M.W., 1989, IFN-γ resistant mutants are defective in the induction of indoleamine 2,3-dioxygenase. *Proc Natl Acad Sci USA.* 86:7144.

Gupta, S.L., Carlin J.M., Pyati P., Dai W., Pfefferkorn E. R., and Murphy M. J., Jr., 1994. Antiparasitic and antiproliferative effects of indoleamine 2,3-dioxygenase enzyme expression in human fibroblasts. *Infect Immun.* 62:2277.

Radzioch D. and Varesio L., 1991, c-*fos* mRNA expression in macrophages is downregulated by interferon-γ at the post-transcriptional level. *Mol Cell Biol* 11: 2718.

Taylor, M.W., and Feng G., 1991, Relationship between interferon-γ, indoleamine 2,3-dioxygenase, and tryptophan catabolism. *FASEB J.* 5:2516.

Thomas, S.M., Garrity L.F., Brandt C.R., Schobert C.S., Feng G.S., Taylor M.W., Carlin J.M, and Byrne G.I., 1993, IFN-γ-mediated antimicrobial response: indoleamine 2,3-dioxygenase-deficient mutant host cells no longer inhibit intracellular *Chlamydia* spp. or *Toxoplasma* growth. *J Immunol.* 150:5529.

Varga, J., Olsen A., Herhal J., Constantine G., Rosenbloom J., and Jimenez S.A.., 1990, Interferon-γ reverses the stimulation of collagen but not fibronectin gene expression by transforming growth factor β in normal human fibroblasts. *Eur J Clin Invest.* 20:487.

Varga J., Li L., Mauviel A., Jeffrey J., and Jimenez S.A., 1994, L-tryptophan in supraphysiologic concentrations stimulates collagenase gene expression in human skin fibroblasts. *Lab Invest.* 70:183

Vilcek J, Oliveira I.C., 1994, Recent progress in the elucidation of interferon-gamma actions: molecular biology and biological functions. *Int Arch Allergy Immunol.* 104:311.

# 24

# ACTIVATED CELLULAR IMMUNITY AND DECREASED SERUM TRYPTOPHAN IN HEALTHY PREGNANCY

D. Fuchs, H. Schröcksnadel, G. Baier-Bitterlich, O. Dapunt, and H. Wachter

Institute of Medical Chemistry and Biochemistry and
Department of Obstetrics and Gynecology
University of Innsbruck, Austria

## ABSTRACT

We analyzed plasma samples from 45 randomly selected healthy pregnant women and from 20 healthy nonpregnant female controls of corresponding age. Tryptophan and kynurenine concentrations were measured by reverse-phase high performance liquid chromatography. The kynurenine per tryptophan ratio was calculated to allow a more accurate estimate of tryptophan degradation. In addition, neopterin was measured by radioimmunoassay. There were significantly lower tryptophan concentrations in pregnant women compared to nonpregnant controls. Kynurenine per tryptophan ratios were increased in pregnant women in the 3rd trimester. Significant correlations existed between neopterin increase and tryptophan decrease as well as kynurenine increase and tryptophan decrease. Also significant correlations between week of pregnancy and lower tryptophan and higher kynurenine per tryptophan ratio were found ($p<0.01$). From our data, cellular immune activation is likely to be the cause of enhanced tryptophan degradation during pregnancy.

## INTRODUCTION

Two enzymic pathways of tryptophan catabolism are well established: tryptophan-5-hydroxylase converts tryptophan to 5-hydroxy-tryptophan to finally form 5-hydroxy-tryptamine (serotonin), and tryptophan pyrrolase and indoleamine (2,3)-dioxygenase (IDO) respectively catalyze the degradation of tryptophan via the kynurenine pathway leading among other products to the formation of N-formyl-kynurenine and kynurenine (Fernstrom and Wurtman, 1972; Byrne et al., 1986; Freese et al., 1990; Fuchs et al., 1993). IDO activity is stimulated by cytokines like interferon-γ and likewise, therapeutic administration of cytokines like interferons in patients is associated with a decrease of tryptophan and an increase of kynurenine concentrations (Brown et al., 1989). In situations when the formation

*Recent Advances in Tryptophan Research*, edited by Graziella Allegri Filippini et al.
Plenum Press, New York, 1996

of endogenous interferon-γ is increased, significant degradation of tryptophan with enhanced production of metabolites such as kynurenine and quinolinic acid can be observed (Freese et al., 1990; Fuchs et al., 1990; 1993; Gasse et al., 1994; Heyes et al., 1992). In parallel, increased formation of neopterin occurs, which sensitively indicates activation of cell-mediated immunity (Fuchs et al., 1988). Since neopterin levels are increased in healthy pregnant women compared to nonpregnant female controls of same age (Fuith et al., 1991) we examined the behavior of tryptophan and of kynurenine in uncomplicated pregnancy.

## PATIENTS AND METHODS

Plasma was analyzed from 45 randomly selected healthy pregnant women with singleton pregnancies and 20 healthy nonpregnant female controls of corresponding age. From the 45 healthy pregnant women 15 were in each trimester. None of the women included in the study had any signs of infection at the time of plasma sampling. Tryptophan and kynurenine concentrations were measured by reverse-phase high performance liquid chromatography using on-line deproteinization of samples without precipitation (Gasse et al., 1994; Fuchs et al., 1992). Relating kynurenine to tryptophan levels the kynurenine per tryptophan ratio was calculated, to allow a more accurate estimate of the tryptophan degradation rate (Fuchs et al., 1990; Gasse et al., 1994). Neopterin was measured by radioimmunoassay (Henning, Berlin, Germany) as described (Fuchs et al., 1992).

## RESULTS

Tryptophan concentrations were similar in the nonpregnant controls (mean ± S.D.: 54.5 ± 13.9 μmol/l) and in pregnancy (54.5 ± 13,9). Levels in the 1st trimester were higher than those of controls (U = 2.45, p= 0.01; Figure 1). In the 2nd and 3rd trimester tryptophan medians were lower differing statistically significantly from the nonpregnant controls (U = 2.12, p = 0.03 and U = 4.05, p < 0.0001, respectively). Values of the trimesters differed significantly amongst each other (Figure1). The correlation of tryptophan to the gestational week was inverse and highly significant (Table 1).

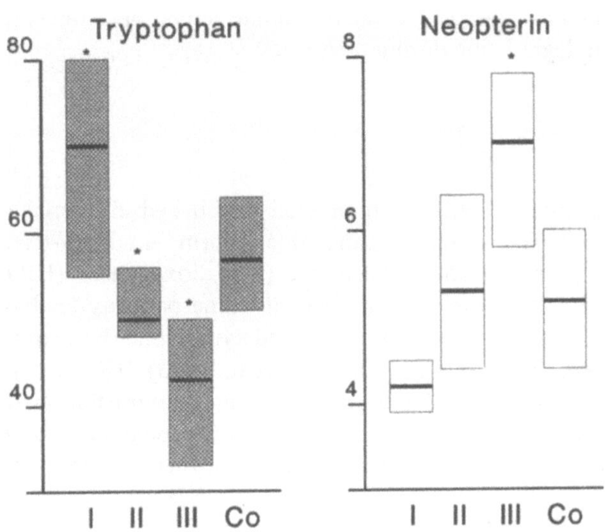


**Figure 1.** Plasma tryptophan (μmol/l) and neopterin (nmol/l) in healthy pregnancy (trimesters I, II, III) and nonpregnant controls (Co), medians (horizontal bars) and inter-quartile ranges (boxes) are shown (* = p<0.01).

**Table 1.** Association between gestational age and plasma concentrations of tryptophan, kynurenine, kynurenine per tryptophan ratio and neopterin in healthy pregnancy (Spearman rank correlation coefficients and p-values are shown)

| | rs | p |
|---|---|---|
| Tryptophan | −0.771 | <0.0001 |
| Kynurenine | −0.096 | n.s. |
| Kynurenine per tryptophan ratio | −0.714 | <0.0001 |
| Neopterin | 0.654 | <0.0001 |

Kynurenine concentrations in nonpregnant controls were indentical to those of pregnant women and no significant differences existed to any trimester and there was no correlation between kynurenine levels and the duration of pregnancy.

Kynurenine per tryptophan ratios in the 1$^{st}$ trimester ($18 \pm 5.3$ µmol/nmol) were lower than those in nonpregnant controls ($25.2 \pm 7.6$; $U = 2.94$, $p = 0.003$). Values in the 3$^{rd}$ trimenon were higher than in the nonpregnant controls ($U = 2.33$, $p = 0.02$). There existed a significant correlation between the kynurenine per tryptophan ratio and the gestational week (Table 1).

Neopterin values of all pregnant women ($5.7 \pm 1.9$ nmol/l) were only slightly higher than those of nonpregnant controls ($5.1 \pm 1.0$). Neopterin levels in the 3$^{rd}$ trimester were highest with a statistical difference to nonpregnant controls ($U = 3.40$, $p = 0.0007$; Fig.1). Neopterin levels in the 2$^{nd}$ trimester were significantly higher than those in the 1$^{st}$, and those in the 3$^{rd}$ were higher than those in the 2$^{nd}$ (Fig.1). There was a significant correlation of neopterin levels to the week of pregnancy (Table 1).

There were significant correlations between neopterin and tryptophan and the kynurenine per tryptophan ratio (Table 2). No significant correlation was found between neopterin and kynurenine levels.

## DISCUSSION

In healthy pregnant women increase of neopterin concentrations was paralleled by a decrease of tryptophan values as is demonstrated by a significant inverse correlation. Kynurenine levels were not higher compared to the nonpregnant controls. However, increased endogenous interferon-γ appears to be the cause of tryptophan degradation because the kynurenine per tryptophan ratio was increased in the 3$^{rd}$ trimester and correlated to the gestational week. Thus, IDO is activated and the decrease of tryptophan may explain, why kynurenine does not exceed the normal range. The increased kynurenine per tryptophan ratio clearly indicates that kynurenine is increased relatively to the decreased concentrations of tryptophan available but does not exceed the normal range.

**Table 2.** Association between parameters of tryptophan metabolism (tryptophan, kynurenine and kynurenine per tryptophan ratio) and neopterin in healthy pregnancy (Spearman rank correlation coefficients and p-values are shown)

| | rs | p |
|---|---|---|
| Tryptophan | −0.566 | p<0.001 |
| Kynurenine | 0.246 | 0.107 |
| Kynurenine per tryptophan ratio | 0.660 | p<0.0001 |

In the 1st trimester there was a trend towards lower neopterin and kynurenine per tryptophan ratio associated with higher tryptophan levels compared to controls possibly indicating a depression of baseline immunoactivity. Only in the further course of pregnancy immune activation increases until gestation. In the postpartum period falling neopterin concentrations indicate a rapid normalization of macrophage activation after pregnancy. Tryptophan degradation seems to decline at that time as shown by higher tryptophan levels on the one hand and a decreasing kynurenine per tryptophan ratio on the other hand.

Alterations of the tryptophan metabolism may indirectly effect the 5-hydroxy-tryptamine (serotonin) pathway which depends from tryptophan concentrations in blood and brain (Fernstrom and Wurtman, 1972; Freese et al., 1990). 5-Hydroxytryptamine is a neurotransmitter, implicated in a variety of neural functions such as pain and sleep, and also causes dose-dependent constriction of smooth muscle. In HIV infection low tryptophan concentrations are associated with impairment of nocturnal sleep (Wiegand et al., 1991), and neopterin and tryptophan concentrations correlate with the degree of cognitive impairment (Fuchs et al., 1990; 1993;). In an investigation about memory loss during pregnancy (Sharp et al., 1993) it was found that 80% of pregnant women subjectively rated their current memory as being impaired. It could be shown in objective tests, that during pregnancy there is a significant impairment of memory as tested by recall or priming but not by recognition. Our data suggest that disturbed metabolism of serotonin is one underlying mechanism by which pregnancy may affect the central nervous system.

## ACKNOWLEDGMENT

This work was supported by the Austrian Ministry of Science and Research, "Sektion Forschung".

## REFERENCES

Brown R.R., Lee C.M., Kohler P.C., et al., 1989, Altered tryptophan and neopterin metabolism in cancer patients treated with recombinant interleukin 2. *Cancer Res* 49:494.

Byrne G., Lehmann L.K., Kirschbaum J.G., et al., 1986, Induction of tryptophan degradation in vitro and in vivo: a gamma interferon stimulated activity. *J Interferon Res* 6:389.

Fernstrom J.D. and Wurtman R.J., 1972, Brain serotonin content: physiological regulation by plasma neutral amino acids. *Science* 178:414.

Freese A., Swartz K.J., During M.J., and Martin J.B., 1990, Kynurenine metabolites of tryptophan: Implications for neurologic diseases. *Neurology* 40: 691.

Fuchs D., Hausen A., Reibnegger G., Werner E.R., Dierich M.P. and Wachter H., 1988, Neopterin as a marker for activated cell-mediated immunity: Application in HIV infection. *Immunol Today* 9:150.

Fuchs D., Möller A.A., Stöckl E., et al., 1990, Decreased serum tryptophan in patients with HIV-1 infection correlates with increased serum neopterin and neurologic/ psychiatric symptoms. *J Acquir Immune Defic Syndr* 3:873.

Fuchs D., Werner E.R. and Wachter H., 1992, Soluble products of immune activation: Neopterin. In: Manual of Clinical Laboratory Immunology, 4th edition. R.R. Rose, E.C. deMacario , J.L. Fahey, H. Friedman and G.M. Penn, eds., American Society for Microbiology 251.

Fuchs D., Möller A.A., Weiss G., Dierich M.P. and Wachter H., 1993, Immune activation, impaired tryptophan metabolism and neurologic/psychiatric symptoms in HIV infection. In: S. Weiss, H. Hippius (eds.), HIV-1 Infection of the Central Nervous System, Seattle: Hogrefe & Huber Publishers 1039.

Fuith L.C., Fuchs D., Hausen A., et al., 1991, Neopterin, a marker of cell-mediated immune activation in human pregnancy. *Int J Fertil* 36:372.

Gasse T., Murr C., Meyersbach P., et al., 1994, Neopterin production and tryptophan degradation in acute lyme neuroborreliosis vers"us late Lyme encephalopathy. *Eur J Clin Chem Clin Biochem* 32:685.

Heyes M.P., Saito K., Crowley S., et al., 1992, Quinolinic acid and kynurenine pathway metabolism in inflammatory and non-inflammatory neurological disease. *Brain* 115:1249.

Sharp K., Brindle P.M., Brown M.W. and Turner G.M., 1993, Memory loss during pregnancy. *Br J Obstet Gynaecol* 100:209.

Wiegand M., Möller A.A. and Schreiber W., et al., 1991, Nocturnal sleep in patients with HIV infection. *Psych Clin Neurosci* 240:153.

# 25

# TRYPTOPHAN STARVATION IS INVOLVED IN HUMAN INTERFERON-GAMMA MEDIATED APOPTOSIS

M. W. Taylor, V. K. Konan, and D. Yu

Department of Biology
Indiana University
Bloomington, Indiana 47405

The tryptophan catabolic enzyme, indoleamine 2,3 dioxygenase (IDO), has been shown to be induced in vivo in lungs of mice by bacterial lipopolysaccharide (Yoshida and Hayaishi, 1978), by influenza virus (Yoshida et al., 1979) and by IFN-gamma (IFN-γ) (Carlin et al., 1987, Taylor and Feng 1991). Unlike other interferon responsive genes, this enzyme is not appreciably induced by any of the type I interferons. We (Feng and Taylor 1989) have previously reported that the human cell line ME180 undergoes cell death very rapidly following treatment with IFN-γ. This cell death appears to be the result of starvation for the essential amino acid tryptophan due to the induction of IDO (Feng and Taylor 1989, Ozaki et al., 1988, Aune and Pogue 1989). In ME180 mutants resistant to the antiproliferative effects of IFN-γ, there is no induction of IDO (Feng and Taylor 1989). Moreover the antiproliferative response to IFN-γ can be reversed by the addition of tryptophan back to the medium (Feng and Taylor, 1989, Ozes and Taylor 1993). In this paper we report that the antiproliferative response to IFN-γ results in apoptosis of treated cells, and that tryptophan starvation by itself appears to be sufficient to drive the cell to apoptosis through the same or similar mechanism.

## RESULTS AND DISCUSSION

Two criteria, the formation of a ladder of DNA on agarose gels as a result of endonucleolytic cleavage, and morphological changes in the cells were used to define programmed cell death or apoptosis. As is shown in Fig. 1, treatment of ME180 cells with IFN-γ, for either a short time (12 hours) or over an extended period results in the formation of a DNA ladder, with bands of approximately 180 bp or mutiples thereof. Ladder formation increases in intensity up to 48-72 hours after treatment. Fig. 2 shows the parallel increase in IDO activity following treatment with IFN-γ over the same time period, and the decrease in cell viability. Addition of tryptophan following IFN-γ treatment completely blocks this endonucleolytic cleavage of DNA.

*Recent Advances in Tryptophan Research*, edited by Graziella Allegri Filippini et al.
Plenum Press, New York, 1996

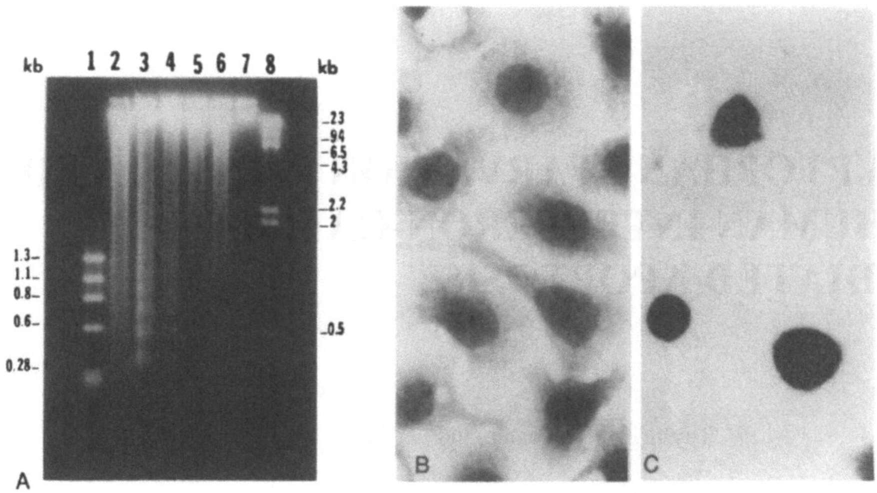


**Figure 1.** A. DNA ladder formed with time following treatment with IFN-γ. Lanes 1,8 size markers, lane 2 VSV infected ME180 cells, lane 7 untreated control, lane 6, cells harvested at 12 hours, lane 5, cells harvested at 24 hours, lane 4, cells harvested at 48 hours, lane 3 cells harvested at 72 hours. B. Untreated cells. C. Morphological changes in treated cells.

In order to ascertain whether tryptophan starvation alone was required for this apoptotic event, cells were grown in medium containing non-essential amino acids and the medium removed at specific times and replaced with tryptophan free medium. When DNA was extracted from such cells and run on a 1% agarose gel, a DNA ladder typical of apoptosis was observed.

The signaling mechanism leading to apoptosis is still unknown. Cells in tissue culture can be induced to undergo apoptosis by a variety of treatments, including withdrawl of growth factors, treatment with gluticorticoids, exposure to ionizing radiation, treatment with cytokines such as TNF and induction of the Fas antigen, an antigen similar to the TNF

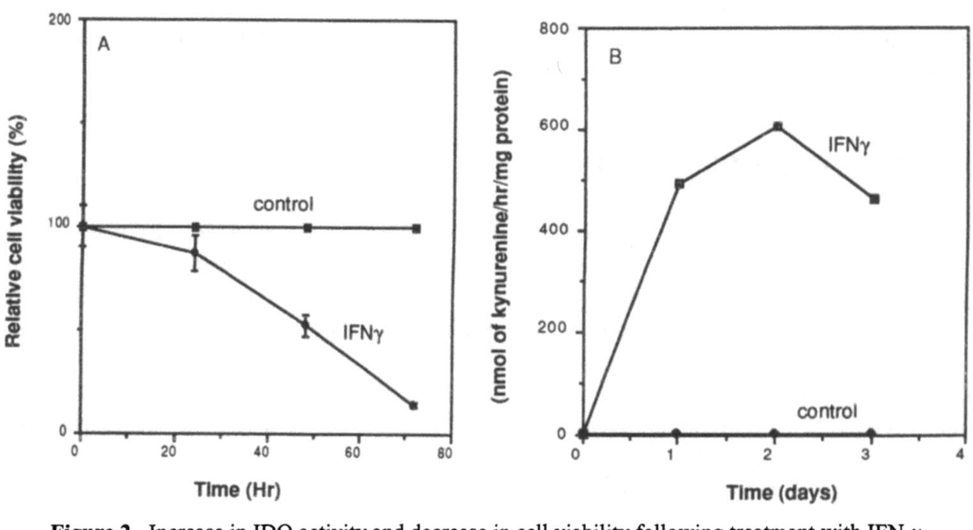


**Figure 2.** Increase in IDO activity and decrease in cell viability following treatment with IFN-γ.

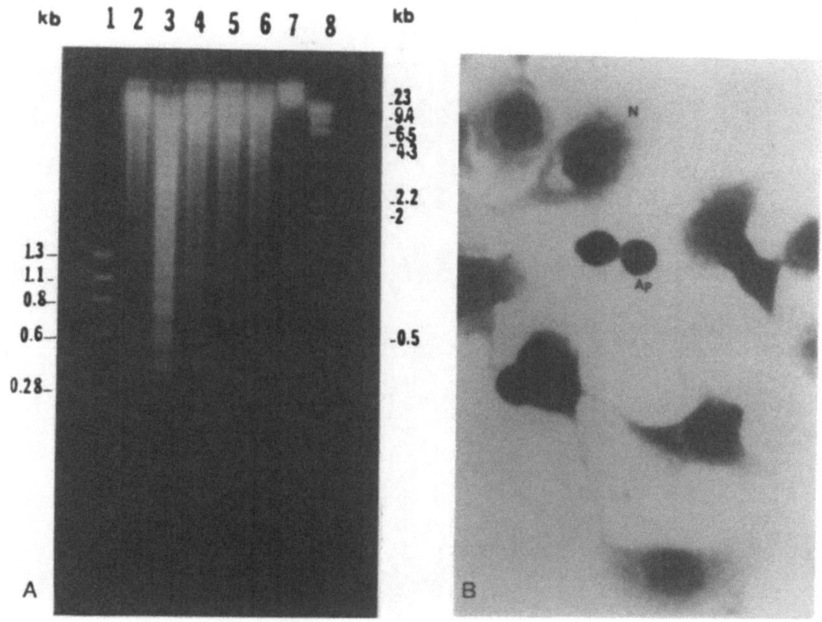


**Figure 3.** A. DNA ladder formed with time in absence of tryptophan from the medium.Lanes 1 and 8, size markers. DNA was collected at 24 hours (lane 6),48 hours (lane 5) amd 72 hours (lane 4).Lane 3. HL-60 cells treated with $10^{-4}$ calcium inonophore-A23187 known to cause apoptosis. B. Changes in cell morphology at 72 hours following starvation.

receptor. It has been proposed that apoptosis is inhibited by the induction of Bcl-2 and a related gene Bcl-X (Chen et al., 1995). Recent studies have proposed that TNF-α and ionizing radiation induction of apoptosis occurs by activating hydrolysis of sphingomyelin to ceramide. It has also been suggested that apoptosis may be mediated through the production of free radicals via oxidative pathways. However recent evidence argues against this hypothesis (Muschel et al., 1995) . A large number of genes known to be involved as tumor suppressors have been reported to be involved in the molecular events leading to apoptosis. These include p53 (Yonish-Rouarch et al., 1991) and p68 kinase (PKR)(Lee and Esteban 1994). It has been suggested that apoptosis is a normal cellular event, and that cancer in fact, may result from a defect in apoptosis. We have previously reported that treatment of ME180 cells with IFN-γ results in induction of PKR activity (Ozes and Taylor 1993). This induction is the result of both enhanced mRNA synthesis and elevated enzyme activity. ME 180 cell mutants, resistant to the antiproliferative effects of IFN-γ, are defective in the IDO response and do not show inducible PKR activity following treatment with IFN-γ (Ozes and Taylor, 1993). Thus it was of interest to examine whether PKR was induced following starvation for tryptophan.

As can be seen from Fig. 4 there was enhanced autophosphorylation when $^{32}\gamma$ P-ATP was mixed with immunoprecipitated PKR from tryptophan starved ME180 cells. There is a linear increase in autophosphorylation and histone phosphorylation between 16-48 hours after removal of tryptophan from the medium. There was also increased mRNA detectable by northern analysis following starvation of ME180 cells for tryptophan. Thus both tryptophan starvation and IFN-γ treatment result in the activation of PKR. However whereas IFN-γ induction appears to involve a dsRNA activation event (McCormack et al., 1994, Samuel 1993), this does not appear to be the case following starvation for tryptophan. PKR autophosphorylation occurs in the absence of dsRNA.

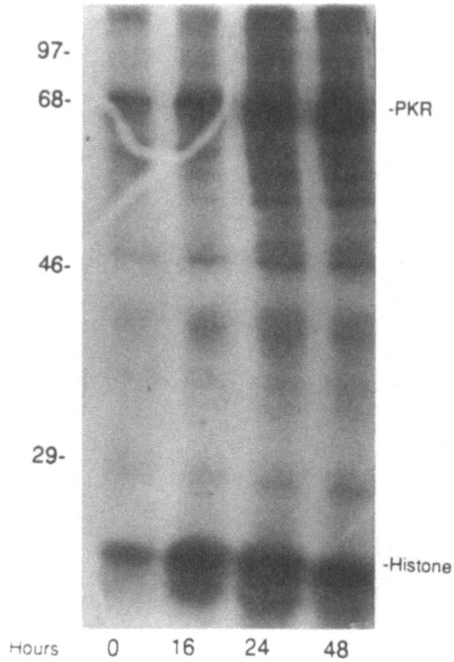


**Figure 4.** Autophosphorylation of PKR and histone phosphorylation following starvation for tryptophan.

The substrate for PKR is the alpha subunit of the eukaryotic initiation factor EIF-2α. The phosphorylation of this factor by PKR lead to the cessation of new protein synthesis from capped mRNAs in vitro (Farrell et al., 1977). That PKR might be a tumor suppressor gene has been proposed by Meurs et al., 1993, and transfection of the PKR gene into mouse 3T3 cells leads to apoptosis (Lee and Estaban 1994).

Chong et al., (1993) from studies with the bakers yeast, *Saccharomycetes cerevisiae* presented a model in which PKR is activated during environmental stress by dsRNA or perhaps tRNA, resulting in inhibition of protein synthesis. It is possible that starvation for a specific amino acid may alter the conformation of the cognate tRNA, or leave it uncharged so that it now interacts with PKR directly and results in autophosphorylation and inhibition of protein synthesis. Our data is in agreement with this hypothesis, since autophosphorylation and eIF-2 phosphorylation occur without the addition of dsRNA following tryptophan starvation .

We present a model, in which IFN-γ acts on cells by induction of IDO which in turn depletes the cell for tryptophan. This tryptophan depletion could lead to the accumulation of uncharged $tRNA^{trp}$, which binds to a specific region of the PKR triggering autophosphorylation. The autophosphorylation may in turn lead to kinase activity on eIF-2α with the blocking of protein synthesis and phosphorylation of a yet unidentified transcription factor, thus enhancing mRNA production for more PKR.

## REFERENCES

Aune, T.M., and Pogue, S.L., 1989, Inhibition of tumor cell growth by interferon gamma is mediated by two distinct mechanism dependent upon oxygen tension: induction of tryptophan degradation in vitro and depletion of intracellular nicotinamide adenine dinucleotide. *J. Clinical Invest.* 84:863.

Carlin, J.M., Borden, E.C., Sondel, P.M. and Byrne, G.I., 1987, Biological Response modifier-induced indoleamine 2,3 dioxygenase activity in human peripheral blood mononuclear cell cultures. *J. Immunol.* 139:2414.

Chen, M., Quintans, J., Fuks, Z., Thompson, C., Kufe, D.W., and Weichselbaum, R.R., 1995, Suppression of Bcl-2 messenger RNA production may mediate apoptosis after ionizing radiation, tumor necrosis factor -a, and ceramide. *Cancer Research* 55:991.

Chong, K.L., Feng, L., Schappert, K., Meurs, E., Donahue, T.F., Freisen, D.J., Havanessian, A.G., and Williams, B.R., 1992, Human p68 kinase exhibits growth suppression in yeast and homology to the translational regulator GCN2. *EMBO-J.*, 11:1553.

Feng, G., and Taylor, M.W., 1989, Interferon-γ resistant mutants are defective in the induction of indoleamine 2,3 dioxygenase. *Proc. Natl. Acad. Sci. USA* 86:7144

Lee, S.B., and Esteban, M., 1994, The interferon induced double-stranded RNA activated protein kinase induces apoptosis. *Virology* 199:491.

McCormack, S.J., Ortega, L.G., Doohan, J.P., and Samuel, C.E., 1994, Mechanisms of interferon action motif I of the interferon induced, RNA-dependent protein kinase (PKR) is sufficient to mediate RNA-binding activity. *Virology*, 198:92.

Muschel, R.J., Bernhard, E.J., Garza, L., McKenna, W.G., and Koch, C.J., 1995, Induction of apoptosis at different oxygen tensions. Evidence that oxygen radicals do not mediate apoptotic signaling. *Cancer Research*, 55:995.

Ozaki, Y., Edelstein, M.P., and Duch, D.S., 1988, Induction of indoleamine 2,3 dioxygenase: a mechanism of the antitumor activity of interferon-γ. *Proc. Nat. Acad. Sci USA* 85:1242.

Ozes, O.N., and Taylor, M.W., 1993, Reversal of IFN-γ resistant phenotype by polyI:C. Possible involvement of double stranded RNA activated kinase in IFN-γ mediated induction of the IDO gene. *J. Interferon Research*. 13:283

Samuel, C.E., 1993, The eIF-2 alpha protein kinases, regulators of translation in eukaryotes and yeasts to humans. *J. Biol. Chem.* 268:7603.

Taylor, M.W., and Feng, G., 1991, relationship between interferon-γ, indoleamine 2,3 dioxygenase and tryptophan catabolism. *FASEB J.* 5:2516.

Wagner, A.J., Small, M.B., and Hay, N., 1993, Myc-mediated apoptosis is blocked by ectopic expression of Bcl-2. *Mol. Cell Biol.* 13:2432.

Yoshida, R., and Hayaishi, O., 1978, Induction of pulmonary indoleamine 2,3 dioxygenase by intraperitoneal injection of bacterial lipopolysaccharide. *Proc. Nat. Acad. Sci. USA* 75:3998.

Yoshida, R., Urade, Y., Tokuda, M., and Hayaishi, O., 1979, Induction of indoleamine 2,3 dioxygenase in mouse lung during virus infection. *Proc. Nat. Acad. Sci. USA* 76:4084..

Yonish-Rouach, E. Resnitzky, D., Lotem, J., Sachs, L., Kimchi, A., and Oren, M., 1991, Wild type p53 induces apoptosis of myeloid leukemic cells that is inhibited by interleukin-6. *Nature 352:345.*

# CYTOKINE AND DRUG MODULATION OF KYNURENINE PATHWAY METABOLISM BY BLOOD MONONUCLEAR CELLS

K. Saito,[1] M. Seishima,[1] A. Noma,[1] S. P. Markey,[2] and M. P. Heyes[2]

[1] Department of Laboratory Medicine
Gifu University School of Medicine
Gifu 500, Japan,
[2] Section on Analytical Biochemistry
National Institute of Mental Health
Bethesda, Maryland 20892

## INTRODUCTION

Quinolinic acid (QUIN), kynurenic acid (KYNA) and L-kynurenine (L-KYN) are neuroactive L-tryptophan metabolites that are synthesized via the kynurenine pathway. These metabolites accumulate in CNS following immune stimulation (particularly those with inflammatory lesions and monocyte infiltration into the CNS) and are attributed to induction of indoleamine-2,3-dioxygenase (IDO) and other kynurenine pathway enzymes (Saito et al., 1993a; Heyes et al., 1992a; 1993).

In fact, direct conversion of L-tryptophan into QUIN by CNS tissue does occur in models of brain inflammation and injury (Saito et al., 1993b; Heyes et al., 1992b). Recently, we have also demonstrated that while various human cells derived from brain and systemic tissue convert L-tryptophan to L-KYN, only macrophage derived cells and certain liver cells convert L-tryptophan to QUIN (Saito et al., 1993c). Because macrophage derived-cells are an important source of QUIN (Heyes et al., 1992a,b; Saito et al., 1993b; Saito et al., 1993d), cytokine and drug modulation of QUIN formation from [$^{13}C_6$]-L-tryptophan (50 μM) by macrophages *in vitro* were investigated.

## MATERIALS AND METHODS

*Reagents:* L-Tryptophan, L-KYN, 3-hydroxykynurenine, anthranilic acid, 3-hydroxyanthranilic acid, QUIN, 1,1,1,3,3,3-hexafluoro-2-propanol, trypan blue, indomethacin, dexamethasone, cyclohexamide, actinomycin-D, LPS *(Saslmonella abortus equii,* sodium ethylenediaminetetraacetic acid, heptanesulphonic acid, methylene blue, ascorbic acid, trichloroacetic acid, MES buffer,. ferrous sulphate and hexafluoroisopropanol

*Recent Advances in Tryptophan Research*, edited by Graziella Allegri Filippini et al.
Plenum Press, New York, 1996

were obtained from Sigma Chemical Co. (St. Louis, MO). Trifluoroacetylimidazole was acquired from Pierce Chemical Co. (Rockford, IL). RPMI-1640, L-glutamine and fetal bovine serum were obtained from Biofluids, Inc. (Rockville, MD). [$^{13}C_6$]-ring-L-tryptophan and [$^{13}C_6$]-anthranilic acid were acquired from MSD Isotopes (Montreal, Canada). Human recombinant cytokines were used and were obtained from Genzyme. 6-Chloro-D,L-tryptophan and 4-chloro-3-hydroxyanthranilate were generous gifts from Dr. J. Parli and the Eli Lilly Company, Indianapolis, IN. The remaining reagents were obtained from Fisher Inc. (Fair Lawn, NJ).

*In Vitro Studies:* THP-1 cells (American Type Culture Collection, Rockville, MD) were grown in RPMI 1640 with 10 % fetal calf serum. Cells were incubated for 48 h in media that contained the experimental reagents as well as 50 μM [$^{13}C_6$]-L-tryptophan, as described previously.

*Biochemical Analyses:* QUIN was quantified by an electron capture negative chemical ionization gas chromatography/mass spectrometry assay. L-KYN was quantified by high performance liquid chromatography and ultraviolet light absorbance spectrometry. Established methods were used to quantify the activities of IDO, kynurenine-3-hydroxylase, kynureninase and 3-hydroxyanthranilate-3,4-dioxygenase.

*Statistical Analyses:* Results were analyzed by one-way analysis of variance with Dunnett's 't'-test for multiple comparisons. Regression analysis was done by the method of least squares. A p value of $<0.05$ was considered statistically significant, and all significant differences are shown at this level. Data are expressed as mean ± one standard error of the mean or as a percent of saline-treated control mice.

## RESULTS

Peripheral blood mononuclear cells stimulated by interferon-g (IFN-γ) or lipopolysaccaride have a large capacity to synthesize L-kynurenine (to >30 μM) and QUIN (to >10 μM) from L-tryptophan by 48 h. In a macrophage cell line (THP-1 cells), IFN-γ induced indoleamine-2,3-dioxygenase (4780 %), kynurenine-3-hydroxylase (125 %), kynureninase (120 %) but not 3-hydroxyanthranilate-3,4-dioxygenase (Figure 1A).

To investigate which enzyme(s) and/or entry of substrate may determine the rate of substrate flux, different substrates were added to the incubation media of unstimulated and IFN-γ stimulated THP-1 cells (Figure 1B). L-Tryptophan and L-KYN were converted into very small quantities of QUIN by unstimulated THP-1 cells, whereas significantly increased QUIN formation was observed following IFN-γ stimulation. In contrast, both 3-hydroxykynurenine and 3-hydroxyanthranilate were readily converted to QUIN in unstimulated cells, and the rate of QUIN formation from 3-hydroxykynurenine and 3-hydroxyanthranilate was not enhanced by IFN-γ stimulation. To identify which cytokines can increase the synthesis of L-KYN and QUIN by THP-1 cells, the effects of IFN-γ, TNF-α, IL-1α, IL-1β, IL-3 and IL-6 alone and in combination were tested. TNF-α, IL-1α, IL-1β, IL-3 and IL-6 alone slightly increased L-KYN production, although these cytokines were not as potent as IFN-γ.

With respect to QUIN synthesis, only IFN-γ., TNF-α and IL-6 stimulation resulted in increased QUIN synthesis. GM-CSF alone had no effect on either L-KYN or QUIN synthesis. In combination however, TNF-α, IL-1β, IL-3 and GM-CSF potentiated the stimulatory effects of IFN-γ, on L-KYN and QUIN accumulation, while IL-6 and GM-CSF potentiated the effects of TNF-α on L-KYN and QUIN accumulation. TNF-α, IL-6 and IL-1β significantly potentiates the effects of IFN-γ. The formation of L-kynurenine and QUIN by lipopolysaccaride activated THP-1 cells is attenuated by IL-4, antibodies to TNF-α or IL-6, but not antibodies to IFN-γ. 4-chloro-3-hydroxyanthranilate (enzyme inhibitor of QUIN

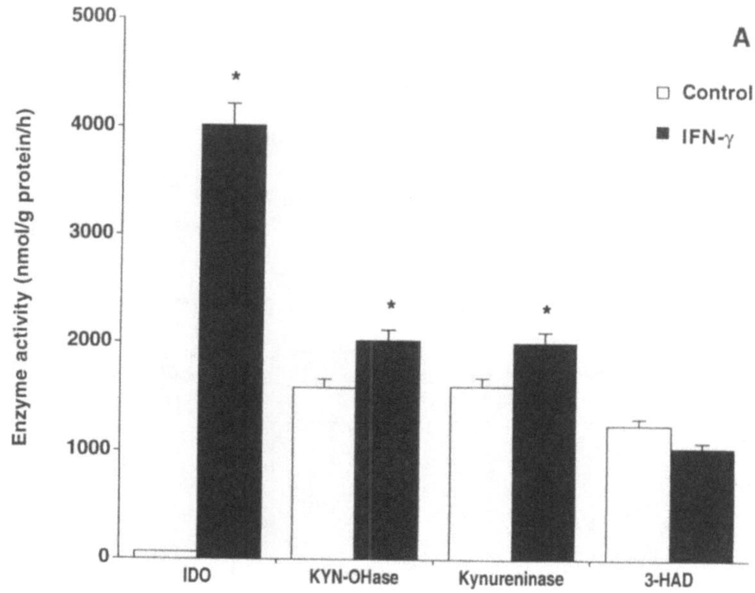


**Figure 1A.** Increased activity of indoleamine-2,3-dioxygenase, kynurenine-3-hydroxylase and kynureninase, but not 3-hydroxyanthranilate-3,4-dioxygenase, in THP-1 cells 48 h following stimulation by interferon-γ (100 IU/mL; n=3) versus unstimulated cells (n=3). Results are expressed as amount of product formed per h per mg protein in the enzyme assay preparation.

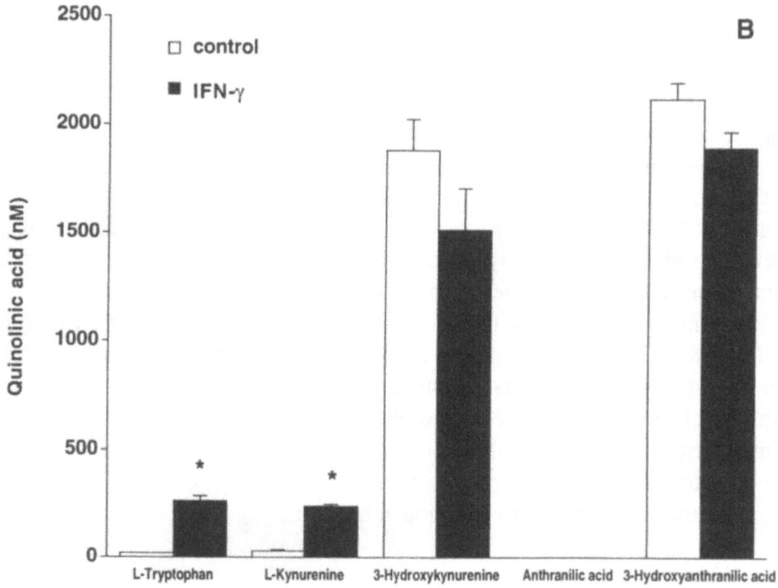


**Figure 1B.** Enhanced conversion of L-tryptophan and L-KYN to QUIN in IFN-γ stimulated THP-1 cells (n=3) versus controls (n=3). THP-1 cells were incubated for 48 h either without or with IFN-γ (100 IU/mL), in media supplemented with either L-tryptophan, L-KYN, 3-hydroxykynurenine, anthranilate or 3-hydroxyanthranilate (all 50 μM).

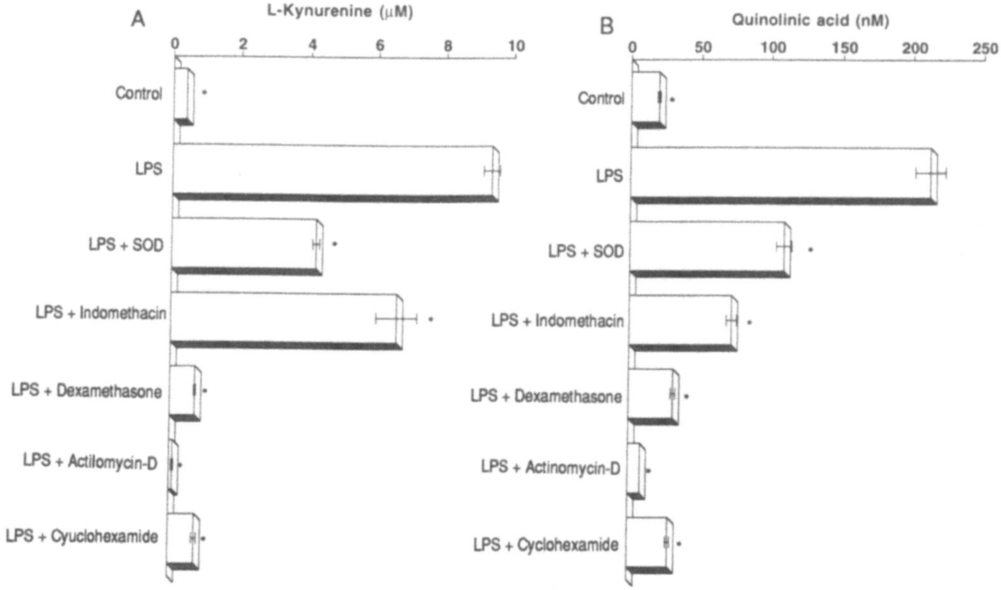


**Figure 2.** Indomethacin (5 μM), dexamethasone, actinomycin-D (0.1 μM) and cycloheximide (1 μM) decrease the production of L-KYN and QUIN production by THP-1 cells in response to LPS (1 ng/mL). Inhibitors were added at the same time as the stimulants, and samples collected 48 h later.

formation), indomethacin, actinomycin-D, cycloheximide, dexamethasone and superoxide dismutase also attenuate QUIN formation (Figure 2).

## CONCLUSIONS

The effects of immune activation on the levels of QUIN and L-KYN in macrophages, blood and brain, demonstrates that an immune stimulation results in increased synthesis of L-KYN and QUIN, that are dependent on a number of intracellular and intercellular events. These events include the release of cytokines, synthesis of prostaglandins, transcription, protein synthesis and the availability of oxygen free radicals and increased activity of selected kynurenine pathway enzymes. This cascade of events offers numerous targets to attenuate the accumulation of QUIN in brain and systemic tissues. While reductions in the synthesis and/or release of physiologically active mediators may have therapeutic benefit following immune activation, such as reducing the concentration of TNF-α in acute septicemia, it would not be possible to discern the role or contribution of QUIN. However, drugs such as an inhibitor of kynurenine pathway enzyme offer tools to attenuate QUIN synthesis in response to immune activation and to investigate what, if any, role QUIN has in mediating detrimental or beneficial responses to immune activation.

## REFERENCES

Heyes M.P., Saito K, Crowley J *et al.*, 1992a, Quinolinic acid and kynurenine pathway metabolism in inflammatory and non-inflammatory neurologic disease. *Brain,* 115:1249.

Heyes M.P., Saito K, Jacobowitz D *et al.*, 1992b, Poliovirus induces indoleamine-2,3-dioxygenase and quinolinic acid synthesis in macaque brain. *FASEB J.*, 6:2977.

Heyes M.P., Saito K, Major E.O. *et al.*, 1993, A mechanism of quinolinic acid formation by brain in inflammatory neurological disease: attenuation of synthesis from L-tryptophan by 6-chlorotryptophan and 4-chloro-3-hydroxyanthranilate. *Brain* 116:1425.

Saito K, Nowak T.S., Suyama K *et al.*, 1993a, Kynurenine pathway enzymes in brain: responses to ischemic brain injury versus systemic immune activation. *J.Neurochem.*, 61:2061.

Saito K, Nowak T.S., Jr., Markey S.P., Heyes M.P., 1993b, Mechanism of delayed increases in kynurenine pathway metabolism in damaged brain regions following transient cerebral ischemia. *J. Neurochem.*, 60:180.

Saito K, Chen C.Y., Masana M *et al.*, 1993c, 4-Chloro-3-hydroxyanthranilic acid, 6-chlorotryptophan and norharmane attenuate quinolinic acid formation by interferon-γ stimulated monocytes (THP-1) cells. *Biochem. J.*, 291:11.

Saito K, Crowley J.S., Markey S.P., Heyes M.P., 1993d, A mechanism for increased quinolinic acid formation following acute systemic immune stimulation. *J. Biol. Chem.*, 268:15496.

# KYNURENINE ENZYMATIC PATHWAY IN HUMAN MONOCYTES-MACROPHAGES

## Effect of Interferon-γ Activation

R. G. Ferrario, S. Barattè, C. Speciale, and P. Salvati

Pharmacia, R&D-CNS
20014 Nerviano (Milano), Italy

## 1. INTRODUCTION

Quinolinic acid (QUIN) and kynurenic acid (KYNA) are two metabolites of the L-tryptophan degradation belonging to the kynurenine pathway and showing opposite effects on the brain. KYNA is a neuroprotectant and acts as an antagonist on the ionotropic excitatory amino acid receptors. In particular KYNA has an high affinity to the glycine site associated to the N-Methyl-D-Aspartate (NMDA) receptor complex (Foster et al., 1984; Kessler et al., 1989). On the contrary, QUIN is a neurotoxin and the observation that QUIN is a potent NMDA receptor agonist as well as a potent lipid peroxidant in brain homogenates has suggested the involvement of this metabolite in many neurodegenerative disorders (Schwarcz et al., 1983; Rios and Santamaria, 1991).

QUIN levels have been found increased in response to immune stimulation in a broad spectrum of inflammatory brain diseases in animal models as well as in human subjects (Heyes, 1993). Furthermore a rise in QUIN concentration was detected in animal models of injury such as spinal cord injury and cerebral ischemia (Saito et al., 1993a).

Cultured human macrophage-derived cells stimulated with interferon-γ (IFN-γ) can synthesized QUIN in concentration (μM) that exceed the levels reported to be neurotoxic (Saito et al., 1993b), suggesting that the recruitment of this cell type at the site of the cerebral insult might contribute to the pathogenesis of neurologic deficits and brain injury in inflammatory diseases.

In the present study human peripheral-blood monocytes/macrophages (MΦ) were isolated and cultured with the following aims: 1- determine the activity of the key enzymes in the kynurenine pathway after stimulation with IFN-γ, a cytokine involved in the regulation of both immune and inflammatory processes; 2- evaluate the extracellular concentrations reached by the neuroactive kynurenine metabolites and in particular KYNA.

*Recent Advances in Tryptophan Research*, edited by Graziella Allegri Filippini et al.
Plenum Press, New York, 1996

# 2. METHODS

## 2.1 Isolation and Stimulation of Human Monocyte/Macrophage Cell Population

The cells obtained from human buffy coats were resuspended in Phosphate Buffered Saline (PBS) and 30 ml of the cell/PBS mixture were slowly layered over 9 ml of Ficoll-Paque solution (Pharmacia LKB, Uppsala, Sweden) into 50 ml tube. After centrifugation (30 min at 900xg, 20°C) the mononuclear cells (60-70% lymphocytes and 30-40% monocytes/macrophages) were transferred into another centrifuge tube, washed twice, seeded ($2x10^6$ cell/ml, 10 ml/flask) and then the MΦ population was isolated from lymphocytes by exploiting their ability to adhere to plastic. After 2 hr-incubation the medium containing nonadherent cells was removed, the cell monolayers washed twice and finally 4.5 ml of complete RPMI 1640 medium (10 % foetal bovine serum and 2 mM glutamine plus antibiotics) were added to each flask. On the 3rd day of culture, MΦ were stimulated with recombinant human IFN-γ (250 U/ml; PharMingen, San Diego, CA) for the next 72hr. Control cells were cultured for 6 days without any treatment.

## 2.2 Enzyme Activities and Extracellular Kynurenines Levels in the Culture Medium

Untreated control and IFN-γ-primed MΦ were scrapered and disrupted by sonication. Cell lysates were centrifuged (20 min, 15,000 g, 4°C), and the supernatants utilized as a source of enzymes for the evaluation of the activities of indoleamine 2,3-dioxygenase (IDO), kynurenine 3- hydroxylase and 3-hydroxy-kynureninase (Saito et al., 1993a).

For each flask 4 culture medium samples (700 µl) were collected and stored at -20°C until HPLC analysis for the quantification of the extracellular levels of L-kynurenine, kynurenic acid, 3-OH-kynurenine, 3-OH-anthranilic acid and anthranilic acid (Saito et al., 1993a). The new production of L-kynurenine upon incubation of the cells was always evaluated prior deduction of the baseline level of L-kynurenine present in the serum added to the incubation medium (approximately 0.5 µM).

# 3. RESULTS

IFN-γ stimulation caused a dramatic induction of IDO activity from not detectable value (detection limit for L-kynurenine was 5 pmoles) to 205 ± 29 nmol/h/mg protein

**Table 1.** Enzyme activities determined in human monocytes/macrophages under control conditions and after 72 hr stimulation with recombinant human IFN-γ (250 U/ml)

| | Enzymes activities (nmol/h/mg protein) | |
|---|---|---|
| | Control | IFN-γ |
| IDO | N.D. (n=4) | 205 ± 29 (n=6) |
| kynurenine-3-hydroxylase | 3.1 ± 1.2 (n=5) | 2.1 ± 0.5 (n=7) |
| 3-OH-kynureninase | 62 ± 8 (n=5) | 66 ± 5 (n=6) |

N.D. = not detectable (below the sensitivity of the assay for L-kynurenine. Data are means ± S.E.M.

**Table 2.** Production of extracellular kynurenine metabolites determined in the culture medium of control and IFN-γ-primed (250 U/ml for 72 hr) human monocytes/macrophages

| | Kynurenine metabolites (nmoles/mg protein) | |
|---|---|---|
| | Control | INF-γ |
| L-kynurenine | N.D. (n=7) | 448 ± 55 (n=13) |
| 3-OH-kynurenine | 0.36 ± 0.08 (n=17) | 4.58 ± 0.71** (n=23) |
| 3-OH-anthranilic acid | 4.9 ± 1.0 (n=17) | 176 ± 16** (n=23) |
| Anthranilic acid | 7.0 ± 1.8 (n=15) | 48 ± 8.0** (n=15) |
| Kynurenic acid | N.D. (n=4) | N.D. (n=4) |

N.D. = not detectable (below the sensitivity of the assay for L-kynurenine and KYNA). See also the Methods. Data are means ± S.E.M.; **$p<0.01$ vs corresponding control cells (t-test).

(Table 1). On the contrary the activity of both kynurenine-3-hydroxylase and 3-hydroxy-kynureninase was not affected by the IFN-γ treatment (Table 1).

IDO induction was paralleled by a massive increase in the production of extracellular L-kynurenine. In fact, even though incubation of the cells per se did not yield to any significant production of L-kynurenine, the exposure to IFN-γ raised the extracellular L-kynurenine to 448 ± 55 nmol/mg protein (Table 2). The increase in L-kynurenine availability triggered also the synthesis and the accumulation in the culture medium of anthranilic acid, 3-OH-kynurenine and -OH anthranilic acid (Table 2). Kynurenic acid extracellular levels were always below the lowest detectable amount (4 pmoles) in control as well as in IFN-γ-primed MΦ cells.

## 4. CONCLUSIONS

The results obtained in this study suggest that in human monocytes/macrophages stimulated with IFN-γ, the IDO induction, and the consequent increase in L-kynurenine availability, might be the key events in the formation of QUIN. This data are in agreement with others previously obtained in different tissues and cell populations. In fact the opening of the L-tryptophan ring by IDO has been described as a critical event which exerts a powerful control on the pathway activity (Takikawa et al., 1986).

In this study we did not attempt to correlate the increased IDO activity with the accumulation of QUIN in the culture media, but Heyes et al.(Saito et al., 1993b) have reported that an increase in L-kynurenine levels in different macrophages-derived cells (U937, THP-1, peripheral-blood monocytes-macrophages) was always paralleled by an increase in QUIN formation, further suggesting that IDO induction play a pivotal role in the activation of the kynurenine pathway in this cell type. Our findings on the lack of an increase in the extracellular levels of kynurenic acid in parallel to the increase in QUIN concentration further support the hypothesis that macrophage infiltration, by causing an imbalance in the endogenous concentrations of the extracellular neuroactive kynurenines, might contribute to the development of the long-term neurologic damage.

## 5. REFERENCES

Foster A.C., Vezzani A., French E.D., and Schwarcz R., 1984, Kynurenic acid blocks neurotoxicity and seizures induced in rats by the related brain metabolite quinolinic acid, *Neurosci. Lett.* 48: 273.

Heyes M.P., 1993, Quinolinic acid and inflammation, *Ann. NY Acad. Sci.* 269: 211.

Kessler M., Terramani T., Lynch G., and Baudry M., 1989, A glycine site associated with N-methyl-D-aspartic acid receptors:characterization and identification of a new class of antagonists, *J Neurochem.* 52: 1319.

Rios C. and Santamaria A., 1991, Quinolinic acid is a potent lipid peroxidant in rat brain homogenates, *Neurochem. Res.* 16: 1139.

Saito K., Nowak T.S., Markey S.P., and Heyes M.P., 1993a, Mechanism of delayed increases in kynurenine pathway metabolism in damaged brain regions following transient global ischemia, *J Neurochem.* 60: 180.

Saito K., Chen C.Y., Masana M., Crowley J.S., Markey S.P., and Heyes M.P., 1993b, 4-chloro-3-hydroxyanthranilate, 6-chlorotryptophan and norharmane attenuate quinolinic acid formation by interferon-γ-stimulated monocytes (THP-1 cells), *Biochem J.* 291:11.

Schwarcz R., Whetsell W.O. Jr, and Mangano R.M., 1983, Quinolinic acid: an endogenous metabolite that produces axon-sparing lesions in rat brain, *Science* 219: 316.

Takikawa O., Yoshida R., Kido R., and Hayashi O., 1986, Tryptophan degradation in mice initiated by indoleamine 2,3-dioxygenase, *J. Biol. Chem.* 261: 3648.

# REGULATION OF THE KYNURENINE PATHWAY BY IFN-γ IN MURINE CLONED MACROPHAGES AND MICROGLIAL CELLS

D. Alberati-Giani,[1] P. Ricciardi-Castagnoli,[2] C. Köhler,[1] and A. M. Cesura[1]

[1] Pharma Division, Preclinical Research, Nervous System Diseases
F. Hoffmann La Roche Ltd.
CH-4002 Basel, Switzerland
[2] CNR Center of Cytopharmacology
University of Milan
Milan, Italy

## 1. INTRODUCTION*

Several pieces of evidence indicate that, in neurological disorders associated with immune activation, brain infiltrating macrophages and microglial cells secrete various neurotoxic factors, such as glutamate and reactive oxygen species, which may greatly influence the survival of neurons (Mallat and Chamak, 1994; Giulian et al., 1993; Stone, 1993). Among the putative endogenous neurotoxins released by phagocytic cells during inflammatory processes, particular attention has been devoted to quinolinic acid (QUIN) (Heyes et al., 1993; Heyes, 1993). This tryptophan metabolite, formed along the kynurenine pathway, exerts its effect through activation of glutamate NMDA receptors and its accumulation within the CNS, observed in brain inflammatory disorders, has been speculatively linked to neuronal dysfunctions (Heyes et al., 1993). The overproduction of QUIN is mainly attributed to the induction of indoleamine-2,3 dioxygenase (IDO), the enzyme converting L-tryptophan into L-kynurenine (Takikawa et al., 1988; Taylor and Feng, 1991), by cytokines, such as IFN-γ. It has been found that IFN-γ-activated human macrophages are able to produce QUIN directly from L-tryptophan (Heyes et al., 1992), whereas neurons and astroglial cells, even if they express inducible IDO activity (Saito et al., 1993a), do not appear to synthesise this neurotoxin directly from L-tryptophan. Whereas most of the available findings suggest that the large increase in QUIN cerebral levels in brain inflammatory conditions is mainly due to infiltrating activated macrophages, the relevance of activated

* The abbreviations used are: IFN-γ, interferon-γ; TNF-α, tumor necrosis factor-α; LPS, lipopolysaccharide; QUIN, quinolinic acid; IDO, indoleamine 2,3 dioxygenase; KAT, kynurenine aminotransferase; KH, kynurenine 3-hydroxylase; 3-HAO, 3-hydroxyanthranilic-acid dioxygenase.

*Recent Advances in Tryptophan Research,* edited by Graziella Allegri Filippini et al.
Plenum Press, New York, 1996

microglial cells in the production of neuroactive kynurenines, has been only partially investigated (Saito et al., 1993a).

In the present work we studied the differences existing in the regulation of the kynurenine pathway enzymes between two phagocytic cloned cell lines from mouse obtained after immortalization of peripheral macrophages and brain microglial cells.

## 2. EXPERIMENTAL SECTION

A microglial cell clone, N11, derived from embryonic mouse brain primary cell cultures immortalised with a recombinant retrovirus (Righi et al., 1989) and a thymic murine macrophage cell clone, MT2, obtained by similar procedure (Pirami et al., 1991), were used. Cells were seeded at a density of $5x10^6$ cell/dish in 10 cm-diameter Petri dishes containing 10 ml of culture medium alone or in the presence of various concentrations of IFN-γ. In some experiments, various additives were given to cell cultures alone or in combination with $10^3$ U/ml of IFN-γ. After treatment, the cells were harvested and stored as pellets at -80°C until analysis.

Cell pellets were homogenized in 0.4 ml of buffer and aliquots of the lysed cells were then used for measuring the activity of IDO (Saito et al., 1993b), kynurenine aminotransferase (KAT) (Alberati-Giani et al., 1995), kynurenine 3-hydroxylase (KH) (Pellicciari et al., 1994), kynureninase (Pellicciari et al., 1994), and 3-hydroxyanthranilic-acid dioxygenase (3-HAO) (Foster et al., 1986).

## 3. RESULTS

### 3.1 Indoleamine 2-3, Dioxygenase

Stimulation by IFN-γ induced the functional expression of IDO activity in both murine MT2 macrophages and N11 microglial cells. This induction was dose- and time-dependent. Fig.1 shows IDO activity upon treatment of the cells with different concentrations of IFN-γ. Although the $ED_{50}$ of IFN-γ for IDO induction were similar in both cell lines

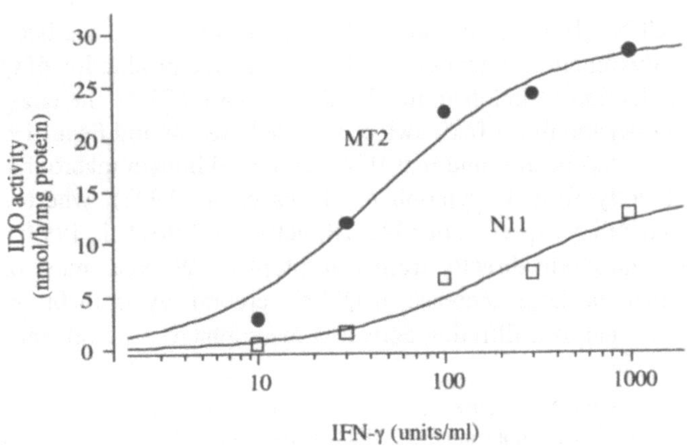


**Figure 1.** Induction of IDO activity in MT2 macrophages and N11 microglial cells by IFN-γ treatment. Cells were cultured for 48 h with medium alone and in the presence of different concentrations of IFN-γ. Results are representative of three experiments.

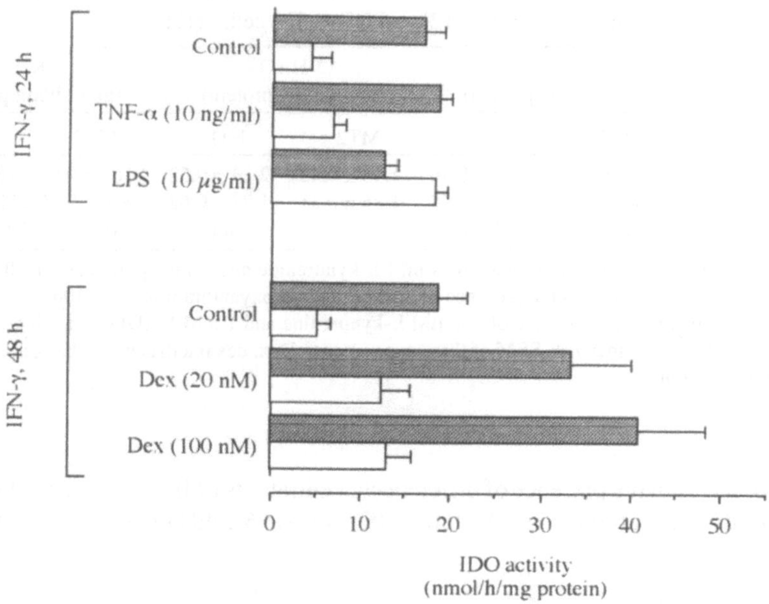


**Figure 2.** Effect of TNF-α, LPS and dexamethasone on IFN-γ-primed MT2 macrophages and N11 microglial cells. Cells were cultured for 24 (TNF-α, LPS) and 48 h (dexamethasone) in the presence of $10^3$ U/ml of IFN-γ. Results are the mean ± SEM of three experiments. *Shaded bars:* MT2 macrophages, *empty bars:* N11 microglial cells.

(43.9±8.4 and 74.2±39.1 U/ml for MT2 and N11 cells, respectively), the maximal activity reached was ~4-fold higher in MT2 macrophages than in N11 microglial cells.

In order to investigate whether other proinflammatory agents were able to modulate the expression of this enzyme, the cells were treated for 24 hours with tumor necrosis factor-α (TNF-α) and lipopolysaccharide (LPS ) alone or in combination with $10^3$ U/ml IFN-γ. Both substances did not affect IDO activity by themselves. Conversely, in the presence of IFN-γ, whereas TNF-α did not modify the induction of the enzyme, LPS differently modulated the induction of IDO by IFN-γ in the two cell lines. Thus LPS slightly reduced IDO activity in MT2 macrophages whereas, in N11 microglial cells, this endotoxin potentiated the effect of the cytokine, leading to a further 3-fold increase in the enzyme activity. Co-treatment of cells with IFN-γ and with the glucocorticoid agent dexamethasone resulted in a further enhancement in the expression of IDO in both cell lines (Figure 2).

*3.1.1. Other Enzymes of the Kynurenine Pathway.* Among the other enzymes of the kynurenine pathway, KAT, 3-HAO and KH appeared to be costitutively expressed in the two cell lines. Whereas KAT and 3-HAO activities did not display any significant changes in both cell lines after exposure to different concentrations of IFN-γ and dexamethasone, IFN-γ induced KH activity in both cell clones, with an approximately 2-fold increase in the activity of this enzyme in the presence of $3x10^2$ U/ml of the cytokine (Table 1).

Regarding the expression of kynureninase activity and its modulation by IFN-γ, major differences were found between the two cell lines. In fact, MT2 macrophages costitutively expressed much higher levels of this enzyme than N11 microglial cells (149±21 and 11.7±0.3 pmol/h/mg protein, respectively). Furthermore, kynureninase activity appeared to be stimulated by INF-γ, in a time- and dose-dependent way, only in MT2 macrophages but not in N11 microglial cells. Figure 3 shows the induction of the kynureninase activity in

**Table 1.** Effect of IFN-γ and dexamethasone on KAT, 3-HAO and KH activities in macrophages (MT2) and microglial cells (N11)

| | KAT (nmol/h/mg protein) | | 3-HAO (nmol/h/mg protein) | | KH (nmol/h/mg protein) | |
|---|---|---|---|---|---|---|
| | MT2 | N11 | MT2 | N11 | MT2 | N11 |
| Control | 2.77 ± 0.12 | 4.14 ± 0.46 | 1.48 ± 0.42 | 2.24 ± 0.52 | 160 ± 0.0 | 118 ± 9.0 |
| IFN-γ ($10^3$ U/ml) | 2.86 ± 0.14 | 4.06 ± 0.48 | 1.86 ± 0.31 | 1.77 ± 0.62 | 358 ± 2.5[a] | 223 ± 2.0[a] |
| Dex (100 nM) | 2.91 ± 0.10 | 3.81 ± 0.35 | n.d. | n.d. | 145 ± 9.0 | 105 ± 9.5 |

KAT activity was measured in the presence of 3 mM L-kynurenine and 5 mM pyruvate for 20 h at 37°C. 3-HAO activity was determined in the presence of 8 μM [$^{14}$C]hydroxyanthranilic acid (30min at 37°C). KH activity was determined in the presence of 100 μM L-kynurenine and 1 mM NADPH for 1 h at 37°C. n.d., not determined. Results are mean ± SEM of three experiments. Dex, dexamethasone. [a]The concentration of IFN-γ was 3 x $10^2$ U/ml.

MT2 macrophages in the presence of different concentrations of IFN-γ. As it can be seen the enzyme activity in the presence of $10^3$ U/ml of IFN-γ was ~6-fold higher than in resting cells.

## 4. DISCUSSION

The purpose of the present study was to characterise the regulation of the kynurenine pathway enzymes by the proinflammatory cytokine IFN-γ in two different populations of phagocytic cells, peripheral macrophages and brain microglial cells. The main conclusions that can be drawn from the present experiments are the following: i) both murine macrophages and microglial cells express the enzymes of the kynurenine pathway which might lead to the formation of QUIN from L-tryptophan; ii) in comparison to microglial cells, activated macrophages may contribute for a much larger extent to the production of QUIN and other kynurenines by virtue of their higher expression levels of IDO and, particularly, of kynureninase activities; iii) the activity of this latter enzyme appears to be stimulated by IFN-γ in macrophages, but not in microglial cells. Our results are in accordance with the available literature suggesting a predominant role of brain invading macrophages in the overproduction of neurotoxic kynurenines in inflammation. In adddition, our data provide

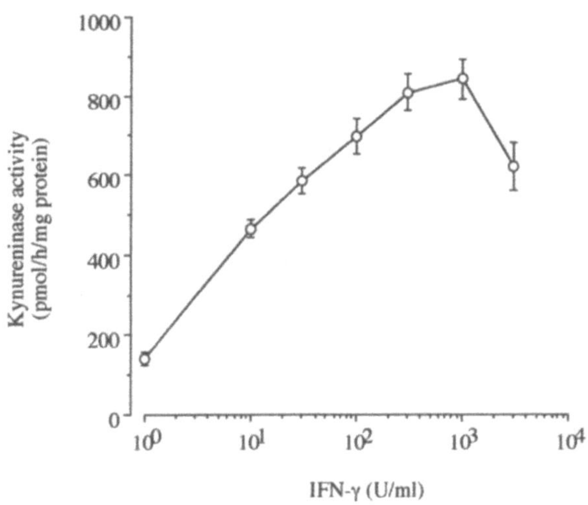


**Figure 3.** Induction of kynureninase activity by IFN-γ in MT2 macrophages. Cells were cultured for 48 h in medium alone or in the presence of different concentrations of IFN-γ. The results are the mean ± SEM of three experiments.

a furher example of the differences existing between brain microglia and macrophages of peripheral origin in their specific functions in response to immune stimuli.

# 5. REFERENCES

Alberati-Giani D., Malherbe P., Köhler C., Lang G., Kiefer V., Lahm H.-W., and Cesura A.M., 1995, Cloning and characterization of a soluble kynurenine aminotransferase from rat brain: identity with kidney cysteine conjugate β-lyase, *J. Neurochem.* 64:1448.

Foster A.C., White R.J., and Schwarcz R., 1986, Synthesis of quinolinic acid by 3-hydroxyanthranilic acid oxygenase in rat brain tissue in vitro, *J. Neurochem.* 47:23.

Giulian D., Vaca K., and Corpuz M., 1993, Brain glia release factors with opposing actions upon neuronal survival, *J. Neurosci.* 13:29.

Heyes M.P., 1993, Metabolism and neuropathologic significance of quinolinic acid and kynurenic acid, *Biochem. Soc. Trans.* 21:83.

Heyes M.P., Saito K., and Markey S.P., 1992, Human macrophages convert L-tryptophan into the neurotoxin quinolinic acid, *Biochem J.* 283:633.

Heyes M.P., Saito K., Major E.O., Milstein S., Markey S.P., and Vickers J.H., 1993, A mechanism of quinolinic acid formation in inflammatory neurological diseases, *Brain* 116:1425.

Mallat M., and Chamak B., 1994, Brain macrophages:neurotoxic or neurotophic effectorcells?, *J.Leuk. Biol.* 56:416.

Pellicciari R., Natalini B., Costantino G., Mahmoud M.R., Mattoli L., Sadeghpour B.M., Moroni F., Chiarugi A., and Carpenedo R., 1994, Modulation of the kynurenine pathway in search of new neuroprotective agents. Synthesis andpreliminary evaluation of (m-Nitrobenzoyl)alanine, a potent inhibitor of kynurenine 3-hydroxylase, *J. Med. Chem.* 37:647.

Pirami L., Stockinger B., Betz-Corradin S., Sironi M., Sassano M., Valsasnini M., Righi M., and Ricciardi-Castagnoli P., 1991, Mouse macrophage clones immortalized by retroviruses are functionally heterogeneous, *Proc. Natl. Acad. Sci. USA* 88:7543.

Righi M., Mori L., De Libero G., Sironi M., Biondi A., Mantovani A., Donini S.D., and Ricciardi-Castagnoli P., 1989, Monokine production by microglial cell clones, *Eur. J. Immunol.* 19:1443.

Saito K., Chen C.Y., Masana M., Crowley J.F., Markey S.P., and Heyes M.P., 1993a, 4-Chloro-3-hydroxyanthranilate, 6-chlorotryptophan and norharmane attenuate quinolinic acid formation by interferon-g stimulated monocytes (THP-1 cells), *Biochem. J.* 291:11.

Saito K., Crowley J.S., Markey S.P., and Heyes M.P., 1993b, A mechanism for increased quinolinic acid formation following acute systemic immune stimulation, *J. Biol. Chem.* 268:1.

Stone T.W., 1993, Neuropharmacology of quinolinic and kynurenic acids, *Pharmacol. Rev.* 45: 309.

Takikawa O., Kuroiwa T., Yamazaki F., and Kido R., 1988, Mechanism of interferon-γ action. Characterization of indoleamine 2,3-dioxygenase in cultured human cells induced by interferon-γ and evaluation of the enzyme-mediated tryptophan degradation in its anticellular activity, *J. Biol. Chem.* 263:2041.

Taylor M.W., and Feng G., 1991, Relationship between interferon-γ, indoleamine 2,3-dioxygenase, and tryptophan catabolism, *FASEB*, 5:2516.

# THE EFFECTS OF INDUCED KYNURENINE PATHWAY ON IMMUNOCYTOCHEMICAL CHANGES IN RAT TISSUES FOLLOWING EXCESSIVE L-TRYPTOPHAN CONSUMPTION

N. Ronen,[1] B. Gross,[2] D. Ben-Shachar,[1] and E. Livne[1]

[1] Bruce Rappaport Faculty of Medicine
[2] Dept. Neurology
Carmel Medical Center and
Bruce Rappaport Faculty of Medicine
Technion-Israel Institute of Technology, Haifa, Israel

## 1. INTRODUCTION

During the past decade it has been reported that diets containing excessive L-tryptophan have caused multisystemic manifestations in humans, known as Eosinophilia Myalgia Syndrome (EMS) (Duffy, 1992). Abnormal metabolism of tryptophan was also reported in clinical syndromes such as carcinoid syndrome (Cooper et al., 1991), scleroderma-like syndrome (Varge et al., 1990; Bruce et al., 1990; Smith et al., 1990), eosinophilic fasciitis (Freis et al., 1973), idiopathic scleroderma (Stenerg et al., 1980) and systemic sclerosis (Freis et al., 1973). Inflammatory responses were elicited in all these syndromes as leukocyte infiltration and fibroblast proliferation were constantly seen in histological sections from lung skin and muscle tissues. It is thus possible that excessive L-tryptophan intake is involved in stimulation of the immune system. However, the mechanism by which tryptophan or its metabolites are involved in these processes is not yet understood. Since it is impossible to precisely follow such effects in humans, it was the purpose of the present study to follow the effects of L- tryptophan and its metabolites in an animal model that demonstrates similar effects to those observed in human tissues.

## 2. METHODS

Three-week-old female rats (Charles River C.D.) were divided into four groups, (6 animals each), that were fed (three weeks) on 10% protein diet derived from casein, with or without the supplementation of 1% tryptophan. At the end of the second week, two groups of animals (with and without 1% L-tryptophan) received two injections of p-chloropheny-

*Recent Advances in Tryptophan Research,* edited by Graziella Allegri Filippini et al.
Plenum Press, New York, 1996

lalanine (p-CPA) (300 mg/kg i.p, 24 h apart), followed by three injections (100 mg/kg) on every alternating day. All animals were kept in conditions of controlled temperature ($24^0$ c) and maintained on a 12h - 12h light-dark cycle. Food and water were supplied ad libitum. At the end of the third week feeding period, animals were scarified, blood was collected for measuring serotonin in serum. Tissue samples including gastrocnemius muscle, spleen, liver, and lung were collected for immunocytochemistry, histology and autoradiography. Brains were collected for measuring serotonin levels. Frozen sections of gastrocnemius muscle and lung were used for demonstration of tumor necrosis factor-α (TNF-α), using polyclonal rabbit anti-rat TNF-α and streptavidin-biotin-peroxidase kit (Zymed Laboratories, South San Francisco,CA). Serotonin levels were calculated in serum samples by HPLC (varian 5000). Brain samples were removed on ice. Serotonin and 5-hydroxy indole acetic acid concentration were determined by HPLC by an electrochemical detector.

## 3. RESULTS

The growth patterns of animals in all experimental groups appeared to be similar except for the period between the second and the third week. The weight gain during this period in animals fed on a diet containing 1% tryptophan+p-CPA and in animals treated with p-CPA alone was significantly ($p < 0.001$) less in comparison with the control group (Figure 1). Injection of p-CPA, an inhibitor of serotonin synthesis, caused significant ($p < 0.001$) reduction in serum serotonin content (Figure 2). Similar significant ($p < 0.001$) reduction was observed in the content of serotonin (5-HT) in the midbrain of treated animals (Figure 3). Cellular infiltration was observed in tissue sections of gastrocnemius muscle from treated animals (Figure 4 a,b). In lungs from animals treated with 1% L-tryptophan and p-CPA, an increased cellular infiltration and fibrosis was also observed (Figure 4 c,d). The

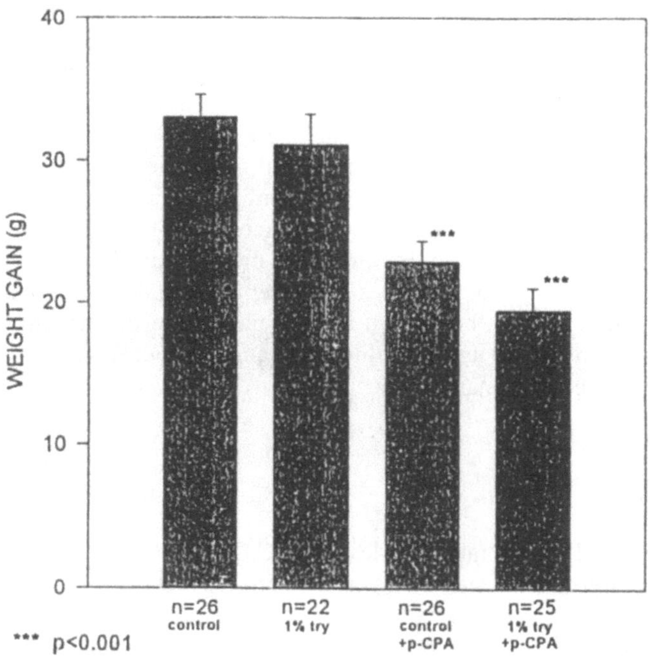


**Figure 1.** Effect of L-tryptophan and p-CPA on weight gain in female rats.

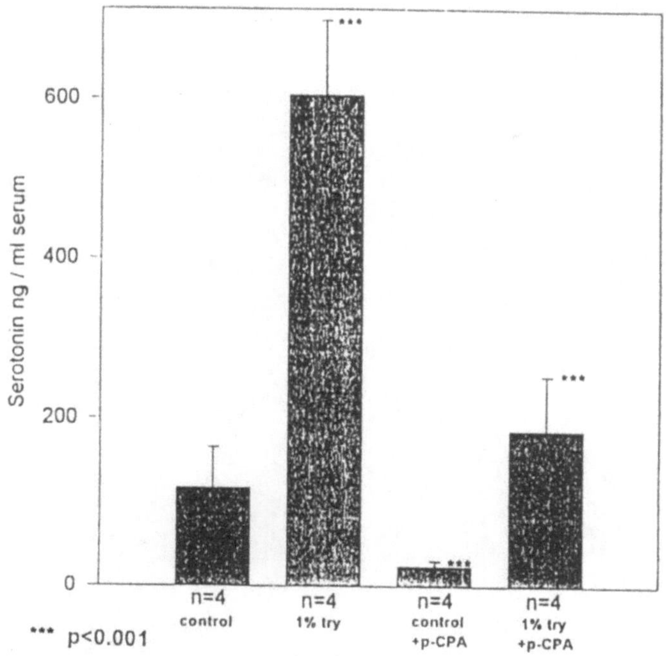


**Figure 2.** The effect of L-tryptophan and p-CPA on the content of serotonin in serum of female rats.

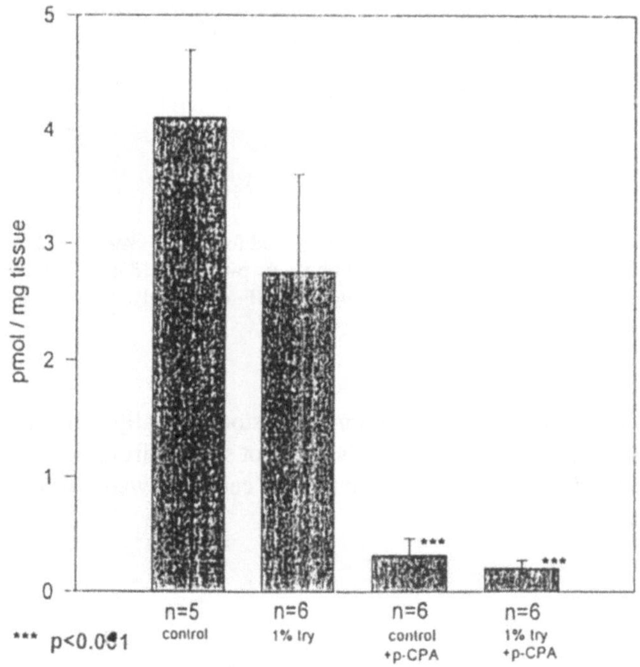


**Figure 3.** The effect of L-tryptophan and p-CPA on the concentration of 5-HT in midbrain of female rats.

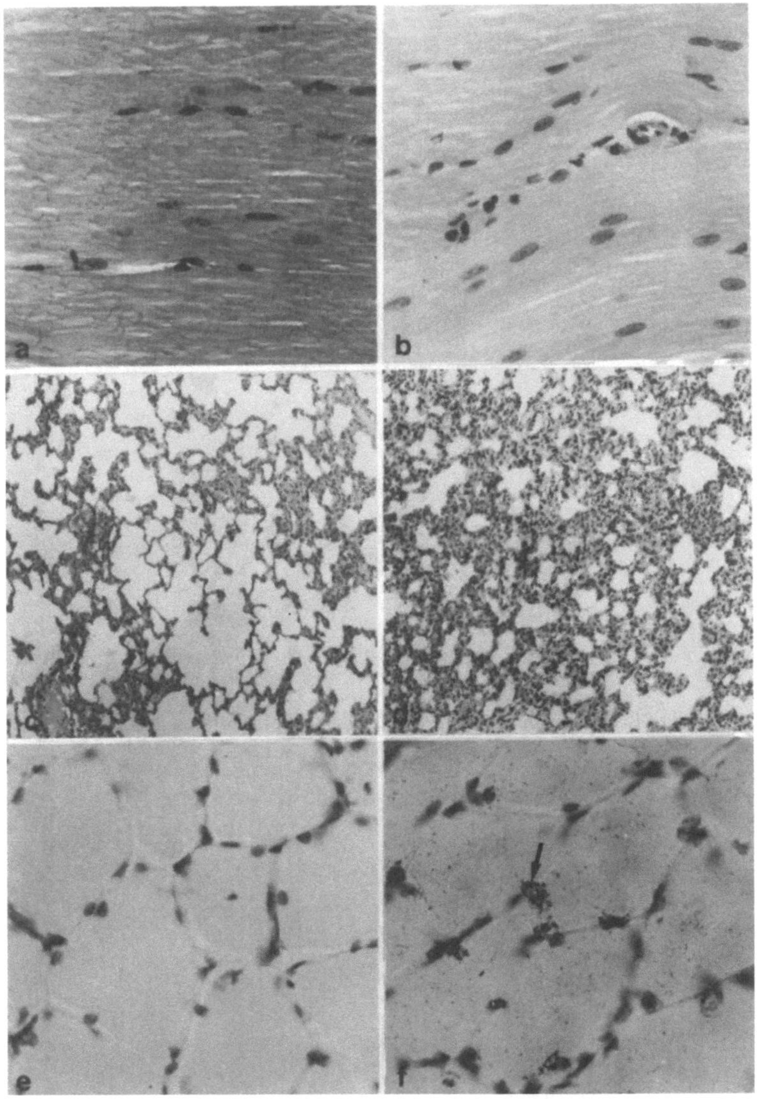


**Figure 4.** Gastrocnemius muscle from rat control animal (a) and following 1%try + p-CPA (b); H&E x 142. Lung from control animal (c) and from animal treated with 1% try p-CPA(d) H&E x 90. Immunohistochemstry for demostration of TNF-α. e - control; f - in the persence of TNF- α antibody. arrow - positive cells x284.

nature of these infiltrating cells was tested immunohistochemically using TNF-α antibody. Both skeletal muscle (Figure 4 e,f) and lung tissues (not shown) from L-tryptophan +p-CPA treated groups, appeared to contain a large number of cells that were positive for TNF-α, a feature indicative of macrophages.

## 4. DISCUSSION

Abnormalities in metabolism of tryptophan have been reported to be associated with tissue fibrosis and inflammatory connective tissue diseases. In the present study the possibility that the kynurenin pathway is involved in tissue fibrosis and in inflammatory processes in rat muscle and lung following excessive L-tryptophan consumption was investigated. The mechanisms that link immune stimulation by the kynurenine pathway metabolism, which includes the production of cytokines, such as interferon-γ and tumor necrosis factor-α are not well understood. It has shown however that indole-amine-2,3-dioxygenase is induced in brain and systemic cells by interferon-γ. It has been previously observed that significant correlations exist between the high concentration of L-kynurenine, quinolinic acid (Leonhardt et al., 1986), and kynurenic acid (Freese et al., 1990) in inflammatory diseases of the central nervous system and in rheaumatic inflammatory diseases such as raumatic arthritis and scleroderma. This supports a role for these metabolites in etiology of inflammatory diseases. Further, the rate of substrate flux through thekynurenine pathway appears to be dependent on activation of indoleamine-2,3-dioxygenase within the systemic tissues. In addition interferon-γ induced indoleamine-2,3-dioxygenase, with concomitant depletion of L-tryptophan has been proposed as a possible mechanism for limitation of replication of intracellular parasites. Picolinic acid, which is also produced, is a co-factor in macrophage activation (Heyes et al., 1992; Varesio et al., 1990). The present study demonstrated that excessive L-tryptophan diet is a possible activator of indoleamine-2,3-dioxygenase and of the immune system. It indicated that kynurenin pathway induction by p-CPA resulted in fibrosis in lungs and cellular infiltration in perimysia of skeletal muscle. Some of the infiltrating cells were shown to be TNF-α positive, indicative of macrophages. We conclude that L-tryptophan diet and p-CPA increased activation of kynurenine pathway in various tissues, in conjunction with macrophage infiltration to affected tissues. This provides a possible mechanism that accounts for increased kynurenine pathway (Saito et al., 1993).

## ACKNOWLEDGMENT

This study was supported by Technion V.P.R. grant no. 181-557.

## 5. REFERENCES

Bruce F., Werth V.P., Rook A.H., O'Connor C.R., Schumacher H.R., 1990, L-Tryptophan ingestion associated with eosinophilic fasci but not progressive systemic sclerosis. *Ann. Int. Med.* 112: 758.

Cooper J.R.., Bloom F.E., Roth R.H., 1991, In: Carcinoid syndrome. The Biochemical Basis of Neuropharmacology. Serotonin (5- hydroxytryptamine) and Histamine. 6th Edition. New York, Oxford. Oxford University Press. p.p.338 - 280.

Duffy J., 1992, The lessons of Eosinophilia-Myalgia Syndrome. *Hospital Practice* 30: 65.

Freese B.A., Kenton J., Swartz B.A., Mattew J., 1990, Kynurenine metabolites of tryptophan: Implications for neurologic diseases. *Neurology* 40: 691.

Freis J.F., Lingren J.A., Bull M.J., 1973, Scleroderma-like lesions and Carcinoid syndrome. *Arch. Intern. Med.* 131: 550.

Heyes M.P., Saito K., Crowley J.S., 1992, Quinolinic acid and kynurenine pathway metabolism in inflammatory and non-inflammatory neurolgic disease. *Brain.* 115: 1249.

Leonhardt A.B., Neal J.W., Kuln J.A., Schwarz M., Plimmer J.R., 1986, Quinolinic acid: An endogenous metabolite that produces axon-sparing lesions in rat brain. *Science.* 219: 316.

Saito K., Crowley J.S., Markry S.P., Heyes M.P., 1993, A mechanism for increased quinolinic acid formation following acute systemic immune stimulation. *J. Biol. Chem.* 268: 15496.

Smith S.A., Roelofs R.I., Gertner E., 1990, Microangiopathy in eosinophilia-myalgia syndrome. *J.Rheumatol.* 17: 1544.

Stenerg E.M., Van Woert M.H., Young S.N., Magnussen I.B., 1980, Development of a scleroderma-like illness during therapy with L-5-hydroxytryptophan and carbidopa. *New Engl. J. Med.* 303: 782.

Varesio L., Calyton M., Ruffman E., Radzioch D., 1990, Picolonic acid, a catabolite of tryptophan, as the second signal in the activation of interferon-γ-primed macrophages. *J. Immunol.* 145: 4265.

Varge J., Peltonen J., Uitto J., Jimenez S., 1990, Development of diffuse fasciitis with eosinophilia during L-tryptophan treatment: Demonstration of Elevated Type I Collagen gene expression in affected tissues. *Annals Int. Med.* 112: 344.

# SERUM TRYPTOPHAN, KYNURENINE, AND NEOPTERIN IN PATIENTS WITH GUILLAIN-BARRE-SYNDROME (GBS) AND MULTIPLE SCLEROSIS (MS)

V. Rudzite,[1] J. Berzinsh,[2] I. Grivane,[2] D. Fuchs,[3] G. Baier-Bitterlich,[3] and H. Wachter[3]

[1] Laboratory of Biochemistry
Latvia Institute of Cardiology, Riga, Latvia
[2] Department of Neurology
Latvia Medicine Academy, Riga, Latvia
[3] Institute of Medical Chemistry and Biochemistry
University of Innsbruck, Innsbruck, Austria

## INTRODUCTION

GBS is a multifocal inflammation of peripheral nerves followed by demyelination as well as lymphocyte and macrophage cell infiltration throughout the peripheral nervous system (Donaghy, 1993, Murray, 1993). MS is a demyelinating disease, specific to the central nervous system. The lesion is located in the white matter of the brain and spinal cord followed by the degeneration of oligodendroglia and perivascular and periadventitial infiltration with mononuclear cells /lymphocytes/ (Ebers, 1995). The etiology of GBS as well as that of MS is unclear, however, pathogenesis can be connected with viral infection or with immunization (Ramsohoff et al.,1994, Hartung et al., 1992, McFarlin, 1990, Stoll et al., 1993).

Both, viral infection and immunization are followed by the activation of immune system and a release of specific cytokines (Whicher and Evans, 1990, Evans and Whicher 1993), as well as by the elevation of kynurenine and some pteridines level (Werner E.R. et al., 1991, Wachter et al., 1992). The purpose of this study is to investigate patients with GBS and with MS by examination of tryptophan, kynurenine and neopterin concentration in blood serum.

## PATIENTS AND METHODS

10 Patients with GBS, 9 patients with MS and 10 healthy persons were studied. 5 Patients with GBS were examined before any treatment (I group). The remaining 5 patients with GBS were observed 1/2 - 8 months after an ineffective treatment with voltaren, analgin, proserin, dimedrol, nicotinic acid, vitamin B1, B6 and B12 in usual therapeutic doses (II group).

*Recent Advances in Tryptophan Research*, edited by Graziella Allegri Filippini et al.
Plenum Press, New York, 1996

Neither patients with GBS (I and II group) and with MS nor healthy persons received corticosteroids, other immunosuppressive therapy or plasmapheresis at the time of any testing.

The determination of tryptophan and kynurenine were carried out in Laboratory of Biochemistry in Latvia Institute of Cardiology, while the determination of neopterin was realised in the institute of Medical Chemistry and Biochemistry University of Innsbruck, Austria. The method of Spice and Chambers (1949) was used for the determination of tryptophan in blood serum, while serum kynurenine content was examined by the method of Joseph and Risby (1975). Blood samples were tested to tryptophan and kynurenine at fasting state (on an empty stomach) as well as 1,2,3,4 and 24 h after an oral dose of 30 mg L-tryptophan / kg of body weight. Neopterin concentration in blood serum samples at fasting state was measured by radioimmunoassay (Wachter et al., 1989).

Statistical analysis were done using the Student's t-test. P-values below 0.05 were considered to indicate significant difference (Weber, 1957).

## RESULTS

Figure 1 shows the tryptophan concentration of blood serum in healthy persons and patients with GBS (I and II group) and with MS on an empty stomach as well as after L-tryptophan loading. We have observed an increase of total tryptophan content in patients with GBS and with MS at fasting state ($p<0.01 - 0.001$) ,as well as 24 h after L-tryptophan

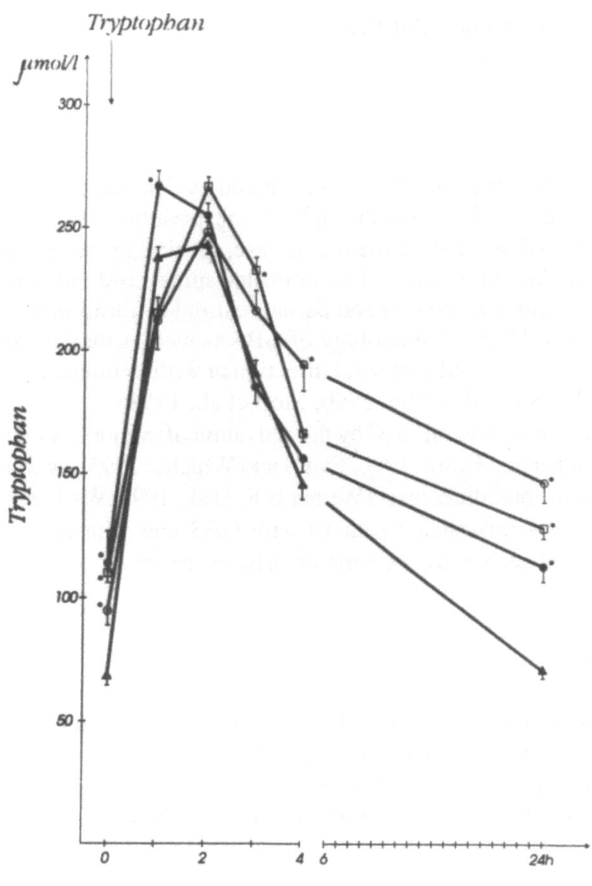


**Figure 1.** Serum concentration of total tryptophan (mean ± S.D.) in healthy persons (- ▲ -) as well as in patients with Guillain-Barre syndrome (GBS) of I group (- O -) or II group (- ● -) and in patients with Multiple sclerosis (MS) (- □ -) before and after oral load of 30 mg L-tryptophan/kg body weight. *Statistically significant ($p<0.05 - 0.002$).

loading (p<0.001) . At 1st , 2nd, 3rd and 4th h after L- tryptophan loading the changes of serum tryptophan content in patients with GBS and with MS were the same as observed in healthy persons.

Figure 2 points out the concentrations of serum kynurenine in healthy persons as well as in patients with GBS (I and II group) and with MS. We have observed an increase of kynurenine content in patients with GBS of the II group ,on an empty stomach (p<0.05) and 24 h after L-tryptophan loading, while changes at 1st, 2nd, 3rd and 4th h after L-tryoptophan loading do not alter from deviations received in healthy individuals. In patients with GBS of the I group as well as in patients with MS we have found a decrease of kynurenine content in blood serum at 1st, 2nd, 3rd and 4th h after L-tryptophan loading (p< 0.05 - 0.001), while at fasting state it was the same as observed in healthy persons.

The results of blood serum neopterin concentrations are seen in Table 1.

We have observed a statistically significant increase of serum neopterin content only in patients with GBS of the II group, while neopterin content in patients with GBS of the I group as well as with MS was normal.

## DISCUSSION AND CONCLUSION

We have not found data in literature about serum total tryptophan content in patients with GBS, while in patients with MS there are different findings: normal (Demish et al.,

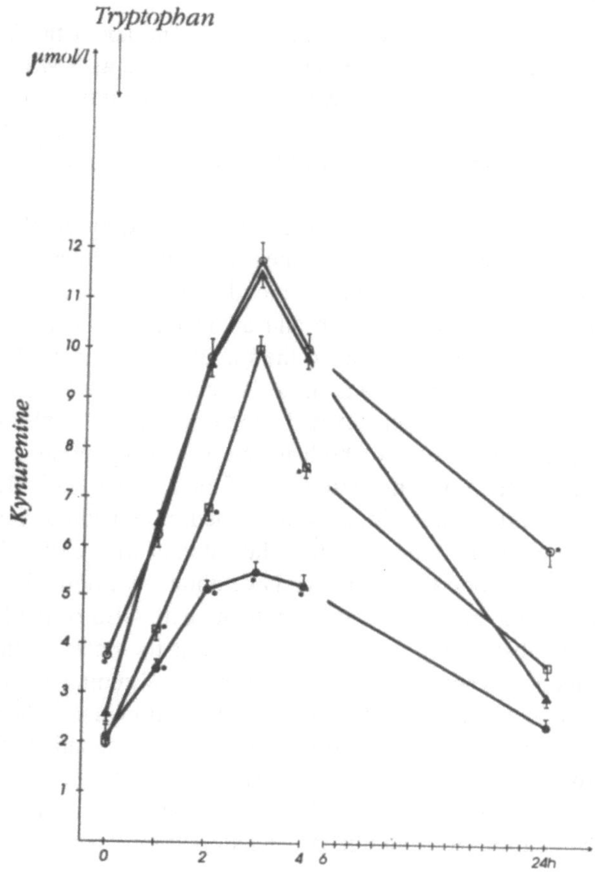


**Figure 2.** Serum concentration of kynurenine (mean ± S.D.) in healthy persons (- ▲ -), as well as in patients with Guillain-Barre syndrome I group (- ○ -), or II group (- ● -) and in patients with Multiple sclerosis (- □ -) before and after oral load of 30 mg L-tryptophan/kg of body weight. *Statistically significant (p<0.05 - 0.002).

**Table 1.** Serum neopterin concentrations (mean ± S.D.) in healthy individuals as well as in patients with Guillain-Barre-syndrome and multiple sclerosis

| Patients | Neopterin in nmol/l | |
|---|---|---|
| GBS ( I group) | 6.8 ± 1.9 | $p>0.25$ |
| GBS ( II group ) | 12.4 ± 3.2 | $p<0.05$ |
| MS | 4.9 ± 0.6 | $p>0.5$ |
| Healthy | 5.3 ± 0.14 | |

1994), or decreased (Milovanovic and Stevanovic, 1994). In contrary to these findings we have observed some increase of serum total tryptophan amount in patients with GBS (I and II group) as well as in patients with MS. We suggest that these different estimations dependent on values found in the groups of healthy persons. For example, Opienska-Blauth et al., (1964) have found 69.3 µmol/l plasma total tryptophan in healthy individuals, Denkla and Dawey (1967) - 54.3 ± 4.9 µmol/l , but Eccleston (1973) - from 42.5 to 106.9 µmol/l . The control group of Demisch et al. (1994) contained 53.4 ± 20.1 µmol/l plasma total tryptophan, while the control group of Milanovic and Stevanocic (1994) - 83.34 ± 29.26 µmol/l, but healthy persons observed by us - 68.6 ± 14.2 µmol/l serum total tryptophan. After loading the increasing of tryptophan content in patients with GBS (I and II group) and with MS was the same as in healthy persons.

In literature we have not found data about the kynurenine content of blood after L-tryptophan loading in patients with GBS and with MS. The results received by us showed a decrease of kynurenine accumulation in patients with MS and with GBS of the I group, while in patients with GBS of the II group the accumulation of kynurenine after L-tryptophan loading was the same as in healthy individuals. Further studies are needed for interpretation of these results.

In viral or bacterial infection followed by an elevation of interferon-gamma level (Demisch et al., 1994) or after the treatment with interferon-gamma (Brown et al., 1987) the amount of total tryptophan in blood plasma decreases. It is followed by an activation of indolamine 2,3-dioxygenase and an increase of kynurenine content in blood and urine (Brown et al., 1994) as well as the stimulation of macrophages and an elevation of neopterin concentration in blood and urine (Wachter et al., 1989, 1992). In patients with GBS some authors have found an elevated neopterin content in plasma (Bansil et al., 1992). We have observed an increase of both neopterin and kynurenine concentration in blood serum at fasting state in patients with GBS of the II group, i.e. after 1/2 - 8 months treatment them with voltaren , proserin, analgin, dimedrol and some vitamins in usual therapeutic doses. However, at the time of testing the patients with GBS of the II group did not receive any therapy. In contrary of that,we have obtained a normal kynurenine and neopterin level in blood serum at fasting state in patients with GBS of the I group and with MS. It allows to suggest that at MS as well as at GBS at the onset of the disease, the interferon-gamma production is not activated. So the process of immunization in such patients is impaired and demyelination appears due to the inflammation in peripheral or central nervous system. It is also observed that the relapses of MS are accompanied with an increase of tumor necrosis factor-α in 85% of such patients (Rieckmann et al., 1995). Moreover, it is also found that tumor necrosis factor-α induces the release of variety of locally acting mediators of inflammation (Rosenblum and Donato,1989).

# REFERENCES

Bansil S., Mithen F.A., Singhal B.S., Cook S.D., Rehewsky-Kochan Ch.,1992, Elevated neopterin levels in Guillain-Barre syndrome, *Arch. Neurology*, 49 : 1277.

Brown R.R., Borden E.C., Sondel P.M., Lee C.M. ,1987, Effect of interferons and interleukin-2 on tryptophan metabolism in humans. In: Progress in Tryptophan and Serotonin research 1986, Bender D.A., Joseph M.H., Kochen W., Steinhart H. Eds., Walter de Gruyter, Berlin-New York, p. 19-26.

Demisch L., Proescholdt M., Ott M., Engelhardt W., Baas H., Enzensberger W., Fischer P.A.,1994, CSF tryptophan in inflammatory neurological diseases. In: L-tryptophan. Current Prospects in Medicine and Drug Safety, Kochen W. and Steinhart H. Eds., Walter de Gruyter, Berlin- New York, p. 139-149.

Dencla W., Dewey M.K.,1967, The determination of tryptophan in plasma, liver and urine, *J. Lab. Clin. Med.*, 69 : 160.

Donaghy M., 1993, Acute idiophatic polyneuritis. In: Brain Diseases of Nervous System, Walton J. Eds., Oxford University Press, p.599 - 603.

Ebers G.C., 1995, Multiple Sclerosis. MS slide lecture series, Produced as an educational service from Schering A.G., Canada, p. 21 - 23.

Eccleston E.G.,1973,A method for estimation of free and total acid-soluble plasma tryptophan using an ultrafiltration technique, *Clin.Chim.Acta*, 48: 269.

Evans S.W., Whicher J.T., 1993, The cytokines: physiological and pathophysiological aspects, *Advances in Clin. Chemistry*, 30 : 1.

Hartung H.P.,Jung S., Stoll G.,Zielasek J., Schmidt B., ArchelosJ.J., Toyka K.V. , 1992, Inflammatory mediators in demyelinating disorders of the CNS and PNS, *J.Neuroimmunology*, 40; 197.

Honavar M., Thanakan J.K.J.,Hugher R.A.C.,Leibovitz S.,Winer J.B., 1991, A clinicopathological study of the Guillain-Barre syndrome, *Brain*, 114: 1245.

Joseph M.H., Risby D., 1975, The determination of kynurenine in plasma, *Clin. Chim.Acta*, 63: 197.

Mc Farlin D.E., 1990, Immunological parametrs in Guillain-Barre syndrome, *Annals of Neurology*, Supplement to 27 : 25.

Milovanovic D.D., Stevatovic H.L., 1992, Cerebrospinal fluid serotonin and 5-hydroxyindolacetic acid and plasma tryptophan in demyelinating diseases and some neurological disorders. In: Advances in Tryptophan Research 1992, Ishigure I., Kido R., Nagatsu Y.,Nagamura Y., Ohta Y. Eds., Fujita Health University Press, Toyoake, p.101 - 106.

Murray D.P., 1993, Impaired mobility: Guillain -Barre syndrome, *J.Neuroscience and Nursing*, 25 : 100.

Opieñska-Blauth J., Chareziñski M.,Berbeč H., 1964, Tryptophan in erytrocytes and plasma in man and some animals, *Acta Biochemica Polonica*, 11 : 147.

Ramsohoff R.M., Tuohy V., Lehman P., 1994, The immunology of multiple sclerosis : new intricacies and new insights, *Current Opinion in Neurology*, 7 : 242.

Rieckmann P., Albrecht M., Kitze B., Weber T., Tumani H., Brooks A., Lüer W., Helvig A.,Poser S.,1995,Tumor necrosis factor-α messenger RNA expression in patients with relapsing-remitting multiple sclerosis is associated with disease activity, *Annals of Neurology*, 37 : 82.

Rosenblum M.G., Donato N.J., 1989, Tumor necrosis factor-α : a multifaced peptide hormone, *CRC Crit.Rev. Immunol.*, 9: 21.

Stoll G., Jung S., Jander S., Meide P., Hartung H.P., 1993, Tumor necrosis factor -α in immunomediated demyelination and Wa llerain degeneration of rat peripheral nervous system, *J.Neuroimmunology*, 45: 175.

Spies J.R., Chambers D.C., 1949, Chemical determination of tryptophan. Study of color formation reaction of tryptophan, p-dimethylaminobenzaldehide and sodium nitrite in sulfuric acid solution, *Analytical Chemistry*, 20: 30.

Wachter H., Fuchs D., Hausen A., Reibnegger G.,Werner E.R.,1989, Neopterin as a marker for activation of cellular immunity: immunologic basis and clinical application, *Advances in Clinical Chemistry*, 27: 81.

Wachter H., Fuchs D.,Hausen A.,Reibnegger G.,Weiss G., Werner E.R., Werner- Felmayer G., 1992, Neopterin: Biochemistry - Methods - Clinical Application, Walter de Gruyter, Berlin- New York.

Werner E.R.,Werner-Felmayer G.,Fuchs D.,Hausen A.,Reibnegger G.,Wachter H., 1991, Relationships between pteridine synthesis and tryptophan degradation, *Advances in Experimental Medicine and Biology*, 294 : 177.

Weber E., 1957, Grundriss der biologischen Statistik für Naturwissenschaftler, Landvirte und Mediziner, Gustav Fischer Verlag, Jena.

Whicher J.T., Evans S.W., 1990, Cytokines in diseasee, *Clin. Chem.*, 36: 1269.

# *KYNURENINES AND DERIVATIVES*

# KYNURENINES AND ANXIETY

I. P. Lapin

Department of Clinical and Experimental Psychopharmacology
Bekhterev Psychoneurological Research Institute
St. Petersburg 193019, Russia

The neuroactivities of kynurenines - the endogenous metabolites of tryptophan have been studied in our laboratory since the begin of 70-ies ( ref. see Lapin, 1989). Initially, the major subjects of our pharmacological experiments on mice and rats were the interaction with monoamines (Lapin, 1972-1976) and the convulsant effect of quinolinic acid (QUIN) and kynurenine (KYN) as well as their excitatory action (Lapin, 1978a,1978b,1980,1981) and the anticonvulsant effect of kynurenic (KYNA), picolinic and xanthurenic acids (Lapin, 1985, Lapin et al., 1986)). Later several experimental evidences have promted the studies of the relevance of kynurenines to anxiety (ref. see - Lapin, 1989):

Initially, those evidences were concerning only KYN itself. They were as follows.

## KYNURENINE (KYN)

### Experimental Data (Ref. See - Lapin, 1989)

1. Similarities in the chemical structures and stereoconfiguration of KYN, diazepam and ligands of the benzodiazepine receptors.
2. Direct radioligand binding data about the action of KYN on the picrotoxin/barbiturate subunit of the benzodiazepine receptor complex.
3. Synergism with caffeine in rats.
4. Diminution of the anticaffeine effect of diazepam in mice.
5. Anxiogen-like activity of KYN in animal models of anxiety in mice: social interaction test and darkness preference (a dark-light chamber).

### Clinical Data

In healthy volunteers blood plasma KYN concentration was markedly increased at the peak of anxiety induced by caffeine. This rise returned to normal after anxiety id ablated (Orlikov and Ryzov, 1991). There was a correlation between KYN concentration and indices of state and trait anxiety (Spielberger-Khanin scale) and anxiety (Hamilton scale) at baseline.

*Recent Advances in Tryptophan Research,* edited by Graziella Allegri Filippini et al.
Plenum Press, New York, 1996

**Table 1.** Relation between the convulsant and the anxiogenic effects of standard and putative endogenous anxiogens (in mice)

| Drug route | Doses (mg/kg - i.p., μg - i.c.v.) Convulsant | Anxiogenic | Ratio |
|---|---|---|---|
| Standard exogenous anxiogens | | | |
| Pentylenetetrazole, i.p. | 80-100 | ca 10 | ca 10 |
| Caffeine, i.p. | 150-200 | ca 20 | ca 10 |
| Yohimbine, i.p. | 100-150 | ca 10 | ca 10-15 |
| Putative endogenous anxiogens | | | |
| Kynurenine, i.c.v. | 50 | 5-10 | ca 5-10 |
| Phenylethylamine*, i.p. | 100 | 5-10 | ca 10-20 |

*(Lapin,1990)

In patients with affective states it was found (Orlikov et al.,1994) that in endogenous anxiety the plasma KYN concentration is increased (1.94 versus 0.9 ug/ml in control). After treatment this concentration became normal. In that study KYN concentration correlated with the severity of anxiety.

## A Proposal about Other Kynurenines

Later one idea has turned our interest to other convulsant kynurenines. We paid attention (Orlikov and Lapin,1993) to the fact that the standard anxiogens which have been proved in humans: pentylenetetrazole, caffeine and yohimbine belong to the group of convulsants (Table 1). In the animal models of anxiety the anxiogenic doses of those drugs relate to their convulsant doses roughly as 1 to 10 (Table 1). Why not to suggest the same or the similar for the endogenous metabolites possessing the stimulant and convulsant actions, e.g. 3-hydroxykynurenine (3-HKYN) or QUIN?

As for anxiolytic action, the standard drugs of this group, benzodiazepines and beta-adrenoblockers, in experiments on animals have the anticonvulsant effect against those anxiogens (pentylenetetrazole, caffeine, yohimbine) used in the convulsant doses. It seemed intriguing to check whether the kynurenines, e.g. KYNA and its precursor indole-3-pyruvic acid (IPA), which possess the anticonvulsant action against the convulsant kynurenines, are active as anxiolytics in the animal models of anxiety.

In the further studies we tested other, than KYN, kynurenines in the animal models of anxiety.

**Table 2.** Anxiogenic activity of 3-hydroxykynurenine (3-HKYN) and quinolinic acid (QUIN) in animal models of anxiety in mice

| Kynurenines | Dose, μg, i.c.v. | Model of anxiety |
|---|---|---|
| 3-HKYN | 0.5-1.5 | Elevated plus-maze* |
| | | Social interaction** |
| QUIN | 1.0 | Socialinteraction** |

*(Lapin and Politi, 1993); ** (Lapin, in press)

**Table 3.** Anxiolytic activity of kynurenic acid (KYNA) and indole-3-pyruvic acid (IPA) in animal models of anxiey in mice

| Kynurenines | Dose, μg/kg, i.p. | Model of anxiety |
|---|---|---|
| KYNA | 40-50 | Elevated plus-maze* |
| IPA | 50-100 | Elevated plus-maze** |

*unpublished data; **(Lapin and Politi, 1993)

## 3-HKYN AND QUIN

It has been found that in mice 3-HKYN and QUIN possess an activity typical of anxiogens (Table 2).

Studies on the anxiolytic effect of the anticonvulsant kynurenines has been started recently.

## KYNURENINC ACID AND INDOLE-3-PYRUVIC ACID

KYNA and IPA appeared to be the compounds possessing the anxiolytic action (Table 3). Their study is the purpose of further experiments.

Thus, the only conclusion which could be made from above is that it is worth to study the neuroactive kynurenines, and not only those mentioned in this paper, in an attempt to find out whether there are compounds with both anxiogenic and anxiolytic activity among them.

## ACKNOWLEDGMENT

This study was supported by two grants: ISF N9L 000 and the Russian Foundation of Fundamental Research N 9 L300

## REFERENCES

Lapin, I.P., 1972, Interaction of kynurenine and its metabolites with tryptamine, serotonin and its precursors and oxotremorine, *Psychopharmacologia* 26:237.

Lapin, I.P., 1973a, Neurotropic activities of kynurenine and its metabolites, in : "Symposium on Pharmacological Agents and Biogenic Amines in the Central Nervous System, J.Knoll and K.Magyar, ed., Akademiai Kiado, Budapest, 149-164.

Lapin, I.P., 1973b, Kynurenines as probable participants of depression, *Pharmakopsychiatrie-Neuropsychopharmakologie* 6: 273.

Lapin, I.P., 1974, Antagonism of nicotinic acid and other kynurenines to antidepressants: one of probable reasons of the therapy-resistance in depression? *Activitas nerv. Super.*(Prague) 16: 260.

Lapin, I.P., 1976, Depressor effect of kynurenine and its metabolites in rats, *Life Sci.* 19: 1479.

Lapin,I.P.,1978a, Stimulant and convulsive effects of kynurenines injected into brain ventricles in mice, *J.Neural Transm* 42:37.

Lapin, I.P.,1978b, Convulsions and tremor in immature rats after intraperitoneal injection of kynurenine and its metabolites, *Pharmacol.Res. Commun.* 10:81.

Lapin, I.P.,1980, Experimental studies on kynurenines as neuroactive metabolites: past, present and future, *Trends in Pharmacol Sci.* 1: 410.

Lapin, I.P.,1981, Kynurenines and seizures, *Epilepsia* 22:257.

Lapin, I.P.,1985, Endogenous antagonists of quinolinic acid and kynurenine as links of the defensive mechanism in epilepsy, *Acta Neurologica* 40:203.

Lapin,I.P., Prakhie, I.V., and Kiseleva, I.P., 1986, Antagonism of seizures induced by the administration of the endogenous convulsant quinolinic acid into rat brain ventricles. *J.Neural Transmis.* 65:177.

Lapin, I.P.,1989, Behavioral and convulsant effects of kynurenines, in: "Quinolinic Acid and the Kynurenines," T.W.Stone, ed., CRC Press, Boca Raton, 193-211.

Lapin, I.P.,1990, Beta-phenylethylamine (PEA): an endogenous anxiogen? Three series of experimental data, *Biol. Psychiatry* 28:997.

Lapin, I.P.,1995, Anxiogenic activity of quinolinic acid and kynurenine in the social interaction test in mice, *J. of Psychopharmacology* (in press).

Lapin, I.P., and Politi, V.,1993, Anxiolytic effect of indole-3-pyruvic acid in mice, *Pharmacol. Res.* 28:129.

Orlikov, A.B., and Lapin, I.P., 1993, Examining the mechanisms of anxiety by anxiogens: clinical and expreimental evidence (Review, 86 ref.), *Exper. and clinical Pharmacology* (Moscow) 56:59.

Orlikov, A.B., Prakhie, I.B., and Ryzov, I.V.,1994, Kynurenine in blood plasma and DST in patients with endogenous anxiety and endoxenous depression, *Biol. Psychiatry* 36:97.

Orlikov, A.B., and Ryzov, I.V., 1991, Caffeine-induced anxiety and increase of kynurenine concentration in plasma of healthy subjects:a pilot study, *Biol. Psychiatry* 29:391.

# BASIC MECHANISMS OF KYNURENINE ACTIONS IN THE CENTRAL NERVOUS SYSTEM

T. W. Stone, D. G. MacGregor, R. A. Smith, P. Jones, W. M. H. Behan, and D. I. Graham

Institute of Biomedical and Life Sciences
University of Glasgow
Glasgow, G12 8QQ, Scotland

## 1. THE KYNURENINE PATHWAY

The kynurenine pathway represents the major biochemical route for the metabolism of L-tryptophan in mammals. The pathway is particularly important in the central nervous system (CNS), since the 'precursor responsiveness' of the pathway exceeds that of tryptophan metabolism via 5-hydroxytryptamine (Freese et al. 1990; Stone, 1993).. Any procedure which alters the systemic availability of tryptophan will therefore have important consequences for CNS function. In particular, it is now well established that within the CNS, the kynurenine pathway lead to the formation of two compounds - quinolinic acid and kynurenic acid - with activity at receptors for neurotransmitter amino acids.

## 2. QUINOLINIC ACID

Quinolinic acid was known to be a convulsant after central administration in several animal species, but it was not until single cell studies of its electrophysiological actions that it was found to be a selective agonist at a population of receptors activated by the excitatory amino acid neurotransmitter, glutamate (Stone & Perkins, 1981). Glutamate normally acts upon several subtypes of ion-channel-linked (ionotropic) receptors sensitive respectively to kainic acid, an isoxazole compound, AMPA, and N-methyl-D-aspartate (NMDA). The last of these is believed to mediate longer latency actions of glutamate at synapses, and to be involved in plastic phenomena such as long-term potentiation, epileptiform bursting activity, and excitotoxicity (Stone & Burton, 1988). It is at NMDA receptors that quinolinic acid is s selective agonist, as originally demonstrated in vivo (Stone & Perkins, 1981), and subsequently confirmed by patch-clamp studies (McLarnon & Curry, 1990).

*Recent Advances in Tryptophan Research,* edited by Graziella Allegri Filippini et al.
Plenum Press, New York, 1996

## 2.1 Excitotoxicity

Excessive stimulation of glutamate receptors, especially those sensitive to NMDA or kainate, can lead to the accumulation of calcium in neuronal cytoplasm, with the consequent activation of cellular proteases which cause cell damage or even death. It remains unclear whether cell death is necrotic or apoptotic (Ignatowicz et al 1991), but it may involve lipid peroxidation within the membranes of susceptible cells. The production of free radicals is therefore likely to occur, although systemic administration of several antioxidants was reported not to affect quinolinate-induced cell damage (Beal et al. 1988).

The ability of the endogenous compound quinolinic acid to act at NMDA receptors therefore raised the possibility that it might be involved in neurodegenerative disorders of the CNS. To date, much evidence has been gathered to support this view from correlations between the amount of quinolinate occurring in a variety of disease states and the degree of neuronal damage (e.g. Heyes et al 1992; Reinhard et al 1994; see Stone 1993), from reports of behavioural and other similarities between quinolinate lesions and Huntington's disease (Popoli et al., 1994) and from neurochemical evidence that the cellular distribution of damage after quinolinate can reproduce that pattern encountered in neurological disorders (Beal et al 1991a,b; Huang et al., 1995). Some of this evidence remains controversial, however (Davies & Roberts 1987; Forloni et al. 1992; Figueredocardenas et al, 1994) and quinolinate's role in neurodegenerative disease remains correspondingly unclear.

# 3. KYNURENIC ACID

The second tryptophan kynurenine metabolite of serious interest in the CNS is kynurenic acid. This compound was first described as a non-selective antagonist at all three of the ionotropic glutamate receptors noted earlier (Perkins & Stone, 1982), and has subsequently received much attention as the lead compound used by several pharmaceutical companies in the development of new glutamate antagonists for use in cerebral ischaemia, stroke or epilepsy.

## 3.1 Kynurenate and Neuroprotection

The systemic injection of kynurenate in high dosage or its more potent analogues such as 7-chloro and 5,7-dichlorokynurenic acids, have been reported to prevent the amino acid release, cerebral damage and behavioural sequelae which follow a period of ischaemia (Germano et al., 1987; Chen et al., 1993; Wood et al., 1993; Ghribi et al., 1994). It has also been observed that kynurenate will prevent neuronal damage induced by traumatic insults (fluid-percussion injury) (Hicks et al., 1994). In contrast, we have found it difficult to obtain significant protection using a model of kainic acid induced neuronal damage. When injected parenterally, kainate induces a pattern of neuronal loss throughout the CNS which closely resembles that which follows repeated epileptic seizures, and some types of ischaemia. The systemic injection of kynurenate, however, in a dose regime which had given protection against ischaemia, did not prevent this damage. Similarly, the injection of L-tryptophan at 200 mg/kg did not modify kainate damage.

It has been demonstrated by several groups that the parenteral injection of L-kynurenine is sufficient to raise the levels of kynurenic acid in the CNS, sufficient to have behavioural and neuroprotective activity against cerebral ischaemia (Nozaki & Beal, 1992). Systemic doses of 100 or 200 mg/kg have been reported to raise cerebral levels of kynurenate from resting values of around 17 nM (Swartz et al 1990) to values near 630 nM (Swartz et al., 1990) or 300 nM (Jauch et al., 1993) respectively. We have therefore also examined this

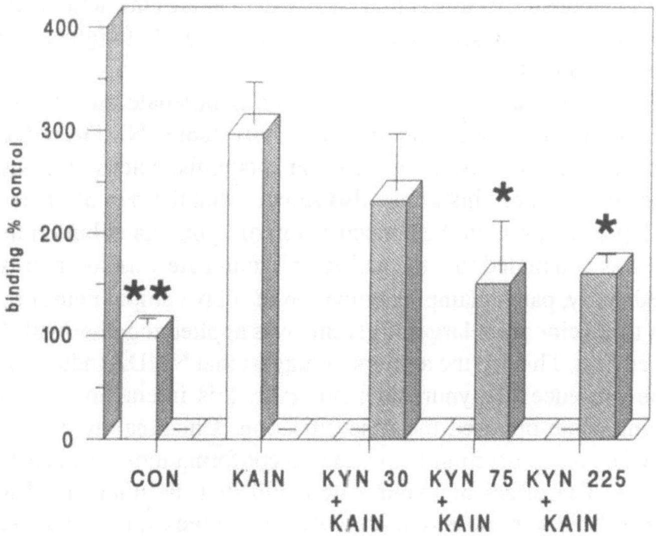


**Figure 1.** Histogram depicting the degree of hippocampal damage measured by the binding of [3H]PK11195 to synaptosomal membranes. The columns show mean ± s.e.m. for rats treated with saline (CON), kainate alone at 10 mg/kg (KAIN), or combinations of kainate with L-kynurenine at the doses indicated (30, 75 and 225 mg/kg, i.p.). Significance is indicated by *P<0.05 and **P<0.01 compared with the kainate group.

route of increasing central kynurenate. The administration of L-kynurenine at 75-220 mg/kg was adequate to protect significantly against kainic acid toxicity, even though the effects did not show clear dose-dependency (Fig. 1).

## 3.2. Indolepyruvic Acid

It has been proposed that an alternative synthetic route to kynurenic acid in the CNS might be via indolepyruvic acid, a normal endogenous compound which can dramatically elevate cerebral kynurenate levels after peripheral administration (Russi et al., 1989. Accordingly we have also examined the effects of systemic administration of this agent. At a dose of 500 mg/kg, indolepyruvic acid was able to protect against kainate neurotoxicity in the rat hippocampus.

Although kynurenate is known to protect against quinolinate toxicity when co-injected into areas such as the nucleus basalis (Boegman et al., 1985), it is not possible to be certain that either L-kynurenine or indolepyruvate are exerting their protective activity in our experiments after metabolism to kynurenate. Nevertheless, there is abundant evidence to indicate that kynurenate has a variety of actions at glutamate receptors, all of which result in blockade or inhibition of receptor function.

## 3.3. Mechanisms of Kynurenate Action

The NMDA receptor complex is subject to several molecular influences, including allosteric modulation via glycine, zinc, and polyamine acceptor sites. The glycine sites are located on the NR1 subunit of the NMDA receptor so that, for example, monomeric receptors consisting entirely of NR1 subunits still show full glycine modulatory properties. Soon after the first report of the glutamate antagonist activity of kynurenate, it was realised that kynurenate could act as an antagonist at the glycine sites, displacing glycine with an IC50 of around 11 μM. The potency of this action clearly demarcates the actions of kynurenate at this site from its weaker and competitive action at the binding site for glutamate itself, where the IC50 is at least 10-fold higher.

The precise mechanistic relationship between kynurenate and the glycine site has attracted considerable attention. Henderson et al. (1990) studied NMDA induced currents in patch clamped cortical neurones, showing that the antagonistic activity of kynurenate was lost in the presence of glycine. This group also showed that the rate of onset of kynurenate antagonism was dependent, not on the concentration of kynurenate, but on the dissociation rate of glycine. It was concluded that the action of kynurenate was competitive.

In a related study, patch clamp examination of hippocampal neurones revealed that current responses to glycine were larger if glycine was applied together with NMDA than if NMDA was applied first. This led the authors to suggest that NMDA induced desensitisation to glycine. In the presence of kynurenate, however, this interaction was not observed, indicating that kynurenate opposed the desensitisation (Chizhmakov et al., 1990). It was suggested that NMDA receptors might exist in two conformations, one sensitive to glycine facilitation, one not. The effect of kynurenate could then be understood as shifting the equilibrium towards the glycine sensitive form. An alternative explanation could be devised, however, involving two states of the glycine site, with NMDA shifting the equilibrium towards a desensitised form and kynurenate acting in the opposite sense.

The complexity of the kynurenate / glycine interaction is emphasised by the fact that another group of modulatory compounds, the polyamines such as spermine, reduce binding to kynurenate but not to glycine. In the cerebellum, there are four times more binding sites for glycine than for kynurenate, suggesting the existence of at least two forms of the glycine binding domain with different ligand selectivity profiles.

The kynurenate binding site is quite complex. Electrophysiological studies of amino acid pharmacology on slices of rat cerebral cortex revealed that kynurenic acid (and quinolinic acid) generated Hill plots with a slope of 2, indicating that two molecules of the agonist and antagonist were required for binding in order to elicit a pharmacological response (Burton et al., 1988). In another quantitative study, by Pullan & Cler (1989), the slope of the Schild plot for kynurenate antagonism of NMDA-induced sodium influx was 0.6 in the absence of any added glycine. This would not be compatible with a competitive action of kynurenate against NMDA. However, when glycine was added to concentrations of 100 and 400 μM, the Schild slopes were increased to 1.0 and 1.9 respectively. This suggests that at least two sites exist for kynurenate, one of which conforms to the pattern expected for an antagonist acting at the agonist recognition site. The result is consistent with the view that kynurenate acts as a competitive antagonist at the glycine site via which the activity of NMDA receptors are modulated, and at a separate site which may be unrelated to NMDA receptors but from which kynurenate can be displaced by glycine.

It may be that with more detailed study these complex patterns of interaction may help to explain some aspects of kynurenate pharmacology which are only poorly understood. For example, in the CA3 region of the hippocampus, both kynurenate and 2-amino-5-phosphono-pentanoic acid (2AP5) are relatively selective antagonists of NMDA applied directly to neurones by microiontophoresis, but only kynurenate is able to prevent the spontaneous bursting discharges of neurones in magnesium-free medium, which are largely the result of activity at NMDA receptor channels (Stone, 1988). Similar results were obtained independently by Brady & Swann (1988). The ability to block bursts was not correlated with antagonism at kainate or AMPA receptors either.

There are, therefore, several unexplained facets of kynurenate activity. Some of them may be explained by newer observations, such as the finding that kynurenate will block high-voltage activated calcium channels in some cells (Dixon & Copenhagen, 1992).

## 4. L-KYNURENINE

Finally, it may be noted that other components of the kynurenine pathway may be involved in neuronal function, other than quinolinate and kynurenate. It has been shown, for example, that L-kynurenine has activity on some hippocampal neurones similar to that of glycine. Neurones in the CA1 region of the hippocampus are normally excited readily by localised applications of NMDA, but this can be suppressed by superfusion with kynurenate. Addition of glycine into the medium at millimolar concentrations is then able to reverse the antagonistic action of kynurenate by reversing at least the competitive component of its action at the glycine site. However, perfusion with L-kynurenine can also reverse kynurenate antagonism, possibly by virtue of the similarity of structure between glycine and the side-chain of L-kynurenine (Stone, 1991). This finding may imply a physiological role for L-kynurenine in the modulation of NMDA sensitivity.

## 5. CONCLUSION

Both quinolinate and kynurenate are important metabolites of tryptophan in the CNS, acting as agonist and antagonist respectively at the NMDA sensitive population of glutamate receptors. Both compounds have the potential for major involvement in physiological and neuropathological situations, yet the details of their mechanisms of action remain unsolved. Particularly in the case of kynurenate, analogues of which are likely to become candidates for therapeutic trials, future research could profitably be targetted at identifying that processes underlying its anticonvulsant and neuroprotectant properties, and at identifying other sites of action which may be clinically relevant.

## 6. ACKNOWLEDGMENTS

We are grateful to the Scottish Home and Health Dept. and the Scottish Epilepsy Research Fund for research funding.

## REFERENCES

Beal, M.F., Ferrante, R.J., Swartz, K.J. and Kowall, N.W., 1991, Chronic quinolinic acid lesions in rats closely resemble Huntington's disease. *J. Neurosci.* 11: 1649.

Beal, M.F., Swartz, K.J., Finn, S.F., Mazurek, M.F. and Kowall, N.W., 1991, Neurochemical characterization of excitotoxin lesions in the cerebral cortex. *J. Neurosci.* 11:147.

Beal., M.F., Kowall, N.K., Swartz, K.J., Ferrante, R.J. and Martin, J.B., 1988, Systemic approaches to modifying quinolinic acid striatal lesions in rats. *J. Neurosci.* 8:3901.

Boegman, R.J., El-Defraqy, S.R., Jhamandas, K., Beninger, R.J. and Ludwin, S.K., 1985, Quinolinic acid neurotoxicity in the nucleus basalis antagonized by kynurenic acid. *Neurobiol. Aging*, 6:331.

Brady, R.J. and Swann, J.W., 1988, Suppression of ictal-like activity by kynurenic acid does not correlate with its efficacy as an NMDA receptor antagonist, *Epilepsy Res.* 2:232.

Burton, N.R., Smith, D.A.S. and Stone, T.W., 1988, A quantitative pharmacological analysis of some excitatory amino acid receptors in mouse neocortex, *Brit. J. Pharmacol.* 93: 693.

Chen, J., Graham, S., Moroni, F. and Simon, R., 1993, A study of the dose dependency of a glycine receptor antagonist in focal ischemia, *J. Pharm. exp. Ther*, 267:937.

Chizhmakov, I.V., Kiskin, N.I., Tsyndrenko, A.Y. and Krishtal, O. A., 1990, Desensitisation of NMDA receptors does not proceed in the absence of kynurenate. *Neurosci. Lett.* 108:88.

Davies, S.W. and Roberts, P.J., 1987, No evidence for preservation of somatostatin-containing neurons after intrastriatal injections of quinolinic acid. *Nature*, 327:326.

Dixon, D.B. and Copenhagen, D.R., 1992, Suppression of a calcium current by CNQX and kynurenate. *Neurosci. Lett.* 147:147.

Figueredocardenas, G., Anderson, K.D.., Chen, Q., Veenman, C.L. and Reiner, A., 1994, Relative survival of striatal projection neurons and interneurones after intrastriatal injection of quinolinic acid in rats. *Exp. Neurol.* 129:37.

Forloni, G.L, Angeretti, N., Rizzi, M. and Vezzani, A., 1992, Chronic infusion of quinolic acid in rat striatum: effects on discrete neuronal populations. *J. Neurol. Sci.* 108:129.

Freese, A., Swartz, K.J., During, M.J. and Martin, J.B., 1990, Kynurenine metabolites of tryptophan: implications for neurologic diseases. *Neurology* 40:691.

Germano,I.M., Pitts, L.H., Meldrum, B.S., Bartkowski, H.M. and Simon, R.P., 1987, Kynurenate inhibition of cell excitation decreases stroke size and deficits. *Ann. Neurol.* 22: 730.

Ghribi, O., Callebert, J., Plotkine, M. and Boulu, R.G., 1994, Effect of kynurenic acid on the ischaemia-induced accumulation of glutamate in rat striatum. *Neuroreport* 5:435.

Henderson, G., Johnson, J.W. and Ascher, P., 1990, Competitive antagonists and partial agonists at the glycine modulatory site of the mouse NMDA receptor., *J. Physiol.* 430:189.

Heyes, M.P., Jordan,E.K., Lee, K., Saito, K., Frank, J.A., Snoy, P. J., Markey, S.P. and Gravell, M., 1992, Relationship of neurologic status in macaques infected with the simian immunodeficiency virus to CSF quinolinic acid and kynurenic acid. *Brain Res.*, 570: 237.

Hicks, R.R., Smith, D.H., Gennarelli, T.A. and McIntosh, T., 1994, Kynurenate is neuroprotective following experimental brain injury in the rat, *Brain Res.* 655:91.

Huang, Q., Zhou, D., Sapp, E., Aizawa, H., Ge, P., Bird, E.D., Vonsattel, J.-P. and DiFiglia, M., 1995, Quinolinic acid-induced increases in calbindin $D_{28k}$ immunoreactivity in rat striatal neurons in vivo and in vitro mimic the pattern seen in Huntington's disease. *Neurosci.* 65: 397.

Ignatowicz, E., Vezzani, A.M., Rizzi, M. and D'Incalci, M., 1991, Nerve cell death induced in vivo by kainic acid and quinolinic acid does not involve apoptosis. *Neuroreport*, 2:651.

Jauch, D.A., Sethy, V.H., Weick, B.G., Chase, T.N. and Schwarcz, R. ,1993, Intravenous administration of L-kynurenine to rhesus monkeys: effects on quinolinate and kynurenate levels in serum and CSF. *Neuropharmacology* 32: 467.

Leeson, P.D. and Iversen, L.L., 1994, The glycine site on the NMDA receptor: structure-activity relationships and therapeutic potential. *J. Med. Chem.*, 37:4053.

McLarnon, J.G. and Curry, K., 1990, Quinolinate activation of NMDA ion channels in rat hippocampal neurons. *Neurosci. Lett.* 116:341.

Nozaki, K. and Beal, M.F., 1992, Neuroprotective effects of L-kynurenine on hypoxia-ischemia and NMDA lesions in neonatal rats. *J. Cereb. Blood Flow M*etab., 12:400.

Perkins, M.N. and Stone, T.W., 1982, An iontophoretic investigation of the actions of convulsant kynurenines and their interaction with the endogenous excitant quinolinic acid. *Brain Res.* 247: 184.

Popoli, P., Pexxola, A., Domenici, M.R., Sagratella, S., Diana, G., Caporali, M.G., Bronzetti, E., Vega, J. and Scotti de Carolis, A., 1994, Behavioural and electrophysiological correlates of the quinolinic acid rat model of Huntington's disease in rats. *Brain Res. Bull.*, 35:329.

Pullan, L.M. and Cler, J.A., 1989, Schild plot analysis of glycine and kynurenic acid at the NMDA excitatory amino acid receptor, *Brain Res.* 497:59.

Reinhard, J.F.jnr., Erickson, J.B. and Flanagan, E.M., 1994, Quinolinic acid in neurological disease: opportunities for novel drug discovery. *Adv. Pharmacol.* 30:85.

Russi, P., Carla, V. and Moroni, F., 1989, Indolepyruvic acid administration increases the brain content of kynurenic acid. Is this a new avenue to modulate excitatory amino acid receptors in vivo? *Biochem. Pharmacol.* 38: 2405.

Stone, T.W., 1993, Neuropharmacology of quinolinic and kynurenic acids, *Pharmacol. Revs.* 45:309

Stone, T.W., 1988, Comparison of kynurenic acid and 2-APV suppression of epileptiform activity in rat hippocampal slices. *Neurosci. Lett.* 84:234.

Stone, T.W., 1991, Kynurenine and glycine enhance neuronal sensitivity to NMDA. *Life Sci.*, 48:765.

Stone, T. W. and Burton, N.R., 1988, NMDA receptors and their endogenous ligands in vertebrate CNS. *Progr. Neurobiol.* 30: 333.

Stone, T.W. and Perkins, M.N., 1988, Quinolinic acid: a potent endogenous excitant at amino acid receptors. *Europ. J. Pharmacol.* 72: 411.

Swartz, K.A., During, M.J., Freese., A. and Beal, M.F., 1990, Cerebral synthesis and release of kynurenate : an endogenous antagonist of excitatory amino acid receptors. *J. Neurosci.* 10: 2965.

Williams, T.L., Smith, D.A.S., Burton, N.R. and Stone, T.W., 1988, Amino acid pharmacology in neocortical slices: evidence for bimolecular actions from an extension of the Hill and Gaddum-Schild equations. *Brit. J. Pharmacol.* 95:805.

Wood, E.R., Bussey, T.J. and Phillips, A.G., 1993, A glycine antagonist 7-chlorokynurenic acid attenuates ischemia-induced learning deficits. *Neuroreport* 4:151.

# 33

# KYNURENINE HYDROXYLASE AND KYNURENINASE INHIBITORS AS TOOLS TO STUDY THE ROLE OF KYNURENINE METABOLITES IN THE CENTRAL NERVOUS SYSTEM

F. Moroni, R. Carpenedo, and A. Chiarugi

Department of Preclinical and Clinical Pharmacology
University of Florence
Viale G.B. Morgagni 65, 50134 Firenze, Italy

## 1. INTRODUCTION

The brain contains a number of tryptophan (TRP) metabolites able to affect neuronal function. A few of them, such as 5-HT, tryptamine and melatonin, have been widely investigated in pharmacological and psychiatric studies during the last 40 years. Others, such as the kynurenine (KYN) metabolites, quinolinic (QUIN) and kynurenic (KYNA) acids, have received much less attention from the scientific community.

Kynurenines originate from tryptophan as a consequence of the action of the rate limiting step enzyme indoleamine-2,3- dioxygenase (IDO; EC.1.13.11.17), an ubiquitously present, induceable protein able to cleave the indol ring. QUIN is a potent NMDA agonist and an excitotoxin, while KYNA competitively antagonises the action of glycine at the NMDA receptor level. Therefore, they may modify neuronal function in an opposite manner (Stone, 1993).

Both QUIN and KYNA have been identified and studied in mammalian biological fluid and in the central nervous system, but the possible role they play in physiology or in pathology is far from clear. Ontogenetic studies show that the concentration of these compounds is quite low (in the nanomolar range) in the brain of new-born animals and that it rapidly increases during development and the normal aging process (Moroni et al., 1986; Moroni et al., 1984b). Brain QUIN and KYNA content changes also in relation to the availability of circulating TRP. The administration of TRP or the reduction of its metabolism in the liver results in a large increase in the brain content of both compounds (Connick et al., 1988; Moroni et al., 1988; Moroni et al., 1986; Moroni et al., 1984a). However, the most effective manner to increase brain KYNA and QUIN content is probably systemic injection of agents able to induce IDO, such as bacterial endotoxins (LPS) (Saito et al., 1992; Moroni

*Recent Advances in Tryptophan Research,* edited by Graziella Allegri Filippini et al.
Plenum Press, New York, 1996

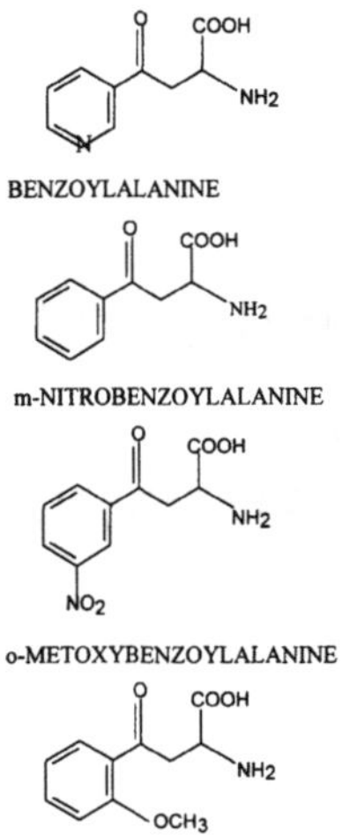

| | IC50 (μM) KYNURENINASE | KYNURENINE HYDROXYLASE |
|---|---|---|
| NICOTINYLALANINE | 800 | 900 |
| BENZOYLALANINE | 300 | 50 |
| m-NITROBENZOYLALANINE | 100 | 0.8 |
| o-METOXYBENZOYLALANINE | 3 | >1000 |

**Figure 1.** Structure activity studies on KYN analogues as inhibitors of kynureninase and kynurenine-hydroxylase. Data are $IC_{50}$ values (μM) and were obtained by using 100 μM KYN as a substrate. Confidential limits were within 20% of the reported values.

et al., 1991). This observation and a series of studies indicating that the content of QUIN increases by several orders of magnitude in the cerebrospinal fluid of patients affected by AIDS dementia and in the rodent brain after ischemic or inflammatory injuries (Saito et al., 1993a; Saito et al., 1993b; Heyes et al., 1991) prompted us to study several possible approaches aimed at selectively changing the brain content of either KYNA or QUIN.

In order to selectively increase KYNA neosynthesis, we attempted to inhibit the metabolism of KYN toward QUIN. KYN is a substrate of several enzymes and, in particular, it may be metabolised into anthranilic acid and into 3-hydroxy-kynurenine by the action of the enzymes kynureninase (EC 7.3.1.3) and kynurenine-hydroxylase (EC 1.14.13.9) respectively. Inhibition of these enzymes may be obtained with nicotynilalanine (NAL), a KYN analogue.

Systemic administration of NAL increased KYNA synthesis in the brain and showed interesting pharmacological properties (Connick et al., 1992; Moroni et al., 1991; Decker et al., 1963); however, NAL potency and selectivity were quite low (Russi et al., 1992). With the aim of understanding the relative role of each of the two above mentioned enzymes involved in KYN metabolism and to further investigate the possibility of significantly increasing KYNA concentration in biological fluids, we tested a number of NAL analogues *in vitro* and *in vivo* (Chiarugi et al., 1995; Carpenedo et al., 1994; Pellicciari et al., 1994). Two of these were selected and further studied. The first was m-nitrobenzoyl-alanine (mNBA), an inhibitor of kynurenine hydroxylase and the second was o-methoxy-benzoyl-alanine (oMBA), an inhibitor of kynureninase (Chiarugi et al., 1995). In this presentation, the biochemical and behavioural actions of these enzyme inhibitors will be discussed.

## 2. BIOCHEMICAL STUDIES ON mNBA AND oMBA

Structure activity studies on NAL analogues synthesised by Prof. R. Pellicciari and by Dr. M.T. Molina were performed by utilising preparations of kynurenine hydroxylase and kynureninase from rat brain or liver (Chiarugi et al., 1995; Carpenedo et al. 1994; Pellicciari et al., 1994). As reported in Fig. 1, mNBA had an acceptable IC50 on kynurenine hydroxylase while oMBA was selectively active as an inhibitor of kynureninase. When these compounds were administered to rats, we observed that the inhibition of kynurenine-hydroxylase, obtained by intraperitoneal administration of mNBA (200-400 mg/kg), resulted in a larger accumulation of KYN and KYNA in the brain, blood, liver and kidney than the inhibition of kynureninase, obtained by intraperitoneal administration oMBA (200-400 mg/kg). Solubility problems prevented the use of larger doses oMBA. The simultaneous administration of 200 mg/kg of both mNBA and of oMBA had additive effects (Chiarugi et al., 1995).

Similarly, the administration of oMBA or of mNBA (50-400 mg/kg i.p.) to rats having microdialysis tubing in their hippocampi significantly increased the concentration of KYNA in the dialysate. The inhibitor of kynurenine-hydroxylase (mNBA) increased KYNA concentration in brain extracellular spaces more efficiently than the inhibitor of kynureninase (oMBA). When both oMBA and mNBA were simultaneously administered, the concentration of KYNA in the dialysate increased in an additive manner (Fig. 2).

The present results thus confirm that hydroxylation is the preferred route of kynurenine metabolism in rodents (Saito et al., 1993b; Bender & McCreanor, 1982) and show the possibility of increasing the concentration of KYNA in brain extracellular fluid by inhibiting kynurenine-hydroxylase or kynureninase with intraperitoneal administrations of either mNBA or oMBA. The simultaneous inhibition of both enzymes resulted in additive effects. The inhibition of kynurenine-hydroxylase, obtained *in vivo* in mice by the administration of mNBA, also resulted in a decrease of 3-OH-kynurenine in blood and brain, while, as expected, a significant increase of this metabolite was observed after the administration of oMBA and the consequent inhibition of kynureninase (Fig. 3). 3-OH-kynurenine may be neurotoxic (Eastman & Guilarte, 1989) and therefore the use of selective kynureninase inhibitors could result in toxic events.

When other kynurenine metabolites were measured in the brain of mice treated with these enzyme inhibitors, we noticed that mNBA was able to decrease the brain concentration of 3OH-anthranilic acid and to increase that of anthranilic acid, while, unexpectedly, oMBA was not able to decrease brain 3OH-anthranilic acid concentration (Fig. 4).

In order to find an explanation for this surprising observation, we investigated the effects of oMBA on kynureninase activity *in vitro* by utilising both 3-OH-kynurenine and

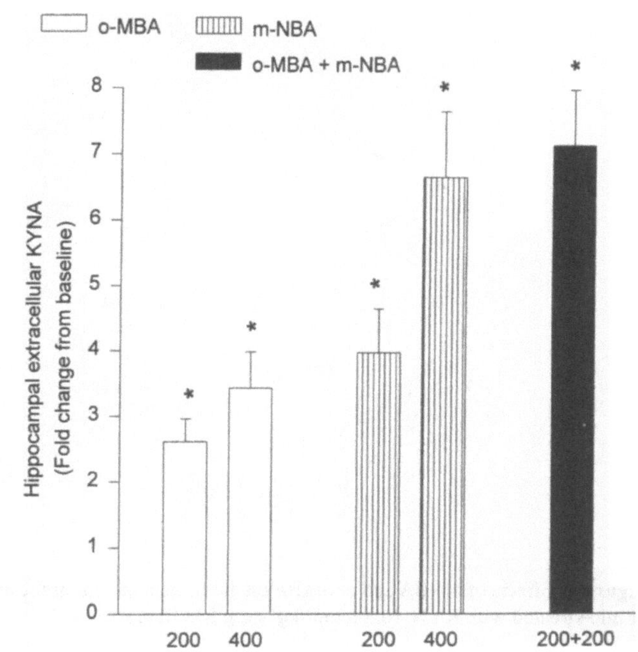


**Figure 2.** Effects of oMBA and of mNBA on the concentration of KYNA in dialysis fluid obtained from the hippocampus of freely moving rats bearing a transverse dialysis probe. The concentration of KYNA found in the control dialysates was 11.1±0.8 nM. Recovery from the dialysis probe was approximately 10%. Values on the abscissa are mg/kg. i.p. Vertical bars are S.E.

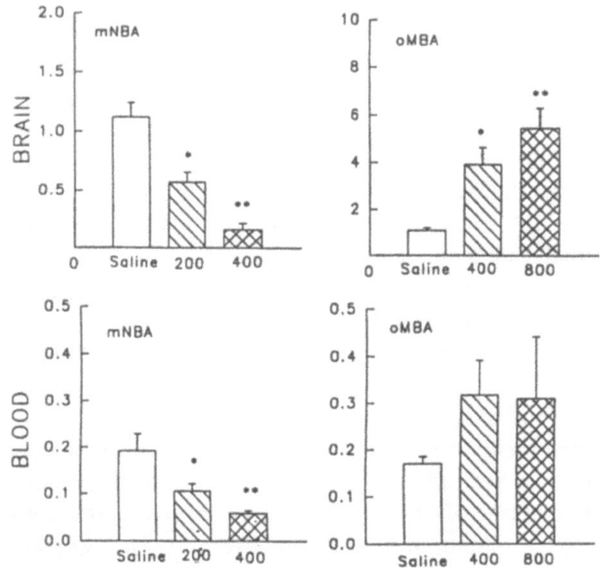


**Figure 3.** Effects of mNBA and oMBA on the content of 3-OH-kynurenine in the blood or brain of mice treated with kynurenine (0.3 mmol/kg. s.c. 1 h. before the sacrifice). oMBA and mNBA were administered at the reported doses (mg/kg i.p.) 90 min. before sacrifice. Columns represent nmoles/g tissue and are means of at least 6 mice. Vertical bars are S.E. * P<0.05; ** P<0.01.

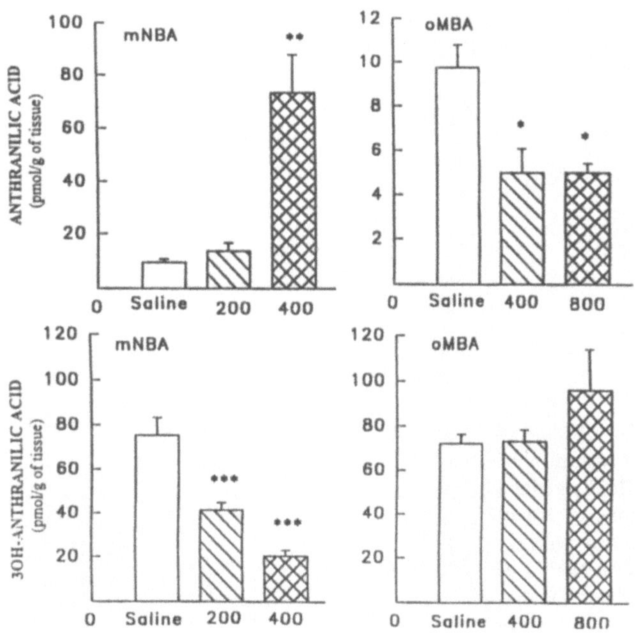


**Figure 4.** Effects of mNBA and of oMBA on the content of anthranilic and 3-OH-anthranilic acids in the brain of mice treated with KYN. (0.3 mmol/kg. s.c.). See figure 3.

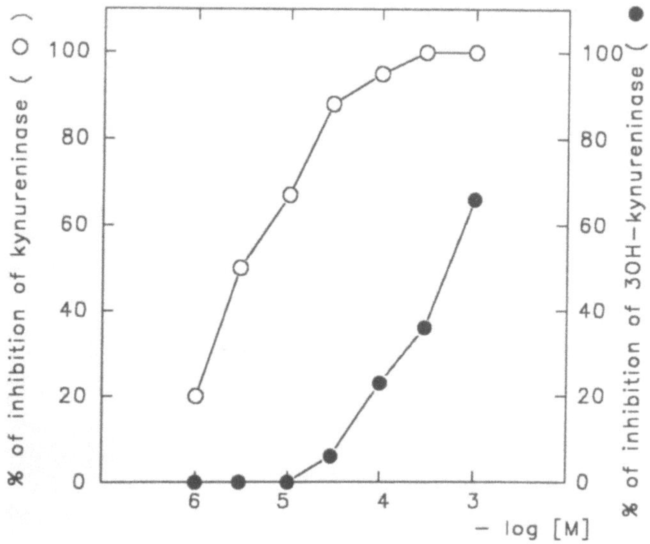


**Figure 5.** Inhibition of kynureninase and 3-OH-kynureninase activity by mNBA. Enzyme activities were evaluated in liver homogenates by utilising either KYN or 3-OH-KYN (100 μM) as a substrate (see: Carpenedo et al. 1994). Each point is the mean of at least 4 experiments conducted in triplicate. S.E. were within 10% of the reported values.

kynurenine as substrates. As reported in Fig. 5, oMBA was two orders of magnitude less potent in inhibiting 3-OH-kynurenine metabolism than in inhibiting kynurenine metabolism.

This observation could explain the reason whereby oMBA is not able to reduce brain 3-OH-anthranilic acid content, and could also open up new and interesting perspectives on the possibilities of modulating KYN metabolism. Experiments on the mechanism of the interactions between oMBA and kynureninase are in progress.

In order to clarify whether or not the 3-6 fold increase of KYNA concentration in brain extracellular spaces, obtained after the administration of the above mentioned enzyme inhibitors caused behavioural actions or neurochemical modifications compatible with an antagonism of the NMDA receptors, further experiments were performed in rats and mice.

## 3. BEHAVIORAL AND NEUROCHEMICAL EFFECTS OF mNBA

Animals (rats and mice) treated with mNBA (50-400 mg/kg i.p.) had a significantly reduced locomotor activity and appeared calm and easy to manage.

As shown in Fig. 6, mNBA not only reduced spontaneous locomotor activity, but was particularly potent in reducing amphetamine-induced hyperlocomotion. Furthermore, mice treated with mNBA had a dose-dependent analgesia (Fig. 7) comparable to that obtained with 5 mg/kg of morphine. Finally, the inhibition of kynurenine hydroxylase and the resulting increase of the extracellular concentration of KYNA changed the output of ACh from the striatum of rats having microdialysis tubing in their brain (Fig. 8).

## 4. CONCLUSION

The increased concentration of KYNA we observed in the brain extracellular spaces after mNBA administration seems associated with decreased locomotor activity, analgesia and inhibition of ACh output. Since similar effects have also been obtained by administering compounds (including KYNA) able to antagonise the glycine recognition site of the NMDA

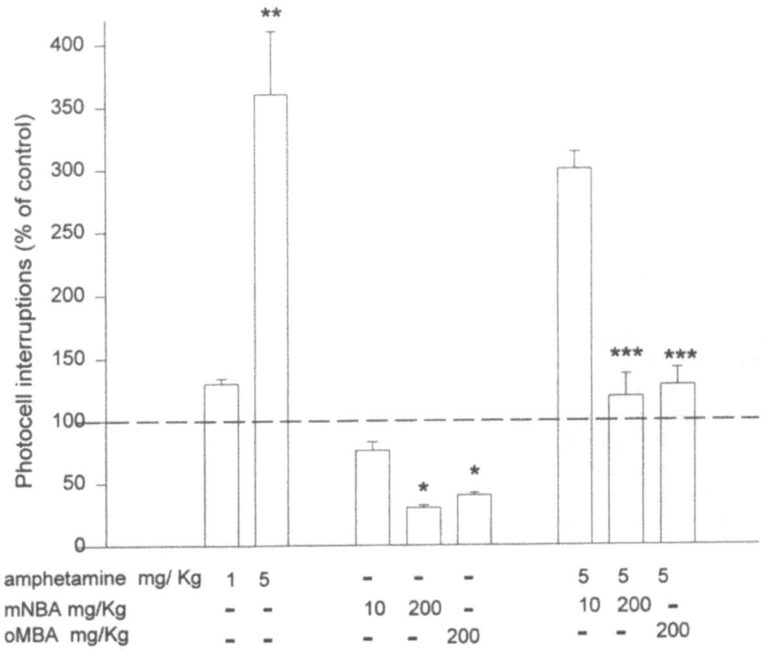


**Figure 6.** Effects of mNBA and oMBA on spontaneous and amphetamine-induced locomotor activity in mice. Columns are mean changes of interruptions of photocells of an Animex activity meter. Vertical bars are S.E.M.; * P<0.05; ** P<0.01; *** P<0.001.

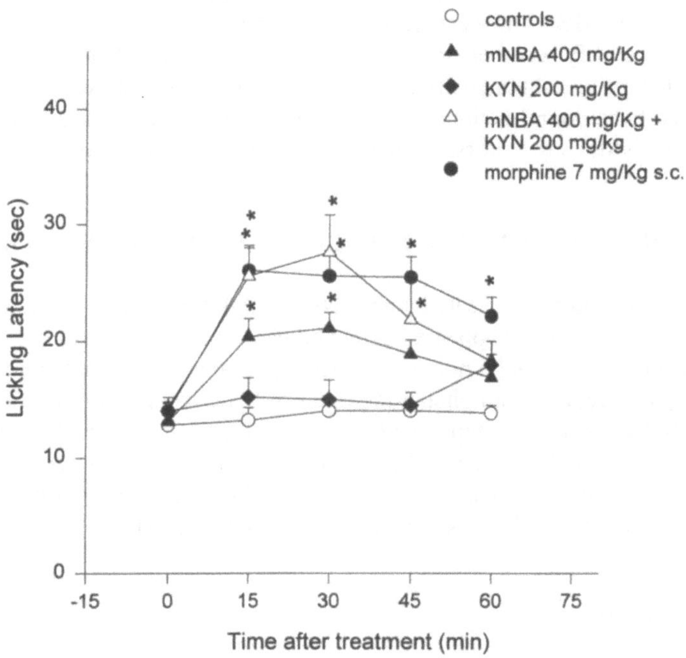


**Figure 7.** Time course and intensity of mNBA-induced analgesia in mice. The licking latency in the hot plate test was evaluated in groups of 10 mice which received mNBA alone or mNBA plus KYN. The compounds were dissolved in saline and administered i.p. For comparison purposes the effects of morphine were also reported.

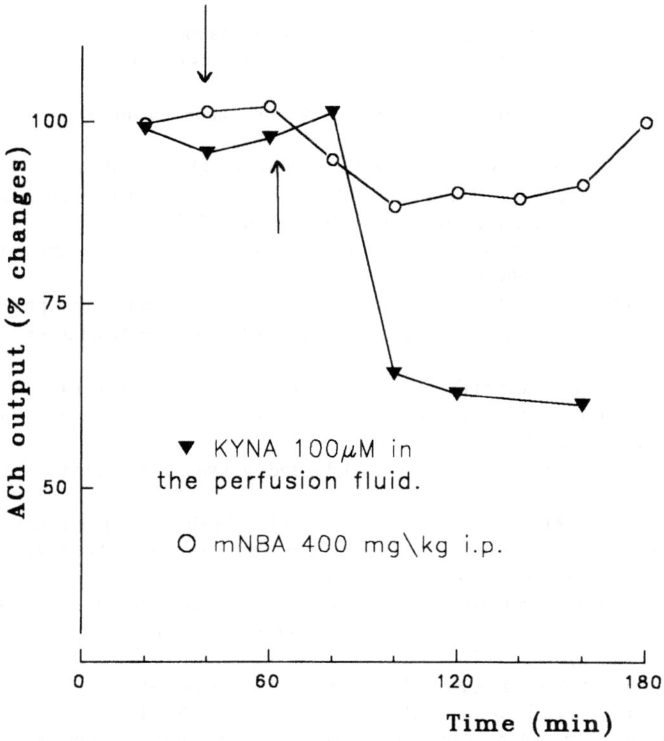


**Figure 8.** The effects of KYNA (100 μM added to the perfusion fluid) or of mNBA (400 mg/kg i.p.) on the output of ACh from the rat striatum. Freely moving animals bearing a transverse microdialysis tubing in the dorsal striatum were used. Values are means obtained in 5 rats for each group.

receptor ion channel complex (Dennison et al. 1992; Rogawski, 1992; Sing et al. 1990), it is reasonable to suggest that the concentrations of KYNA reached after mNBA administration are active and reduce the function of glutamate neurotransmission.

Inhibitors of kynurenine hydroxylase and kynureninase seem useful tools to understand the physiological role of QUIN and KYNA in brain function and pathology. A pharmacological use of such compounds may be envisaged in the numerous situations in which an abnormal function of glutamate receptors play a pathogenetic role.

## 5. REFERENCES

Bender DA, McCreanor GM (1982) The preferred route of kynurenine metabolism in the rat. *Biochim Biophys Acta* 717:56.

Carpenedo R, Chiarugi A, Russi P, Lombardi G, Carlà V, Pellicciari R, Mattoli L, Moroni F 1994, Inhibitors of kynurenine hydroxylase and kynurenine increase cerebral formation of kynurenic acid and have sedative and anticonvulsant activities. *Neuroscience* 61:237.

Chiarugi A, Carpenedo R, Molina MT, Mattoli L, Pellicciari R, Moroni F (1995) Comparison of the neurochemical and behavioral effects resulting from the inhibition of kynurenine hydroxylase and/or kynureninase. *J Neurochem* in press.

Connick J, Lombardi G, Beni M, Moroni F (1988) Decrease in rat cerebral quinolinic acid concentration following hydrocortisone treatment. *Neurosci Lett* 88:216.

Connick J, Heywood GC, Sills GJ, Thomson GG, Brodie MJ, Stone TW (1992) Nicotinylalanine increases cerebral kynurenic acid content and has anticonvulsant activity. *Gentile Pharmacol* 23:235.

Decker RH, Brown RR, Price JM (1963) Studies on the biological activity of nicotynilalnine, an analogue of kynurenine. *J Biol Chem* 238:1049.

Dennison Z, Ossenkop KP, Cain DP (1992) Effects of central administration of kynurenic acid on spontaneous locomotor activity in kindled rat: a multivariate approach of using the automated Digiscan monitoring system. *Pharmacol Biochem Behav* 43:807.

Eastman CL, Guilarte TR (1989) Citotoxicity of 3-hydroxykynurenine in a neuronal hybrid cell line. *Brain Res* 495:225.

Heyes MP, Brew BJ, Martin A, Price RW, Salazar AM, Sidtis JJ, Yergey JA, Mouradian MM, Sadler AE, Keilp J, Rubinov D, Markey SP (1991) Quinolinic acid in cerebrospinal fluid and serum in HIV-1 infection: relationship to clinical and neurological status. *Ann Neurol* 29:202.

Moroni F, Lombardi G, Carlà V, Moneti G (1984a) The excitotoxin quinolinic acid is present and unevenly distributed in the rat brain. *Brain Res* 295:352.

Moroni F, Lombardi G, Moneti G, Aldinio C (1984b) The excitotoxin quinolinic acid is present in the brain of several animal species and its cortical content increases during the aging process. *Neurosci Lett* 47:51.

Moroni F, Lombardi G, Carlà V, Pellegrini D, Carassale GL, Cortesini C (1986) Content of quinolinic acid and of other tryptophan metabolites increases in brain regions of rats used as experimental models of hepatic encephalopathy. *J Neurochem* 46:869.

Moroni F, Russi P, Lombardi G, Beni M, Carlà V (1988) Presence of kynurenic acid in the mammalian brain. *J Neurochem* 51:177.

Moroni F, Russi P, Gallo-Mezo MA, Moneti G, Pellicciari R (1991) Modulation of quinolinic and kynurenic acid content in the rat brain: effects of endotoxins and nicotinylalanine. *J Neurochem* 57:1630.

Pellicciari R, Natalini B, Costantino G, Mahmoud MR, Mattoli L, Sadeghpour BM, Moroni F, Chiarugi A, Carpenedo R (1994) Modulation of the kynurenine pathway in search for new neuroprotective agents. Synthesis and preliminary evaluation of (*m*-nitrobenzoyl)alanine, a potent inhibitor of kynurenine-3-hydroxylase. *J Med Chem* 37:647.

Rogawski M (1992) The NMDA receptor, NMDA antagonists and epilepsy therapy. *Drugs* 44:279.

Russi P, Alesiani M, Lombardi G, Davolio P, Pellicciari R, Moroni F (1992) Nicotinylalanine increases the formation of kynurenic acid in the brain and antagonizes convulsions. *J Neurochem* 59:2076.

Saito K, Markey SP, Heyes MP (1992) Effects of immune activation on quinolinic acid and kynurenine pathway metabolism in the mouse. *Neuroscience* 51:25.

Saito K, Crowley JS, Markey SP, Heyes MP (1993a) A mechanism for increased quinolinic acid formation following acute systemic immune stimulation. *J Biol Chem* 268:15496.

Saito K, Nowak TSJ, Markey SP, Heyes MP (1993b) Mechanism of delayed increase in kynurenine pathway metabolism in damaged brain regions following transient cerebral ischemia. *J Neurochem* 60:180.

Sing L, Oles RJ, Tricklebank MD (1990) Modulation of seizure susceptibility in the mouse by the strychnine-insensitive glycine recognition site of the NMDA receptor/ion channel complex. *Br J Pharmacol* 99:285.

Stone TW (1993) Neuropharmacology of quinolinic and kynurenic acids. *Pharmacol Rev* 45:309.

# KYNURENINE METABOLISM IN THE RAT BRAIN IN VIVO

## Effect of Acute Excitotoxic Insults

R. Schwarcz, G. Ceresoli, and P. Guidetti

Maryland Psychiatric Research Center
University of Maryland School of Medicine
Baltimore, Maryland 21228

## INTRODUCTION

The neuroactive kynurenine pathway metabolites quinolinic acid (QUIN) and kynurenic acid (KYNA) have been speculatively linked to the pathogenesis of several neuro-psychiatric diseases (Schwarcz, 1992; Heyes, 1993; Reinhard et al., 1994). The idea that QUIN or KYNA may be causally involved in brain disorders derived originally from the discovery that intracerebral application of QUIN results in excitotoxic neurodegeneration which resembles the lesions seen in Huntington's disease, temporal lobe epilepsy and cerebral ischemia, that KYNA can serve as a potent neuroprotectant against excitotoxic injury, and that both compounds are normal constituents of the mammalian brain. Subsequently, individual enzymes of the kynurenine pathway were shown to be present in the brain, and abnormal enzyme activities and metabolite levels were found in brain tissue in a number of human brain diseases (Schwarcz and Du, 1991; Stone, 1993).

Although the kynurenine pathway is responsible for the large majority of tryptophan degradation in the periphery, and in spite of efforts to incorporate radiolabelled or unlabelled tryptophan into kynurenine pathway metabolites in the brain, the existence of a functional cascade from tryptophan to QUIN or KYNA has so far not been demonstrated to exist in the normal brain (Speciale et al., 1989; Heyes, 1993; Saito et al., 1993a). This may be due to the very low cerebral activity of indoleamine-2,3-dioxygenase, the enzyme which is responsible for the oxidative cleavage of tryptophan (Saito et al., 1992). Notably, the *de novo* synthesis of QUIN and KYNA from tryptophan can be readily demonstrated after induction of this enzyme under pathophysiological conditions (Saito et al., 1993a).

In view of the possibility that L-kynurenine (KYN), rather than tryptophan, serves as the primary precursor of both QUIN and KYNA in the brain (cf. Fig. 1), we have recently begun to examine if intracerebrally injected KYN can serve as a bioprecursor of kynurenine pathway metabolites *in vivo* (Guidetti et al., 1995). In the following, we describe the results of some of our studies, using rapid and prolonged infusions of 5-$^{3}$H-KYN into the rat striatum.

*Recent Advances in Tryptophan Research,* edited by Graziella Allegri Filippini et al.
Plenum Press, New York, 1996

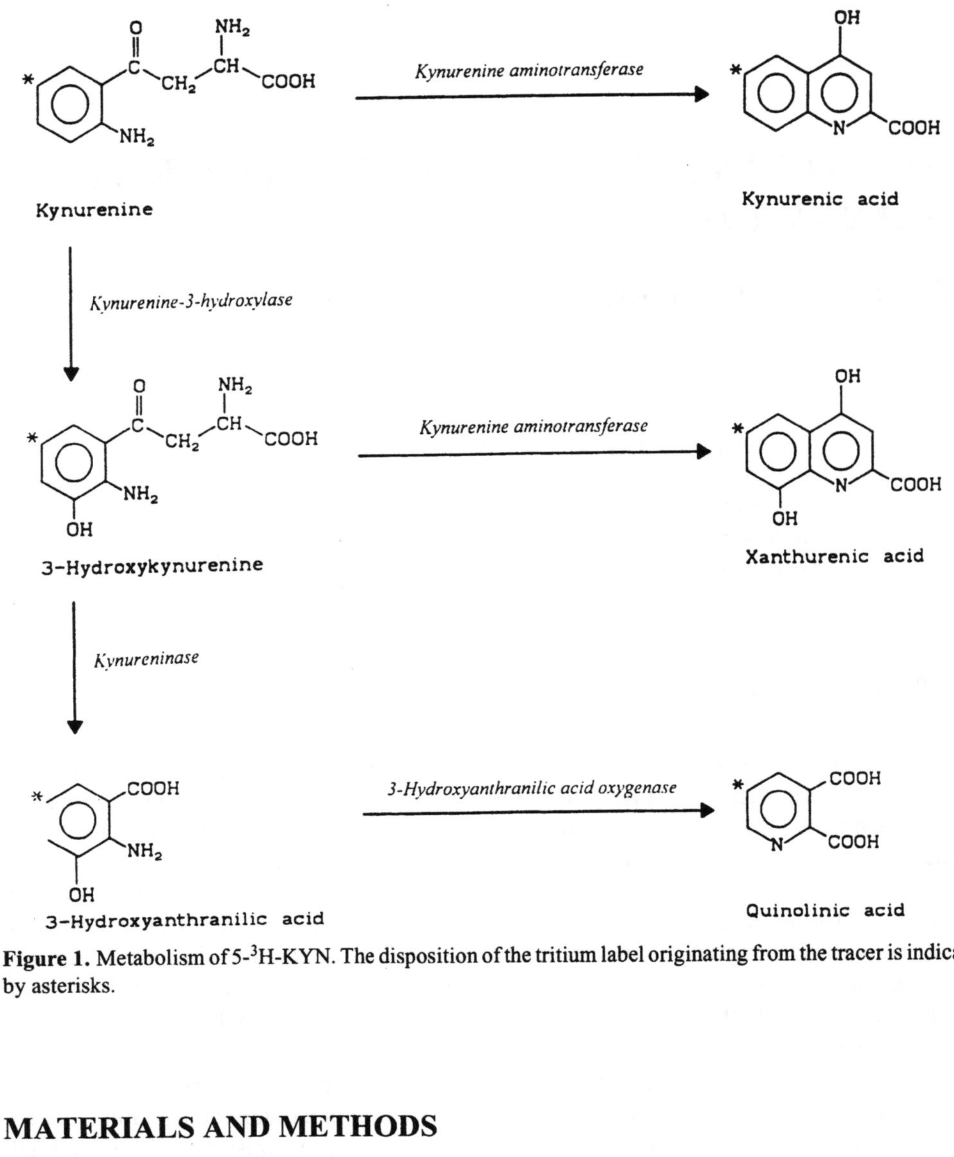


**Figure 1.** Metabolism of 5-$^3$H-KYN. The disposition of the tritium label originating from the tracer is indicated by asterisks.

## MATERIALS AND METHODS

*Animals.* Adult male Sprague-Dawley rats (200-220 g) or male rat pups [postnatal days (PND) 7 and 14] of the same strain were used in all experiments. The animals were housed in a light- and temperature-controlled animal facility with free access to food and water.

*Materials.* L-KYN sulfate, DL-3-hydroxykynurenine (3HK),3-hydroxyanthranilic acid (3HANA), QUIN, KYNA, xanthurenic acid (XA) and ibotenic acid were purchased from Sigma Chemical Co. (St. Louis, MO). 5-$^3$H-KYN (28 Ci/mmol) was synthesized by Amersham (Arlington Heights, IL) from L-5-chlorokynurenine (kindly provided by Dr. M. Varasi, Milan, Italy). The tracer was purified by HPLC and dried under vacuum prior to use (Speciale and Schwarcz, 1990). DL-$^3$H-3HK (spec. activity: 3.6 Ci/mmol) was a generous gift from Dr. T. Guilarte (Baltimore, MD). $^3$H-QUIN (29 Ci/mmol) and $^3$H-KYNA (9 Ci/mmol) were obtained from Amersham, and 1-$^{14}$C-3HANA (6 mCi/mmol) from NEN-Dupont (Wilmington, DE).

*Intrastriatal Infusion of 5-$^3$H-KYN and Excitotoxins.* Animals were anesthetized with chloral hydrate (350 mg/kg, i.p.), placed in a stereotaxic frame, and injection needles were lowered bilaterally into the striatum. Stereotaxic coordinates were: A: 0.7, L: 2.2, V: 4.2 (PND 7); A: 0.8, L: 2.3, V: 4.3 (PND 14); and A: 0.8, L: 2.7, V: 4.8 (adult rats). 15 μM 5-$^3$H-KYN (2.5 μCi in 2 or 6 μl) was then infused for 10 min or 2 hours using a microinfusion pump. In some experiments, the excitotoxins ibotenic acid (40 μg) or QUIN (20 μg) were co-infused with 5-$^3$H-KYN. Animals were killed 2 hours after the beginning of the infusion.

*Tissue Preparation.* Animals were killed by decapitation, and their striata were dissected out rapidly at 4°C. The tissue was frozen on dry ice and kept at -80°C until assayed. For the determination of kynurenine pathway metabolites, the tissue was first sonicated in ultrapure water (1:10, w/v). Recovered radioactivity was measured in a 10 μl aliquot of the original homogenate. 50 μl of 6% perchloric acid were then added to 100 μl of the homogenate, and the precipitated proteins were removed by centrifugation in a microfuge. The resultant supernatant was diluted with 100 μl of the mobile phase and subjected to HPLC analysis (cf. below). In pilot experiments, labeled standards (1-1.5 nCi each) were added to separate brain tissue samples after sonication for the determination of extraction efficiency (> 90% in all cases).

*HPLC Determination of KYN and Its Metabolites.* 200 μl of the tissue extract were applied to a $C_{18}$ reverse-phase column (Lichrosorb, 250 x 4.6 mm; Alltech Assoc., Deerfield, IL) and eluted isocratically at a flow rate of 1.2 ml/min with a mobile phase consisting of 100 mM $(NH_4)H_2PO_4$ and 100 mM acetic acid (titrated to pH 3.2 with phosphoric acid), containing 1.25 mM octanesulphonic acid and 6% (v/v) acetonitrile. The eluate was collected with the aid of a fraction collector. The volume per fraction varied between 0.4 and 1.2 ml and was determined by the elution pattern of standard samples which were detected by UV spectroscopy at 280 nm (QUIN and picolinic acid) or 340 nm (KYN, 3HK, 3HANA, KYNA and XA) (cf. Fig. 2A). A background chromatogram was generated in every experiment by HPLC analysis of a suitable quantity of 5-$^3$H-KYN.

The identity of the radioactive peaks recovered from HPLC analyses was confirmed in separate experiments either by re-analysis using two additional chromatographic systems or, in the case of QUIN, by re-analysis using one additional chromatographic system and a specific radioenzymatic procedure (Guidetti et al., 1995).

# RESULTS

## Metabolism of $^3$H-KYN in the Young and Adult Rat Striatum

Radioactive metabolites were positively identified in tissue extracts obtained from animals infused for 10 min with the radiolabelled tracer (Fig. 2B). Thus, all radiochemical chromatograms from 5-$^3$H-KYN-injected striata showed an intense peak of radioactivity which was readily identifiable as unmetabolized KYN. This peak co-migrated with authentic 5-$^3$H-KYN and unlabelled KYN. 4.2%, 3.8% and 5.8% of the injected radioactivity was recovered in PND 7, PND 14 and adult rats, respectively, and peaks corresponding to authentic KYN, 3HK, KYNA, XA, 3HANA and QUIN were isolated. In addition to unmetabolized KYN (82.4%, 80.4% and 74.7% of recovered radioactivity PND 7, PND 14 and adult rats), the greatest proportion of radioactivity was recovered as $^3$H-3HK at all ages examined (Fig. 3). Less than half that amount was recovered as $^3$H-KYNA. Tritiated 3HANA, QUIN and XA each accounted for less than 1% of the recovered radioactivity at all age groups (Table 1 and Fig. 3). In no case was there an indication that 5-$^3$H-KYN was metabolized to picolinic acid, quinaldic acid or anthranilic acid.

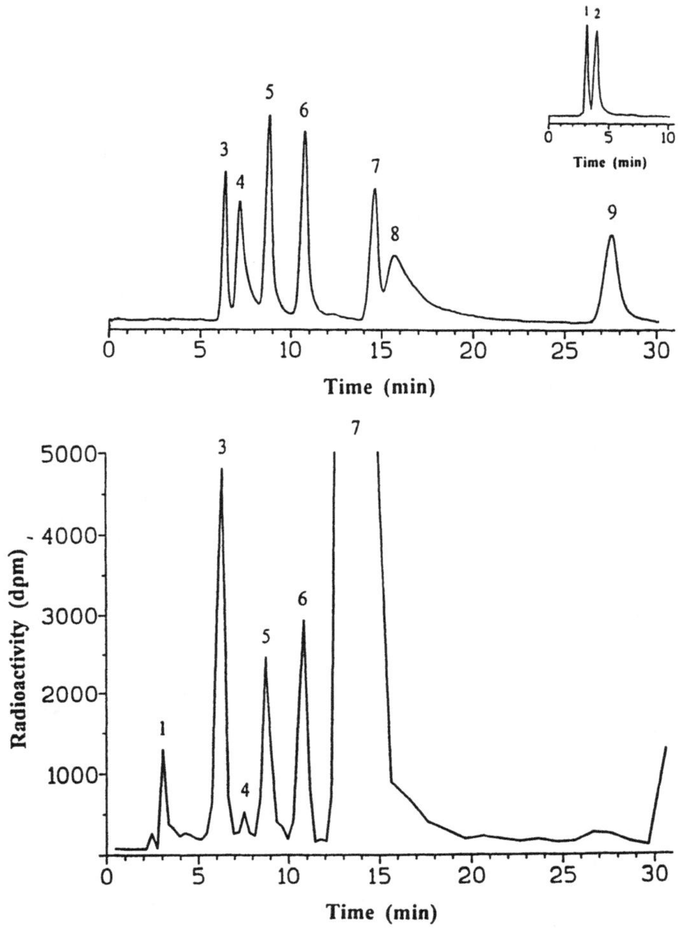


**Figure 2.** Resolution of kynurenine pathway metabolites. Top: HPLC trace showing the approximate retention times of non-radioactive standards detected by UV absorbance at 340 nm or 280 nm (inset). Bottom: Typical radiochemical HPLC trace obtained from a normal rat striatum 2 hours following a focal injection of 2.5 μCi 5-$^3$H-KYN. 1: QUIN; 2: Picolinic acid; 3: 3HK; 4: XA; 5: KYNA; 6: 3HANA; 7: KYN; 8: Quinaldic acid; 9: Anthranilic acid.

Analysis of the developmental pattern of cerebral KYN metabolism revealed increases in the production of all measured pathway metabolites between PND 7 and PND 14 and, with the exception of the proportional recovery of $^3$H-3HK, further increases between PND 14 and adult rats (Fig. 3). Notably, developmental changes in the QUIN branch of the pathway occurred mainly during the second postnatal week and preceded the increase in the proportional recovery of $^3$H-KYNA which was particularly pronounced between PND 14 and adulthood.

## Striatal Metabolism of $^3$H-KYN Is Dependent on the Speed of Tracer Infusion

A separate set of experiments was designed to examine possible qualitative differences in brain KYN metabolism between animals receiving rapid and prolonged intrastriatal

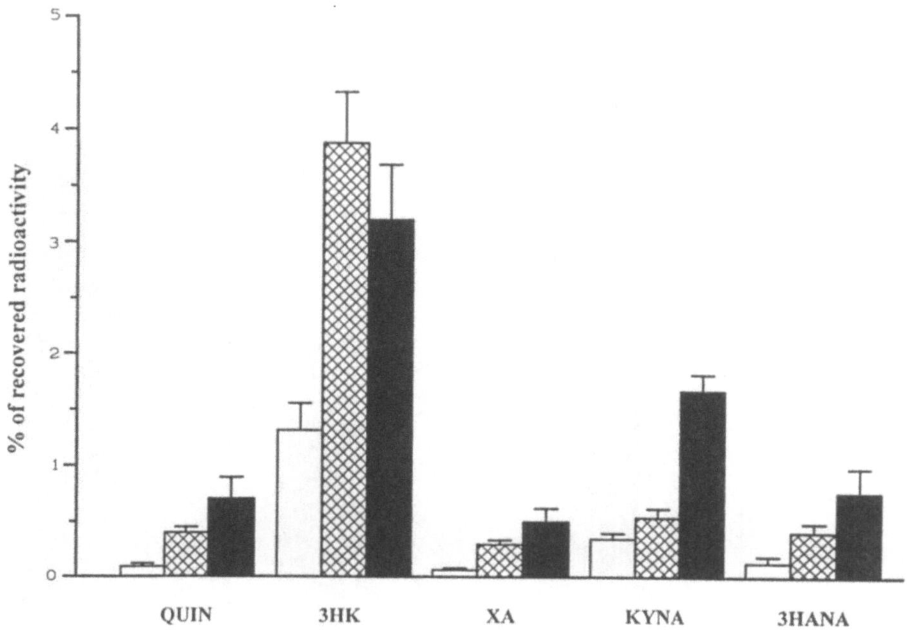


**Figure 3.** Proportional recovery of kynurenine pathway metabolites 2 hours after a rapid (10 min) intrastriatal infusion of 2.5 μCi 5-$^3$H-KYN into PND 7 (open bars; n = 14), PND 14 (cross-hatched bars; n = 11) and adult (solid bars; n = 10) rats. Data are the mean ± SEM.

tracer infusions. As shown in Table 1, the proportional recoveries of $^3$H-3HK, $^3$H-QUIN and $^3$H-XA, in particular, were substantially reduced in animals which were infused over a period of 2 hours as compared to animals which received a 10 min infusion (and were killed after 2 hours). In contrast, the *de novo* production of $^3$H-KYNA from $^3$H-KYN was unaffected by the duration of the tracer infusion.

**Table 1.** Striatal 5-$^3$H-KYN metabolism: dependency on the period of tracer infusion

| | Infusion Period | |
|---|---|---|
| Recovered metabolite | 10 Minutes | 2 Hours |
| $^3$H-3HK | 3.19 ± 0.49 | 0.93 ± 0.15 |
| $^3$H-XA | 0.50 ± 0.12 | 0.08 ± 0.01 |
| $^3$H-3HANA | 0.76 ± 0.22 | 0.35 ± 0.09 |
| $^3$H-QUIN | 0.70 ± 0.19 | 0.13 ± 0.02 |
| $^3$H-KYNA | 1.67 ± 0.15 | 1.47 ± 0.13 |
| $^3$H-KYN | 74.68 ± 3.81 | 87.9 ± 1.74 |

Animals were infused with 2.5 μCi 5-$^3$H-KYN as described in Materials and Methods. Data are expressed as a percentage of the radioactivity recovered from the tissue 2 hours after the beginning of the infusion, and are the mean ± SEM of 10 (10 minute infusion) or 15 (2 hour infusion) animals.

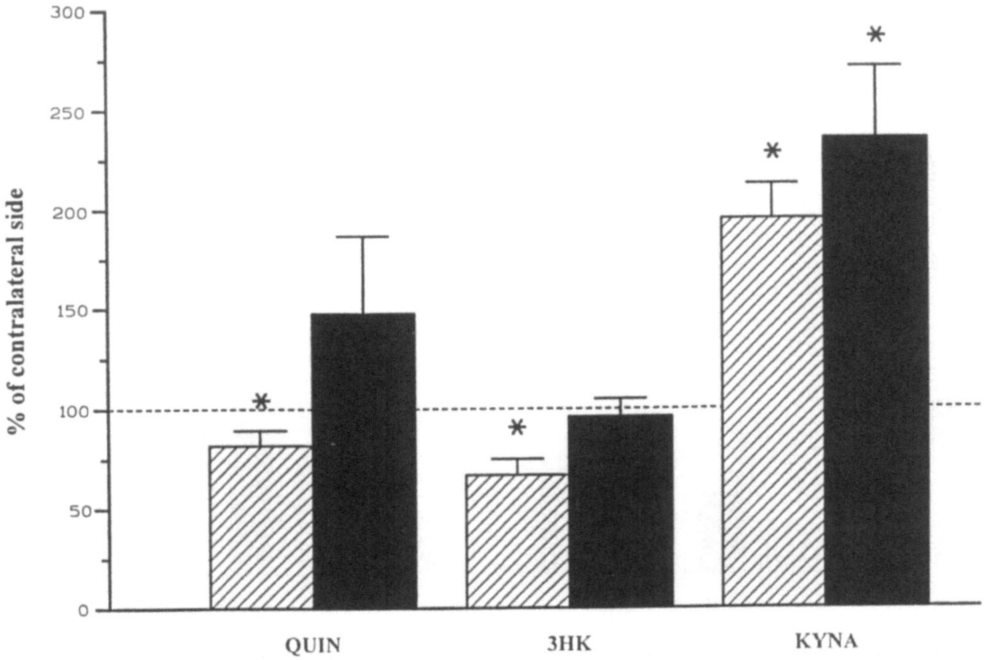


**Figure 4.** Effect of intrastriatal infusions of ibotenic acid (40 µg; hatched bars) or QUIN (20 µg; solid bars) on the local *de novo* synthesis of tritiated 3HK, QUIN and KYNA from 5-$^3$H-KYN. Tracer and toxin were co-infused for 2 hours. Data (mean ± SEM of 6 animals per group) are expressed as a percentage of the proportional metabolite recovery on the contralateral side (cf. Table 1 for control values). * $P < 0.05$ as compared to the contralateral striatum (paired Student's t-test).

## Striatal $^3$H-KYN Metabolism: Effect of Excitotoxin Infusion

Co-infusion of the excitotoxins ibotenic acid or QUIN with 5-$^3$H-KYN caused a rapid increase (to 235 and 195%, respectively, as compared to the contralateral control striatum) in the *de novo* production of $^3$H-KYNA (Fig. 4). In the same tissues, no increases were observed in the proportional recovery of $^3$H-QUIN (Fig. 4), $^3$H-XA and $^3$H-3HANA (data not shown). The synthesis of $^3$H-3HK, while unchanged in response to a QUIN infusion, was in fact found to be slightly reduced after ibotenate (Fig. 4).

# DISCUSSION

The present data, together with the recent more extensive characterization of the dynamics of 5-$^3$H-KYN metabolism in the rat brain (Guidetti et al., 1995), provide the first evidence for the existence of a functional kynurenine pathway in the normal mammalian brain. Previous attempts to detect the cerebral *de novo* production of QUIN from its remote precursors had failed because of the use of tryptophan as the putative parent compound (cf. Introduction) or because of insufficient assay sensitivity for the determination of relatively small changes in QUIN concentration (Speciale et al., 1989). The use of intracerebral infusions of $^3$H-KYN and of a newly developed HPLC methodology permitted the monitoring of newly synthesized tritiated 3HK, 3HANA, QUIN, KYNA and XA in a single run, and enabled us to examine both qualitative and quantitative features of brain KYN metabolism.

This methodology has physiological relevance since approximately 60% of brain KYN derives from the periphery (Gál and Sherman, 1978), i.e. the majority of KYN needs to enter the extracellular compartment before being actively transported into brain cells. Notably, extracellular KYN is rapidly accumulated by astrocytes (Speciale and Schwarcz, 1990) which have been shown to contain several KYN pathway enzymes (Schwarcz and Du, 1991; Stone, 1993).

As described in detail elsewhere (Guidetti et al., 1995), the synthesis of all identified KYN metabolites (expressed as a percentage of the total radioactivity recovered from the injected striatum) was critically dependent both on the concentration of KYN used and on the time that had elapsed from the beginning of the tracer infusion. For example, KYNA synthesis, but not 3HK, 3HANA and QUIN production, increased linearly when KYN concentration were raised beyond 15 μM. This is probably due to the different kinetic characteristics of kynurenine aminotransferase, the enzyme responsible for the production of KYNA (Okuno et al., 1991), and kynurenine-3-hydroxylase, which catalyzes the formation of 3HK from KYN (Saito et al., 1993b). Probably for the same reason, KYNA synthesis, but not the formation of all other identified KYN metabolites, was found to be maximal within a few minutes after the beginning of tracer infusion. Taken together, these data indicated that the activities of kynurenine aminotransferase and kynurenine-3-hydroxylase, and the availability of KYN for these two enzymes, are important determinants of the relative flux along the two branches of the kynurenine pathway, and hence of the production of KYNA and QUIN in the brain.

The three novel findings reported here were obtained using intracerebral infusions of 15 μM KYN, a concentration chosen to minimize changes in the ambient concentration of KYN in the adult rat brain (Gál and Sherman, 1978). The first series of experiments demonstrated that a functional cerebral KYN pathway exists in the rat striatum during the early postnatal period, and that the respective production of the various identified KYN pathway metabolites increases with increasing age, albeit with different developmental profiles. Thus, the formation of 3HK reached adult proportions by PND 14, probably reflecting a preferential rapid maturation of kynurenine-3-hydroxylase (G.C., unpublished data). The lack of proportional increases in the cerebral synthesis of the downstream metabolites 3HANA, QUIN, and XA in PND 14 rats suggests that in the perinatal period kynurenine-3-hydroxylase activity must not be considered the sole rate-limiting step in the cascade leading to QUIN production. This could be of relevance for future attempts to identify appropriate targets for influencing cerebral QUIN function in young animals.

The respective activities of kynurenine aminotransferase and kynurenine-3-hydroxylase probably also hold the key to the unexpected observation that a slow infusion of $^3$H-KYN, designed to duplicate the physiological situation of low stable extracellular KYN concentrations, substantially decreased the proportional recovery of metabolites along the QUIN branch of the KYN pathway without affecting the *de novo* production of KYNA. This result, which has recently been confirmed using intrastriatal 5-$^3$H-KYN infusions for up to 16 hours (C. Eastman, unpublished data), suggests that small amounts of KYN, upon transport into the cell, are preferentially converted by kynurenine aminotransferase, a cytosolic enzyme (Roberts et al., 1992), but do not adequately reach the outer mitochondrial membrane which harbors kynurenine-3-hydroxylase (Okamoto and Hayaishi, 1969). Thus, KYNA production appears to outweigh the formation of 3HK and its metabolites in the normal adult brain *in vivo*. This finding has implications for the supposition that the ratio of KYNA and QUIN may be of significance for excitatory amino acid receptor function (Stone, 1993). In other words, an excess of QUIN over KYNA production is likely to occur only in pathophysiological situations where KYNA synthesis is impaired or QUIN formation is selectively enhanced.

The final set of experiments was designed to examine if, in fact, qualitative changes in KYN metabolism occur when pathological events take place. To this end, the classic excitotoxins ibotenic acid or QUIN were infused intrastriatally over 2 hours together with $^{3}$H-KYN. At the end of the infusion period, the *de novo* production of $^{3}$H-KYNA, alone among all identified metabolites, was increased. Since excitotoxic damage is still reversible within the first 2 hours (Foster et al., 1988), the selective increase in KYNA synthesis, which is probably accompanied by an increased release of KYNA into the extracellular compartment (Wu et al., 1992), can be interpreted as the brain's (futile) endogenous anti-excitotoxic defense strategy. Notably, this acute response is qualitatively different from the chronic effects of an excitotoxic striatal lesion, which are characterized by a massive up-regulation of the QUIN branch of the KYN pathway (Guidetti et al., 1995).

Taken together, the data reported by Guidetti et al. (1995) and in the present study clearly indicate that a functional KYN pathway, originating with KYN as the parent compound, exists in the young and adult rat brain. Moreover, the complexities of cerebral KYN metabolism are amenable to experimental scrutiny by the intracerebral infusion of 5-$^{3}$H-KYN and the subsequent analysis of newly produced tritiated pathway metabolites. Substantial qualitative differences were noticed between rapid and prolonged tracer infusion, and the new method was successfully used to investigate the reaction of cerebral KYN pathway metabolism to acute excitotoxic insults. The results demonstrate that the novel methodology will be applicable to further studies of the *in vivo* physiology, pharmacology and pathology of the brain's KYN pathway. In appropriately modified form, this tracer technique may also be used for *in vivo* brain imaging studies in experimental animals and humans.

## ACKNOWLEDGMENTS

We thank Ms. Joyce Burgess for excellent secretarial assistance. This work was supported by USPHS grants NS 16102 and NS 28236 and a grant from the Huntington's Disease Society of America (to P.G.).

## REFERENCES

Foster, A.C., Gill, R., and Woodruff, G.N., 1988, Neuroprotective effects of MK-801 in vivo: selectivity and evidence for delayed degeneration mediated by NMDA receptor activation. *J. Neurosci.* 8: 4745.

Gál, E.M., and Sherman, A.D., 1978, Synthesis and metabolism of L-kynurenine in rat brain. *J. Neurochem.* 30: 607.

Guidetti, P., Eastman, C.L., and Schwarcz, R., 1995, Metabolism of [5-$^{3}$H]kynurenine in the rat brain in vivo: evidence for the existence of a functional kynurenine pathway. *J. Neurochem.* in press.

Heyes, M.F., 1993, Metabolism and neuropathologic significance of quinolinic acid and kynurenic acid. *Biochem. Soc. Trans.* 21: 83.

Okamoto, H., and Hayaishi, O., 1969, Solubilization and partial purification of kynurenine hydroxylase from mitochondrial outer membrane and its electron donors. *Arch. Biochem. Biophys.* 131: 603.

Okuno, E., Schmidt, W., Parks, D.A., Nakamura, M., and Schwarcz, R., 1991, Measurement of rat brain kynurenine aminotransferase at physiological kynurenine concentrations. *J. Neurochem.* 57: 533.

Reinhard, J.F. Jr., Erickson, J.B., and Flanagan, E.M., 1994, Quinolinic acid in neurological disease: opportunities for novel drug discovery. *Adv. Pharmacol.* 30: 85.

Roberts, R.C., Du, F., McCarthy, K., Okuno, E., and Schwarcz, R., 1992, Immunocytochemical localization of kynurenine aminotransferase in the rat striatum: a light and electron microscopic study. *J. Comp. Neurol.* 326: 82.

Saito, K., Markey, S.P., and Heyes, M.P., 1992, Effects of immune activation on quinolinic acid and neuroactive kynurenines in the mouse. *Neuroscience* 51: 25.

Saito, K., Nowak, T.S. Jr., Markey, S.P., and Heyes, M.P., 1993a, Mechanism of delayed increases in kynurenine pathway metabolism in damaged brain regions following transient cerebral ischemia. *J. Neurochem.* 60: 180.

Saito, K., Quearry, B.J., Saito, M., Nowak, T.S. Jr., Markey, S.P., and Heyes, M.P., 1993b, Kynurenine 3-hydroxylase in brain: species activity differences and effect of gerbil cerebral ischemia. *Arch. Biochem. Biophys.* 307: 104.

Schwarcz, R., 1992, Excitotoxins, kynurenines and neuropsychiatric diseases: implications for drug development, *in*: "Drug Research Related to Neuroactive Amino Acids", A. Schousboe, N.H. Diemer, and H. Kofod, eds., Munksgaard, Copenhagen.

Schwarcz, R., and Du, F., 1991, Quinolinic acid and kynurenic acid in the mammalian brain, *in*: "Kynurenine and Serotonin Pathways: Progress in Tryptophan Research", R. Schwarcz, S.N. Young, and R.R. Brown, eds., Plenum, New York.

Speciale, C., and Schwarcz, R., 1990, Uptake of kynurenine into rat brain slices. *J. Neurochem.* 54: 156.

Speciale, C., Ungerstedt, U., and Schwarcz, R., 1989, Production of extracellular quinolinic acid in the striatum studied by microdialysis in unanesthetized rats. *Neurosci. Lett.* 104: 345.

Stone, T.W., 1993, Neuropharmacology of quinolinic and kynurenic acids, *Pharmacol. Rev.* 45: 310.

Wu, H-Q., Baran, H., Ungerstedt, U., and Schwarcz, R., 1992, Kynurenic acid in the quinolinate-lesioned rat hippocampus: studies in vitro and in vivo. *Eur. J. Neurosci.* 4: 1264.

# KYNURENIC ACID-ENHANCING AND ANTI-ISCHEMIC EFFECTS OF THE POTENT KYNURENINE 3-HYDROXYLASE INHIBITOR FCE 28833 IN RODENTS

C. Speciale,[1] M. Cini,[1] H.-Q. Wu,[2] P. Salvati,[1] R. Schwarcz,[2] A. Molinari,[1] M. Calabresi,[1] and M. Varasi[1]

[1] Pharmacia, R&D CNS
Nerviano, Italy
[2] Maryland Psychiatric Research Center
University of Maryland School of Medicine
Baltimore, Maryland 21228

## 1. INTRODUCTION

When injected into the brain of experimental animals, quinolinic acid (QUIN) and kynurenic acid (KYNA) act as a neurotoxin and neuroprotectant, respectively, and these effects are mediated by excitatory amino acid (EAA) receptors (Stone, 1993). Since their identification in the mammalian brain, QUIN and KYNA have therefore been proposed to play roles as endogenous modulators of EAA receptors. Both compounds are metabolites of the kynurenine pathway in the periphery and in the brain, and a large number of studies have been performed to examine their source, metabolism and disposition in the central nervous system in physiological and pathological conditions (see Stone, 1993, for review). In particular, an increased formation of QUIN was observed in several models of nerve cell damage, such as excitotoxin lesions (Speciale et al., 1987), spinal cord trauma (Popovich et al., 1994) and global ischemia (Heyes and Nowak, 1990; Saito et al., 1993), and in humans infected with the HIV virus (Heyes et al., 1991). This enhancement of QUIN production was suggested to contribute to the loss of neurological function. On the other hand, KYNA was found to be likely involved in cerebral self-defence (Schwarcz et al., 1992). This is supported both by occurrence of cell death following the pharmacologically induced decrease in brain KYNA (McMaster et al., 1991) and by the increase in cerebral KYNA production following acute excitotoxic (Schwarcz et al., this volume) or convulsive (Wu and Schwarcz, 1994; Baran et al., 1995) insults.

Since it contains two neuroactive molecules with opposing effects, the kynurenine pathway offers unique opportunities for pharmacological intervention (Schwarcz, 1992; Reinhard et al., 1994). Thus, a shift in the balance between the QUIN and KYNA branches

*Recent Advances in Tryptophan Research*, edited by Graziella Allegri Filippini et al.
Plenum Press, New York, 1996

of the pathway may have a significant impact on brain function. To date, attempts to develop appropriate pharmacological tools have largely focused on a decrease in QUIN formation by blocking its synthesising enzyme 3-hydroxyanthranilic acid oxygenase (3HAO) (Parli et al., 1980), and on an increase in KYNA formation by supplying more kynurenine (KYN) as substrate for its synthesis (Miller et al., 1992). Indeed, both approaches have been linked to decreased neuronal vulnerability (Blight et al., 1995; Nozaki and Beal, 1992).

Using the KYN analogue nicotinylalanine (NAL) as an experimental tool, Decker and collaborators (1963) proposed that inhibition of kynurenine 3-hydroxylase (KYN 3-OH-ase) and kynureninase (KYN-ase) is likely to increase KYNA production. More recently, Moroni, Pellicciari and collaborators, as well as Reinhard's group (this meeting) have described the synthesis and the metabolic effects of other KYN 3-OH-ase inhibitors (Pellicciari et al., 1994). In line with our long-standing interest in kynurenine neurobiology, we now present data obtained with a novel, potent and selective KYN 3-OH-ase inhibitor, FCE 28833 [(R,S)-2-amino-4-oxo-4-(3'-4'-dichlorophenyl) butanoic acid] (Giordani et al., this meeting). In particular we shall describe the ability of this compound to increase brain KYNA levels and to provide neuroprotection in the gerbil model of global ischemia.

## 2. METHODS

### 2.1 in Vivo Metabolic Studies

Male Wistar rats (Charles River, Italy) weighing 200-250 g were used for all the metabolic studies. Experiments were performed after a single oral treatment with FCE 28833 (suspended in methocel). Control animals were treated with the vehicle alone. For the study of time-related effects, the dose of 400 mg/kg was used. Dose-response relationships were tested by using 50 to 400 mg/kg of FCE 28833. Animals were sacrificed by decapitation and their plasma was recovered. Brains were rapidly excised, frozen on dry ice and stored at -80°C. On the day of the assay, tissues were thawed and homogenised in 2.5-3 volumes of 0.5 M $HClO_4$ with a Polytron homogeniser (setting 8, twice for 15 sec). After sonication, the tissues were centrifuged at 12,500 x g for 15 min, and the supernatants were subjected to HPLC analysis. Plasma was mixed with 2 volumes of $HClO_4$ (final concentration: 0.5 M) and then processed as the tissues. Hippocampal microdialysis for KYNA measurements (recovery from the dialysis probe: approximately 20%) was performed in unanesthetized, cannulated animals (Speciale et al., 1989b). In separate animals, aminooxyacetic acid (AOAA) was administered through the dialysis probe at a concentration of 1 mM for three hours, starting three hours after FCE 28833 treatment.

### 2.2 Assay of Enzyme Activities and Metabolite Levels

Brain KYN 3-OH-ase activity was measured according to Saito et al. (1993). The activity of KYN-ase was tested in liver homogenates by following the conversion of KYN to anthranilic acid (ANA) (Saito et al., 1993). Brain kynurenine aminotransferase (KAT) activity was measured according to Okuno et al. (1991). KYN content was determined by HPLC using a 10 cm $C_{18}$ reversed phase column (3 μm) with UV detection (340 nm). The mobile phase consisted of 3.6 mM heptanesulphonic acid in 50 mM ammonium acetate buffer:methanol (97:3) at pH 6, and was used at a flow-rate of 1 ml/min. ANA was determined by HPLC with fluorescence detection (excitation 313 nm, emission: 420 nm) using a 30 cm $C_{18}$ reversed phase column (5 μm) and a mobile phase consisting of 80 mM potassium phosphate buffer:acetonitrile (83:17) at pH 2.5, and a flow-rate of 1 ml/min. KYNA was determined by HPLC with zinc acetate post-column derivatization and fluorescence detection (excitation: 254 nm, emission: 404 nm) (Swartz et al., 1990).

## 2.3 Global Ischemia in Gerbils

Male Mongolian gerbils (50-70 g body weight) were anaesthetised with a mixture of 2% halothane, 75% nitrogen and 30% oxygen. Sham-operated and ischemic gerbils were pre-treated 3 hours before the beginning of the surgery with either vehicle (methocel) or FCE 28833 (400 mg/kg, p.o., 5 ml/kg). This schedule was chosen on the basis of previous results obtained in the gerbil after a single administration of 400 mg/kg, p.o. showing that 2-4 hours are needed to reach elevated KYNA levels in the brain (Cini et al., this meeting). For the surgery, a ventral midline incision was made in the neck, both common carotid arteries were exposed, and, subsequently, occluded for 5 min using non-traumatic aneurysm clips. At the end of the occlusion period, the clips were released to restore the blood flow, and the incision sutured. Sham-operated controls were anaesthetised and subjected to the neck incision but carotid arteries were not clamped. Throughout the occlusion time, the electroencephalogram (EEG) was recorded. Animals not showing isoelectric EEG were excluded from the study. Seven days after surgery, gerbils were sacrificed and brains processed for histology (hematoxylin/eosin staining of 4 μm coronal sections). Three sections (approximately 150 μm distance) of each brain were analysed blindly. The size of hippocampal injury was measured by counting the number of viable CA1 pyramidal cells. Thus, neurons showing signs of irreversible ischemic damage, defined by their shrunken appearance, dark pyknotic nuclei, or poor stainability, were excluded from cell counting. Neurons without such changes were classified as normal and counted.

# 3. RESULTS

The selectivity and potency of FCE 28833 to inhibit some enzymes of the kynurenine pathway was established *in vitro*. The $IC_{50}$ of FCE 28833 was 0.2 μM for KYN 3-OH-ase, and >200 μM for KYN-ase and KAT, respectively. This makes the compound the most potent selective KYN 3-OH-ase known to date. After a single oral administration, in a dose range varying between 50 and 400 mg/kg, FCE 28833 caused a prominent and sustained increase in KYN and KYNA in plasma and brain (Table 1). The two metabolites were found to be significantly elevated at 50 mg/kg, and no saturation of the effect was reached at the highest dose tested. Time-related changes of the metabolites were investigated up to 8 hours after the p.o. treatment with 400 mg/kg (Table 1). The compound showed a prompt and long lasting ability to evoke KYN and KYNA increases in both plasma and brain tissue. In particular, KYN and KYNA levels were still rising 8 hours after the treatment.

A better understanding of the time-related effect of FCE 28833 on brain KYNA was obtained in experiments using intrahippocampal microdialysis for up to 18 hours post-treatment. In these rats, extracellular KYNA was found to increase approximately 10-fold from 4 to 9 hours after the administration of the KYN 3-OH-ase inhibitor (Figure 1), and significant changes were obtained until 14 hours following the administration of FCE 28833. In separate rats, the non-specific transaminases inhibitor AOAA was delivered through the intrahippocampal dialysis probe to prevent the local formation of KYNA (Speciale et al., 1990). These experiments demonstrated that most of the FCE 28833-induced *de novo* synthesis of KYNA occurred in the brain (Figure 1).

Experiments designed to examine the neuroprotective potential of FCE 28833 were conducted in the gerbil global ischemia model. Histological analysis was performed in animals sacrificed 7 days after the ischemic insult. Given once at 400 mg/kg, p.o., 3 hours

**Table 1.** Time- and dose-related effects of oral FCE 28833 treatment on KYN and KYNA content in plasma and brain tissue from Wistar rats

| | Kynurenine | | Kynurenic acid | |
|---|---|---|---|---|
| | Plasma | Brain | Plasma | Brain |
| Time (h) | | | | |
| Control | 5217 ± 764 | 319 ± 14 | 261 ± 50 | 5 ± 2 |
| 2 | 14184 ± 978** | 2067 ± 362** | 2451 ± 830** | 23 ± 8** |
| 4 | 18169 ± 621** | 3560 ± 119** | 2942 ± 577** | 33 ± 5** |
| 6 | 18818 ± 2264** | 3912 ± 324** | 4534 ± 1415** | 56 ± 12** |
| 8 | 22231 ± 1936** | 5711 ± 508** | 7606 ± 1259** | 69 ± 5** |
| Dose (mg/kg) | | | | |
| Control | 4940 ± 341 | 411 ± 34 | 97 ± 10 | 4 ± 1 |
| 50 | 10049 ± 1037** | 1699 ± 239** | 580 ± 89** | 22 ± 2** |
| 100 | 17354 ± 1141** | 3857 ± 264** | 1154 ± 146** | 33 ± 3** |
| 200 | 32599 ± 1001** | 8281 ± 515** | 4143 ± 516** | 86 ± 6** |
| 400 | 36898 ± 1650** | 20499 ± 710** | 7578 ± 982** | 142 ± 26** |

Time-related effects were studied by treating the animals with 400 mg/kg, FCE 28833, p.o.. For dose-dependency experiments, rats were sacrificed 8 hours after the administration of the compound. Values related to plasma and brain tissue are expressed as pmoles/ml and pmoles/g, respectively. Data are the means ± S.E.M. of at least five individual animals per group. **P < 0.01 as compared to untreated control rats (Dunnett's test).

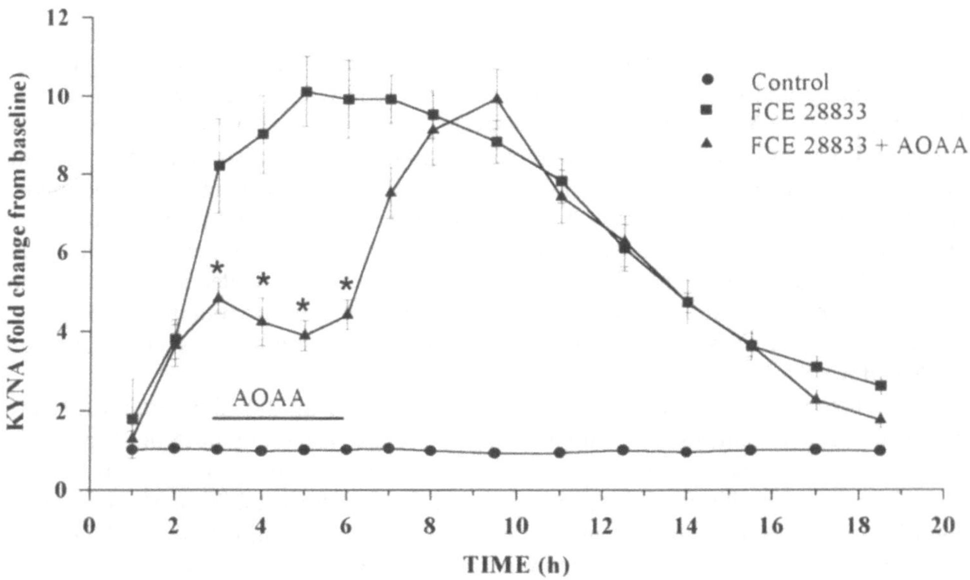


**Figure 1.** Effect of FCE 28833 (400 mg/kg, p.o.) on extracellular KYNA concentrations in the rat hippocampus (see Methods for details). In separate animals, AOAA (1 mM) was included in the infusion solution for a period of 3 hours (bar). Data are expressed as a percentage of baseline values (145.4±3.4 fmoles KYNA/60 μl dialysate) and are the means ± S.E.M. of six animals for each experimental group. The effect of FCE 28833 was evaluated by two way ANOVA followed by Newman-Keuls post hoc analysis [$F(17,170) = 24.85$, $P<0.01$]. *$P< 0.05$ as compared to FCE 28833 alone at the corresponding time point (Bonferroni's t test).

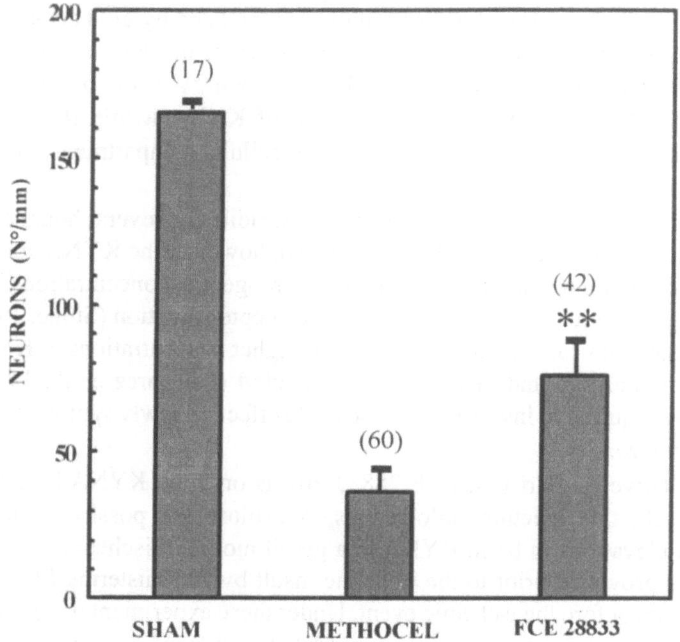


**Figure 2.** Neuroprotective effect of FCE 28833 in the gerbil model of global ischemia. FCE 28833 was given at 400 mg/kg, p.o., 3 hours before 5 minutes bilateral carotid artery occlusions. Gerbils were sacrified 7 days later (see Methods for details). Bar graphs are the means ± S.E.M. of neuronal density (surviving neurons /mm CA1 in the dorsal hippocampus) of 4 different experiments. The number of animals per group is given in parenthesis. **$P<0.01$ as compared to untreated control animals (Mann-Whitney's non parametric test).

before 5 min ischemia, FCE 28833 had significant neuroprotective effects as measured by counting of apparently viable CA1 pyramidal cells (Figure 2). Notably, binding studies demonstrated that FCE 28833 does not interact directly with the ionotropic EAA receptors (data not shown).

## 4. DISCUSSION

Modulation of kynurenine pathway metabolism in the brain offers novel opportunities for pharmacological intervention in brain pathologies. m-Nitrobenzoylalanine, an inhibitor of KYN 3-OH-ase ($IC_{50}$: 3 μM), the first enzyme involved in the formation of QUIN from KYN, was the first compound shown to increase the content of KYNA in the rat brain (Carpenedo et al., 1994), but the effect of this treatment has not been studied in animal models of nerve cell death. The present results suggest that the enhancement of brain KYNA by KYN 3-OH-ase inhibition may be linked to neuroprotection in the global ischemia model in the gerbil. Thus, the novel KYN 3-OH-ase inhibitor FCE 28833 ($IC_{50}$: 0.2 μM) caused a >10 fold increase in cerebral KYNA concentrations and significantly attenuated the death of hippocampal CA1 neurons *in vivo*. Notably, administration of neuroprotective quantities of FCE 28833 did not cause any apparent behavioral changes in rats or gerbils (data not shown).

After the application of FCE 28833, rises in KYNA content closely paralleled the changes in KYN concentrations in both plasma and brain. This suggests that the KYN 3-OH-ase inhibitor had acted by blocking not only the cerebral but also the peripheral enzyme, leading to a large increase in circulating KYN and KYNA. Indeed, KYN readily crosses the blood-brain barrier, and is then rapidly accumulated by glial cells for sequestration to KYNA (Speciale et al., 1989a). Confirmation of these mechanisms was provided by microdialysis experiments (Wu et al.; 1992). Thus, the ability of the non-specific KYNA synthesis inhibitor AOAA to block the production of KYNA *within the brain* strongly suggests that most of the KYNA measured in the extracellular compartment was synthesised locally from blood-derived KYN.

In normal rats, brain KYNA levels increased steadily for several hours after a single oral administration of FCE 28833. Although elevated, however, the KYNA content in both tissue and microdialysate remained in the nanomolar range, i.e. concentrations below those reported to be necessary for the antagonism of EAA receptor function (Stone, 1993). On the other hand, neuroanatomical evidence suggests that higher concentrations of KYNA may be reached in close vicinity to and inside the synaptic cleft (Schwarcz et al., 1992). Further studies are clearly required to investigate the possible effect of newly synthesised KYNA on EAA receptors *in vivo*.

The qualitative similarity of FCE 28833 effects on brain KYNA levels in rats and gerbils (Cini et al., this meeting) allowed us to explore the possible neuroprotective consequences of elevations in brain KYNA in a gerbil model of ischemia. The increase in brain KYNA was provoked prior to the ischemic insult by administering FCE 28833 (400 mg/kg, p.o.) 3 hours before the ischemic event. Under these experimental conditions, and in the absence of any change in body temperature (data not shown), a modest, yet significant attenuation of CA1 nerve cell loss was observed in drug-treated animals. These findings support the hypothesis that endogenously synthesised KYNA can exert a neuroprotective effect. However, there is currently no information on the effect of KYN 3-OH-ase inhibition on the levels of brain QUIN. Further studies will need to investigate the effects of FCE 28833, and of functionally related compounds, on the *ratio* of QUIN and KYNA in both the normal and the injured brain over time. Such data will be helpful for the design of studies examining the possible role of endogenous KYNA and QUIN in the development of ischemic nerve cell loss. Thus, prolonged KYN 3-OH-ase inhibition may provide neuroprotection acutely by increasing brain KYNA levels, but may also prevent the late occurring ischemia-induced increase in brain QUIN (Heyes and Nowak, 1990; Saito et al., 1993) and thus abolish delayed nerve cell death. More generally, compounds capable of influencing the activity of various kynurenine pathway enzymes may provide not only useful experimental tools, but also turn out to be of value in the treatment of brain disorders which are linked to a dysfunction of cerebral EAA receptors.

## 5. REFERENCES

Carpenedo, R., Chiarugi, A., Russi, P., Lombardi, G., Carlà, V., Pellicciari, R., Mattioli, L., and Moroni, F., 1994, Inhibitors of kynurenine hydroxylase and kynureninase increase cerebral formation of kynurenate and have sedative and anticonvulsant activities. *Neurosci.* 61:237.

Baran, H., Gramer, M., Hönack, D., and Löscher, W., 1995, Systemic administration of kainate induces marked increases of endogenous kynurenic acid in various brain regions and plasma of rats. *Eur. J. Pharmacol.*,286:167.

Blight, A.R., Cohen, T.I., Saito, K., and Heyes, M.P., 1995, Quinolinic acid accumulation and functional deficits following experimental spinal cord injury, *Brain* 118:735.

Decker, R.H., Brown, R.R., and Price, J.M., 1963, Studies on the biological activity of nicotinylalanine, an analogue of kynurenine, *J.Biol.Chem.* 238:1049.

Heyes, M.P., and Nowak Jr., T.S., 1990, Delayed increases in regional brain quinolinic acid follow transient ischemia in the gerbil. *J. Cereb. Blood Flow Metab.* 10:660.

Heyes, M.P., Brew, B.J., Martin, A:, Price, R.W., Salazar, A.M., Sidtis, J.J., Yergey, J.A., Mouradian, M.M., Sadler, A.E., Keilp J., Rubinow, D., and Markey, S.P., 1991, Quinolinic acid in cerebrospinal fluid and serum in HIV-1 infection: reationship to clinical and neurologic status. *Ann. Neurol.* 29:202.

McMaster, O.G., Du, F., French, E.D., and Schwarcz, R., (1991) Focal injection of aminooxyacetic acid produces seizures and lesions in rat hippocampus: evidence for mediation by NMDA receptors. *Exp. Neurol.* 113:378.

Miller, J.M., MacGarvey, U., and Beal, M.F., 1992, The effect of peripheral loading with kynurenine and probenecid on extracellular striatal kynurenic acid concentrations. *Neurosci. Lett.* 146:115.

Nozaki, K., and Beal, M.F., 1992, Neuroprotective effects of L-kynurenine on hypoxia-ischemia and NMDA lesions in neonatal rats. *J. Cereb. Blood Flow Metab.* 12:400.

Okuno, E., Schmidt, W., Parks, D.A., Nakamura, M., and Schwarcz, R., 1991, Measurement of rat brain kynurenine aminotransferase at physiological kynurenine concentrations. *J. Neurochem.* 57:533.

Parli, C.J., Krieter, P., and Schmidt, B., Metabolism of 6-chlorotryptophan to 4-chloro-3-hydroxyanthranilic acid: a potent inhibitor of 3-hydroxyanthranilic acid oxidase. *Arch. Biochem. Biophys.* 203:161.

Pellicciari, R., Natalini, B., Costantino, G., Mahmoud, M.R., Mattioli, L., Sadeghpour, B.M., Moroni, F., Chiarugi, A., and Carpenedo, R., 1994, Modulation of the kynurenine pathway in search for new neuroprotective agents. Synthesis and preliminary evaluation of (*m*-nitrobenzoyl)alanine, a potent inhibitor of kynurenine-3-hydroxylase. *J. Med. Chem.* 37:647.

Popovich, P.G., Reinhard Jr., J.F., Flanagan, E.M., and Stokes, B.T., 1994, Elevation of the neurotoxin quinolinic acid occurs following spinal cord trauma. *Brain Res.* 633:348.

Reinhard Jr., J.F., Erickson, J.B., and Flanagan, E.M., 1994, Quinolinic acid in neurological disease: opportunities for novel drug discovery. *Adv. Pharmacol.* 30:85.

Saito, K., Nowak, T.S., Markey, S.P., and Heyes, M.P., 1993, Mechanism of delayed increases in kynurenine pathway metabolism in damaged brain regions following transient cerebral ischemia. *J. Neurochem.* 60:180.

Swartz, K.J., Matson, W.R., Mac Garvey, U., Ryan, E.A., and Beal, M.F., 1990, Measurement of kynurenic acid in mammalian brain extracts and cerebrospinal fluid by high-performance liquid chromatography with fluorimetric detection and coulometric electrode array detection. *Anal. Biochem.* 185:363.

Schwarcz, R., 1992, Excitotoxins, kynurenines and neuropsychiatric diseases: implications for drug development, *in:* "Drug Research Related to Neuroactive Amino Acids", A. Schousboe, N.H. Diemer, and H. Kofod, eds., Munksgaard, Copenhagen.

Schwarcz, R., Du, F., Schmidt, W., Turski, W.A., Gramsbergen, J.P.B., Okuno, E., and Roberts, R.C., 1992, Kynurenic acid: a potential pathogen in brain disorders. *in:* "Neurotoxins and Neurodegenerative Disease", J.W. Langston and A. Young eds., Ann. N.Y. Acad. Sci. Vol. 648, New York, p. 140.

Speciale, C., Okuno, E., and Schwarcz, R., 1987, Increased quinolinic acid metabolism following neuronal degeneration in the rat hippocampus. *Brain Res.* 436:18.

Speciale, C., Hares, K., Schwarcz, R., and Brookes, N., 1989a, High-affinity uptake of L-kynurenine by $Na^+$ independent transporter of neutral amino acids in astrocytes. *J. Neurosci.*, 9:2066.

Speciale, C., Ungerstedt, U., and Schwarz, R., 1989b, Production of extracellular quinolinic acid in the striatum: studies by microdialysis in unanesthetized rats. *Neurosci. Lett.* 104:345.

Speciale, C., Wu, H.-Q., Gramsbergen, J.B., Turski, W.A., Ungerstedt, U., and Schwarcz, R., 1990, Determination of extracellular kynurenic acid in the striatum of unanesthetized rats: effects of aminooxyacetic acid. *Neurosci. Lett.* 116:198.

Stone, T.W., Neuropharmacology of quinolinic and kynurenic acids, 1993, *Pharmacol. Rev.* 45:309.

Wu, H.-Q, Baran, H., Ungerstedt, U., and Schwarcz, R., 1992, Kynurenic acid in the quinolinate-lesioned rat hippocampus: studies in vitro and in vivo. *Eur. J. Neurosci.* 4: 1264.

Wu, H.-Q., and Schwarcz, R., 1994, Seizure-related increases in extracellular kynurenic acid levels in the rat hippocampus. *Soc. Neurosci.*, Abstr. 20, 408.

# REGULATION OF QUINOLINIC ACID IN THE NORMAL RAT BRAIN BY KYNURENINE PATHWAY PRECURSORS

J. Luthman,[1*] E. Vänerman,[1] G. Fredriksson,[2] and B. Fornstedt-Wallin[1]

[1] Behavioral and Biochemical Pharmacology
Preclinical R&D
[2] Bioanalysis, Pharmaceutical R&D
Astra Arcus AB
S-151 85 Södertälje, Sweden

## 1. INTRODUCTION

The kynurenine pathway constitutes the major route of L-tryptophan (TRY) metabolism in mammalians in which essential products are nicotinamide and its nucleotide conjugates (see Stone, 1989). However, several intermediates of the kynurenine pathway have also been found to possess significant biological effects. In particular, there is an increasing awareness in the role kynurenines may play in the central nervous system (CNS); an interest that has been spurred by the finding that certain kynurenines may act on excitatory amino acid (EAA) receptors (see Stone 1993). One important kynurenine in this regard is quinolinic acid (QUIN), a dicarboxylic acid that has been shown to act as an agonist on the N-methyl-D-aspartate (NMDA) EAA receptor in the CNS (Stone and Perkins, 1981; Ganong and Cotman, 1986). When administered into the brain, QUIN causes convulsions (Lapin, 1978) and neurodegeneration (Schwarcz et al., 1983) that can be counteracted by NMDA antagonists (see Stone 1993). Furthermore, in a number of studies, increased levels of QUIN in cerebral tissue or cerebrospinal fluid (CSF) have been found in various conditions of brain injury, encephalitis, systemic immune stimulation and liver dysfunction both in humans (Moroni et al., 1986; Schwarcz et al., 1988; Heyes et al., 1991; see also Heyes et al., 1992) and in animal models (see Heyes et al., 1992; Stone, 1993; Reinhard et al., 1994). Based on these findings, it has been suggested that QUIN under certain circumstances could act as a neurotoxin and participate in neuropathological processes (Schwarcz et al., 1986).

In order to better understand the possible role of QUIN as an endogenous pathogen in different cerebral disorders, it is important to have a well-developed understanding of the

* Correspondence: Johan Luthman, Assistant professor, Behavioural and Biochemical Pharmacology, Preclinical R&D, Astra Arcus AB, S-151 85 Södertälje, Sweden. Fax 46-8-55328891.

origin of QUIN in the CNS. One important aspect is that the concentration of QUIN is substantially higher in plasma than in the brain. Therefore in some conditions an increase in cerebral QUIN could solely reflect an impaired blood brain barrier (BBB) for QUIN. In other conditions, altered kynurenine pathway metabolism in the liver or other peripheral tissues may lead to enhanced plasma levels of cerebral precursors of QUIN, or QUIN itself, with subsequently increased BBB transport. Potentiated QUIN synthesis may also occur within the brain itself due to increased enzyme activities; particularly in conditions of brain injury and gliosis (Heyes et al., 1992; Saito et al., 1993a, b; see Reinhard et al., 1994). QUIN is formed from 3-hydroxyanthranilic acid (3-HANA) through an initial step involving the 3-hydroxyanthranilic acid 3,4-dioxygenase (3-HAO) enzyme (Okuno et al., 1987; Malherbe et al., 1994) and a second step in which the product, 2-amino-3-carboxy muconic acid semialdehyde (Long et al., 1954), is transformed to QUIN non-enzymatically. The 3-HAO enzyme appears to be present in the normal rat brain (Foster et al., 1986; Köhler et al., 1988; Okuno et al., 1987) and cerebral administration of 3-HANA leads to major increases in QUIN (Heyes et al., 1988; Speciale et al., 1989). On the other hand, TRY and kynurenine (KYN) do not enhance cerebral QUIN when given into the normal rat brain (Speciale et al., 1989). In agreement with these observations the activities of several other enzymes of the kynurenine pathway, e,g. idoleamine dioxygenase (IDO), kynurenine hydroxylase and kynureninase, appear to be low in the brain (see Stone, 1993). On the other hand. it has been reported that 3-HANA can be formed from anthranilic acid (ANA) in rat brain slices (Baran and Schwarcz, 1990), indicating that ANA could serve as a cerebral precursor of QUIN.

Nonetheless, there are still major uncertainties concerning which kynurenines that are able to act precursors of cerebral QUIN and which key enzymes of the kynurenine pathway that are active in the brain as well as how these factors may be involved in regulation of QUIN under normal and pathological conditions (see Stone, 1993). In an attempt to obtain some further insight into the production and regulation of cerebral QUIN in rats under normal conditions, different kynurenine precursors were given either via intracerbroventricular (i.c.v.) injections or by systemic administrations and QUIN was measured in brain and plasma. In addition, a 3-HAO enzyme inhibitor of the 3-HANA analog class (Todd et al ., 1989) was used as a tool to further characterize cerebral QUIN production.

## 2. EXPERIMENTAL PROCEDURES

### 2.1 Chemicals

TRY (L-tryptophan; Fluka AG, Switzerland; batch 307857 or Sigma Chemical Company, St Louis, U.S.A.; batch 10F-0134; MW: 204.2), KYN (L-kynurenine sulfate; Sigma; batch 110H0531; MW 306.3), 3-hydroxy-DL-kynurenine (3-OH-KYN, Sigma; batches 21H4021 and 124H4062; MW: 224.2), anthranilic acid (ANA; Merck AG, Darmstadt, Germany; batch 123764; MW: 137.1), 3-HANA (Aldrich Chemical Company, Milwaukee, MI, U.S.A.; batch 02820KY; MW: 153.1) and QUIN (Sigma; batch 50H3418; MW: 167.1), 4,6-bromo-3-hydroxyanthranilic acid (NCR-631; Astra Arcus AB; batch 205/94).

### 2.2 Animals Treatments

Male Sprague-Dawley rats (170-180 g) were used (B&K Universal AB, Sollentuna, Sweden). The rats were housed under controlled conditions with free access of food and water. Systemic administrations were performed by either subcutaneous (s.c.; 5 ml/kg) or intraperitoneal (i.p.; 5 ml/kg) injections. The compounds were dissolved immediately before use in NaOH, diluted in saline and the pH was adjusted to 9 - 10 by HCl. The rats given i.c.v. injections were anaesthestized with a mix of enflurane (Efrane; Abbott, Campoverde, Italy), $O_2$ and $N_2O$, and placed in a stereotaxic frame. A unilateral hole on the right side of the skull bone was made and an i.c.v. injection of 10 µl of a 10 mM phosphate buffer, containing 105 mM NaCl, 2.5 mM KCl, 1.18 mM $MgCl_2$, 1.26 mM $CaCl_2$ and 0.1% ascorbic acid, and various doses of kynurenines were performed for one minute using a Hamilton microliter syringe (gauge 22S, 25 µl). In one experiment, osmotic pumps (Alzet mini-osmotic pump model 2002; Alza Corporation, Palo Alto, CA, U.S.A; flow rate of 0.5 µl/h, 14 days) were implanted s.c. and ANA at a concentration of 80 mM was infused i.c.v. for two weeks via a cannula fixed with dental cement to the skull bone. Before sacrifice, the animals were anaesthestized with enflurane and cardiac blood was collected in heparinized tubes followed by transcardial perfusion with heparinized saline (1 IE/l). The brains were rapidly removed and the hippocampus from the right side was dissected out for analysis. The blood samples were centrifuged at 2600 g for 10 min and the supernatants were used for analysis.

## 2.3 Analysis of Quinolinic Acid

Analysis of QUIN was performed essentially according to Heyes and Markey (1988) using gas chromatography coupled to a mass spectrometer (GC/MS). In brief, the tissue was homogenized in 0.3 M formic acid with $[^{18}O]_4$-quinolinic acid added as internal standard. After filtration, the samples were applied to pretreated anion-exchanger column and thereafter evaporated. The carboxylic groups were esterfied with hexafluoroisopropanol in the presence of trifluoroacetylimidazole, water and n-heptane were then added. Extraction of the derivatives to the organic layer was performed by whirl mixing followed by centrifugation. The organic phase was washed with water and after a second centrifugation the water was discarded and the samples evaporated with a stream of nitrogen at room temperature. The GC (Carlo Erba MFC 500; Fisons Instruments, Milano, Italy) was operating in splitless injection mode and a 25 m capillary column (NB-225) with an i.d. of 0.32 mm was used. Helium was employed as carrier gas. The oven was run at an initial temperature of 70°C up to 135°C (20°C/min). The MS (VG, TS-250; Fisons Instruments, Manchester, England) was adjusted to record m/e 467 QUIN and m/e 471 (internal standard) by negative chemical ionization. Methane was used as reagent gas. Statistical calculations were performed on the numerical data or on the natural logarithm of the data by means of one-way analysis of variance (ANOVA) followed by Scheffe's post hoc test.

# 3. CENTRAL ADMINISTRATION OF KYNURENINES

In order to examine the cerebral QUIN production from kynurenines, several metabolites of this pathway were administered into the rat brain by i.c.v. injections and the QUIN levels in the hippocampus determined. Only the immediate substrate for the 3-HAO enzyme, i.e. 3-HANA, and QUIN itself were able to significantly elevate QUIN (Fig. 1). The distribution of 3-HANA into the brain parenchyma and the conversion to QUIN was obviously rapid and efficient, as suggested by time response and dose response studies (data not shown). The effects seen after 3-HANA administration is in agreement with previous studies, suggesting that the 3-HAO enzyme exists in the brain and that it has a large capacity to metabolize 3-HANA (Foster et al., 1986; Heyes et al 1988, Speciale et al., 1989). Furthermore, the observed accumulation of QUIN supports the reported low activity of QUIN's degradative enzyme, quinolinic acid phosphoribosyltransferase (QPRT), in the rat brain (Foster et al., 1985). The importance of cerebral 3-HAO for de novo synthesis of QUIN was further studied by co-administration of 3-HANA with the 3-HAO inhibitor NCR-631,

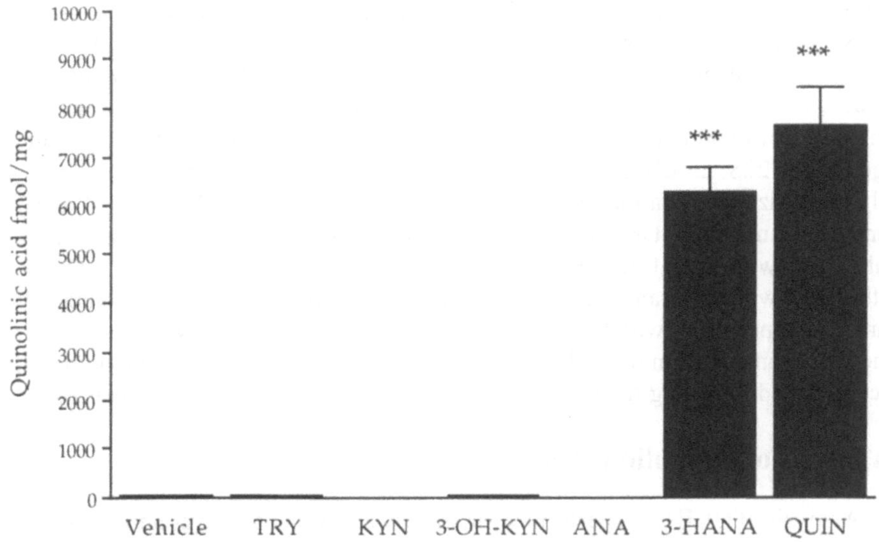


**Figure 1.** The effect of L-tryptophan (100 nmol), L-kynurenine (100 nmol), 3-hydroxykynurenine (300 nmol), anthranilic acid (100 nmol), 3-hydroxyanthranilic acid (100 nmol) or quinolinic acid (100 nmol) on quinolinic acid levels in the hippocampus. All kynurenines were administered i.c.v. 2 h before the rats were killed. Shown are the means ± s.e.m. expressed as fmol/mg wet weight (n=6-12). Statistical analysis was performed using ANOVA followed by Scheffe's post hoc test. ***$p<0.001$.

which potently inhibited the QUIN accumulation in hippocampus (Fig. 2), as previously shown with another 3-HAO inhibitor of the same chemical class (Heyes et al., 1988).

Neither TRY, KYN, or 3-OH-KYN had any significant effect on hippocampal QUIN levels when administered i.c.v. (Fig. 1). This finding is in agreement with previous studies where no detectable transformation to QUIN was seen in normal rat brain after dialysis

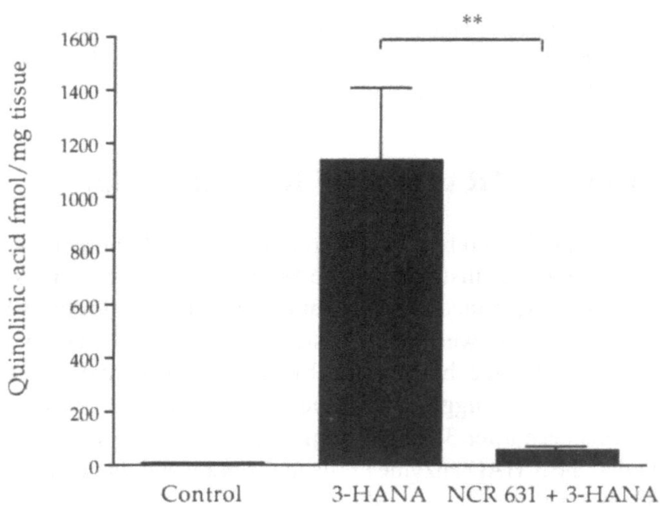


**Figure 2.** The effect of NCR-631 given i.c.v. at a doses of 300 nmol against quinolinic acid production induced by concomitant i.c.v. administration of 3-HANA at a dose of 10 nmol. The animals were sacrificed 2 h after injection and the quinolinic acid levels were determined in the hippocampus. Shown are the means ± s.e.m. expressed as fmol/mg wet weight. Statistical analysis was performed using Student's t-test. **$p<0.01$.

infusion of TRY or KYN (Speciale et al., 1989), or in gerbil brain after intracisternal TRY administration (Saito et al., 1993b). It has been reported that 3-HANA can be formed from anthranilic acid (ANA) in rat brain slices, indicating that ANA thereby could serve as a cerebral precursor of QUIN (Baran and Schwarcz 1990). However, ANA given i.c.v acutely failed to increase cerebral QUIN, but since only one time point was studied in the initial experiment, an additional study was performed where ANA was infused i.c.v. for two weeks via osmotic minipumps (40 nmol/h). Still, ANA was unable to increase QUIN levels in hippocampus in spite of the extended infusion time (ANA group: 44.0 ± 9 fmol/mg; Saline group: 64 ± 15 fmol/mg). A similar observation has previously been reported in gerbils, where no synthesis of QUIN was seen after intracisternal administration of labelled ANA (Saito et al., 1993b).

## 4. SYSTEMIC ADMINISTRATION OF KYNURENINES

### 4.1 Brain Levels of QUIN after Systemic Administration

To study the influence of systemic kynurenines on cerebral QUIN levels, different kynurenines were given s.c. or i.p. at a dose of 250 mg/kg, which corresponds to a relatively narrow dose range when expressed as micromolar doses. It is therefore unlikely that any of the major differences observed between the kynurenines related to the particular dose given. However, to compensate for the relatively low dose of KYN, which was used as a sulfate salt, a dose of 500 mg/kg was also given. Another methodological issue of concern is that the 3-OH-KYN used was a racemic mixture. Therefore, it is possible that some of the effects seen could relate to a relatively lower dose and possibly even unwanted effects, e.g. enzyme inhibition, of the D-enantiomer.

A s.c. administration of TRY (1.22 mmol/kg) or KYN (1.63 mmol/kg) increased hippocampal QUIN approximately 10 fold, while 3-HANA (1.63 mmol/kg) and QUIN itself (1.50 mmol/kg) enhanced cerebral QUIN about 500 fold, when analyzed 2 h after administration (Fig. 3). In contrast, no significant effect on hippocampal QUIN was detected following s.c. administration of ANA (1.82 mmol/kg) or 3-OH-KYN (1.12 mmol/kg). Also i.p. injections of TRY, 3-HANA or QUIN resulted in significant QUIN elevations in the hippocampus, although the increases seen were considerably lower as compared to the s.c. administrations (Fig. 4). This could relate to a more efficient liver metabolism and elimination after the i.p. route of administration, reducing the available levels of QUIN precursors, or QUIN itself, for BBB transport. Also, there was one obvious difference in the relative effects of the various precursors when given i.p. compared to s.c.; 3-HANA was not as effective as TRY in enhancing cerebral QUIN. Moreover, only a slight tendency towards increased QUIN levels was found following the i.p. injection of KYN. Again, this may relate to a more efficient metabolism in the liver, even by enzymes distal to QUIN such as QPRT. The finding that TRY given i.p. induces an increase of QUIN in the brain is in agreement with a previous report in which a major and persistent increase of extracellular QUIN was observed in the striatum (During et al., 1989).

To study the possibility that 3-HANA acted as a cerebral precursor of QUIN when after i.p. administration of TRY, the 3-HAO inhibitor was given i.c.v. in a dose of 300 nmol before teh TRY treatment. However, no reduction on the QUIN accumulation in hippocampus could be detected (Fig. 5), suggesting that the increase in cerebral seen after TRY depends entirely on extracerebral metabolism. The massive effect of QUIN itself on hippocampal QUIN after either s.c. or ip, administration is in line with this hypothesis. On the other hand, there appear to be important species differences in the effects of peripheral kynurenine loading. In mice, 3-OH-KYN, when given i.p. at a dose of 0.3 mmol/kg, is an efficient

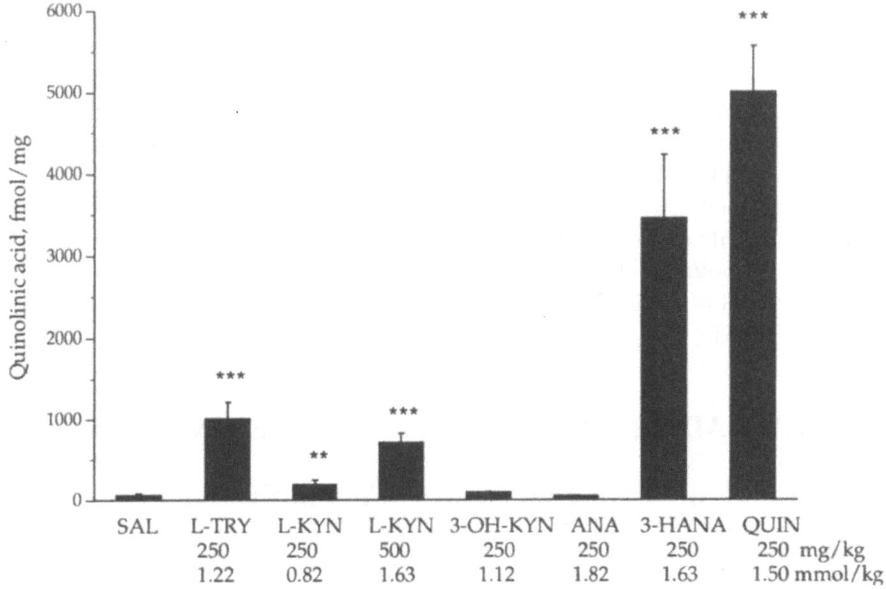


**Figure 3.** The effect of s.c. administeration of L-tryptophan, L-kynurenine, 3-hydroxykynurenine, anthranilic acid, 3-hydroxyanthranilic acid or quinolinic acid on quinolinic acid levels in the hippocampus. Determinations at 2 h after the kynurenine administrations are presented. Shown are the means ± s.e.m. expressed as fmol/mg wet weight (n=6-10). Statistical analysis was performed using ANOVA followed by Scheffe's post hoc test. **p<0.01, ***p<0.001.

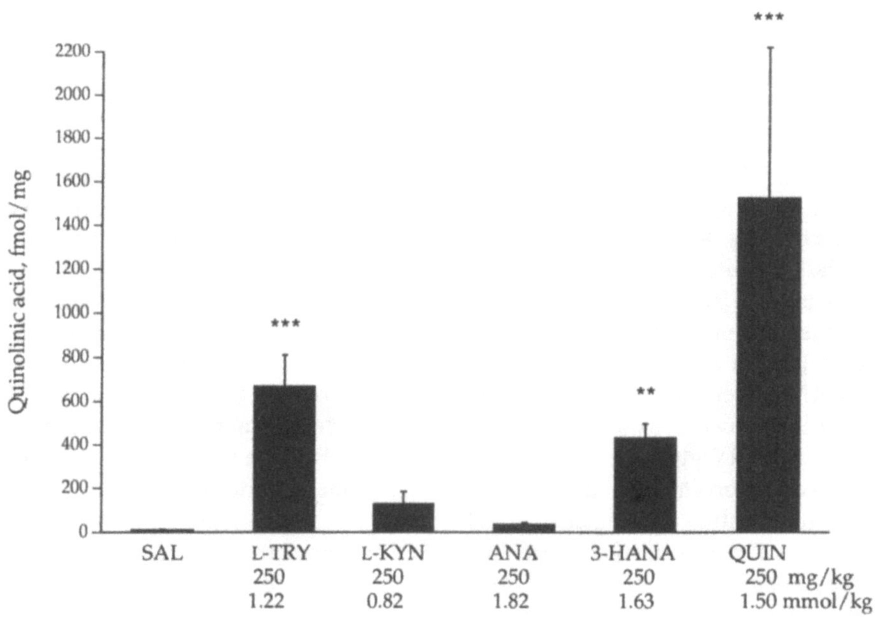


**Figure 4.** The effect of i.p. administration of L-tryptophan, L-kynurenine, anthranilic acid, 3-hydroxyanthranilic acid or quinolinic acid on quinolinic acid levels in the hippocampus. All kynurenines were given at a dose of 250 mg/kg and killed 2 h after treatment. Shown are the means ± s.e.m. expressed as fmol/mg wet weight (n=4-7). Statistical analysis was performed using ANOVA followed by Scheffe's post hoc test. **p<0.01, ***p<0.001.

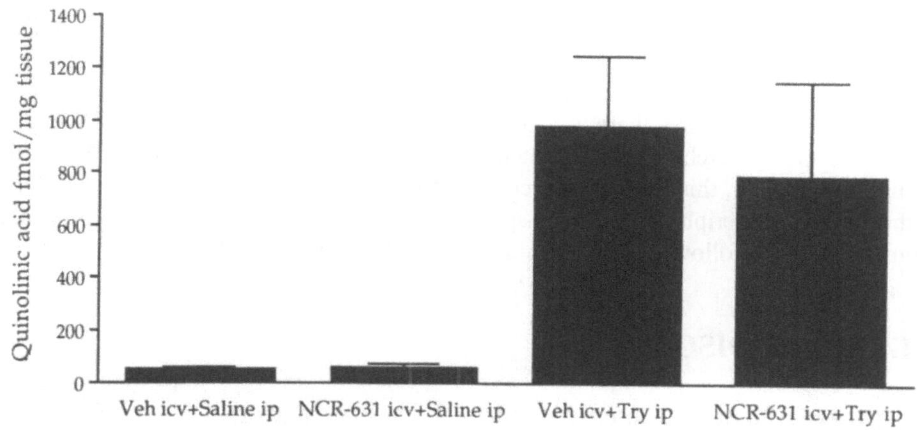


**Figure 5.** The effect of NCR-631 given i.c.v. at a dose of 300 nmol against quinolinic acid production induced by i.p. administration of 250 mg/kg L-tryptophan given 5 min after the NCR-631 injection. The animals were sacrificed 2 h after the L-tryptophan injection and the quinolinic acid levels were determined in the hippocampus. Shown are the means ± s.e.m. expressed as fmol/mg wet weight (n=5-6).

precursor of cerebral QUIN, even compared to 3-HANA, while KYN appears to raise QUIN more than TRY (Reinhard et al., 1994). Interestingly, ANA given i.p. in mice did not have any notable effect on QUIN, in agreement with the present results.

## 4.2 Plasma Levels of QUIN after Systemic Administration

The plasma levels of QUIN were studied after s.c. administration of kynurenines (Table 1). At 2 h following s.c. administration of TRY or QUIN, the plasma levels of QUIN increased 245-fold and 448-fold, respectively.

Large increases in plasma concentrations of QUIN were also found following s.c. injections of 500 mg/kg of KYN or 250 mg/kg of 3-HANA, but the levels did not reach those

**Table 1.** Effects of s.c. administration of different kynurenine metabolites on the quinolinic acid concentration in blood plasma 30 min, 2 h and 6 h following administration

| Treatment | 0.5 h | 2 h | 6 h |
|---|---|---|---|
| Control | 1.07 ± 0.08.08 | 0.67 ± 0.10.10 | 0.72 ± 0.03.03 |
| L-Tryptophan 250 mg = 1.22 mmol/kg | N.d. | 163.00 ± 18.90*** | N.d. |
| L-Kynurenine 250 mg =0.82 mmol/kg | 3.35 ± 0.64.64 | 7.80 ± 2.57.57*** | 1.70 ± 0.07.07 |
| L-Kynurenine 500 mg=1.64 mmol/kg | N.d. | 57.49 ± 16.66.66*** | N.d. |
| 3-Hydroxykynurenine 250 mg=1.12 mmol/kg | 2.40 ± 0.50.50 | 2.42 ± 0.12.12*** | 2.31 ± 0.16.16 |
| 3-Hydroxyanthranilate 250 mg=1.63 mmol /kg | 73.03 ± 15.54.54*** | 73.10 ± 15.51.51*** | 14.70 ± 5.51.51*** |
| Quinolate 250 mg=1.50 mmol/kg | N.d. | 299.50 ± 16.66.66*** | N.d. |

Data are presented as means ± s.e.m in nmol/ml quinolinic acid in blood plasma (n=4-10). N. d.= not determined. Statistical analysis was performed using ANOVA, followed by Scheffe's post hoc test. ***p<0.001.

seen after administration of TRY, or QUIN itself. A very modest effect was seen following 250 mg/kg KYN or 3-OH-KYN, mirroring the restricted effects seen in the brain. Plasma QUIN levels were also measured at 2 h after a single i.p. dose of TRY (250 mg/kg). This treatment resulted in a 350-fold increase of plasma QUIN (211 ± 49 nmol/ml) at the 2 h time-point studied. Hence, there was generally a correspondence between QUIN in plasma and the resulting cerebral tissue content of QUIN after administration of the different kynurenines. Again, this finding is in contrast to findings in mice, where 3-OH-KYN only slightly enhanced peripheral QUIN, in spite of a major increase in cerebral QUIN, while the opposite was seen following TRY administration (Reinhard et al., 1994).

# 5. GENERAL DISCUSSION

## 5.1 Intracerebral Kynurenines and QUIN Production

The present experiments demonstrate an inability of the normal rat brain to produce QUIN from several kynurenine metabolites. Indeed, only 3-HANA appears to act as an important intracerebral precursor of QUIN in the normal rat brain. This supports the notion that some of the upstream enzymes in the kynurenine pathway are either missing in the normal rat brain, or at least have a very low capacity (see Stone, 1993). In the normal rat brain, KYN appears to readily convert to 3-OH-KYN, but not to 3-HANA (Vecsei et al., 1992). Therefore, a likely reason for the inability of TRY, KYN or 3-OH-KYN to act as precursors of QUIN in the brain is that the kynureninase enzyme is missing, or has a very low activity. An additional factor that may limit the production of QUIN from TRY in the CNS under normal condition is the low IDO activity (see Stone, 1993). The inability of ANA, given either acutely or for two weeks i.c.v., to enhance QUIN suggests that hydroxylation of ANA to 3-HANA does not occur in vivo in the rat brain. This finding is in agreement with a previous in vivo study in gerbils (Saito et al., 1993b), but in apparent contrast to an in vitro study on rat brain tissue where 3-HANA was reported to be produced from ANA (Baran and Schwarcz, 1990).

Hence, it seems that there is an incomplete kynurenine metabolism in the rat brain, restricting the metabolism of several kynurenines and their flow to QUIN through the kynurenine pathway. Indeed, this may constitute an important protective mechanism against build-up of neuroactive concentrations of QUIN. This becomes particularly important considering that TRY is required in normal brain functioning, e.g. as a precursor of serotonin (see Stone, 1993), and that it actively passes over the BBB by the neutral amino acid transporter (Fukui et al., 1991). Also, KYN, 3-OH-KYN and ANA may cross the BBB relatively easily in rats, while 3-HANA, an efficient cerebral precursor of QUIN, as well as QUIN itself do not readily enter the brain (Fukui et al., 1991). This suggests that an obstruction of both kynureninase activity and ANA hydroxylation is important in preventing a rise in cerebral QUIN and may constitute a critical regulatory mechanism. On the other hand, it is crucial to take into account that a minor conversion of the kynurenines to QUIN may still occur in the normal brain, even though the present study failed to detect any such activity. Hence, i.c.v. injections may not deliver the precursors to the cellular or subcellular compartments where the conversion takes place. Moreover, the mechanisms involved in suppressing cerebral QUIN appear to be susceptible to pathological changes, and a complete kynurenine pathway has been described to develop in the injured or immune-stimulated brain (Heyes et al., 1992).

## 5.2 Systemic Kynurenines and QUIN Production

The characterization of possible precursors of cerebral QUIN becomes more complicated after systemic administration of kynurenines. The resulting plasma levels as well as an enhanced passage over the BBB have to be considered. In contrast to the intracerebral injections, several kynurenines were shown to give rise to significant increases in brain QUIN, particularly after s.c. administration. One obvious explanation for this difference is that the systemically administered kynurenines enhanced plasma QUIN, and thereby seemed to enhance QUIN transport over the BBB. Indeed, a marked increase in plasma QUIN was observed following systemic TRY loading, suggesting that the rise in cerebral QUIN was due to enhanced translocation of QUIN into the brain. An alternative explanation is that peripheral TRY metabolism produced a metabolite that entered the brain as a precursor of QUIN. If so, a possible candidate is 3-HANA, but this is unlikely since i.c.v. administration of the 3-HAO inhibitor, which potently inhibited QUIN accumulation after intracerebral 3-HANA, did not affect the increase of brain QUIN after systemic TRY administration. Moreover, while the BBB transport of both QUIN and 3-HANA is very restricted (Fukai et al., 1991), systemic QUIN loading has been shown to lead to an increase in cerebral QUIN (During et al., 1989). On the other hand, 3-HANA given s.c. was the most potent QUIN precursor of cerebral QUIN, while it was not as effective in enhancing peripheral QUIN. This indicates that an enhanced BBB transport of 3-HANA, and subsequent intracerebral metabolism, occurred after systemic 3-HANA loading. In fact, this is supported by the observation that 3-HANA given i.p, when a more pronounced liver metabolism along the kynurenine pathway can be expected, was less efficient in increasing cerebral QUIN. Taken together, this suggests that the increase in cerebral QUIN seen after systemic administration of TRY, and possibly KYN, was solely related to an increase in plasma QUIN, while the effect of peripheral 3-HANA was dependent on the route of administration and consequently a possible penetration directly into the brain. Measurements of 3-HANA in plasma and brain as well as further studies using 3-HAO inhibitors may provide further information concerning the site of QUIN production after systemic administration of kynurenines.

The contribution of peripheral 3-OH-KYN in increasing cerebral QUIN is difficult to decipher at the present time, since rather small effects were observed. If a complete kynurenine pathway was involved in the production of QUIN from TRY in peripheral tissues, e.g. the liver, as assumed, administration of 3-OH-KYN should have lead to a significant increase of QUIN in plasma and brain. However, as discussed earlier, the 3-OH-KYN used for the present study may not have been optimal. Other possible explanations for this apparent discrepancy are insufficient distribution of 3-OH-KYN from the injection sites, competition for transport mechanisms, shunts of kynurenines into enzymatic side-tracks of the kynurenine pathway, or rapid elimination before metabolism. A less likely explanation would be that the peripheral kynurenine pathway is deficient in the rat, while still TRY can be metabolized to QUIN through other pathways. Importantly, there seem to be major species differences in the kynurenine pathway and its regulation of cerebral QUIN, which become particularly apparent after TRY and 3-OH-KYN administration.

The inability of ANA, given peripherally, to enhance QUIN, suggests that hydroxylation of ANA to 3-HANA does not constitute a significant metabolic pathway in the rat at all. This finding is in agreement with previous in vivo studies in the mouse (Reinhard et al., 1994). However, in the present study, a very weak trend towards increased hippocampal QUIN was actually observed after i.p. administration of ANA at a dose of 1.82 mmol/kg. Furthermore, in mice it seems that a dose of 1 mmol/kg given i.p. may elevate cerebral QUIN (Reinhard et al., 1994). This suggests that a minor peripheral metabolism of ANA to 3-HANA, and thereby QUIN, could occur in rodents, at least after high doses and when a major proportion of the ANA passes the liver. However, measurements of both 3-HANA and QUIN in different tissues will be required to further characterize the possibility of a any significant 3-hydroxylation of ANA in vivo.

## 5.3 Conclusions

The present experiments support and extend the observation that the kynurenine pathway to QUIN appears to be incomplete in the normal rat brain; only 3-HANA metabolized by the 3-HAO enzyme may enhance cerebral QUIN after intracerebral injection. On the other hand, the systemic availability of several intermediates of the kynurenine pathway, including TRY, may affect the plasma concentration of QUIN, and thereby regulate cerebral QUIN.

# 6. REFERENCES


Baran, H., and Schwarcz, R., 1990, Presence of 3-hydroxyanthranilic acid in rat tissues and evidence for its production from anthranilic acid in the brain, *J. Neurochem.* 55:738.

During, M.J., Heyes, M.P., Freese, A., Markey, S.P. Martin, J.B., and Roth R.H., 1989, Quinolinic acid concentrations in striatal extracellular fluid reach potentially neurotoxic levels following systemic tryptophan loading, *Brain Res.* 476:384.

Foster, A.C., Zinkand, W.C., and Schwarcz, R., 1985, Quinolinic acid phosphoribosyltransferase in rat brain, *J. Neurochem.* 44:446.

Foster, A.C., White, R.J., and Schwarcz, R., 1986, Synthesis of quinolinic acid by 3-hydroxyanthranilicacid oxygenase in ratbrain tissues in vitro, *J. Neurochem.* 47:23.

Fukui, S., Schwarcz, R., Rapoport, S.I., Takada, Y., and Smith, Q.R., 1991, Blood-brain barrier transport of kynurenines: Implications for brain synthesis and metabolism, *J. Neurochem.* 56:2007.

Ganong, A.H., and Cotman, C.W., 1986, Kynurenic acid and quinolinic acid act at N-methyl-D-aspartate receptors in the rat hippocampus, *J. Pharmacol Exp. Ther.* 236:293.

Heyes, M.P., Brew, B.J., Martin, A., Price, R.W., Salazar, A.M., Sidtis, J.J., Yergey, J.A., Mourdian, M.M., Sadler, A.E., Keilp, J., Rubinow, D., and Markey, S.P., 1991, Quinolinic acid in cerebrospinal fluid and serum in HIV-1 infection: Relationship to clinical and neurological status, *Ann. Neurol.* 29:202.

Heyes, M.P., Hutto, B., and Markey, S.P., 1988, 4-Chloro-3-hydroxyanthranilate inhibits brain 3-hydroxyanthranilate oxidase, *Neurochem. Int.* 13:405.

Heyes, M.P., and Markey, S.P., 1988, Quantification of quinolinic acid in rat brain, whole blood, and plasma by gas chromatography and negative chemical ionization mass spectrometry: Effects of systemic tryptophan administration on brain and blood quinolinic acid concentrations, *Anal. Biochem.* 174:349.

Heyes, M.P., Saito, K., Crowley, J.S., Davis, LE., Demitrack, M.A., Der, M., Dilling, L.A., Elia, J., Kruesi, M.J.P., Lackner, A., Larsen, SA., Lee, K., Leonard, H.L., Markey, S.P., Martin, A., Milstein, S., Mouradian, M.M., Pranzatelli, M.R., Quearry, B.J., Salazar, A., Smith, M., Strauss, S.E., Sunderland, T., Swedo, S.W., and Tourtellotte, W.W., 1992. Quinolinic acid and kynurenine pathway metabolism in inflammatory and non-inflammatory neurological disease, *Brain* 115:1249.

Köhler, C., Eriksson, L.G., Okuno, E., and Schwarcz R., 1988, Localization of quinolinic acid metabolizing enzymes in the rat brain. Immunohistochemical studies using antibodies to 3-hydroxyanthranilic acid oxygenase and quinolinic acid phosphoribosyltransferase, *Neuroscience* 27:49.

Lapin, I.P., 1978, Stimulant and convulsant effects of kynurenines injected into brain ventricles in mice, *J. Neuronal. Trans.* 42:37.

Long, C.L., Hill, H.N., Weinstock, I.M., and Henderson, L.M., 1954, Studies of the enzymatic transformation of 3-hydroxyanthranilate to quinolinate, *J. Biol. Chem.* 211:405.

Malherbe, P., Köhler, C., Da Prada, M., Lang, G., Kiefer, V., Schwarcz, R., Lahm, H.-W., and Cesura, A.M., 1994, Molecular cloning and functional expression of human 3-hydroxyanthranilic-acid dioxygenase, *J. Biol. Chem.* 269:13792.

Moroni, F., Lombardi, G., Carlà, V., Lal, S., Etienne, P., and Nair, N.P.V., 1986, Increase in the content of quinolinic acid in cerebrospinal fluid and frontal cortex of patients with hepatic failure, *J. Neurochem.*, 47:1667.

Okuno, E., Köhler, C., and Schwarcz, R., 1987, Rat 3-hydroxyanthranilic acid oxygenase: Purification from the liver and immunocytochemical localization in the brain, *J. Neurochem.* 49:771.

Reinhard, J.F. Jr., Erickson, J.B., and Flanagan, E.M., 1994, Quinolinic acid in neurological disease: opportunities for novel drug discovery, *Adv. Pharmacol.* 30:85.

Saito, K., Nowak, T.S. Jr, Markey, S.P., and Heyes, M.P., 1993a, Mechanism of delayed increases in kynurenine pathway metabolism in damaged brain regions following transient cerebral ischemia, *J Neurochem*. 60: 180.

Saito, K., Nowak, TS. Jr., Suyama, K., Querry, BJ, Saito, M., Crowley, J.S., Markey, S.P., and Heyes, M.P., 1993b, Kynurenine pathway enzymes in brain: responses to ischemic brain injury versus systemic immune activation, *J. Neurochem*., 61:2061.

Schwarcz, R., Whetsell, W.O. Jr., and Mangano, R.M., 1983, Quinolinic acid: An endogenous metabolite that produces axon-sparing lesions in rat brain, *Science* 219:316.

Schwarcz, R., Okuno, E., and Köhler, C., 1986, Endogenous excitotoxins: Focus on quinolinic acid, in: "Excitatory Amino Acids", Roberts, P.J., Storm-Mathisen, J., and Bradford, H.F. eds., McMillan, London.

Schwarcz, R., Okuno, E., White, R.J., Bird, E.D., and Whetsell, W.O. Jr., 1988, 3-Hydroxyanthranilate oxygenase activity is increased in the brains of Huntington disease victims, *Proc. Natl. Acad. Sci. U.S.A*. 85:4079.

Speciale, C., Ungerstedt, U., and Schwarcz, R., 1989, Production of extracellular quinolinic acid in the striatum studied by microdialysis in unanesthetized rats, *Neurosci. Lett*. 104:345.

Stone, T.W., 1989, "Quinolinic acid and the Kynurenines", CRC Press, Boca Raton, Florida.

Stone, T.W., 1993, Neuropharmacology of quinolinic and kynurenic acids, Pharm. Rev. 45:309.

Stone, T.W. ,and Perkins, M.N., 1981, Quinolinic acid: A potent endogenous excitant at amino acid receptors in CNS, *Eur. J. Pharmacol*. 72:411.

Todd, W.P., Carpenter, B.K.,and Schwarcz, R., 1989, Preparation of 4-halo-3-hydroxyanthranilates and demonstration of their inhibition of 3-hydroxyanthranilate oxygenase activity in rat and human brain tissue, *Prep. Biochem*. 19:155.

Vecsei, L., Miller, J., MacGarvey, U., and Beal, F., 1992, Effects of kynurenine and probenicide on plasma and brain tissue concentrations of kynurenic acid, *Neurodegeneration* 1:17.

37

# THE NEUROTOXIN QUINOLINIC ACID IS INCREASED IN SPINAL CORDS OF MICE WITH HERPES SIMPLEX VIRUS ENCEPHALITIS

J. F. Reinhard, Jr. and E. M. Flanagan

The Wellcome Research Laboratories
3030 Cornwallis Rd., ResearchTriangle Park, North Carolina

## ABSTRACT

Following retroperitoneal, intradermal inoculation of mice with HSV-1, signs of encephalomyelitis (hind-limb paralysis, flaccid tail and loss of bladder control) appeared 6-7 days later. Levels of quinolinic acid (QUIN; determined by gas chromatography with mass-spectrometry), rose approximately 40-fold in mice with encephalomyelitis, primarily in the spinal cord. Live virus could also be grown from homogenates of the affected spinal cords. Time-course studies, demonstrated that the increase in QUIN coincided with the appearance of encephalomyelitis. Large increases in indoleamine dioxygenase activity were observed in spinal cords from the affected mice, suggesting that the QUIN was synthesized within the spinal cord. It is, therefore, possible that QUIN may contribute to the pathology of HSV-1 encephalomyelitis.

## INTRODUCTION

Quinolinic acid, a neurotoxin formed *in vivo* from tryptophan, is associated with a number of infectious and inflammatory diseases of the nervous system(Reinhard et al., 1994). While the sites of formation of quinolinic acid (QUIN) are not entirely clear, macrophages, activated by cytokines, may account for the increased QUIN (Heyes et al., 1992b).

Introduction: Increased levels of QUIN have been reported for a number of conditions in which macrophage activation is a prominent feature. Examples include spinal cord injury (Blight et al., 1993; Popovitch et al., 1994), retroviral infection (Heyes et al., 1989), polioviral infection (Heyes et al., 1992a), autoimmune encephalitis (Flanagan et al., 1995) and ischemic injury (Saito et al., 1993b).

*Recent Advances in Tryptophan Research*, edited by Graziella Allegri Filippini et al.
Plenum Press, New York, 1996

Herpes simplex virus-1 (HSV-1) is a DNA virus which is neurotropic and neurotoxic (Meyer et al., 1960). It is possible that HSV-1, like the retroviruses and polio virus, increase QUIN formation. We tested this possibility by determining QUIN in the spinal cords of mice infected with HSV-1 and now report increases in both the QUIN content and QUIN biosynthetic enzyme activities in the affected spinal cords.

## MATERIALS AND METHODS

Male balb/c mice were purchased from the Charles River Laboratories (Wilmington MA) and maintained on a 12-hour light dark cycle with *ad libitum* access to food and water.

The reagents were obtained from the Sigma Chemical Co. (St. Louis MO) unless otherwise noted. Mice were inoculated in the depilidated, abraded, retroperitoneal dermal flank with $10^5$ plaque forming units (pfu) of the SC16 strain of HSV-1, in a volume of 100 µl. Approximately 80% of the inoculated animals developed signs of encephalomyelitis (hindlimb paralysis, flaccid tails, and loss of urinary bladder control) within 7-8 days post-inoculation at which time they were killed (unless otherwise noted).

Quinolinic acid was measured by gas chromatography with mass spectrometry (GC/MS), operated in the negative chemical ionization mode as described (Heyes and Markey, 1988), with minor modifications (Flanagan et al., 1995). Assay of the QUIN biosynthetic enzymes was performed on cell free homogenates of tissues which were sonically disrupted in 10-20 volumes of TRISoHCl (pH 7.7, 50 mM). Indoleamine dioxygenase (IDO), and kynureninase activities were assayed using HPLC as described (Saito et al., 1993a; Okuno and Kido, 1991) while kynurenine hydroxylase (KH) and hydroxyanthranilate 2,3-dioxygenase (HAD) were assayed radiometrically (Erickson et al., 1992; Okuno et al., 1987).

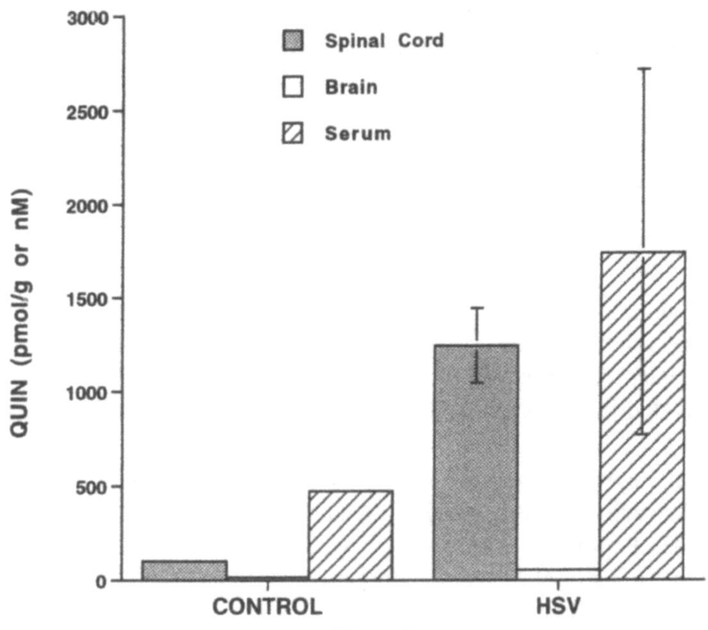


Figure 1

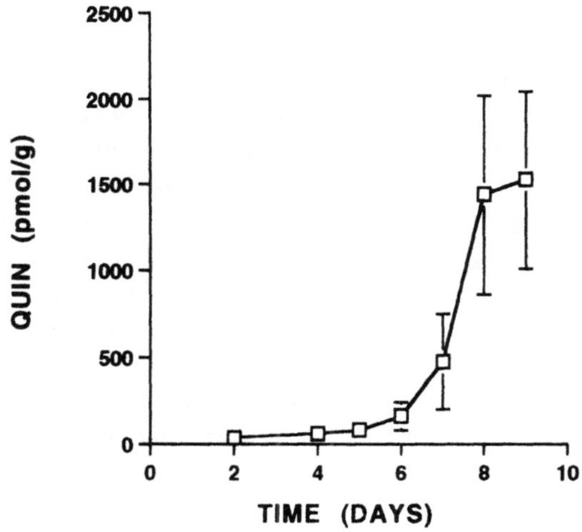


Figure 2

Statistical analyses were performed using the JMP® statistical software for the Macintosh computer. A probability of 0.05 was considered to be significant.

## RESULTS

Following inoculation of animals with HSV-1 approximately 80% developed signs of encephalomyelitis (hindlimb paralysis, flaccid tails, and loss of urinary bladder control). Tissue QUIN contents in forebrain, spinal cord and serum are presented as Figure 1. HSV-encephalomyelitis was associated with ~40-fold elevations in spinal cord QUIN while smaller increases were observed in the forebrain and the blood.

In a time-course study, spinal cord QUIN was not elevated until day 6 after which time QUIN rose rapidly to a level ~42-times that of the un-infected, control animals (Figure 2).

The activity of the QUIN biosynthetic enzymes, indoleamine 2,3-dioxygenase (IDO), kynurenine 3-monooxygenase (KH), kynureninase (Kase) and hydroxyanthranilate 3,4-dioxygenase (HAD) are presented as Table 1. Of the 4 enzymes studied, only IDO showed marked alterations in activity in the infected animals. IDO activity was ~ 20-times that of the control IDO activity. A small increase in KH activity was also observed.

**Table 1.** Kynurenine pathway enzymes in HSV encephalitis

| Treatment | IDO | KH | Kase | HAD |
|---|---|---|---|---|
| Control | 0.56 ± 0.33 | 48.02 ± 5.37 | 217 ± 7 | 21600 ± 7800 |
| HSV | 11.04 ± 0.40* | 60.96 ± 5.16* | 217 ± 34 | 45000 ± 25800 |

Data from mouse spinal cords, are the means x SEM of 3 animals per group expressed as pmol/hour *mg protein. Asterisks denote significant differences ($p<0.05$) from the corresponding control groups.

## DISCUSSION

The present data demonstrate that HSV-1 encephalomyelitis is associated with increases in tissue QUIN content. Neurological signs following HSV-1 have been observed following either intravenous, intracerebral and intranasal (Anderson and Field, 1983) or intradermal (Dillard et al., 1995) inoculation with the virus. The present data, with retroperitoneal dermal infection, demonstrate large increases in QUIN which are largely restricted to the spinal cord. Smaller increases in QUIN were observed in the brain and in serum. The result is not surprising since the spinal cord is the initial site of viral entry into the CNS following intradermal inoculation. The first neurological signs observed were consistent with spinal cord involvement and are reminiscent of experimental autoimmune encephalomyelitis (EAE). Indeed, in EAE, tissue QUIN content was increased (Flanagan et al., 1995).

The time-course for QUIN formation was very similar to the time-course for the development of encephalomyelitis. Thus, it is possible that some of the neurologic signs are a result of increased QUIN in the tissue. The time required for the appearance of neurological signs and QUIN may be due to axonal transport of the virus to the spinal cord or the time for the appearance and activation of peripheral macrophages in the spinal cord.

The present data don't identify which cell (or cells) make the quinolinate. The increases in tissue IDO and KH activities suggest that QUIN is being formed within the CNS, most likely in macrophages. However, the data do not establish that other cells may contribute to the increase in tissue QUIN. Since macrophages contain all of the enzyme activities necessary to form QUIN, it is highly likely that some, if not all, of the QUIN arises from these cells. Additionally, astrocytes could also contribute to the increased QUIN by metabolizing 3-hydroxyanthranalic acid to QUIN, since they contain HAD (Okuno et al., 1987) and since activated monocytoid cells also release 3-hydroxyanthranalic acid (Werner-Felmayer et al., 1989). Neuronal synthesis of QUIN has not been demonstrated and, thus, appears unlikely to be the source of the spinal cord QUIN. Rather, infiltrating and resident macrophages, activated by HSV-1, are the most likely source of the toxin.

Herpes viruses differ significantly from the retroviruses. The herpes viruses are neurotropic DNA viruses which may enter cells through the basic fibroblast growth factor receptor (Kaner et al., 1990) and are transported trans-synaptically throughout nerve pathways (Kristensson et al., 1978). Herpes viruses, dividing within cells, can cause cell death through lysis. In contrast, retroviruses are not neurotropic *per se* but appear to enter the CNS either through infected monocytes. To date, no compelling evidence has been presented to document retroviruses in neurons. However, neurotoxicity does not have to be the direct result of virus in the neuron but could be explained through the release of soluble factors which, in turn, mediate the neurotoxicity. Monocyte cell lines, infected in culture with HIV-1, were reported to release neurotoxic activity whose apparent molecular weight was <500, was resistant to proteolysis and heat and whose effects were blocked by antagonists to the NMDA receptor (Giulian et al., 1993). The identity of the neurotoxic substance(s) released from the macrophages has not been established but the authors of the study claimed that it was not QUIN. However, subsequent studies have shown that both cytokines and HIV-1 can induce QUIN formation in macrophages and in THP-1 cells (Heyes et al., 1993). Moreover, the concentrations of QUIN released are higher than was reported to cause toxicity to neurons in culture (Whetsell and Schwarcz, 1989).

It is possible that, in HSV, some of the neuropathological changes result from the increased QUIN. However, since herpes viruses kill cells through lysis, it is difficult to determine which neurological signs are caused by the increased QUIN. In cases of forebrain involvement, it is possible that QUIN might corrupt the otherwise orderly transmission of information through the NMDA receptor. Unlike glutamate, QUIN is not readily inactivated through uptake or enzymatic processes in the brain (Foster et al., 1984). Thus, QUIN effects at the receptor are likely to persist much longer than those of glutamate. The resulting alterations in NMDA receptor sensitivity could have manifold, adverse actions on excitatory amino acid transmission and may, thus explain some of the neurological complications of herpes viral infection.

## REFERENCES


Anderson J.R., Field H.J., 1983, The distribution of herpes simplex virus type 1 antigen in mouse central nervous system after different routes of inoculation. *J Neurol Sci* 60:181.

Blight A.R., Saito K., Heyes M.P., 1993, Increased levels of the excitotoxin quinolinic acid in spinal cord following contusion injury. *Brain Res* 632:314.

Dillard S.H., Cheatham W.J., Moses H.L., 1995, Electron microscopy of zosteriform herpes simplex virus infection in the mouse. *Lab Invest* 1972:391.

Erickson J.B., Flanagan E.M., Russo S., Reinhard J.F., Jr., 1992, A radiometric assay for kynurenine 3-hydroxylase based on the release of $^3H_2O$ during hydroxylation of L-[3,5-$^3$H]Kynurenine. *Anal Biochem* 205:257.

Flanagan E.M., Erickson J.B., Viveros O.H., Chang S.Y., Reinhard J.F.,Jr., 1995, Neurotoxin quinolinic acid is selectively elevated in the spinal cords of rats with experimental allergic encephalomyelitis. *J Neurochem* 64:1192.

Foster A.C., Miller L.P., Oldendorf W.H., Schwarcz R., 1984, Studies on the disposition of quinolinic acid after intracerebral or systemic administration in the rat. *Exp Neurol* 84:428.

Giulian D., Corpuz M., Chapman S., Mansouri M., Robertson C., 1993, Reactive mononuclear phagocytes release neurotoxins after ischemic and traumatic injury to the central nervous system. *J. Neuroscience Res.* 36:681.

Heyes M.P., Markey S.P., 1988, Quantification of quinolinic acid in rat brain, whole blood, and plasma by gas chromatography and negative chemical ionization mass spectrometry: effects of systemic L-tryptophan administration on brain and blood quinolinic acid concentrations. *Anal Biochem* 174:349.

Heyes M.P., Rubinow D., Lane C., Markey S.P., 1989, Cerebrospinal fluid quinolinic acid concentrations are increased in acquired immune deficiency syndrome. *Ann Neurol* 26:275.

Heyes M.P., Saito K., Markey S.P., 1992a, Human macrophages convert L-tryptophan into the neurotoxin quinolinic acid. *Biochem J* 283:633.

Heyes M.P., Saito K., Jacobowitz D., Markey S.P., Takikawa O., Vickers J.H., 1992b, Poliovirus induces indoleamine-2,3-dioxygenase and quinolinic acid synthesis in macaque brain. *FASEB J* 6:2977.

Heyes M.P., Saito K., Major E.O., Milstien S., Markey S.P., Vickers J.H., 1993, A mechanism of quinolinic acid formation by brain in inflammatory neurological disease. Attenuation of synthesis from L-tryptophan by 6-chlorotryptophan and 4-chloro-3-hydroxyanthranilate. *Brain* 116:1425.

Kaner R.J., Baird A., Mansukhani A., et al., 1990, Fibroblast growth factor receptor is a portal of cellular entry for herpes simplex virus type 1. . *Science* 248:1410.

Kristensson K., Vahlne A., Persson L.A., Lycke E., 1978, Neuronal spread of herpes simplex virus types 1 and 2 after corneal or subcutaneous (footpad) inoculation. *J Neurol Sci* 35:331.

Meyer H.M., Jr, Johnson R.T., Crawford I.P., Dascomb H.E., Rogers N.G., 1960, Central nervous system syndromes of "viral" aetiology-A study of 713 cases. *Am J Med* 29:334.

Okuno E., Kido R., 1991, Kynureninase and Kynurenine 3-Hydroxylase in Mammalian Tissues. In: Kynurenine and Serotonin Pathways. Schwarcz R. ed. Plenum Press, New York, 167-176.

Okuno E., Kohler C., Schwarcz R., 1987, Rat 3-hydroxyanthranilic acid oxygenase: purification from the liver and immunocytochemical localization in the brain. *J Neurochem* 49:771.

Popovitch P.G., Reinhard J.F.,Jr., Flanagan E.M., Stokes B.T., 1994, Elevation of the neurotoxin quinolinic acid occurs following spinal cord trauma. *Brain Res* 633: 348

Reinhard J.F.,Jr., Erickson J.B., Flanagan E.M.., 1994, Quinolinic acid in neurological disease: opportunities for novel drug discovery. *Adv Pharmacol* 30:85.

Saito K., Crowley J.S., Markey S.P., Heyes M.P., 1993a, A mechanism for increased quinolinic acid formation following acute systemic immune stimulation. *J Biol Chem.* 268:15496.

Saito K., Nowak T.S.,Jr., Markey S.P., Heyes M.P., 1993b, Mechanism of delayed increases in kynurenine pathway metabolism in damaged brain regions following transient cerebral ischemia. *J Neurochem* 60:180.

Werner Felmayer G., Werner E.R., Fuchs D., Hausen A., Reibnegger G., Wachter H., 1989, Characteristics of interferon induced tryptophan metabolism in human cells in vitro. *Biochim Biophys Acta* 1012:140.
Whetsell W.O.J., Schwarcz R., 1989, Prolonged exposure to submicromolar concentrations of quinolinic acid causes excitotoxic damage in organotypic cultures of rat corticostriatal system. *Neurosci Lett* 97:271.

# 38

# QUINOLINIC ACID AND ACTIVE OXYGENS

## Possible Contribution of Active Oxygens during Cell Death in the Brain

K. Goda, R. Kishimoto, S. Shimizu, Y. Hamane, and M. Ueda

Department of Nutrition
Kobe Gakuin University
Kobe 651-21, Japan

## INTRODUCTION

Quinolinic acid (QA, pyridine-2,3-dicarboxylic acid) is an intermediate of the kynurenine pathway in the tryptophan metabolism, which has been detected in the central nervous system (Gal and Sherman, 1978; Speciale and Schwarcz, 1993), and behaves as an excitotoxin (Lapin, 1978; Schwarcz et al., 1983). Recently, it has been proposed that QA and other metabolites of tryptophan may be involved in the brain pathology accompanying neuroinflammatory conditions, and also implicated in the pathophysiology of brain ischemia (Jhamandas and Boegman, 1994). The neuroexcitatory and neurotoxic actions have been suggested to be mediated by NMDA receptor (recently, its subtypes) (Schwarcz et al., 1984; Nakanishi, 1992),. However, the proximate cause of cell death at metabolic level has remained elusive. In the brain, there are numerous sources of oxygen-derived free radicals and they may exert a large variety of effects upon importnat central nervous system functions. Several studies have pointed to the role of metal ions, especially iron, in forming oxygen-derived free radicals (Barber, 1966; Vladimirov et al., 1980), which introduce several reactions such as lipid peroxidation, DNA chain breakage and others in the brain. The purpose of this experiment was to see the interaction of QA with iron ion (complex formation), the electron transfer to oxygen molecules from the complex (superoxide formation) and superoxide-mediated reactions such as lipid peroxidation and DNA chain breakage *in vitro*. Also, tryptophan metabolites which are known to be antagonist for QA were examined in the QA-iron system.

## MATERIALS AND METHODS

Quinolinic acid (QA), kynurenic acid (KA) and picolinic caid (PA) as antagonists for QA were obtained from Wako Pure Chemicals. Cytochrome c (Type VI, horse heart) and catalase were from Sigma, superoxide dismutase (SOD, bovine erythrocyte) from Boehrin-

Recent Advances in Tryptophan Research, edited by Graziella Allegri Filippini et al.
Plenum Press, New York, 1996

ger Mannheim. Methyl linoleate and DNA (calf thymus) were used as the substrate for lipid peroxidation and DNA chain breakage for each other. Ethidium bromide was used as an intercalator for DNA, deferoxamine mesylate (DEF) as a chelator and 2,2'-Azobis [2-(5-methyl-2-imidazolin-2-yl) propane] dihydrochloride (AIPD) as a free radical source, which is a water-soluble polymerization initiator at low temperature, were used. Other chemicals were commercially available and used without further purification.

The electron transfer to oxygen molecules from quinolinato-iron (II), QA-iron(II) complex , was evaluated by the cytochrome c reduction. The reduction was monitored by following the increase of absorbance at 550 nm. The determination of TBARS formed during lipid peroxidation was conducted spectrophotometrically, in which methyl linoleate was used as an unsaturated fatty acid of the substrate. DNA chain breakage was determined by the fluorometric measurements (Morgan and Pulleyblank, 1974), which is based on the change with fluorescence intensity of ethidium bromide intercalated into DNA molecules.

## RESULTS AND DISCUSSION

Certain regions of the brain contain unusually high concentrations of iron (Francois et al., 1981). The iron ion has been suggested to participate in the initiation of lipid peroxidation even in the central nervous system. In this study, several reactions mediated by the interaction of QA with iron ion in the presence of biological substances were investigated as a mimic of neuronal cell lesions induced by QA as an excitotoxin.

Superoxide Generation from QA-iron Complex: The reduction of cytochrome c in the aerobic condition was confirmed in the presence of QA-iron (II) complex as shown in Table 1. The reduction was ascertained to be mediated by superoxide, which was proved by the inhibition by SOD as seen in the table . This indicates the electron transfer from QA-iron (II) complex to oxygen molecules, resulting in the formation of superoxide. Superoxide spontaneously dismutates to hydrogen peroxide, and hydrogen peroxide interacts with superoxide and/or electron donors such as QA-iron (II) complex to form hydroxyl radical (Haber-Weiss reaction) (Haber and Weiss, 1934). In this system, several reactions such as lipid peroxidation and DNA chain breakage might be expected.

### 1. Lipid Peroxidation Induced by QA-iron (III) Complex and Effect of Antagonists for QA

To estimate the lipid peroxidation. TBARS was measured during the reaction by TBA method. The lipid peroxidation was greatly accelerated in the presence of QA-iron (III) complex as seen in Figure 1. To evaluate actual radical species in this reaction, several radical scavengers for active oxygens were addded to the reaction mixture. Table 2 shows the reaction was prevented by catalase, catalase + SOD, and t-butanol, an effective hydroxyl

Table 1. Reduction of cytochrome c during 5 minutes

| | Reduced Cyt.c (μM) |
|---|---|
| QA-iron (II) | 0.29 |
| QA-iron (II) + SOD | 0.18 |
| Iron (II) | 0.15 |
| Iron +.SOD | 0.12 |

Cyt.c, 2.96 μM; QA-iron (II), 8 μM; iron (II), 8 μM; SOD, 1.6 μg/ml

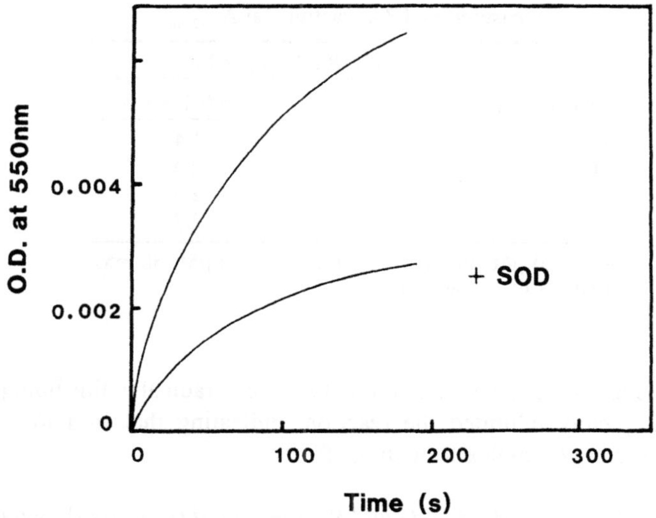


**Figure 1.** Reduction of cytochrome c mediated by superoxide in the presence of QA-iron (II) complex. Cyt. c, 2.96 μM; iron (II), 8 μM; QA-iron (II) complex, 8μM; SOD, 1.6 μg/ml.

Table 2. Prevention of TBARS formation by QA-iron (III)-stimulated lipid peroxidation

| Additives | TBARS/$\Delta A_{532\ nm}$ |
|---|---|
| None | 0.040 |
| SOD, 13.3 μg/ml | 0.040 |
| Catalase, 13.3 μg/ml | 0.015 |
| SOD + Catalase | 0.018 |
| t-BuOH, 10 mM | 0.020 |
| DEF, 167 μM | ~0 |

Methyl linoleate, 300 μM; QA-iron (III), 167 μM

$$R\text{-}OOH + (QA\text{-}Fe^{3+}) \longrightarrow R\text{-}OO\cdot + H^{+} + (QA\text{-}Fe^{2+})$$

$$(QA\text{-}Fe^{2+}) + O_2 \longrightarrow (QA\text{-}Fe^{3+}) + O_2^{-}$$

$$H_2O_2 + O_2^{-} \longrightarrow \cdot OH + OH^{-} + O_2$$

$$H_2O_2 + (QA\text{-}Fe^{2+}) \longrightarrow \cdot OH + OH^{-} + (QA\text{-}Fe^{3+})$$

Scheme 1. Active oxygens formation mediated by QA-iron complex.

Table 3. Metal-ion stimulated lipid peroxidation measured as TBA reactivity at $A_{532\ nm}$

| | TBARS/$A_{532\ nm}$ x $10^3$ | |
|---|---|---|
| Time (h) | Iron (III) | Iron (III) + KA |
| 0.5 | 1.3 | 1.4 |
| 1.0 | 2.5 | 2.0 |
| 2.0 | 3.2 | 2.6 |
| 3.0 | 3.3 | 2.7 |

ML, 300 μM; iron (III), 33 μM; KA, 100 μM; 50 mM TAPS, pH 8.3; Temp., 40°.

radical scavenger, suggesting the participation of hydroxyl radical in this lipid peroxidation. Deferoxamine completely inhibited the reaction, indicating that iron ion is absolutely necessary for this system (complex formation effect).

*1.1 Effect of Kynurenic Acid (KA) and Picolinic Acid (PA).* As shown in Figure 1, it is interesting that KA inhibited the lipid peroxidation in this system. Its function to prevent tha reaction was investigated, which is a chelator and has no catalytic activity (electron transfer ability) or a radical scavenger. Their results are shown in Table 3 and 4. KA does not form a complex with iron ion (data not shown). Table 3 is an indirect evidence, in which KA plays a role as an inhibitor without forming the complex, because TBARS formation will not be expected if KA forms a complex. Table 4 shows the inhibitory effect of KA on the radical-initiated lipid peroxidation, which clearly shows KA is a radical scavenger.

As described in the above, KA inhibits the peroxidation as a radical scavenger. Also, PA shows a marked inhibition of the QA-iron complex mediated lipid peroxidation (Table 5). However, the inhibitory manner of PA is different from that of KA. PA is an iron ion-chelator, which complex has no catalytic activity to initiate lipid peroxidation (electron transfer ability), and is not a radical scavenger (Table 6 and Table 7). In Table 6, PA-iron complex is shown to be unable to initiate peroxidation reaction, and in Table 7, PA has no radical scavenging ability, which means PA chelates iron of QA-iron complex in the lipid peroxidation system and, as the result, inhibits the peroxidation.

## 2. DNA Chain Breakage Induced by QA-Iron (II) Complex and Effect of KA and PA

DNA chain breakage during the reaction was monitored by measuring the fluorescence intensity of ethidium intercalated into DNA molecules. Figure 2 shows the time course

Table 4. Effect of KA on the AIPD-initiated peroxidation of methyl linoleate

| KA (μM) | TBARS $\Delta A_{532\ nm}$/h x $10^2$ | Inhibition % |
|---|---|---|
| 0 | 1.09 | 0 |
| 33 | 0.93 | 13 |
| 66 | 0.77 | 29 |
| 99 | 0.50 | 54 |

ML, 300 μM; AIPD, 1mM; 50 mM TAPS, pH 8.3; Temp., 40°

Table 5. Effect of Picolinic acid (PA) on the peroxidation of methyl linoleate

| Additives | TBARS (μmoles/ml/3h) |
|---|---|
| None | 0.43 |
| PA | |
| 30 μM | 0.29 |
| 167 μM | 0.03 |

ML, 300 μM; QA-iron (III), 167 μM; 50 mM TAPS, pH 8.3; Temp., 37°

Table 6. Effect of PA-iron complex on the peroxidation of methyl linoleate

| Additives | TBARS (μmoles/ml/3h) |
|---|---|
| None | 0.03 |
| PA-iron (III) | 0.03 |
| PA-iron (II) | 0.03 |
| PA-iron (II) + Ascorbate | 0.03 |

ML, 300 μM; PA-iron (II) and PA-iron (III), 167 μM; Ascorbate, 500 μM; 50 mM TAPS, pH 8.3; Temp., 37°

Table 7. Effect of PA on the AIPD-initiated peroxidation of methyl linoleate

| Additive | TBARS ($\Delta A_{532\ nm}$/h) |
|---|---|
| None | 0.015 |
| PA, 30 μM | 0.015 |

ML, 300 μM; AIPD, 1 mM; 50 mM TAPS, pH 8.3; Temp., 40°

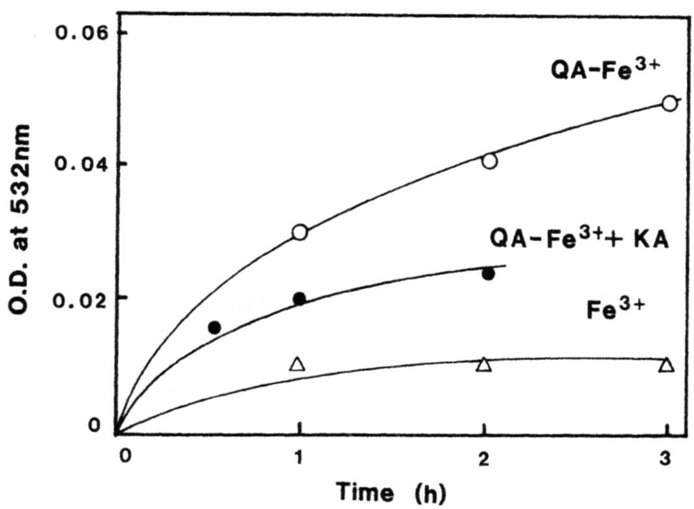


**Figure 2.** Time course of DNA chain breakage mediated by QA-iron (II) complex. Calf thymus DNA, $2A_{260}$; QA-iron (III) complex, 1 mM; 100mM TAPS, pH 7.6; Temp., 40°. (O) Ascorbic acid, 3mM.

Table 8. DNA chain breakage mediated by QA-iron complex

| Additives | DNA breakage (%/h) | Ratio (%) |
|---|---|---|
| None | 4.8 | 100 |
| SOD, 20 μg/ml | 4.6 | 95 |
| Catalase, 20 μM | 2.0 | 42 |
| α-Tocopherol, 20 μM | 1.5 | 31 |
| Formate, 5 mM | 2.8 | 58 |
| Mannitol, 10 mM | 2.7 | 56 |
| Thiourea, 10 mM | 0.7 | 15 |
| DEF, 1 mM | 0.4 | 8 |

Calf thymus DNA, $2A_{260}$ ; QA-iron (III), 1 mM; Ascorbate, 3 mM; 100 mM TAPS; ph7.6; Temp., 40°

of DNA chain breakage in the presence of QA-iron (II) complex. Table 8 reveals that actual radical species participated in the reaction is hydroxyl radical as like as in the lipid peroxidation.

*2.1 Effect of Kynurenic Acid (KA) and Picolinic Acid (PA).* As seen in Table 9, KA has an inhibitory effect on the chain breakage of DNA in the presence of QA-iron (II) complex. Even in the chain breakage reaction, KA plays a role as a radical scavenger, not as a chelator (data not shown), which is found in the results of KA during a radical-initiated reaction in Table 10. PA in the chain breakage reaction like as the lipid peroxidation has inhibitory action as shown in Table 11. PA acts as a chelator for iron in QA-iron (II) complex, and does not as a radical scavenger (Table 11).

As mentioned in the above, several reactions are induced by active oxygen species, actually hydroxyl radical, derived from superoxide, which is produced by the electron transfer to oxygen molecules from QA-iron complex. Such reactions may give deleterious

Table 9. Effect of KA to DNA breakage

| Additives | DNA chaine scission (%) | Inhibition (%) |
|---|---|---|
| None | 6.5 | 0 |
| KA | | |
| 20 μM | 4.1 | 37 |
| 100 μM | 1.8 | 72 |

The reaction condition is same to that in Table 8 except the reaction time, 3h.

Table 10. Effect of KA on the AIPD-initiated DNA breakage

| Additives | DNA chain scission (%) | Inhibition (%) |
|---|---|---|
| None | 8.2 | 100 |
| KA | | |
| 20 μM | 6.4 | 78 |
| 100 μM | 3.6 | 44 |

The reaction condition is same to that in Table 9 except the reaction time, 2h.

Table 11. Effect of PA to DNA breakage

| Additives | DNA chain scission (%) | Inhibition (%) |
|---|---|---|
| QA-iron (II) | 5.3 | 100 |
| plus PA, 0.1 mM | 2.6 | 49 |
| PA-iron (II) | 0 | 0 |
| Initiator | 5.4 | 100 |
| plus PA, 30 μM | 5.4 | 100 |

Calf thymus DNA, $2A_{260}$ ; QA-iron complez, 1 mM; PA-iron complex, 1 mM; Ascorbate, 3 mM; 100 mM TAPS, pH 7.6; Temp., 40°; Reaction time, 3h.

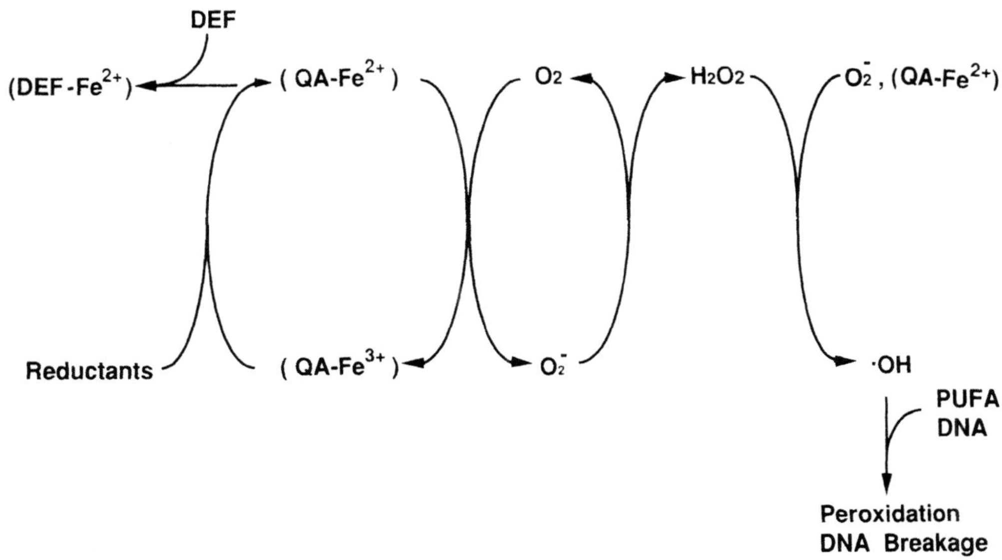


Scheme 2. Formation of active oxygens and induced reactions.

effects on cell components, resulting in damage of membrane structure by lipid peroxidation and others. There is the possibility that the structure change of excitotoxin receptors, might be, occur, and leads to cell death.

## REFERENCES

Barber A.A., 1966, Lipid peroxidation in rat tissue homogenates: Interaction of iron and ascorbic acid as the normal catalytic mechanism, *Lipids* 1:146.

Francois C., Nguyen-Legros J., Pecheron G., 1981, Topographical and cytological localization of iron in rat and monkey brains, *Brain Res.*, 215:317.

Gal E.M. and Sherman A.D., 1978, Synthesis and metabolism of L-kynurenine in rat brain, *J.Neurochem.* 30:607.

Haber F., Weiss J., 1934, The catalytic decomposition of hydrogen peroxide by iron salts, *Proc.Soc.Lond Series A*, 147:332.

Jhamandas K.H. and Boegman R.J., 1994, Quinolinic acid induced brain neurotransmitter deficits: modulation by endogenous excitotoxin antagonists, *Can.J.Physiol.Pharmacol.* 72:1473.

Lapin I.P., 1978, Stimulant and convulsive effect of kynurenines injected into brain ventricles in mice, *J.Neural Trasm.* 42:37.

Morgan A.R. and Pulleyblank D.E., 1974, Native and denaturated DNA, cross-linked and palindromic DNA and circular covalently-closed DNA analysed by a sensitive fluorometric procedure, *Biochem. Biophys. Res.Commun.*, 61: 346.

Nakanishi S., 1992, Molecular deversity of glutamate receptors and implications for brain function, *Science* 258:597.

Schwarcz R., Whetsell W.O., and Mangano R.M., 1983, Quinolinic acid: an endogenous metabolite that produces axon-sparing lesions in rat brain, *Science* 219: 316.

Schwarcz R., Foster A.C., French E.D., Whetsell W.O. and Kohler C., 1984, Excitotoxic models of neurodegenerative disorders, *Life Sci.* 35:19.

Speciale C. and Schwarcz R., 1993, On the production and disposition of quinolinic acid in rat brain slices, *J. Neurochem.* 60:212.

Vladimirov Y.U.A., Olenev V.I., Suslova T.B., 1980, Lipid peroxidation in mitochondrial membrane, *Adv.Lipid Res.*, 17:173.

39

# KYNURENINE AND PTERIDINES IN CHANGES OF MEMBRANE FLUIDITY

V. Rudzite,[1] E. Jurika,[1] D. Fuchs,[2] G. Baier-Bitterlich,[2] and H. Wachter[2]

[1] Laboratory of Biochemistry Latvia Institute of Cardiology
Riga, Latvia
[2] Institute of Medical Chemistry and Biochemistry
University of Innsbruck
Innsbruck, Austria

## INTRODUCTION

Examination of blood serum lipid and kynurenine content in patients with ischemic heart disease and patients with essential hypertension showed that the elevated whole cholesterol and decreased HDL-cholesterol content were not dependent on kynurenine accumulation in blood. The difference was found out in triglycerol and phospholipid amount. Kynurenine accumulation was followed by an increase of triglycerol and a decrease of phospholipid, especially lecitin amount as well as an increased saturated and decreased unsaturated fatty acid content of phospholipids (Rudzite et al.,1988, 1994). The same changes of lipids were also found in white rats with experimentally induced kynurenine accumulation in blood serum (Rudzite and Jurika, 1991). Moreover, kynurenine addition to incubation medium for phospholipid biosynthesis in vitro was followed by an increase of palmitic and oleic acid, a decrease of linoleic and arachidonic acid incorporation into phospholipids. Further studies showed, that similar deviation in fatty acid incorporation into phospholipids was seen after neopterin addition to incubation medium for phospholipid biosynthesis in vitro. The difference was only in oleic acid incorporation: neopterin decreased, while L-kynurenine increased the oleic acid incorporation into phospholipids (Rudzite et al., 1993,1994).

The aim of this study was to investigate the effect of other pteridines to cholesterol and phospholipid content as well as to the process of fatty acid incorporation into phospholipids. In this context some experiments with L- kynurenine and D-erythroneopterin were repeated.

## METHODS

Homogenized liver tissue samples of 32 male Wistar rats in Krebs-Ringer phosphate buffer (pH = 7.4) containing 0.3 % albumin were used for these experiments (Nilsson, 1970,

*Recent Advances in Tryptophan Research,* edited by Graziella Allegri Filippini et al.
Plenum Press, New York, 1996

Rudzite et al. 1993, 1994). Samples were incubated at 37 ° C two times for 4 hours. To investigate kynurenine and different pteridines effect L-kynurenine (4 nmol/g wet weight), erythro - D - neopterin, sepiapterine , 7,8 - dihydrobiopterine, 5,6,7,8 - tetrahydrobiopterin (5 and 30 pmol/g wet weight) and xanthopterin (4 and 20 nmol/g wet weight) were added to experimental test tubes before the first incubation . We have used L-kynurenine sulfate (SERVA) and pteridine derivatives (Schircks Laboratories, Jova Schwitzerland) in these experiments. We thanks prof. H. Wachter , doz. E.Werner and assistent prof. D. Fuchs (Institute of Medical Chemistry and Biochemistry University of Innsbruck) for kindly assistance by handing the pteridine derivatives used by us in these experiments.

All the metabolites were diluted with bidestilled water. The same volume of bidestilled water (0.1 ml) without addition of metabolites was added to control test tubes. After the incubation at 37°C for 4 hours 60 µl of 20% fatty acid solution in chloroform (palmitic, stearic, linoleic and arachidonic acid) in equivalent amount (SIGMA, Münich, Germany) and 20 µl of glycerol were added to all test tubes. Then the test tubes were incubated again at 37°C for 4 hours. After this second incubation 75 ml of chloroform / methanol mixture (2:1) was added to each sample. The samples were mixed and left for 40 min. at room temperature. Then the chloroform/methanol extract was filtered and the filtrate was carefully treated according the method of Folch and others (Folch et al., 1967). The whole cholesterol concentration was determinated by the method of Engelhard and Smirnova (Predtetchenski et al., 1950) and the amount of phospholipids by the method of Urbach - Raabe (Homolka, 1964). The method of Kates (Kates, 1978) was used to prepare the phospholipid samples for fatty acid examination by gas chromatography. Helium was used as a carrier gas and the sorbent was chromosorb covered with DEAG. Separation of fatty acid methylesters were monitored at 210°C, using flame ionisation detection and an adequate flow of helium. The separation was controlled by internal methylester standard substances (SERVA). Statistical analyses were done using Student's t -test. P - values bellow 0.05 were considered to indicate significant differences (Weber , 1957).

## RESULTS

Figure 1 shows the changes of phospholipid biosynthesis after addition of kynurenine (4 nmol/g wet weight).We observed an increase of palmitic acid, oleic acid incorporation

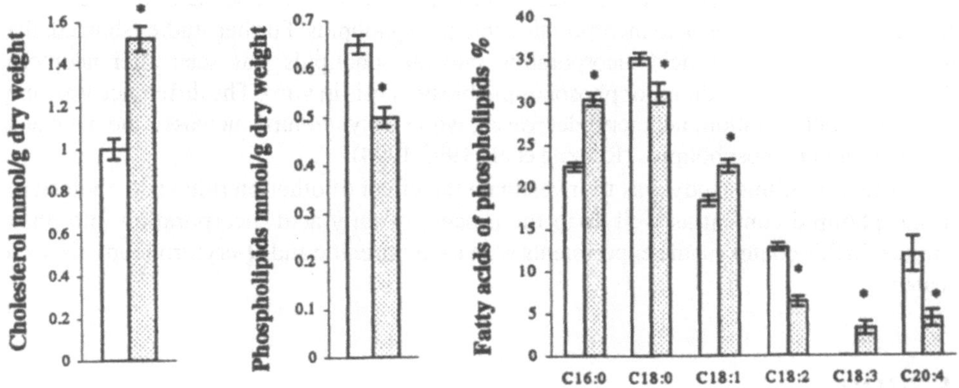


**Figure 1.** Concentrations (mean ± S.D.) of cholesterol and phospholipids as well as fatty acid content (in percent: mean ± S.D.) of phospholipids in controls (open bars) and in samples with 4 nmol/g wet weight (dotted bars) L - kynurenine. Samples of rat liver tissue homogenate were used to test for phospho- lipid biosynthesis in vitro. Incubation medium: Krebs-Ringer-phosphsate buffer (pH=7.4), fatty acid mixture and glycerol. Samples were incubated twice for 4 h at 37°C. * The changes are statistically significant (p<0.001).

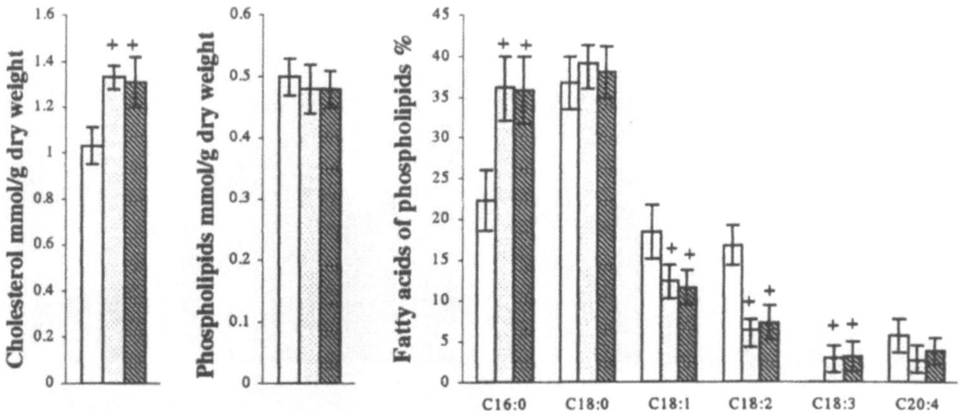


**Figure 2.** Concentrations (mean ± S.D.) of cholesterol and phospholipids as well as fatty acid content (in percent : mean ± S.D.) of phospholipids in controls (open bars) and in samples with 5 pmol/g wet weight (dotted bars) or with 30 pmol/g wet weight (hatched bars) D- neopterin. Samples of rat liver tissue homogenate were used to test for phospholipid biosynthesis in vitro. Incubation medium: Krebs-Ringer-phosphate buffer (pH=7.4), fatty acid mixture and glycerol. Samples were incubated twice for 4 h at 37°C. + The changes are statistically significant ($p<0.01$ - $0.001$).

($p< 0.001$; $p< 0.001$) and decrease of linoleic and arachidonic acid incorporation ($p< 0.05$; $p< 0.001$) into phospholipids. At time of biosynthesis the amount of phospholipids decreased ($p< 0.01$) while the amount of cholesterol increased ($p< 0.001$) in samples containing kynurenine.

Figure 2 points out the changes of phospholipid biosynthesis after addition of neopterin to samples. With both concentrations (5 and 30 pmol/g wet weight) neopterin promotes palmitic acid incorporation into phospholipids ($p < 0.001$) and reduces the incorporation of oleic, linoleic and arachidonic acid ($p<0.01$ - $p<0.001$) into phospholipids. The cholesterol content increased in neopterin treated samples ($p < 0.001$), while the amount of phospholipids was not altered.

Figure 3 shows the changes of phospholipid biosynthesis after addition of sepiapterin (5 and 30 pmol/g wet weight) to samples. We observed a decrease of stearic acid ($p<0.05$) and increase of arachidonic acid ($p<0.001$) incorporation into phospholipids. Chromatogramms of experimental samples showed the appearance of usually observed 5,8,11,14 - arachidonic acid as well as 8,11,14,17 - arachidonic acid, while control samples contained only 5,8,11,14 - arachidonic acid. Addition of sepiapterin did not change the phospholipid content in samples while cholesterol content significantly decreased ($p<0.001$).

Figure 4 points out the deviation of phospholipid biosynthesis after addition of 7,8 - dihydrobiopterin to incubation medium. We observed the same changes as in the cases with sepiapterin : a decrease of stearic acid ($p<0.001$) and an increase of two types of arachidonic acid. (5,8,11,14; $p<0.001$ and 8,11,14,17 - arachidonic acid; $p<0.001$). The cholesterol content in samples containing 7,8 - dihydrobiopterin decreased ($p<0.001$), but the amount of phospholipids remained unchanged.

Figure 5 shows the changes of phospholipid biosynthesis after addition of 5,6,7,8 - tetrahydrobiopterin (5 and 30 pmol/g wet weight) to samples. We have observed the same changes as in the cases with sepiapterin and with 7,8 - dihydro- biopterin: a decrease of stearic acid ($p< 0.05$) and an increase of two types of arachidonic acid (5,8,11,14 - $p<0.001$

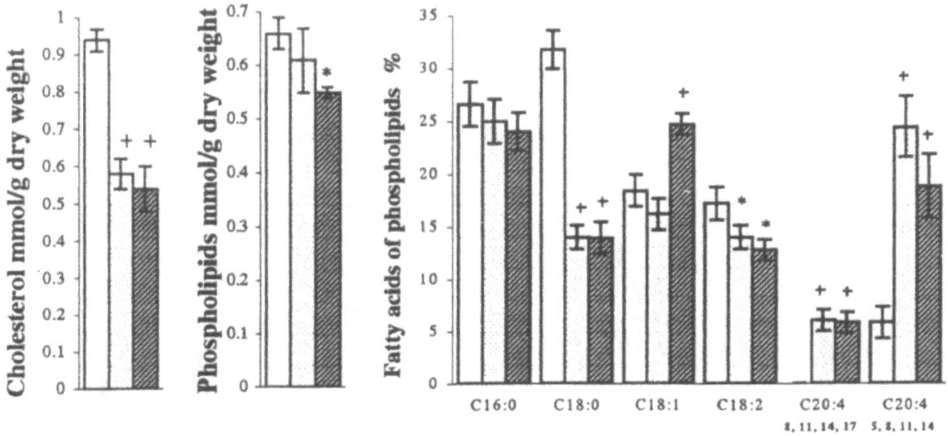


**Figure 3.** Concentrations (mean ± S.D.) of cholesterol and phospholipids as well as fatty acid content (in percent : mean ± S.D.) of phospholipids in controls (open bars) and in samples with 5 pmol/g wet weight (dotted bars) or with 30 pmol/g wet weight (hatched bars) sepiapterin. Samples of rat liver tissue homogenate were used to test for phospholipid biosynthesis in vitro. Incubation medium: Krebs-Ringer-phosphate buffer (pH=7.4), fatty acid mixture and glycerol. Samples were incubated twice for 4 h at 37°C. * The changes are statistically significant ($p<0.05$). + The changes are statistically significant ($p<0.001$).

and 8,11,14,17 - $p<0.001$) into phospholipids. The amount of cholesterol decreased in samples, but the content of phospholipids was not altered.

Figure 6 reflects an influence of xanthopterin (4 and 20 nmol/g wet weight) on phospholipid biosynthesis in vitro. Addition of xanthopterin to samples for phospholipid biosynthesis in vitro altered neither incorporation of fatty acids into phospholipids nor the content of cholesterol and phospholipids.

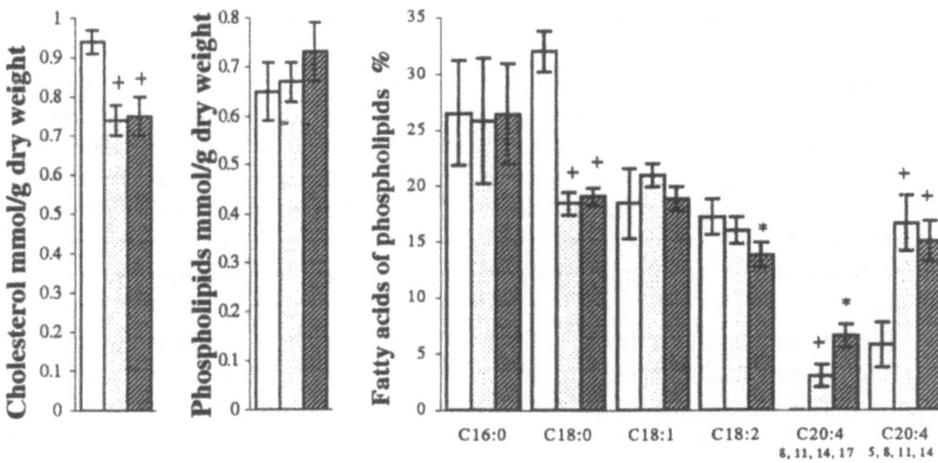


**Figure 4.** Concentrations (mean ± S.D.) of cholesterol and phospholipids, as well as fatty acid content (in percent : mean ± S.D.) of phospholipids in controls (open bars) and in samples with 5 pmol/g wet weight (dotted bars) or with 30 pmol/g wet weight (hatched bars) 7,8-dihydrobiopterin. Samples of liver tissue homogenate were used to test for phospholipid biosynthesis in vitro. Incubation medium: Krebs-Ringer-phosphate buffer (pH=7.4), fatty acid mixture and glycerol. Samples were incubated twice for 4 h at 37°C. * The changes are statistically significant ($p<0.05$). + The changes are statistically significant ($p<0.001$).

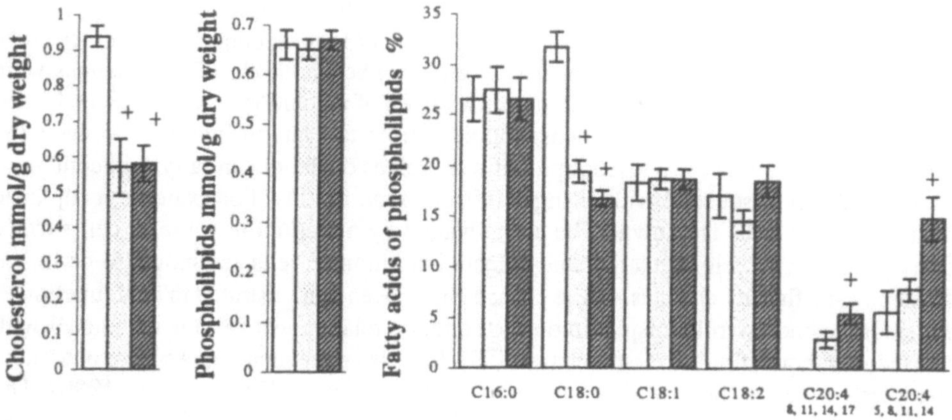


**Figure 5.** Concentrations (mean ± S.D.) of cholesterol and phospholipids, as well as fatty acid content (in percent : mean ± S.D.) of phospholipids in controls (open bars) and in samples with 5 pmol/g wet weight (dotted bars) or with 30 pmol/g wet weight (hatched bars) 5,6,7,8- tetrahydrobiopterin. Samples of rat liver tissue homogenate were used to test phospholipid biosynthesis in vitro. Incubation medium: Krebs-Ringer-phosphate buffer (pH=7.4), fatty acid mixture and glycerol. Samples were incubated twice for 4 h at 37°C. + The changes are statistically significant ($p<0.001$).

## DISCUSSION AND CONCLUSIONS

We have observed an increase of saturated fatty acids (palmitic acid) and a decrease of polyunsaturated fatty acid (linolic, arachidonic acid) incorporation into phospholipids after the addition of either L-kynurenine or D-neopterin to samples. L-kynurenine promoted, while D-neopterin reduced oleic acid incorporation into phospholipids. The cholesterol content elevated in samples contained either kynurenine or neopterin. Otherwise, sepiap-

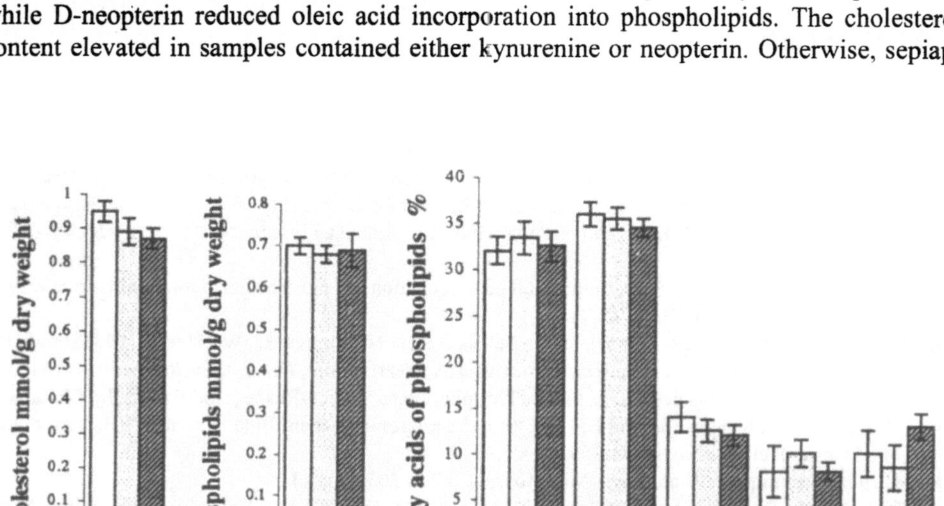


**Figure 6.** Concentration (mean ± S.D.) of cholesterol and phospholipids, as well as fatty acid content (in percent: mean ± S.D.) of phospholipids in controls (open bars) and in samples with 4 nmol/g wet weight (dotted bars) or with 20 nmol/g wet weight (hatched bars) xanthopterin. Samples of rat liver tissue homogenate were used to test for phospholipid biosynthesis in vitro. Incubation medium: Krebs - Ringer - phosphate buffer (pH=7.4), fatty acid mixture and glycerol. Samples were incubated twice for 4 h at 37°C.

terin, 7,8 - dihydrobiopterin and 5,6,7,8 - tetrahydrobiopterin induced contrary effect both to cholesterol content and to fatty acid incorporation into phospholipids due to phospholipid biosynthesis in vitro: we have observed a decrease of cholesterol content in samples and a decrease of saturated (stearic) acid incorporation into phospholipids. These changes were followed by especially high incorporation of two tipes of arachidonic acids (5,8,11,14 and 8,11,14,17 - arachidonic acid) into phospholipids. From the data of literature we have the information that cholesterol/phospholipid ratio as well as deviations in fatty acid content of phospholipids are associated with changes in membrane fluidity. For example, neoplastic transformation of cells is known to be accompanied by a marked increase in cell surface fluidity. Both - cell cycle regulation and contact inhibition of cells are shown to be linked with membrane fluidity changes. These changes have been demonstrated to be a function of the membrane cholesterol/phospholipid ratio so that a relative low ratio is associated with increased membrane fluidity (Ostro et al., 1980). It is also dependent on cholesterol content in samples as well as fatty acid composition of phospholipids (Ferber et al., 1980). The increased cholesterol content in membranes and decreased polyunsaturated fatty acid content of membrane phospholipids have been observed in patients with arterial hypertension, myocardial infarction as well as at aging (Cloete et al., 1993, Montila et al.,1993, Munitz et al.,1993, Olivieri et al., 1993).

Moreover,such diseases as arterial hypertension (Rudzite et al., 1987, 1994), cardiomyophaty (Fuchs et al., 1993, Rudzite et al., 1987,1994, Samsonov et al., 1992), atherosclerosis (Tatzber et al., 1991, Weiss et al., 1994) as well as aging (Reibnegger et al., 1988) are accompanied with elevated kynurenine and neopterin level in blood serum. Therefore, the results received in our experiments allow us to suggest that kynurenine and neopterin depress, while 5,6,7,8 - tetra - hydrobiopterin and its precursors (sepiapterin, 7,8 - dihydrobiopterin) stimulate the cell cycle both by changing cholesterol / phospholipid ratio and fatty acid incorporation into phospholipids .

## REFERENCES

Cloete F., de Waal J.M. , Badenhorst P.W., Kelderman M. 1993,Abnormal fatty acid - composition of plasma and platelets in patients with acute myocardial infarction. In: Abstracts of 1st International Congress "Fatty Acids and Lipids from Cell Biology to Human Disease. Lugano, Switzerland, p.109.

Ferber E., Kröner E., Schmidt B., Ficher H., Peskar B,A., Anders C. 1980, Dynamics of membrane fatty acids during lymphocyte stimulation by mitogens. In: Membrane Fluidity .Biophysical Techniques and Cellular Regulation. M. Kates and A. Kuksis Eds. The Humana Press Inc. Clifton, New Jersey. p. 239 - 263.

Folch J., Lees M., Stanley G.H.1957, A simple method for isolation and purification of total lipids from animal tissues , *J. Biol Chem.* 226: 495.

Fuchs D., Samsonov M., Titz G.P., Reibnegger G., Belenkov J.N., Nassonov E.L., Wachter H. 1993, Stimulated cellular immune systems in patients with congestive heart failure, *Eur. J. Clin. Biochemistry*, 31: 111

Homolka J., 1964, Bestimmung des anorganischen Phosphors nach Urbach-Raabe (modifiziert). In: Chemische Diagnostik im Kindersalter mit Bevorzugung der mikrometrischen Blutanalysen. WEB, Verlag Volk und Gesundheit , Berlin, p.354-355.

Kates M. 1978, Technique of Lipidology. Mir, Moscow. p.201-203. (Russ.)

Montilla C., Muniz O., Villar J., Ruiz-Gutierrez V., Stiefer P., Serreras J.L. 1993, Erythrocyte membrane cholesterol and phospholipids composition in hypertensives vs normotensives. In: Abstracts of 1st International Congress" Fatty acids and Lipids from Cell Biology to Human Disease" Lugano, Switzerland, p.116.

Muniz O., Montilla C., Villar J., Ruiz-Guitierrez V., Stiefer P., Serreras J.L., 1993, Erythrocute membrane fatty acids composition in hypertensives vs normotensives. Abstracts of 1st International Congress "Fatty Acids and Lipids From Cell Biology to Human Disease" Lugano, Switzerland, p.116.

Nilsson S., 1970, Synthesis and Secretion of biliary phospholipids by man, *Acta Chirurg. Scand.* Suppl.405: 1.

Olivieri O.,Stanzial A.M.,Girelli D.,Trevisan M.T.,Guarini P., Terzi M., Caffi S., Fontana F., Casaril M ., Ferrari S., Corrocher R., 1993 , Selenium, status fatty acids, antioksidant vitamins and aging. The nove study. In: Abstract of 1st International Congress"Fatty Acids and Lipids from Cell Biology to Human Disease "Lugano, Switzerland, p.136.

Ostro M.J., Bessinger B., Summers J., Dray S., 1980, Effect of membrane lipid composition on mobility of lymphocyte surface immunoglobulins. In: Membrane Fluidity . Biochemical Techniques and Cellular Regulation. M. Kates and A Kuksis Eds.The Humana Press Inc.Clifton, New Jersey, p.105-117.

Predtetchenski V.J., Borovskaja V.M., Margolina L.T. Handbook of Clinical and Laboratory Investigation, Medgiz, Moscow. (Russ.)

Reibnegger G., Huber L.A., Jürgens G., Schönitzer D., Werner E.R., Wachter H., Wick G., Traili K. N., 1988, Approach to define "normal aging" in man. Immune function , serum lipids , lipoproteins and neopterin levels. *Mech . Aging Developm.*, 46: 67.

Rudzite V., Jurika E.,Arajs J.,Andrejev N., 1987,The relationship between kynurenine and arterial blood pressure. In: Progress in Tryptophan and Serotonin Research 1986. Bender D.A., Joseph M.H., Kochen W., Steinhart H. Eds. Walter de Gruyter & Co, Berlin - New York, p.127 - 130.

Rudzite V., Jirgensons J., Jurika E., Sileniece G., Zirne R., Jirgensone S., 1988, Besonderheiten der Nikotinsäurebildung bei koronarer Herzkrankheit unter Berücksichtigung von Herzrhythmusstörungen. *Z. gesamte inn. Med* .,43: 60

Rudzite V., Jurika E., 1991,Kynurenine and Lipid Metabolism. *Advances in Expe rimental Medicine and Biology.* 294: 463.

Rudzite V., Sileniece G., Liepina D., Dalmane A., 1991 ,Impairment of kynurenine metabolism in cardiovascular disease. *Advances in Experimental Medicine and Biology*, 294: 663.

Rudzite V ., Skards J.I., Fuchs D ., Reibnegger G., Wachter H., 1992 , Serum kynurenine and neopterin concentrations in patients with cardiomyopathy. *Immunol. Lett.* 32 : 125.

Rudzite V., Jurika E., Reibnegger G., Weiss G., Wachter H., 1993, Influence of Kynurenine , Neopterin , Noradrenaline and Pyridoxal - 5 - Phosphate on Cholesterol Content and Phospholipid Biosynthesis in vitro. *Pteridines*, 4: 126

Rudzite V., Jurika E., Fuchs D., Reibnegger G., Wachter H., 1994, The influence of kynurenine, neopterin and pyridoxal -5-phosphate /P-5-P/ on cholesterol content and phospholipid biosynthesis in vitro. In: L-Tryptophan Current Prospects in Medicine and Drug Safety. Kochen W. and Steinhart H. Eds. Walter de Gruyter, Berlin - New York, 165-170.

RudziteV., Sileniece G.,Jirgensons J., Dalmane A., Zirne R., Jurika E., Skards J.I., 1994, The significance of increased kynurenine levels in blood to the development of cardiovascular pathology. *Acta Medica Baltica*, 1: 4.

Samsonov M.,Fuchs D.,Reibnegger G.,Belenkov J.N.,Nassonov E.L.,Wachter H., 1992, Patterns of serological markers for cellular immune activation in patients with dilated cardiomyopathy and chronic myocarditis. *Clin. Chem* ., 38 : 678.

Tatzber F.,Rabl H.,Koriska K.,Erhart U., Puhl H.,Waeg G., Krebs A., Esterbauer H., 1991,Elevated serum neopterin levels in atherosclerosis.*Atherosclerosis* 89: 203.

Weber E.,1957, Grundri der biologischen Statistik für Naturwissenschaftler, Landwirte und Mediziner. Gustav Fischer Verlag, Jena.

Weiss G.,Fuchs D.,Hausen A.,Reibnegger G., Werner E.R., Werner-Felmayer G., Semenitz E ., Dierich M.P., Wachter H., 1993, Neopterin modulates toxicity mediated by reactive oxygen and chloride species. *FEBS Lett.*, 89.

# THE KYNURENINES AND THE SEIZURES

V. I. Guzeva, I. B. Mikhailov, and N. V. Melnikova

St. Peterburg State Pediatric Medical Academy
Russia

At present, patients with epilepsy are resistant against all antiepileptic drugs in 15-25 per cent of cases. Hence the search for new pathogenetically sound approaches to treatment of this disease. Different neuroactive substances have been studied, of them kynurenines attracting a particular attention.

The objective of the present work involves studying of the influence of quinolinic acid (QUIN), anthranilic acid (AA), kynurenic acid (KYNA) kynurenine (KYN) upon the activity of experimental epileptogenic foci (EF), i.e. upon the function of neurons with pathologically high excitability.

Chronic experiments were carried out with implanting electrochemotrodes into the hippocampus. With the help of micromanipulator, an EF was created with Benzylpenicillin-natrium injection (100 units in 1 mcl). 5 min prior to the injection, solutions of kynurenines had been injected into the EF region (KYNA 1, 5, 10, 20 mcg, i.c., QUIN 1 and 10 mcg, i.c., AA 20 mcg, i.c., and KYN 10 mcg, i.c.).

The EEG control of pathological activity was used. After the penicillin injection, the EEG was recorded during 120 min in order to count the number of interictal epileptogenic discharges (IED) per 1-min epoch of the EEG recording at 10-min intervals, and the average number of electrographic seizure correlates (ESC) at 10-min intervals.

The 1st experiment with the EEG recording was conducted on the 6th day after the operation, with subsequent experiments in the same rat at 3-day intervals.

Instead of kynurenines solutions, control rats were injected with the distilled water which had served as a vehicle for Benzylpenicillin-natrium.

The results were statistically processed with the Mann-Lithy "U" criterion.

In all, 73 experiments were performed in white non-linear male rats of 120-170 g body weight, 42 of them with KYNA, 15 with QUIN, 5 with AA and 10 with KYN.

Before the operation, all rats were anesthetised with Nembutal (50 mg/kg, i.p.).

## RESULTS

The data presented in the Table 1 show that the preliminary injection KYNA into the EF region results in lowering the intensity of the IED and ESC. THE KYNA injection in all

*Recent Advances in Tryptophan Research,* edited by Graziella Allegri Filippini et al.
Plenum Press, New York, 1996

**Table 1.** The change of the EF activity

| Injected substance and its dose, control | | Number of experiments | In per cent with respect to results of penicillin injection into the hippocampus | |
|---|---|---|---|---|
| | | | Intensity of the IED | Intensity of the ESC |
| Distilled water | (control) | 16 | 90.5* | 221.9* |
| KYNA | 1 | 11 | 26.8* | 66.2* |
| | 5 | 10 | 10.4* | 65.8* |
| | 10 | 12 | 2.3* | 5.6* |
| | 20 | 7 | 0* | 0* |
| QUIN | 1 | 6 | 120.0 | 563.0 |
| | 10 | 9 | 209.0** | 1157.6** |
| KYN | 10 | 5 | 94.4 | 308.0 |
| AA | 20 | 10 | 213.3** | 471.1 |

*Significance of differences $p<0.01$ with respect to control.
**$p<0.05$. The intensity of pathological activity was measured during the 2nd-6th creation of the EF in the same point of the hippocampus.

the doses induced slow diffuse theta-rhythm in the EEG. Antiepileptic drugs are known to cause the same effect sometimes.

AA did not affect the EF activity, QUIN and KYN produced an effect of activation.

KYNA seems to play a protective role in the pathogenesis of epilepsy. We were interested therefore, whether the injection of KYNA changes the latency of the seizures (i.e. the time elapsing from the injection of penicillin into the hippocampus until the occurrence of the first IED (Table 2). The latency was found to be proportional to an increase in the KYNA dose and to be always a longer one than in the control.

The data obtained show a direct depressing effect of the KYNA upon the excitability of the CNS neurons, and an activating effect of the QUIN and KYN. On the basis of our previous research, we suggested allopurinol, the inhibitor of tryptophan-pyrrolase, for shifting the tryptophan intermediate metabolism towards the serotonine way by means of inhibiting the kynurenine way. Serotonine inhibits the pathological activity of the EF. However, our present data show the problem to be more complicated.

Apparently, not only an absolute contents of the kynurenines is important, but their ratio as well.

A search for new pathogenetic approaches to treatment of epilepsy is the most promising when considering the balance among different kynurenines, endogenous convulsants and anticonvulsants.

**Table 2.** The latency

| Injected substance and its dose, mcg | The latency in min |
|---|---|
| Distilled water (control) | 1 |
| KYNA | |
| 1 | 8 |
| 5 | 15 |
| 10 | 52 |
| 20 | |

# SYSTEMIC ADMINISTRATION OF KAINATE INDUCES MARKED INCREASES OF ENDOGENOUS KYNURENATE IN VARIOUS BRAIN REGIONS AND PLASMA OF RATS

H. Baran, M. Gramer, D. Hönack, and W. Löscher

Department of Pharmacology, Toxicology, and Pharmacy
School of Veterinary Medicine
Bünteweg 17, 30559 Hannover, Germany

The brain metabolite kynurenic acid (KYNA, Moroni et al., 1988 J.Neurochem. 51, 177), is a broad spectrum antagonist of excitatory amino acid receptors, can antagonize the strychnine-insensitive glycine site associated with the N-methyl-D-aspartate receptor (Carter 1992 Drug Future, 595; Stone, 1993, Pharmacol. Rev. 45, 309) and has neuroprotective and anticonvulsant properties (Foster et al., 1984 Neurosci. Lett. 48, 273; Andine et al., ,1988 Neurosci. Lett., 90, 208). KYNA has been speculatively linked to the pathogenesis of epilepsy and several other brain disorders (Schwarcz et al., 1992 Ann. N.Y. Acad. Sci., 648, 140; Stone 1993 Pharmacol Rev. 45, 309).

In the present study, alterations in KYNA levels were examined in the kainic acid (KA) epileptic model of temporal lobe epilepsy. KA was sistematically injected in female Wistar rats, weiging 210-230 g, at a dose (10 mg/kg) which induces a wet dog shakes (WDS), characteristic behavioural syndrome with stereotypies and focal (limbic) and generalized seizures, eventually progressing into seveve status epilepticus. KYNA was determined 3 h after injection in various brain regions (pons/medulla, olfactory bulb, hippocampus, frontal cortex, nucleus accumbens, cerebellar cortex, piriform cortex, thalamus, superior and inferior colliculus, amygdala, caudate/putamen) and plasma, using a sensitive HPLC method (Shibata, 1988 J.Chromatogr. 430, 376; Turski et al., 1988 Brain Res. 454, 164). In control, KYNA levels showed an uneven regional distribution in the brain, between 1-5 pmol/g wet weight tissue. In plasma, KYNA levels were more than 10 times higher (95 pmol/ml) than in brain regions. KA induced symptoms were accompanied by marked increases of KYNA in the brain and plasma. When data were analysed irrespective of individual seizures severity, significant increases in KYNA were determined in plasma and all brain regions examined except hippocampus, nucleus accumbens and pons/medulla. The most marked (300-600% of control) increases were seen in pyriform cortex, amygdala, and cereberal cortex. In plasma, levels of KYNA increased by 210% above controls. When KYNA levels were determined in subgroups of rats with different behavioural alterations in response to KA, the

*Recent Advances in Tryptophan Research*, edited by Graziella Allegri Filippini et al.
Plenum Press, New York, 1996

most marked KYNA increases in the brain regions and in the plasma were seen in rats with status epilepticus, but also in rats which developed mild (focal) seizures or stereotype behaviours, WDS. In plasma, the highest KYNA levels (415% above control) were determined in rats with status epilepticus, whereas in animals with focal seizures or WDS were 131% and 55%.

1. Present data demonstrate that systemic administration of KA (10mg/kg) markedly enhances endogenous levels of KYNA in several brain regions and plasma.
2. The extent of KYNA increase seemed to relate to seizure severity: the increase in KYNA was not only due to severe seizures but also due to focal convulsions and WDS.
3. Our data suggest that plasma KYNA determination may be used to indirectly monitor alterations in the brain KYNA levels in KA epileptic model.

## ACKNOWLEDGMENT

Supported by the Deutsche Fund (Lo 274/2-3) and by Austrian Fund (Charlotte Bühler Habilitationsstipendium No HO17-MED and H00-43 to H.Baran (Present address: Department of Pediatrics, University of Vienna, Währinger Gürtel 18-19, 1090 Vienna, Austria).

# 42

# EFFECT OF ROUTE OF ADMINISTRATION ON THE METABOLISM OF QUINOLINIC ACID-6-$^{*14}$C IN ANIMALS

L. V. Hankes and M. Schmaeler

Med. Res. Center
Brookhaven Natl. Lab.
Upton, New York 11973-5000

## INTRODUCTION

After a study of quinolinic acid metabolism in patients with pellagra (Hankes et al., 1992), a study was done to determine the effect of route of administration of the labeled compound on its metabolism in an animal species which had an intestinal synthesis in a ceacum.

## EXPERIMENTAL

The quinolinic acid-6-$^{*14}$C used in the study was from the same batch that had been used in the study with patients with pellagra ((Hankes et al., 1992). Rats were interperitoneally injected or stomach tubed with measured amounts of a solution of the labeled compound. The animals were housed in a metabolics system arranged to continuously collect breath and urine. Quinolinic acid, picolinic acid and nicotinic acid were isolated from the urines by methods used in other studies ((Hankes et al., 1977). At the end of the studies, tissues were taken from the animals and they were analyzed for total $^{14}C$ activity (Mahin and Lofberg, 1966).

## RESULTS AND COMMENTS

Table 1 shows that at 12 hrs the interperitoneally injected groups expired an average 2.45% activity in the $CO_2$. The first 12 hr urines contained 75.62% of the dose activity with an additional 5% coming out in the second 12 hr period. The stomach-tubed animals expired relatively no activity in the first 3 hrs and then a rapid production of $^{14}CO_2$ beginning at the 4.5 hour period. The activity level then jumped from 4.5% to 3818% over the next 7.5%/hrs (4.5%/hr). The high activity in the early urine of the injected group was probably due to the

*Recent Advances in Tryptophan Research,* edited by Graziella Allegri Filippini et al.
Plenum Press, New York, 1996

**Table 1.** Percent excretion of $^{14}C$ from quinolinic acid-6-$^{14}C$ in $^{14}CO_2$ and urine of rats

| | | $CO_2$ Time intervals | | | | | | Urine Time intervals | |
|---|---|---|---|---|---|---|---|---|---|
| Group number | Animals | 1.5 | 3.0 | 4.5 | 6.0 | 9.0 | 12 | 0–12 | 12–24 |
| Intraperitoneally injected | | | | | | | | | |
| 112 | two | 0.02 | 0.05 | 0.14 | 0.43 | 1.11 | 1.33 | 80.16 | 4.95 |
| 117 | two | 0.012 | 0.037 | 0.30 | 1.23 | 2.92 | 3.57 | 71.08 | 4.59 |
| Average | | 0.016 | 0.044 | 0.22 | 0.83 | 2.02 | 2.45 | 75.62 | 4.77 |
| Stomach tubed | | | | | | | | | |
| 119 | three | 0.00 | 1.25 | 7.18 | 17.25 | 40.48 | 53.47 | 5.62 | 3.27 |
| 121 | three | 0.00 | 0.48 | 3.81 | 8.44 | 19.93 | 32.35 | 6.03 | 3.61 |
| 122 | two | 0.00 | 0.14 | 1.69 | 4.26 | 11.61 | 24.00 | 8.49 | 11.40 |
| Average*+ | | 0.00 | 0.68 | 4.54 | 10.70 | 25.56 | 38.18 | 6.49 | 5.43 |

*Added the sum of (119 x 3) plus (121 x 3) and 122 x 2) for total values.
+Divided the total values by 8 for averages of Stomach-tubed.

rapid absorption in the tissues and excretion in the urine. The latent high level of activity in the $CO_2$ of the stomach-tubed group was probably due to the action of the bacterial colonies in the ceacum of the rats breaking down the quinolinic acid molecule.

To determine which compounds of the tryptophan-niacin pathway contained $^{14}C$ activity, quinolinic acid, nicotinic acid and picolinic acid were isolated from the urines of groups 112 and 121 by carrier techniques. The data of Table 2 shows that in the interperitoneally injected group 77.2% of the dose came out in the quinolinic acid in the first 12 hrs, while in the stomach-tubed group 4.99% of the dose came out as quinolinic acid in the same period of time, with only an additional 2.65% coming out in the 2nd 12 hr urine for a total of 7.64% of the dose. The 24 to 53.4% of the dose appearing in the $CO_2$ of the stomach-tubed

**Table 2.** Route of administration effect on metabolism of quinolinic acid-6-$^{14}C$ in the rat

| | | Group | | | |
|---|---|---|---|---|---|
| | | 112 | | 121 | |
| No. of animals | | Two | | Three | |
| Route | | Intraperitoneally | | Stomach tube | |
| Carbon-14 administered μCi | (a) | 0.5588 | | 0.6171 | |
| | | Urine collection period | | | |
| | | 1st 12hrs | 2nd 12hrs | 1st 12hrs | 2nd 12hrs |
| Carbon-14 excreted μCi | (b) | 0.4479 | 0.0277 | 0.0372 | 0.0223 |
| Quinolinic acid | | | | | |
| $^{14}C$ excreted in QA μCi | (c) | 0.4318 | 0.0192 | 0.0308 | 0.0164 |
| Percent of dose | (c/a) | 77.27 | 3.44 | 4.99 | 2.65 |
| Percent urinary | (c/b) | 96.40 | 69.39 | 82.76 | 73.50 |
| Nicotinic acid | | | | | |
| $^{14}C$ excreted in NA μCi | (d) | 0.0026 | 0.0013 | 0.0004 | 0.0006 |
| Percent of dose | (d/a) | 0.46 | 0.23 | 0.06 | 0.10 |
| Percent of urinary | (d/b) | 0.58 | 4.67 | 0.97 | 2.76 |
| Picolinic acid | | | | | |
| $^{14}C$ excreted in PA μCi | (e) | 0.0001 | <0.0001 | 0.0003 | <0.0001 |
| Percent of dose | (e/a) | 0.02 | <0.01 | 0.05 | 0.01 |
| Percent of urinary | (e/b) | 0.02 | 0.04 | 0.80 | 0.17 |

**Table 3.** Percent carbon-14 in rat tissues 24 hours after administration of quinolinic acid-6-$^{14}$C

| Group | Route | Dose (μc) | Kidney | Liver | Spleen | Heart | Testes | Brain | Serum | Cells |
|---|---|---|---|---|---|---|---|---|---|---|
| 112A | I.P. | .2444 | 7.61 | 3.03 | 0.12 | 0.09 | 0.13 | 0.03 | 0.00 | 0.00 |
| 112B | I.P. | .3139 | 5.83 | 2.19 | 0.07 | 0.09 | 0.20 | 0.03 | 0.00 | 0.00 |
| 117A | I.P. | .3515 | 5.83 | 2.68 | 0.06 | 0.09 | 0.00 | 0.05 | 0.00 | 0.00 |
| 117B | | .3186 | 6.10 | 3.12 | 0.10 | 0.08 | 0.00 | 0.02 | 0.00 | — |
| Ave | | | 6.34 | 2.76 | 0.09 | 0.09 | 0.08 | 0.03 | 0.00 | 0.00 |
| ±S.D. | S.E. | | 0.43 | 0.21 | 0.01 | 0.00 | 0.05 | 0.01 | 0.00 | 0.00 |
| 119A | Oral | .1783 | 0.59 | 0.46 | 0.02 | 0.02 | 0.06 | 0.04 | 0.27 | 0.00 |
| 119B | Oral | .2172 | 0.89 | 0.79 | 0.02 | 0.01 | 0.05 | 0.01 | 0.05 | 0.00 |
| 119C | Oral | .1899 | 0.66 | 0.31 | 0.01 | 0.00 | 0.00 | 0.00 | 0.00 | 0.00 |
| 121A | Oral | .1967 | 0.42 | 0.30 | 0.01 | 0.00 | 0.00 | 0.01 | 0.00 | 0.00 |
| 121B | Oral | .2156 | 0.79 | 0.52 | 0.01 | 0.00 | 0.00 | 0.00 | 0.00 | 0.00 |
| 121C | Oral | .2093 | 0.61 | 0.93 | 0.02 | 0.01 | 0.42 | 0.03 | 0.00 | 0.00 |
| 122B | Oral | .2119 | 0.57 | 0.96 | 0.02 | 0.01 | 0.00 | 0.03 | 0.98 | 0.00 |
| 122C | Oral | .2219 | 0.67 | 0.61 | 0.01 | 0.00 | 0.00 | 0.00 | 0.41 | 0.00 |
| Ave | | | 0.65 | 0.61 | 0.015 | 0.006 | 0.07 | 0.015 | 0.21 | 0.00 |
| ±S.D. | S.E. | | 0.05 | 0.09 | 0.002 | 0.003 | 0.05 | 0.006 | 0.12 | 0.00 |

group and just an additional 7.64% coming out in the urine suggests that the dose must have been partially bound in the intestinal contents. Very little quinolinic acid-6-$^{14}C$ was decarboxylated to form nicotinic acid and none was converted into picolinic acid.

The tissue activity data in Table 3 suggests that very little of the activity from the quinolinic acid-6-$^{14}C$ crosses over the blood-brain barrier into the brain, when it is administered by either route. Although large quantities of quinolinic acid are produced in the body after a 2 gr. load of L-tryptophan, little, if any crosses the blood-brain barrier.

After a study of quinolinic acid-6-14C metabolism in pellagra patients, a study was done with rats given the compound in solution by stomach tube (ST) or by interperitoneal injection (IP). The breath was controlled continuously for 12 hours and the urines for two 24 hour periods. The IP group showed about 2.5% $^{14}C$ activity in the $CO_2$ of the first 12 hours, 75% in the first 12 hour urine and 4.8% in the 2$^{nd}$ 12 hour urine. The ST group showed 38% $^{14}C$ activity in the $CO_2$ of the 1$^{st}$ 12 hours, 6.5% in the 1$^{st}$ 12 hr urine and 5.4% in the 2$^{nd}$ 12 hr urine. The 1$^{st}$ 12 hr urine had 77% of the dose in quinolinic acid (QA), 0.46% in nicotinic acid (NA) and 0.02% in picolinic acid (PA) from the IP group; and 5% of the dose in QA, 0.06 in NA and 0.05% in PA from the ST group. The 2$^{nd}$ 12 hour urine showed 3.44% in QA, 0.23% in NA and 0.17% in PA from the IP group; and 2.65% in QA, 0.1% in NA and 0.01% in PA from the ST group. The tissues of the rats were taken at 24 hours and the total $^{14}C$ activity present determined by scintillation counting, after perchloric acid $H_2O_2$ digestion. In the IP group the kidney collected 6.34%, liver 2.76%, spleen 0.17%, heart 0.09%, testes 0.08%, brain 0.03%, serum 0.00% and cells 0.00% of the dose. In the ST group, the kidney collected 0.65%, liver 0.61%, spleen 0.015%, heart 0.006%, testes 0.07%, brain 0.015%, serum 0.21% and cells 0.00% of the dose.

## CONCLUSIONS

1. Very little QA-6-$^{14}C$ activity crosses the blood-brain barrier.
2. High activity present in the early urine of the IP group was due to the rapid absorption of the compound into the tissues and excretion into the urine.
3. The high level of activity in the breath of the ST group due to the activity of bacterial colonies in the ceacums of the rats breaking down the QA-6-$^{14}C$ molecule.
4. Very little QA was decarboxylated to form NA.
5. None of the dose was converted into PA.

## ACKNOWLEDGMENTS

The authors wish to thank Jean Hani and Catherine A. Schindewolf of the Institut für Biochemie, Genzentrum der Ludwig-Maximillian Universität, Würmtalstr. 221, München, Germany for assistance in the preparation of the article.

Supported by U.S. Dept. of Energy under Contract DE-AC02-76CH00016

## REFERENCES

Hankes, L.V., De Bruin, E., Jansen, C. R., Vorster, L., and Schmaeler, M., 1977, *South African Med. Jour.* 51:383.

Hankes, L.V., Hugo, N., Jansen, C. R., and Schmaleler, M., 1992, Metabolism of quinolinic acid-6-$^{14}$C in Bantu with pellagra, in: Advances in Tryptophan Research, Ishiguro I., Kido R., Nagatsu T., Nagamura Y., Ohta Y., eds., Fujita Health University Press, Toyoake, pp 269-272.

Mahin, D.T., and Lofberg, R.T., 1966, A simplified method of sample preparation for determination of tritium, carbon-14, or sulfur-35 in blood or tissue by liquid scintillation counting, *Anal. Biochem.* 16: 500.

# MODULATION OF EXTRACELLULAR KYNURENIC ACID CONTENT BY EXCITATORY AMINO ACIDS IN PRIMARY CULTURES OF RAT ASTROCYTES

L. Curatolo,[1] C. Caccia,[1] C. Speciale,[1] L. Raimondi,[1] M. Cini,[1] M. Marconi,[1] A. Molinari,[1] and R. Schwarcz[2]

[1] Pharmacia, R&D-CNS
20014 Nerviano, Milano, Italy
[2] Maryland Psychiatric Research Center
University of Maryland School of Medicine
Baltimore Maryland 21228

## 1. INTRODUCTION

Kynurenines, i.e. tryptophan metabolites along the kynurenine pathway, have recently received attention because of their possible role in the immunocompromised brain and in the pathogenesis of human neurodegenerative diseases (Stone, 1993). In particular, high levels of the endogenous excitotoxin quinolinic acid and low levels of the excitatory amino acid (EAAs) antagonist kynurenate (KYNA) have been speculatively linked to the occurrence of seizures and nerve cell death.

Since a decreased formation of KYNA, which has a particularly high affinity for the glycine co-agonist site of the N-methyl-D-aspartate (NMDA) receptor, could favor excitotoxic damage (Schwarcz et al., 1992), several attempts have been made to examine the production and possible function of KYNA in the normal and impaired brain.

Immunocytochemical studies have localised the majority of KYNA's biosynthetic enzyme kynurenine aminotransferase (KAT) to astrocytes (Du et al., 1992). Biochemical experiments have demonstrated an increased formation of KYNA in brain tissue which shows astrogliosis secondary to nerve cell loss (Turski et al., 1989). These data suggest that in vivo most KYNA in the brain derives from astrocytes.

In preliminary experiments performed in rat brain tissue slices treated with KYNA's bioprecursor L-kynurenine (L-KYN) we had noted the ability of certain EAAs to decrease the amount of KYNA recovered from the incubation medium (Gramsbergen et al., 1989). Additional studies suggested that events related to neuron-glia interactions may regulate the disposition of KYNA in the brain's extracellular compartement. Thus, veratridine, a commonly used tool for probing $Na^{+}$ channels, significantly decreased KYNA liberation into the

*Recent Advances in Tryptophan Research,* edited by Graziella Allegri Filippini et al.
Plenum Press, New York, 1996

extracellular milieu in intact tissue but not after experimental neuronal ablation (Wu et al., 1992). In order to further delineate the likely role of astrocytes in the production and release of KYNA, we now used primary astrocyte cultures to examine the ability of EAAs and veratridine to modulate KYNA disposition in vitro.

## 2. METHODS

### 2.1 Preparation of Enriched Rat Cortical Astrocytic Cultures

Brains from 1-2 day old Sprague-Dawley rats were used to prepare primary astrocyte cultures. After cortex isolation and peeling off of the meninges, the tissue was mechanically dissociated in order to obtain a cell suspension. Cells were vitally counted and plated at $1\text{-}1.4\times10^5$ cells/cm$^2$ in T175 pre-coated plastic flasks. The average yield was $10\text{-}15\times10^6$ cells/cortex. After primary seeding, the cells were incubated at 37°C in 5% $CO_2$ for at least two days without disturbance to avoid clump formation. Confluent monolayers were split every 7 days. Culture medium (Dulbecco's Modified Eagle Medium) supplemented with 33 mM glucose, 2 mM glutamine and 10% fetal bovine serum was changed every 2-3 days. The purity of the astrocyte population was checked by glial fibrillary acidic protein immunostaining. All experiments were carried out using Type-1 astrocytes completely differentiated after two transplantations (21 DIV).

### 2.2 KYNA Assay

After removal of the culture medium and addition of fresh Locke's solution, astrocytes were incubated up to 4 h at 37°C in 5% $CO_2$ in the presence of 50 μM of L-KYN. Following incubation, 900 μl of the medium were added to 100 μl 5 N perchloric acid. The mixture was centrifuged, and the supernatant was used for the determination of KYNA. KYNA was assessed by HPLC with a fluorescence detection (Swartz et al., 1990).

## 3. RESULTS

Exposure of the cells to 50 μM L-KYN yielded a linear increase in the extracellular content of KYNA up to 4 h of incubation. Linearity with time was not seen at concentrations exceeding 100 μM (calculated $K_m$ and $V_{max}$ values were 89 μM and 2189 pmoles KYNA/h/mg cell protein, respectively).

In routine experiments, the cells were incubated with 50 μM L-KYN for 2 h at 37°C in 5% $CO_2$ . Several test compounds reduced the extracellular KYNA content, with quisqualate ($EC_{50}$= 3 μM) being the most active (Table 1). Incubation with the endogenous lysine metabolite L-α-aminoadipate resulted in a pronounced decrease of KYNA release ($EC_{50}$= 8 μM), while the D-enantiomer was ineffective up to 500 μM. L-glutamate had an $EC_{50}$ of 81 μM. AMPA, kainate, NMDA, the prototypical agonists at ionotropic EAA receptors, and L-homocysteate did not show any effect up to a concentration of 500 μM. Notably, incubation of astrocytes with up to 100 μM veratridine did not affect the extracellular level of KYNA.

## 4. DISCUSSION

The present study demonstrates that primary cultures of rat cortical astrocytes are able to produce and release KYNA, and that astrocytic KYNA formation is controlled by regulatory mechanisms involving EAAs. The absence of an effect of veratridine on astro-

**Table 1.** Effects of EAA receptor agonists and veratridine on KYNA release from rat cortical astrocytes

| | $EC_{50}$ (μM) |
|---|---|
| Endogenous compounds | |
| L-α-Aminoadipate | 8 |
| D-α-Aminoadipate | >500 |
| L-Glutamate | 81 |
| L-Homocysteate | >500 |
| Exogenous compounds | |
| Quisqualate | 3 |
| AMPA | >500 |
| Kainate | >500 |
| NMDA | >500 |
| Veratridine | >100 |

Rat cortical astrocytes were incubated for 2 hours in the presence of 50 μM L-KYN at 37°C in 5% $CO_2$. KYNA was measured in the incubation medium prior to deproteinization. $EC_{50}$ values were calculated by least square method by application of non-linear regression on values from at least three different experiments. Control astrocytes produced 2189 pmoles KYNA/h/mg protein.

cytes, which contrasts with the ability of this compound to reduce the extracellular levels of KYNA in vivo (Wu et al., 1992), supports the idea that in vivo veratridine acts indirectly by evoking the neuronal release of endogenous agent(s) which in turn affect glial KYNA production. In contrast, endogenous EAAs such as L-α-aminoadipate or glutamate, which also decrease KYNA levels in vivo (Wu et al., 1992; 1995), appear to act directly via astrocytes, presumably either through membrane receptors or by interaction with KAT following active transport into the cell.

The EAA-induced decrease in KYNA may remove this endogenous neuroprotective agent from the glutamatergic synapse and may thus precipitate excitotoxic neuropathology (Schwarcz et al., 1992). Our data suggest that cultured primary astrocytes can be used to study some, but not all, of the mechanisms which determine the liberation of KYNA into the brain's extracellular milieu, and which therefore may play a role in the pathogenesis of neurodegenerative and seizure disorders.

## 5. REFERENCES

Du F., Schmidt W., Okuno E., Kido R., K"hler C and Schwarcz R., 1992, Localization of kynurenine aminotransferase immunoreactivity in the rat hippocampus, *J. Comp. Neurol.* 321:477.

Gramsbergen J.B.P., Turski W.A. and Schwarcz R., 1989, Sodium-dependent inhibition of kynurenate synthesis by quisqualate in rat brain. *Soc. Neurosci. Abstr.* 15:328

Schwarcz R., Du F., Schmidt W., Turski W.A., Gramsbergen J.B.P., Okuno E. and Roberts R.C., 1992, Kynurenic acid: a potential pathogen in brain disorders, *in*: "Neurotoxins and Neurodegenerative Disease", J.W. Langston and A. Young eds., *Ann. N.Y. Acad. Sci.*, New York 648:140

Stone T.W., 1993, Neuropharmacology of quinolinic and kynurenic acids, *Pharmacol. Rev.* 45: 310

Swartz K.J., Matson W.R., MacCarvey U., Ryan E.A. and Beal M.F., 1990, Measurement of kynurenic acid in mammalian brain extracts and cerebrospinal fluid by high-performance liquid chromatography with fluorimetric and coulometric electrode array detection, *Anal. Biochem.* 185:363.

Turski W.A., Gramsbergen J.B.P., Traitler H. and Schwarcz R., 1989, Rat brain slices produce and liberate kynurenic acid upon exposure to L-kynurenine, *J. Neurochem.* 52:1629

Wu H.-Q., Baran H., Ungerstedt U. and Schwarcz R., 1992, Kynurenic acid in the quinolinate- lesioned rat hippocampus: studies in vitro and in vivo, *Eur. J. Neurosci.* 4:1264

Wu H.-Q., Ungerstedt U. and Schwarcz R., 1955, L-α-Aminoadipic acid as a regulator of kynurenic acid production in the hippocampus: a microdialysis study in freely moving rats, *Eur. J. Pharmacol. 281: 55.*

# *INDOLE DERIVATIVES AS ANTIOXIDANTS*

# 44

# EVOLUTIONARY ASPECTS OF INDOLEAMINES AS RADICAL SCAVENGERS

## Presence and Photocatalytic Turnover of Indoleamines in a Unicell, *Gonyaulax polyedra*

R. Hardeland,[1] G. Behrmann,[1] B. Fuhrberg,[1] B. Poeggeler,[2] S. Burkhardt,[1] H. Uría,[3] and B. Obst[1]

[1] I. Zoologisches Institut, Georg-August-Universität
D-37073 Göttingen, Germany
[2] Department of Cellular and Structural Biology
The University of Texas Health Science Center
San Antonio, Texas 78284-7762
[3] Departamento de Morfología y Biología Celular
Universidad de Oviedo
E-33006 Oviedo, Spain

## INTRODUCTION

Indolic compounds can act as potent radical scavengers, a finding which has been particularly well studied in the case of melatonin and which has led to the conclusion that this function may be of physiological relevance, in addition to the roles of indolecamines in signaling (Reiter et al., 1993, 1994; Hardeland et al., 1993; Hardeland and Rodríguez, 1995). Among substances which are capable of terminating radical reaction chains, melatonin is the most efficient hydroxyl radical scavenger found to date in biological material (Tan et al., 1993). Along with the observation that melatonin accumulates in nuclei of various cell types (Menendez-Pelaez et al., 1993; Menendez-Pelaez and Reiter, 1993), this property suggests a protective role at DNA level.

Melatonin is not restricted to vertebrates. It has also been found in various taxa of invertebrate animals (reviews: Vivien-Roels and Pévet, 1993; Hardeland and Rodríguez, 1995), in 16 families of higher plants (Van Tassel et al., 1993; Dubbels et al., 1995; Hattori et al., 1995; Balzer and Hardeland, 1995), in a sea-weed, *Pterygophora californica* (Fuhrberg et al., unpubl.), and even in a unicell, the dinoflagellate *Gonyaulax polyedra* (Poeggeler et al., 1991). Melatonin and other indolic compounds may, therefore, be regarded as evolutionarily old molecules, which may contribute to antioxidative protection in many organisms.

*Recent Advances in Tryptophan Research*, edited by Graziella Allegri Filippini et al.
Plenum Press, New York, 1996

# MATERIALS AND METHODS

*Gonyaulax polyedra* was grown unialgally, in a modified f/2 sea-water medium, at 20°C, under a light-dark cycle of 12:12 h, as previously described (Balzer and Hardeland, 1991). Effects of lowered temperature were studied after cells had been transferred to 15°C at zeitgeber time (ZT) 3:00 h (Balzer and Hardeland, 1991). Melatonin and 5-methoxytryptamine were extracted under special preservative conditions, and determined by HPLC with electrochemical detection, as described earlier (Poeggeler et al., 1991). Tryptophan hydroxylase activity was determined by a fluorometric assay specially developed for *Gonyaulax* homogenates (Burkhardt and Hardeland, 1995).

Photooxidation of melatonin, 5-methoxytryptamine and 5-methoxytryptophol was studied using *Gonyaulax* extracts as a source of photocatalysts. For this purpose, *Gonyaulax* cells were collected by filtration (Poeggeler et al., 1991) until falling dry. The pellet was homogenized in DMSO, using a Potter-Elvehjem homogenizer with a tightly fitting glas pestle. Homogenates were centrifuged for 5 min at 3,000 × *g*. The supernatant was diluted with DMSO to give 2 O.D. at 440 nm. The photooxidation system consisted of 1.2 ml DMSO, 0.15 ml of the respective indolic compound, dissolved in DMSO at a concentration of 20 mM, and 0.15 ml of extract. Samples were irradiated by means of a fluorescent lamp, at 8,000 lx, for 2 h. Educts and products were studied fluorometrically.

# RESULTS AND DISCUSSION

The dinoflagellate *Gonyaulax polyedra* produces all metabolites of the melatonin biosynthetic pathway, such as 5-hydroxytryptophan, serotonin, N-acetylserotonin (Poeggeler, 1992), and melatonin (Poeggeler et al., 1991), as well as the catabolites 5-methoxytryptamine (Poeggeler, 1992) and 5-methoxytryptophol (Fuhrberg and Hardeland, 1995). Serotonin is predominantly formed during photophase, as judged from the circadian pattern of tryptophan hydroxlase activity (Fig. 1a), whereas the maxima of melatonin and 5-methoxytryptamine appear during night (Fig. 1b,c).

The 5-methoxylated indoles mentioned attain remarkably high concentrations in the unicell; at 20 °C, the circadian maxima exceed 1 or even 2 ng/mg protein. Moreover, the levels of these substances can be dramatically increased upon exposure to a temperature of 15 °C. In this case, melatonin content rises to 50 ng/mg protein (Fig. 1d). Under same conditions, 5-methoxytryptophol levels can exceed 35 ng/mg protein (Fuhrberg and Hardeland, 1995), and initial evidence indicates similar stimulations in 5-methoxytryptamine (Fuhrberg and Hardeland, unpubl.).

The response to a decreased temperature is seen already after a few hours, and is not prevented by light (cf. Fig. 1d). If the strong augmentation of 5-methoxylated indoles should be related to antioxidative protection, this might be seen in connection with the fact that the cells' enzymatic management of highly reactive oxygen species, based primarily on superoxide dismutatases and catalases/peroxidases, looses capacity in an ectothermic organism when temperature drops. This can later be partially compensated in *Gonyaulax*, since glutathione peroxidase is induced by low temperature in this organism (Obst and Hardeland, unpubl.). Nevertheless, there may still remain an additional requirement for antioxidative protection, accomplished by increased levels of indoleamines, especially as a photoautotrophic bright-light organism such as *Gonyaulax* is continually exposed to high light intensities during photophase.

Protection from photooxidative radical reactions may be one of the important functions of indoles in an organism such as *Gonyaulax*. This unicell contains cellular

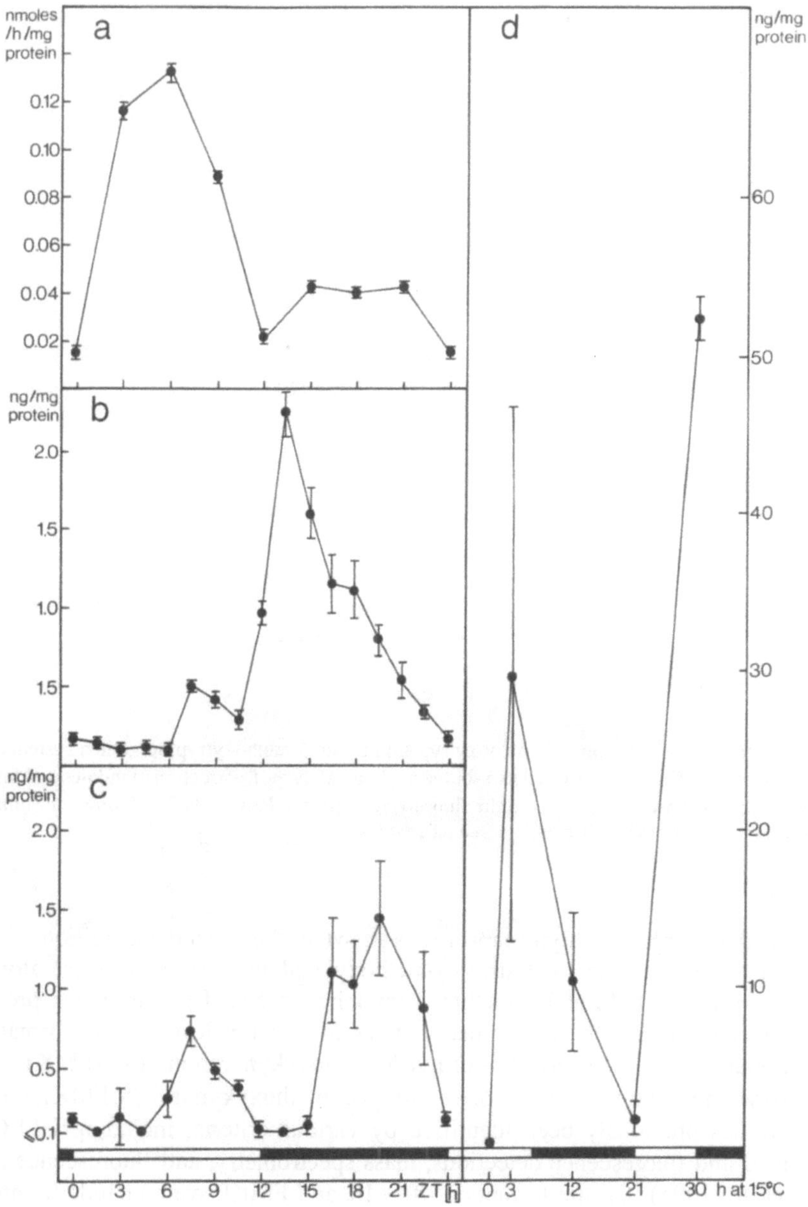


**Figure 1.** Indoleamines in the dinoflagellate *Gonyaulax polyedra*. Temporal patterns and the effect of lowered temperature on melatonin. Circadian rhythms of (a) tryptophan hydroxylase activity, (b) melatonin concentration, and (c) 5-methoxytryptamine concentration, at 20 °C. (d) Dramatic increase of melatonin concentration upon transfer to 15 °C. ZT: zeitgeber time [h]. Black and white bars: light-dark cycle. Vertical lines: s.e.m.

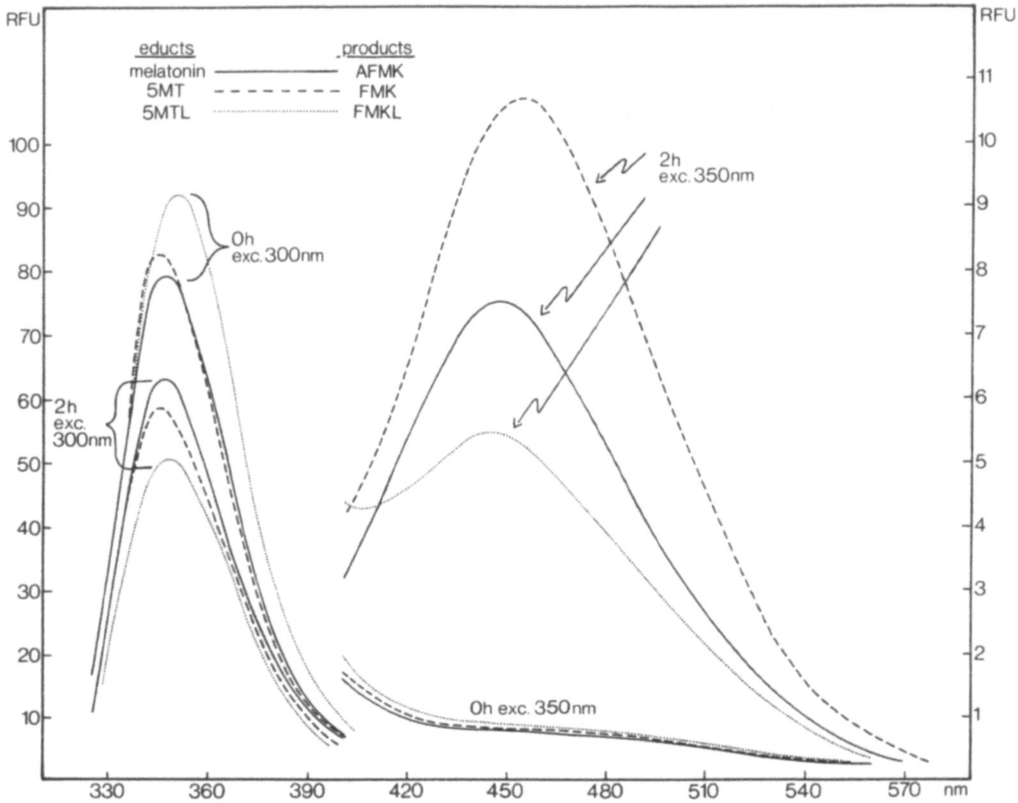


**Figure 2.** Photooxidation of melatonin, 5-methoxytryptamine, and 5-methoxytryptophol in a system containing DMSO as solvent and *Gonyaulax* extract as a source of photocatalysts. Destruction of indoles and formation of kynuric products are shown by changes in the fluorescence spectra. RFU: relative fluorometer units; exc.: excitation wavelength. For other abbreviations see current text.

constituents with high photocatalytic capacity, as shown in Fig. 2. Under irradition by visible light, the cell extract is able to destroy 5-methoxylated indoles such as melatonin, 5-methoxytryptamine (= 5MT) and 5-methoxytryptophol (= 5MTL), even when present in millimolar concentrations. One of the major types of reaction leads to the formation of kynuric compounds, i.e., $N^1$-acetyl-$N^2$-formyl-5-methoxykynuramine (= AFMK), $N^2$-formyl-5-methoxykynuramine (= FMK), or N-formyl-5-methoxykynurol (= FMKL), respectively. AFMK has previously been identified by various criteria, including HPLC with electrochemical and fluorescence detections, mass spectrometry, and fluorescence studies (Behrmann et al., 1995). On the formation of FMK and FMKL we conclude by analogy, according to similar fluorescence spectra of products. In the case of 5MTL oxidation, a large proportion of the products is, however, different from FMKL, as indicated by the shoulder at lower wavelengths in the emission spectrum. Basicly, the reactions are identical with those obtained in the presence of protoporphyrin IX, the photocatalytic action of which is well understood, and whose interaction with melatonin has been studied in detail using various inhibitors and scavengers (Hardeland et al., 1995). The formation of AFMK or analogs indicates a two-step mechanism, in which first an electron is abstracted from the indole body, leading to the formation of the respective indolyl cation radical, which afterwards combines with a superoxide anion, a step neutralizing the charges and allowing electron pairing, thereby terminating the radical reaction chain. As compared to the reaction mechanism described, interaction with light-induced singlet oxygen does not play a major role in photooxidation of 5-methoxylated indoles, since inhibitors such as DABCO or azide (1 mM) only weakly interfere with indole destruction and formation of kynuric products (Hardeland et al., 1995). Moreover, photooxidation does not require hydroxyl radicals, since the reactions take place at a high rate even when using the hydroxyl radical scavenger DMSO as the only solvent (Fig. 2).

The fact that 5-methoxylated indoles are present in high concentrations already in a unicell, and that this organism contains substantial amounts of photocatalysts, may provide a hint for the original role of these indoles in evolution. Their radical-scavenging capacity might have been the primary advantage for their formation in the remarkable quantities found. With regard to the photocatalytic activity of cellular constitutents, protection from photooxidative radical reaction chains may be particularly important in the light-exposed unicells. Protection from free radicals, especially those induced by irradation, may have been the primary function of 5-methoxylated indoles.

## ACKNOWLEDGMENT

Supported by the Deutsche Forschungsgemeinschaft.

## REFERENCES

Balzer, I., and Hardeland, R., 1991, Photoperiodism and effects of indoleamines in a unicellular alga, *Gonyaulax polyedra*. *Science* 253:795.

Balzer, I., and Hardeland, R., 1995, Melatonin in green plants, *in*: "Cellular Rhythms and Indoleamines", R. Hardeland, ed., University of Göttingen, Göttingen, p. 152.

Behrmann, G., Uría, H., Fuhrberg, B., Poeggeler, B., and Hardeland, R., 1995, Melatonin terminiert photooxidative radikalische Reaktionsketten. *Verh. Dtsch. Zool. Ges*. 88.1:94.

Burkhardt, S., and Hardeland, R., 1995, Tryptophan hydroxylase in the dinoflagellate *Gonyaulax polyedra*. I. Differences to the vertebrate enzyme, *in*: "Cellular Rhythms and Indoleamines", R. Hardeland, ed., University of Göttingen, Göttingen, p. 33.

Dubbels, R., Reiter, R.J., Klenke, E., Goebel, A., Schnakenberg, E., Ehlers, C., Schiwara, H.W., and Schloot, W., 1995, Melatonin in edible plants identified by radioimmunoassay and by high performance liquid chromatography-mass spectrometry. *J. Pineal Res*. 18:28.

Fuhrberg, B., and Hardeland, R., 1995, 5-Methoxytryptophol in *Gonyaulax polyedra*, *in*: "Cellular Rhythms and Indoleamines", R. Hardeland, ed., University of Göttingen, Göttingen, p. 25.

Hardeland, R., Balzer, I., Poeggeler, B., Fuhrberg, B., Uría, H., Behrmann, G., Wolf, R., Meyer, T.J., and Reiter, R.J., 1995, On the primary functions of melatonin in evolution: Mediation of photoperiodic signals in a unicell, photooxidation, and scavenging of free radicals. *J. Pineal Res*. 18:104.

Hardeland, R., Reiter, R.J., Poeggeler, B., and Tan, D.-X., 1993, The significance of the metabolism of the neurohormone melatonin: antioxidative protection and formation of bioactive substances. *Neurosci. Biobehav. Rev*. 17:347.

Hardeland, R., and Rodríguez, C., 1995, Versatile melatonin: A pervasive moleciule serves various functions in signaling and protection. *Chronobiol. Int.* 12, in press.

Hattori, A., Migitaka, H. Iigo, M., Itoh, M., Yamamoto, K., Ohtani-Kaneko, R., Hara, M., Suzuki, T., and Reiter, R.J., 1995, Identification of melatonin in plants and its effects on plasma melatonin levels and binding to melatonin receptors in vertebrates. *Biochem. Mol. Biol. Int.*, in press.

Menendez-Pelaez, A., Poeggeler, B., Reiter, R.J., Barlow-Walden, L., Pablos, M.I., and Tan, D.-X., 1993, Nucelar localization of melatonin in different mammalian tissues: immunocytochemical and radioimmonoassay evidence. *J. Cell Biochem*. 53:373.

Menendez-Pelaez, A., and Reiter, R.J., 1993, Distribution of melatonin in mammalian tissues: The relative importance of cytosolic versus nuclear localization. *J. Pineal Res*. 15:159.

Poeggeler, B., 1992, "Vergleichende Untersuchungen über Melatonin und strukturverwandte Tryptophan-metabolite. Zur Rolle von Melatonin und 5-Methoxytryptamin bei einem Dinoflagellaten, *Gonyaulax polyedra*, sowie pinealen und extrapinealen 5-methoxylierten Indolaminen bei Vertebraten", Ph.D. thesis, Göttingen.

Poeggeler, B., Balzer, I., Hardeland, R., and Lerchl, A., 1991, Pineal hormone melatonin oscillates also in the dinoflagellate *Gonyaulax polyedra*. *Naturwissenschaften* 78:268.

Reiter, R.J., Poeggeler, B., Tan, D.-X., Chen, L.-D., Manchester, L.C., and Guerrero, J.M., 1993, Antioxidant capacity of melatonin: A novel action not requiring a receptor. *Neuroendocrinol. Lett.* 15:103.

Reiter, R.J., Tan, D.-X., Poeggeler, B., Menendez-Pelaez, A., Chen, L.-D., and Saarela, S., 1994, Melatonin as a free radical scavenger: Implications for aging and age-related diseases. *Ann. N.Y. Acad. Sci.* 719:1.

Tan, D.-X., Chen, L.-D., Poeggeler, B., Manchester, L.C., and Reiter, R.J., 1993, Melatonin: a potent, endogenous hydroxyl radical scavenger. *Endocr. J.* 1:57.

Van Tassel, D.L., Li, J., and O'Neill, S.D., Melatonin: Identification of a potential dark signal in plants. *Plant Physiol.* 102, Suppl. 1, p. 659.

Vivien-Roels, B., and Pévet, P., 1993, Melatonin: presence and formation in invertebrates. *Experientia* 49:642.

# UTILIZATION OF SUPEROXIDE ANION BY INDOLEAMINE OXYGENASE-CATALYZED TRYPTOPHAN AND INDOLEAMINE OXIDATION

O. Hayaishi

Osaka Bioscience Institute
6-2-4 Furuedai, Suita, Osaka 565, Japan

## I. DISCOVERY OF INDOLEAMINE DIOXYGENASE

In 1963, we isolated from rabbit intestine a new enzyme that catalyzed the formation of D-kynurenine from D-tryptophan (Higuchi et al., 1963). In contrast to the classical tryptophan dioxygenase, this enzyme exhibits broad substrate specificity and acts upon various indoleamine derivatives, including tryptophan. Therefore it was termed indoleamine 2,3-dioxygenase, abbreviated as IDO. Typtophan dioxygenase, abbreviated as TPO, is found only in the liver, and acts only on L-tryptophan. In contrast, IDO is widely distributed in various tissues and organs, except the liver. It has a broad substrate specificity, and acts upon not only L- and D-tryptophan but also 5-hydroxy-tryptophan, tryptamine, serotonin, etc, although tryptophan is a far better substrate in terms of Km and Vmax. TPO utilizes molecular oxygen, whereas IDO requires and utilizes superoxide anion and molecular oxygen for its activity. Although both IDO and TPO are hemoproteins, their molecular properties are clearly different with regard to their molecular weight, subunit structure, carbohydrate and heme contents, and the turnover number. They are also different immunologically. Thus, there are two distinctly different enzymes both catalyzing essentially identical reactions, namely the formation of formylkynurenine from tryptophan. However, unlike TPO, IDO is not induced by either tryptophan or glucocorticoid, but is induced under various pathological conditions.

## II. REQUIREMENT FOR $O_2^-$

One of the most striking features of IDO is the fact that the purified enzyme is by itself totally inactive, even in the presence of oxygen, and requires the superoxide anion. Evidence for the participation of the superoxide anion in the IDO-catalyzed reaction has already been published in detail (Hayaishi, 1975) and may be briefly summarized as follows.

*Recent Advances in Tryptophan Research*, edited by Graziella Allegri Filippini et al.
Plenum Press, New York, 1996 

First, IDO requires and utilizes the superoxide anion for its activity. The superoxide anion can be supplied as a relatively stable potassium salt, $KO_2$ or can be generated either chemically or enzymatically. The superoxide is required for the initiation of the reaction as well as for maintenance of the steady state of the reaction. Second, the reaction is inhibited by scavengers of the superoxide anion, such as superoxide dismutase, tyron, and so forth, but scavengers of singlet oxygen, $H_2O_2$ or hydroxyl radicals are totally without effect, even at high concentrations. Third, the superoxide anion labeled with a heavy oxygen isotope, oxygen-18, is incorporated into the product of the reaction. However, when the reaction is carried out in the presence of heavy oxygen-containing air, oxygen-18 is also incorporated into the product of the reaction to some extent. These results indicate that IDO is a unique oxygenase and requires both molecular oxygen and the superoxide anion for its activity.

An explanation for this dual requirement for $O_2$ and $O_2^-$ is given in the reaction scheme shown in Fig. 1. This scheme illustrates a possible reaction sequence based on the kinetic experiments in our laboratory. It is obviously an oversimplification but may be useful as a point of orientation for discussing the reaction mechanism. Let us start with the native ferric enzyme. First, it binds with the substrate, tryptophan, which is probably a very fast process; then the ES complex reacts with superoxide to form a ternary complex of [$Fe^{+3} \times O_2^- \times$ Try] which is identical with the [$Fe^{+2} \times O_2 \times$ Try] complex. This charge transfer complex is also produced by the binding of molecular oxygen with the ferrous form of enzyme and substrate. The unstable binary complex of [$Fe^{2+} \times$ Try] can be produced experimentally by photolysis of the [$Fe^{2+} \times$ Try $\times$ CO] complex. Once the ternary complex is produced, it decomposes to yield the reaction product, generating a ferrous form of enzyme. The first-order rate constant for this reaction is estimated to be 2.0/sec. Thus, during the steady state of the reaction cycle, $K_4$ is the rate-determining step and is in good agreement with the turn-over number of the enzyme. During the catalytic cycle, however, IDO is slowly autooxidized with the rate constant of 0.03/sec to the ferric form which, in absence of the superoxide anion, remains inactive. In other words, once in every 60 or 70 cycles, the enzyme undergoes autooxidation to become a ferric form of enzyme, which then binds with tryptophan and the superoxide anion. In fact, when the supply of the superoxide anion is cut off during the steady state of the reaction, the reaction continues for several seconds until all the enzyme molecules are converted to the inactive ferric form. Thus the enzyme utilizes both molecular oxygen and the superoxide anion. But in order to initiate the reaction and

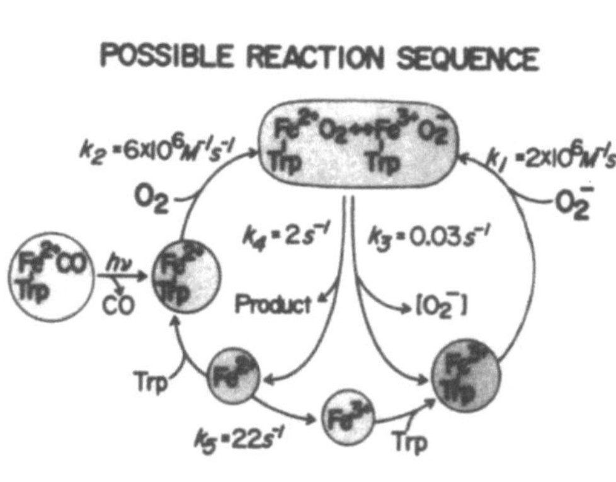


Figure 1

maintain the steady state of the reaction, the superoxide anion is absolutely essential, and the supply of $O_2^-$ is the rate-determining step in the overall process.

## III. EXPERIMENTS WITH CELL SUSPENSIONS

Now, even though IDO utilizes the superoxide anion in vitro, the possibility still remains that these observations are an in vitro artifact and that the enzyme utilizes only molecular oxygen in vivo. To carry the point further, one may ask: "Does this enzyme actually exist and function in vivo?", because the superoxide anion is very unstable and the steady state concentration in the cell is presumably extremely low. In an attempt to answer these questions, we isolated mucosal cells from rabbit intestine and determined the intracellular IDO activity of suspensions of dispersed enterocytes as a model of in vivo experiments.

Because superoxide dismutase, SOD, is active in enterocytes, it is reasonable to assume that the supply of superoxide anion is low and must be one of the rate-limiting factors of the intracellular IDO activity. Now how can one regulate the level of $O_2^-$ in the cell? In order to approach this problem we did the following two types of experiments: First, a number of enzyme systems may generate $O_2^-$ in the enterocytes, but since xanthine oxidase generates $O_2^-$ in vitro and is abundant in the intestine, the effect of inosine, a substrate of xanthine oxidase, on the intracellular IDO activity was examined. Second, diethyldithiocarbamate (DDC), a copper chelator and a potent in vivo inhibitor of superoxide dismutase, was used to inhibit the intracellular dismutase activity. Both of these reagents should increase the concentration of the superoxide anion inside the enterocytes and thereby stimulate the endogenous IDO activity in the cell. The results of these experiments show that the addition of inosine increased the IDO activity nearly 5 fold. It was dose dependent. The amount of inosine required for the maximum stimulation was in the order of 0.1 mM, and this effect was abolished by the addition of allopurinol, a specific inhibitor of xanthine oxidase. Neither inosine nor allopurinol had any effect on the purified IDO in vitro. These results indicate that the superoxide anion generated by the xanthine oxidase system is utilized by IDO and that it is the rate-determining step in intact cells.

In the same manner, the intracellular IDO activity in the dispersed enterocytes was markedly enhanced by the addition of 5 mM DDC, an inhibitor of superoxide dismutase. The diethyldithiocarbamate, however, had no activating effect on the purified IDO in vitro. These results indicate that DDC probably inhibits superoxide dismutase, increases the intracellular concentration of superoxide anion, and stimulates the apparent IDO activity, again indicating that $O_2^-$ is the rate-limiting factor, in the intact cells.

## IV. PHYSIOLOGICAL FUNCTION OF IDO

So far, I have talked about the role of superoxide in the IDO-catalyzed reaction. Now let us look at the physiological function of the superoxide anion in vivo.

The superoxide anion is an active form of oxygen and is produced in leukocytes when bacteria are ingested by leukocytes. During phagocytosis, NADPH oxidase on the membrane of phagosomes is activated and molecular oxygen is reduced by one electron to produce the superoxide anion. The superoxide anion alone, or together with $H_2O_2$ that is generated from the superoxide, kills the ingested bacteria. Therefore the superoxide anion is implicated in the defense mechanism against bacterial infection. Indoleamines are also implicated in the inflammatory process, as it is a well-known fact that tryptophan is degraded under inflammatory and other pathological conditions. So, how do we put all these pieces of information together to elucidate the biological function of this enzyme ? As the lung is an aerobic organ and probably generates a fair amount of superoxide anion, we decided to subject mice to viral infection to cause viral pneumonia and determined the IDO activity in the lung.

When we exposed mice to influenza virus, the virus replication in the lung began within 24 hours after the infection, reached a peak by the 3rd day and persisted until the 9th day, causing viral pneumonia. Thereafter, the virus was rapidly eliminated from the lung, and by the 14th day almost completely disappeared. The IDO activity in the lung increased almost linearly from the 5th day after infection. More than a 120-fold increase was observed on the 11th day, after which the activity gradually decreased to the normal value over a two-week period. These results indicate that, with viral infection, the IDO activity increases tremendously and the breakdown of tryptophan and other indoleamine derivatives in the lung is accelerated to a large extent (Yoshida and Hayaishi, 1978).

In an attempt to determine the mechanism of enzyme induction, we then switched to an in vitro system and exposed lung slices to LPS, or a lipopolysaccharide fraction. When LPS was given at time zero, IDO was gradually induced but with a sizable lag period. As expected, the interferon titer increased almost instantaneously after LPS administration. When interferon itself was administered at time zero, there was no lag in the induction of IDO. These time courses indicate that when LPS enters the cell, interferon is produced first and it then induces the IDO activity.

Our initial studies were carried out with mice and murine α- and β- interferon but later with γ-interferon. More recently, Yasui and coworkers in our laboratory extended these observations to human lung with human γ-interferon.

Even more recently, Pfefferkorn reported that γ-interferon blocks the growth of Toxoplasma gondii in human fibroblasts by inducing the host cells to degrade tryptophan (Pfefferkorn, 1984).

When human fibroblasts were treated with γ- interferon, the growth of Toxoplasma gondii, an obligate intracellular protozoan parasite, was blocked, and at the same time tryptophan in the host cells was degraded to produce a large amount of formylkynurenine and kynurenine. He concluded that the enzyme induced by physiological doses of γ-interferon described in his experiment is probably "indoleamine dioxygenase." Similar induction of IDO by γ- interferon in a mixed culture of lymphocytes was reported by Werner and co-workers.

## V. SUMMARY

The following is our current working hypothesis concerning the biological significance of IDO induction. When tissues are invaded by virus, bacteria, or parasites, leukocytes and lymphocytes will accumulate at the site and interferon will be produced by these cells in the inflammatory loci. The interferon thus produced is released and interacts with the cell surface to trigger IDO induction in the same or other types of cells. As a consequence of inflammation, superoxide anion is liberated and serves as a substrate for IDO. Although it is possible that some tryptophan metabolites may activate the immune system or act as bacteriostatic agents, available evidence does not support this hypothesis. We therefore tentatively conclude that tryptophan is degraded by IDO and depleted, whereby the growth of viruses, bacteria and certain parasites is inhibited, because tryptophan is the least available and therefore most important essential amino acid for their growth.

## REFERENCES

Hayaishi O., 1975, Indoleamine 2, 3-dioxygenase a new vista in tryptrophan metabolism, in : "Acta VItaminologica et Enzymologica" [Proceedings of the First ISTRY Meeting, Padua, May 2-4, 1974].

Higuchi K., Kuno S., and Hayaishi O., 1963, Enzymic formation of D-kynurenine, *Federation Proc.*, 22: 242

Pfefferkorn E.R., 1984, Interferon γ blocks the growth of *Toxoplasma gondii* in human fibroblasts by inducing the host cells to degrade tryptophan, *Proc Natl. Acad. Sci. USA.*, 81: 908.

Yoshida R., and Hayaishi O., 1978, Induction of pulmonary indoleamine 2,3-dioxygenase by interaperitoneal injection of bacterial lipopolysaccharide, *Proc. Natl. Acad.Sci. USA. 75: 3998*

# ANTIOXIDANT PROPERTIES OF INDOLE-3-PYRUVIC ACID

V. Politi, S. D'Alessio, G. Di Stazio, and G. De Luca

Polifarma Res.Centre
Via Tor Sapienza 138 - 00155 Roma, Italy

## 1. INTRODUCTION

Oxidative stress is believed to be involved in the pathogenesis of many important degenerative diseases, especially in the Central Nervous System, where oxygen is utilized in very high amount, and scavengers of toxic substances might be insufficient when neurons are over-stimulated or ageing processes accumulate peroxide lipids (Halliwell, 1992).

To counteract oxidative stress, living organisms selected several biomolecules, able to scavenge oxygen free-radicals, or to transform them to less injurious agents: Glutathion, Vitamins C and E, Enzymes (e.g. Superoxide Dismutase and Catalase), and so on. Indole compounds constitute one of the most important classes of substances interacting with toxic oxygen derivatives and, because of their widespread presence in any kind of living cell, they can be considered within the first defense systems developed during evolution. Among the others, Indole-3-pyruvic acid (IPA), the keto-analogue of tryptophan that can be easily obtained from the aminoacid by means of several aromatic aminotransferases or aminoacid oxidases, is arising as a pre-eminent indole in living organisms because it replenishes cells with important biochemicals and behaves like a strong antioxidant.

## 2. IPA IN THE VEGETAL KINGDOM

Indoles are extensively utilized by plants as "auxins", or growth-inducers, and indole-3-acetic acid (IAA) is recognized as the most important compound of this class. The biochemical pathway of IAA synthesis from tryptophan, including IPA as the central intermediate, has been established in several plants and microorganisms, and enzymes like aromatic aminotransferase and indolpyruvate decarboxylase have been described in details (El-Abyad et al., 1994; Koga et al., 1994), so that this is considered the usual mean utilized by plants for their growth, even if some indole side-chain reactions have been recently proposed as alternate pathways (Normanly et al., 1995).

When plants stop growing, however, IAA production is strongly reduced, and IPA accumulates in senescent leaves up to very high concentrations (Atsumi and Hayashi, 1979).

*Recent Advances in Tryptophan Research*, edited by Graziella Allegri Filippini et al.
Plenum Press, New York, 1996

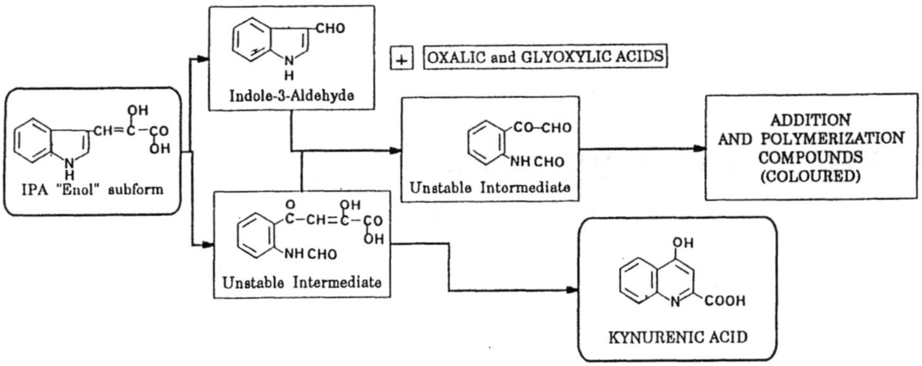


**Figure 1.** IPA main Pathways after oxygen attack. Development of coloured compounds is believed to prevail in plants, while Kynurenic acid is more important for animal cells.

Bright colours, going from yellow to red and brown, are often observed in dying leaves of higher plants during the autumn: this reflects reactions of IPA with oxygen species (Figure 1), and the same colours are easily obtained in test tubes when IPA is dissolved in aerated solutions in the presence of light. On the other hand IPA, as well as other indoles found in plants, are well recognized as good electron donors, and intense colours develop when they are put in contact with electron acceptors (Hutzinger et al., 1972), suggesting a direct participation also to colour development in flowers.

## 3. IPA AS A TARGET FOR RADICAL ATTACK

It is believed that the strong antioxidant properties of indoles present in living cells are especially due to a radical attack on carbon atoms in positions 2 and 3, after abstraction of hydrogen from the N-H bond (Lissi et al., 1991). When considering IPA molecule, however, reactivity toward oxygen and oxygen radicals is increased, because in its "enol" configuration the ketoacid has a planar structure, with another double bond in the side chain, conjugated with that present in the indole nucleus.

Some time ago, it has been shown that oxygen attack on IPA induces release of chemiluminescence, attributed to several open-ring species (Zinner et al., 1974). The discovery that oxydation of IPA induces also formation of kynurenic acid both in the presence and absence of enzymatic systems (Russi et al., 1989; Politi et al., 1991), authorizes some conclusions about chemicals formed when IPA reacts with oxygen free-radicals. As described in Fig.1, after an attack on the side-chain, compounds like oxalic and glyoxylic acid are easily found in solutions, while the opening of indole ring by oxygen radicals is followed by formation of unstable compounds, further transformed to kynurenic acid or to addition and polymerization products (most of them coloured).

As a strong free-radical scavenger, IPA was found to inhibit radical damages in abiotic systems as well as in organ homogenates (Politi et al., 1991). Furthermore, the ketoacid showed powerful des-mutagenic properties in the Ames test (Table 1): using non-toxic concentrations of IPA, dramatic reductions in the mutagenicity of 4-Nitroquinoline-N-oxide (4NQO), 2-Nitrofluorene (2NF), Captan (CAP) and Folpet (FOL) were observed. These effects were not slight diminutions in activity, but substantial reductions in the mutagenic potency of the products, recognized as free-radical-forming agents. Instead, no reductions were observed in the mutagenicity of other substances, like Ethylmethanesulphonate, Methylmethanesulphonate or Sodium Azide (data not shown), whose mechanism of toxicity

**Table 1.** Reverse mutation in S. typhimurium TA 100

| | Number of Revertant Colonies/Plate | | |
|---|---|---|---|
| Agent | Dose/plate (mcg) | Mutagen | ±IPA(1 mg) |
| Control | — | 200 | 200 |
| 4NQO | 0.035 | 380 | 190 |
| 4NQO | 0.070 | 590 | 200 |
| 4NQO | 0.125 | 950 | 210 |
| 4NQO | 0.250 | 1500 | 260 |
| 2NF | 1 | 280 | 200 |
| 2NF | 2 | 400 | 210 |
| 2NF | 4 | 650 | 240 |
| 2NF | 8 | 1150 | 260 |
| CAP | 1.25 | 350 | 180 |
| CAP | 2.50 | 500 | 200 |
| CAP | 5.0 | 850 | 230 |
| CAP | 10 | 1200 | 250 |
| FOL | 12.5 | 400 | 190 |
| FOL | 25 | 700 | 200 |
| FOL | 50 | 950 | 240 |
| FOL | 100 | 1300 | 320 |

Values express the mean of two independent experiments.
For abbreviations. see text.

does not include formation of free-radicals, thus confirming the anti-oxidant properties of IPA.

## 4. IPA EFFECTS IN MAMMALIAN BODIES

Aromatic aminotransferases are diffused in all living cells, and therefore part of the tryptophan taken up by cells from circulation is surely transformed to IPA. However, it is difficult to detect the presence of the ketoacid in tissues under normal situations, because IPA is trapped by radical-forming systems. For the same reasons, transamination of tryptophan cannot be considered as truly reversible, but as a steady, albeit slow, flux from the amino- to the keto-acid, triggering normal defense systems in mammalian bodies, and assuming higher importance when aromatic transaminases increase in concentration in cells.

### 4.1 IPA in the Liver

In liver IPA inhibits tryptophan-2,3-dioxygenase (TPO, EC 1.13.11.11), an enzyme induced by glucocorticoid, strongly involved in the regulation of circulating levels of tryptophan, and believed to be the primary responsible of tryptophan depletion during stress in mammalians. Inhibition has been shown to be dependent on interference with linkage of the coenzyme to the apoenzyme (Lavaggi et al., 1987); moreover, like melatonin and other indoles, the mechanism of inhibition might be at the level of oxygen binding to the coenzyme (Walsh et al., 1994). As a consequence of IPA activity against TPO, more tryptophan is available in blood, and can reach brain structures, where it follows pathways leading to serotonin and melatonin biosynthesis (Bacciottini et al., 1987; Ruggeri et al., 1990; Ferretti et al., 1990; Squadrito et al., 1990).

Furthermore, the presence in liver of aromatic aminotransferase activities three times more elevated at night than during the day (Chia et al., 1984) suggests that IPA is involved in physiological mechanisms favouring induction of sleep and rest in brain: in fact, tryptophan is rendered more available to CNS during the evening, when are at their peak also other important enzymes of indole pathways, like Tryptophan hydroxylase and serotonin N-acetyltransferase, crucial for serotonin and melatonin biosynthesis (Green et al., 1995).

## 4.2 IPA in the Brain

IPA can cross Blood-brain barrier, even if in limited amount, as demonstrated with the use of a labeled compound (unpublished results). However, most of IPA normally present in brain structures comes from endogenous synthesis, because aromatic aminotransferases are very much diffused and play a significant role in nervous tissue (Benuck et al., 1971). This means that brain cells, through transamination of tryptophan, have the chance to use a powerful system for the scavenging of oxygen free-radicals. Moreover, because under radical attack IPA is partially transformed to Kynurenic acid (KYNA), a good inhibitor of excitatory aminoacids (Perkins & Stone, 1985), this pathway leads finally to decreased brain excitation and to rest of neurons.

It has been shown that KYNA content in brain is roughly two times higher during sleep than during awakening, and this strongly suggests a role for KYNA in reducing brain activity at bedtime (Moroni et al., 1990). Classical pathways for KYNA synthesis in the brain, however, emphasize the intervention of kynurenine aminotransferase, a well characterized enzyme transforming kynurenine to KYNA (Baran et al., 1994): if this was the only mean for KYNA formation, we should conclude that kynurenine, a compound known for its pro-convulsant activity (Lapin, 1980), is always balancing the anti-excitatory effects of KYNA in the brain, therefore marginalizing global interventions of this compound on physiological biorhythms. We believe, instead, that kinurenine is the key KYNA precursor only in specific pathologic situations, i.e. when Indoleamine-2,3-dioxygenase in brain is increased as a consequence of damages due to inflammations or immunological responses (Taylor and Feng, 1991), while IPA is a preferred precursor in healthy brains, when neurons need to decrease excitatory activity.

It is interesting to underline that IPA effects against excitatory aminoacids are not only mediated by KYNA formation, but are also exerted directly on neurons (Bruno et al., 1991), as a consequence of antagonism on oxygen free-radicals: in fact, the role of free-radicals in neurotoxicity induced by excitatory aminoacids has been well established (Fagni et al., 1994). Unfortunately, however, the level of IPA protection in brain is not enough to antagonize brain damages observed in models of stroke in rats (Zoli et al., 1993).

# 5. IPA IN SLEEP AND STRESS

The finding that IPA increases melatonin turnover in the pineal gland of rats, acting as a precursor of both tryptophan and serotonin, and increasing the activity of Serotonin-N-acetyl-transferase (Ferretti et al., 1990) has important consequences in the drawing of a biochemical pathway leading to physiological sleep induction (Figure 2).

In this picture, IPA plays a pivotal role in liver, reducing TPO activity and therefore increasing free tryptophan available to the brain. Here, IPA has again a fundamental place, detoxicating cells from free-radicals, reducing activity of excitatory aminoacids and stimulating biosynthesis of melatonin. Because melatonin is now well recognized as a primary endogenous indole, whose deficiency is implicated not only in sleep disorders, but also in ageing (Reiter, 1995), radical damages (Poeggeler et al., 1993), tumours (Das, 1994) and

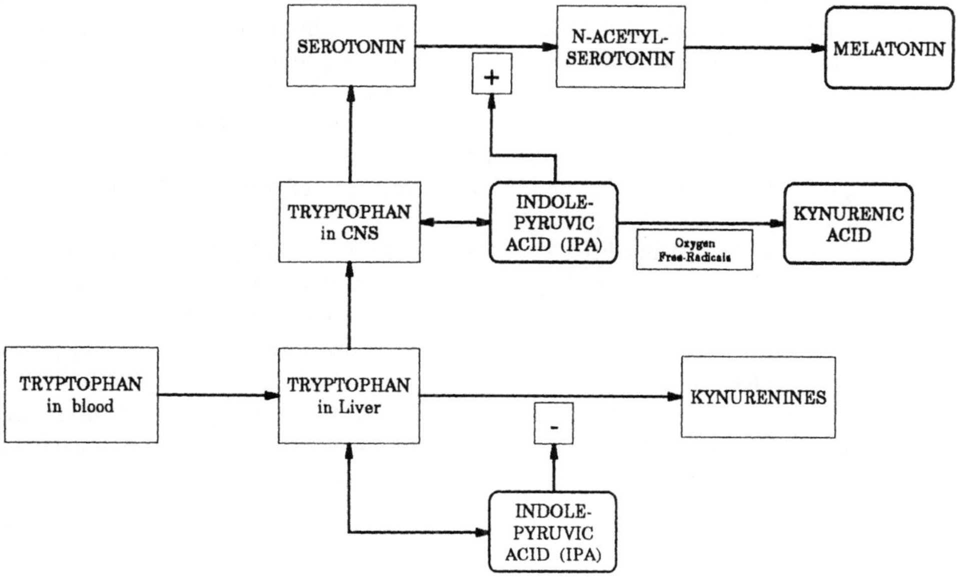


**Figure 2.** IPA effects on Tryptophan metabolism in liver and in brain. Reducing tryptophan degradation in liver, IPA increases aminoacid concentration in brain and enhances formation of kynurenic acid and melatonin.

psychiatric disorders (Brown, 1995), the pathway drawn in fig.2 has important consequences, suggesting that, throughout evolution, IPA maintained the function of cell defender against oxygen damages, especially during ageing.

That this is the case, has been demonstrated also using the Katz test on unpredictable stress in rats (Biagini et al., 1993): IPA reduced serum corticosterone levels of stressed rats, and protected their brains from degeneration induced by glucocorticoids.

## 6. IPA AS A DRUG

Based on the recognized strong anti-oxidant properties of IPA, the ketoacid has been subjected to extensive studies in mice, rats and finally in humans, with the aim to detect the most suitable indications for human diseases. Table 2 shows some of the most relevant findings obtained up to now.

**Table 2.** Activities of IPA used as a drug

| Pharmacological effect | Species | Reference |
|---|---|---|
| Increased SW sleep | Rat | Merlo Pich et al, 1990 |
| Increased total sleep | Human | Silvestri et al., 1991 |
| Anti-stress | Rat | Biagini et al., 1993 |
| Anxiolytic | Mouse | Lapin and Politi, 1993 |
| Sedation | Rat | Bacciottini et al., 1987 |
| Anorectic | Rat | Squadrito et al., 1991 |
| Anti-alcohol | Mouse | Lapin and Politi, 1994 |
| Anti-depressant | Rat | Biagini et al., 1993 |
| Analgesia | Rat | Bacciottini et al., 1987 |

Results on sleep, both in rats and humans, suggest that the drug cannot be defined as a hypnotic, because sleep latency is not reduced after its administration: instead, a physiological sleep is recovered, indicating a mechanism of activity through the melatonin pathway. Furthermore, IPA effects on sleep and anxiety are clearly different from those of benzodiazepines: in fact, IPA does not induce withdrawal signs nor drowsiness in patients, and pharmacological activity is not diminished by its chronic use. Besides, mood of patients appears highly improved after IPA administration, recalling effects described with the use of melatonin.

Studies in animals, using different stressors, indicate that IPA protects brain structures from glucocorticoid damages: according to a well-grounded theory on brain ageing (Sapolsky, 1992), this means that IPA should be especially useful to aged persons undergoing stressful events in life, chiefly because in elderly people other important detoxicating agents, like melatonin, are lacking.

Effects of IPA in obese rats and against alcohol damages are significant and suggest further uses of the ketoacid in other disfunctions (e.g. alcohol abuse and eating disorders): however, further studies are needed before the potential use in humans is validated.

## 7. CONCLUSIONS

Indoles are at present subjected to accurate studies, in order to understand their intervention in living cells, as powerful inhibitors of damages induced by oxygen free-radicals. In fact, as shown in fig. 3, it is now proved that stress events induce release of a cascade of biochemical agents, like glucocorticoids, oxygen free-radicals and excitatory aminoacids, that stimulate brain activities and, in the long period, dissipate metabolic resources, inducing irreversible neuron damages. On the other hand, compounds originating from tryptophan counteract effects of stress, reducing energetic requirements and facilitating neuronal rest and recovery.

For their characteristics, a number of tryptophan derivatives should be good candidates as drugs for CNS diseases leading to neuronal decay: unfortunately, kynurenic acid does not seem strong enough to antagonize devastating effects of excitatory aminoacids during stroke, probably because it cannot reach in high concentrations the damaged neuronal structures. However, a much brighter future is likely both for IPA and melatonin as promoters of physiological sleep in insomniac subjects and as preservers of brain from pathologic consequences of ageing.

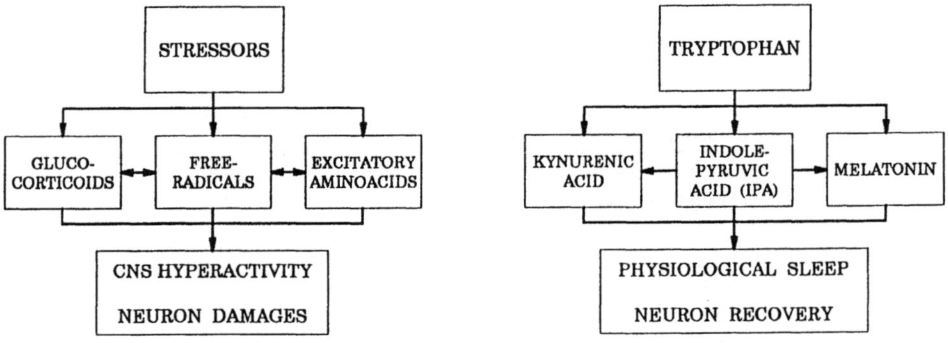


**Figure 3.** Balancing systems in CNS cells. Chronic stress events damage neuron structures; some tryptophan derivatives induce rest and recovery of neurons.

# REFERENCES


Atsumi,S., and Hayashi,T., 1979, Examination of the pronounced increase in auxin content of senescent leaves. *Plant & Cell Physiol.* 20: 861.

Bacciottini,L., Pellegrini-Giampietro,D.E., Bongianni,F., De Luca,G.,Beni,M., Politi,V., and Moroni,F., 1987, Biochemical and behavioural studies on indole-pyruvic acid: a keto-analogue of tryptophan. *Pharmacol.Res.Commun.* 19: 803.

Baran,H., Okuno,E., Kido,R., and Schwarcz,R., 1994, Purification and characterization of kynurenine aminotransferase I from human brain. *J.Neurochem.* 62: 730.

Benuck,M., Stern,F., and Lajtha,A., 1971, Transamination of amino acids in homogenates of rat brain. *J.Neurochem.* 18: 1555.

Biagini,G., Merlo-Pich,E., Carani,C., Marrama,P., Gustafsson,J.A., Fuxe,K., and Agnati,L.F., 1993, Indolepyruvic acid, a tryptophan ketoanalogue, antagonizes the endocrine but not the behavioral effects of repeated stress in a model of depression. *Biol.Psychiat.* 33: 712.

Brown,G.M., 1995, Melatonin in psychiatric and sleep disorders. CNS *Drugs* 3: 209.

Bruno,V., Canonico,P.L., Aronica,E., Nicoletti,F., and Scapagnini,U., 1991, Modulation of excitatory amino acid receptors by indolpyruvic acid. *Neurosci.Res.Commun.* 8: 137.

Chia,Y.C., Smith,G.W., and Lees,G.J., 1984, Differences in properties between aromatic amino acid : aromatic keto acid aminotransferases and aromatic aminoacid: alfa-ketoglutarate aminotransferases. *Life Sci.* 34: 2443.

Das,U.N., 1994, Melatonin: is it an endogenous anti-mutagenic, oncostatic and anti-ageing molecule? *Med.Sci.Res.* 22: 463.

El-Abyad,M.S., El-Sayed,M.A., El-Shanshoury,A.R., Farid,M., 1994, Optimization of culture conditions for indole-3-pyruvic acid production by Streptomices Griseoflavus. *Can.J.Microbiol.* 40: 754.

Fagni.L., Lafon-Cazal,L., Rondouin,G., Manzoni,O., Lerner-Natoli,M., and Bockaert,J., 1994, The role of free radicals in NMDA-dependent neurotoxicity. *Progr.Brain Res.* 103: 381.

Ferretti,C., Blengio,M., Ghi,P., and Genazzani,E., 1990, Differential effects of indolepyruvic acid and 5-hydroxytriptophan on indole metabolism in the pineal gland of the rat during the light-dark cycle. *Europ.J. Pharmacol.* 187: 345.

Green,C.B., Cahill,G.M., Besharse,J.C., 1995, Regulation of tryptophan hydroxylase expression by a retinal circadian oscillator in vitro. *Brain Res.* 677: 283.

Halliwell,B., 1992, Reactive oxygen species and the Central Nervous System. *J.Neurochem.* 59: 1609.

Hutzinger,O., Heacock,R.A., Macneil,J.D., Frei,R.W., 1972, Indoles and auxins. XIII. Identification and analysis of naturally occurring indoles via electron donor-acceptor complexes. *J.Chromatogr.* 68: 173.

Koga,J., Syono,K., Ichikawa,T., Adachi,T., 1994, Involvement of L-Tryptophan aminotransferase in indole-3-acetic acid biosynthesis in Enterobacter Cloacae. *Biochim.Biophys.Acta* 1209: 241.

Lavaggi,M.V., Politi,V., De Luca,G., Gorini,A., 1987, Effects of indole-3-pyruvic acid on tryptophan pyrrolase activity "in vitro" and "in vivo". In: Progress in Tryptophan and Serotonin Research 1986, W.de Gruyter & Co., Berlin, 51-54.

Lapin,I.P., 1980, Experimental studies on kynurenines as neuroactive tryptophan metabolites: past, present and future. *TIPS* 1: 410.

Lapin,I.P., and Politi,V., 1993, Anxiolytic effect of indole-3-pyruvic acid (IPA) in mice. *Pharmacol.Res.* 28: 129.

Lapin,I.P., and Politi,V., 1994, Antiethanol effects of indol-3-ylpyruvic acid (IPA) in mice. *Alcohol and Alcoholism* 29: 265.

Lissi,E.A., Faure,M., Montoya,N., Videla,L.A., 1991, Reactivity of indole derivatives towards oxygenated radicals. *Free Rad.Res.Commun.* 15: 211.

Merlo-Pich,E., Solfrini,V., Biagini,G., Fuxe,K., and Agnati,L.F., 1990, Effects of indole-pyruvic acid on sleep and food intake in the rat. *Acta Physiol.Scand.* 139: 583.

Moroni,F., Russi,P., Lombardi,G., Carlà,V., Alesiani,M., Cherici,G., and Moneti,G., 1990, Tryptophan metabolites and excitatory amino acid receptors: studies in physiology and pathology. In: Neurotoxicity of Excitatory Amino Acids, Raven Press, New York, 203-216.

Normanly,J., Slovin,J.P., Cohen,J.D., 1995, Rethinking auxin biosynthesis and metabolism. *Plant Physiol.* 107: 323.

Perkins,M.N., and Stone,T.W., 1985, Actions of kynurenic acid and quinolinic acid in the rat hippocampus in vivo. *Exptl.Neurol.* 88: 570.

Poeggeler,B., Reiter,R.J., Tan,D-X, Chen,L-D, Manchester,L.C., 1993, Melatonin, hydroxyl radical-mediated oxidative damage, and ageing: a hypothesis. *J.Pineal Res.* 14: 151.

Politi,V., Lavaggi,M.V., Di Stazio,G., and Margonelli,A., 1991, Indole-3-pyruvic acid as a direct precursor of kynurenic acid. In: Kynurenine and Serotonin Pathways, Plenum Press, New York, 515-518.

Reiter.R.J., 1995, Oxidative processes and antioxidative defense mechanisms in the ageing brain. *FASEB J.* 9: 526.

Ruggeri,M., Merlo-Pich,E., Zini,I., Fuxe,K., and Agnati,L.F., 1990, Indole-pyruvic acid increases 5-hydroxyindoleacetic acid levels in the cerebrospinal fluid and fronto-parietal cortex of the rat. A microdyalysis study. *Acta Physiol.Scand.* 138: 97.

Russi,P., Carlà,V., Moroni,F., 1989, Indolepyruvic acid administration increases the brain content of kynurenic acid. Is this a new avenue to modulate excitatory aminoacid receptors in vivo? *Biochem.Pharmacol.* 38: 2405.

Sapolsky,R.M., 1992, Stress, the Aging Brain, and the Mechanisms of Neuron Death. MIT Press, Cambridge.

Silvestri,R., Mento,G., Raffaele,M., De Luca,G., Buttini,G., Casella,C., Tisano,A., De Domenico,P., Di Rosa,A.E., Di Perri,R., 1991, Indole-3-pyruvic acid as a possible hypnotic agent in insomniac subjects. *J.Intl. Med.Res.* 19,:403.

Squadrito,F., Sturniolo,R., Trimarchi,G.R., Campo,G.M., De Luca,G., Gorini,A., and Caputi,A.P., 1990, Antihypertensive activity of indolpyruvic acid: a keto analogue of tryptophan. *J.Cardiovasc. Pharmacol*: 15: 102.

Squadrito,F., Calapai,G., Campo,G.M., Zingarelli,B., Arcoraci,V., Ioculano,M., Buemi,A.L., Cucinotta,D., Saitta,A., De Luca,G., Gorini,A., and Caputi,A.P., 1991, Indolepyruvic acid, a tryptophan keto analogue, reduces food intake in obese Zucker rats. *Neurosci.Res.Commun.* 9: 27.

Taylor,M.W., and Feng,G., 1991, Relationship between interferon-gamma, indolamine 2,3-dioxygenase, and tryptophan catabolism. *FASEB J.* 5: 2516.

Walsh,H.A., Daya,S., Whiteley,C.G., 1994, Mode of inhibition of rat liver tryptophan 2,3-dioxygenase by melatonin. *J.Pineal Res.* 16: 188.

Zinner,K., Casadei de Baptista,R., and Cilento,G., 1974, Indole-3-pyruvic acid as a potential luciferin. *Biochem.Biophys.Res.Commun.* 61: 889.

Zoli,M., Merlo-Pich,E., Ferraguti,F., Biagini,G., Fuxe,K., and Agnati,L.F., 1993, Indole-pyruvic acid treatment reduces damage in striatum but not in hippocampus after transient forebrain ischemia in the rat. *Neurochem.Int.* 23: 139.

# PLATELET SEROTONIN ACTS AS A LOCALLY RELEASABLE ANTIOXIDANT

G. Huether[1] and P. Schuff-Werner[2]

[1] Department of Psychiatry
[2] Department of Internal Medicine
University of Göttingen
von Siebold-Str. 5, D 37075 Göttingen, Germany

## 1. INTRODUCTION

It has long been noticed that serotonin (5-HT) is readily oxidized under various conditions to give colored, but poorly identified products. Examples are the enzymatic oxidation of 5-HT by peroxidaseoxidases (Napolitano et al., 1988; Methodiewa et al., 1989; Svensson, 1989) and the nonenzymatic oxidation of 5-HT by oxyhaemoglobin (Blum and Ling, 1959), by ferricytochrome (Alivisatos and Williams-Ashman, 1964) and by inorganic oxidants such as copper and $AgNO_3$ (Eriksen et al., 1960), ammonium persulfate, potassium ferricyanide and sodium periodate (Napolitano et al., 1988). 5-HT has also found to be readily oxidized by human serum and erythrocyte haemolysates (Blum and Ling, 1959; Martin et al., 1960) by liver and brain microsomes (Uemura et al., 1979; 1980) and by spleen homogenates (Nelson and Huggins, 1975) but the products of these oxidation process have only poorly been identified.

The oxidation of the indole nucleus of 5-HT is thought to generate radical intermediates although the fate of these radicals is unknown. Wrona and Dryhurst (1989) investigated the electrochemical oxidation of 5-HT and proposed an oxidative pathway involving the formation of phenoxyl and nitrogen centred radicals which could either couple to give a series of oligomeric products, or loose another electron to generate an unstable quinone imine. The intermediate formation of the same quinone imine was also suggested to occur under the more biomimetic conditions of hydroxyindole oxidation in the presence of peroxidase and $H_2O_2$, resulting in the ultimate formation of a complex mixture of oligomeric products (Napolitano et al., 1988).

The physiological conditions which may lead to the formation of such oxidation products, the kinds of 5-HT oxidation products which are formed under such conditions, their possible role and their potential pathological implications are presently unclear. Circulating 5-HT is stored in high concentrations in thrombocytes. It is released and may reach local extracellular concentrations of 100 μM when platelets are activated at sites of endothelial damage, inflammation of bacterial infiltration. This local release of 5-HT from

*Recent Advances in Tryptophan Research*, edited by Graziella Allegri Filippini et al.
Plenum Press, New York, 1996

**Table 1.** Effect of different concentrations of 5-HT on the formation of reactive oxygen metabolites formed in the presence of peroxidase/$H_2O_2$ and during the respiratory burst of activated phagocytes

| Source of reactive oxygen species | Concentration of 5-HT [mol/l] | | | | |
|---|---|---|---|---|---|
| | 0 | $10^{-3}$ | $10^{-4}$ | $10^{-5}$ | $10^{-6}$ |
| $H_2O_2$/peroxidase | | | | | |
| peak [cpm x$10^{-4}$] | 184.6 | 0.7 | 5.3 | 41.0 | 121.4 |
| integral [cpm x $10^{-6}$] | 120.9 | 0.6 | 3.7 | 25.5 | 120.8 |
| Monocytes | | | | | |
| peak [cpm x$10^{-4}$] | 65.6 | 21.4 | 26.8 | 28.5 | 36.8 |
| integral [cpm x $10^{-6}$] | 106.8 | 22.0 | 33.2 | 37.1 | 71.3 |
| Granulocytes | | | | | |
| peak [cpm x$10^{-4}$] | 276.3 | 93.7 | 157.5 | 209.1 | 237.7 |
| integral [cpm x $10^{-6}$] | 801.4 | 240.9 | 457.9 | 638.6 | 691.1 |

Suppression of radial formation was measured by the decrease of luminol-enhanced chemiluminescence. The cell free assay was started by the addition of $H_2O_2$ as substrate. Monocytes were activated by PMA and granulocytes by opsonized S. aureus. Data show the results of one representative experiment.
Peak: maximal light emission, integral - total light emission

platelets is additionally stimulated by humoral immunological reactions (Henson, 1970), by bacterial endotoxins (Davis et al., 1961) and by activated phagocytes (Clark et al., 1980). Therefore, the local concentration of free 5-HT is particularily high at sites of inflammation and local irritation, i.e. at sites were the oxidative pressure and the likelihood of the formation of reactive oxygen metabolites is particularily high. Encouraged by this coincidence we performed a series of experiments in order to study the possible role of platelet derived serotonin as a locally releasable antioxidant and radical scavenger.

## 2. ANTIOXIDATIVE PROPERTIES OF 5-HT IN A CELL-FREE, RADICAL GENERATING MODEL SYSTEM

The horseradish peroxidase/superoxide (HRP/$H_2O_2$)-system was used as a model for the investigation of the antioxidative properties of 5-HT and for the characterization of the major oxidation products formed under these conditions. In the presence of HRP and $H_2O_2$, 5-HT was almost completely oxidized during the first minute of incubation. and could no longer be measured by HPLC and ECD. With the addition of increasing amounts of 5-HT to the HRP/$H_2O_2$ system, a dose-dependent suppression of the chemiluminescence signal was observed (Table 1). As seen in autoradiograms of TLC-separations, the loss of 5-HT coincided with the appearance of several oxidation products of 5-HT with smaller Rf-values. The pattern of TLC separations of the reaction products was dependent on the $H_2O_2$ concentration used. Separation of the reaction products on a Sephadex LH 20 column revealed a major peak representing about 50 % of total radioactivity at about half the elution volume of 5-HT. This major oxidation product was identified by plasma desorption mass spectrometry as a 5-HT-dimer which was formed under abstraction of 2 protons. The most likely structure of this dimer, as suggested in Fig. 1, is that of a 5,5′dihydroxy-4,4′-bitryptamine. Almost no oxidation of 5-HT was observed during incubation in the presence of either HRP or $H_2O_2$ alone. The oxidation of 5-HT in the presence of HRP and $H_2O_2$ was prevented by the addition of ascorbic acid or $NADH_2$. Glutathione was less effective and gave rise to a conjugate characterized by a very small Rf-value in TLC-separations. In the presence of plasma or homogenates of liver and brain tissue, the rate of 5-HT oxidation was reduced, and the

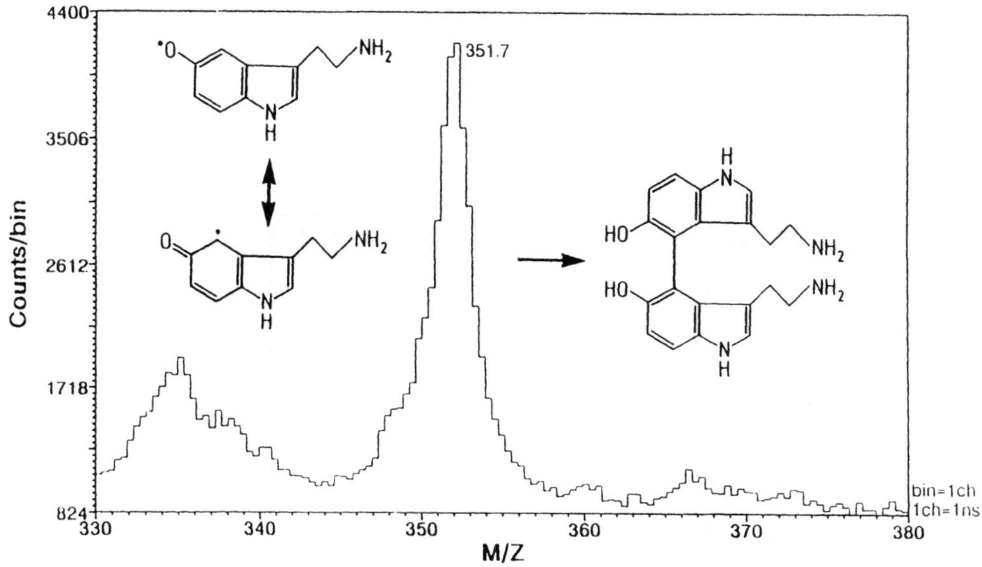


**Figure 1.** Plasma desorption time-of-flight mass sepctrum of the major oxidation product formed in the presence of peroxidase/$H_2O_2$ acid in the presence of activated phagocytes. Most likely the 5-HT dimer is a 5,5′ dihydroxy-4,4′-bitryptamine which is formed by a free radical mechanism involving the one-electron oxidation of 5-HT and the formation of semiquinone imine or phenoxy radicals in which the unpaired electron is located at Ċ (4).

formation of soluble oxidation products was diminished on the expense of considerable binding of oxidized 5-HT to tissue proteins (Huether et al., 1990).

The formation of dimeric and polymeric oxidation products of 5-HT has long been thought to imply the initial formation of 5-HT radicals (Perez-Reyes and Mason, 1981; Wrona and Dryhurst, 1988). There is some evidence that 5-HT can be oxidized to a radical, e.g., by caeruloplasmin, (Walaas and Walaas, 1961), oxidation of pyridine nucleotides by cytochrome C (Alivisatos and Williams-Ashman, 1964), during permanganate oxidation (Borg, 1965) and during autoxidation in alkaline solutions (Perez-Reyes and Mason, 1981). Free radical reactions of the serotonin free radicals with themselves, with authentic serotonin and its oligomeric oxidation products, and with nucleophilic sites of tissue proteins may be important in the formation of melanin like material observed upon incubation of microsomes in the presence of 5-HT and NADPH (Uemura et al., 1980). The suppression of chemiluminescence seen during coincubation of 5-HT with HRP and $H_2O_2$ indicates that 5-HT, like other hydroxylated indoles and catechols, prevent the formation reactive oxygen metabolites during $H_2O_2$ cleavage. The generation of reactive oxygen metabolites during peroxidatic degradation of $H_2O_2$ seem to be responsible for the substrate induced inactivation of peroxidase (Arnao et al., 1990). That 5-HT, as a radical scavenger and reductant cosubstrate, can prevent this inactivation process is indicated by the increased consumption of $H_2O_2$ by HRP in the presence of 5-HT (Fig. 2).

## 3. SEROTONIN INHIBITS THE OXIDATIVE MODIFICATION OF LOW DENSITY LIPOPROTEINS

Oxidative modification of low density lipoproteins (LDL) is thought to play a special role in the pathogenesis of atherosclerosis. Factors that affect oxidative LDL modification

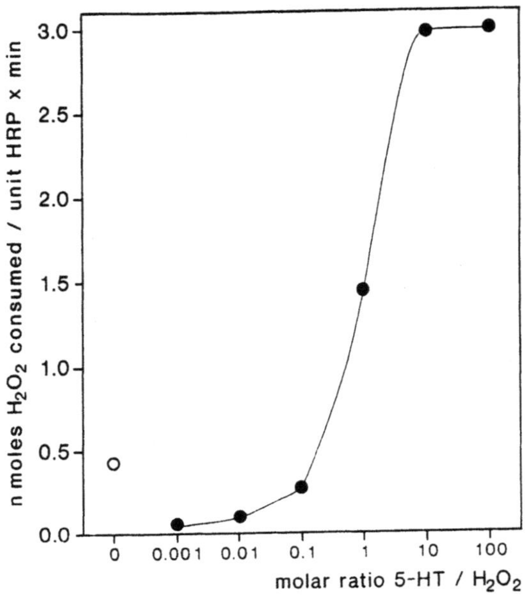


**Figure 2.** Stimulation of $H_2O_2$-degradation by 5-HT in the peroxidase/$H_2O_2$-system. 50 µl peroxidase (25 mU/µl), 80 µl 0,1 M phosphate buffer (pH 7.0) and 20 µl buffer containing increasing concentrations of 5-HT (to give final concentrations between 0.1 µM and 1.0 mM) were mixed and the reaction was stated by the addition of 50 µl $H_2O_2$ (0.2 mM). Even at the highest concentration, 5-HT was completely oxidized during the 15 min of incubation.

**Table 2.** Effect of different indole and catechol derivatives (100 µmol/l) on the oxidation of LDL lipids by copper ions (10 µmol/l)

| Compound added | % MDA formed |
|---|---|
| Cu-LDL without addition | 100 |
| tryptamine | 98 |
| tyramine | 97 |
| L-tryptophan | 94 |
| 5-hydroxy-indole acetic acid | 67 |
| 5-hydroxy-tryptophan | 21 |
| 3-hydroxy-tyramine | 15 |
| 3-hydroxy-anthranilic acid | 15 |
| epinephrine | 13 |
| L-DOPA | 10 |
| 5-hydroxy-tryptamine | 9 |
| norepinephrine | 4 |
| No Copper Control (NCC) | 5 |

Values are means of duplicate determinations from 4 representative experiments. 100 % MDA correspond to 46.7 ± 6.4 nmol MDA/mg LDL protein (n=4).

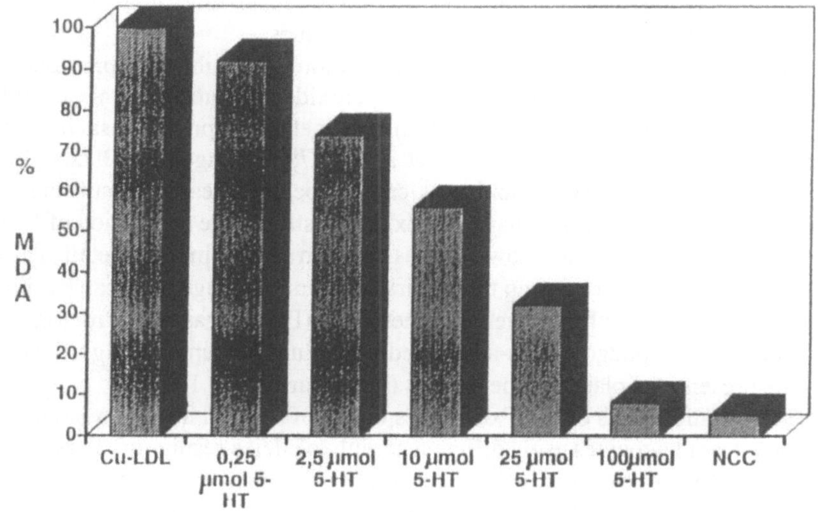


**Figure 3.** Effect of increasing 5-HT-concentrations on the LDL lipid peroxidation by copper ions. 100% TBARS corresponds to 83 nmol MDA/mg LDL protein. Values are means of duplicate determinations from a representative experiment.

may therefore influence atherogenesis. The ability of serotonin to inhibit the oxidative modification of (LDL) was tested in a system in which $Cu^{++}$ was used as a trigger for the oxidative modification of LDL. Serotonin, like indole and catechol derivatives inhibited the $Cu^{++}$ induced peroxidation of LDL-lipids (Table 2). This effect was dose-dependent and a half-maximal protection was seen at a 5-HT concentration of about 10 µmol/l (Fig. 3).

Serotonin, like other hydroxylated aromatic amines has been reported to inhibit lipid peroxidation also in liver microsomes, liposomes and arachidonic acid micelles (Reyftman et al., 1990). Since circulating platelets release their stored 5-HT at sites of endothelial injury, at sites of inflammation and during thrombus formation, local concentrations of 10 µmol/l may easily be reached under these conditions (Benedict et al., 1986). Such concentrations would be sufficient for a significant antioxidative protection of LDL against oxidative modification.

## 4. SEROTONIN ACTS AS A RADICAL SCAVENGER AND IS OXIDIZED TO A DIMER DURING THE RESPIRATORY BURST OF STIMULATED PHAGOCYTES

Polymorphonuclear and mononuclear phagocytes constitute a powerful defence system against microorganisms invading the host's organism. Upon activation either by opsonized bacteria or by soluble mediators they respond immediately by generating a so-called "respiratory burst" in the course of which reactive oxygen species are formed which contribute directly or indirectly to their microbicidal activity (Jonston et al., 1975; Babior, 1978). The reactive oxygen species produced by activated phagocytes are involved in inflammatory processes and may contribute to cell and tissue damage either directly or through inactivation of serum anti-proteases (Weiss, 1989). In addition to inflammation, reactive oxygen species appear to be implicated in the pathogenesis of cancer, arteriosclerosis, and other diseases (Halliwell, 1987). Obviously, a delicate balance exists between activation of circulating or marginated phagocytes directed toward defence and those inflammatory responses that ultimately damage host tissues.

Protection against overshooting radical formation is thought to be provided by a set of antioxidant defences, including proteins (superoxide dismutase, catalase and metal binding proteins), vitamins (vitamin C and E) and metabolic products such as reduced glutathione, uric acid or bilirubin (Stocker et al., 1987; DeLange and Glazer, 1989). In contrast to 5-HT, none of these antioxidants can be locally released in sufficiently high amounts under conditions of inflammation or oxidative stress. The interaction of 5-HT with the respiratory burst of phagocytes, however, has not been studied in some detail. The release of reactive oxygen metabolites during the respiratory burst of phagocytes can be monitored by luminol or lucigenin amplified chemiluminescence (Prendergast and Proctor, 1980). It has been reported that phagocytosis-associated chemiluminescence is significantly suppressed in the presence of platelet supernatants (Nannmark et al., 1992).

We studied the effects of 5-HT on the responses of isolated human mononuclear and polymorphonuclear phagocytes using different stimuli to elicit a respiratory burst. As shown in Table 1 the PMA-induced, luminol-enhanced chemiluminescence (peak and total light generation) of both granulocytes and monocytes is inhibited in a dose-dependent fashion in the presence of 5-HT. This effect was reproducibly seen in all experiments independent of whether or not the actual initial values were high or low. At a 5-HT concentration of 1 mmol/l the chemiluminescence signal was reduced by about 60 to 70 %, whereas at a 5-HT concentration of 100 μmol/l the chemiluminescence decreased by more than one half of the amplitude reached in the absence of 5-HT. This effect was independent of the stimulus used to elicit the chemiluminescence response.

Thin-layer chromatography and autoradiographic analysis of the reaction products of 5-HT revealed that 5-HT is only minimally metabolised in the presence of unstimulated human phagocytes. During the PMA-elicited respiratory burst of these cells, 5-HT is converted into several degradation products. In the presence of antioxidants ($NADH_2$, ascorbic acid, tocopherol) this reaction is prevented, indicating that 5-HT is oxidized during the respiratory burst. The stimulation of granulocytes with opsonized staph. aureus leads to quantitative but not to qualitative differences in the formation of 5-HT metabolites. Activated monocytes and granulocytes produce an identical pattern of at least three major oxidation products with different Rf-values (for original data see Schuff-Werner et al., 1995).

Using preparative gel chromatography on Sephadex LH 20, sufficient quantities of the major 5-HT metabolite formed in the presence of activated granulocytes could be isolated for further characterization by HPLC and mass spectrometry. HPLC separation and electrochemical detection of the major oxidation product revealed a distinct signal at about half the retention time of authentic 5-HT. The mass spectrometric analysis of the major oxidation product of 5-HT indicated that this was the same 5-HT dimer as that found in the cell-free radical generating system.

Both, the radical scavenging property of 5-HT derived from activated platelets and the resulting formation of a novel 5-HT oxidation product extend the spectrum of hitherto known interactions between leucocytes and thrombocytes in inflammatory, immunological and thrombotic processes (as recently reviewed by Cerletti et al., 1992, and Bazzoni et al., 1992). Several inhibitory effects of platelet products, e.g., $O_2$-generation (Naum et al., 1991), response to stimulation (Del Principe et al., 1990) or other functions of human neutrophils may now be explained by the actions of platelet-derived 5-HT on the respiratory burst of activated phagocytes. Until now, interactions of 5-HT with leucocytes have almost exclusively been discussed on the basis of an assumed activation of 5-HT receptors. However, 5-HT receptors have not yet been detected on polymorphonuclear phagocytes, and the concentrations at which 5-HT was found to influence certain functions of these cells are in the micromolar range. As shown in this study, the $IC_{50}$ of 5-HT on the luminol amplified chemiluminescence of activated phagocytes is about 100 µmol/l. The concentration of free 5-HT in the plasma is in the picomolar range. Circulating 5-HT almost completely stored in the secretory granules of thrombocytes where it reaches concentrations of 100-300 mmol/l (Reimers, 1985). Whole blood serotonin concentrations are about 1 µmol/l. During thrombus formation, due to the massive release of 5-HT from thrombocytes, 5-HT concentrations of 85 µmol/l have been reported at thrombotic sites (Benedict et al., 1986). Consequently, local concentrations of 100 µmol/l may well be reached at the immediate sites of 5-HT release by activated thrombocytes, e.g. at sites of inflammation and/or endothelial damage. Such concentrations would indeed be sufficient to profoundly attenuate the damage of surrounding cells by the reactive oxygen metabolites released during the respiratory burst of phagocytes.

Therefore the release of 5-HT at the sites of action of phagocytes appears to be a very efficient mechanism to improve the power of the cellular defence system and to minimize damage to the host tissue. This adds an interesting new aspect to the physiological role of 5-HT stored in circulating platelets, a role which is still not fully understood (for review see Pletscher, 1970). The slow turnover of 5-HT and the absence of a biosynthetic pathway for 5-HT (platelets are loaded during their passage through the gut and the portal blood with 5-HT synthesized by the enterochromaffine cells of the intestinal mucosa) are hardly compatible with a major regulatory function under physiological conditions. It has therefore long been speculated that 5-HT is stored in circulating platelets and will be released in case of emergencies in large quantities at sites of irritation. Because such sites are always sites of increased radical formation (Halliwell, 1987) the primary function of the platelet stored 5-HT may be that of an unique, i.e. locally releasable, antioxidant and radical scavenger.

## ACKNOWLEDGMENTS

This research was supported by a grant of the Deutsche Forschungsgemeinschaft (Hu 351/10-1).

## REFERENCES

Alivisatos, S.G.A., Williams-Ashman, H.G., 1964, Serotonin-mediated oxidation of dihydronicotinamide derivates by cytochrome C. *Biochim. Biophys. Acta* 86:392.

Arnao, M.B., Acosta, M., Del Rio, J.A., Garcia-Canovas, F., 1990, Inactivation of peroxidase by hydrogen peroxide and its protection by a reductant agent. *Biochim. Biophys. Acta* 1038:85.

Babior, B.M., 1978, Oxygen-dependent microbial killing by phagocytes. *N. Engl. J. Med.* 298:659.

Baehner, R.L., Boxer, L.A., Allen, J.M., Davis, J., 1977, Autooxidation as a basis for altered function by polymorphonuclear leukocytes. *Blood* 50:327.

Bazzoni, G., Defana, E., Del Maschio, A., 1992, Polymorphonuclear leucocyte modulation of platelet function: Relevance to the pathogenesis of thrombosis. *Pharmacol. Res.* 26:269.

Benedict, C.R., Mathew, B., Rex, K.A., Cartwright, J., jr., Sordahl, L.A., 1986, Correlation of plasma serotonin changes with platelet aggregation in an in vivo dog model of spontaneous acclusive coronary thrombus formation. *Circ. Res.* 58:58.

Blum, J.J., Ling, N.S., 1959, Oxidation of serotonin and 5-hydroxyindoles during the denaturation of oxyhaemoglobin. *Biochem. J.* 73:530.

Borg, D.C., 1965, Transient free radical forms of hormones. EPR spectra from iodothyrosins indoles, estrogens, and insulin. *Prog. Natl. Acad. Sci.* USA, 53:829.

Cerletti, Ch., Evangelista, V., De Gaetano, G., 1992, Polymorphonuclear leucocyte-dependent modulation of platelet function: Relevance to the pathogenesis of thrombosis. *Pharmacol. Res.* 26:261.

Clark, R.A., Klebanoff, S.J., 1980, Neutrophil-platelets interaction mediated by myeloperoxidase and hydrogen peroxide. *J. Immunol.* 124:399.

Davis, R.B., Meeker, W.R. jr., Bailey, W.L. 1961, Serotonin release by bacterial endotoxin. *Proc. Soc. Exp. Biol. Med.* 108:775.

Del Principe, D., Menichelli, A., DiGiulio, S., De Matteis, W., Giordani, M., Pentassuglio, A.M., and Finazzi-Agro, A., 1990, Stimulated platelets release factor(s) affecting the in vitro response of human polymorphonuclear cells. *J. Leuk. Biol.* 48:7.

DeLange, R.J., and Glazer, A.N., 1989, Phycoerythrin in fluorescence-based assay for peroxy radials: a screen for biologically relevant protective agents. *Anal. Biochem.* 177:300.

Eriksen, N., Martin, G.M., Benditt, E.P., 1960, Oxidation of the indole nucleus of 5-HT and the formation of pigments: isolation and partial characterization of a dimer of 5-HT. *J. Biol. Chem.* 235:1662.

Halliwell, B., 1970, Oxidants and humans disease: some new concepts. *FASEB J.* 1:358.

Henson, P.M., 1970, Mechanisms of release of constituents from rabbit platelets by antigen-antibody complexes and complement. *J. Immunol.* 105: 476

Huether, G., Reimer, A., Schmidt, F., Schuff-Werner, P., and Brudny, M.M., 1990, Oxidation of the indole nucleus of 5-hydroxytryptamine and formation of dimers in the presence of peroxidase and $H_2O_2$. *J. Neural. Transmission* 32:249.

Johnston, R.B. Jr., Keele, B.B. Jr., Misra, H.P., Lehmeyer, J.E., Webb, L.S., Baehner, R.L., and Rajagopolan, K.V., 1975, The role of superoxide anion generation in phagocytic bactericidal activity. Studies with normal and chronic granulomatous disease leukocytes. *J. Clin. Invest.* 55:1357.

Martin, G.M., Benditt, E.P., Erikson, N., 1960, Enzyme oxidation of the indole nucleus of 5-HT: properties of an enzyme in human serum and the products of oxidation. *Arch. Biochem. Biophys.* 90:208.

Methodiewa, D., Reszka, K., Dunford, M.B., 1989, Evidence for a peroxidatic oxidation of norepinephrine, a catecholamine, by lactoperoxidase. *Biochem. Biophys. Res. Commun* 160:1183.

Nannmark, U., Sennerby, L., Bjursten, L.M., Skolnik, G., and Bagge, U., 1992, Inhibition of leukocyte phagocytosis by serotonin and its possible role in tumour cell distruction. *Cancer Letters* 62:83.

Napolitano, A., D'Ischia, M., Prota, G., 1988, A profile of the oxidation chemistry of 5-hydroxyindole under biomimetic conditions. *Tetrahydron* 44:7265.

Naum, C.C., Kaplan, S.S., Basford, R.E., 1991, Platelets and ATP prime neutrophils for enhanced $O_2$-generation at low concentrations but inhibit $O_2$-generation at high concentration. *J. Leuk. Biol.* 49:83.

Nelson, D.R., Huggins, A.G., 1975, The interaction of 5-HT and related hydroxyindoles with horseradish and mammalian peroxidase systems. *Biochem Pharmacol.* 24:181.

Perez-Reyes, E., Mason, R.P., 1981, Characterization of the structure and reactions of free radicals from serotonin and related indoles. *J. Biol. Chem.* 256:2427.

Pletscher, A., 1987, The 5-Hydroxytryptamine system of blood platelets: physiology and pathophysiology. *Int. J. Cardiol.* 14:177.

Prendergast, E., and Proctor, R., 1981, Chemiluminescence of neutrophils in presence of luminol. *J. Clin. Microbiol.* 13:390.

Read, N.G., 1985 The accumulation of an aryloxya/kylamidine (501C) an 5-hydroxytryptamine in human polymorphonuclear leukocytes: a quantitative electron microscopy study. *J. Pharm. Pharmacol.* 37:96.

Reimers, H.J. Adenine nucleotides in blood platelets. In the Platelets: Physiology and pharmacology edited by Longenecken G., Orlando, Florida, 1985, 85-112.

Reyftmann, J.P., Santus, R., Mazière, J.C., Morlière, P., Salmon, S., Candide, Ch., Mazière, C., and Haigle, J., 1990, Sensitivity of tryptophan and related compounds to oxidation induced by lipid autoperoxidation. Application to human serum low- and high-density lipoproteins. *Biochim. Biophys. Acta* 1042:159.

Schuff-Werner, P., Splettstösser, W., Schmidt, F., and Huether, G., 1995, Serotonin acts as a radical scavenger and is oxidized to a dimer during the respiratory burst of human mononuclear and polymorphonuclear phagocytes. *Europ. J. Clin. Invest.* 25, in press.

Stocker, R., Yamamoto, Y., McDonagh, A.F., Glazer, A.N., and Ames, B.N., 1987, Bilirubin is an antioxidant of possible physiological inportance. *Science* 235:1043.

Svensson, B.E., 1989, Involvement of cysteine, serotonin and their analogues in peroxidase-oxidase reactions. *Chem. Biol. Interact.* 70: 305.

Uemura, T., Shimazu, T., Miura, R., Yamano, T., 1980, NADPH-dependent melanin pigment formation from 5-OH-indolealkylamines by hepatic and cerebral microsomes. *Biochem. Biophys. Res. Commun* 93:1074.

Walaas, E., Walaas, O., 1961, Oxidation of reduced phosphopyridine nucleotides by p-phenylenediamines, catecholamines and serotonin in the presence of ceruloplasmin. *Arch. Biochem. Biophys.* 95:151.

Weiss, S.J., 1989, Tissue destruction by neutrophils. *N. Engl. J. Med.* 320:365.

Wrona, M.Z., Dryhurst G., 1988, Further insights into the oxidation chemistry of 5-HT. *J. Pharmacol. Sci.* 77:911.

# THE INDOLEAMINE MELATONIN AS A FREE RADICAL SCAVENGER, ELECTRON DONOR, AND ANTIOXIDANT

## In Vitro and in Vivo Studies

R. J. Reiter

Department of Cellular and Structural Biology
The University of Texas Health Science Center
7703 Floyd Curl Drive, San Antonio, Texas 78294-7762

## 1. INTRODUCTION

Following its discovery in bovine pineal tissue in the late 1950s (Lerner et al, 1958) melatonin was functionally linked to the reproductive system (Hoffman and Reiter, 1965). Soon after its discovery, the pineal gland, which is a primary source for circulating melatonin, was shown to be functionally dependent on the photoperiodic cycle (Quay, 1964; Snyder et al., 1964), found to regulate photoperiod-mediated reproductive physiology (Czyba et al., 1964; Hoffman and Reiter, 1965, Reiter, 1973), and shown to depend on its sympathetic innervation for its activity (Reiter and Hester, 1966).

Although it had been strongly suspected for years, it was not until 1976, however, that the hormone melatonin was proven to be the actual pineal envoy responsible for the reproductive changes that had been observed (Reiter et al., 1976; Tamarkin et al., 1976). These endocrine aspects of the indoleamine melatonin were the subject of a previous conference of this society (Reiter and Richardson, 1980). Because of these finding research on the pineal gland and melatonin was highly invested in neuroendocrine-reproductive physiology for decades, a trend that continues to the present time (Goldman and Nelson, 1993; Malpaux et al., 1993).

Simultaneous with these observations, however, were findings which linked the pineal gland and melatonin to circadian physiology (Arendt et al., 1987). As a result, it is now well accepted that melatonin, in addition to mediating the seasonal reproductive events described above, also influences circadian changes in physiology thereby linking some functional aspects of the organism with the prevailing light:dark cycle (Armstrong and Redman, 1993).

Despite these well characterized functions of melatonin, data accumulated from studies in which the indoleamine was given suggested that melatonin's actions were far more ubiquitous than merely its demonstrated effects of reproductive and circadian physiology

*Recent Advances in Tryptophan Research,* edited by Graziella Allegri Filippini et al.
Plenum Press, New York, 1996

(Reiter, 1991). About 4 years ago we made a set of observations which led us to propose a new functional role for melatonin. While studying cardiovascular biochemistry, we observed a pineal-dependent circadian rhythm in the activity of $Ca^{2+}$ + $Mg^{2+}$-dependent ATPase (the calcium pump) in vivo and furthermore we found that when cardiac muscle cells were incubated with melatonin the activity of this enzyme increased (Chen et al., 1994). These novel observations had several important implications which alerted us to the possibility that melatonin may have other basic functions within the cell. Firstly, melatonin stimulated the activity of the calcium pump in an organ in which melatonin receptors had not been identified, and this effect also did not rely on specific receptors located elsewhere since even in culture the cardiomyocytes responded to the melatonin signal.

Among several potential explanations for the stimulatory effect of melatonin on calcium pump activity, we surmised that melatonin may have altered the redox state of the cell and then the activity of the enzyme passively increased. Soon thereafter we showed that melatonin is in fact a highly efficient scavenger of the hydroxyl radical (Tan et al., 1993a). This observation opened a new field of melatonin research which has linked this hormone with oxygen metabolism in all aerobic organisms.

## 2. OXYGEN ACTIVATION AND FREE RADICALS

Free radicals, which are molecules with unpaired electrons, are generally unstable and highly reactive. To achieve a level of stability free radicals rid themselves of their unpaired electron. This can be done in two ways. A free radical can contribute its unpaired electron to another molecule, in which case the radical is a reducing species [such as the peroxyl radical ($ROO^\bullet$) which is produced during lipid peroxidation], or it can accept an electron, thereby serving as an oxidizing agent [such as the hydroxyl radical ($^\bullet OH$) which is formed by the addition of three electrons to molecular oxygen ($O_2$)]. The $^\bullet OH$ is formed from $O_2$ as follows:

$$O_2 \xrightarrow{e^-} O_2^{\bar{\bullet}} \xrightarrow[SOD]{e^-} H_2O_2 \xrightarrow{e^-} {}^\bullet OH \qquad \text{Eq. 1}$$

Depending on the specific molecule that encounters a radical, the radical can be either a reducing or oxidizing agent. A radical can also autoxoxidize in a reaction known as dismutation. An example is the union of two superoxide anion radicals ($O_2^{\bar{\bullet}}$) as follows:

$$O_2^{\bar{\bullet}} + O_2^{\bar{\bullet}} \xrightarrow{2H^+} H_2O_2 + O_2 \qquad \text{Eq. 2}$$

In this reaction one $O_2^{\bar{\bullet}}$ is reduced while the other is oxidized.

Some of chief by-products of $O_2$ metabolism within cells which interest free radical biologists are $O_2^{\bar{\bullet}}$, hydrogen peroxide ($H_2O_2$) and $^\bullet OH$ (Floyd, 1993). They are related as shown in Eq. 1. The monovalent reduction of $O_2$ generates the $O_2^{\bar{\bullet}}$. Although the $O_2^{\bar{\bullet}}$ is not especially toxic to cellular molecules, in vivo its destructive potential may relate to the fact that, in the presence of $H_2O_2$, the $O_2^{\bar{\bullet}}$ can generate the highly toxic $^\bullet OH$ in what is known as the Haber-Weiss reaction:

$$O_2^{\bar{\bullet}} + H_2O_2 \longrightarrow O_2 + OH^- + {}^\bullet OH \qquad \text{Eq. 3}$$

$O_2^{\bar{\bullet}}$ can also undergo dismutation (Eq. 2 above) to generate $H_2O_2$. This reaction in vivo is catalyzed by one of several isoforms of an enzyme identified as superoxide dismutase (SOD). $H_2O_2$, rather than being a free radical (it has no unpaired electrons), is generally referred to as a reactive oxygen intermediate (ROI); $H_2O_2$ is a potent oxidizing agent. In the presence of a transition metal, most often $Fe^{2+}$ but also $Cu^{1+}$, $H_2O_2$ decomposes into a harmless $OH^-$ ion and the highly reactive •OH; this conversion involving the transition metal is known as the Fenton reaction (Meneghini and Martin, 1993; Koppenal, 1994) and is the final step shown in Eq. 1. The reactivity of the •OH is high in biological media with the radical indiscriminantly attacking virtually any organic molecule. These interactions produce secondary radicals due to the loss of a $H^+$ or the transfer of an electron. Of all the oxygen-based radicals produced it is generally concluded that the •OH is the most reactive, therefore the most toxic.

Another radical of interest to the present discussion is the ROO•. This radical is generated by the interaction of an oxygen-based radical, either $O_2^{\overline{\cdot}}$ or •OH, with a polyunsaturated fatty acid (PUFA) (Pryor, 1994). The ROO• is somewhat less reactive and more selective than the •OH; however, it can re-initiate (propagate) the oxidation of a second PUFA once produced. Thus, it is responsible for the chain reactions associated with the peroxidation of lipids. As a result, once the process of lipid peroxidation is underway, the chain reaction could theoretically result in the destruction of all the lipid in an organism if the process is not interrupted, e.g., by a chain breaking antioxidant.

## 3. MELATONIN AS AN ELECTRON DONOR, FREE RADICAL SCAVENGER AND ANTIOXIDANT

### 3.1 In Vitro Evidence

The ability of melatonin to scavenge the •OH was initially shown in a series of simple in vitro experiments. In the first study to test this possibility, we merely exposed a solution of $H_2O_2$ to ultraviolet light (uv); this results in the decomposition of the $H_2O_2$ to produce the •OH. However, because of its extremely short half-life (due to recombination reactions), it is not possible to directly quantify the •OH generated. Thus, a spin trap also is added to the mixture. The spin trapping reagent (in this case 5, 5-dimethyl-pyrroline-N-oxide or DMPO) forms a adduct with the •OH to provide an index of the number of radicals generated according to the following equation:

$$H_2O_2 + \text{DMPO} \xrightarrow[254m]{uv} \text{DMPO-}^{\bullet}\text{OH} \qquad \text{Eq. 4}$$

The DMPO-•OH has a relatively long half life and, in the case of Tan and co-workers (1993a), they were identified by means of high performance liquid chromatography and electron spin resonance spectroscopy. By additionally adding melatonin (or other potential radical scavengers) to the system, the reduction in the formation of the DMPO-•OH was used as an index of the ability of the scavenger to neutralize the •OH. Using this system, Tan et al (1993a) went on to show that melatonin was actually a more efficient scavenger of the •OH than was glutathione, a well known intracellular scavenger. Certainly, melatonin's ability in this regard far exceeded the expectations of the authors of the report. In the same paper, it was shown that indoles structurally similar to melatonin lacked the efficient scavenging ability of melatonin and that both the methoxy group at position 5 of the indole nucleus as well as the acetyl group on the side chain are important for the scavenging activity of the pineal hormone.

The non-enzymatic degradation of melatonin when it scavenges a •OH is believed to occur when melatonin contributes and electron to the free radical as follows:

$$\text{melatonin} + {}^{\bullet}\text{OH} \longrightarrow \text{indolyl cation radical} \qquad \text{Eq. 5}$$

In so doing, melatonin itself becomes a free radical because of its unpaired electron (Poeggeler et al., 1994). The reactivity of the indolyl cation radical is very low compared to that of the •OH; therefore, the reaction greatly reduces the overall toxic environment of a cell. Once formed, the indolyl cation radical is believed to scavenge the $O_2^{\cdot-}$, which is a precursor of the •OH.

indolyl cation radical + $O_2^{\cdot-}$ ⟶ $N^1$-acetyl-$N^2$-formyl-5-methoxy kynuramine (AFMK) Eq. 6

This reaction further reduces •OH generation by removing an important precursor of this high toxic entity. AFMK is excreted in the urine. Finally, we are currently looking into the possibility that the indolyl cation radical is recycled back to melatonin allowing the product to neutralize yet another •OH.

Shortly after it was shown that melatonin neutralizes the •OH, Pieri et al (1994) found, also in an in vitro system, that melatonin also scavenged the ROO• and it did so more efficiently than does vitamin E (tocopherol). Vitamin E is a well known chain breaking antioxidant which plays a very important role in cellular membranes where it interrupts the process of the oxidative destruction of PUFA. Preliminary studies from our laboratory have subsequently shown that melatonin may act syngeristically to enhance the scavenging actions of acorbate, the vitamin E analog trolox, and glutathione (Poeggeler et al., 1995). The mechanisms of these interactions have yet to be defined.

A large number of other in vitro studies which used homogenized tissue samples or cells have provided indirect evidence that melatonin is an efficient scavenger of toxic free radicals (Pierrefiche and Laborit, 1995; Reiter et al., 1995). These studies revealed that melatonin is highly protective of DNA against ionizing radiation (Vijayalaxmi et al., 1995), a process known to involve the highly toxic •OH.

Also, melatonin was found to inhibit the oxidation of membrane lipids in cellular homogenates of various brain regions (Pierrefiche et al., 1993; Melchiorri et al., 1995b; Sewerynek et al., 1995a). Furthermore, melatonin reduces excitatory neurotransmitter toxicity in neurons (Giusti et al., 1995) and inhibits apoptosis in nerve cells following their exposure to singlet oxygen (Cagnoli et al., 1995). Oxidative stress is purported to be a mediator of apoptosis (Burke and Sandstrom, 1994). As already stated, while these measurements are all indirect evidence of the antioxidant potential of melatonin, collectively they present a rather compelling argument that melatonin is a noteworthy antioxidant and free radical scavenger in vitro.

## 3.2 In Vivo Evidence

Whereas free radicals are difficult to measure in in vitro situations, their quantification is even more problematic in vivo although there are ways of assessing their production using, for example, spin trapping agents. At this point there have been no such studies illustrating melatonin's ability to scavenger free radicals in whole animals. However, the damage inflicted on macromolecules has been measured as an index of melatonin's antioxidant capability (Reiter et al., 1994; Reiter, 1995). While melatonin certainly exhibits this capability in vitro, the effects in vivo are at least as dramatic.

Following the demonstration by Tan and colleagues (1993a) that in a cell free system melatonin quenched the •OH, they soon tested the indole in vivo to determine whether it would limit the toxicity of a chemical carcinogen, safrole, which is believed to damage DNA due to its ability to generate free radicals. In a series of two reports in which safrole-treated rats were either co-treated with pharmacological doses of melatonin or not given supplemental quantities of the indole, melatonin was found to be highly protective of DNA in reducing safrole toxicity (Tan et al., 1993b). In the non-melatonin supplemented animals, the nocturnal rise in endogenous melatonin levels were likewise shown to reduce DNA damage following safrole administration (Tan et al., 1994). This was further emphasized by the observation that otherwise untreated rats which were pinealectomized, to deprive them of their circadian melatonin rhythm (Reiter, 1991), exhibited augmented safrole-induced damage to hepatic DNA. The protection of nuclear DNA from oxidative attack by melatonin is consistent with the high intranuclear concentrations of the indole (Reiter et al., 1995).

Melatonin's ability to protect macromolecules from the toxic effects of free radicals is, however, not limited to the nucleus. When Abe and colleagues (1994) treated newborn rats that were pharmacologically depleted of their antioxidant glutathione with supplemental melatonin, cataract development normally associated with glutathione depletion was overcome. Thus, melatonin had apparently substituted for glutathione in reducing the oxidative damage that is believed to be responsible for cataracts. It is generally accepted that lenticular proteins, when oxidized, cause clouding of the lens in animals including man. Thus, the implication of the findings of Abe and co-workers is that melatonin acts in the aqueous cytosol to reduce oxidative stress to proteins much like it does in the nucleus to protect DNA. Major disruptions of cellular physiology occur when the integrity of cellular membranes are compromised. A frequent cause of membrane damage relates to the ready oxidizability of the PUFA in the membranes.

A great deal of effort has been put forth to identify molecules that will protect membrane lipids from the damage caused by free radicals. The ability of melatonin to reduce this damage in vivo seems to be highly significant. One of the first of these studies was that of Melchiorri et al (1995a). This group treated rats with the highly toxic herbicide, paraquat, a compound that is activity taken up in the lungs where it causes massive destruction of lipids in cell membranes. Not surprisingly the treatment caused large increases in malonaldehyde and 4-hydroxyalkenal levels in the lungs associated with a reduction of reduced glutathione levels. When paraquat-injected animals were also given melatonin, however, the changes indicative of lipid peroxidation were prevented. Considering the great potency of this herbicide in damaging membrane lipids, the prevention of these changes by melatonin is certainly quite remarkable.

Other in vivo models have been used to test for melatonin's antioxidant ability. A commonly used method for inducing lipid peroxidation is to treat animals with the bacterial endotoxin, lipopolysaccharide (LPS). This toxin not uncommonly causes mulitple organ failure because of the damage it inflicts on macromolecules throughout the organism; its destructive effects relate in part to its ability to generate free radicals. When injected into rats, we (Sewerynek et al., 1995b, 1995c) found LPS to induce large increases in hepatic malonaldehyde and 4-hydroxyalkenal levels within 6 hours. These products are reflective of the level of lipid peroxidation. When the rats were concurrently treated with melatonin the rise in lipid peroxidation products as a consequence of LPS administration was prevented. The protective effect of the indole was believed to relate to its antioxidant activity.

Another highly toxic compound that works via free radicals is carbon tetrachloride ($CCl_4$). Its induction of hepatic damage has been thoroughly investigated and when Daniels and co-workers (1995) injected it into rats the expected rise in hepatic lipid peroxidation products was observed. As in the LPS study, if the rats were given both $CCl_4$ and melatonin, however, the degree of hepatic lipid peroxidation was highly significantly reduced. While glucose-6-phosphatase activity was also inhibited by $CCl_4$, this change was not reversed by melatonin administration.

## 3. CONCLUDING REMARKS

Besides those summarized above, other studies have shown melatonin to counteract free radical damage in vitro and vivo under a variety of experimental conditions (Reiter, 1995; Reiter et al., 1995). It is clear that this protection against free radical damage extends to DNA, proteins and lipids. This suggests that melatonin readily enters each subcellular compartment where it actively reduces the toxicity of free radicals. Whereas melatonin's protective actions may be directly related to its scavenging ability (Tan et al., 1993a; Pieri et al., 1994) the indole also acts indirectly to stimulate at least one antioxidative enzyme, glutathione peroxidase (Barlow-Walden et al., 1995).

## REFERENCES


Abe, M., Reiter, R.J., Orhii, P.B., Hara, M., and Poeggeler, B., 1994, Inhibitory effect of melatonin on cataract formation in newborn rats: Evidence for an antioxidative role for melatonin. *J. Pineal Res.* 17:94.

Arendt, J., Aldhous, M., English, J., Marks, V., and Arendt, J.H., 1987, Some effects of jet lag and their alleviation by melatonin. *Ergonomics* 30:1379.

Armstrong, S.M., and Redman, J.R., 1993, Melatonin and circadian rhythmicity, in: Melatonin, H.S. Yu and R.J. Reiter, Eds., CRC Press, Boca Raton, pp. 187-224.

Barlow-Walden, L.R., Reiter, R.J., Abe, M., Pablos, M.I., Menendez-Palez, A., Chen, L.D., and Poeggeler, B., 1995, Melatonin stimulates brain glutathione peroxidase activity. *Neurochem Int.* 26:497.

Burke, T.H., and Sandstrom, P.A., 1994, Oxidative stress as a mediator of apoptosis. *Immunol. Today*, 15:7.

Cagnoli, C., Atabay, C., Kharlamova, E., and Manev, H., 1995, Melatonin protects neurons from singlet oxygen-induced apoptosis. *J. Pineal Res.*, in press.

Chen, L.D., Kumar, P., Reiter, R.J., Tan, D.X., Manchester, L.C., Chambers, J.P., Poeggeler, B., and Saarela, S., 1994, Melatonin prevents the suppression of cardiac $Ca^{2+}$ - stimulated ATPase activity. *Amer. J. Physiol.* 267:E57.

Czyba, J.C., Girod, C., and Durant, M., 1964, Sur l' antagonism épiphyseo-hypophysaire et les variations saisonnières de la spermatogénèse chez le Hamster doré (*Mesocricetus auratus*). *C.R. Soc. Biol.* 158:742.

Daniels, W.M.V., Reiter, R.J., Melchiorri, D., Sewerynek, E., Pablos, M.I. and Ortiz, G.G., 1995, Melatonin counteracts lipid peroxidation induced by carbon tetrachloride but does not restore glucose-6-phosphatase activity. *J. Pineal Res.*, in press

Floyd, R.A., 1993, Basic free radical biochemistry, in: Free Radicals in Aging, B.P. Yu, Ed., CRC Press, Boca Raton, pp. 39-56.

Giusti, P., Gusella, M., Lipartiti, D., Milam, D. Zhu, W., Vicini, S., and Manev, H., 1995, Melatonin protects primary cultures of cerebellar granule neurons from kainate but not from N-methyl aspartate toxicity. *Exp. Neurol.* 131:39.

Goldman, B.D., and Nelson, R.J., 1993, Melatonin and seasonality in mammals, in: Melatonin, H.S. Yu and R.J. Reiter, Eds., CRC Press, Boca Raton, pp. 225-252.

Hoffman, R.A. and Reiter, R.J., 1965, Pineal gland: Influence on gonads of male hamsters. *Science* 148:1609.

Koppenal, W.H., 1994, Chemistry of iron and copper in radical reactions, in: Free Radical Damage and Its Control, C.A. Rice-Evans and R.H. Burdon, Eds., Elsevier, Amsterdam, pp. 3-24.

Lerner, A.B., Case, J.D., Takahashi, Y., Lee, Y., and Mori, W., 1958, Isolation of melatonin, a pineal gland factor that lightens melanocytes. *J. Amer. Chem. Soc.* 8:2587.

Malpaux, B., Chemineau, P., and Pelletier, J., 1993, Melatonin and reproduction in sheep and goats, in: Melatonin, H.S. Yu and R.J. Reiter, Eds., CRC Press, Boca Raton, pp. 253-288.

Melchiorri, D., Reiter, R.J., Attia, A.M., Hara, M., Burgos, A., and Nistico, G., 1995a, Potent protective effect of melatonin on in vivo paraquat-induced oxidative damage in rats. *Life Sci.* 96:83.

Melchiorri, D., Reiter, R.J., Sewerynek, E., Chen, L.D., and Nistico, G., 1995b, Melatonin reduces kainate-induced lipid peroxidation in homogenates of different brain regions. *FASEB J.*, in press.

Meneghini, R., and Martin, E.L., 1993, Hydrogen peroxide and DNA damage, in: DNA and Free Radicals, B. Halliwell and O.I. Aruoma, Eds., Ellis Harwood, London, pp. 83-94.

Pieri, C., Maria, M., Moroni, F., Decchioni, R., and Marcheselli, 1994, Melatonin: A peroxyl radical scavenger more potent than vitamin E. *Life Sci.* 55:271.

Pierrefiche, G., and Laborit, H., 1995, Oxygen free radicals, melatonin, and aging. *Exp. Gerontol.* 30:213.

Pierrefiche, G., Topall, G., Courboin, G., Henriet, I., and Laborit, H., 1993, Antioxidant activity of melatonin in mice. *Res. Commun. Chem. Pathol. Pharmacol.* 80:211.

Poeggeler, B., Saarela, S., Reiter, R.J., Tan, D.X., Chen, L.D., Manchester, L.C., and Barlow-Walden, L.R., 1994 Melatonin-a highly potent endogenous radical scavenger and electron donor: New aspects of the oxidation chemistry of this indole assessed in vitro. *Ann. N.Y. Acad. Sci.* 738:419.

Poeggeler, B., Reiter, R.J., Hardeland, R., Sewerynek, E., Melchiorri, D., and Barlow-Walden, L.R., 1995, Melatonin, a mediator of election transfer and repair reactions, acts synergistically with the chain-breaking antioxidants ascorbate, trolox and glutathione. *Neuorendocrinol. Lett.*, in press.

Pryor, W.A., 1994, Free radicals and lipid peroxidation, in: Natural Antioxidants in Human Health and Disease, B. Frei, Eds., Academic Press, San Diego, pp. 1-24.

Quay, W.B., 1964, Circadian and estrous rhythms in pineal melatonin and 5-hydroxyindole-3-acetic acid. *Proc. Soc. Exp. Biol. Med.* 114:718.

Reiter, R.J., 1973, Comparative physiology: Pineal gland. *Annu. Rev. Physiol.* 35:305.

Reiter, R.J., 1991, Melatonin: That ubiquitously acting pineal hormone. *News Physiol. Sci.* 6:223.

Reiter, R.J., 1995, Oxidative processes and antioxidative defence mechanisms in the aging brain. *FASEB J.* 9:526.

Reiter, R.J., and Hester, R.J., 1966, Interrelationships of the pineal gland, the superior cervical ganglia and the photoperiod in the regulation of the endocrine system of hamsters. *Endocrinology* 79:1168.

Reiter, R.J., and Richardson, B.A., 1980, The physiology of melatonin, in: Biochemical and Medical Aspects of Tryptophan Metabolism. O. Hayaishi, Y. Ishimura, and R. Kido, eds., Elsevier/North Holland, Amsterdam, pp. 247-256.

Reiter, R.J., Blask, D.E., Johnson, L.Y., Rudeen, P.K., Vaughan, W.K., and Waring, P.J., 1976. Melatonin inhibition of reproduction in the male hamster: Its dependency on time of day of administration and on an intact and sympathetically innervated pineal gland. *Neuroendocrinology* 22:107.

Reiter, R.J., Tan, D.X., Poeggeler, B., Menendez-Pelaez, A., Chen, L.D., and Saarela, S., 1994, Melatonin as a free radical scavenger: Implications for aging and age-related diseases. *Ann. N.Y. Acad. Sci.* 719:1.

Reiter, R.J., Melchiorri, D., Sewerynek, E., Poeggeler, B., Barlow-Walden, L.R., Chuang, J.I., Ortiz, G.G., and Acuña-Castroviejo, D., 1995, A review of the evidence supporting melatonin's role as an antioxidant. *J. Pineal Res.* 18:1.

Sewerynek, E., Melchiorri, D., Ortiz, G.G., Poeggeler, B., and Reiter, R.J., 1995a, Melatonin reduces $H_2O_2$-induced lipid peroxidation in homogenates of different rat brain regions. *J. Pineal Res.*, in press.

Sewerynek, E., Melchiorri, D., Chen, L.D., and Reiter, R.J., 1995b, Melatonin reduces both basal and bacterial lipopolysacchoride-induced lipid peroxidation in vivo. *Free Rad. Biol. Med.*, in press

Sewerynek, E., Melchiorri, D., Reiter, R.J., Ortiz, G.G., and Lewinski, A., 1995, Lipopolysacchoride-induced hepatotoxicity is inhibited by the antioxidant melatonin. *Eur. J. Pharmacol.*, in press.

Snyder, S.H., Zweig, M., and Axelrod, J., 1964, Control of the circadian rhythm in serotonin content of the rat pineal gland. *Life Sci.* 3:1175.

Tamarkin, L., Westrom, W.K., Hamill, A.I., and Goldman, B.D. 1976, Effect of melatonin on the reproductive systems of male and female Syrian hamsters. Diurnal rhythm in sensitivity to melatonin. *Endocrinology* 101:631.

Tan, D.X., Chen, L.D., Poeggeler, B., Manchester, L.C., and Reiter, R.J., 1993a, Melatonin: A potent, endogenous hydroxyl scavenger. *Endocrine J.* 1:57.

Tan, D.X., Poeggeler, B., Reiter, R.J., Chen, L.D., Chen, S., Manchester, L.C., and Barlow-Walden, L. R., 1993b, The pineal hormone melatonin inhibits DNA adduct formation induced by the chemical carcinogen safrole in vivo. *Cancer Lett.*, 70:65.

Tan, D.S., Reiter, R.J., Chen, L.D., Poeggeler, B., Manchester, L.C., Barlow-Walden, LR., 1994, Both physiological and pharmacological levels of melatonin reduce DNA adduct formation induced by the carcinogen safrole. *Carcinogenesis* 15:615.

Vijayalaxmi, Reiter, R.J., and Meltz, M.L., 1995, Melatonin protects human blood lymphocytes from radiation induced chromosome damage. *Mutat. Res.* 346:23.

# NEUROPROTECTIVE EFFECTS OF MELATONIN

M. Lipartiti,[1] D. Franceschini,[1] R. Zanoni,[2] M. Gusella,[1] P. Giusti,[1] C. M. Cagnoli,[3] A. Kharlamov,[3] and H. Manev[3]

[1] Department of Pharmacology, University of Padova
35131 Padova, Italy
[2] Fidia Research Laboratory
35031 Abano Terme, Italy
[3] Department of Psychiatry, Medical College of Pennsylvania and Hahnemann University and
Allegheny- Singer Research Institute
Pittsburgh, Pennsylvania 15212

## ABSTRACT

The full range of physiological actions of melatonin is not completely known. In mammals, it modulates gonadal function and regulates biological rhythms. Furthermore, it has also been reported to have anxyolitic, sedative, and anticonvulsant properties, both in human and animals. Recently it has been shown that melatonin is a potent, endogenous hydroxyl radical scavenger suggesting that it might interfere with neurodegenerative processing involving free-radical formation and excitatory amino acid release. Using primary cultures of rat cerebellar neurons and *in vivo* models of brain injury in rats, we demonstrate that melatonin might be considered an endogenous neuroprotective factor useful for the pharmacological treatment of neurological disorders and neural degeneration produced by glutamate excitotoxicity and oxidative stress.

## INTRODUCTION

Melatonin, 5-methoxy-*N*-acetyl-tryptamine, is the only indoleamine derivative other than 5-hydroxy-tryptamine that is known to have a physiological function. In mammals it is produced predominantly by the pineal gland. The retina, the Harderian gland, the intestines, and some regions of the brain are other sources of this hormone (Brown *et al.*, 1983). Melatonin is released into both the blood and cerebral spinal fluid in a circadian manner, with a peak occurring during the night (Wurtman *et al.*, 1964); and, then, it is delivered to specific target sites.

*Recent Advances in Tryptophan Research*, edited by Graziella Allegri Filippini et al.
Plenum Press, New York, 1996

At this time, the full range of physiological actions of melatonin is not known. In mammals, its major effect seems to be modulating gonadal function, generally in an inhibitory way (Tamarkin *et al.*, 1985). Melatonin has also been shown to play a major role in the regulation of biological rhythms, e.g. the synchronization of mammalian reproduction with the seasons (Reiter *et al.*, 1993). Melatonin has also been reported to have anxyolitic, sedative (Golombek *et al.*, 1933), and anticonvulsant properties, both in human and animals (Romijin, 1978; Champney & Champney, 1992). Recently it has been shown that melatonin is a potent, endogenous hydroxyl radical scavenger (Tan *et al.*, 1993).

Reactive oxygen species (ROS) may cause a rapid proliferation of free radical reactions which are involved in damage to many cellular elements, such as lipids, nucleic acids, and proteins. This effect is particularly evident in the brain, due to its low content of protective antioxidant compounds (Phillis, 1994). Furthermore, free-radical formation increases the release of the excitatory neurotransmitter glutamate (Pellegrini-Giampietro *et al.*, 1990). Glutamate binds and activates two main groups of central nervous system receptors, ionotropic glutamate receptors and metabotropic glutamate receptors (Nakanishi, 1992). The protracted binding of glutamate to ionotropic receptors causes protracted neuronal excitation, which ultimately leads to neuronal damage and death (Choi, 1988). In addition, glutamate may increase ROS production, creating a feedback loop that causes progressive brain damage (Dykens *et al.*, 1987). Kainate (KA) is an agonist for a subtype of glutamate receptors, the non-*N*-methyl-D-aspartate (non-NMDA) ionotropic glutamate receptors. A central role in the mechanisms of KA-induced neuronal death in cerebellar granule cells has been attributed to the generation of ROS (Puttfarcken *et al.*, 1993).

The ROS-scavenging action of melatonin suggests that it also might interfere with neurodegenerative processes that involve free radicals and excitotoxicity. The aim of our studies is to characterize the putative neuroprotective role of melatonin. To this end, we have used primary cultures of rat cerebellar neurons and *in vivo* models of brain injury in rats to investigate the effects of melatonin on both oxidative-stress- and KA-induced neuronal damage.

## NEUROPROTECTIVE ACTION OF MELATONIN *IN VITRO*

We recently demonstrated that melatonin has neuroprotective action in an *in vitro* model of glutamate excitotoxicity (Giusti *et al.*, 1995; Gusella *et al.* 1995). In particular, the exposure of primary cultures of cerebellar granule neurons to different doses of glutamate resulted in dose-dependent excitotoxicity. To assess the putative neuroprotective action of melatonin, a concentration of 50 μM glutamate was selected. This dose reduced cell viability by about 40%. Co-treatment with melatonin was neuroprotective, with a maximal effect attained at 500 μM melatonin. In addition, melatonin was able to shift the dose-response curve of glutamate to the right. When two selective agonists for ionotropic glutamate receptors, NMDA and the non-NMDA agonist KA, were used to learn whether melatonin protected from excitotoxicity regardless of which subtype of glutamate receptor was stimulated, we observed complete protection from KA-induced excitotoxicity. On the other hand, melatonin was ineffective in protecting neurons against NMDA-induced neurotoxicity (Table 1). Thus, the neuroprotective effect of melatonin against glutamate-mediated excitotoxicity seems to be specific for KA-mediated toxicity.

We also tested the effect of melatonin on apoptotic cell death unrelated to glutamate in primary cultures of cerebellar granule neurons. In this model of oxidative stress, apoptosis was triggered by singlet oxygen that was produced by the interaction of light with a photosensitive dye, rose bengal (Manev *et al.*, 1995). The data in Table 2 demonstrate that melatonin was active against the neurotoxic effect of oxidative stress (Cagnoli, *et al.*, 1995).

**Table 1.** Protective effect of melatonin against KA-, but not NMDA-, induced neurotoxicity in cerebellar granule neurons

| | Neuronal death (%) | |
|---|---|---|
| Treatment (μM) | Vehicle (ethanol 0.5%) | + Melatonin (500 μM) |
| Control | 1 + 5.2 | 3 + 4.6 |
| KA (100) | 10 ± 4.6 | 1 + 1.8 |
| KA (250) | 26 ± 3.2 | 4 ± 3.6* |
| KA (500) | 47 ± 2.3 | 7 ± 4.6* |
| KA (1000) | 58 ± 5.9 | 9 ± 8.2* |
| NMDA (10) | 51 ± 4.8 | 51 ± 2.3 |
| NMDA (100) | 54 ± 2.1 | 55 ± 3.2 |
| NMDA (500) | 63 ± 1.9 | 62 ± 6.1 |

KA and NMDA were dissolved in Locke's buffer. Melatonin (500 μM) was dissolved in ethanol and administered to cultures in Locke's buffer containing final 0.5% ethanol. The same concentration of ethanol was added to cultures in all corresponding control conditions. Incubation with drugs was performed at 22-24°C in Locke's buffer: KA (30 min), NMDA (60 min). To allow for the full activation of NMDA-sensitive ionotropic glutamate receptors, incubation with NMDA was carried out in a buffer without $MgCl_2$. Melatonin was added during the exposure to glutamate agonists. At the end of the treatments, cultures were washed in Locke's buffer and returned to the culture-conditioned medium for an additional 18 h; then an MTT (3-(4,5-dimethylthiazol-eyl)-2,5-diphenyl tetrazolium bromide) assay was performed. Results are means ± S.D. of different cultures *P>0.01 in comparison to the corresponding vehicle-treated group.

## NEUROPROTECTIVE ACTION OF MELATONIN *IN VIVO*

In addition to the above-described *in vitro* experiments, we also found melatonin capable of reducing KA-neurotoxicity *in vivo*. As already reported (Sperk, 1994), in adult male Wistar rats, a single dose of KA (10 mg/kg/ml) administered i.p. resulted in a set of behavioral responses consisting of several distinct phases. The first period was characterized mainly by ataxia. This behavior was then replaced by head nodding and masticatory

**Table 2.** Melatonin prevents neuronal death induced by photochemical oxidative stress (POS) in cerebellar granule neurons

| Treatment | Neuronal death (%) |
|---|---|
| Control | 10 ± 5 |
| POS | 85 ± 3 |
| Melatonin | 12 ± 6 |
| POS + Melatonin | 55 ± 6* |

Cerebellar granule neurons were prepared from 7-day-old rat pups and used between 8-10 days *in vitro*. Cells were incubated for 15 min with 2 μg/ml rose bengal at 24 °C. After the removal of the dye, the cultures were washed three times and melatonin (1 mM) was added 10 min prior to the exposure to light (green light, 1 sec). Cell viability was assessed by trypan blue exclusion (0.4% trypan blue solution, 4 min). P<0.05 in comparison with POS.

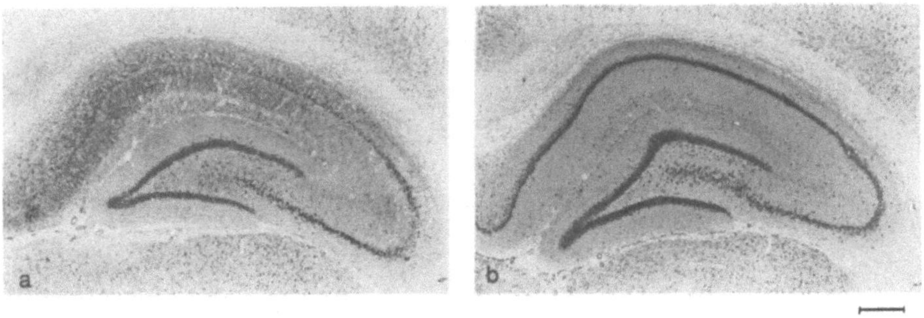

**Figure 1.** Melatonin reverses KA-induced neurotoxicity in the hippocampus. **a**) Rats received KA (10 mg/kg i.p.); **b**) Animals were injected with melatonin (2.5 mg/kg i.p.) 20 and 0 min before, and 20 and 40 min after the administration of KA. Both groups stained with cresyl violet. *Scale bar* represents 200 μm.

movements. At the same time, numerous intermittent, wet-dog-shaking movements were noticed. In the next phase, rearing and myoclonic twitches of the forelimbs were apparent, followed by generalized clonic convulsions with rearing and falling-down episodes. In the last phase, recurrent tonic-clonic convulsions were observed. This behavioral pattern was associated with mortality, and in the survivors, with loss of body weight. Massive neuronal degeneration was observed in the brains of surviving animals 11 days after KA administration. The areas most severely affected were the amygdala and pyriform cortex, where cellular and tissue colliquative necrosis were present. Tissue damage was associated with intense gliosis. A consistent pattern of neuronal damage was also observed in the dorsal hippocampus (Fig. 1, a).

The administration of four doses of melatonin (2.5 mg/kg/ml, i.p.) did not block KA-induced convulsions. Conversely, melatonin was able to counteract the high rate of acute mortality observed within 24 h from KA administration and body weight loss evaluated 11 days after the excitotoxin. Almost complete neuroprotection was observed in dorsal hippocampus after melatonin treatment (Fig 1, b).

We also tested the effects of melatonin in an *in vivo* model of photochemically induced oxidative stress and apoptosis (Manev *et al.*, 1994). In female Sprague-Dawley rats, stroke was induced by the rose bengal and irradiation with light, a procedure that generates singlet oxygen. Melatonin (2.5 mg/kg/ml, i.p., 20 min before and 1 and 4 h after the light exposure) significantly reduced the total area of injury (Table 3).

**Table 3.** Melatonin reduces lesion size following photothrombotic brain injury in the rat

| Treatment | Number of animals | Lesion size (arbitrary units) |
|---|---|---|
| Vehicle | 7 | 33.6 ± 2.2 |
| Melatonin | 7 | 25.3 ± 3.4* |

Photochemical brain injury (stroke) was induced in female Sprague-Dawley rats using rose bengal and light as described by Manev *et al.* (1994). Melatonin (2.5 mg/kg/ml, i.p.) was injected into the animals 20 min before and 1 and 4 h after light exposure. The rats were sacrificed 24 h after injury, and the brains were processed for TTC (2,3,5-triphenyltetrazolium bromide) staining.* $P<0.05$

## MECHANISM OF THE NEUROPROTECTIVE ACTION OF MELATONIN

The antineuronotoxic effects of melatonin occurred in the absence of any interference with glutamate recognition sites (Giusti *et al.*, 1995). In fact, our binding experiments did not show a direct inhibitory action of melatonin on glutamate receptors. It is important to note that our electrophysiological studies excluded a direct inhibitory action of melatonin on ionotropic KA receptors. Finally, melatonin did not interfere with basal free intracellular calcium concentration and did not affect the KA-induced $[Ca^{2+}]_i$ elevation. Thus, we propose that the antiexcitotoxic action of melatonin might be attributed to its action on mechanisms located downstream from KA receptors.

Although melatonin's effects on KA-induced neurotoxicity *in vivo* could be attributed to a general depressant action on nerve cell activity, e.g., an effect on $GABA_A$ receptors (Niles, 1991), our electrophysiological studies demonstrated that melatonin combined with GABA failed to affect whole cell current in primary cultures of cerebellar granule cells (Giusti *et al.*, 1995).

Typically, the neurotoxic action of KA can be blocked with selective antagonists for KA-sensitive glutamate receptors. However, KA-induced neurotoxicity can also be prevented by scavengers of ROS, such as mannitol, superoxide dismutase, or xanthine-oxidase inhibitors (Dykens 1987). Thus, the formation of neurotoxic ROS appears to be an important mechanism of KA-triggered neuronal death (Puttfarcken *et al.*, 1993). Moreover, ROS are central mediators of apoptosis, a type of cell death that can be induced by KA and is described in acute and chronic neurodegerative diseases (Serafian & Bredesen, 1994). Because melatonin is a potent scavenger of hydroxyl ROS (Tan *et al.*, 1993), it is likely that its neuroprotective action resides in this antioxidant activity. The anti-apoptopic action of melatonin we found in a previous study (Cagnoli *et al.*, 1995) suggests that the antioxidant action of melatonin could be a more general neuroprotective mechanism that is not only limited to protecting against the glutamate-receptor-mediated neurotoxicity.

## THERAPEUTIC PERSPECTIVES

Several excellent review articles have summarized the importance of glutamate-receptor-mediated signaling and the role of ROS in the pathophysiology of neurologic disorders and neural degeneration (Gotz *et al.*, 1994; Lipton & Rosenberg, 1994). Studies that describes glutamate excitotoxicity and oxidative stress imply two types of brain pathologies: a) acute, such as stroke, cerebral ischemia, brain injury, epileptic attacks; and b) chronic and progressive disorders, such as Alzheimer's disease, Parkinson's disease, amyotrophic lateral sclerosis. Although these studies point to endogenous systems (i.e., glutamatergic neurotransmission and endogenously formed ROS) as the jeopardizing factors for the integrity of neural cells, it should be stressed that in physiological conditions these factors are kept in balance by endogenous protective systems which include endogenous antioxidants and neurothrophins (Mattson & Scheff, 1994; Reiter, 1995). It has been suggested that in the aging brain the efficacy of endogenous defense mechanisms declines, leaving the brain more susceptible to excitotoxicity and ROS-mediated damage (Reiter, 1995).

Different neuroprotective strategies are currently being considered for the treatment of pathologies described above. One promising direction is the development of selective glutamate receptor antagonists. However, the side effects associated with the inhibition of vitally important neurotransmitter systems, such as the glutamatergic, may remain a serious obstacle. One proposed alternative is to utilize endogenous neuroprotective factors (Mattson & Scheff, 1994; Reiter, 1995). The pineal hormone melatonin appears to be one of these. Our *in vitro* and *in vivo* studies suggest that melatonin might be considered for the pharmacological treatment of glutamate excitotoxicity and ROS-triggered neurodegeneration.

In our hands, *in vivo* neuroprotection was obtained with repetitive administration of 2.5 mg/kg; the cumulative effective dose did not exceed 10 mg/kg. Melatonin treatment is known to have wide margin of safety: in rats, the administration of melatonin up to 30 mg/kg failed to produce any detectable behavioral effects (Chuang & Lin, 1994), whereas in one human study, patients who received 1,000 mg/die (approximately 14 mg/kg/die) for 30 days exhibited only minimal adverse effects (Norlund & Lerner, 1977). Thus, we envisage that the early administration of a pharmacological dose of melatonin in acute human pathologies such as stroke or brain trauma could be attempted in the future. It is also worth investigating whether chronic melatonin administration could slow down the progression of chronic neurodegenerative disorders such as Alzheimer's disease and Parkinson's disease.

## ACKNOWLEDGMENT

This work was supported in part by Allegheny-Singer Research Institute Grant 96-006-IR (H.M.)

## REFERENCES

Brown, G.M., Pulido, O., and Niles, L.P., 1983, Differential localization of melatonin and *N*-acetyl serotonin in brain, *in*: "The pineal gland and its endocrine role", J. Axelrod, F. Fraschini, and G.P. Velo, eds., Plenum Press, New York.

Cagnoli, C.M., Atabay, C., Kharlamov, E., and Manev, H., 1995, Melatonin protecs neurons from singlet oxigen-induced apoptosis, *J. Pineal Res.*, in press.

Champney, T. H., & Champney, J.-A. C., 1992, Novel anticonvulsant action of chronic melatonin in the gerbil, *NeuroReport*, 3: 1152.

Choi, D. W., 1988, Glutamate neurotoxicity and diseases of the nervous system. *Neuron,* 1: 623.

Chuang, J.I., & Lin M.T., 1994, Pharmacological effects of melatonin treatment on both locomotor activity and brain serotonin release in rats, *J. Pineal. Res.*, 17: 11.

Dykens, J.A., Stern, A., Trenker, E., 1987, Mechanism of kainate toxicity to cerebellar neurons in vitro is analogous to reperfusion tissue injury. *J. Neurochem.*, 49: 1222.

Giusti, P., Gusella, M., Lipartiti, M., Milani, D., Zhu, W., Vicini, S., and Manev, H., 1995, Melatonin protects primary cultures of cerebellar granule neurons from kainate but not from *N*-methyl-d-aspartate excitotoxicity, *Exp. Neurol.*, 131: 39.

Golombek, D.A., Martini M., and Cardinali, D.P., 1993, Melatonin as an anxiolytic in rats: time dependence and interaction with the central GABAergic system., *Eur. J Pharmacol.*, 237: 231.

Gotz, M.E., Kunig G., Riederer, P., and Youdim, M.B.H., 1994, Oxidative stress: free radical production in neural degeneration, *Pharmac. Ther.*, 63: 37.

Gusella, M., Lipartiti, M., Franceschini, D., Manev, H., and Giusti, P., 1995, Melatonin protects from kainate-induced neurotoxicity in cerebellar granule cells, *Behav. Pharmacol.*, 6 (suppl.1): 98.

Lipton, S.A., and Rosemberg, P.A., 1994, Excitatory amino acids as a final common pathway for nerological disorders, *New England J. Med.*, 330:613.

Manev., H., Kharlamov, A., and Armstrong, D., 1994, Photochemical brain injury in rats triggers DNA fragmentation, p53 and HSP72, *NeuroReport*, 5:2661.

Manev, H., Cagnoli, C.M., Atabay, C., Kharlamov, E., Ikomonovic, M.D., and Grayson, D.R., 1995, Neuronal apoptosis in an in vitro model of photochemically induced oxidative stress, *Exp. Neurol.*, 133: in press.

Mattson, M.P., & Scheff, S.W., 1994, Endogenous neuroprotection factors and traumatic brain injury. Mechanisms of action and implications for therapy, *J. Neurotrauma*, 11: 3.

Nakanishi, S., 1992, Molecular diversity of glutamate receptors and implications for brain function, *Science*, 258: 597.

Niles L., 1991, Melatonin interaction with the benzodiazepine-GABA receptor complex in the CNS, *in*: "Kynurenine and serotonin pathways", R. Schwarcz, ed, Plenum Press, New York.

Norlund, J.J., Lerner, A.B., 1977, The effects of oral melatonin on skin color and on the release of pituitary hormones, *J. Clin. Endocrinol. Metab.*, 45: 768.

Pellegrini-Giampietro, D. E., Cherici, E. G., Alesiani, M., Carla, V., Moroni, F., 1990, Excitatory amino acid release and free radical formation may cooperate in the genesis of ischemia-induced neuronal damage, *J. Neurosci.*, 10:1035.

Phillis, J.W., 1994,: A "radical" view of cerebral ischemic injury, *Prog. Neurobiol.* 42: 441

Puttfarcken, P. S., Getz, R. L., Coyle, J. T., 1993, Kainic acid-induced lipid peroxidation: Protection with butylate hydroxytoluene and U78517F in primary cultures of cerebellar granule cells. *Brain Res.*, 624: 223.

Reiter, R. J., Poeggeler, B., Tan, D.-X., Chen, L. D., Manchester, L. C., and Guerrero, J. M., 1993, Antioxidant capacity of melatonin: A novel action not requiring receptors. *Neuroendocrinol. Lett.* 15: 103.

Reiter, R. J., 1995, Oxidative processes and antioxidative defense mechanisms in the aging brain. *FASEB J.*, 9: 526.

Romijin, H. J., 1978, The pineal, a tranquilizing organ?, *Life Science*, 23: 2257.

Serafian, T.A., & Bredesen, D.E., 1994, Is apoptosis mediated by reactive oxygen species?, *Free Rad. Res.*, 21:1.

Sperk, G., 1994, Kainic acid seizure in the rat, *Progress in Neurobiology*, 42: 1.

Tamarkin, L., Baird, C.J., and Almeida, O.F.X., 1985, Melatonin: A coordinating signal for mammalian reproduction?, *Science* 227:714.

Tan, D.-X., Chen, L.-D, Poeggeler, B., Manchester, L.C., and Russel, J. R., 1993, Melatonin: a potent, endogenous hydroxyl radical scavenger. *Endocrine J.*, 1: 57.

Wurtman, R. J., Axelrod, J., and Fisher, J. E., 1964, Melatonin synthesis in the pineal gland: Effect of the light mediated by the sympathetic nervous system, *Science*, 143: 1328.

# *METABOLIC AND CLINICAL ASPECTS*

50

# PATHOGENESIS OF L-TRYPTOPHAN EOSINOPHILIA MYALGIA SYNDROME

E. M. Sternberg*

National Institute of Mental Health
Bethesda, Maryland

## INTRODUCTION

Illnesses which develop in association with exposure to environmental chemicals can be classified according to the nature of the chemical inducing agent or according to the pattern of disease which develops in the host. In the case of eosinophilia myalgia syndrome, the initial definition of the illness was a clinical one (Hertzman et al., 1990; Swygert et al., 1990), defined by the presence of eosinophilia and myalgias, and evolving into a chronic inflammatory and fibrotic disease (Silver et al., 1990) involving fascia, lungs, heart, skin and nerves. Since L-tryptophan ingestion was associated with virtually every case which developed during the 1989 epidemic, the syndrome was called L-tryptophan eosinophilia myalgia syndrome (L-TRP EMS).

However, clinically similar syndromes existed prior to the 1989 epidemic, and many studies have subsequently addressed the role of pure or impure L-tryptophan in causing this syndrome. Overall, the evidence suggests that many etiologic agents may together or alone cause a clinical syndrome of eosinophilia, myalgias and fasciitis. Among the probable causal agents are certain impurities found in case-associated L-tryptophan, L-5-hydroxytryptophan (L-5-HTP) and chemicals related to aniline, such as those found in toxic oil syndrome (TOS) case-associated oils (see below). L-tryptophan itself may also play a role, and appears to contribute to some of the scarring manifestations of the illness. Once these identified etiologic agents are accounted for, there are still a proportion of patients who develop the syndrome with no known exposures λ-idiopathic EMS. This syndrome is clinically very similar to eosinophilic fasciitis, or Schulman's syndrome, (Shulman, 1990), although it differs primarily in that eosinophilic fasciitis exhibits fewer systemic manifestations and often has its onset after strenuous exercise.

* Address reprint requests to: Esther M. Sternberg, M.D., NIMH/NIH Bldg. 10, Rm. 3S-231.10 Center Dr., MSC 1284, Bethesda, MD 20892.

Recent Advances in Tryptophan Research, edited by Graziella Allegri Filippini et al.
Plenum Press, New York, 1996

# ETIOLOGY AND PATHOGENESIS

## Distinction of Terms and Definitions

Development of disease after exposure to environmental agents is dependent not only on exposure to the causal, or etiologic agent, but also on the subsequent pathologic responses of the host. Thus, the pathogenesis of disease depends not only on the characteristics of the inducing chemical but also on the host's responses. Identifying the etiologic environmental agent and proving that it causes disease requires different studies than those designed to elucidate the mechanism of disease induction, and the cascade of host responses after exposure to the etiologic agent which lead to full expression of pathology, i.e. the pathogenesis of the illness.

## Proof of Cause and Effect

It is difficult to prove that a cause and effect relationship exists between exposure to an environmental chemical and the development of illness. Ultimately, proof of cause and effect depends on weighing the evidence from studies using a variety of different approaches, including case reports associating exposure and onset of illness; large scale epidemiological studies associating exposure to the putative agent and development of illness; chemical studies defining the etiologic agent; in vitro studies showing biological effects of the agent; in vivo animal studies showing development of characteristic pathologic features of the illness in animals treated with the putative agent. In this three dimensional approach, no one line of evidence provides full proof of cause and effect, and some lines of evidence provide stronger proof than others.

Case reports provide the weakest circumstantial evidence for cause and effect, but are suggestive and may provide potential clues for further study. Chemical data is critical in defining and synthesizing a pure putative agent to test alone or in combination with others. Furthermore, since variations in chemical characteristics, route, dose and duration of exposure to putative etiologic agents can contribute to variable disease outcomes, availability of pure chemical toxicants for testing is essential to define such dose and time course effects. In vitro studies are important both in understanding mechanisms of pathogenesis and providing evidence that a putative agent has biological activity.

However, ultimately proof of cause and effect requires reproducing the characteristic features of the illness in animals after exposure to the putative etiologic agent. The drawbacks of animal studies are that pathological responses vary in different species, and thus animal models are rarely perfect replicas of human disease. Animal models which consistently, across different species, exhibit some or many similar characteristic pathological features seen in humans nonetheless provide strong proof that a particular agent caused disease.

It is this weight of evidence, in which findings from multiple lines of study point in the same direction, which ultimately builds a body of work proving that a putative environmental exposure caused disease. In this regard, studies of the etiology and pathogenesis of L-tryptophan EMS exemplify the range of studies necessary to prove cause and effect, and elucidate the pathogenic mechanism of induction of the clinical constellation of symptoms known as eosinophilia myalgia syndrome.

# ETIOLOGIC AND PATHOGENIC STUDIES OF L-TRYPTOPHAN EMS

## Epidemiological and Chemical Studies

Epidemiological and chemical studies showed an association of EMS cases with certain lots of L-TRP manufactured by a single company (Swygert et al., 1990). Two of the over 60 different impurities detected on high performance liquid chromatography (HPLC) (Belongia et al., 1990; Smith et al., 1991; Goda et al., 1992; Adachi et al., 1995): 1,1' ethylidene bis [L-tryptophan], (EBT) and 3-(phenylamino)-L-alanine (PAA) have been identified, synthesized and studied in animals. Epidemiological studies suggest a strong EMS case-association with EBT (Belongia et al., 1990), but a weaker association with EBT, and stronger association with other impurities when time factors are included (Hill et al., 1993). The association with PAA is interesting, since this impurity structurally resembles one of the possible contaminants in the TOS, 3-phenylamino-1,2-propanediol.

### Animal Studies

Animal studies in rats (Crofford et al., 1990; Love et al., 1993) and mice (Silver et al., 1994) support the hypothesis that impure L-TRP and one or more of the impurities, EBT, are important etiologic agents in the fascial and inflammatory features of the syndrome. These studies also suggest that while pure L-TRP does not alone cause the syndrome, it may contribute to the scarring features. Evidence supporting these conclusions includes the findings that treatment of animals with doses equivalent to the median doses taken by patients, is associated with severe pancreatic acinar hyperplasia, and some mild increased collagen and adhesion molecule expression in tissues. Animal studies with PAA show a particular accumulation in brain tissue, Adachi et al., 1995. The remarkable concordance of findings in rats and mice (Crofford et al., 1990; Love et al., 1993; Silver et al., 1994) in separate studies, using different dosing regimens is strongly supportive of the role of tryptophan and its impurities in causing the fibrotic and inflammatory features of this syndrome.

## INTERACTIONS BETWEEN ENVIRONMENTAL AGENTS AND HOST RESPONSES

Interactions between environmental agents and host responses are numerous in EMS. These include inflammatory mediator induced activation of tryptophan metabolism; resultant quinolinic acid elevation; suppression of neuroendocrine inflammatory counter-regulatory responses; and tissue effects of eosinophil degranulation products. It is important to note that although these host inflammatory responses amplify certain pathogenic features of the disease, they are not themselves responsible for initiating disease. Furthermore, differences in host responses play a role in variable expression of disease, ranging from absence to severe symptoms or death, and contribute not only to differences in severity of acute symptoms, but also to variations in progression to chronicity. The existence of variable host responses does take away from the importance of the environmental agent in the initiating the disease. Nonetheless, it is important to define the full range of host pathologic responses, both to enhance understanding of pathogenic mechanisms of similar idiopathic diseases, and to provide avenues for treatment once these host response mechanisms have been set into motion by the initiating etiologic agent(s).

### Tryptophan Metabolism

Tryptophan metabolism is altered in patients with L-TRP EMS compared to non-inflamed controls. The pattern of tryptophan metabolites of elevated plasma quinolinic acid and plasma kynurenine, is consistent with activation of the enzyme indoleamine 2,3-dioxygenase by inflammatory mediators, such as gamma interferon, and initiation of the syndrome by an inflammatory stimulus, rather than an inborn error of tryptophan metabolism. Regardless of the mechanism of induction of inflammation, subsequent elevation of tryptophan metabolites could still play a role in amplifying some of the pathological features of the syndrome. Quinolinic acid, a neurotoxic excitatory amino acid, could contribute to some of the neuropathic features. Genetic HLA factors play only a weak role, showing only a trend to association with progression to chronic disease (Kaufman et al., 1991).

## Suppression of Neuroendocrine Inflammatory Counter-Regulatory Responses

Impure L-TRP and EBT also affect neuroendocrine host responses important in modulating severity of inflammation (Crofford et al., 1990; Brady et al., 1994). Case-associated impure L-TRP or EBT, but not pure L-TRP or vehicle control, were associated with suppression of hypothalamic CRH mRNA expression. The mechanism of suppression is unclear, whether a direct neurotoxic or pharmacologic effect, or secondary to chronic inflammation. In vitro studies suggest that neurotoxicity is possible, since one of the tetrahydrobetacarboline isomers of the acid breakdown product of EBT is toxic to mature murine spinal cord neurons in culture, in a maturation and interleukin-1 dependent manner (Brenneman et al., 1993). Since interruptions of the hypothalamic-pituitary-adrenal axis are associated with exacerbations of inflammatory disease (Sternberg et al., 1992), this could represent another mechanism for enhancement of severity of inflammation in EMS.

## Eosinophil Degranulation Products

Although EMS is associated with initial blood eosinophilia, eosinophils are rarely found in tissues, with instead lymphocytes and mononuclear cells predominating. CD8+ lymphocytes have been reported in lung biopsies, suggesting a possible immune cytotoxic/suppressor lymphocyte role in the pathogenesis of EMS.

Nonetheless, eosinophil degranulation products, eosinophil-derived neurotoxin (EDN) and major basic protein (MBP), are increased in affected EMS tissues (Hertzman et al., 1992), and could potentially amplify some toxic features of the syndrome. Whether the initial eosinophilia plays a role in the later fibrotic features of the syndrome, or whether the etiologic agent induces both early eosinophilia and later fibrosis independently is not clear. Animal studies in which fibrosis occurs in the absence of eosinophilia suggest that eosinophils do not play an important etiologic role in the development of the fibrosis.

Cytokines such as IL-8 (Reitamo et al., 1993), and growth factors such as TGFß (Tazelaar et al., 1993) are also elevated. Increases in the eosinophil and mast cell maturation factor IL-5 have also been shown (Owen et al., 1990; Yamaoka et al., 1994), and are consistent with the increased mast calls, eosinophilia and bone marrow eosinophil precursors (Philen et al., 1991). The absence of elevation of IgE suggests that classic allergic phenomena are unlikely to play a role since in most cases, although mast cells have been reported in some patients (Kaufman et al., 1990).

## L-5-Hydroxytryptophan Related Scleroderma: A Closely Related Clinical Syndrome

Since 1980, several cases of a clinically and biochemically closely related syndrome have been reported in association with L-5-hydroxytryptophan (L-5-HTP) (Sternberg et al., 1980, reviewed in Michelson et al., 1994). Most recently, members of a family in Canada developed full-blown EMS, or partial manifestations after exposure to L-5-HTP (Michelson et al., 1994). The case-associated L-5-HTP contained several unidentified peak. Substitution of a source of L-5-HTP not containing these peaks was associated with resolution of clinical signs in the children.

## SUMMARY

Taken together, these studies suggest that many different etiologic agents alone or together may initiate the common final pathways of tissue pathologic response resulting in the clinical syndrome of eosinophilia, myalgias and fasciitis. Tryptophan itself may contribute to some of the scarring features of the illness, while impure L-tryptophan, and one or more of the impurities cause the characteristic features of the illness. The altered tryptophan metabolism in EMS is secondary to inflammation.

## REFERENCES

Adachi J., Gomez M., Smith C.C., Sternberg. E.M., 1995, Accumulation of 3-(phenylamino)alanine, a constituent in L-tryptophan products implicated in eosinophilia-myalgia syndrome, in blood and organs of Lewis rats. *Archives Toxicol.* 69(4):266.

Belongia E.A., Hedberg C.W., Gleich G.J., White K.E., Mayeno A.N., Loegering D.A., 1990, An investigation of the cause of the eosinophilia-myalgia syndrome associated with tryptophan use. *N Engl J Med* 323:357.

Brady L.S., Page S.W., Thomas F.S., Rader J.L., Lynn A.B., Polterak B., Zelazowski E., Crofford L.J., Zelazowski P., Smith C., Raybourne R.B., Love L.A., Gold P.W., Sternberg E.M.,1994, 1,1'-ethylidenebis (L-tryptophan) (EBT), a contaminant implicated in L-tryptophan eosinophilia myalgia syndrome, suppresses mRNA expression of hypothalamic corticotropin-releasing hormone in Lewis (LEW/N) rat brain. *Neuroimmunomodulation* 1:59.

Brenneman D.E., Page S.W., Schultzberg M., Thomas F.S., Zelazowski P., Burnet P., Avidor R., Sternberg E.M., 1993, A decomposition product of a contaminant implicated in L-tryptophan eosinophilia myalgia syndrome affects spinal cord neuronal cell death and survival through stereospecific, maturation and partly IL-1-dependent mechanisms. J *Pharmacol Exp Ther* 266:1029.

Crofford L.J., Rader J.I., Dalakas M.C., Hill R.H.J., Brady L.S., Page S.W., Needham L.L., Heyes M.P., Wilder R.L., Gold P.W., Illa I., Smith C., Sternberg E.M., 1990, L-tryptophan implicated in human eosinophilia -myalgia syndrome causes fascitis and perimyositis in the Lewis rat. *J Clin Invest* 86:1757.

Goda Y., Suzuki J., Maitani T., 1992, 3-Anilino-L-alanine, structural determination of UV-5, a contaminant in EMS-associated L-tryptophan samples. *Chem Pharm Bull* 40:2236.

Hertzman P.A., Blevins W.L., Mayer J., Greenfield B., Ting M., Gleich G.J., 1990, Association of the eosinophilia-myalgia syndrome with the ingestion of tryptophan. *N Engl J Med* 323:357.

Hertzman P.A., Maddoux G.L., Sternberg E.M., Heyes M.P., Mefford I.N., Kephart G.M., Gleich G.J., 1992, Repeated coronary artery spasm in a young woman with the eosinophilia myalgia syndrome. *JAMA* 267:2932.

Hill R.H.J., Caudill S.P., Philen R.M., 1993, Contaminants in L-tryptophan associated with eosinophilia-myalgia syndrome (EMS). *Arch Environ Contam Toxicol* 25:134.

Kaufman L.D., Seidman R.J., Phillips M.E., 1990, Cutaneous manifestations of the L-tryptophan-associated eosinophilia-myalgia syndrome: A spectrum of sclerodermatous skin disease. *J Am Acad Dermatol* 23:1063.

Kaufman L.D., Gruber B.L., Gregersen P.K., 1991, Clinical follow-up and immunogenetic studies of 32 patients with eosinophilia-myalgia syndrome. *Lancet* 337:1071.

Love L.A., Rader J.L., Crofford L.J., Raybourne R.B., Principato M.A., Page S.W., Trucksess M.W., Smith M.J., Dugan E.M., Turner M.L., Zelazowska E., Zelazowski P., Sternberg E.M., 1993, Pathological and immunological effects of ingesting L-tryptophan (L-TRP) and 1,1'-ethylidene bis[tryptophan] in Lewis rats. *J Clin Invest* 91:804.

Michelson, D., Page, S.W., Casey, R., Trucksess, M., Love, L., Milstein, S., Wilson, C., Massaquoi, S., Crofford, L., Hallett, M., Gold, P.W., Sternberg, E.M., 1994, Eosinophilia myalgia syndrome and related disorders associated with exposure to L-5-hydroxytryptophan, *J. Rheumatol.* 21: 2261.

Owen W.F.J., Petersen J., Scheff D.M., Folkerth R.D., Anderson R.J., Corson J.M., A.L. S, Austen K.F., 1990, Hypodense eosinophils and interleukin 5 activity in the blood of patients with eosinophilia-myalgia syndrome. *Proc Natl Acad Sci USA* 87:8647.

Philen R.M., Eidson M., Kilbourne E.M., 1991, Eosinophilia-myalgia syndrome: A clinical case series of twenty-one patients. *Arch Intern Med* 151:533.

Reitamo S., Remitz A., Varga J., 1993, Demonstration of interleukin-8 and autoantibodies to interleukin-8 in the serum of patients with systemic sclerosis and related disorders. *Arch Dermatol* 129:189.

Shulman L.E., 1990, The eosinophilia myalgia syndrome associated with ingestion of L-tryptophan. *Arthritis Rheum* 33:913.

Silver R.M., Heyes M.P., Maize J.C., Quearry B., Vionnet F.M., Sternberg E.M., 1990, Scleroderma, fasciitis, and eosinophilia associated with the ingestion of tryptophan [see comments]. *N Engl J Med* 322:874.

Silver R.M., Ludwicka A., Hampton M., Ohba T., Bingel S.A., Smith T., Harley R.A., Maize J., Heyes M.P., 1994, A murine model of eosinophilia-myalgia syndrome induced by 1,1'-ethylidenebis(L-tryptophan). *J. Clin. Invest.* 93:1473.

Smith M.J., Mazzola E.P., Farrell T.J., 1991, 1, 1'-Ethylidenebis (L-tryptophan), structure determination of contaminant "97" - implicated in the eosinophilia-myalgia syndrome (EMS). *Tetrahedron Lett* 32:991.

Sternberg E.M., Chrousos G.P., Wilder R.L., Gold P.W., 1992, The stress response and the regulation of inflammatory disease. *Ann Intern Med* 117:854.

Sternberg E.M., Van W.M., Young S.N., Magnussen I., Baker H., Gauthier S., Osterland C.K., 1980, Development of a scleroderma-like illness during therapy with L-5-hydroxytryptophan and carbidopa. *N Engl J Med* 303:782.

Swygert L.A., Maes E.F., Sewell L.E., Miller L., Falk H., Kilbouren E.M., 1990, Eosinophilia-myalgia syndrome. *JAMA* 264:1698.

Tazelaar H.D., Myers J.L., Strickler J.G., 1993, An immunophenotypic, immunofluorescent, and electron microscopic study. *Modern Pathol* 6:56.

Yamaoka K.A., Miyasaka N., Inuo G., Saito I., Kolb J.P., Fujita K., Kashiwazaki S., 1994, 1,1'-Ethylidenebis(tryptophan) (Peak E) induces functional activation of human eosinophils and interleukin 5 production from T lymphocytes: association of eosinophilia-myalgia syndrome with a L-tryptophan contaminant. *J Clin Immunol* 14:50.

51

# CLINICAL FEATURES OF EOSINOPHILIA MYALGIA SYNDROME AND RELATED DISORDERS

D. J. Clauw

Georgetown University Medical Center
3800 Reservoir NW, Washington, D.C. 20007

## I. INTRODUCTION

This chapter will review the clinical features of Eosinophilia Myalgia Syndrome (EMS). Since it is likely that this disorder falls within a spectrum of eosinophilic connective tissue diseases that also includes Eosinophilic Fasciitis (EF) and the Toxic Oil Syndrome (TOS), these other disorders will also be discussed. To better understand these disorders, it is helpful to also review the epidemiology of these disorders, as well as potential pathogenic mechanisms.

## II. EPIDEMIOLOGY

EF is the first of the diseases in this spectrum to be described, initially by Shulman in 1974, and later by Rodnan and colleagues (Shulman, 1974; Rodnan et al., 1975). The early descriptions of EF were of a disorder characterized by peripheral eosinophilia and fasciitis, which could be differentiated from scleroderma by the distinctive pattern of skin involvement (sparing the digits and involving fascia rather than dermis). In these early reports of EF, this disorder was considered to be clinically and histopathologically localized to the fascia.

However, larger series of cases of EF revealed a widening spectrum of findings (Michet et al., 1981; Lakhanpal, et al.,1988). Clinical and histopathological evidence of muscle involvement was noted, as evidenced by an increased frequency of myalgias, as well as evidence of myopathy (documented by EMG), and of inflammation of the epimysium and perimysium. A number of other symptoms were noted to occur frequently, including arthritis, entrapment neuropathies, and occasional visceral involvement and concurrent lymphoproliferative conditions.

In 1989 an epidemic of an eosinophilic connective tissue disease was recognized in the United States, and was termed Eosinophilia Myalgia Syndrome (Belongia et al., 1990; Hertzman et al., 1991; Philen and Posada, 1993). Because of the rapid work of public health

*Recent Advances in Tryptophan Research,* edited by Graziella Allegri Filippini et al.
Plenum Press, New York, 1996

officials, it became clear that most of these individuals were consuming L-tryptophan, an amino acid that was available in the United States as an over-the-counter nutritional supplement. Subsequent epidemiologic studies suggested that most individuals who developed EMS during the epidemic had consumed L-tryptophan manufactured by a single Japanese company (Slutsker et al., 1990; Belongia et al., 1992). Although only approximately 1700 individuals were reported to the Centers for Disease Control and Prevention (CDC) who met surveillance criteria established for this diagnosis (an absolute eosinophil count > 1,000 cells/mm$^3$, severe incapacitating myalgias, and the absence of other conditions to explain these findings), it has been estimated that the true number of cases was several-fold greater. The clinical spectrum of EMS includes fasciitis indistinguishable from that seen in EF, but neuromuscular and visceral involvement is more common, and the rapid response to corticosteroid therapy initially reported in EF is not typically observed in EMS (Culpepper et al., 1991; Silver, 1992).

The EMS epidemic led to a re-examination of the Toxic Oil Syndrome (TOS), a similar epidemic of an eosinophilic connective tissue disorder that afflicted approximately 20,000 Spaniards in 1981 (Kilbourne et al., 1991; Philen and Posada, 1993). TOS was clinically and histopathologically similar to EMS, and in this case the inciting agent appeared to be adulterated rapeseed oil (Aldridge, 1992; Alonso-Ruiz et al., 1993). Chemically similar trace constituents have been found in the implicated samples of rapeseed oil and tryptophan, raising the possibility that a single toxin, or a group of related substances, may have been responsible for these two epidemics (Hill et al., 1993).

The recognition that both EMS and TOS were likely induced by toxins also led to a re-examination of earlier cases of EF. Several studies retrospectively examined EF cases that occurred prior to the EMS epidemic, and found that a significant percentage of these individuals had been consuming L-tryptophan (Blauvelt and Falanga, 1991; Hibbs et al., 1992). These early cases of EMS associated with L-tryptophan ingestion, which were characterized as EF until 1989, likely contributed to the widening of the spectrum of EF in the late 1980's. Further reports have demonstrated that EMS or clinically indistinguishable conditions can occur in association with exposure to several substances other than L-tryptophan or rapeseed oil, including other nutritional supplements or chemicals (see Table 1) (Farinelli et al., 1991; Patmas, 1992; Waller et al., 1994). Although only the associations with rapeseed oil and L-tryptophan have been confirmed by epidemiologic studies, the distinct clinical features of EF as well as the rarity of this condition strengthen the weight of case reports as proof of potential causation.

Epidemiologic studies into EMS have given some insight into potential mechanisms by which toxin exposure may lead to the development of a connective tissue disease (Kilbourne, 1992). Most such studies in EMS suggested that only a small percentage of those who ingested implicated L-tryptophan became ill, thus suggesting that host factors may be important in the development of disease (Belongia et al., 1990; Clauw et al., 1990; Hertzman et al., 1991). Although several studies suggest that the likelihood of developing EMS was related to the quantity of L-tryptophan ingested (and perhaps increased age), it is clear that many afflicted individuals took a very small quantity, and that many individuals who consumed large quantities of implicated L-tryptophan did not become ill (Kamb et al., 1992;

**Table 1.** Environmental agents associated with development of EF or EMS-like disorders

| | |
|---|---|
| Rapeseed oil | Lysine |
| 5 - hydroxytryptophan | Niacin |
| L - tryptophan | Trichloroethylene |

Back et al., 1993). Similarly low attack rates are seen in most other types of environmentally-induced connective tissue diseases. There are data to suggest that individuals with abnormal xenobiotic (P450) metabolism may have been predisposed to develop of EMS, either because these individuals could not properly eliminate a toxin, or perhaps because they converted a "pro-toxin" to a toxin (Flockhart et al., 1994). This is analogous to that seen with other drug- or toxin- induced connective tissue diseases, such as drug -induced SLE and idiopathic scleroderma (May et al., 1990; Hess, 1991).

To summarize, the accumulated evidence suggests that there is a continuum between EF, which is a *localized* inflammation of the fascia, and EMS and TOS, where the inflammation (and perhaps other pathogenic features) extend into other organs and tissues. It is likely that most of the diseases within this spectrum may be associated with the ingestion or exposure to environmental agents. Because the discrete toxin(s) which may have triggered these outbreaks has not yet been identified, it is likely that similar epidemics will occur in the future. Therefore, when confronted with a patient with EF or EMS, a careful history of potential environmental exposure(s) is critical.

## III. CLINICAL FEATURES AND POTENTIAL PATHOGENIC MECHANISMS

In this section, the clinical features of EF and EMS are discussed concurrently with potential pathogenic mechanisms. This approach allows for a better understanding of the potential therapy of these disorders, which will not be specifically discussed, but is symptom-directed and based on the organ or tissue involved.

Table 2 lists the clinical features that are typically associated with EF, as well as the additional features that may occur in association with EMS. For simplicity, in this chapter

**Table 2**

| Clinical features of EF | Other features seen in EMS |
|---|---|
| *Cutaneous* | *Cutaneous* |
| Fasciitis / induration | Urticaria |
| Edema | Maculopapular rash |
| Pigmentation changes | Mucinosis |
| *Rheumatic* | Dermatographism |
| Joint contractures | Alopecia |
| *Neurologic* | *Rheumatic* |
| Carpal tunnel syndrome | Myalgias / myopathy |
| *Other* | *Neurologic* |
| Fevers | Distal sensorimotor neuropathy |
| Lymphoproliferative disorders | Mononeuritis multiplex |
| *Laboratory* | Cognitive symptoms |
| Eosinophilia | *Cardiopulmonary* |
| Hypergammaglobulinemia | Pneumonitis |
| | Respiratory muscle dysfunction |
| | Pulmonary hypertension |
| | *Other* |
| | Severe fatigue |
| | Sicca symptoms |
| | Gastrointestinal |
| | *Laboratory* |
| | Elevated serum aldolase |
| | Antinuclear antibodies |

EF will be considered to be a disorder localized to the fascia, whereas individuals with systemic features will be considered to have EMS. In reality, this semantic distinction between EF and EMS may be relatively unimportant, since the treatment of these entities is symptom-directed, and no global treatments can be recommended for all patients with either entity.

## 1. Skin and Fascia

Skin and fascial involvement can be a significant cause of morbidity in these conditions (Martin et al., 1990; Oursler et al., 1992). Although peau d'orange ("orange peel") skin, along with venous furrowing of the uplifted arm ("groove sign"), are the classically described manifestations of the fasciitis seen in EF or EMS, most individuals with these entities have slightly less impressive findings. In many cases, individuals with fasciitis may present with either induration or non-pitting edema of the affected area. This involvement is typically most prominent in the distal leg and forearm, but can occur anywhere in the body, with prominent truncal involvement seen relatively frequently. Pigmentation changes and joint contractures are late consequences of fasciitis. Other cutaneous findings, seen more frequently in EMS than EF, include early maculopapular pruritic eruptions, morphea, urticaria, alopecia, mucinous papules, dermatographism, and subcutaneous nodules (Kaufman et al., 1990; Varga et al., 1993). It is important to note that neither Raynaud's phenomenon nor digital skin involvement (e.g. sclerodactyly) occur in EMS or EF, and these clinical features remain the best manner to distinguish these conditions from scleroderma.

Common histopathological features seen within the skin of both EF and EMS patients are inflammatory infiltrates involving the dermis (especially prominent in the perivascular regions), dermal fibrosis, and increased deposition of mucin/glycosaminoglycans (Kaufman et al., 1990; Valicenti et al., 1991; Oursler et al., 1992; Rongioletti and Rebora, 1992; Feldman et al., 1992). The inflammatory infiltrates in these regions contain primarily mononuclear cells and eosinophils, and mast cells may also be increased in number. Qualitatively similar inflammatory infiltrates extend into the fascia, and in EMS the inflammation may also involve the connective tissue surrounding muscle fibers (epimysium, perimysium, and perivascular regions) and nerves (Rongioletti and Rebora, 1992). It is important to note that the primary location of the inflammatory infiltrates in these entities differs from those seen in inflammatory muscle disease, where more prominent endomysial inflammation is seen.

Considerable effort has been expended in further delineating the nature of the inflammatory response in EF and EMS. There is good evidence for a central role for T-cells in these disorders, as this is the predominant cell type in the affected tissues, and a significant percentage of the T cells are activated (Emslie-Smith et al., 1991; McKinley et al., 1993). There are also data suggesting that fibroblasts play a prominent role in the pathogenesis of these conditions, as one would expect in a fibrosing illness. Varga and colleagues have extensively studied the fibroblast in EMS, and have established that these cells exhibit increased collagen synthesis, as well as increased expression of both $\alpha$ 1(1) procollagen and transforming growth factor $\beta_1$ genes (Peltonen et al., 1991; Varga et al., 1993). Although both peripheral and tissue eosinophilia are frequently seen in these disorders, the role of these cells in the pathogenesis of these entities is less certain. It is clear that even when eosinophils are not detected in tissues, there is frequently evidence of eosinophil degradation, as measured by elevations in concentrations of major basic protein and/or eosinophil-derived neurotoxin in serum and tissues (Martin et al., 1990).

## 2. Neuromuscular Symptoms

Neuromuscular symptoms arguably lead to the greatest morbidity in these disorders, and the pathogenesis of these symptoms is poorly understood. Neurological and muscular symptoms are discussed concurrently because it is likely that there are overlapping mechanisms responsible for these symptoms.

Muscle pain is common, and likely has several different causes. Constant diffuse myalgias that are present at rest and worsen with exertion may be indicative of concurrent fibromyalgia. Many individuals who developed EMS were consuming L-tryptophan as self-medication for insomnia, pre-menstrual syndrome, or depression. It should not be surprising, then, that many of these individuals had fibromyalgia prior to developing EMS, and continue to experience fibromyalgia symptoms after the inflammatory phase of EMS has resolved (Hudson et al., 1993). It is important to emphasize that the symptom of muscle pain should not be interpreted as evidence for active fascial or muscle inflammation, since myalgias are not prominent in either "pure" EF or inflammatory muscle disease (Shulman, 1974; Rodnan et al., 1975; Slutsker et al., 1990).

Diffuse muscle pain and weakness in EMS may also be due to the development of a metabolic myopathy. EMS patients display evidence of a unique and selective lowering of intramuscular ATP levels, as determined $P^{31}$ magnetic resonance spectroscopy (Clauw et al., 1994a). There are additional data that suggest that the administration of parenteral magnesium can lead to both improvement in symptoms and ATP levels (Clauw et al., 1994b).

There are other potential causes of neuromuscular symptoms in both EF and EMS. Entrapment neuropathies, particularly carpal tunnel syndrome, occur commonly in both of these conditions, presumably because of fibrosis or inflammation causing compression of the median nerve. A distal sensorimotor neuropathy also can occur in EMS, and may be partly responsible for symptomatology, especially involving the distal extremities (Donofrio et al., 1992; Satsangi et al., 1992; Freimer et al., 1992). Other potential causes for chronic muscle pain and weakness have been postulated, but are less well established. Some authors have suggested the term "downslope claudication" to describe a type of compartment syndrome which may result from fascial fibrosis and the resultant restricted blood flow to the extremities in these conditions (Naschitz et al., 1991). In addition to the potential restriction of blood flow that might occur at the level of the large vessels from fascial inflammation and/or fibrosis, ischemia may also result from impeded microvascular flow from the occlusive microangiopathy seen pathologically in EMS. Although rare cases of a Guillian-Barre-like illness have been reported in EMS, it is also possible that a more proximal motor neuropathy, perhaps at the spinal cord level (as is seen in anterior horn cell disorders e.g. Amyotrophic Lateral Sclerosis), may occur in EMS and contribute to neuromuscular dysfunction (Donofrio et al., 1992; Satsangi et al., 1992; Slotkoff et al., 1994).

## 3. Hematologic

In nearly all cases of EF and EMS, peripheral eosinophilia resolves with time, and is typically quite corticosteroid-responsive (Wallace, 1993). When eosinophilia persists or is unresponsive to steroid therapy, then it is more likely that the underlying disorder falls within the spectrum of Hypereosinophilic Syndrome (HES).

## 4. Cardiopulmonary

There are a variety of cardiopulmonary symptoms that occur in these conditions. Early pulmonary involvement, most commonly characterized by a transient pneumonitis that resolved spontaneously or with corticosteroid therapy, was commonly seen in EMS (Tazelaar et al., 1993). The early pulmonary involvement seen in TOS was more fulminant, and is one of the principal clinical differences between these two entities (Kilbourne et al., 1991; Philen and Posada, 1993). Chronic pulmonary symptoms are common in EMS. The most common cause of these chronic symptoms appears to be respiratory muscle dysfunction (Read et al., 1992). Pulmonary function testing including maximal static inspiratory and expiratory pressures should be performed, with the latter determinations frequently the sole abnormality when the dyspnea is due to pain or dysfunction of the chest wall musculature. In more severe cases of neuromuscular chest wall disease (due to neuropathy, weakness, or fasciitis of the chest wall region), a restrictive pattern may be seen on pulmonary function testing. A low diffusing capacity that does not correct for decreased alveolar ventilation, or infiltrates or fibrosis on imaging studies (chest radiograph or high resolution chest computed tomography), may suggest pulmonary hypertension or interstitial lung disease, both of which may occur in EMS (Tazelaar et al., 1993; Chauhan and Mengel, 1993; Cheng, 1993).

Direct cardiac inflammation, as well as coronary artery spasm, have been noted as rare complications of EMS (Hertzman et al., 1993; James et al., 1994). A much more frequent symptoms in EMS is non-cardiac chest pain. This may be due to muscular chest wall involvement, or esophageal dysmotility. Another frequent symptom in EMS is resting tachycardia and postural hypotension; this is likely due to unrecognized autonomic neuropathy (unpublished observation).

### 5. Other Symptoms

Severe debilitating fatigue, cognitive dysfunction, recurrent fevers, and xerostomia are among the other symptoms that are reported frequently in EMS (Crofford and Clauw, 1995). The pathogenic mechanisms remain unclear. There are several potentially identifiable causes, all of which may to occur in a higher than expected frequency in patients with these entities. By far the most common is concurrent fibromyalgia, which may not only contribute to fatigue and cognitive dysfunction, but may also be accompanied by headaches, irritable bowel and sicca symptoms, and esophageal dysmotility (Slotkoff et al., 1994). Other potential concurrent conditions which may contribute to these symptoms include depression, hypothyroidism, and sleep apnea.

## IV. SUMMARY

EMS, EF, and TOS are all relatively rare disorders. There are considerable data to suggest that most (if not all) of these cases may be due to toxin exposure, although the precise etiologic agent(s) has yet to be identified. It is likely that the pathogenic mechanisms responsible for disease are similar in these entities, and thus the distinctions between these illnesses may be largely semantic. Rational therapy includes the removal of an inciting agent if identified, and the application of symptom-based treatment based on the organ or tissue involved, and whether there is evidence of active inflammation is present.

## REFERENCES

Aldridge WN., 1992, The toxic oil syndrome (TOS, 1981): from the disease towards a toxicological understanding of its chemical aetiology and mechanism. *Toxicol Lett* 59.

Alonso-Ruiz A, Calabozo M, Perez-Ruiz F, Mancebo L., 1993, Toxic oil syndrome. A long-term follow-up of a cohort of 332 patients. *Medicine (Baltimore)*; 72:285.

Back EE, Henning KJ, Kallenbach LR, Brix KA, Gunn RA, Melius JM., 1993, Risk factors for developing eosinophilia myalgia syndrome among L-tryptophan users in New York. *J Rheumatol.* 20:666.

Belongia EA, Hedberg CW, Gleich GJ, et al., 1990, An investigation of the cause of the eosinophilia-myalgia syndrome associated with tryptophan use. *N Engl J Med* 323(6):357.

Belongia EA, Mayeno AN, Osterholm MT., 1992, The eosinophilia-myalgia syndrome and tryptophan. *Ann Rev Nutr* 12:235.

Blauvelt A, Falanga V., 1991, Idiopathic and L-tryptophan-associated eosinophilic fasciitis before and after L-tryptophan contamination. *Arch Dermatol* 127:1159.

Chauhan D., Mengel C.E., 1993, Eosinophilia-myalgia syndrome and fasciitis with an active alveolitis, *Kans. Med.* Jun. 94: 175.

Cheng T.O., 1993, Pulmonary hypertension in patients with eosinophilia-myalgia syndrome ortoxic oil syndrome [letter, comment]. Mayo Clin. Proc., 1993 Aug., 68: 823.

Clauw DJ, Nashel DJ, Umhau A, Katz P., 1990, Tryptophan-associated eosinophilia connective tissue disease. A new clinical entity? *JAMA* 263:1502.

Clauw DJ, Hewes B, Nelson M, Katz P, Rajan S., 1994a, 31-P magnetic resonance spectroscopy of skeletal muscle in the Eosinophilia Myalgia Syndrome. *J Rheumatol.* 21:654.

Clauw DJ, Ward K, Wilson B, Katz P, Rajan S., 1994b, Magnesium deficiency in the eosinophilia myalgia syndrome: report of clinical and biochemical improvement with repletion. *Arthritis Rheum.* 37(9):1331.

Crofford L., Clauw D.J., 1995, Eosinophilic rheumatic disorders, *Rheum. Dis. Clin. N.Amer.*, 21(1): 231.

Culpepper RC, Williams RG, Mease PJ, Koepsell TD, Kobayashi JM., 1991, Natural history of the eosinophilia-myalgia syndrome. *Ann Intern Med* 115:437.

Donofrio PD, Stanton C, Miller VS, et al., 1992, Demyelinating polyneuropathy in eosinophilia-myalgia syndrome. *Muscle Nerve.* 15:796.

Emslie-Smith AM, Engel AG, Duffy J, Bowles CA., 1991, Eosinophilia myalgia syndrome: I. Immunocytochemical evidence for a T-cell-mediated immune effector response. *Ann Neurol* 29:524.

Farinelli S, Mariani A, Grimaldi A, Mariani M, Iannessi A, De Rosa F., 1991, Eosinophilia-myalgia syndrome associated with 5-OH-tryptophan. Description of a case]. *Recenti Prog Med.* 82:381.

Feldman SR, Silver RM, Maize JC., 1992, A histopathologic comparison of Shulman's syndrome (diffuse fasciitis with eosinophilia) and the fasciitis associated with the eosinophilia-myalgia syndrome. *J AM Acad Dermatol.* 26:95.

Flockhart DA, Clauw DJ, Buchert E, Harris J, Woosley RL., 1994, Pharmacogenetic characteristics of the eosinophilia myalgia. *Clin Pharm Ther.* 56:398.

Freimer ML, Glass JD, Chaundhry V, et al., 1992, Chronic demylelinating polyneuropathy associated with eosinophilia-myalgia syndrome. *J Neurol Neurosurg Psychiatry* 55:352.

Hertzman PA, Falk H, Kilbourne EM, Page S, Shulman LE., 1991, The eosinophilia-myalgia syndrome: the Los Alamos Conference. *J Rheumatol* 18:867.

Hertzman P.A., Maddoux G.L., Sternberg E.M., Heyes M.P., Mefford I.N., Kephart G.M., Gleich G.J., 1993, Repeated coronary artery spasm in a young wuman with the eosiniphilia-myalgia syndrome, *JAMA* 1992 Jun. 3, 267:2932.

Hess EV., 1991, Drug-related lupus, *Curr Opin Rheumatol* 3:809.

Hibbs JR, Mittleman B, Hill P, Medsger TA, Jr., 1992, L-tryptophan-associated eosinophilic fasciitis prior to the 1989 eosinophilia-myalgia syndrome outbreak. *Arthritis Rheum* 35:299.

Hill RH, Jr., Caudill SP, Philen RM, et al., 1993, Contaminants in L-tryptophan associated with eosinophilia myalgia syndrome. *Arch Environ Contam Toxicol* 25:134.

Hudson JI, Pope HG, Jr., Daniels SR, Horwitz RI., 1993, Eosinophilia-myalgia syndrome or fibromyalgia with eosinophilia? *JAMA* 269:3108.

James T., Kamb M., Sandberg G.A., Silver R.M., Kilbourne E.M., 1994, Postmortem studies of the heart in three fatal cases of the eosinophilia-myalgia syndrome, *Ann. Intern, Med.* 115: 102.

Kamb ML, Murphy JJ, Jones JL, et al., 1992, Eosinophilia-myalgia syndrome in L-tryptophan-exposed patients. *JAMA* 267:77.

Kaufman LD, Seidman RJ, Phillips ME, Gruber BL., 1990, Cutaneous manifestations of the L-tryptophan-associated eosinophilia-myalgia syndrome: A spectrum of sclerodermatous skin disease. *J AM Acad Dermatol.* 23:1063.

Kilbourne EM, Posada de la Paz M, Abaitua Borda I, Diez Ruiz-Navarro M, Philen RM, Falk H., 1991, Toxic oil syndrome: a current clinical and epidemiologic summary, including comparisons with the eosinophilia-myalgia syndrome. *J Am Coll Cardiol* 18:711.

Kilbourne EM., 1992, Eosinophilia-myalgia syndrome: coming to grips with a new illness. *Epidemiol Rev.* 14:16.

Lakhanpal S, Ginsburg WW, Michet CJ, Doyle JA, Moore SB., 1988, Eosinophilic fasciitis: clinical spectrum and therapeutic response in 52 cases. *Seminars Arth Rheum* 17 (4):221.

Martin RW, Duffy J, Engel AG, et al., 1990, The clinical spectrum of the eosinophilia-myalgia syndrome associated with L-tryptophan ingestion. *Ann Intern Med.* 113:124.

May DG, Black CM, Olsen NJ, et al., 1990, Scleroderma is associated with differences in individual routes of drug metabolism: a study with dapsone, debrisoquin, and mephenytoin. *Clin Pharmacol Ther.* 48:286.

McKinley KL, Harati Y, Schneider LW., 1993, Chronic immune activation in the eosinophilia-myalgia syndrome. *Muscle Nerve* 16:947.

Michet CJ, Doyle JA, Ginsburg WW., 1981, Eosinophilic fasciitis: report of 15 cases. *Mayo Clin Proc* 56:27.

Naschitz JE, Yeshurun D, Misselevitch I, Boss JH., 1991, Downslope claudication in eosinophilic fasciitis: a fourth type of claudication-case presentations. *Angiology* 167.

Oursler JR, Farmer ER, Roubenoff R, Mogavero HS, Watson RM., 1992, Cutaneous manifestations of the eosinophilia-myalgia syndrome. *Br J Dermatol.* 127:138.

Patmas MA., 1992, Eosinophilia-myalgia syndrome not associated with L-tryptophan. *N J Med J.* 89:285.

Peltonen J, Varga J, Sollberg S, Uitto J, Jimenez SA., 1991, Elevated expression of the genes for transforming growth factor-beta 1 and type VI collagen in diffuse fasciitis associated with the eosinophilia-myalgia syndrome. *J Invest Dermatol.* 96:20.

Philen RM, Posada M., 1993, Toxic oil syndrome and eosinophilia-myalgia syndrome: May 8-10, 1991, World Health Organization meeting report. *Sem Arthritis Rheum* 23(2):104.

Read C.A., Clauw D., Weir C., Da Silva A.T., Katz P., 1992, Dyspnea and pulmonary function in the L-tryptophan-associated eosinophilia-myalgia syndrome, Chest 101: 1282.

Rodnan GP, DiBartolomeo A, Medsger TA, Barnes EL., 1975, Eosinophilic fasciitis: report of six cases of a newly recognized scleroderma-like syndrome. *Arthritis Rheum* 18:525.

Rongioletti F, Rebora A., 1992, Cutaneous toxic mucinoses. *J Am Acad Dermatol.* 26:789.

Satsangi J, Donaghy M., 1992, Multifocal peripheral neuropathy in eosinophilic fasciitis. *J Neurol.* 239:91.

Shulman LE., 1974, Diffuse fasciitis with hypergammaglobulinemia and eosinophilia: A new syndrome? *J Rheumatol* 1(S):46.

Silver RM., 1992, Eosinophilia-myalgia syndrome, toxic-oil syndrome, and diffuse fasciitis with eosinophilia. *Curr Opin Rheumatol* 4:851.

Slotkoff A, Woodward M, Manz H, Katz P, Clauw DJ., 1994, Eosinophilia myalgia syndrome: case report of possible spinal cord involvement. *Arthritis Rheum.* 37(6S):25.

Slutsker L, Hoesley FC, Miller L, Williams LP, Watson JC, Fleming DW., 1990, Eosinophilia-myalgia syndrome associated with exposure to tryptophan from a single manufacturer. *JAMA* 264:213.

Tazelaar H.D., Myers J.L., Strickler J.G., Colby T.V., Duffy J., 1993, Tryptophan-induced lung disease: an immunophenotypic, immunofluorescent, and electron microscopic study, *Mod. Pathol.* 6: 56.

Valicenti JM, Fleming MG, Pearson RW, Budz JP, Gendleman MD., 1991, Papular mucinosis in L-tryptophan-induced eosinophilia-myalgia syndrome. *J Am Acad Dermatol.* 25:54.

Varga J, Jimenez SA, Uitto J., 1993, L-tryptophan and the eosinophilia-myalgia syndrome: current understanding of the etiology and pathogenesis. *J Invest Dermatol.* 100:97S.

Wallace D.J., 1993, Persistent eosinophilia in 2 patients with eosinophilia-myalgia syndrome, *Arthritis Rheum.* 36: 1488.

Waller PA, Clauw DJ, Cupps T, Metcalf JS, Silver RM, LeRoy EC., 1994, Fasciitis (not scleroderma) following prolonged exposure to an organic solvent (trichloroethylene). *J Rheumatol* 21(8):1567.

# 52

# THE EOSINOPHILIA-MYALGIA SYNDROME AND THE TOXIC OIL SYNDROME

## Pursuing Parallels

P. A. Hertzman

Los Alamos, New Mexico

In May 1981 one of the largest epidemics of an intoxication ever recorded occurred in Spain in the provinces surrounding Madrid. The toxic oil syndrome (TOS) affected more than 20,000 patients and resulted in more than 12000 hospitalizations (Toxic Epidemic Study Group, 1982).

In October 1989 a cluster of patients with a similar, but apparently new and unusual syndrome was identified in the United States (Hertzman et al., 1990). Eventually more than 1500 cases of the Eosinophilia myalgia syndrome (EMS) were reported to the Centers for Disease Control (CDC), and more than 30% of these patients required hospitalization. Both outbreaks were associated with food related exposures. The culprit in the Spanish illness was identified as a cheap, unlabelled cooking oil sold in unmarked 5-liter plastic containers at outdoor markets and door to door by itinerant salesmen. In the US Case control studies demonstrated a association between EMS and the use of L-tryptophan (LT) containing products. The causative agent, however has never been proven for either illness.

HPLC studies demonstrated more than 60 trace contaminants in implicated batches of LT. While early studies showed a strong statistical association with 1,1-ethylidene-bis-tryptophan (Belongia et al., 1990), more recent studies have identified 3-Phenyl(aminoalanine), another substance which is also statistically associated with EMS (Mayeno et al., 1992). The similarity of this substance to 3-phenylamino-1,2-propanediol, one of the contaminants found in the toxic Spanish cooking oil, suggests an etiologic link between the two illnesses. Unfortunately, no convincing animal model has yet been developed for either TOS or EMS.

## CLINICAL FEATURES

The more common clinical features of these two illnesses involve lung, skin and subcutaneous tissue, and nerve (Hertzman and Abaitua-Borda, 1993).

*Recent Advances in Tryptophan Research,* edited by Graziella Allegri Filippini et al.
Plenum Press, New York, 1996

## PULMONARY FEATURES

There appears to be a significant difference between the early stage respiratory illness observed in the two diseases. The acute onset of cough and severe SOB seen in the majority of TOS patients was typically associated with radiographic and pathologic changes consistent with a distinctive non-inflammatory, non-cardiogenic pulmonary edema. While the majority of patients recovered uneventfully, there were approximately 250 TOS deaths due to this "mechanical" complication.

In contrast acute respiratory involvement in EMS was inflammatory, occurred less frequently and was less severe. Interstitial pneumonitis usually either resolved spontaneously or was highly responsive to glucocorticoids.

Pulmonary arterial hypertension occasionally developed in the intermediate stage of both diseases. It usually improved spontaneously, but a small subset of patients in both diseases followed a slowly progressive fatal course.

## CUTANEOUS FEATURES

A variety of cutaneous manifestations were seen in both diseases. In the acute phase these included polymorphous eruptions, edema, and pruritus. Later, the characteristic skin and subcutaneous tissue physical findings in both illnesses were those of fasciitis.

Edema was described in both illnesses, however the pattern may have varied in each. Morphea and linear scleroderma like changes were seen in both illnesses. Contractures seen in EMS were usually secondary to fasciitis (rather than neuropathy).

We examined a number of TOS patients with a history of skin involvement nearly eleven years after the onset of illness. There was a suggestion of subcutaneous tissue atrophy in some individuals, however no active inflammatory process was apparent. In EMS patients the skin also gradually improved in most patients and new lesions are not being reported. Thus, skin involvement in both illnesses is considered to be uniformly in remission.

## NEUROMUSCULAR MANIFESTATIONS

Neuromuscular involvement in TOS was generally more severe than in EMS, but the clinical and pathological manifestations of both diseases are qualitatively very similar.

In the acute and intermediate phases myalgias, muscle cramps, weakness, arthralgias, paresthesias, and dysaesthesias were reported. Objective evidence of neuropathy or myopathy was sometimes present. Carpal tunnel syndrome occurred in both illnesses.

In the later phases a severe, sensory-motor polyradiculopathy with progressive weakness at times resulting in respiratory muscle weakness and respiratory failure developed. This complication occurred more frequently in TOS (30%) than in EMS (<5%). Some TOS patients developed severe muscle wasting with atrophy. In others deformities of the hands and wrists with neuropathic contractures occurred, but such findings were unusual in EMS.

Distinctive muscle cramping, with or without involuntary muscle activity, occurred commonly and has persisted (albeit less severe) in many patients with both diseases. In addition, a significant proportion of patients continue to complain of persistent fatigue and/or myalgias but have no identifiable objective clinical abnormalities.

## PSYCHOLOGICAL AND NEUROCOGNITIVE MANIFESTATIONS

Many (20 - 30%) TOS and EMS patients have complained of intellectual or psychological impairment. Some adults showed reactivation of previous psychiatric conditions.

In EMS the prevalence of anxiety and major depression appears to be higher than would be expected in a general population of patients with medical illness. The presence of significant psychiatric comorbidity in some patients makes it difficult to assess the significance of certain complaints like memory loss or fatigue. The significance of neurocognitive complaints remains controversial. In TOS patients EEGs, CT scans, and MRIs were normal. At autopsy the brain has not revealed pathologic abnormalities. Currently a multicenter study designed to better understand neurocognitive aspects of EMS is in progress.

## HISTOPATHOLOGIC ABNORMALITIES

The histopathology in both illnesses was qualitatively similar showing perivascular, perimysial, and perineural inflammation.

## NATURAL COURSE OF ILLNESS

In the majority of patients with both diseases, the development of new symptoms ceased by about 8 to 12 months after disease onset and was followed by slow, spontaneous improvement.

Currently most TOS patients are living normal lives or have only minor limitations. In a small proportion of TOS patients, residual tissue damage persists (Alonso-Ruiz et al., 1993). In EMS, most objective disease manifestations continue to slowly improve or have resolved (Hertzman et al., 1995).

Many patients still have a variety of subjective complaints without objective abnormalities. The basis for continued symptoms of fatigue, myalgia, muscle cramps and cognitive changes in the absence of objective findings in both diseases is currently unexplained.

## DIFFERENCES BETWEEN TOS AND EMS

Although TOS and EMS are clinically remarkably similar there are some notable differences. In a study among a group of EMS and TOS experts, we attempted to determine whether experienced physicians could discriminate clinically between these 2 illnesses (Hertzman, 1993). 21 physicians expert in EMS(7) or TOS(14) were shown 35mm photographic slides of patients with clinical features of EMS, TOS, and eosinophilic fasciitis(EF) in a random sequence. Many of the features represented were uniformly judged to be consistent with both syndromes. These included tight shiny skin, contractures, and mild pulmonary infiltrates. Characteristics which were invariably determined to be more representative of EMS included macular rash, mottled hyperpigmentation, indurated skin, linear scleroderma, and morphea. Features more often associated with TOS included morbilliform rash, swollen distal digits, extreme emaciation, and marked pulmonary edema. The results of this small study suggest that although EMS and TOS are clinically similar, experienced physicians are able to discern recognizable features that separate the two illnesses.

## CONCLUSION

TOS and EMS are two unique conditions which are strikingly similar. They affect the same organs, share many unusual features and pathophysiological mechanisms, and appear to have the potential for common long-term sequelae. Although TOS in the early stage with acute pulmonary involvement and later with neuropathic involvement, was more severe than EMS, the "average patient" with either disease has remarkably similar clinical manifestations and outcome.

Within an eight year time period we have witnessed epidemics of two very similar illnesses, both associated with food related products. It is likely that in the future new food-associated, drug-associated or environmentally-associated diseases will occur or may even be occuring unrecognized at this moment.

There is a clear need to prepare for these events in advance.

We should develop systems for early recognition and generate action plans today so that we are equipped for future episodes. One example of an early warning system could employ regional medical centers, utilizing computer systems that may already be in place, to monitor for unusual trends or the occurence of clusters of laboratory abnormalities such as eosinophilia or liver function abnormalities. Second, with current powerful technology available such as the capability for looking at gene expression there are exciting new avenues for the investigation into disease etiology and pathogenesis. A coordinated system with capability for centralized tissue banking is essential to ensure the collection of adequate quantities of blood and tissue samples early in an epidemic. Finally, the development of patient data banks with the capacity for long term followup would permit more efficient tracking with better definition of a disease's natural course and could also facilitate the early institution of therapeutic trials.

## REFERENCES

Alonso-Ruiz A., Calabozo M., Perez-Ruiz F., Mancebo L., 1993, Toxic Oil Syndrome: A Long-term follow-up of a Cohort of 332 *Patients. Medicine* 72:285.

Belongia E.A., Hedberg C.W., Gleich G.J., et. al., 1990, An investigation of the cause of the eosinophilia-myalgia syndrome associated with tryptophan use. *NEJM* 323:357.

Hertzman P.A., Blevins, W.L., Mayer J., Greenfield B., Ting M., Gleich G.J., 1990, "Association of the Eosinophilia-Myalgia Syndrome with the Ingestion of Tryptophan". *N Eng J Med* 322:869.

Hertzman P.A., 1993, Are the eosinophilia-myalgia syndrome (EMS) and the Toxic Oil syndrome (TOS) Clinically distinguishable? *Arthrit Rheum* 36;(9s):s115.

Hertzman P., Abaitua-Borda I., 1993, The Toxic Oil Syndrome and the Eosinophilia-myalgia syndrome: Pursuing Clinical Parallels. *J Rheumatol* 20:1707.

Hertzman P.A., Clauw D.J., Kaufman L.D., Varga J., Silver R.M., et. al., 1995, The Eosinophilia-myalgia syndrome: Status of 205 patients and results of treatment 2 years after onset. *Ann Internal Med* 122:851.

Mayeno A.N., Belongia E.A., Lin F., Lundy S.K., Gleich G.J., 1992 3-Phenyl(aminoalanine), a novel aniline derived amino acid associated with the eosinophilia-myalgia syndrome: a link to the toxic oil syndrome? *Mayo Clin Proc* 67:1134.

Toxic Epidemic Study Group. Toxic epidemic syndrome, Spain, 1981. *Lancet* 1982;326:697-702.

# EFFECTS OF TRYPTOPHAN RELATED COMPOUNDS ON NUCLEAR REGULATORY CONTROL

## Possible Role in the Eosinophilia-Myalgia Syndrome

H. Sidransky, E. Verney, P. Latham, and A. Schwartz

Department of Pathology
George Washington University Medical Center
Washington, DC 20037

## 1. INTRODUCTION

For many years our laboratory has been concerned with the unique effect of L-tryptophan on rat hepatic protein synthesis (Sidransky, 1985). Tryptophan administered alone, but not an administration of other single essential amino acids, induced a stimulatory response of protein synthesis in mice and rats as measured in vivo or in vitro(Sidransky, 1985; Sydransky et al., 1967; Sidransky et al., 1968). Other laboratories have confirmed our findings (Pronezuk et al., 1968; Cammarano et al., 1968; Rothschild et al., 1969; Oravec and Sourkes, 1970; Park et al., 1973; Jorgensen and Majumdar, 1975; Majumdar, 1982). Subsequent studies from our laboratory have enabled us to gain much insight into the mechanism/s by which L-tryptophan stimulates hepatic protein synthesis. Many of these findings will be reviewed in a subsequent section.

In view of our continuing interest in how L-tryptophan acts on regulatory controls within the liver under normal conditions, we became intrigued with the recent occurrence of the eosinophilia-myalgia syndrome (Belongia et al., 1990; Slutsker et al., 1990; Varga et al., 1992; Belongia et al., 1992). Epidemiologic evidence indicated that the syndrome was related to the ingestion of L-tryptophan from a single manufacturer, Showa Denko of Japan (Belongia et al., 1990; Slutsker et al., 1990; Varga et al., 1992; Belongia et al., 1992). Analysis of the L-tryptophan implicated in cases of the syndrome revealed the presence of a number of contaminants (revealed by HPLC), especially the presence of 1,1'-ethylidene-bis(tryptophan) (EBT) (Mayeno et al., 1990; Smith et al., 1991) and 3-phenylamine-L-alanine (PAA) (Goda et al., 1992; Mayeno et al., 1992).

Our first thoughts related to the eosinophilia-myalgia syndrome focused upon determining whether the L-tryptophan implicated in the syndrome, or its contaminants, may affect or alter the regulatory control mechanisms that pertain to protein synthesis, and which are attributed to L-tryptophan's effects under normal conditions. This paper will review a

*Recent Advances in Tryptophan Research*, edited by Graziella Allegri Filippini et al.
Plenum Press, New York, 1996

**Table 1.** Acute effects of the administration of L-tryptophan on liver

| |
|---|
| Increases in: |
| 1. DNA-independent RNA polymerase activity |
| 2. Ribosomal and DNA-like RNA synthesis in nuclei |
| 3. Polyribosomal RNA synthesis |
| 4. Nuclear poly(A) polymerase activity |
| 5. Specific nuclear envelope L-tryptophan receptor binding |
| 6. Nucleocytoplasmic translocation of mRNA |
| 7. Cytoplasmic mRNA (poly (A)-mRNA) |
| 8. Heavier polyribosomal aggregates |
| 9. Protein synthesis (in vivo and in vitro) |

number of our findings that relate to nuclear tryptophan receptor binding, considered to play a regulatory role in nucleocytoplasmic translocation of mRNA, and how L-tryptophan and/or certain contaminants implicated in the eosinophilia-myalgia syndrome affect this important process. Based upon these data, we are able to speculate how such a disturbance may possibly be related to the induction of the syndrome.

Before dealing with the latter subject, it becomes important to briefly review and summarize our earlier data pertaining to how L-tryptophan acts under normal conditions in affecting hepatic protein synthesis. An understanding of these findings is necessary before moving on to the studies relating to L-tryptophan implicated in the eosinophilia-myalgia syndrome.

## 2. STUDIES WITH L-TRYPTOPHAN (PURE, NON-IMPLICATED)

Table 1 summarizes data pertaining to actions of L-tryptophan that affect RNA and protein metabolism in the liver cells.

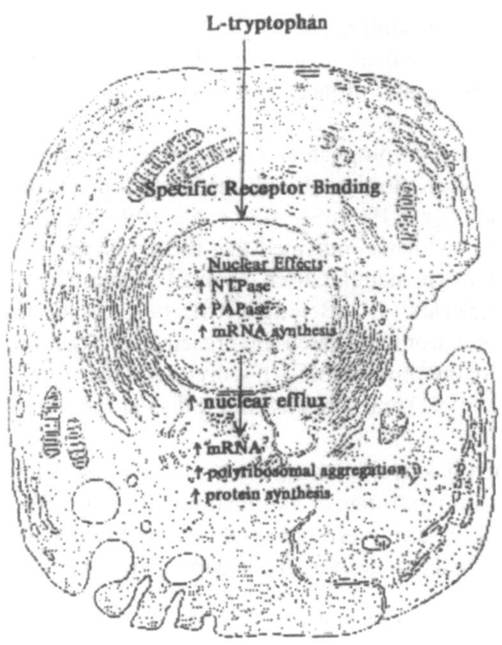


**Figure 1.** Schematic chart of postulated early events in the hepatic cell following the administration of L-tryptophan, (NTPase, nucleo-side triphosphatase; PAP, poly(A)polymerase).

**Table 2.** Essential sites of L-tryptophan molecule and of nuclear tryptophan receptor protein for binding effect

| | |
|---|---|
| L-tryptophan | |
| | Stereoisometric configuration (L-not D-) |
| | Carbonyl oxygen available for dipolar or hydrogen bind interactions |
| | Amino group (steric considerations important) |
| | R region permits a variety of compounds that mimic indole group to compete (contains hydrophobic protein interactions). |
| Nuclear receptor | |
| | Sulfhydryl groups |

Figure 1 presents a schematic chart pertaining to the early events occurring in hepatic cells following the administration of L-tryptophan and the projected sequence of events.

## 2.1. The Nuclear Receptor for L-Tryptophan

In 1987 our laboratory described and characterized a nuclear envelope receptor for L-tryptophan (Kurl et al., 1987; Kurl et al., 1988). Subsequently, we have reported upon how a number of tryptophan-related compounds compete for the nuclear tryptophan receptor (Sidransky et al., 1990; 1992a; 1994a). Based upon the latter findings we have been able to reconstruct the vital areas of the L-tryptophan molecule which are essential for special binding to the nuclear receptor (Table 2). Also, in a recent study based upon in vitro experiments using sodium selenite (Sidransky and Verney, submitted), we have reported data suggesting that the receptor protein contains sulfhydryl groups which are essential binding sites within the receptor.

# 3. STUDIES WITH L-TRYPTOPHAN AND TWO CONTAMINANTS IMPLICATED IN THE EOSINOPHILIA-MYALGIA SYNDROME

## 3.1. Implicated L-Tryptophan

Our first report, using the in vitro binding assay for L-tryptophan and rat hepatic nuclei or nuclear envelopes, revealed that significant differences were observed in the binding assay when L-tryptophan obtained from Showa Denko (implicated in cases of the eosinophilia-myalgia syndrome) was compared to non-implicated L-tryptophan from other suppliers (Table 3) (Sidransky et al., 1992b). The implicated L-tryptophan revealed less specific $^{3}H$-tryptophan binding to hepatic nuclear envelopes upon in vitro assay than did non-implicated L-tryptophan. This suggested that the implicated L-tryptophan contained one or more contaminants that affected the binding assay.

In another study (Sydransky and Verney, 1994) we investigated whether a strain difference of rats could influence the animals' responses to the administration of control or implicated L-tryptophan. Specifically, we compared the responses of Sprague-Dawley and of Lewis female rats. The Lewis strain has been reported to be suspectable to many inflammatory diseases and has been used in attempts to develop an experimental model of the eosinophilia-myalgia syndrome (Crofford et al., 1990; Love et al., 1993). In our comparative study we observed that there was a decrease in the affinity of in vitro $^{3}H$-tryp-

**Table 3.** Inhibition of total *in vitro* $^3$H-tryptophan binding using an excess ($10^{-4}$M) of unlabeled test compounds

| | Hepatic nuclear envelopes | |
|---|---|---|
| Compound/s | % Inhibition of binding | % Change |
| L-TRP (control) | (20) 63.7 ± 1.6[a] | |
| L-TRP (implicated) | ( 7) 55.8 ± 1.8[b] | 12.4 |
| L-TRP (control) | (13) 67.4 ± 2.3 | |
| 1,1′-ethylidene-bis-(TRP) (EBT) | (12) 20.8 ± 3.7[b] | –69.1 |
| L-TRP + EBT | ( 8) 59.6 ± 3.5 | –11.6 |
| L-TRP (control) | ( 8) 66.9 ± 2.0 | |
| 3-phenylamino-L-alanine (PAA) | ( 8) 0.03 ± 0.8[b] | –99.96 |
| L-TRP + PAA | ( 7) 44.8 ± 7.3[b] | –33.0 |

[a]Mean ± SEM; number of experiments in parentheses.
[b]$P < 0.01$

tophan binding to hepatic nuclei or nuclear envelopes of Lewis compared with Sprague-Dawley rats. In both strains, implicated L-tryptophan revealed less specific binding than the control, non-implicated L-tryptophan. However, stimulation of hepatic protein synthesis due to tube-feeding control L-tryptophan was similar in both strains.

## 3.2. 1,1′-Ethylidenebis(Tryptophan)(EBT)

Next, we investigated whether EBT, the first reported contaminant of implicated L-tryptophan from Showa Denko, would affect nuclear tryptophan binding (Table 3). Our results indicated that EBT alone or when added to control L-tryptophan had little inhibitory effect compared with that of control L-tryptophan alone on hepatic nuclei (Sidransky et al., 1994b). However, using nuclei of cultured murine macrophages (WLG5), we found that the addition of EBT plus control L-tryptophan caused less inhibition of $^3$H-tryptophan binding than did the addition of L-tryptophan alone.

Further investigation with EBT revealed that when it was added to an in vitro hepatic protein synthesis system in which $^3$H-tryptophan was incorporated into acid-precipitable proteins, EBT competed in a manner similar to that observed with equimolar concentrations of unlabeled L-tryptophan (Sidransky et al., 1994b), EBT did not affect $^{14}$C-leucine incorporation into proteins. $^{14}$C-EBT became incorporated in vitro into proteins (acid-precipitable) and this incorporation was diminished in the presence of equimolar concentrations of unlabeled EBT or L-tryptophan. The latter finding with $^{14}$C-EBT has been confirmed by others (Buss et al., 1995).

## 3.3. 3-Phenylamino-L-Alanine (PAA)

PAA has been observed to affect in vitro $^3$H-tryptophan binding to rat hepatic nuclei or nuclear envelopes (Sidransky et al., 1994c). PAA alone has no inhibitory effect on binding. However, when PAA is added to excess unlabeled, control L-tryptophan, the combination inhibits labeled tryptophan binding to a lesser extent than unlabeled L-tryptophan alone (Table 3). This effect is like that found earlier with L-tryptophan implicated in the eosinophilia-myalgia syndrome (Sidransky et al., 1992b) but not with EBT plus control unlabeled L-tryptophan (Sidransky et al., 1994b). Using nuclei of cultured murine macrophages or of rat brain, addition of PAA and control L-tryptophan did not affect $^3$H-tryptophan binding in comparison to control L-tryptophan alone (Sidransky et al., 1994c).

Further studies with PAA revealed that when it was added to an in vitro hepatic protein synthesis system in which $^3$H-tryptophan or $^3$H-alanine was incorporated into acid-precipitable proteins, PAA competed similarly but somewhat less, than did equimolar concentrations of unlabeled L-tryptophan or L-alanine (Sidransky et al., 1994c). This suggest that PAA or a breakdown compound becomes incorporated into proteins.

## 3.4. Effect on Other Receptors

In light of our reported findings with implicated L-tryptophan or its contaminants on the hepatic nuclear receptor for L-tryptophan, we decided to investigate whether other receptors may become affected by implicated L-tryptophan or its contaminants. We decided to test whether peripheral-type benzodiazepine receptors of nuclei of rat liver or of cultured murine macrophages may become affected by implicated L-tryptophan or its contaminants. A number of investigations have stressed the possible importance of peripheral benzodiazepine receptors on leucocytes in relation to immunological responses (Rocca et al., 1993; Drugan, 1994; Cahard et al., 1994). Conceivably induction of altered receptor responses of immunocompetent blood cells by implicated L-tryptophan and benzodiazepine (frequently used in patients who developed the eosinophilia-myalgia syndrome) could induce disturbances in immunomodulary actions which could possibly be of importance in the pathogenesis of the eosinophilia-myalgia syndrome.

Preliminary studies using in vitro ($^3$H)PK11195 binding to nuclei of rat liver or cultured murine macrophages have provided the following data summarized in Table 4 (Latham et al., unpublished data). It is apparent that while control L-tryptophan has little effect on in vitro ($^3$H)PK11195 binding to nuclear preparations, the addition of implicated L-tryptophan does have an appreciable inhibitory effect on binding. PAA alone, but not EBT alone, appears to act as does implicated L-tryptophan when added to the in vitro assay system. The possible significance of this finding merits further investigations which are being pursued in our laboratory.

# 4. EFFECT OF SELECTED DRUGS ON NUCLEAR TRYPTOPHAN BINDING

## 4.1 Studies with Non-implicated L-Tryptophan

The use of antidepressant or psychotropic medications was common in many patients who developed the eosinophilia-myalgia syndrome and was considered as possibly increas-

**Table 4.** Inhibition of total in vitro ($^3$H)PK 11195 binding using an excess ($10^{-4}$M) of unlabeled test compounds

| Compound | Liver | Macrophages (clone 5) |
|---|---|---|
| pK11195 | (9) 78.5 ± 4.13$^a$ | (7) 96.8 ± 0.52$^a$ |
| L-TRP (control) | (9) 8.8 ± 3.77$^b$ | (7) + 1.6 ± 4.92$^b$ |
| L-TRP (implicated) | (4) 32.9 ± 5.97$^{b,c}$ | (5) 33.3 ± 5.23$^{b,c}$ |
| 1,1′-ethylidene-bis-(TRP) (EBT) | (4) 10.8 ± 6.80$^b$ | (5) 12.3 ± 9.31$^b$ |
| 3-phenylamino-L-alanine (PAA) | (4) 30.3 ± 7.82$^{b,d}$ | (5) 29.1 ± 7.11$^{b,c}$ |

$^a$Mean ± SEM; number of experiments in parentheses.
$^b$P< 0.01, compared with pK 11195 group.
$^c$P< 0.01, compared with L-TRP (control) group.
$^d$0.05 >P > 0.01, compared with L-TRP (control) group.

ing the risk of developing the syndrome (Back et al., 1993; Henning et al., 1993). Therefore, we studied whether some drugs such as benzodiazepines (Sidransky et al., 1992c; 1992d) or metyrapone, an inhibitor of endogenous adrenal corticosteroid synthesis (Verney and Sidransky, 1994) would affect in vitro $^{3}$H-tryptophan binding to rat hepatic nuclei or nuclear envelopes. Indeed each group of drugs, (Sidransky et al., 1992c; 1992d; Verney and Sidransky, 1994) demonstrates an interactive effect with unlabeled L-tryptophan to diminish specific binding when compared to the effect of excess, unlabeled L-tryptophan alone. The effect is not present with drug alone.

### 4.2 Studies with Implicated L-Tryptophan

Preliminary studies as outlined above using implicated L-tryptophan instead of control L-tryptophan revealed similar responses to each drug as indicated above.

## 5. CONCLUDING REMARKS

Although the pathogenesis of the eosinophilia-myalgia syndrome remains unresolved, a number of reports have speculated as to its cause. Based mainly on clinical observations, some have suggested that an autoimmune mechanism may be involved (James et al., 1991; Kaufman et al., 1991; Varga et al., 1993; Illa et al., 1993). The ingestion of contaminants present in implicated L-tryptophan, which possibly may become incorporated into proteins (forming abnormal proteins) as suggested by experimental studies with EBT (Sidransky et al., 1994b; Buss et al., 1995) and also with PAA (Sidransky et al., 1994c), are supportive of this concept.

Although certain drugs used simultaneously with the ingestion of implicated L-tryptophan have been described in the case of eosinophilia-myalgia syndrome, no definitive correlation between the two has been unraveled based upon epidemiologic studies. Yet based upon experimental studies with selected drugs, it appears that certain drugs, especially benzodiazepines, may influence the effects of L-tryptophan such as nuclear L-tryptophan receptor binding (Sidransky et al., 1992d; Verney and Sidransky, 1994). Our preliminary findings with liver cells and macrophages suggest that peripheral-type benzodiazepine receptors may become affected by implicated L-tryptophan or by a contaminant such as PAA. Such alterations possibly involving leukocytes could affect immunologic responses as suggested by others (Rocca et al., 1993; Drugan, 1994; Cahard et al., 1994). Current investigation into this possible relationship is in progress.

The pathogenesis of the eosinophilia-myalgia syndrome appears to be multifactorial (Sidransky, 1994). Although the ingestion of implicated L-tryptophan is an essential requirement, only a few of the many individuals who ingested the implicated L-tryptophan developed the eosinophilia-myalgia syndrome. Therefore, other risk factors appear to be required and many have been considered in reports (Back et al., 1993; Henning et al., 1993). Intrinsic factors, such as how individuals metabolize L-tryptophan (Silver et al., 1992) or how they are susceptible to effects (toxic and carcinogenic-related) of L-tryptophan may be of importance. Several experimental studies do suggest differences in responses by different species or strains of animals to the administration of L-tryptophan (Sidransky, 1986; Sidransky and Verney, 1994). Extrinsic factors, such as the concurrent ingestion of certain drugs as discussed earlier (section 4) must also be considered for a possible role.

Our own experimental approach has been to investigate whether a disturbance in the effects of L-tryptophan by implicated L-tryptophan itself and/or related compounds (contaminants) on nuclear L-tryptophan receptor binding occurs. If present, as indicated by our studies to date, these abnormalities may conceivably play a role in the pathogenesis of the eosinophilia-myalgia syndrome. Our current experimental findings encourage us to probe further.

## ACKNOWLEDGMENT

This work was supported by United States Public Health Service Research Grant DK-45353 from the National Institute of Diabetes and Digestive and Kidney Diseases.

## REFERENCES

Back E.E., Henning K.J., Kallenbach L.R., Briz K.A., Gunn R.A., and Melius J.M., 1993, Risk factors for developing eosinophilia-myalgia syndrome among L-tryptophan users in New York. *J Rheumatol* 20:666.

Belongia E.A., Hedberg C.W., Gleich G.J., White K.E., Mayeno A.N., Loegering D.A., Diennetti S.L., Pirie P.L., and MacDonald K.L., and Osterholm M.T., 1990, An investigation of the cause of the eosinophilia-myalgia syndrome associated with tryptophan use. *N Engl J Med* 323:357.

Belongia E.A., Mayeno A.N., and Osterholm M.T., 1992, The eosinophilia-myalgia syndrome and tryptophan. *Annu Rev Nutr* 12:235.

Buss W.C., Bankhurst A.D., Stephanek J., Pastussn A., and Mayen A.N., 1995, The tryptophan contaminant EBT associated with eosinophilia-myalgia syndrome is an amino acid analog. *FASEB J* 9:A1028.

Cahard D., Canat X., Carayon P., Rogue C., Casella P., and LeFur G., 1994, Subcellular localization of peripheral benzodiazepine receptors on human leukocytes. *Lab. Invest* 70:23.

Cammarano P., Chinali G., Gaetini S., and Spandoni M.A., 1968, Involvement of adrenal steroids in the changes of polysome organization during feeding of imbalanced amino acid diets. *Biochem Biophys Acta* 155:302.

Crofford L.J., Rader J.I., Dalakas M.C., Hall R.H. Jr., Page S.W., Needham L.L., Brody L.S., Heyes M.P., Wilder R.L., and Gold P.W., Illa I, Smith C, Sternberg E.M., 1990 L-tryptophan implicated in human eosinophilia-myalgia syndrome causes fasciitis and perimyositis in the Lewis rat. *J Clin Invest* 86:1757.

Drugan R.C., 1994, Are the nonmitochondrial peripheral benzodiazepine receptors on leukocytes a novel intermediary of brain, behavior and immunity? *Lab Invest* 70:1

Goda Y., Suquki J., Maitani T., Yohihira K., Takeda M., and Uchiuama M., 1992, 3-Anilino-L-alanine, structural determination of UV-5, a contaminant in EMS-associated L-tryptophan sample. *Chem Pharm Bull* (Tokyo) 40:2236

Henning K.J., Jean-Baptiste E., Singh T., Hill R.H., and Friedman S.M., 1993, Eosinophilia-myalgia syndrome in patients ingesting a single source of L-tryptophan. *J Rheumatol* 20:273.

Kaufman L.D., Gruber B.L., and Gregerson P.K., 1991, Clinical follow-up on immunogenetic studies of 32 patients with eosinophilia-myalgia syndrome. *Lancet* 337:1071.

Kurl R.N., Verney E., and Sidransky H., 1987, Tryptophan binding sites on nuclear envelopes of rat liver. *Nutr Rep Inter* 36:669.

Kurl, R.N., Verney E., and Sidransky H., 1988, Identification and immunohistochemical localization of a tryptophan binding protein in nuclear envelopes of rat liver. *Arch Biochem Biophys* 265:286.

Illa I., Dinsmore S., and Dalakas M.C., 1993, Immunomedicated mechanisms and immune activation of fibroblasts in the pathogenesis of eosinophilia-myalgia syndrome induced by L-tryptophan. *Hum Pathol* 24:702.

James T.N., Kamb M.L., Sandber G.A., Silver R.M., and Kilbourne E.M., 1991, Postmortem studies of the heart in three fatal cases of the eosinophilia-myalgia syndrome. *Ann Intern Med* 115:102.

Jorgensen A.J.F., and Majumdar A.P.N., 1975, Bilateral adrenalectomy: Effect of a single tube-feeding of tryptophan on amino acid incorporation into plasma albumin and fibrinogen in vivo. *Biochem Med* 13:231.

Latham P., Verney E., and Sidransky H., (Unpublished data).

Love L.A., Rader J.I., Crofford L.J., and Raybourne R.B., Principato M.A., Page S.W., Trucksess M.W., Smith Dugan M.J., Turner M. L., Zelasonski E., Zelazonski P., and Sternberg E.M., 1993, Pathological and immunological effects of ingesting L-tryptophan and 1,1'-ethylidenebis (tryptophan) in Lewis rats. *J Clin Invest* 91:804.

Majumdar A.P.N., 1982, Tryptophan requirement for protein synthesis. A review. *Nutr Rep Int* 26:509.

Mayeno A.N., Lin F., Foot C.S., Loegering D.A., Ames M.M., Hedberg C.W., and Gleich G.J., 1990, Characterization of "Peak E", a novel amino acid associated with eosinophilia-myalgia syndrome. *Science* 250:1707.

Mayeno A.N., Belongia E.A., Lin F., Lundy S.K., and Gleich G.J., 1992, 3-(phenylamino)alanine, a novel aniline-derived amino acid associated with the eosinophilia-myalgia syndrome: a link to the toxic oil syndrome. *Mayo Clin Proc* 67:1134.

Oravec M., and Sourkes T.L., 1970, Inhibition of hepatic protein synthesis by a-methyl-DL-tryptophan in vivo.Further studies on the glyconeogenic action of a-methyltryptophan. *Biochemistry* 9:4458.

Park O.J., Henderson M., and Swan P.B., 1973, Effects of the administration of single amino acids on ribosome aggregation in rat liver. *Proc Soc Exp Biol Med* 142:1023.

Pronezuk A.W., Baliga B.S., Triant J.W., and Munro H.S., 1968, Comparison of the effect of amino acid supply on hepatic polysome profiles in vivo and in vitro. *Biochem Biophys Acta* 157:204.

Rocca P., Bellone G., Benna P., Bergamasco B., Ravizza L., and Ferrero P., 1993, Peripheral-type benzodiazepine receptors and diazepam binding inhibitor-like immunoreactivity distribution in human peripheral blood mononuclear cells. *Immunopharmacology* 25:163.

Rothschild M.A., Oratz M., Mongelli J., Fishman L., and Schreiber S.S., 1969, Amino Acid regulation of albumin synthesis. *J Nutr* 98:395.

Sidransky H., 1985, Tryptophan, unique action by an essential amino acid,in: Nutritional Pathology. Pathobiochemistry of Dietary Imbalances. H. Sidransky, ed, Marcel Dekker Inc New York p 1.

Sidransky H., 1986, Role of tryptophan in carcinogenesis. In Advances in Experimental Medicine and Biology, Vol 206, Essential Nutrients in Carcinogenesis, edited by L.A. Poirier, P.M. Newberne, N.W. Pariza. p 187, New York, Plenum.

Sidransky H., 1994, Eosinophilia-myalgia syndrome: a recent syndrome serving as an alert to new diseases ahead. *Mod Pathol* 7:806.

Sidransky H., Bongiorno M., Sarma D.S.R., and Verney E., 1967, The influence of tryptophan on hepatic polyribosomes and protein synthesis in fasted mice. *Biochem Biophys Res Comm* 27:242.

Sidransky H., Sarma D.S.R., Bongiorno M., and Verney E., 1968, Effect of dietary tryptophan on hepatic polyribosomes and protein synthesis in fasted mice. *J Biol Chem* 243:1123.

Sidransky H., Verney E., and Kurl R.N., 1990, Comparison of effect of L-tryptophan and a tryptophan analog, D,L-a-(1-naphthy)alanine on processes relating to hepatic protein synthesis in rats. *J Nutr* 120:1157.

Sidransky H., Verney E., Cosgrove J.W., Schwartz A.M., 1992a, Studies with compounds that compete with tryptophan binding to rat hepatic nuclei. *J Nutr* 122:1085.

Sidransky H., Verney E., and Cosgrove J.W., 1992b, Competitive studies relating to tryptophan binding to rat hepatic nuclear envelopes as a sensitive assay for unknown compounds. *Toxicology* 76:89.

Sidransky H., Verney E., Cosgrove J.W., and Schwartz A.M., 1992c, Effect of benzodiazepines on tryptophan binding to rat hepatic nuclei. *Toxicol Pathol* 20:350.

Sidransky H., Verney E., Cosgrove J.W., and Schwartz A.M., 1992d, Inhibitory effect of demoxepam of tryptophan binding to rat hepatic nuclei. *Biochem Med Metab Biol* 47:270.

Sidransky H., and Verney E., 1994, Comparative studies on tryptophan binding to hepatic nuclear envelopes in Sprague-Dawley and Lewis rats. *Am J Physiol* 267:R502.

Sidransky H., Verney E., Cosgrove J.W., Latham P. and Schwartz A.M., 1994a, Indolic compounds affect tryptophan binding to rat hepatic nuclei *J Nutr* 124:779.

Sidransky H., Verney E., Cosgrove J.W., Latham P.S., and Mayeno A.N., 1994b, Studies with 1,1'-ethylidenebis (tryptophan), a contaminant associated with L-tryptophan implicated in the eosinophilia-myalgia syndrome. *Toxicol Appl Pharmacol* 126:108.

Sidransky H., Verney E., Cosgrove J.W., and Latham P.S., 1994c, Effect of 3-phenylamino-L-alanine on tryptophan binding to rat hepatic nuclear envelopes. *Toxicology* 86:135.

Sidransky H., and Verney E., Effect of selenite in rat hepatic nuclei tryptophan receptor. *Proc Soc Exp Biol Med* (Submitted).

Silver R.M., McKinley K., Smith E.A., Quearry B., Harati Y., Sternberg E.M., and Heyes M.P., 1992, Tryptophan metabolism via the kynurenine pathway with the eosinophilia-myalgia syndrome. *Arthritis Rhem* 35:1097.

Slutsker L., Hoesly F.C, Miller L., Williams L.P., Watson J.C., and Fleming D.W., 1990, Eosinophilia-myalgia syndrome associated with exposure to tryptophan from a single manufacturer. *JAMA* 264:213.

Smith M.J., Mazzola E.P., Farrell T.J., Page S.W., Ashley D., Sirimanne S.R., Hill R.H., and Needham L.L., 1991, 1,1'-ethylidenebis(tryptophan) structure determination of contaminant "97" implicated in the eosinophilia-myalgia syndrome (EMS) *Tetrahedron Lett* 32:991.

Varga J., Uitto J., and Jimenez S.A., 1992, The cause and pathogenesis of the eosinophilia-myalgia syndrome.*Ann Intern Med* 116:140.

Varga J., Jimenez S.A., and Uitto J., 1993, L-tryptophan and the eosinophilia-myalgia syndrome: current understanding of the etiology and pathogenesis. *J Invest Dermatol* 100:975.

Verney E., and Sidransky H., 1994, Effect of metyrapone on tryptophan binding to rat hepatic nuclei. *Metabolism* 43:79.

# EOSINOPHILIA MYALGIA SYNDROME

## The Role of Contaminants, the Role of Serotonergic Metabolism Set Up

M. Nicolodi and F. Sicuteri

Interuniversity Centre of Neurochemistry and
Clinical Pharmacology of Idiopathic Headache
Florence University
Unit- 18 V.le Pieraccini, Florence, Italy

## INTRODUCTION

The serotonin theory of migraine (M) we proposed in the seventies indicated a pivotal role of serotonin in mechanism of M (Sicuteri et al., 1974). The analgesizing role of the indolamine was then stressed (Yaksh et al., 1977, Fields and Anderson, 1978) . We began to study the serotonin mechanism in primary headache in the sixties. At that time we hypothesized that serotonin could have had a mainly excitatory effect on polymodal receptors. Their serotonin-mediated activation was supposed to evoke headache pain. This theory re-emerged in recent years, in spite of the fact that substantial data showing have been obtained indicating that M, as well as other related primary pains, are caused chiefly by a failure of serotonergic analgesizing systems in the CNS (Sicuteri et al., 1973, Sicuteri et al., 1978, Del Bianco et al., 1982, Rajiv et al., 1989, Lance, 1992). As a matter of fact, serotonin is a transmitter characterized by a double-faceted effect, in that it is a pain-sensitising agent at the periphery (by acting on polymodal receptors) (Sicuteri et al., 1965), whereas it displays an evident antinociceptive effect by activating serotonergic analgesizing systems in the CNS. The centrally mediated analgesizing effect seems to largely overcome the pain-sensitising peripheral effect of the amine (Kimball et al., 1960). Among the more convincing outcomes supporting this assumption we can cite the effect of fenclonine, which we tested first in the late sixties (Sicuteri et al., 1973, Sicuteri et al., 1978). Fenclonine is a potent and irreversible inhibitor of tryptophan (Trp) hydroxylase, the rate-limiting enzyme in serotonin biosynthesis. When fenclonine is administered, enzyme activity is reduced in both cell bodies and nerve terminals. Recovery of the enzyme activity takes 1-2 weeks depending on the different area of the brain (Park et al., 1994). At that time, our hypothesis was that this compound was able to counteract the peripheral effect of serotonin, which was at the time hypothesized to promote headache by exciting nociceptors. The administration of this drug in 18 headache sufferers resulted in a very severe, widespread myalgic pain showing all the features of fibromyalgia (Sicuteri et al., 1973, Sicuteri et al., 1978).

*Recent Advances in Tryptophan Research*, edited by Graziella Allegri Filippini et al.
Plenum Press, New York, 1996

The effect we obtained was exactly the opposite of what we expected. We were thus compelled to completely revise our hypotheses and suggest that M sufferers were subjects characterized by a defective serotonergic/serotonin-dependent analgesia. Indeed, treatment with fenclonine, provoked severe pain involving the whole body. In other words, we obtained a farmacologically-induced fibromyalgia (Sicuteri et al., 1973, Sicuteri et al., 1978). The fenclonine-induced fibromyalgia cannot be considered the only pharmacologically induced fibromyalgia since it is well-known that the sensory syndrome of opiate-abstinence is related to a fibromyalgia-like syndrome (Tilson et al., 1983). On the other hand, fenclonine-induced fibromyalgia is, to our knowledge, the only pharmacologically-induced syndrome clearly overlapping spontaneous fibromyalgia.

A derangement of serotonin receptor activity in fibromyalgia was indicated by the dramatic inversion of physiological tachyphylaxis of the vein constrictor activity of serotonin we observed in the dorsal hand vein of subjects affected by this type of painful syndrome (Sicuteri et al., 1973, Sicuteri et al., 1978, Del Bianco et al., 1982, Sicuteri et al., 1983). Therefore, the experiment involving a treatment with fenclonine and the clinical pharmacological test evidencing the occurrence of serotonin receptor supersensitivity as well as the inversion of serotonin tachyphylaxis remain the only substantial experimental data providing a breakthrough in the neurobiochemistry and in the receptor dysfunction of fibromyalgia (Sicuteri et al., 1973, Sicuteri et al., 1978, Del Bianco et al., 1982). The very first cases of pharmacologically-induced eosinophilia fibromyalgia syndrome associated with fever were those we described following the administration of the serotonin depleting agent, fenclonine (Sicuteri et al., 1973). This agrees with the hypothesis that primary pain depends on an abnormal functioning of analgesia systems permitting hyperactivity of sensory pathway neurones (Wall, 1992).

There are data suggesting that the deficit of analgesia chiefly concerns serotonergic systems in both headache and fibromyalgia (Sicuteri et al., 1974, Rajiv et al., 1989, Lance, 1992, Moldofsky and Warsh, 1978). Since the seventies we have used Trp to cure headache and fibromyalgia syndrome (Sicuteri, 1972, Sicuteri, 1973). We have taken this approach for three considerations: a) the synthesis of brain serotonin depends on the availability of plasma Trp (Knott and Curzon, 1972, Messing et al., 1976) b) animal experiments have shown that altered availability of Trp can lead to a hyperalgesia state (Messing et al., 1976) , c) free plasma-Trp seems inversely related to the severity of primary, central musculoskeletal pain syndrome ( 15). Free Trp significantly increases during M attack, thus suggesting a decrease in 5-HT synthesis in the brain (Salmon et al., 1982).

In this paper we point out that, in spite of the very large number of subjects of different ages treated with L-Trp, none subjects reported eosinophilia myalgia syndrome (EMS). This information supports the acknowledged assumption indicating EMS, the newly diagnosed disorder characterized by eosinophilia and myalgia, is likely to be due to contaminants (Mayeno et al., 1992, Silver, 1992, Varga et al., 1993, Adachi et al., 1993, Kurihara et al., 1993, Philen et al., 1993, Taylor and McNeil, 1993, Donofrio et al., 1992)

In this paper we discuss mainly the hypothesis that Trp by itself is incapable of provoking fibromyalgia or EMS. EMS never appeared in the patients of ours who received only uncontaminated Trp or 5-hydroxytryptophan (5-HTP), so stressing once again the role of contaminants at least in provoking the auto-immune aspects of EMS. Indeed, even the long lasting (1-2 years) oral treatment with L-Trp or 5-HTP never induced EMS.

We also discuss the possible side-effects of excessive doses of Trp.

## MATERIAL AND METHODS

A retrospective study was carried out on 18 M sufferers (**Group A**) treated with fenclonine as well as on 319 subjects suffering from both fibromyalgia and M ( **Group B**).

All the subjects were diagnosed as migraine sufferers in agreement with the criteria for classification set forth by the International Headache Society in 1988. Subjects included in **Group B** also fulfilled the criteria for fibromyalgia (Wolfe et al., 1990). All the subjects included in **Group B** underwent 1-month placebo treatment; thereafter 219 of them were treated with uncontaminated L-Trp (Marxer Co. Pharmaceutics, Italy), the remainder 100 patients were given 5-HTP (Sigma-Tau Pharmaceutics, Italy), the hydroxylated form of Trp capable of crossing easily BBB in view of the fact that it is unbounded to plasmaproteins and does not compete with other aminoacids in this passage.

The incidence and clinical course of pain, eosinophilia, fever and the characteristics (age and sex) of the treated patients as well as the severity of the symptoms affecting the treated subjects and the characteristic of intake (i.e. dose and duration) have been observed.

## RESULTS AND COMMENT

A syndrome including myalgia, eosinophilia and wide-spreaded hyperalgesia occurred in 33.3% of the fenclonine-treated subjects; in one half of these patients mild fever was also present In the remainder of the subjects, only a spontaneous-like, severe fibromyalgia was reported. Both the fenclonine-induced EMS and fibromyalgia syndrome disappeared 1-2 weeks (mean 9 ± 4.9 STD days) after drug discontinuation. Fenclonine was able to induce fibromyalgia or EMS in headache sufferers only, whereas headache-exempt control subjects having also a family history negative for headache did not report any clinical signs. None of the subjects we treated were affected by skin involvement. Fenclonine combined with testosterone also failed to induce any sign of EMS in a group of subjects who were absolutely exempt from any type of primary pain but severely ailed by sexual performance deficiency or impotence that was independent of anatomical and/or endocrine defects (Sicuteri, 1974). The fact that fenclonine is capable of inducing fibromyalgia pain in headache sufferers only is seemingly supported by the fact that fenclonine fails to induce this type of syndrome when administered to cure intestinal carcinoid syndrome. When applied to cure the latter pathology, fenclonine is reported to sometimes provoke allergic phenomena associated to eosinophilia, which is acknowledged to be a marker of allergic diathesis (Samiy et al., 1987) Trp (2 g/day/ orally during 12 months or 1 g/ day/i.v. 2 periods of 4 months with an interval of 1 month ) never induced EMS in our patients. Also 5-HTP never induced EMS in our patients, both when given orally in the dose of 200-400 mg. for 12 months and when administered in the dose of 600 mg/i.v./day for 1 month. We can also observe that both the age and sex of the treated patients were not influent in evoking a different pattern in adverse reactions. In fact, no side-effect, except a randomly occurring somnolence during the infusion of Trp (but never during the administration of 5-HTP), was reported by any of the three different age groups: 9 to 18 years (Group 1), 20 to 37 years (Group 2), 38 to 58 years (Group 3). The only side-effect we observed during the treatment with 5-HTP consisted in very mild stomach-ache (11%) which never occurred in the group including younger patients. Group 1 reported the greatest benefit from L-Trp and 5-HTP therapy as regards sleep pattern and pain relief. The oldest subjects (Group 3) obtained the poorest results from this therapeutic approach. With reference to this, we must underline that the least responsive subjects were always those having a longer and more severe fibromyalgia/ headache history. L-Trp and 5-HTP-induced benefit on FS that was significantly greater than that induced by placebo and accounts for less than 30% of cases characterized by a partial refractoriness (benefit less than 30% decrease of pain score versus baseline) to the therapy.

Benefit from Trp or 5-HTP therapy was observed in a large part of the M/fibromyalgia patients after long term therapy. This seemingly suggests that remission of painful symptoms may not depend on short-term availability of serotonin.The therapeutic effectiveness of serotonin precursors well agrees with the "serotonin-theory" of both M and fibromyalgia.

It seems also noteworthy that in a large number of patients treated over a long period of time by using uncontaminated Trp, given i.v., or oral 5-HTP EMS was never provoked. This result seems to represent a further support to the hypothesis that EMS is caused by contaminated Trp, as stressed by other authors (Mayeno et al., 1992, Silver, 1992, Varga et al., 1993, Adachi et al., 1993, Kurihara et al., 1993, Philen et al., 1993, Taylor and McNeil, 1993, Donofrio et al., 1992). Contaminants can be responsible for triggering an immune response as suggested by other studies (McKinley et al., 1993). The immune, and probably auto-immune reaction seems to largely explain the presence of eosinophilia, characterizing the acute phase, followed by progressive neuropathy and myopathy. In particular, the discovery of an aniline-derived contaminant supported the hypothesis that EMS and toxic-oil syndrome may share a common trigger (Mayeno et al., 1992).

Increases or decreases in blood Trp as well as varied serotonin receptor sensitivity (due to a chronic exposure to abnormal levels of the specific transmitter) do not seem to be related to EMS onset, as stressed in the following points:

## A) Spontaneous or Induced Changes in Serotonin Availability

i. Long lasting treatments (3 weeks) of 5-HT creatinine sulphate (10-15 mg/i.v./day) improved M syndrome. The only side-effect consisted in a vasoconstriction and irritation of the injected vein.

ii. In two cases of carcinoid, we observed a clear-cut increase of plasma and platelet Trp, but never EMS symptoms (Salmon et al., 1986). It is also noteworthy that the treatment of fenclonine, while evoking a benefit particularly evident on diarrhoea, induced eosinophilia, thus it was suspended in spite of the fact that the carcinoid syndrome largely benefits from the above treatment (Samiy et al., 1987).

iii. Long-lasting treatments with reserpine, employed to treat anxiety or hypertension, obviously largely depleted platelet serotonin content but never induced any one of the symptoms characterizing EMS.

## B)Lack of EMS Signs in Conditions Characterized by Changes in the Sensitivity of Serotonin Receptors

i. During chronic treatment with MAOI-A/B a significant increase in both 5-HT and NA receptor sensitivity always occurred. The changed serotonin receptor sensitivity that never was associated to EMS.

ii. In the two above mentioned carcinoid syndromes, a hyporesponsivity to 5-HT was observed that was never associated to EMS. The sensitivity to 5-HT regained baseline levels when fenclonine treatment decreased diarrhoea characterizing carcinoid.

iii. In fibromyalgia syndrome, a dramatic hypersensitivity to serotonin was present that exceeded 100-500 times the reactivity of control subjects (Sicuteri et al., 1973, Sicuteri et al. 1978, Del Bianco et al., 1982); anyway, eosinophilia, fever or cutaneous-nervous pathological effect were never complained or observed.

These observations agree with the outcomes of the majority of researchers in the field of EMS. These authors disregarded the possibility that Trp by itself can give origin to EMS. It seems possible to reach some conclusions by using exclusion criteria. With regard to this it seems also possible to stress that long-term therapy with Trp or 5-HTP cannot induce EMS symptoms through a receptor metabolic change. It seems proved by the fact that in intestinal carcinoid, where serotonin increase, reaching levels up to 100 times those of baseline, is associated to specific receptor down-regulation but never to EMS. This observation seems to make inconsistent the hypothesis of a Trp or 5-HTP-induced EMS. Even large doses of uncontaminated L-Trp or 5-HTP given as a long-lasting, chronic daily treatment failed to induce EMS. In our experience, as well as in that of other pain specialists, this type of treatment can largely improve the painful picture of primary FS (Sicuteri, 1972, Sicuteri, 1973, Bono et al., 1984).

Apparently the problem of EMS could be ascribed to contaminants, nevertheless it seems noteworthy to underline that also some data are intriguing. Indeed the first pharmacologically-induced EMS was observed following treatment with fenclonine, which, contrary to Trp, acts by inhibiting serotonin metabolism. In spite of their opposite modes of action fenclonine and Trp act on serotonin metabolism lending to receptor and metabolic changes which are opposite in nature. Thus it may be possible to propose that Trp overflow can induce an abnormal set-up of enzymes and metabolism in serotonergic pathways.

With regard to this topic it seems noteworthy that patients suffering from EMS seem to show an abnormal Trp metabolism which was hypothesized to consist in increased activity of indoleamine 2,3-dioxygenase (Varga et al., 1993). It seems also that, as reported in the literature, sometimes the patients use too large doses of Trp which are acknowledged to lead to a decrease of cerebral serotonin since on one hand, the affinity of Trp for hydrxylase is lower than that for pyrrolase, on the other hand formylkinurenine inhibits Trp in crossing BBB; moreover, the excess of Trp inhibits Trp hydroxylase and activates the enzyme Trp-2,3-dioxygenase. In other words an excess of Trp inhibits the increase of serotonin content in the brain; conversely, it results in an augmented function of the metabolic pathway leading to kynurenines. The metabolites of kynurenines can display a puzzling effect on nociceptive transmission since its metabolites, represented by kynurenic acid, shown to be present in mammalian brain (Moroni et al., 1988) and quinolinic acid are agonists or antagonists of excitatory amino acids receptor, respectively (Foster et al., 1984, Moroni, 1984, Ascher et al., 1988). Indeed, neurogenic, painful input transmission might be impaired by inhibition of excitatory amino acids action, whereas an hyperalgesia state might be promoted by a sensory neuron firing due to quinolinic acid.

## ACKNOWLEDGMENTS

This work was partly supported by a grant from FATMA, National Council of Researches, Rome

## REFERENCES

Adachi, J., Naito, T., Ueno, Y., Ogawa, Y., Ninomiya, I., Tatsuno, Y., 1993, Metabolism and distribution in the rat of the peak E substance, a constituent in L-trytophan product implicated in eosinophilia-myalgia syndrome. *Arch Toxicol* 67(4): 284

Ascher, P., Henderson, G., Johnson, J.W., 1988, Dual inhibitory actions of kynurenate on the N-methyl-o-aspartate (NMDA)-activated response of cultured mouse cortical neurons. *J Physiol* 406:141

Bono, G., Micieli, G., Sances, G., Calvani, M., Nappi, G., 1984, L-5HTP treatment in primary headaches: an attempt at clinical identification of responsive patients. *Cephalalgia* 4:159

Del Bianco, P.L., Franchi, G., Anselmi, B., Sicuteri, F., 1982, Monoamine sensitivity of smooth muscle in vivo in nociception disorders, in: "Advances in Neurology", M. Critchley, AP. Friedman, , S. Gorini, F. Sicuteri , eds, Raven Press, New York

Donofrio, P.D., Stanton, C., Miller, V.S., Oestreich, L., Lefkowitz, D.S., Walker, F.O., Ely, E.W., 1992, Demyelinating polyneuropathy in eosinophilia-myalgia syndrome. *Muscle Nerve* 15 (7): 796

Fields, H.L., Anderson, S.D., 1978, Evidence that raphe-spinal neurons mediate opiate and midbrain stimulation-produced analgesias. *Pain* 5:33

Foster, A.C., Vezzani, A.M., French, E.D., Schwarcz, R., 1984, Kynurenic acid blocks neurotoxicity and seizures in rats by the related brain metabolite quinolinic acid. Neurosci Lett 48:273 International Headache Society : Classification and diagnostic criteria for headache disorders cranial neuralgias and facial pain. 1988 *Cephalalgia* 8(7)

Kimball, R.W., Friedman, A.P., Valleyo E., 1960, Effect of serotonin in migraine patient. *Neurology* 10:107

Knott, P.J,. Curzon, G., 1972, Free tryptophan in plasma and brain tryptophan metabolism. *Nature* 239:452

Kurihara, N., Yanagisawa, H., Jin , Z., Wada, O., 1993, Production of polyclonal antibodies against 1,1-3-ethyliden bis(L-tryptophan (EBT), a potential contaminant causing eosinophilia-myalgia syndrome. *Toxicol Lett* 66(3):231

Lance, J.W., 1992, Possible role of serotonin in migraine syndrome. in, "Advances in Pain Research and Therapy", F Sicuteri, L Terenius, L Vecchiett, CA Maggi ,eds, Vol 20, Raven Press, New York

McKinley, K.L., Harati, Y., Schneider, L.W., 1993, Chronic immune activation in the eosinophilia-myalgia syndrome. *Muscle Nerve* 16(9): 947

Mayeno, A.N., Belongia, E.A., Lin, F., Lundy, S.K., Gleich, G.J., 1992, 3-(Phenylamino)alanine, a novel aniline-derived amino acid associated with the eosinophilia-myalgia syndrome: a link to the toxic oil syndrome? *Mayo Cli Proc* 67 (12): 1134

Messing, R.B., Fisher , L.A., Phebus, L., Lytle, L.D., 1976, Interaction of diet and drugs in the regulation of brain 5-hydroxyindoles and the response to painful electric shock. *Life Sci* 18 :707

Moldofsky, H., Warsh, J.J., 1978, Plasma tryptophan and musculoskeletal pain in non-articular rheumatism ("fibrositis syndrome") *Pain* 5:65

Moroni, F., Lombardi, G., Carlà, V., Moneti, G., 1984, The excitotoxin quinolinic acid is present and unevenly distributed in the rat brain. *Brain Res* 295: 352

Moroni, F., Russi, P, Carlà, V., Lombardi, G., 1988, Kynurenic acid is present in the rat brain and its content increases during development and aging process. *Neurosci Lett* 94:145

Park, D.H., Baker, H., Kim, K.S, John T.H. Early induction of rat brain tryptophan hydroxylase (TPH) mRNA following paracholorophenylalanine (PCPA) treatment. *Brain Res Mol Brain Res*. 22 (1-4): 20-8, 1994

Philen, R.M., Hill, R.H., Flanders, W.D., Caudill, S.P., Needham, L., Sewell, L., Sampson, E.J., Falk, H., Kilboune, E.M., 1993, Tryptophan contaminants associated with eosinophilia-myalgia syndrome. The Eosinophilia-Myalgia Studies of Oregon, New York and New Mexico. *Am J Epidemiol* 138(3): 154

Rajiv, J., Welch, K.M.A., D'Andrea, G., 1989, Serotonergic hypofunction in migraine:a syntesis of evidence based on platelet dense body dysfunction. *Cephalalgia* 9: 18

Salmon, S., Bonciani, M., Fanciullacci, M., Marianelli, L., Michelacci, S., Sicuteri, F., 1982, A putative 5-HT central feedback in migraine and cluster headache attacks.in, "Advances in Neurology", M. Critchley et al., , eds, Raven Press, New York

Salmon, S., Curradi, C., Geppetti, P., Sicuteri, F., 1986, Tryptophan, 5-hydroxytryptophan and substance P in carcinoid patients. in, "Progress in Tryptophan and Serotonin Research", Walter de Gruyter & Co., Berlin

Samiy , A.H., Douglas, R.G., Barondess, J.A., 1987, Diagnostic Medicine. Lea & Febiger, Philadeplhia

Sicuteri, F., 1972, 5-Hydroxytryptophan in the prophylaxis of migraine. *Pharmacol Res Comm* 4:213

Sicuteri F., 1973, The ingestion of serotonin precursors (L-5-hydroxytryptophan and L-tryptophan). *Headache* 13:19

Sicuteri, F., 1974, The influence of tryptophan and parachlorophenylalanine on the sexual activity in man. Abstract book "First International Meeting on Tryptophan metabolism: Biochemistry , Pathology and Regulation, Padova, 2-4 May, 1974

Sicuteri, F., Fanciullacci, M., Franchi, G., Del Bianco, P.L., 1965, Serotonin-bradykinin potentiation on the pain receptors in man. *Life Sci* 4: 409

Sicuteri, F., Anselmi, B., Del Bianco, P.L., 1973, 5-hydoxytryptamine supersensitivity as a new theory of headache and central pain. A clinical pharmacological approach with p-chlorophenylalanine. *Psychopharmacology* 29: 347

Sicuteri, F., Anselmi, B., Fanciullacci, M., 1974, The serotonin (5-HT) theory of migraine. *Adv Neurol* 4: 383

Sicuteri, F., Anselmi, B., Del Bianco, P.L., 1978, Systemic non-organic central pain : a new syndrome with decentralization supersensitivity. *Headache* 18: 133

Sicuteri, F., Panconesi, A., Franchi, G., Del Bianco, P.L., Anselmi, B., 1983, Impaired 5-hydroxytryptamine tachyphylaxys in migraine. *Cephalalgia* 3: 139

Silver, R.M., 1992, Eosinophilia-myalgia syndrome, toxic-oil syndrome, and diffuse fascitis with eosinophilia. *Curr Opin Rheumatol* 4(6): 851

Taylor, R., McNeil, J.J., 1993, Eosinophilia-myalgia syndrome: lessons for public health researchers. *Med J Aust* 158 (1): 51

Tilson, H.A., Rech, R.H., Stolman, S., 1983, Hyperalgesia during withdrawal as a means of measuring the degree of dependence in morphine dependent rats. *Psychopharmacology* 28: 287

Varga, J., Jmenez, S.A., Uitto, J., 1993, L-tryptophan and eosinophilia-myalgia syndrome; current understanding of the etiology and pathogenesis. *J Invest Dermatol* 100(1): 97S

Wall P.D., 1992, The biological function and dysfunction of differerent pain mechanism. in, "Advances in Pain Research and Therapy", F. Sicuteri, L. Terenius, L. Vecchiett, CA. Maggi ,eds, Vol 20, Raven Press, New York

Wolfe, F., Smythe, H.A., Yunus, M.B., 1990, The american college of rheumatology 1990 criteria for the classification of fibromyalgia, report of the multicenter criteria committee. *Arthritis Rheum* 33: 160

Yaksh , T.L., Plant, R.L., Rudy, T.A., 1977, Studies on the antagonism by raphe lesions of antinociceptive action of systemic morphine. *Eur J Pharmacol 41:399*

# PERSISTENT MICROVASCULOPATHY IN CHRONIC EOSINOPHILIA-MYALGIA SYNDROME

S. A. Smith

University of Minnesota Hennepin Faculty Associates Neuromuscular Center
825 South 8th Street, Suite LL-05
Minneapolis, Minnesota 55404

## 1. ABSTRACT

Acute eosinophilia-myalgia syndrome (EMS) due to contaminated L-tryptophan (LT) exposure is an inflammatory microangiopathy of the dermis, fascia, and muscle. Select individuals evolve from acute EMS to have persistence of myalgia, fatigue, cramps, and skin changes for years. Many develop memory dysfunction and confusion. The objective of this study is to delineate the pathology in individuals with chronic EMS. Seventeen patients with ongoing symptoms representing chronic EMS are studied by skin, fascia, and muscle biopsies four to five years after exposure to contaminated LT and initial onset of EMS. All have microvascular disease. Most have lymphocytic inflammatory infiltrates. Several have dermal sclerosis. The findings indicate that persistent microvascular disease is present in chronic EMS. The pathologic changes are similar to those of acute EMS but with notable differences. Tissue eosinophil infiltration is rare in the chronic state as compared to acute EMS. The degree of inflammation in chronic EMS is markedly reduced compared to acute EMS. The persistence of endothelial pathology indicates continuing microvascular dysfunction.

## 2. INTRODUCTION

Contaminated LT was linked with EMS in 1989 (Eosinophilia-myalgia syndrome, 1989; Glickstein et al., 1990; Hertzman et al., 1990). Acute EMS pathologically is a microangiopathy of dermis, fascia, and skeletal muscle accompanied by an inflammatory infiltrate of lymphocytes, histiocytes, and eosinophils (Silver et al., 1990; Smith et al., 1990; Smith, 1992;). Follow-up clinical studies indicate evolution of acute EMS in select individuals into a chronic condition with persistent symptoms involving skin and muscle (Kaufman, 1994). Clinical findings include myalgia, cramps, fatigue, neuropathy, arthralgia, dyspnea,

*Recent Advances in Tryptophan Research,* edited by Graziella Allegri Filippini et al.
Plenum Press, New York, 1996

**Table 1.** Clinical manifestations

| | |
|---|---|
| Myalgia | 14 |
| Excessive fatigue | 9 |
| Thickened skin or hair loss | 8 |
| Muscle cramping | 5 |
| Confusion and dizziness | 5 |
| Word finding difficulty | 3 |

**Table 2.** Biopsy results—17 cases

| | |
|---|---|
| Endothelial dysfunction | 17 |
| Perivasculitis | 16 |
| Muscle necrosis/atrophy | 7 |
| Sclerosis | 6 |
| Inflammatory neuropathy | 4 |
| Necrotic vessels | 2 |
| Tissue eosinophils | 2 |

skin thickening, alopecia, and neurocognitive dysfunction. The persistence of EMS may result in significant morbidity (Kaufman, 1993) and mortality (Swygert et al., 1993). This study delineates the pathology of chronic EMS in individuals with persistent symptoms four and five years after disease onset.

# 3. METHODS

## 3.1 Subjects

Seventeen individuals had onset of symptoms indicative of EMS after exposure to contaminated LT in 1989. All continued to be symptomatic indicating evolution to chronic

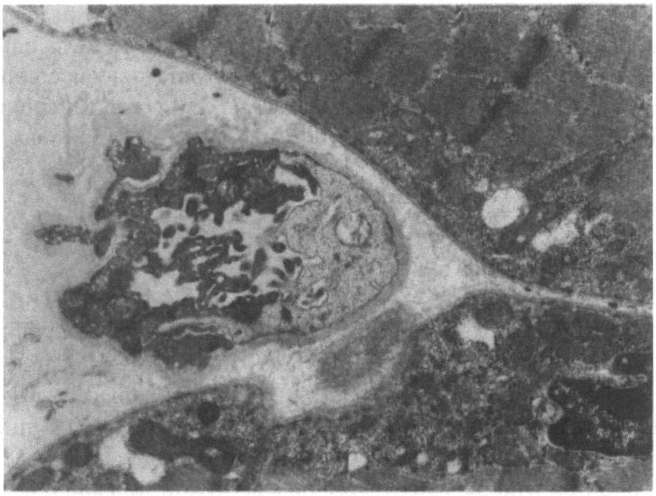

**Figure 1.** Electron micrograph from skeletal muscle biopsy illustrating microangiopathy in a capillary with swollen endothelium on the right and redundant luminal endothelium on the left. (18,500X).

EMS. They were studied by skin, fascia, and muscle biopsies four to five years after exposure to contaminated LT and initial onset of EMS. The most common persistent complaint present in 14 patients is myalgia. Nine have bothersome and excessive fatigue most noticeable after modest activity. Eight patients have thickened skin, that developed during the acute phase of illness often accompanied by hair loss. Five experience intermittent muscle cramping. Five have mental confusion and/or dizziness. Three have word finding difficulty or short term memory dysfunction.

## 3.2 Pathology

Subjects had skin, fascia, and skeletal muscle biopsies, either from the deltoid or biceps region in the arm or the quadriceps in the leg. Muscle was quickly frozen for light microscopy and histochemistry and preserved in fixative for electron microscopy. Skin and fascia were prepared for light microscopy and electron microscopy.

The results from biopsies show that all 17 have microvascular disease in one or more tissues. This typically consists of capillary and small vessel endothelial swelling or thickening accompanied by redundancy of luminal endothelial surface as noted in figure 1. Sixteen biopsies demonstrate mononuclear cell perivasculitis in one or more tissues, typically the muscle as seen in figure 2. In addition, four have evidence for vasculitis with invasion of the adventitia of small blood vessels by lymphocytes illustrated in figure 3. Two subjects have necrotic capillaries seen in figure 4. Lymphocytic infiltrate in one or more tissues is present with eosinophils found in two cases in separate tissues as observed in figure 5. Secondary muscle necrosis or atrophy is present in six. Skin sclerosis is found in four. Infiltration of nerve perineurium is noted in four biopsies as illustrated in figure 2.

Fourteen of the subjects have been symptomatic for four years at the time of study. Three others are five years into their illness at biopsy. The findings are similar between the two groups, although the longer term study individuals do not have evidence for inflammatory neuropathy. One patient from each group has eosinophils along with lymphocytes in their biopsies. Tables 3 and 4 detail the pathologic findings in each group.

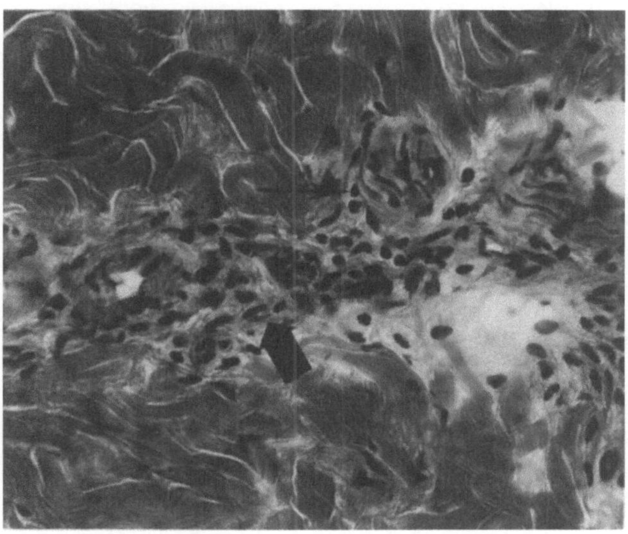

**Figure 2.** Skin biopsy showing perivascular, lymphocytic inflammatory infiltrate (big arrow) and invasion of nerve perineurium by the infiltrate (small arrow). (H&E stain, 490X).

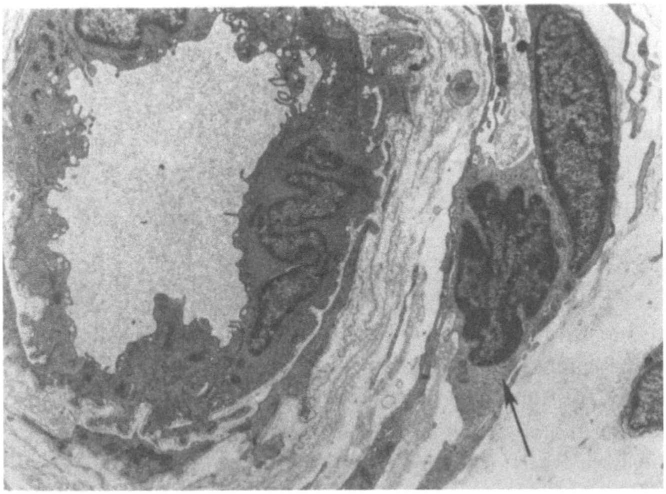

**Figure 3.** Electron micrograph from skin biopsy showing an arteriole with lymphocytic invasion (arrow) beneath the adventitia. (7,800X).

## 4. DISCUSSION

The pathologic changes in acute EMS are distinctive (Silver et al., 1990; Smith et al., 1990; Smith,1992; Emslie-Smith et al., 1991). They are different from idiopathic eosinophilic fasciitis, toxic oil syndrome, dermatomyositis, lupus myositis, and polymyositis. The phenotypic nature of the inflammatory infiltrate and the microvascular involvement define the unique pathology of EMS (Smith et al., 1990; Smith, 1992). The main pathologic abnormality is a microangiopathy with endothelial swelling and necrosis plus sub-adventitial

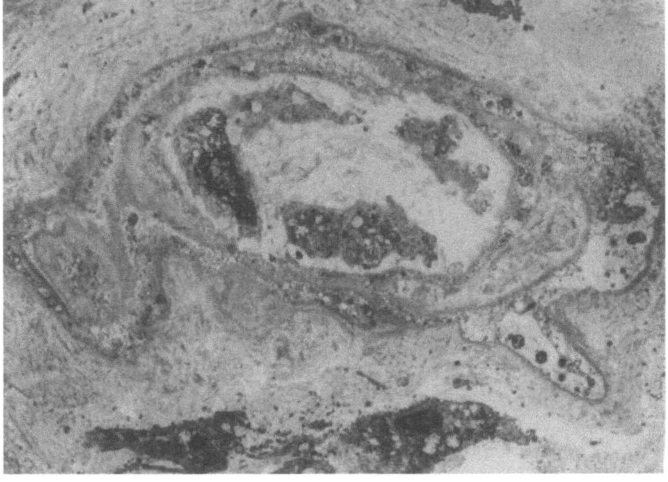

**Figure 4.** Electron micrograph from skin biopsy illustrating endothelial necrosis and loss of vessel lumen. (6,600X).

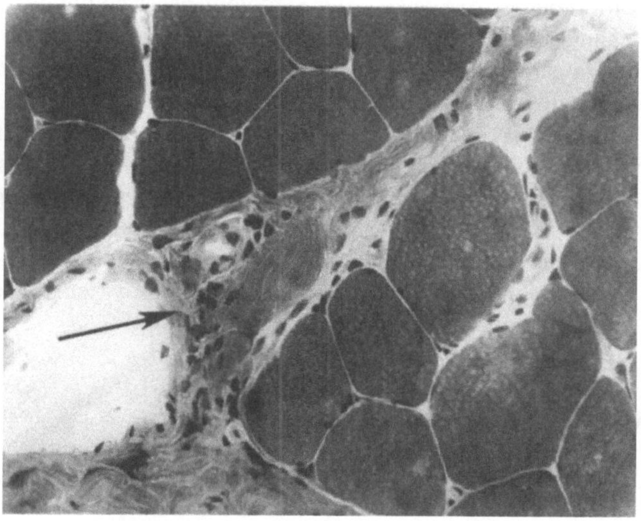

**Figure 5.** Muscle biopsy showing perivascular and diffuse lymphocytic and eosinophilic (arrow) infiltrates. (H&E, 340X).

**Table 3.** Four year biopsy results—14 cases

| | Muscle | Fascia | Dermis |
|---|---|---|---|
| Endothelial dysfunction | 9 | 4 | 12 |
| Perivasculitis | 12 | 3 | 1 |
| Muscle necrosis/atrophy | 6 | | |
| Sclerosis | | 1 | 3 |
| Inflammatory neuropathy | 1 | 1 | 2 |
| Necrotic vessels | | | 1 |
| Tissue eosinophils | 1 | | 1 |

**Table 4.** Five year biopsy results—3 cases

| | Muscle | Fascia | Dermis |
|---|---|---|---|
| Endothelial dysfunction | 2 | 2 | 3 |
| Perivasculitis | 2 | 1 | 2 |
| Muscle necrosis/atrophy | 1 | | |
| Sclerosis | | 0 | 3 |
| Inflammatory neuropathy | | | |
| Necrotic vessels | 1 | 0 | 0 |
| Tissue eosinophils | 1 | | 1 |

invasion of small vessels by lymphocytes. Perivascular lymphocytic and eosinophilic infiltrates in the dermis, fascia, and perimysium of the muscle are present. A more diffuse lymphocytic and histiocytic infiltrate with eosinophils is found in the fascia. The vascular disease progresses to necrosis of the endothelium. Repeat biopsies two and one-half years after disease onset show findings identical to those seen in acute phase biopsies with continuation of the microvasculopathy, in particular the swelling and necrosis of the endothelium associated with adventitial invasion of small vessels by lymphocytes (Kaufman, 1994; Smith, 1994).

Chronic EMS four to five years after symptom onset continues to show similar, although fewer, changes compared to those seen in earlier stages of the condition. The continuing microangiopathy is recognized by capillary endothelial swelling, invasion of adventitia by lymphocytes, and capillary endothelial necrosis. The degree of inflammation in chronic EMS is markedly reduced compared to acute EMS. Muscle remains the dominant site for residual microvascular and inflammatory pathology with the skin less involved. The fascia is least involved in the longer term, as compared to significant pathologic involvement early on. Tissue eosinophils, so prominent initially, are rarely seen four and five years into the illness. Invasion of adventitia of small vessels is present but less apparent than in the acute phase. Of greatest significance is the persistence of microvascular endothelial alterations indicating continuing microvascular dysfunction.

## 5. REFERENCES

Emslie-Smith A.M, Engel A. G., Duffy J., et al., 1991, Eosinophilia myalgia syndrome, I:immunocytochemical evidence for a T-cell-mediated immune effector response. *Ann Neurol* 29:524.

Eosinophilia-myalgia syndrome and L-tryptophan-containing products — New Mexico, Minnesota, Oregon, and New York. *MMWR* 38:785-88(1989).

Glickstein S.L., Gertner E., Smith S.A., et al, 1990, Eosinophilia-myalgia syndrome associated with L-tryptophan. *J Rheumatol* 17:1534.

Hertzman P.A., Blevins W.L., Mayer J., et al, 1990, Association of the eosinophilia-myalgia syndrome with the ingestion of tryptophan. *N End J Med* 322:869.

Kaufman L.D., 1993, Eosinophilia-myalgia syndrome: morbidity and mortality [editorial]. *J Rheumatol* 20:1644.

Kaufman L.D., 1994, The evolving spectrum of eosinophilia myalgia syndrome. *Rheum Dis Clin North Amer* 20:973.

Silver R.M., Heyes M.P., Maize J.C., et al, 1990, Scleroderma, fasciitis, and eosinophilia associated with the ingestion of tryptophan. *N Eng J Med* 322:874.

Smith S.A., 1992 Pathology of eosinophilia-myalgia syndrome, in: "Advances in Tryptophan Research," I. Ishiguro, R. Kido, T. Nagatsu, et al, eds., Fujita Health University Press, Toyoake, p. 9375-78.

Smith S.A., 1994, Pathology of the microangiopathy and immunohistochemistry of the inflammatory infiltrate in eosinophilia-myalgia syndrome, in: "L-Tryptophan Current Prospects in Medicine and Drug Safety," W. Kochen, H. Steinhart, eds., Walter de Gruyter, Berlin 223-33.

Smith S.A., Roelofs R.I., Gertner E., 1990 Microangiopathy in the eosinophilia-myalgia syndrome. *J Rheumatol* 17:1544.

Swygert L.A., Back E.E., Auerbach S.B., et al., 1993, Eosinophilia-myalgia syndrome: mortality data from the U.S. national surveillance system. *J Rheumatol* 20:1711.

56

# A COMPARATIVE STUDY OF TISSUE DISTRIBUTION AND EXCRETION AMONG THREE SUBSTANCES IMPLICATED IN EOSINOPHILIA-MYALGIA SYNDROME

J. Adachi,[1] Y. Ueno,[1] Y. Tatsuno,[1] M. Gomez,[2] C. C. Smith,[2] and E. M. Sternberg[2]

[1] Department of Legal Medicine
Kobe University School of Medicine
Kobe 650 Japan
[2] Clinical Neuroendocrinology Branch
National Institute of Mental Health
Bethesda, Maryland 20892

## 1. INTRODUCTION

Epidemiological studies have suggested that L-tryptophan(L-TRP) produced by a specific manufacturer may be implicated in eosinophilia-myalgia syndrome(EMS) in the United States and other countries(Belongia et al., 1990). The implicated L-TRP has been shown to contain more than 60 impurities(Hill et al., 1993), some of which might contribute to EMS(Centers for Disease Control, 1990; Crofford et al., 1990). Of these, the first impurity statistically associated with EMS was identified as 1,1'-ethylidenbis [tryptophan] (EBT) (Mayeno et al., 1990). EBT was vulnerable to acid and readily decomposed to 1-methyl-tetrahydro-ß-carboline(MTCA) in artificial gastric fluid(CDC 1990). In addition MTCA was detected in blood and urine of rats treated with EBT(Adachi et al., 1993). A second impurity has now been isolated(Toyo'oka et al., 1991), and its chemical structures has been determined to be 3-(phenylamino)alanine(PAA) (Goda et al., 1992). We administered EBT, MTCA and PAA to rats to compare their distributions and excretions. We then investigated the effect of chronic treatment of PAA on the tissue concentration and the retention of PAA in their tissues following the animals'return to a standard control diet. Finally, we investigated its metabolites in rat urine.

*Recent Advances in Tryptophan Research*, edited by Graziella Allegri Filippini et al.
Plenum Press, New York, 1996 

**Table 1.** Concentrations(ng/g tissue) at 5 h

| | Dose (mg/kg) | Blood | Brain | Kidney | Liver |
|---|---|---|---|---|---|
| MTCA | 1.6 | 3.8 | 0.2 | 75.4 | 12.9 |
| PAA | 5 | 899 | 2139 | 2225 | 1290 |
| MTCA* | 5 | 11.8 | 0.6 | 233 | 40 |
| PAA/MTCA* | | 76 | 3500 | 10 | 32 |

*Calculated by dose.

# 2. TISSUE DISTRIBUTION AND EXCRETION

## 2.1 Comparison of Tissue Distribution between the Rats Administered MTCA and PAA

After overnight fast, rats were given by gastric gavage a single 1.6 mg/kg dose of MTCA(n=5)(Ogawa et al., 1993) and a single 5 mg/kg dose of PAA(n=5)(Adachi et al., 1995), respectively. They were sacrificed at 5 h and blood, brain, kidney and liver were collected. MTCA was analyzed by HPLC with fluorometric detection, while PAA by HPLC with electrochemical detection.

Table 1 shows levels of MTCA and PAA in blood and organ tissues. MTCA was scarcely seen in brain, whereas very high levels of PAA was observed in brain indicating the passage of PAA through the blood-brain barrier. The concentration of PAA in the kidney (2225 ng/g) was highest, and similar to that of brain(2139 ng/g), with a lower concentration seen in the liver(1290 ng/g). The smallest level of PAA was seen in the blood(899 ng/g), however the level was 76 times higher than that of MTCA in blood calculated by dose. The level of PAA in the liver was 32 times higher, and that in the kidney was 10 times higher than those of MTCA calculated by dose, respectively. Thus, an accumulation of PAA in each tissue was observed.

## 2.2 Comparison of Excretion among the Rats Administered MTCA, EBT and PAA

After overnight fast, rats were perorally administered the following drugs: (A) a single 1.6 mg/kg dose of MTCA(n=15); (B) a single 50 mg/kg dose of EBT(n=14)(Adachi et al., 1993); (C) a single 100 mg/kg dose of PAA(n=12) (Adachi et al., 1994). They were sacrificed at 5 h and blood and contents of the large intestine were collected. Urine was

**Table 2.** Blood levels and excretions

| | Dose (mg/kg) | Blood 5 h (ng/ml) | Urine 0–24 h | | Large intestine | |
|---|---|---|---|---|---|---|
| | | | μg | % | μg | % |
| MTCA | 1.6 | 4 | 111 | 35 | 198 | 61 |
| EBT | 50 | 83 | 11 | 0.1 | 3136 | 31 |
| PAA | 100 | 4220 | 106 | 0.5 | <1 | 0 |
| MTCA* | 100 | 237 | 6950 | | 12375 | |
| EBT* | 100 | 166 | 22 | | 6272 | |

*Calculated by dose.

collected separately in glass containers at the bottom of each cage for 5 h and 5 to 24 h. EBT was measured by HPLC with UV detection.

Table 2 shows blood levels and excretions of three substances. The blood level of PAA was 17 times higher than that of MTCA and 25 times higher than that of EBT. In addition, 35 % of MTCA was excreted into urine and 61 % of MTCA was seen in the large intestine. Thirty-one percent of EBT was seen in the large intestine, while 0.1 % of EBT was excreted into urine. In contrast, only 0.5 % PAA was excreted into urine within 24 hrs and PAA remained scarcely in the large intestine. Thus, within 24 hrs most of MTCA was excreted with or without absorption, whereas PAA was scarcely excreted.

## 3. EFFECT OF CHRONIC TREATMENT ON PAA CONCENTRATIONS IN TISSUE

Twenty-six rats were used. Six rats were administered PAA incorporated in a solid diet(50 mg/kg chow) and killed following 16 and 24 h ad libitum exposure to this diet. A chronic feeding study of 6 weeks duration using solid PAA enriched diet was performed on 20 rats. Of these, 14 rats were killed at the end of 6 weeks, and the remaining six rats were fed an identical control diet(no PAA) for an additional 12 days and then killed.

PAA tissue concentrations are represented in Table 3. There were no significant increases in PAA content in any of the tissue between 16-and 24-h time points of the acute study. A comparison of tissue levels following 42 days on a PAA diet with those following 24 h showed a 10-fold increase in blood levels. Brain, liver and kidney levels increased approximately 2-fold. Following removal of PAA for 12 days, drug levels decreased by 94-98%, however detectable levels of the drug were present in all tissues.

## 4. IDENTIFICATION OF PAA METABOLITES

Rats were given perorally a single 100 mg/kg dose of PAA and urine was collected. The samples of trimethylsilyled(TMS) derivatives were analyzed by GC-MS(Adachi et al., 1994).

Figure 1 shows a gas chromatogram (TIC) of TMS derivatives obtained from rat urine extract (upper column) and the mass spectra of those derivatives. The figure also shows the molecular ion chromatogram (m/z 223,325 and 311) of each standard N-phenylglycine, 3-(phenylamino)lactic acid, and N-(hydroxyphenyl)glycine (lower column). The ion peaks originating from TMS derivatives of standard N-phenylglycine, 3-(phenylamino)lactic acid, and N-(hydroxyphenyl)glycine were seen at RTs of 7.8 min, 11.8 min, and 14.2 min, respec-

**Table 3.** PAA concentrations(ng/g ± SEM) in rats which received PAA as supplement in a solid diet(50 mg/kg chow)

| | Blood | Brain | Kidney | Liver |
|---|---|---|---|---|
| (A) | 146 ± 17 | 598 ± 180 | 1051 ± 109 | 683 ± 88 |
| (B) | 121 ± 25 | 939 ± 102 | 1094 ± 60 | 659 ± 46 |
| (C) | 1272 ± 75* | 1998 ± 75* | 2764 ± 105* | 1356 ± 36* |
| (D) | 10 ± 2 | 26 ± 3 | 68 ± 11 | 84 ± 9 |

(A),(B),(C) and (D) represent the groups divided by durations of administration.
A; acute 16 h(n=3) B; 24 h(n=3) C; chronic 42 d(n=14) D; 42+12 d(n=6)
*Significantly greater than the corresponding acute value ($P < 0.05$).

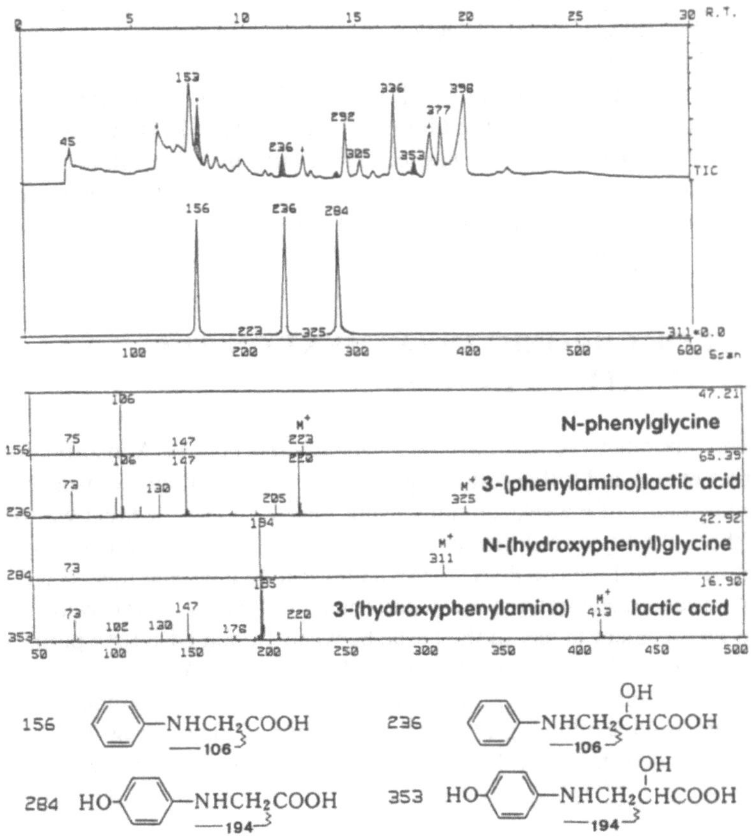


**Figure 1.** Gas chromatograms(TIC) and mass spectra of trimethylsilyled derivatives: (upper column) rat urine extract; (under column) standard specimens.

tively. The ion peak obtained from the urine extract appearing at an RT of about 7.8 min had molecular ion(m/z 223) and fragment ions similar to those for standard N-phenylglycine. Similarly, the ion peaks from the urine extract appearing at RTS of about 11.8 min and 14.2 min had molecular ion(m/z 325 and m/z 311) and fragment ions similar to those for standard 3-(phenylamino)lactic acid and N-(hydroxyphenyl)glycine, respectively. Therefore, these peaks extracted from urine were identified based on the similarity of their RTs and mass spectra with those of standard specimens. The ion peak from the urine extract appearing at an RT of 17.65 min had molecular ion(m/z 413) and a base peak ion at m/z 194. N-phenylglycine and 3-(phenylamino)lactic acid had base peak at m/z 106, while N-(hydroxyphenyl)glycine had one at m/z 194, corresponding to TMS derivative of hydroxy group. Considering that the base peak (m/z 194) was the same as that of N-(hydroxyphenyl)glycine, the chemical structure of the ion peak was determined to 3-(hydroxyphenylamino)lactic acid. Fragmentation patterns of these four compounds are shown under their mass spectra.

## 5. DISCUSSION

We have shown for the first time the metabolism and distribution of PAA in the rat. Comparison of tissue levels between animals administered PAA and MTCA, respectively, showed much higher levels of PAA than MTCA. A further comparison of excretion of PAA, EBT and MTCA showed that the amount of PAA excreted was the lowest of these three compounds suggesting an accumulation of PAA in rat tissues. In addition, we confirmed the accumulation of PAA in each tissue following chronic treatment with solid food diet enriched with PAA.

We have also found four metabolites of PAA in the urine of rats that received PAA perorally. Since GC/MS showed that each metabolite had the same RT and fragmentation pattern as each standard compound, we were able to identify the three metabolites. The fourth metabolite was elucidated by its similar fragmentation pattern to N-(hydroxyphenyl)glycine. This suggested a common fragmentation at the amino acid residue of these compounds. The presence of these metabolites in urine suggests a possible metabolic pathway for these impurities. The first step may be hydroxylation of PAA to hydroxy-PAA, after which a reaction may be catalyzed by phenylalanine hydroxylase (monooxygenase). The reductant may be tetrahydrobiopterine. The next step may be the transamination of PAA to 3-(phenylamino)pyruvic acid. The α-keto acid must be unstable, so it would be immediately reduced by aldehyde reductase to 3-(phenylamino)lactic acid. The last step may be decarboxylation by aromatic-L-amino acid decarboxylase to phenylethylenediamine, which may subsequently be oxidized by monoamineoxidase to phenylaminoaldehyde, immediately resulting in N-phenylglycine.

We have showed for the first time an outline of PAA biotransformation, which suggests a possible degradation pathway with various metabolic intermediates at each step. These secondary metabolites could play a role in toxicity, however, the involvement of each substance in toxicity is not known and must be elucidated in the future.

## 6. ACKNOWLEDGMENT

The authors would like to thank Ms Sachiko Fujimoto for technical assistance.

## 7. REFERENCES

Adachi, J., Naito, T., Ueno, Y., Ogawa, Y., Ninomiya, I., Tatsuno, Y., 1993, Metabolism and distribution in the rat of peak E substance, a constituent in L-tryptophan product implicated in eosinophilia-myalgia syndrome, *Arch Toxicol.* 67: 284.

Adachi, J., Mio, T., Ueno, Y., Naito, T., Nishimura, A., Fujiwara, S., Sumino, K., Tatsuno, Y., 1994, Identification of four metabolites of 3-(phenylamino)alanine, a constituent in L-tryptophan product implicated in eosinophilia-myalgia syndrome, in rats, *Arch Toxicol* 68: 500.

Adachi, J., Gomez, M., Smith, C.C. and Sternberg, E.M, 1995, Accumulation of 3-(phenylamino)alanine, a constituent in L-tryptophan products implicated in eosinophilia-myalgia syndrome, in blood and organs of the Lewis rats, *Arch Toxicol.* 69:266.

Belongia, E.A., Hedberg, C.W., Gleich, GJ., White, K.E., Mayeno, A.N., Loegering, D.A., Dunnette, S.L., Pirie, P.L., MacDonald, K.L., Osterholm, M.T., 1990, An investigation of the cause of the eosinophilia-myalgia syndrome associated with tryptophan use, *N Engl J Med* 323: 357.

Centers for Disease Control, 1990, Analysis of L-tryptophan for the etiology of eosinophilia-myalgia syndrome, New Mexico MMWR. 39: 589.

Crofford, L.J., Rader, J.I., Dalakas, M.C., Hill, R.H. Jr., Page, SW., Needham, L.L., Brady, L.S., Heyes, M.P., Wilder, R.L.,Gold, P.W., Illa I., Smith, C.C., Sternberg, E.M., 1990, L-tryptophan implicated in human eosinophilia-myalgia syndrome causes fascitis and perimyositis in the Lewis rat, *J Clin Invest.* 86: 1757.

Goda, Y., Suzuki, J., Maitani, T., Yoshihira, K., Takeda, M., Uchiyama, M., 1992, 3-Anilino-L-alanine, structural determination of UV-5, a contaminant in EMS-associated L-tryptophan samples, *Chem Pharm Bull.* 40: 2236.

Hill, R.H Jr., Caudilli, S.P., Philen, R.M., Bailey, S.L., Flanders, W.D., Driskell, W.J., 1993, Containants in L-tryptophan associated with eosinophilia-myalgia myalgia syndrome(EMS), *Arch Environ Contam Toxicol.* 25: 134.

Mayeno, A.N., Lin, F., Foote, C.S., Loegering, D.A., Ames, M.M., Hedberg, C.W., Gleich, G.J. 1990, Characterization of "peak E", a novel amino acid associated with eosinophilia-myalgia syndrome, *Science.* 250: 1707.

Ogawa, Y., Adachi, J., Tatsuno, Y., 1993, Accumulation of 1-methyl-tetrahydro-b-carboline-3-carboxylic acid in blood and organs of rat. A possible causative substance of eosinophilia-myalgia syndrome associated with ingestion of L-tryptophan, *Arch Toxicol.* 67: 290.

Toyo'oka ,T., Yamazaki, T., Tanimoto, T., Sato, K., Sato, M., Toyoda, M., Ishibashi, M., Yoshihira, K., Uchiyama, M., 1991, Characterization of contaminants in EMS-associated L-tryptophan samples by high-performance liquid chromatography, *Chem Pharm Bull*, 39: 820.

# 57

# EOSINOPHILIA-MYALGIA SYNDROME

## Status of Patients at Onset and after Four Years of Disease

L. A. Criswell and T. Pincus

University of California San Francisco
San Francisco, California and
Vanderbilt University School of Medicine
Nashville, Tennessee

In order to characterize long-term outcomes of the eosinophilia-myalgia syndrome (EMS), the 3,500 members of the American College of Rheumatology (ACR) were contacted to identify patiens with EMS. Among 1,535 ACR members who responded, 440 reported seeing 896 patients with EMS. When asked about the course of these patients, these rheumatologists indicated that 4% had died, 31% had severe functional disability, 37% had mild functional disability, and 5% had minimal residual disabilitity but required corticosteroids. 46% had physical findings. 14% had completely recovered. Most patients took L-tryptophan (LT) as a result of a self-medication, but 30% took LT upon recommendation of a health professional. The 440 rheumatologists who reported cases of EMS, and the physicians who had reported 1,400 cases to the Centers for Disease Control and Prevention (CDC), were asked to present a self-report questionnaire to their patients with EMS in 1993, four years after onset. The questionnaire include question regarding demographic data, functional status, status of EMS at onset, status prior to EMS onset, pain scores, fatigue scores, self-report of somatic symptoms, and self-report of cognitive symptoms. Among the initial 344 patients who returned questionnaires, 251 indicated that they had been reported to State Health Departments. Mean age at diagnosis was 57 years, 88% were female, and 96% were Caucasian. Among these 344 patients, four years after disease onset 29 reported themselves as "fully recovered", 121 as "much better", 4 as "better", 4 as "the same", 23 as "worse", 129 as "better in some ways and worse in others", and 33 "other". The status of EMS at onset was only minimally predictive of status 4 years later. A severe or very severe initial acute course of EMS was reported by 52% of patients whose follow-up status was "recovered", 67% "much better", 56% "worse", and 67% of those whose status was "better in some ways and whorse in others". Four years after disease onset, symptoms reported by more than 50% of patients included fatigue, muscle pain, poor memory, poor sleep, anxiety, muscle cramps, numbness, muscle weakness, depression, joint pain and headaches. Self-report scores for functional status, global status, pain, fatigue, and learned helplessness all indicated significantly better status in patients who described themselves as "recovered" or "much better" versus those who described themselves as "the same", "worse", or "better in some ways and worse in others". Scores for the cognitive symptoms inventory were correlated significantly with the number of "purely somatic" symptoms. These data indicate that although most patients with EMS have shown clinical improvement, most report persistent symptoms.

# 58

# FIBROMYALGIA AND MIGRAINE, TWO FACES OF THE SAME MECHANISM

## Serotonin as the Common Clue for Pathogenesis and Therapy

M. Nicolodi and F. Sicuteri

Interuniversity Centre of Neurochemistry and
Clinical Pharmacology of Idiopathic Headache-Florence
University Unit, 18
V.le Pieraccini, I-50139 Florence, Italy

## SUMMARY

In this study fibromyalgia sufferers were randomly administered a combination of monoamine -oxidase inhibitors (MAOIs) -A/B with 5-HTP, 5-HTP alone, MAOIs-A/B alone, or the tricyclic drug amitriptyline in order to compare the efficacy of these treatments. The benefits on the painful syndrome were assessed by using Visual Analogic Scale score rating from 0 to 4. The combination of MAOIs with 5-HTP significantly improved fibromyalgia syndrome as determined by Visual Analogic Scale whereas the other treatments yielded poorer benefits. No subject withdrew from the trial due to adverse effects, even if some sleep disturbances and mild stomach-ache were reported. The tolerability of the association MAOIs/5-HTP was good, although a transient *cheese effect* occurred in one of the patients treated with MAOIs as well as in a patient treated with the association MAOIs and 5-HTP. No one of these two cases was due to pharmacological dietetic mistake of the patient. In both the cases the transient hypertension was associated to very dramatic emotional events.

The benefits obtained by using the combination of MAOIs with 5-HTP can be explained with a treatment-induced enhancement of aminergic and serotonergic transmission. The recently shown high prevalence of migraine in the population of fibromyalgia sufferers, suggests a common ground shared by fibromyalgia and migraine. Migraine has been demonstrated to be characterized by a defect in the serotonergic and adrenergic systems. A parallel dramatic failure of serotonergic systems and a defect of adrenergic transmission have been evidenced to affect fibromyalgia sufferers too. Enhancing serotonergic analgesia while increasing adrenergically mediated analgesia seems to be an important tool in fibromyalgia. Treatment consisting with the association MAOIs/5-HTP is aimed at enhancing serotonergic/adrenergic transmission by inducing an up-regulation of serotonergic/adrenergic receptors and a simultaneous increase of serotonin levels in the central nervous system.

*Recent Advances in Tryptophan Research,* edited by Graziella Allegri Filippini et al.
Plenum Press, New York, 1996

# INTRODUCTION

Today there is large agreement that fibromyalgia syndrome (FS) depends on a central imbalance between the painful transmission and the controlling systems (Simons, 1990, Zimmerman, 1991, Nicolodi and Sicuteri, 1992).

At the beginning of the seventies, studies carried out in our laboratory indicated that fenclonine can evoke a systemic FS which is reverted by discontinuation of the drug and appears again with new administration of fenclonine (Sicuteri, 1971, Sicuteri et al., 1973). It was also proved that fenclonine treatment induced FS in migraine (M) sufferers and not in headache-exempt subjects (Cremata and Koe, 1966). The experiment highlighted that an impairment of serotonergic pathways can lead to the appearance of FS .

This is supported by the dramatic supersensitivity to serotonin which has been shown by using the original technique of computerized vein constriction in FS patients (Sicuteri et al., 1978). This supersensitivity is due to an up-regulation of serotonin receptors caused by chronic exposure to low doses of the receptor agonists, as established by a pharmacological rule involving also serotonin receptor (Creese and Sibley, 1981). Till today these studies are among the few available on this topic and represent substantial receptor/biochemical breakthrough in the mechanism of FS.

It seems noteworthy to underline that the clinical investigation stressed that fenclonine can promote FS in M sufferers. This points out a peculiar fragility of serotonergic pathways in M sufferers. In fact, M sufferers were sensitive to fenclonine-induced decrease in serotonin in the CNS whereas control subjects, undergoing the same fenclonine-induced block of serotonin metabolism, showed no clinical feature of such impairment. Taken together, these observations and experimental data suggest that: i) a positive anamnesis for headache can be considered as the pivotal factor predisposing to suffer from FS; ii) primary headache and FS share a common pathogenic background since they both share the common feature of a baseline fragility of serotonergic pain controlling systems (Sicuteri et al., 1974). In a recent study, we observed that the prevalence of headache in FS is dramatically higher than in healthy subjects who are exempt from FS. In a retrospective study, we estimated that primary headache affects 97% of the subjects in a group of 201 FS patients versus the 14% - 40% of the general population, as observed by other authors (Bastos et al., 1993, Munoz et al., 1993). It is important to consider that headache always preceded FS. since it seems possible to indicate headache as the chief predisposing factor for developing FS. Therefore, it seemed suitable to us to consider headache's neurochemical mechanisms and the biochemical data available on the topic of FS in our therapeutic approach to FS. Thus, a defect in the serotonergic system, largely stressed to underlie headache (Rajiv et al., 1989, Sicuteri and Nicolodi, 1991, Lance, 1992) and also evidenced as a factor predisposing to FS, seems to be a rationale starting point for planning a therapeutic approach to FS, acknowledged as a disease where therapy can induce little changes in time (McCain, 1990).

A therapeutic approach enhancing serotonergic transmission seems to represent a rational tool in FS therapy. Furthermore, the fact that headache is the factor predisposing to FS, leads to evaluate the possibility that FS sufferers may share with headache sufferers, in addition to serotonin defects, also defects affecting other neuronal pathways such as the adrenergic. A low availability of adrenergic transmitters has been indirectly shown to affect M sufferers (Sicuteri et al., 1988). Moreover, an adrenergic receptor supersensitivity has been shown to characterize FS patients (Sicuteri et al., 1973, Sicuteri et al., 1978). These outcomes seem to testify an adrenergic transmitters' low level which results, as established by a pharmacological rule (Creese and Sibley, 1981), in an up-regulation and hyper-sensitivity of the specific receptors.

Both serotonergic and adrenergic pathways play a pivotal role in maintaining analgesia. Therefore, drugs which enhance these pathways can be considered suitable for the treatment of FS. Animal investigations suggest that brain serotonin content physiologically depends on the levels of free serum tryptophan (Trp) (Curzon, 1979, De Montis, 1977). The administration of the appropriate substrate in association with monoamine oxidase inhibition, known to increase aminergic transmission, should greatly improve FS. Trp availability represents the fundamental mechanism for controlling serotonin synthesis in the central nervous system . In the present study, we compared the beneficial effect induced by the association MAOIs- A/B and 5-hydroxytryptophan (5-HTP), the hydroxylated Try capable to easily cross BBB, with the benefits induced by amitriptyline or by MAOIs-A/B or 5-HTP alone, already proved to be active in controlling primary painful syndromes (Sicuteri, 1975, King, 1980, Seltzer, 1985).

## MATERIAL AND PROCEDURE

We enrolled 200 subjects who were diagnosed to be suffering from FS in agreement with the criteria established by the American College of Rheumatology (ACR) in 1990 (Wolfe et al., 1990). All the subjects were also suffering from M without aura which was diagnosed in agreement with the criteria proposed by International Headache Society in 1988. Examines and visits were performed to discard all subjects affected by systemic diseases, hypertension, liver diseases or dysfunction, stomach diseases, ocular diseases, clear-cut mood disturbances, pregnant or breast-feeding women. Monitoring of psychological reaction was performed by using self-administered questionnaires regarding anxiety (Wang et al., 1976) and depression (Zung et al., 1965), respectively. Routine laboratory examines and neuropsychiatric visit were performed each month during the 12-month period of treatment. It was previously established to interrupt the therapy in case same clearly abnormal value or clinical sign should be present. Ethical Committee of the Interuniversity Headache Centre gave it consent. All the enrolled subjects were informed of the plan, of the purposes of the study and gave their formal consent. After a wash-out period lasting 1 month, all the subjects were treated with placebo for 1 month. Thereafter, the enrolled patients were divided in 4 groups and randomly assigned to: a) amitriptyline, b) pargyline or phenelzine, c) 5-HTP, d) association of pargyline( or phenelzine) and 5-HTP, as indicated in Table 1.

The four groups of patients were matched according to sex and age. Moreover, there was no significant difference in their baseline visual analogical scale (VAS) score of pain.

**Table 1.** Research plan

| | | |
|---|---|---|
| | | 5-HTP<br>(400 mg/day, oral administration) |
| | | Tricyclic drugs<br>(amitriptyline 10–50 mg/day, oral administration) |
| Wash-out | Placebo | |
| | | MAOIs-A/B<br>(pargyline 5–10 mg/day or phenelzine 10–15 mg/day, orally) |
| | | MAOIs-A/B associated to 5-HTP<br>(200 mg/day, oral administration) |
| 0 30 days | 0 30 days | 1 2 3 4 5 6 7 8 9 10 11 12 months |

**Table 2.**

| VAS (Mean ± STD) Values at the End of 12 Months of Treatment<br>Treatment | VAS (Mean ± STD) at the End of 1 Month of Placebo Treatment<br>Treatment | |
|---|---|---|
| | Placebo | 67.4 ± 6.33 |
| MAOI-A/B (Pargyline/Phenelzine) | 39.6 ± 10.5 p>0.0001 vs PL | |
| Tricyclics (Amitriptyline) | 44.6 ± 6.6 p>0.0001 vs PL | |
| 5-HTP | 48.5 ± 6.0 p>0.0001 vs PL | |
| MAOI-A/B plus 5-HTP | 14.8 ± 18.1 p> 0.0001vs PL | unpaired Students' T-test |

During the wash out period and during the pharmacological treatment all patients under investigation were asked to fulfil a pain diary where they daily scored their pain on a VAS rating from 0 to 4. Mean pain levels were calculated. Mann Whitney U test and Student's T-test for unpaired data were used to compare the benefits induced by the different regimens. The same tests were also used to compare the mean VAS pertaining the wash-out period and the placebo-treatment period with the VAS recorded after 12 months each treatment. The patients were also asked to indicate all side-effects occurring during the treatment.

# RESULTS

Mean VAS pain levels following wash-out and the four different pharmacological treatments are summarised in Table 2. Side-effects were indicated in Table 3.

No one of the patients who experienced side-effects withdrew of the study. On the other hand, none of the listed side-effects were severe. As regards the transient hypertension occurring in two subjects during MAOIs treatment or MAOIs plus 5-HTP treatment, respectively, it seems noteworthy that they did not depend upon a dietetic or pharmacological mistake of the patients. In fact all the patients undergoing this type of therapy were very carefully instructed as regards what they have to avoid. Indeed, both the transient hypertension episodes were concomitant to exceptionally dramatic stressful event. Routine laboratory and clinical examines performed monthly to observe the stemming off of some clear sign of liver dysfunction/neuropsychiatric disorder never underlined significant changes. Self administered tests for anxiety never evidenced any sign of increased anxiety. The administered tests also indicated an amelioration of the mood. This improvement , more evident when using the association of MAOIs and 5-HTP can not be finally ascribed to a direct effect of the used drugs; in fact it can also depend on the induced pain relief. According to this study,

**Table 3**

| Treatment | Side-Effects |
|---|---|
| Placebo | dizziness(1.5%,) stiffness(9.5%), somnolence(0.5%), insomnia(3.5%), nausea(9.5%), stomach-ache(1.5%) |
| MAOI-A/B (Pargilyne/Phenelzine) | insomnia 6%, nausea 2%, somnolence 4%, stomach-ache (8%) mild transient hypertension (2%) |
| Tricyclics (Amitryptiline) | drowsiness 20%, dizziness (6%), nausea (4%), stomach-ache (4%) |
| 5-HTP | stomach-ache (8%) |
| MAOI-A/B plus 5-HTP | nausea (2%), mild transient hypertension ( 2%), stomach-ache(8%) |

the association of MAOIs A/B and 5-HTP induced significantly greater benefits than the other pharmacological regimens.

## DISCUSSION

The association MAOIs A/B and 5-HTP induced much greater benefit than those obtained with amitriptyline, the conventional treatment of FS therapy. The benefits induced by the combination of MAOIs A/B with 5-HTP was also much greater than those induced by either 5-HTP or MAOIs A/B given alone. We established to administer moderate dose of 5-HTP and not L-Trp to avoid, at least in part, side-effects due to an interference of Trp on excitatory aminoacids (Perkins and Stone, 1982, Foster et al., 1984, Moroni et al., 1984). In particular, by quinolinic acid, a potentially toxic metabolite of Trp might induce a further impairment of the hyperalgesia state characterizing M and FS, by enhancing the excitatory amino acids-induced neuronal hyperactivity which seems to be the functional background of hyperalgesia (Handworker, 1993).

It is important to notice that MAOIs A/B treatment alone induces markedly lower benefits than the association MAOIs A/B and 5-HTP. This stresses the key role of a serotonergic defect in FS mechanism. It is known that a serotonin-dependent syndrome can be evoked by 5-HTP receptor inhibitors and Trp in animals pre-treated with MAOI A/B (Koshikawa, 1985). It is also acknowledged that the increase in cortical 5-HT levels shortly (3 h) after the administration of 5-HTP is significantly greater than after serotonin uptake inhibitors administration (Koshikawa, 1985). Therefore, the simultaneous administration of 5-HTP and MAOIs A/B is expected to induce a positive "circle" where 5-HTP enhances the action of MAOIs-A/B while they increase serotonergic and catecholaminergic activity (Scheme1) . We can underline the occurrence of a positive "circle" between 5-HTP, 5-HT and MAOIs A/B in view of the following outcomes: i) Experiments carried out in our laboratory evidenced a motor hyper-activity following reserpine administration in rodents pre-treated with MAOIs A/B; ii) Our group showed that an hyper-responsiveness to both NA and 5-HT as well as to tyramine, a NA releaser, occurred in subjects pre-treated with MAOIs A/B; iii) It was also demonstrated that a long-lasting (some hours) hypertension reaction occurs in humans pre-treated with MAOIs A/B when given metaraminol, which is a "false NA transmitter" refractory to monoamine oxidase. This effect can aid to explain the "cheese effect" of MAOIs. These results indicate that MAOIs A/B treatment does not induce aminergic receptor down-regulation. On the contrary, this regimen is able to evoke a post-synaptic super-sensitivity as shown by the above cited behavioural and pharmacological investigations (Sicuteri et al., 1962, Sicuteri et al., 1965) The hyper-response to NA and 5-HT shown in both animals and humans pre-treated with MAOIs A/B, is consistent with the induction of a post-synaptic up-regulation resulting from the chronically poor level of infra-synaptic specific transmitter. The induced receptor up-regulation can explain the increased effect of 5-HTP when administered to MAOI treated patients. iiii) moreover, other authors showed that the largest effect of monoamine oxidase activity was observed with serotonin as substrate (Panagiotidis, 1993), thus indicating the dependency of serotonin effect on monoamine oxidase set up.

| | 5-HT | |
|---|---|---|
| 5-HTP increased availability | ⟶ | MAOIs induced 5-HT receptor supersensitivity |

**Scheme 1.** Strengthening serotonergic transmission by means of the combination of MAOIs with 5-HTP

In conclusion, by using MAOIs A/B plus 5-HTP, we can expect a reduction of the defect in serotonergic/catecholaminergic systems which affects FS sufferers. Post-synaptic up-regulation seems fundamental to enhance serotonergic transmission, which is shown to be defective in FS as well as in headache which has lately been proved to be the chief risk factor for developing FS. The minor benefit evoked by tricyclic drugs could be explained in view of the receptor down-regulation which they can induce when given chronically. Down-regulation can indeed, at least in part, counteract the benefit induced by the increase in the synaptic levels of the specific transmitter.

The beneficial effect induced by 5-HTP in subjects treated with MAOIs A/B could depend also on other variables than those listed above. Indeed, we can mention some other conditions capable of influencing catecholaminergic system which is known to play a chief role in analgesia. The impairment of the adrenergic system has been shown in primary headache, that has been recently shown to be the *chief risk factor* for developing FS. The fact that primary headache is the ground for FS leads us to consider that headache and FS are likely to share a common pathogenic pattern. Therefore, restoring adrenergic analgesia can co-operate in curing FS. The action of MAOIs on adrenergic activity are well known. Recently, it has been demonstrated that MAOI-B inhibition could induce an increment of phenylethylamine, a "trace amine", having a neuromodulatory effect on catecholaminergic transmission (Boulton et al., 1990, Heinonen and Lammintausta, 1991). The expected benefit from strengthening adrenergic pain controlling systems is confirmed by the present study highlighting the significant amelioration induced even by MAOIs-A/B regimen given alone. This suggests that in FS an adrenergic impairment occurs, in addition to the demonstrated dramatic defect of serotonergic transmission.

## ACKNOWLEDGMENTS

This work was partly supported by a grant from FATMA-Stress, National Council of Researches, Rome

## REFERENCES

Bastos, S.B., de Almeida-Filho, N., Santana, V.S., 1993, Prevalence of headache as a symptom in an urban area of Salvador, *Bahia. Arq Neuropsiquiatr*, 51 (3): 307

Boulton, A.A., Juorrio, A.V., Peterson, I.A., 1990, Phenylethylamine in the CNS: effects ofmonoamine oxidase inhibiting drugs, deuterium substitution and lesions and its role in the neuromodulation of catecholaminergic neurotransmission. *J Neural Trans* 29: 119

Creese, I., Sibley, D.R., 1981, Receptor adaptations to centrally acting drugs. *Ann Rev Pharmacol Toxicol* 21: 357

Cremata, V.Y., Koe, B.K., 1966, Clinical pharmacological evaluation of p-chlorophenylalanine : a new serotonin-depleting agent. *Clin Pharmacol Ther* 7: 768

Curzon, G., 1979, Relationship between plasma, CSF and brain tryptophan. *J Neural Transm* 15:81

De Montis, M.G., Olianas, M,C,. Mulas, G., Tagliamonte, A., 1977, Evidence that only free serum tryptophan exchanges with the brain. *Phar Res Commun* 9 (2): 215

Foster, A.C., Vezzani, A.M., French, E.D., Schwarcz, R., 1984, Kynurenic acid blocks neurotoxicity and seizures in rats by the related brain metabolite quinolinic acid. *Neurosci Lett* 48:273

Handworker, H.O., 1993, Mechanisms of cutaneous and subcutaneous hyperalgesia, in "New Trends in Referred Pain and Hyperalgesia", Vecchiet L, Albe-Fessard D, Lindblom U, eds, Elsevier, Amsterdam (1993)

Heinonen, E.H., Lammintausta, R.A., 1991, A review of the pharmacology of selegiline. *Acta Neurol Scand* 84 (136): 44

International Headache Society: Classification and diagnostic criteria for headache disorders, cranial neuralgias and facial pain. *Cephalalgia* 1988, 8 (7)

King, R., 1980, Pain and tryptophan. *J Neurosurg* 53: 44

Koshikawa, F., Koshikawa, N., Stephenson, J.D., 1985, Effects of antidepressant drugs and cortical 5-HT2 receptors and wet-dog shakes in rats. *Eur J Pharmacol* 118: 273

Lance, J.W., 1992, Possible role of serotonin in the migraine syndrome. in: "Advances in Pain Research and Therapy" , Sicuteri F., Terenius L., Vecchiet L., Maggi CA. ,eds, Raven Press, New York

Mc Cain, G.A., 1990, Management of fibromyalgia syndrome. In: JR Fricton, A Essam (eds), Advances in Pain Research and Therapy, Vol 17, Raven Press, New York

Moroni, F., Lombardi, G., Carlà, V., Moneti G., 1984, The excitotoxin quinolinic acid is present and unevenly distributed in the rat brain. *Brain Res* 295:352

Munoz, M., Dumas, M., Boutros-Toni , F., Coquelle, D., Nicolas, A., Ndzanga, E., Boa, F., Vallat, J.M., Dartigues, J.F., 1993, Prevalence of headache in a representative sample of the population in a French department (Haute-Vienne-Limousin) *Headache* 33 (10) 521

Nicolodi, M., Sicuteri, F., 1992, Central panalgia as an expression of systemic analgodeficiency. in: "Advances in Pain Research and Therapy", F. Sicuteri, L Terenius, L Vecchiet, CA Maggi, eds, Raven Press, New York

Panagiotidis, G., Stenstrom, A., Lundquist, I., 1993, Effects of adrenergic and cholinergic stimulation on islet monoamine oxidase activity and insulin secretion in the mouse. *Eur J Pharmacol* 233 (3): 285

Perkins, M.N., Stone, T.W., 1982, An iontophoretic investigation of the actions of convulsant kynurenines and the interaction with endogenous excitant quinolinic acid. *Brain Res* 247: 184

Rajiv, J., Welch, K.M.A., D'Andrea, G., 1989, Serotonergic hypofunction in migraine: a synthesis of evidence based on platelet dense body dysfunction. *Cephalalgia* 9: 18

Seltzer, S., 1985, Pain relief by dietary manipulations and tryptophan supplements. *J Endodontics* 11(10): 449

Sicuteri, F., 1971, Pain syndrome in man following treatment with p-chlorophenylalanine. *Pharmacol Res Comm* 3(4): 401

Sicuteri, F., 1975, 5-hydroxytryptophan in the prophylaxis of migraine. *Pharmacol Res Commun* 4(3): 213

Sicuteri, F., Michelacci, S,. Franchi, G., 1962, Contributo alla caratterizzazione della farmacologia della reserpina sul sistema nervoso centrale. antagonismo di un bloccante adrenergico sul fenomeno dell'eccitazione paradossa da reserpina nel topo. *Boll Soc Ital Biol Sper* 38 (8): 381

Sicuteri, F., Fanciullacci, M., Del Bianco, P.L., 1965, Variata reattività alla 5-idrossitriptamina ed alla nor-adrenalina da anti monoamino ossidasici nell' uomo. *Boll Soc Ital Biol Sper* 41 (9): 467

Sicuteri, F., Anselmi, B., Del Bianco, P.L., 1973, 5-hydroxytryptamine supersenstivity as a new theory of headache and central pain: a clinical pharmacological approach with p-chlorophenylalanine. *Psychopharmacology* 29: 347

Sicuteri, F., Anselmi, B., Fanciullacci, M., 1974, The serotonin (5-HT) theory of migraine. *Adv Neurol* 4: 383

Sicuteri, F., Anselmi, B., Del Bianco, P.L., 1978, Systemic non organic central pain with decentralization super-sensitivity. *Headache* 18: 133

Sicuteri, F., Nicolodi, M., Fusco, B.M., 1984, Abnormal sensitivity to agonists, antagonists and releasers. in: "Basic Mechanism of headache", J Olesen, LEdvinnsson ,eds, Elsevier, Amsterdam

Sicuteri, F., Nicolodi, M., 1991, Serotonin: a "Sphynx" in migraine. in, "Headache and depression: serotonin pathways as a common clue", Nappi G et al., eds, Raven Press, New York

Simons, D., 1990, Muscular pain syndromes. in: "Advances in Pain Research and Therapy", JR Fricton, A Essam ,eds, Raven Press, New York

Wang, R.I.H., Wiesen, R.L., Treul, S., Stockdale, S., 1976, Brief anxiety rating test. *J Clin Pharmacol* 16 (2/3): 99-105

Wolfe, F., Smythe, H.A., Yunus, M.B., 1990, The american college of rheumatology 1990 Criteria for the classification of fibromyalgia, report of the multicenter criteria committee. *Arthritis Rheum* 33: 160

Zimmermann, M., 1991, Pathophysiological mechanisms of fibromyalgia. *Clin J Pain* 7(1): 8

Zung, WW. 1965 A self rating depression scale. *Arch Gen Psychiatry* 12: 163-65.

# 59

# MACROPHAGES MODULATE LIVER CELL FUNCTION VIA TRYPTOPHAN METABOLITES

Y. Nagamura,[1] K. Uesugi,[1] J. Naito,[1] K. Saitoh,[2] A. Noma,[2] and I. Ishiguro[1]

[1] Fujita Health University
Toyoake Aichi 470-11, Japan
[2] Gifu University
Tukasamachi Gifu 500, Japan

We have reported that administration of tryptophan (Trp) usually cures hepatic injury caused by carbontetrachrolide but occasionally causes aggravation of it and the serum concentration of kynurenine (Kyn) in hepatic injured mouse increases even though the low activity of tryptophandioxygenase (TDO) in hepatic injury. We will discuss that the acceleration of Kyn generation would be caused by macrophage(MP) and some Trp metabolites may play an important role in both the cure of hepatic injury or aggravation of it.

## MATERIALS AND METHODS

3-hydroxyanthranilic acid (3OA), 3-hydroxykynurenine (3OK), Kyn, anthranilic acid (Anth), quinolinic acid (QA), and kynurenic acid (KA) were purchased from Sigma and collagenase and other analytical reagents from Wako Pure Chemical.

The experiment was carried out as follows. Six weeks old ddy strain mice weighing 30g were used as experimental animals. One group of animals were composed of 7 to 10 mice. At first all mice were fasted in order to make the energy condition even, since the feeding condition of mice is strongly affected to the level of liver injury caused by $CCl_4$. After fasting the control group was fed with glucose and tryptophan administrated group was fed with the mixture of 0.1% tryptophan and glucose. After 24 feeding 0.1% or 10% of $CCl_4$ /300ul /30g mouse were injected intraperitonealy and mice were bred for 24 hours the same feed. After 24hrs blood and liver were collected under anesthesia with pentobarbital. Transaminase AST and ALT activities of blood were assayed as the indicators of liver injury by automated blood analyzer COBAS Mira. TG, LPO and glycogen in liver tissue were also assayed as the chemical histological indicator of liver. TG was assayed by acetyl acetone method and LPO was assayed by thiobarbituric method. Tryptophan metabolites Kyn. 3OK and 3OA were separated with HPLC. Kyn was estimated with UV detector and both 3OK and 3OA were detected electrochemicaly with a graphite electrode. We tried to examine the

*Recent Advances in Tryptophan Research*, edited by Graziella Allegri Filippini et al.
Plenum Press, New York, 1996

effects of Mps and Trp metabolites in liver injury. The experiment was carried out as follows. We planed to inhibit the Mps activity by intra venous injection of fine carbon particles prepared by dialysis of drawing ink. At first both control and Trp groups were fasted as previous experiment. Then carbon particles were injected intra venously. After MP treatment with carbon particles, 0.1% or 10% $CCl_4$ were injected intraperitonealy. After further 24 hrs blodd and liver were collected under anesthesia with pentobarbituricacid. Blood and tissues were analyzed as previous experimaent. We tried to examine the direct effects of Trp metabolites on isolated hepatocytes injury caused by $CCl_4$ vapor. The experiment was carried out as follows. Isolated mouse hepatocytes were harvested by clloragenase perfusion method. Hepatocytes prepared with 4 to 5 mice were mixed, since mouse liver was not so large to get enough amount of hepatocytes. Cell viabilities were confirmed by crystal violet excursion test. 5 ml of 1 million cells/ml cell suspension mixed with various concentration of Trp metabolites were saturated with $O_2$ $CO_2$ gas in 10 ml flask. Cell suspension was exposed with $CCl_4$ vapor and incubated at 37° C. 0.25ml cell suspension were collected into 1.5ml of mini test tube. This tube was centrifuged to separate the medium from hepatocytes. AST and ALT activities of supernatant released from hepatocytes were assayed with autoanalyzer Cobas Mirra.

## RESULTS AND DISCUSSION

Fig.1 shows serum transaminase activity of $CCl_4$ liver injury mouse. In the case of 1 % $CCl_4$, Trp administration restrained the activities of both AST and ALT about 50 %. While in the case of 10% $CCl_4$ dosage, Trp administration made transaminase activities increase about 35%. The serum transaminase activities did not always show the histological state of the liver. From the data of our laboratory, the lipid concentration of liver reflected the histological state well in the case of $CCl_4$ liver injury. In the case of 1% $CCL_4$ dosage, LPO was restrained about 80% by Trp administration. On the contrary, in the case of 10% $CCl_4$ dosage, Trp administration made LPO concentration of liver increase. The same phenomena were observed in TG concentration of liver. Thus the effects of Trp administration were also confirmed histologically. That is, administration of tryptophan acts on cure for low concentration of $CCl_4$ liver injury and worsen the liver injury at high concentration of $CCl_4$.

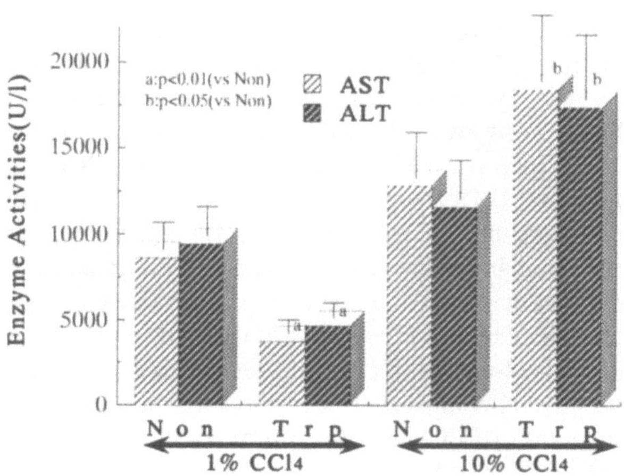


**Figure 1.** Effects of tryptophan on liver injury caused by $CCl_4$.

**Table 1.** Serum kynurenine concentration and TDO activity in liver injury caused by $CCL_4$

| | AST (U ± SD) | Kyn (nmol/ml ± SD) | TDO (μmol Ky/mg Pro) |
|---|---|---|---|
| 1 % $CCl_4$ | | | |
| + Non | 10229 ± 2175 | 1.582 ± 0.457 | 0.363 ± 0.187 |
| + 0.1 % Trp | 8714 ± 1704 | 1.746 ± 0.582 | 0.527 ± 0.295 |
| 10 % $CCl_4$ | | | |
| + Non | 12285 ± 2658 | 2.739 ± 0.838 | 0.131 ± 0.104 |
| + 0.1 % Trp | 17484 ± 3102 | 4.085 ± 1.382 | 0.186 ± 0.135 |
| Control | | | |
| + Non | 117 ± 38 | 0.839 ± 0.295 | 1.483 ± 0.385 |
| + 0.1 % Trp | 122 ± 57 | 4.835 ± 1.730 | 1.967 ± 0.704 |

The effects of Trp concentration on the 1% $CCl_4$ liver injury were examined. The concentration of Trp were chosen from 0 to 5%. Both AST and ALT activities were restrained at the concentration of 0.1% like previous data. But high dosage of Trp worsen the liver injury. This phenomena were confirmed by the LPO and TG analysis of the liver.

These data suggest that Trp metabolites participate in various kinds of way for $CCl_4$ liver injury. Trp in the liver is usually metabolized by TDO which exists only in the liver and its activity falls by liver injury.

Table 1 shows the serum Kyn concentration and TDO activity in liver injury. In 1% $CCl_4$ liver injury AST activity was increased and it was depressed by administration of Trp. Liver TDO activity of non treated control was 1.483 μmol Kyn formed /mg protein and that of injured liver was 0.363 μmol/l. This value was under the quator of controls one. But serum Kyn concentration of control was 0.839 nmol and that of liver injured mouse was 1.582 nmol. The TDO activity of mice administrated Trp was 1.967 μmol Kyn formed / mg protein. This value was higher than that of non treated control, suggesting the induction of TDO by administration of Trp. The TDO induction activity of Trp was more strongly in rat, but not so strongly in mouse.

In the case of 10% $CCl_4$ liver injury the same tendency was more clearly observed. The serum Kyn concentration of 10% $CCl_4$ loaded mice was 2.739. This value is higher than that of 1% $CCl_4$ treated group. The TDO activity of 10% $CCl_4$ injured liver was 0.131. This value was about one tenth of non treated control. The serum Kyn concentration was 2.739 and this value was higher than that of 1% $CCl_4$ injury mouse, even though the low activity of TDO. Trp administration increased the both AST activity and serum kyn concentration, but not TDO activity. The TDO activity was 0.186 and this value was about one tenth of control. The enzyme which is able to decompose Trp to Kyn except TDO was IDO in macrophages. The above data shows that the IDO in MPs which were induced by $CCl_4$ liver injury supplied Kyn. And Kyn thus formed in Mps changed the state of liver injury in several ways.

Fig. 2 shows the effects of intra venous injection of fine carbon particles on $CCl_4$ liver injury. When 1% $CCl_4$ loaded, liver injury became worse by I.V. injection of carbon particles. On the other hand, when 10% $CCl_4$ loaded, the levels of liver injury were strongly restrained by I.V. injection of carbon particles. The restriction of Mps activity by injection of carbon particles acts on cure in low concentration $CCl_4$ liver injury and became worse liver injury in high concentration $CCl_4$. In other words Trp metabolites act on cure or aggravation of liver injury on the condition.

Then the effects of Trp metabolites on liver injury of both 1% and 10% were examined. Fig.3 shows the effects of Trp metabolites on 1% $CCl_4$ liver injury. After injection

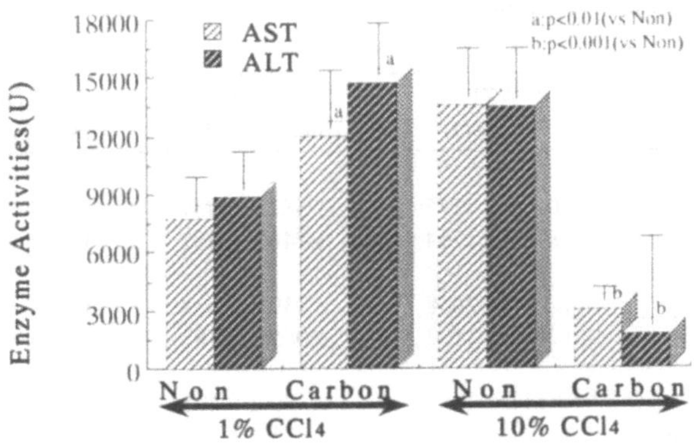


**Figure 2.** Effects of I.V. injection of carbon particles on $CCl_4$ liver injury.

of $CCl_4$, Trp metabolites were injected 4 times every 6 hrs and 24 hrs blood was collected. The serum transaminase activities of Kyn and 3OK administrated mice were restricted compared with control, but not that of 3OA administrated one. 3OK was the most effective agent for depression of transaminase activities by liver injury.

The effects of Trp metabolites on 10% $CCl_4$ liver injury were also examined. In this case, both Kyn and 3OK depressed the transaminase activities, but 3OA did not. The liver injury with 10 % $CCl_4$ was strongly depressed by injection with carbon particle. Therefore we supposed that the effects of Trp metabolites on liver injury might be different between 1% and 10 % $CCl_4$ injection. But these results were not anticipated one suggesting that the difference of the rate of Trp metabolism between 1% and 10% $CCl_4$ liver injury caused the difference of Trp metabolites and some metabolites acted on cure and some became worse.

Fig.4 shows the effects of several tryptophan metabolites on isolated mouse hepatocytes injury. Intracellular transaminase of hepatocytes were released gradually. There are some differences of the releasing pattern between AST and ALT. Many Trp metabolites except 3OK accelerated the enzyme releasing, suggesting Trp metabolites made worse the liver injury. These phenomena were good corresponded to in vivo study.

We assayed the serum concentration of Trp metabolites, since, the Trp metabolites seem to affect on the liver injury in various way. Table 2 shows the serum Trp metabolites concentration in 1 % $CCl_4$ liver injury. The serum Kyn concentration of non treated control was 0.713, while that of 1% $CCl_4$ liver injured mouse was 1.647 nmol/ml. The Kyn

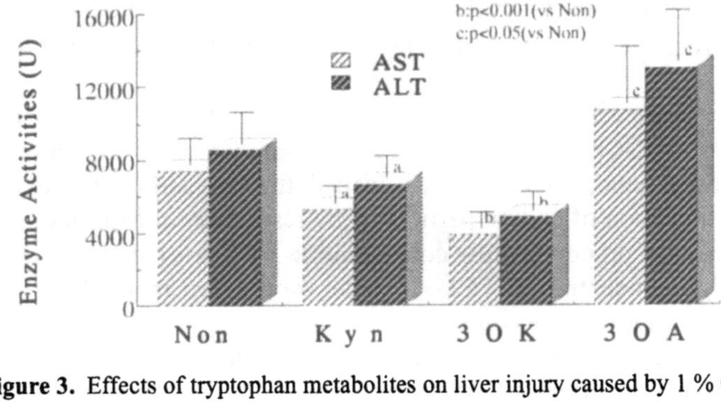


**Figure 3.** Effects of tryptophan metabolites on liver injury caused by 1 % $CCl_4$.

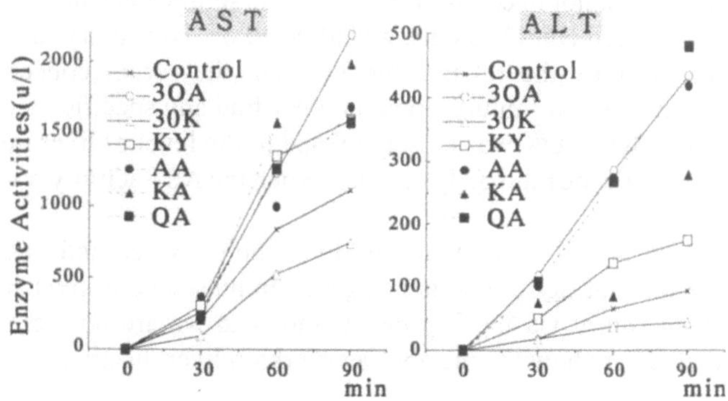


**Figure 4.** Effects of tryptophan metabolites on isolated mouse hepatocytes injury caused by $CCl_4$ vapor.

concentration of carbon injected control mouse was 0.628 nmol/ml, this value was not so decreased from nontreated control. But the Kyn concentration of carbon injected and 1% $CCl_4$ liver injured mouse was a half of 1% $CCl_4$ injured mouse. It means that the depression of MPs activity cause decrease of Kyn formation in 1% liver injured mouse, in other words, MPs are responsible for the increasing of Kyn concentration in liver injured mouse. In the case of Trp administration, the Kyn formation was similar to that without Trp administration. The serum 3OK concentration of non treated control was 0.98, while that of 1% $CCl_4$ liver injured mouse was 0.027 and the difference of serum 3OK concentration between non and carbon injected mouse increased in $CCl_4$ treated mouse. The serum 3OA concentration of non treated control was the same as 1% $CCl_4$ liver injured mouse and carbon injection increased the formation of 3OA. When Trp was administrated carbon particle loading decreased the 3OA formation by one fifth but not in control.

The serum Kyn concentration and the effect of carbon particle loading in 10% $CCl_4$ injured mouse was more clearly observed. The carbon loading decreased the serum Kyn concentration by one 3rd in liver injured mouse, but not in control mouse. The serum 3OK concentration of both control mouse and liver injured one were not so affected by carbon particle injection. While the serum 3OA concentration was affected by carbon particle

**Table 2.** Transaminase activities and the ratio of 3OA/3OK

| | AST (U ± SD) | 3OA/3OK |
|---|---|---|
| 1 % $CCl_4$ | | |
| Non | 7947 ± 1683 | 1.703 |
| Carbon | 11753 ± 2948 | 6.538 |
| Trp | 4260 ± 1327 | 1.051 |
| Trp + Carbon | 5979 ± 1528 | 4.536 |
| 10% $CCl_4$ | | |
| Non | 14038 ± 3629 | 2.125 |
| Carbon | 12746 ± 2855 | 1.318 |
| Trp | 19537 ± 3857 | 3.912 |
| Trp + Carbon | 9276 ± 1745 | 1.116 |
| Control | | |
| Non | 98 ± 27 | 0.459 |
| Carbon | 103 ± 31 | 0.623 |
| Trp | 88 ± 25 | 0.401 |
| Trp + Carbon | 108 ± 38 | 0.336 |

loading in liver injured, but not in control mouse like the serum concentration of Kyn. Thus the Trp metabolites changed complicatedly, and affected on liver injury in various ways.

We tried to examine the relationship between the serum concentration of Trp metabolites and the level of liver injury, and could not find any specific metabolites.

But the ratio of 3OA/3OK was closely related to the level of liver injury. This slide shows the AST activity and the ratio of 3OA/3OK. When the AST activity was high, the ratio was also high.

As I mentioned above, 3OA made the liver injury worse and 3OK act on cure. Therefore, the large ratio of 3OA/3OK is not good circumstance in liver injury. Thus the MPs affected on liver injury via the Trp metabolites and we are not able to specify the metabolites and to elucidate how the metabolites work on liver disease.

# NEUROPSYCHIATRIC IMPLICATIONS OF BRAIN TRYPTOPHAN PERTURBATIONS APPEARING IN HEPATIC ENCEPHALOPATHY

F. Bengtsson and P. Bergqvist

Department of Clinical Pharmacology
Lund University Hospital
S-221 85 Lund, Sweden

## 1. INTRODUCTION

The metabolic encephalopathies exhibit CNS-symptoms that rapidly may alter in location of brain region of engagement, indicating a continuous fluctuation of the level of which brain functions are disturbed. This is particularily obvious in chronic conditions, such as chronic or latent hepatic encephalopathy (HE), where the fluctuating level of consciousness may be a dangerous sudden developing symptom in a person otherwise appearing to have normal brain functions. If these individuals are occupied with e.g. driving vehicle in traffic or working in heavy industrial practice, such chronic portal-systemic encephalopathy (PSE) (Sherlock et al., 1954) may be a particular hazard to both themselves and others. Unfortunately, the pathogenesis of PSE is still unknown. Hence, the delineation of the pathogenesis for HE/PSE, and therapeutic modalities developing therefrom, is highly warrented. Numerous pathogenetic mechanisms and etiological factors for PSE have been suggested (Conn and Liebertahl, 1979).

Elevations in body/blood and brain ammonia content is probably a central event in all conditions of HE/PSE (Butterworth, 1992). Besides the direct toxic effects of ammonia, however, many other metabolic disturbances proven to be associated with PSE have suggested this condition to be of multifactorial origin (Conn and Liebertahl, 1979; Butterworth, 1992).

Some of the factors in recent years that have attracted most interest in PSE pathophysiology/pathogenesis are related to body/blood and brain amino acid disturbances, notably aromatic amino acids in general and L-tryptophan (L-TRP) in particular has been in focus for this discussion. Whether or not, however, increased brain L-TRP concentrations primarily are the consequence of an overall change in concentrations of neutral amino acids in the body in PSE is still an unresolved issue.

*Recent Advances in Tryptophan Research,* edited by Graziella Allegri Filippini et al.
Plenum Press, New York, 1996

From a functional point of view, however, the parallel increments in brain serotonin (5-hydroxytryptamine; 5-HT) and 5-HT metabolite concentrations repeatedly shown in both patients and animal models of various types of hepatic disorders (Tyce et al., 1967, Baldessarini and Fischer, 1973; Curzon et al., 1973; 1975; Knott and Curzon, 1974; Reichle and Reichle, 1975; Cummings et al., 1976a; 1976b; Jellinger and Riederer, 1977; Ono et al., 1978; Bloxam and Curzon, 1978; Tricklebank et al., 1978; 1981; Kamata et al., 1980; Herlin et al., 1982; Denis et al., 1983; Martin et al., 1983; Jonung et al., 1985; Mans and Hawkins, 1986; Bergeron et al.,1989; 1990; Mans et al., 1990; Bengtsson, 1991) made more of a case for suggesting an general overactivity in CNS serotonergic neurotransmission as being essential in the development of all types of HE. It is possible that different forms of HE may have common pathogenic denominators. It is however most likely that these various conditions also depend on a multitude of separate specific factors to be manifested, in each case establishing a partly resembling symptomatology (Conn and Liebertahl, 1979; Bengtsson et al., 1989b; Butterworth, 1992).

## 2. HE/PSE AND BRAIN L-TRP

This issue can be scrutinized by addressing several aspects of the problem, such as neurochemical/pharmacological aspects including L-TRP metabolic processes, L-TRP catabolites exerting specific pharmacological activities, L-TRP catabolites possessing potential biological activities, and in particular 5-HT metabolic processes, 5-HT release, and 5-HT receptors in HE/PSE.

However, also functional aspects such as behavioral disturbances in chronic HE/PSE and the worsening of neurological status progressing from normalcy to latent HE/PSE until hepatic coma should preferably be considered in parallel with the scrutiny on the neurochemistry/neuropharmacology of these conditions in order to search for true causal relations. In the following, some of the major L-TRP derived factors mainly studied by our own research efforts in this area, will be summarized.

## 3. SEROTONIN

Most experimental HE studies have suggested brain L-TRP and 5-HIAA levels to be markedly increased while 5-HT concentrations increments were much less pronounced. Elevations in brain 5-HT do not appear to correlate directly with the degree of HE unless ratios of L-TRP/5-HT/5-HIAA (usually 5-HIAA/5-HT) are calculated to be used as an indicator of brain 5-HT turnover. However, measurements of absolute levels of L-TRP and L-TRP metabolites always have the disadvantage of imaging static events from which the dynamics of the system cannot be inferred. In addition, whole brain analyses of these compounds probably also frequently can obscure important regional variations, especially so since it has been shown that L-TRP hydroxylation activities normally differs in different parts of the brain (Baumgarten et al., 1973). The normal day-night rhythm variations for brain L-TRP and L-TRP metabolite concentrations also has to be considered, especially so in PSE research since there are reasons to believe that the normal circadian pattern may be disturbed in this condition (Bergeron et al., 1991).

In the later part of the 1980ies our research group undertook a series of studies addressing the issue of 5-HT metabolic pathology in the brain in different animal models of HE/PSE. These experiments supported the contention of an existing overall and regional increase in 5-HT synthesis and turnover in the brain of portacaval shunted (PCS) rats ranging from one day up to six months after the operation, as well as in rats with toxin-induced liver cirrhosis (Bengtsson et al., 1985; 1986; 1987a; 1987b; 1987c; 1988; 1991). In these experiments we used an in vivo method with complete decarboxylase inhibition and determination of the accumulation of 5-hydroxytryptophan (5-HTP) as a measure of the activity in the rate-limiting reaction of 5-HT turnover, i.e. the L-TRP hydroxylating activity in brain parenchyma in vivo (Bengtsson et al., 1985; Carlsson and Lindqvist M., 1973). The repeated similar findings from our studies, additional results when using monoamine oxidase inhibition in vivo supporting the same contention (Bengtsson et al., 1987b), and incresed brain histochemical fluorescence of 5-HT in the brain following PCS in the rat (Bengtsson et al., 1988) made us firmly suggest an increased brain 5-HT turnover associated with chronic PSE.

However, several lines of evidence in our studies mentioned above indicated no clear functional relevance as to the alterations in brain 5-HT and the encephalopathy in liver dysfunction (Bengtsson et al., 1986; 1989b; Bugge et al., 1986; 1987a; 1987b; 1987c; 1988; 1989; Bengtsson et al., 1989a; Peroutka, 1990; Robinson et al., 1992; Johansson et al., 1981). In a recent study we have shown that the neuronal release of brain 5-HT in rats with both chronic and acute HE seems, in fact, not to be increased (Bergqvist et al., 1995a). As of yet, however, these findings only encounters neocortical brain dialysis studies in ambient resting (i.e. not activated or manipulated for release) serotonergic neuronal systems. Thus, these results have to be further substantiated before any firm conclusions about the CNS 5-HT release pattern in PSE can be made. This type of functionally relevant studies, as well as the pharmacological approach towards the delineation of 5-HT subreceptor properties in brain in HE/PSE, seem to be the most highly warranted studies in the area today.

Even more puzzeling is the potential uncoupling between neuronal 5-HT metabolism and neuronal 5-HT release in HE if the 5-HT receptors are considered. In fact, our most recent data strongly suggest a major change in the number of 5-HT receptors associated with PCS in the rat (Apelqvist et al., 1995). These alterations show an increased density of 5-HT1A receptors, a decreased density of 5-HT1B receptors, but no changes in the number of 5-HT2 receptors in the brain of PCS rats (Apelqvist et al., 1995). Certainly, these first preliminary findings indicating a great likelyhood for also brain 5-HT receptor changes in HE/PSE has to be followed-up by future research.

Thus, taken altogether it is not unlikely that the brain 5-HT turnover increase primarily occurs because high brain L-TRP levels are present in most situations associated with hepatocerebral dysfunction. Accordingly, in this situation the change in 5-HT metabolism change could refl ect a metabolic "event" or an epiphenomenon rather than being causally related to the development of the encephalopathy. If this is the case (i.e. with no serotoninergic neurotransmission failure in HE/PSE), it will be another important evidence supporting the hypothesis of the existence of two pools of 5-HT in the brain, one functionally active and another non-releaseable pool that both shares the same metabolic pathway (Bengtsson et al., 1989b; Grahame-Smith, 1973). However, complex interactions can be anticipated to exist in HE/PSE that can conceal any straightforward correlation to even a major single pathogenetic factor such as brain 5-HT might still be in this condition.

## 4. SEROTONIN RELATED METABOLISM

With some scrutiny on the L-TRP hydroxylation it should be recalled that L-erythro-5,6,7,8-tetrahydrobiopterin (BH4) is the natural cofactor in vivo for L-TRP hydroxylase (Bengtsson et al., 1991). Thus, during L-TRP hydroxylation one atom of molecular oxygen is used in the 5-hydroxylation to 5-HTP, while the other is reduced to water. Electrons are donated by the reduced BH4 and the unstable quinonoid dihydropterin that results is, depending on the ambient redox state of the tissue, regenerated to the BH4 from the

NADH/NAD(+)-linked quinonoid dihydropyridine reductase. Thus, BH4 acts like a true catalyst in this reaction. Further, it is important to note that it has been shown that the PCS rat exhibits a higher degree than normal of reduction of the cytoplasmic NADH/NAD(+) system in the brain parenchyma although the PCS procedure does not result in any major acid-base changes or disrupted balance in the energy state of the cells (Holmin and Siesjö, 1974). Based on this information it is conceivable that the BH4/quinonoid dihydropterin conversion in the PCS rat brain, due to the increased reduction in the redox system, shifts towards increased availability of BH4. In turn, with increased endogenous levels of L-TRP as in the brain of PCS rats, this could work in concert with the increased availability for the rate-limiting BH4 cofactor to result in the supranormal L-TRP hydroxylating activity and increased 5-HT turnover seen in vivo in this animal model of chronic PSE (Bengtsson et al., 1991). In the extension of this observation (Bengtsson et al., 1991) it could be speculated that the state of the redox systems in the brain may very well play a role in regulation brain 5-HT turnover in vivo. As an example it has been shown that also in experimental status epilepticus the brain 5-HT turnover is increased (Calderini et al., 1978) and here we can also note a reduced redox state in the parenchyma compared to normal (Bengtsson et al., 1990). Conversly, in experimental insulin-induced hypoglycemia the 5-HT turnover in the brain is decreased (Agardh et al., 1979). Here, however, the redox state of the tissue is oxidized compared to normal (Bengtsson et al., 1990). Finally, therefore, one could pose the question if the alteration in redox state is a sensitive mechanism by which a pathological conditions result in changes in brain 5-HT metabolism and, perhaps, a mechanism by which also brain serotonin turnover can be regulated under normal conditions? Although a preliminary study from our group showed that there may most likely be a pertubation in the overall L-TRP hydroxylating cofactor metabolic system, we failed to demonstrate any alteration of the functionally active biopterin (i.e. BH4) levels per se in the brain parenchyma in experimental chronic PSE (Bergqvist et al., 1995b). However, though, a change in the redox state towards more a more reduced system may also promote a relative change in the formation of 5-hydroxytryptophol (5-HTOL), the alcohol otherwise formed only in limited amounts in the breakdown of 5-HT on behalf of the normal major route for catabolizing 5-HT to 5-HIAA. Certainly, these and related issues of altering the metabolism of brain L-TRP and 5-HT have to be addressed in the future.

## 5. OTHER MAJOR L-TRP PERTURBATIONS IMPLICATED IN HE/PSE

In the middle of the 1980's, Dr. Moroni and his distiguished research group described a novel finding of increments in the L-TRP metabolite quinolinic acid in brain of both experimental HE/PSE (Moroni et al., 1986a) as well as in cerebrospinal fluid of patients suffering from HE and in the post mortem brain parenchyma of patients dying in hepato-cerebral dysfunction (Moroni et al., 1984; 1986b). These most exciting reports have since then, however, not been further verified or studied in respect as to the functional relevance of this finding, probably due to the analytical problems involved with the determination of the overall low concentrations of quinolinate in the brain. However, although low in concentration, quinolinate has the interesting physiological/pharmacological potential of being active as a potent N-methyl-D-aspartate (NMDA) receptor agonist. This excitatory glutamate subreceptor type in the brain has attended much interest during many recent years because of its relation to calcium influx into cells, and thereby exerting one important regulator function of intracellular calcium homeostasis, as well as when activated being implicated as a central event in underlying so called excitotoxic lesions in the brain in conditions where excitotoxicity are likely to be a pathogenetic mechanism in cell damage (Siesjö et al., 1989). Moreover, alterations in the NMDA receptor activity has also been suggested to play a role in the pathogenesis of HE (Butterworth, 1992). Unfortunately, the most recent data available in this field does not promote quinolinic acid as a major pathogenetic factor in HE/PSE (Basile et al., 1995a; 1995b; Bergqvist et al., 1995 in press; Bergqvist et al., submitted). The quinolinic acid story in HE is discussed in more detail at another section of the present volume (see Bergqvist et al. *Is quinolinic acid involved in the pathogenesis of hepatic encephalopathy?).*

Another factor to consider might be the amine formation of L-TRP without hydroxylation to yield tryptamine. Tryptamine is normally an endogenous trace amine in the brain in the pico-molar range. However, it has been shown to increase in brain concentrations in some neurological disorders, and has the potential to exert functional CNS-activities in higher concentrations (Mousseau, 1993). In HE there are as of yet only availability for a limited information on the subject of possible changes in the brain tryptamine activities, although some data suggest an increased level of this trace amine to exist in patients suffering from HE (Mousseau, 1993). This trace amine must therefore also be in focus of some future studies on the L-TRP related pathogenesis for HE/PSE.

## 6. BRAIN L-TRP PATHOLOGY IN HE IN SUMMARY

Accepting the high levels of brain L-TRP and increased de novo brain 5-HT synthesis due to the induction of L-TRP hydroxylation but unaltered brain quinolinate concentartion as present in the hyperammonemic condition of PSE, we would suggest the following facts and speculative pathophysiological events to may be come into action in HE:

1) If increases in brain 5-HT turnover in HE/PSE is only the result of increases in brain L-TRP, such a "metabolic" increase in brain 5-HT metabolism per se may not necessarily result in increased 5-HT release or an increased serotonergic "tone" in the CNS in HE. If this is the case, this phenomenon in HE will straighten the likelyhood that 5-HT metabolism clearly always have the potential to not directly relate to the functional state of serotonergic neurons (Lookingland et al., 1986; de Simoni et al., 1987).

2) The 5-HT turnover increase is possibly due to induction of brain L-TRP hydroxylation in HE (Bengtsson et al., 1991). This induction may in some way be caused by changes in the biopterin cofactor metabolism in brain in HE/PSE due to changes in the redox state of the brain parenchyma in this condition (Bengtsson et al., 1991; Bergqvist et al., 1995b).

Furthermore, if the change in biopterin metabolism will lead to greater availability for the pterin cofactor, this can also affect neurotransmitter release of e.g. 5-HT since both L-TRP (Carboni et al., 1989), ammonia (Erecinska et al., 1987), redox changes (Woodward and Blair, 1991) and the pterin BH4 in itself (Wolf et al., 1991) have been shown to exert direct effects on neurotransmitter release under certain circumstances in vitro.

3) Whether or not the biopterin cofactor is involved as stated above, a change in the redox systems of the brain in HE/PSE might promote the formation of the 5-HT derived alcohol 5-HTOL. 5-HTOL may in higher concentrations exert a functional activity in the brain and, hence, a novel potential pathogenetic factor in HE is presented.

4) Increases in brain quinolinate in HE/PSE may, but is not likely, to have a functional impact on the development of HE/PSE (Basile et al., 1995a; 1995b). Other trace amines from L-TRP, though, like tryptamine may however be possible candidates for exerting neuroactive effects in HE/PSE.

5) Finally, we have to add another "confounding" factor of clinical relevance to patients suffering from HE/PSE. In the vast majority of cases with HE/PSE the underlying

etiology for the liver impairment is chronic alcohol abuse. It is therefore with great interest to study in parallel with "plain" PSE-models the direct effect of ethanol administration to brain L-TRP and 5-TRP-related changes. In fact, the effect on brain L-TRP/5-HT may be quite different if ethanol is given acute or chronic, or as an acute intoxication to an individual with chronic use of large amounts of alcohol (as is the case with many patients suffering from PSE (Badawy et al., 1989). We therefore have, at least at some point, to include the findings of ethanol effects on L-TRP metabolism and disposition also in the discussion on brain L-TRP and HE to be clinically relevant.

## ACNOWLEDGMENTS

Grants from the Swedish Medical Research Council (No. 4X-9889, 14V-9919, and 14P-10160), the Swedish Society of Medicine, the Åke Wibergs Foundation, the Wenner-Gren Center Foundation for Scientific Research, the Sandoz-Stiftelsen för Medicinsk-Biologisk Forskning, Stockholm, and the Medical Faculty at Lund University, the Crafoordska Foundation, the Anna Lisa and Sven-Eric Lundgrens Foundation for Medical Research, the Anna och Nanny Blomgrens fond, Lund, Sweden, are all gratefully acknowledged.

## REFERENCES

Agardh C-D., Carlsson A., Lindqvist M,, Siesjö B.K., 1979, The effect of pronounced hypoglycemia on monoamine metabolism in rat brain. *Diabetes* 28:804.

Apelqvist G., Bergqvist P.B.F., Bugge M., and Bengtsson F., 1995, Autoradiographic characterization of regional brain 5-HT subreceptors changes in chronic experimental hepatic encephalopathy. submitted.

Badawy AA-B., Morgan C.J., Lane J., Dhaliwal K., Bradley D.M., 1989, Liver tryptophan pyrrolase. A major determinant of the lower brain 5-hydroxytryptamine concentration in alcohol-preferring C57BL mice. *Biochem. J.* 264:597.

Baldessarini R.J., Fischer J.E., 1973, Serotonin metabolism in rat brain after surgical diversion of the portal venous circulation. *Nature New Biol.* 245:25.

Basile A.S., Saito K., Al-Mardini H., Record C.O., Hughes R.D., Harrison P., Williams R., Li Y., and Heyes M.P., 1995a, The relationship between plasma and brain quinolinic acid levels and the severity of hepatic encephalopathy. *Gastroenterology* 108: 818.

Basile A.S., Saito K., Li Y., and Heyes M.P., 1995b, The relationship between plasma and brain quinolinic acid levels and the severity of hepatic encephalopathy in animal models of fulminant hepatic failure. *J. Neurochem.* 64: 2607.

Baumgarten H.G., Victor S.J., Lovenberg W., 1973, Effect of intraventricular injection of 5,7-dihydroxytriptamine on regional tryptophan hydroxylase of rat brain. *J. Neurochem.* 21:251.

Bengtsson F., 1991, Round table discussion on brain monoamines: some personal reflections. In: Bengtsson F, Jeppsson B., Almdal T., Vilstrup H., eds. Progress in hepatic encephalopathy and metabolic nitrogen exchange. Boca Raton, Florida, CRC Press Inc, , pp. 233-239

Bengtsson F., Gage F.H., Jeppsson B., Nobin A., Rosengren E., 1985, Brain monoamine metabolism and behavior in portacaval-shunted rats. *Exp. Neurol.* 90:21.

Bengtsson F., Nobin A., Falck B., Gage F.H., Jeppsson B., 1986, Portacaval shunt in the rat: selective alterations in behavior and brain serotonin. *Pharmacol. Biochem. Behav.* 24:1611.

Bengtsson F., Bugge M., Vagianos C., Jeppsson B., Nobin A., 1987a, Brain serotonin metabolism and behavior in rats with carbon tetrachloride-induced liver cirrhosis. *Res. Exp. Med.* 187:429.

Bengtsson F., Bugge M., Hansson L., Fyge K., Jeppsson B., Nobin A., 1987b, Serotonin metabolism in the central nervous system following sepsis or portacaval shunt in the rat. *J. Surg. Res.* 43:420.

Bengtsson F., Nobin A., Falck B., Gage F.H., Jeppsson B., 1987c, Effect of oral branched chain amino acids on behavior and brain serotonin metabolism in portacaval shunted rats. *World J. Surg.* 12:246.

Bengtsson F., Ekman R., Nobin A.., 1987d, Substance P-like immunoreactivity in the rat central nervous system following portacaval shunt. *Surg. Res. Comm.* 2:65-69.

Bengtsson F., Bugge M., Brun A., Falck B., Henriksson K-G., Nobin A., 1988, The impact of time after portacaval shunt in the rat on behavior, brain serotonin, and brain and muscle histology. *J. Neurol. Sci.* 83:109.

Bengtsson F., Bugge M., Hall H., Nobin A., 1989a, Brain 5-HT1 and 5-HT2 binding sites following portacaval shunt in the rat. *Res. Exp. Med.* 189:249.

Bengtsson F., Bugge M., Nobin A., 1989b, Hepatocerebral dysfunction and brain serotonin. In: Butterworth R.F., Pomier Layrargues G., eds. Hepatic Encephalopathy. Humana Press, Clifton, New Jersey, pp. 355-387.

Bengtsson F., Boris-Möller F., Hansen A-J., Siesjö B.K., 1990, Extracellular pH in the rat brain during hypoglycemic coma and recovery. *J. Cereb. Blood Flow Metab.* 10:262.

Bengtsson F., Bugge M., Johansen K.H., Butterworth R.F., 1991, Brain tryptophan hydroxylation in the portacaval shunted rat: A hypothesis for the regulation of serotonin turnover in vivo. *J. Neurochem.* 56:1069.

Bergeron M., Reader T.A., Layrargues G.P., Butterworth R.F., 1989, Monoamines and metabolites in autopsied brain tissue from cirrhotic patients with hepatic encephalopathy. *Neurochemical Research* 14:853.

Bergeron M., Swain M.S., Reader T.A., Grondin L., Butterworth R.F., 1990, Effect of ammonia on brain serotonin metabolism in relation to function in the portacaval shunted rat. *J Neurochem.* 55:222.

Bergeron M.., Reader T.A., Butterworth R.F., 1991, Early changes of serotonin turnover in brain following portacaval anastomosis: Relation to altered sleep patterns and diurnal rhythms. In: Bengtsson F., Jeppsson B., Almdal T., Vilstrup H., eds. Progress I hepatic encephalopathy and metabolic nitrogen exchange. Boca Raton, Florida, CRC Press Inc, pp. 219-232.

Bergqvist P.B.F., Vogels B.A.P.M., Bosman D.K., Maas M.A.W., Hjorth S., Chamuleau R.A.F.M., and Bengtsson F., 1995a, Neocortical dialysate monoamines of rats after acute, subacute, and chronic liver shunt, *J. Neurochem.* 64: 1238.

Bergqvist P.B.F., Werner E.R., Apelqvist G., Bugge M., Wachter H., and Bengtsson F., 1995b, Brain biopterin metabolism in rats with chronic experimental hepatic encephalopathy. *Metab. Brain Dis.* 10: 143.

Bergqvist P.B.F., Heyes M.P., Bugge M., and Bengtsson F., 1995, Brain quinolinic acid in chronic experimental hepatic encephalopathy: effects og an exogenous ammonium acetate challenge, *J. Neurochem.* 65: in press.

Bergqvist P.B.F., Heyes M.P., Apelqvist G., Butterwoth R.F., and Bengtsson F., 1995, Brain extracellular quinolinic acid in chronic experimental hepatic encephalopathy as assessed by in vivo microdialysis: effects of pharmacological manipulation of the L-tryptophan availability. submitted.

Bloxam D.L., Curzon G., 1978, A study of proposed determinants of brain tryptophan concentrations in rats after portocaval anastomosis or sham operation. *J Neurochem* 31:1255.

Bugge M., Bengtsson F., Nobin A., Jeppsson B., Herlin P., 1986, Metabolism of monoamines in the brain after total hepatectomy in the rat. *Exp. Neurol.* 94:11.

Bugge M., Bengtsson F., Nobin A., Jeppsson B., Herlin P., 1987a, The turnover of brain monoamines after total hepatectomy in rats infused with branched chain amino acids. *World J. Surg.* 11:810.

Bugge M., Bengtsson F., Nobin A., Jeppsson B., Herlin P., 1987b, The effect of liver ischaemia on brain monoamine synthesis in the rat. *Res. Exp. Med.* 187:119-130.

Bugge M., Bengtsson F., Nobin A., Jeppsson B., Hultberg B., Jonung T., Herlin P., 1988, Serotonin metabolism in the rat brain following ammonia administration. In: Soeters P.B., Wilson J.M.P., Meijer A.J., Holm E., eds. Advances in ammonia metabolism and hepatic encephalopathy. Amsterdam, Exerpta Medica, International Congress Series 761, Elsevier Science Publishers B.B., pp 454-461.

Bugge M., Bengtsson F., Nobin A., Hall H., Wedel I., Jeppsson B., Herlin P., 1989, Serotonin receptors in the brain following total hepatectomy in rats treated with branched-chain amino acids. *J. Parenter. Enteral Nutr.* 13:235-239.

Butterworth R.F., 1992, Pathogenesis and treatment of portal-systemic encephalopathy: an update. *Dig Dis Sci* 37:321.

Calderini G., Carlsson A., Nordström C-H., 1978, Monoamine metabolism during bicuculline-induced epileptic seizures in the rat. *Brain Res.* 157:295.

Carboni E., Cadoni C., Tanda G.L., Di Chiara G., 1989, Calcium-dependent, tetrodotoxin-sensitive stimulation of cortical serotonin release after a tryptophan load. *J. Neurochem.* 53:976.

Carlsson A., Lindqvist M., 1973, In vivo measurements of tryptophan and tyrosine hydroxylase activities in mouse brain. *J. Neural. Transm.* 34:79.

Conn HO, Liebertahl M.M., 1979, The hepatic coma syndromes and lactulose. The Williams & Wilkins Co, Baltimore,

Cummings M.G., James J.H., Soeters P.B., Keane J.M., Foster J., Fischer J.E., 1976a, Regional brain study of indoleamine metabolism in the rat in acute hepatic failure. *J.Neurochem.* 27:741.

Cummings M.G., Soeters P.B., James J.H., Keane J.M., Fischer J.E., 1976b, Regional brain indoleamine metabolism following chronic portacaval anastomosis in the rat. *J Neurochem* 27:501.

Curzon G., Kantamaneni B.D., Winch J., Rojas-Bueno A., Murray-Lyon I.M., Williams R., 1973, Plasma and brain tryptophan changes in experimental acute hepatic failure. *J Neurochem* 21:137.

Curzon G., Kantamaneni B.D., Fernando J.C., Woods M.S., Cavanagh J.B., 1975, Effects of chronic portocaval anastomosis on brain tryptophan, tyrosine and 5-hydroxy-tryptamine. *J Neurochem* 24:1065

Denis J., Delorme M.L., Boschat M., Nordlinger B., Opolon P., 1983, Respective roles of ammonia, amino acids, and medium-sized molecules in the pathogenesis of experimentally induced acute hepatic encephalopathy. *J. Neurochem.* 40:10.

Erecinska M., Patuszko A., Wilson D.F., Nelson D., 1987, Ammonia-induced release of neurotransmitters from rat brain synaptosomes: differences between the effects of amines and amino acids. *J. Neurochem.* 49:1258.

Grahame-Smith D.G. 1973, Does the total turnover of brain 5-HT reflect the functional activity of 5-HT in brain? In: Barchas J., Usdin E., eds. Serotonin and behavior. New York, London, Academic Press, pp 5-7.

Herlin P.M., James J.H., Gimmon Z., Fischer J.E., 1982, Combined or individual administration of branched-chain amino acids following total hepatectomy in the rat: effects on amino acids and indoleamines in brain. *J Parenter. Enter. Nutr.* 6:383.

Holmin T., Siesjö B.K. 1974, The effect of porta-caval anastomosis upon the energy state and upon acid-base parameters of the rat brain. *J. Neurochem.* 22:403.

Jellinger K., Riederer P., 1977, Brain monoamines in metabolic (endotoxic) coma. A preliminary biochemical study in human postmortem material. *J Neural. Transm.* 41:275.

Johansson O., Hökfelt T., Pernow B., Jeffcoate S.L., White N., Steinbusch H.W.M., Berhofstad A.A.J., Emsson P.C., Spindel E., 1981, Immunohistochemical support for three putative transmitters in one neuron: Coexistence of 5-hydroxytryptamine, substance P and thyrotropin releasing hormone-like immunoreactivity in medullary neurons projecting to the spinal cord. *Neuroscience* 6:1857

Jonung T., Ramzy A., Herlin P., James J.H., Edwards L., Fischer J.E., 1985, Indole amines and amino acids in various brain regions after infusion of branched chain amino acids into hepatectomized rats. *Eur. Surg. Res.* 17:83.

Kamata S., Okada A., Watanabe T., Kawashima Y., Wada H., 1980, Effects of dietary amino acids on brain amino acids and transmitter amines in rats with a portacaval shunt. *J.Neurochem.* 35:1190

Knott P.J., Curzon G., 1974, Effects of increased rat brain tryptophan on 5-hydroxytryptamine and 5-hydroxyindoleacetic acid in the hypothalamus and other brain regionons. *J Neurochem* 22:1065.

Lookingland K.J., Shannon N.J., Chapin D.S. Moore K.E., 1986, Exogenous tryptophan increases synthesis, storage, and intraneuronal metabolism of 5-hydroxytryptamine in the rat hypothalamus. *J. Neurochem.* 47:205.

Martin J.R., Dedek J., Driscoll P., 1983, Portacaval anastomosis in rats: effects on behaviour and brain serotonin metabolism. *Pharmacol. Biochem. Behav.* 18:269.

Mans A.M., Hawkins R.A., 1986, Brain monoamines after portacaval anastomosis. *Metabolic Brain Disease* 1:45.

Mans A.M., DeJoseph M.R., Davis D.W., Vina J.R., Hawkins R.A., 1990, Early establishment of cerebral dysfunction after portacaval shunting. *Am. J. Physiol.* 259:E104.

Moroni F., Carlà V., Lombardi G., Pellegrini D., Carassale G.L., Cortesini F., 1984, Excitatory amino acids, tryptophan metabolites, and endogenous neurotoxins in porta-caval shunted rats. In: Kleinberger G., Ferenci P., Riederer P., Thaler H., eds. Advances in hepatic encephalopathy and urea cycle diseases. Basel, Karger, pp 385-393.

Moroni F., Lombardi G., Carlà V., Pellegrini D., Carassale G.L., Cortesini C., 1986a, Content of quinolinic acid and of other tryptophan metabolites increases in brain regions of rats used as experimental models of hepatic encephalopathy. *J. Neurochem.* 46:869.

Moroni F., Lombardi G., Carlà V., Lal S., Etienne P., Nair N.P.V., 1986b, Increase in the content of quinolinic acid in cerbrospinal fluid and frontal cortex of patients with hepatic failure. *J. Neurochem.* 47:1667.

Mousseau D.D., 1993, Tryptamine: a metabolite of tryptophan implicated in various neuropsychiatric disorders. *Metab. Brain Dis.*, 8: 1.

Ono J, Huston D.G., Dombro R.S., Levi J.U., Livingstone A., Zeppa R., 1978, Tryptophan and hepatic coma. *Gastroenterology* 74:196.

Peroutka S.J., 1990, Receptor "families" for 5-hydroxytryptamine. *J. Cardiovasc. Pharmacol.* 16(suppl 3):S8.

Reichle R.M., Reichle F.A., 1975, Effect of portacaval shunt and acute hepatic ischemia on brain and liver serotonin and catecholamines. *Surg. Forum* 26:413.

Robinson M.B., Anegawa N.J., Gorry E., Qureshi I.A., Coyle J.T., Lucki I., Batshaw M.L., 1992, Brain serotonin2 and serotonin1A receptors are altered in the congenitally hyperammonemic sparse fur mouse. *J. Neurochem.* 58:1016.

Sherlock S., Summerskill W.H.J., White L.P., Phear E.A., 1954, Portal-systemic encephalopathy. Neurological complications of liver disease. *Lancet* 11:453.

Siesjö B.K., Bengtsson F., 1989, Calcium fluxes, calcium antagonists, and calcium-related pathology in brain ischemia, hypoglycemia and spreading depression: a unifying hypothesis. *J. Cereb. Blood. Flow Metab.* 9:127.

de Simoni G.M., Sokola A., Fodritto F., Dal Toso G., Algeri S., 1987, Functional meaning of tryptophan-induced increase of 5-HT metabolism as clarified by in vivo voltammetry. *Brain Research* 411:89.

Tyce G.M., Flock E.V., Owen C.A., Stobie G.H.C., David C., 1967, 5-Hydroxyindole metabolism in the rat brain after hepatectomy. *Biochem. Pharmol.* 16:979.

Tricklebank M.D., Smart J.L., Bloxam D.L., Curzon G., 1978, Effects of chronic experimental liver dysfunction and L-tryptophan on behaviour in the rat. *Pharmacol. Biochem. Behav.* 9:181.

Tricklebank M.D., Bloxam D.L., Kantamaneni B.D., Curzon G., 1981, Brain 5-hydroxytryptamine metabolism after portacaval anastomosis: Relationship with ambulation. *Pharmacol. Biochem. Behav.* 14:259.

Woodward J.J., Blair R., 1991, Redox modulation of N-Methyl-D-Aspartate-stimulated neurotransmitter release from rat brain slices. *J. Neurochem.* 57:2059.

Wolf W.A., Ziaja E., Arthur Jr R.A., Anastasiadis Z, Levine R.A., Kuhn D.M., 1991, Effect of tetrahydrobiopterin on serotonin synthesis, releas and metabolism in superfused hippocampal slices. *J. Neurochem. 57:1191.*

61

# IS QUINOLINIC ACID INVOLVED IN THE PATHOGENESIS OF HEPATIC ENCEPHALOPATHY?

P. B. F. Bergqvist,[1] M. P. Heyes,[2] and F. Bengtsson[1]

[1] Department of Clinical Pharmacology
Lund University Hospital
S-211 85 Lund, Sweden
[2] Section on Analytical Biochemistry
Laboratory of Clinical Science, NIMH
Bethesda, Maryland 20892

## 1. INTRODUCTION

The pathogenic mechanisms resulting in the serious syndrome associated with acute and chronic liver diseases, better kown as hepatic encephalopathy (HE), are still unknown. The syndrome is most often referred to as a neuropsychiatric disease of multifactorial origin (Conn and Lieberthal, 1979). Many different hypotheses have been suggested for the patogenesis of HE. No single such hypothesis has so far been proven to be valid. Among these hypotheses are increased brain levels of the compound quinolinic acid (QUIN; pyridine-2,3-dicarboxylic acid) one that suggests this potential neurotoxin to contribute to the development of HE (Moroni et al., 1986*a,b*).

QUIN is an excitotoxic kynurenine pathway metabolite acting as a competetive agonist of the N-methyl-D-aspartate (NMDA) subclass of excitatory amino acid receptors mediating neuronal membrane calcium ion transients and thereby excitatory neurotransmission in the brain (for recent overview, see e.g. Stone, 1993). Under normal circumstances, QUIN concentrations in brain tissue are in the nanomolar range. In general, from micro- to millimolar concentrations of QUIN have been shown to promote an exitotoxic lesion to nerve cells (Stone, 1993). Some studies have, however, reported excitotoxicity to be exerted by QUIN in certain neuronal cell types in vitro by only slightly higher than physiological concentrations (Whetsell and Schwarcz, 1989).

Several factors speak in favor of an involvement of QUIN in the development of HE. Firstly, in patients with varying degree of HE, cerebrospinal fluid (CSF) (Ono et al., 1978; Smith et al., 1978; Hutson et al., 1979; Cascino et al., 1982; Rössle et al., 1984) as well as post mortem brain tissue (Jellinger et al., 1978; Bergeron et al., 1989) levels of the QUIN precursor, L-tryptophan (L-TRP), have been shown to be increased. Elevated L-TRP levels have also been demonstrated in brain tissue (Baldessarini and Fischer, 1973; Curzon et al., 1975; Cummings et al., 1976; Mans et al., 1982, 1990; Bengtsson et al., 1991) and in the brain extracellular fluid (Bergqvist et al., 1995*a*) in portacaval shunted (PCS) rats, the most commonly used and non-toxin induced experimental in vivo model of chronic HE. Secondly, in 1986 Moroni and coworkers reported increased QUIN levels in the cerebral neocortex and cerebellum but in no other brain regions of the PCS rat (Moroni et al., 1986*a*). Following a coma-inducing ammonium acetate (NH4Ac) challenge, an increased neocortical brain QUIN level was reported in acute hyperammonemic rats by the same investigators (Moroni et al., 1986*a*). Thirdly, increased QUIN concentrations were found by the same group in the CSF in patients suffering from HE as well as post-mortem in the frontal neocortical brain tissue in patients dying in hepatic coma (Moroni et al., 1986*b*). The latter results (i.e. Moroni et al., 1986 *a,b*) are schematically summarized in Figure 1. These observations led to the hypothesis that increased L-TRP availability in the brain may be associated with an increase in brain QUIN levels in HE and that this could contribute to the development of encephalopathy (Moroni, 1986*a,b*). A fourth factor speaking in favor of an involvement of QUIN in the pathogenesis of HE may partly be supported by the fact that reduced NMDA receptor binding sites in the brain in experimental HE as has been reported in the PCS rat (Peterson et al., 1990).

However, contrary to these factors speaking in favor of a role of QUIN in the pathogenesis of HE indeed several factors seem to argue against the involvement of QUIN in this condition. Firstly, in experimental HE, glial rather than neuronal degeneration is present (Bengtsson et al., 1988), although patients dying from HE may exhibit indications of neuronal degeneration (Victor et al., 1965). Lack of neuronal damage is usually not the

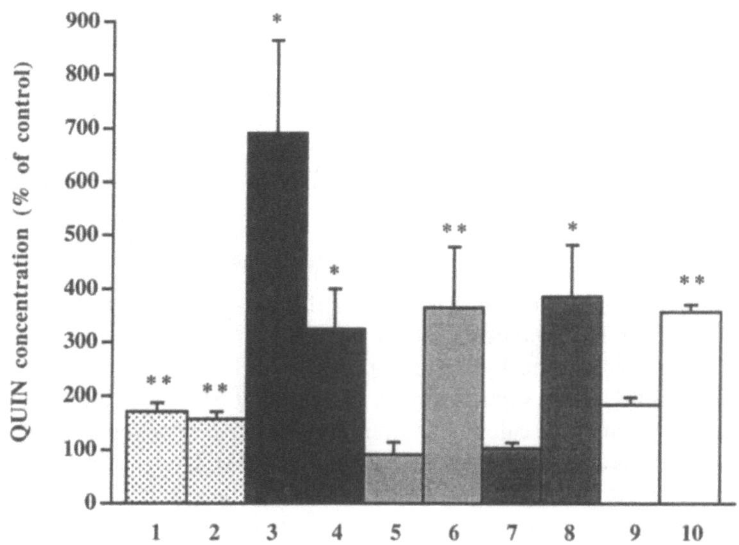


**Figure 1.** A schematic presentation of some QUIN results obtained in previous studies (results adopted and modified from Moroni et al., 1986*a,b* and Basile et al., 1995*a,b*). Numbers below bars indicate type of study and from which reference the data have been obtained; **1.** neocortical brain tissue, PCS rat (Moroni et al., 1986*a*); **2.** neocortical brain tissue, acute hyperammonemic rats (Moroni et al., 1986*a*); **3.** CSF, HE patients (Moroni et al., 1986*b*); **4.** neocortical brain tissue, deseased HE patients (Moroni et al., 1986*b*); **5.** neocortical brain tissue, deseases chronic HE patients (Basile et al., 1995*a*); **6.** neocortical brain tissue, deseased acute HE patients (Basile et al., 1995*a*); **7.** neocortical brain tissue, rats with fulminant hepatic failure (FHF) stage I (Basile et al., 1995*b*); **8.** neocortical brain tissue, rats with FHF stage IV (Basile et al., 1995*b*); **9.** neocortical brain tissue, rabbits with FHF stage I (Basile et al., 1995*b*); **10.** neocortical brain tissue, rabbits with FHF stage IV (Basile et al., 1995*b*). All values are means ± SEM and expressed as percentage of controls. * and ** denote significant differences ($p<0.05$ and 0.01, respectively) as compared to respective control group.

expected event occuring on the basis of common excitotoxic brain lesions. Secondly, in HE an overall depressed neuronal activity including a general EEG slowing commonly appears (Conn and Lieberthal, 1979) which is not the usual event preceeding a potential excitotoxic effect. Thirdly, in a recently published study elevated brain QUIN levels in patients dying in acute but not in chronic HE are shown (Basile et al., 1995*a*). Importantly, the latter study also displayed that the QUIN levels in the patients dying from acute HE were not correlated to the neurological status of the patients at time of death (Basile et al., 1995*a*). In another recent study from Basile and collaborators, they showed in two toxin-induced experimental models of acute HE that no correlation could be shown between the brain levels of QUIN and neurological impairements of the animals (Basile et al., 1995*b*). The latter results (Basile et al., 1995*a,b*) are also summarized in Figure 1.

At present, therefore, there seems to be a general disagreement on possible role for brain QUIN in the pathogenesis of HE. Hence, verifying or escluding brain QUIN as a pathogenetic factor in HE is today one of the major issues in this research field, with certain bearings on the future therapies to be investigated in this condition (Bengtsson, 1992; Butterworth, 1992). The present chapter summarizes two experiments performed by our group in order to further investigate a possible involvement of QUIN in the pathogenesis of HE (Bergqvist et al., 1995*b,c*).

## 2. MATERIALS AND METHODS

The present study constituted of two separate trials. In both trials the chronic PCS rat (Lee and Fischer, 1961) was used as an experimental model of chronic HE. Sham-operated rats were used as controls. The QUIN levels were determined by the use of a gas chromatography mass spectrometry assay with an electron-capture negative chemical ionization detection (Heyes and Markey, 1988*a,b*) .

In trial 1 plasma as well as neocortical brain tissue QUIN levels were determined (Bergqvist et al., 1995*b*). In this trial the PCS rats was also provoked into an acute reversible coma by an intraperitoneal administration of NH4Ac (5.2 mmol/kg body weight; Sigma Chemical Co., St. Louis, MO, U.S.A.). The neurological status of the PCS-NH4Ac rats was evaluated by using a method previously described in great detail by Giguére and Butterworth (1984). Four different stages of neurological status were defined. These stages were referred to as basal, pre-coma, coma, and post-coma, respectively. At these various stages, a number of PCS rats were sacrificed along with rats from the sham-operatedcontrol group.

Since QUIN is such a potent in vivo glutamatergic NMDA-receptor agonist it is most relevant also to determine the brain *extracellular* QUIN levels in HE. Therefore, in a second trial the intracerebral microdialysis technique (Ungerstedt, 1984) was applied to measure the frontal neocortical brain extracellular QUIN levels (Bergqvist et al., 1995*c*). The impact of alterations in the amino acid precursor availability on the extracellular levels of QUIN was also studied in this trial. This was performed by systemic administration of an initial single loading dose of L-TRP (280 mg/kg, i.p.; Sigma). This administration was repeated three hours later by four successive injections of lower doses of L-TRP (4 x 50 mg/kg, i.p.) during the four consecutive hours to follow.

## 3. RESULTS

The results revealed that 5-6 weeks following the PCS procedure the basal plasma, brain tissue as well as brain extracellular fluid QUIN levels were decreased in the PCS rats as compared to the sham-operated controls (Figure 2).

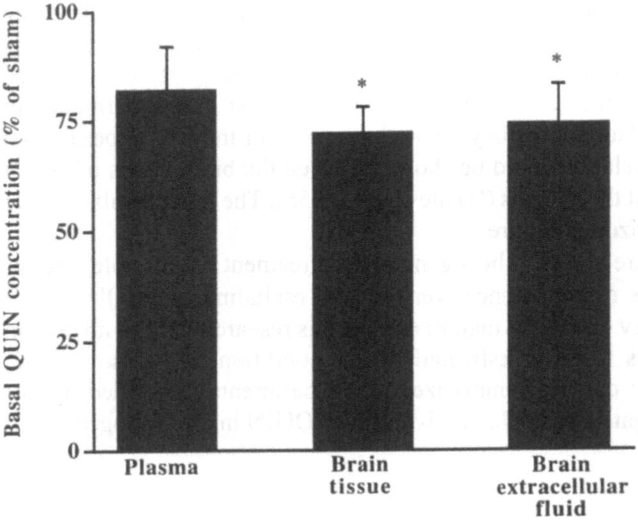


**Figure 2.** Basal QUIN levels in plasma, brain tissue, and brain extracellular fluid in rats 5-6 weeks following the creation of a PCS (results adopted from Bergqvist et al., 1995*b,c*). All values are means ± SEM and expressed as percentage of sham-operated rats. * denotes a significant difference ($p<0.05$) as compared to the sham-operated controls.

The NH4Ac challenge affected the PCS rats in the following manner; At the basal stage i.e. just prior to the ammonia challenge the PCS rats showed no visible symptoms of neurological disturbances as compared to the control rats. After 10-15 min the ammonia administered PCS rats were in the pre-coma stage and had lost their righting reflexes. Following 20-25 min the PCS rats had lost both their righting and corneal reflexes and fell into a deep reversible coma. After 60-75 min following the initiation of the ammonia challange the PCS rats had restored their righting and corneal reflexes and were in the post-coma stage. The administration of NH4Ac to the sham-operated rats did not affect the neurological status of these rats.

The sham-operated rats displayed some increases in plasma QUIN following the ammonia administration although these rats were not neurologically affected by the challange (Figure 3a). No changes in plasma QUIN levels in the PCS rats as compared to the basal value were seen following the ammonium challenge. Hence, neither is there any relationship between the neurological status and plasma levels of QUIN in the PCS rats (Figure 3b). The only change in brain tissue QUIN in the PCS rats was displayed when the coma had cleared. In parallel with the case of plasma QUIN the sham-operated rats also displayed some neurologically-unrelated increases in brain QUIN.

Increased precursor availability by the exogenous L-TRP load was associated with an increased brain extracellular fluid QUIN levels to a similar degree in both PCS and sham rats (Figure 4). Although statistically non-significant, the area under curve (AUC) for the PCS-L-TRP rats were about 30% lower as compared to the sham-L-TRP animals (77.1±7.7 µmol/L x min versus 107±33 µmol/L x min) indicating a generally lower accumulation of QUIN following the L-TRP administration in the PCS rats. It is impossible from the present study to define the more detailed kinetics for the QUIN accumulations under these circumstances. It can be seen, however, that both the PCS and the sham-operated rats reached a steady-state level of QUIN about 3 hours following the initial bolus dose of L-TRP. It can also be seen that iterations with lower L-TRP doses did not affect these steady-state QUIN levels in any major way.

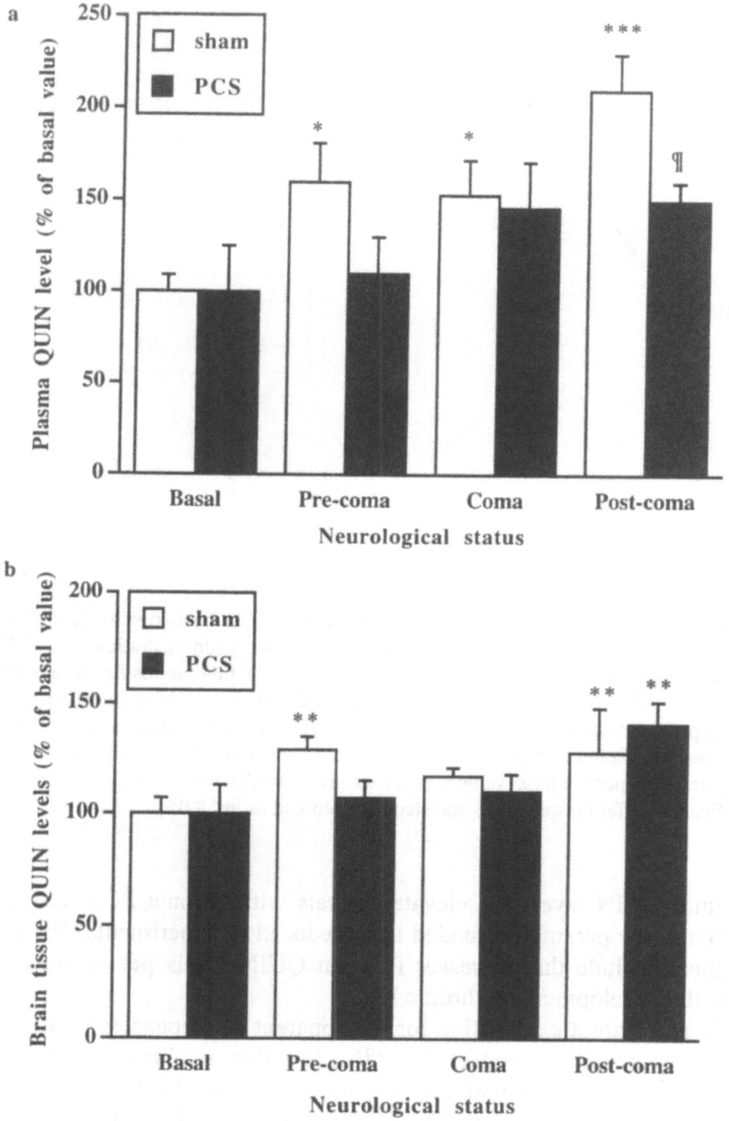


**Figure 3.** The influence of a single parenteral $NH_4Ac$ challenge on the **(a)** plasma and **(b)** brain QUIN levels in PCS and sham rats (results adopted from Bergqvist et al., 1995*b*). All values are expressed as percentage (means ± SEM) of the corresponding basal value. Relative QUIN values at the different time-periods that differ from the basal value within each of the four experimental groups are denoted as $^{*}p<0.05$, $^{**}p<0.01$, and $^{***}p<0.001$. Relative QUIN values that differ between PCS and sham are denoted as ¶p<0.05.

# 4. DISCUSSION

The potentially excitotoxic compound QUIN has been suggested to be involved in several CNS disorders such as epilepsy, the neuropathology associated with Acquired ImmunoDeficiency Syndrome, Huntingtons's disease as well as HE (for overview, see Stone, 1993). In the present chapter two experiments investigating the possible involvement of QUIN in the pathogenesis of HE show results that neither basal plasma nor brain tissue or

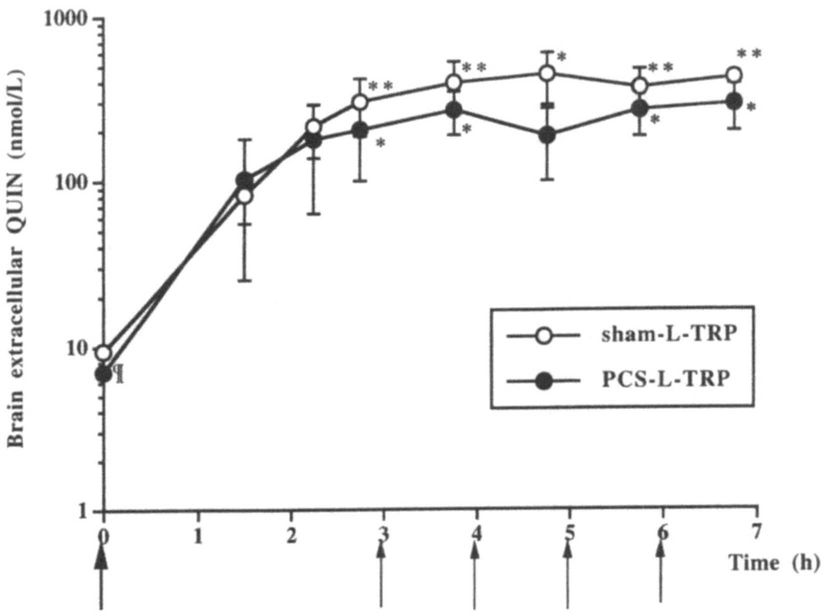


**Figure 4.** The time-dependent effects of an exogenous systemic L-TRP challenge on the brain extracellular QUIN during a 6.5 h period of continuous neocortical dialysis following administrations of L-TRP to PCS and sham-operated rats (results adopted from Bergqvist et al., 1995*c*). The time-points for the administration of a single loading dose of L-TRP (280 mg/kg body weight, i.p.) and of four successive injections of lower L-TRP doses (50 mg/kg, i.p.) are indicated by a bold arrow and by regular arrows, respectively. All values are means ± SEM and expressed as nanomoles per liter dialysate fluid (uncorrected for probe recovery). Significant differences compared to respective basal value within each group are denoted as $*p<0.05$ and $**p<0.01$. QUIN values that significantly differ between PCS and sham are denoted as ¶$p< 0.05$.

extracellular fluid QUIN levels are elevated in rats with chronic PCS (Bergqvist et al., 1995*b,c*). In fact, these parameters tended to be reduced in experimental HE compared to normal. We thus conclude that increases in brain QUIN levels per se are probably not responsible for the development of chronic HE.

There is no clearcut explanation for the apparent discrepancies in the brain QUIN results described previously (Moroni et al., 1986 *a,b*) and those obtained both by Basile and coworkers (Basile et al., 1995 *a,b*) and by us (Bergqvist et al., 1995 *b,c*). Accordingly, the seemingly disagreement on the possible role for brain QUIN in HE is today best referred to as dependent on possible methodological differences between these studies.

A single exogenous NH4Ac challenge slightly elevated the plasma and brain tissue UIN levels in all rats subjected to this treatment, but there were no major differences in the overall magnitude of this response between the PCS and sham-operated rats. Furthermore, no relevant temporal relationship of changes in brain QUIN levels and the severity of the neurological status in the PCS rats following the acute NH4Ac load could be demonstrated. This is in agreement with the results from a recent study by Basile et al. (1995*b*) where no correlation between the plasma and brain QUIN levels and the neurological status in either the thioacetamide-treated rat model or the galactoseamine-treated rabbit model of acute HE was reported.

Another important observation is the rather similar accumulation profile for brain extracellular fluid QUIN in both PCS and sham rats following an increased availability to L-TRP. The only difference observed between sham and PCS rats following the pharmacological manipulations of the L-TRP availability with regard to brain extracellular QUIN levels was a non-significant decrease in AUC in the PCS rats as compared to the sham-operated controls. This phenomenon further call the contention of an involvement of QUIN in the pathogenesis of HE into question. Therefore, if at all, it seems at present that brain QUIN may only play a very minor role in the pathogenesis of HE.

The enzyme demonstrated to be responsible for the major source of kynurenine production in the body is the hepatic enzyme tryptophan-2,3-dioxygenase (TDO). Although QUIN can be produced in other organs in the body than the liver, it has been indicated from the results in several studies that the amount of QUIN found in the brain in normal conditions probably results from a peripheral QUIN production in the liver followed by a passive diffusion (Kitt and Spector, 1987) of QUIN in the blood across the blood-brain barrier (BBB) into the brain (see e.g. Heyes et al., 1993). The hepatic dysfuction previously described in the PCS rats (Schröder et al., 1985) may possibly contribute to a reduced liver TDO-activity with a subsequently reduced peripheral QUIN production as the result. Hence, the mechanism behind the possibly slightly lower basal brain tissue and extracellular fluid QUIN levels (Figure 2) in the PCS rats could be promoted by a reduced peripheral QUIN production by the liver dysfunction of the PCS rats followed by a decreased diffusion of QUIN from the blood into the brain in experimental HE. On the other hand, the trends towards increases in plasma and brain QUIN following the ammonia challenge in both sham (corresponding controls to the pre-coma and post-coma stages for PCS rats) and PCS (in the post-coma stage) rats could be speculated as being the result of an increased peripheral QUIN production by an elevated liver TDO activity e.g. due to stress-released glucocorticoids (Saito et al., 1993) followed by an impaired BBB resistance for QUIN promoted by the very profound hyperammonemia arising after an exogenous ammonia load. This hypothesis could be supported by the increased hepatic TDO activity and the unaltered brain indoleamine-2,3-dioxygenase (the enzyme responsible for the extrahepatic QUIN production) activity recently reported in acute experimental liver failure (Basile et al., 1995*b*). Previous studies have reported an abnormal BBB permeability in acute experimental liver failure (Livingstone et al., 1977; Crinquette et al., 1982; Zaki et al., 1983; Horowitz et al., 1983). In line with the results from the present investigation, however, the ultimate cause for the elevations in brain QUIN following ammonia exposure should result from an augmented diffusion over the BBB in acute hyperammonemic states *without* prior liver dysfunction. At present, therefore, the data available from brain tissue and extracellular fluid do not point out QUIN as a causative agent for the neurological impairments associated with acute hyperammonemia.

In conclusion, based on the results obtained in the two trials described in the present chapter in conjunction with those findings recently published by Basile and coworkers it is concluded that QUIN seems not to be involved in the pathogenesis of either acute or chronic HE. Therefore, other factors of pertubations in the metabolism of L-tryptophan than excessive QUIN formation should be searched for in future HE research.

## 5. ACKNOWLEDGMENTS

Dr. Marcelle Bergeron is gratefully acknowledged for excellent technical assistance. The performance of these investigations was made possible by the generous support by a Research Fellowship for Bsc Peter Bergqvist from the Thorsten and Elsa Segerfalks Foundation. This work was further supported by grants from the Swedish Society of Medicine, the Åke Wibergs Foundation (Stockholm, Sweden), the Medical Faculty of Lund University, the Crafoord Foundation, the Royal Physiographic Society in Lund, the Anna Lisa and Sven-Eric Lundgrens Foundation for Medical Research, and the Maggie Stephens Foundation (Lund, Sweden).

# 6. REFERENCES


Baldessarini R. J. and Fischer J. E., 1973, Serotonin metabolism in rat brain after surgical diversion of the portal venous circulation. *Nature New Biology* 245: 25.

Basile A. S., Saito K., Al-Mardini H., Record C. O., Hughes R. D., Williams R., Li Y., and Heyes M. P., 1995*a*, The relationship between plasma and brain quinolinic acid levels and the severity of hepatic encephalopathy. *Gastroenterology* 108: 818.

Basile A. S., Saito K., Li Y., and Heyes M. P.,1995*b*, The relationship between plasma and brain quinolinic acid levels and the severity of hepatic encephalopathy in animal models of fulminant hepatic failure. *J. Neurochem.* 64: 2607.

Bengtsson F., Bugge M., Brun A., Falck B., Henriksson K. G., and Nobin A., 1988, The impact of time after portacaval shunt in the rat on behavior, brain serotonin, and brain and muscle histology. *J. Neurolog. Sci.* 83: 109.

Bengtsson F., Bugge M., Johansen K. H., and Butterworth R. F.,1991, Brain tryptophan hydroxylation in the portacaval shunted rat: a hypothesis for the regulation of serotonin turnover in vivo. *J. Neurochem.* 56: 1069.

Bengtsson F., 1992, Neurotransmission failure in hepatic encephalopathy involving the combined action of different brain tryptophan-related pathology: a speculative synthesis, in Advances in tryptophan research *1992* (Ishiguro I., Nagatsu T., and Nagamura Y., eds), pp. 303-308. Fujita Health University Press, Toyoake, Japan.

Bergeron M., Pomier Layrargues G., and Butterworth R. F., 1989, Aromatic and branched-chain amino acids in autopsied brain tissue from cirrhotic patients with hepatic encephalopathy. *Metab. Brain Dis.* 4: 169.

Bergqvist P. B. F., Vogels B. A. P. M., Bosman D. K., Maas M. A. W., Hjorth S., Chamuleau R. A. F. M., and Bengtsson F., 1995*a*, Neocortical dialysate monoamines of rats after acute, subacute, and chronic liver shunt. *J. Neurochem.* 64: 1238.

Bergqvist P. B. F., Heyes M. P., Bugge M., and Bengtsson F.,1995*b*, Brain quinolinic acid in chronic experimental hepatic encephalopathy: Effects of an exogenous ammonium acetate challenge. *J. Neurochem. In press.*

Bergqvist P. B. F., Heyes M. P., Apelqvist G., and Bengtsson F., 1995*c*, Brain extracellular quinolinic acid in chronic experimental hepatic encephalopathy as assessed by in vivo microdialysis: Effects of altered L-tryptophan availability. *Submitted.*

Butterworth R. F., 1992, Evidence that hepatic encephalopathy results from a defect of glutamatergic synaptic regulation. *Mol. Neuropharmacol.* 2: 229.

Cascino A., Cangiano C., Fiaccadori F., Ghinelli F., Merli M., Pelosi G., Riggio O., Rossi-Fanelli F., Sacchini D., Stortoni M., and Capocaccia L., 1982, Plasma and cerebrospinal fluid amino acid patterns in hepatic encephalopathy. *Dig.Dis.Sci.* 27:828.

Conn H. O. and Lieberthal M. M., 1979, The Hepatic Coma Syndromes and Lactulose, pp. 1-47. The Williams & Wilkins Co., Baltimore, MD, U.S.A.

Crinquette J. -F., Boschat M., Rapin J. -P., Delorme M. -L., and Opolon P., 1982, Early changes in blood-brain barrier permeability after porto-caval shunt and liver ischaemia. *Clin. Physiol.* 2: 241.

Cummings M. G., Soeters P. B., James J. H., Keane J. M., and Fischer J. E., 1976, Regional brain indoleamine metabolism following chronic portacaval anastomosis in the rat. *J. Neurochem.* 27: 501,.

Curzon G., Kantamaneni B. D., and Fernando J. C., 1975, Effects of chronic porto-caval anastomosis on brain tryptophan, tyrosine and 5-hydroxytryptamine. *J. Neurochem.* 24: 1065.

Giguére J. F. and Butterworth R. F., 1984, Amino acid changes in regions of the CNS in relation to function in experimental portal-systemic encephalopathy. *Neurochem. Res.* 9: 1309.

Heyes M. P. and Markey S. P., 1988*a*, Quantification of quinolinic acid in rat brain, whole blood, and plasma by gas chromatography and negative chemical ionization mass spectrometry: effects of systemic L-tryptophan administration on brain and blood quinolinic acid concentrations. *Anal. Biochem.* 174: 349.

Heyes M. P. and Markey S. P., 1988*b*, ($^{18}$O)Quinolinic acid: its esterification without back exchange for use as internal standard in the quantification of brain and CSF quinolinc acid. *Biomed. Environ. Mass Spectrom.* 15: 291,.

Heyes M. P., Saito K., Major E. O., Milstein S., Markey S. P., and Vickers J. H, 1993, A mechanism of quinolinc acid formation by brain in inflammatory neurological disease. Attenuation of synthesis from L-tryptophan by 6-chlorotryptophan and 4-chloro-3-hydroxyanthranilate. *Brain* 116: 1425.

Horowitz M. E., Schafer D. F., Molnar P., Jones E. A., Blasberg R. G., Patlak C. S., Waggoner J., and Fenstermacher J. D., 1983, Increased blood-brain barrier transfer in a rabbit model of acute liver failure. *Gastroenterology* 84: 1003.

Hutson D. G., Ono J., Dombro R. S., Levi J. U., Livingstone A., and Zeppa R., 1979, A longitudinal study of tryptophan involvment in hepatic coma. *Am. J. Surg.* 137: 235.

Jellinger K., Riederer P., Rausch W. D., and Kothbauer P., 1978, Brain monoamines in hepatic encephalopathy and other types of metabolic coma. *J. Neural Transm.* 14:103.

Kitt T. M. and Spector R., 1987, Transport of quinolinic acid into rabbit and rat brain. *Neurochem. Res.* 12: 625.

Lee S. H. and Fisher B., 1961, Portacaval shunt in the rat. *Surgery* 50: 668.

Livingstone A.S., Potvin M., Goresky C.A., Finlayson M.H., and Hinchey E.J., 1977, Changes in the blood brain barrier in hepatic coma after hepatoectomy in the rat. *Gastroenterology* 73: 697.

Mans A.M., Biebuyck J.F., Shelly K., and Hawkins R.A., 1982, Regional blood-brain barrier permeability to amino acids after portacaval anastomosis. *J. Neurochem.* 38: 705.

Mans A.M., DeJoseph M.R., Davis D.W., Viña J.R., and Hawkins R.A., 1990, Early establishment of cerebral dysfunction after portacaval shunting. *Am. J. Physiol.* 259: E104.

Moroni F., Lombardi G., Carlà V., Pellegrini D., Carassale G. L., and Cortesini C., 1986*a*, Content of quinolinic acid and of other tryptophan metabolites increases in brain regions of rats used as experimental models of hepatic encephalopathy. *J. Neurochem.* 46: 869.

Moroni F., Lombardi G., Carlà V., Lal S., Etienne P., and Nair N. P. V., 1986*b*, Increase in the content of quinolinic acid in cerebrospinal fluid and frontal cortex of patients with hepatic failure. *J. Neurochem.* 47: 1667.

Ono J., Hutson D.G., Dombro R.S., Levi J.U., Livingstone A., and Zeppa R., 1978, Tryptophan and hepatic coma. *Gastroenterology* 74: 196.

Peterson C., Giguére J. F., Cotman C. W., and Butterworth R. F., 1990, Selective loss of *N*-methyl-D-aspartate-sensitive L-[$^3$H]glutamate binding sites in rat brain following portacaval anastomosis. *J. Neurochem.* 55: 386.

Rössle M., Luft M., Herz R., Klein B., Lehmann M., and Gerok W., 1984, Amino acid, ammonia and neurotransmitter concentrations in hepatic encephalopathy: serial analysis in plasma and cerebrospinal fluid during treatment with an adapted amino acid solution. *Klin. Wochenschr.* 62: 867.

Saito K., Markey S. P., and Heyes M. P., 1993, Effects of immune activation on quinolinic acid and neuroactive kynurenines in the mouse. *Neuroscience* 51: 25.

Schröder R., Müller O., and Bircher, J., 1985, The portacaval and splenocaval shunt in the normal rat. A morphometric and functional reevaluation. *J. Hepatol.* 1: 107.

Smith A. R., Rossi-Fanelli F., Ziparo V., James J. H., Perelle B. A., and Fischer J. E., 1978, Alterations in plasma and CSF amino acids, amines and metabolites in hepatic coma. *Ann. Surg.* 187: 343.

Stone T. W., 1993, Neuropharmacology of quinolinic and kynurenic acids. *Pharmacol. Rev.* 45: 309.

Ungerstedt U., 1984, Measurement of neurotransmitter release by intracranial dialysis, in Measurement of Neurotransmitter Release In Vivo (Marsden C. A., ed), pp.81-105. John Wiley & Sons, New York.

Victor M., Adams R. D, and Cole M., 1965, The acquired (non-Wilsonian) type of chronic hepatocerebral degeneration. *Medicine* 44: 345.

Whetsell W.O., Jr. and Schwarcz R., 1989, Prolonged exposure to submicromolar concentrations of quinolinic acid causes excitotoxic damage in organotypic cultures of rat corticostriatal system. *Neurosci. Lett.* 97: 271.

Zaki A.E.O., Ede R.J., Davis M., and Williams R., 1983, Experimental studies of blood brain barrier permeability in acute hepatic failure. *Hepatology* 4: 359.

# 4-CHLORO-3-HYDROXYANTHRANILATE ATTENUATE QUINOLINIC ACID ACCUMULATION IN BRAIN FOLLOWING TRANSIENT CEREBRAL ISCHEMIA IN THE GERBIL

K. Saito,[1] M. Seishima,[1] A. Noma,[1] K. Suyama,[2] S. P. Markey,[3] and M. P. Heyes[3]

[1] Department of Laboratory of Medicine
Gifu University School of Medicine
Gifu 500, Japan
[2] Department of Neurosurgery
Nagasaki University School of Medicine
Nagasaki 852, Japan
[3] Section on Analytical Biochemistry, Laboratory of Clinical Science
National Institute of Mental Health
Bethesda, Maryland 20892

## INTRODUCTION

The kynurenine pathway is the predominant route of L-tryptophan metabolism in most tissues. Two aspects of L-tryptophan catabolism through this pathway are of particular interest. Firstly, the initial enzyme of the kynurenine pathway in extrahepatic tissues, indoleamine-2,3-dioxygenase (IDO), is induced by interferon-γ and other immune stimuli (Saito et al., 1991,1992; Takikawa et al., 1986). Such induction is observed in patients with cancer, infectious disease and following therapeutic interferon administration , and has been implicated in the anti-proliferative and anti-microbial effects of interferon-γ (Yasui et al., 1986; Taylor et al., 1991). Secondly, the neuroactive nature of certain metabolites, including quinolinic acid (QUIN), an excitotoxic agonist of N-methyl-D-aspartate receptors, and kynurenic acid (KYNA) an antagonist of excitatory amino acid receptors, has implicated their involvement in a broad spectrum of neurodegenerative disease (Heyes et al., 1992,1993). In a previous study, accumulation of QUIN in gerbil brain was observed 4 and 7 days following transient cerebral ischemia (Saito et al., see Figure 1). Such increases have been demonstrated to be secondary to increases of IDO and other kynurenine pathway enzymes in infiltrated

*Recent Advances in Tryptophan Research*, edited by Graziella Allegri Filippini et al.
Plenum Press, New York, 1996 

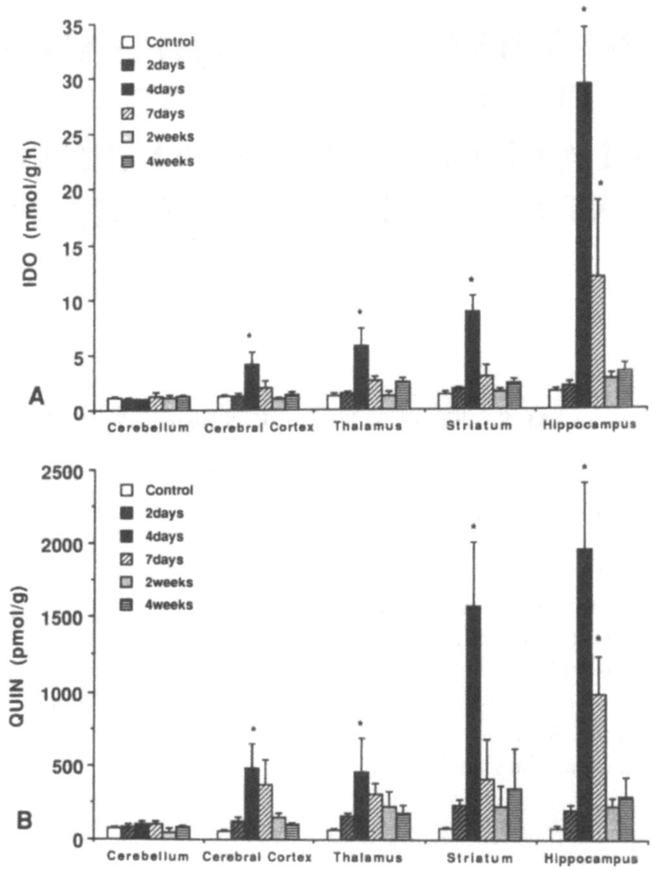


**Figure 1.** Regional brain IDO activity (A) and QUIN concentrations (B) 1, 2, 4, 7, 14, and 28 days following 10 min bilateral carotid artery occlusion (Saito et al., 1993). Note the marked and sustained responses in the hippocampus, which shows the most severe neurodegenerative and inflammatory lesions, compared to the cerebellum which is undamaged. $^{*}p<0.05$.

inflammatory cells. While the neurotoxic effects of QUIN are well established, and circumstances where QUIN levels within the CNS are elevated have been identified, it is not known whether such accumulations are of neuropathogenic or physiologic significance. To investigate this question, the purpose of the present study was to directly examine whether 4-chloro-3-hydroxyanthranilate, an inhibitor of 3-hydroxy-anthranilate-3,4-dioxygenase, or dexamethasone (anti-inflammatory drug) reduces brain QUIN accumulations in that model.

# METHODS

## Experimental Animals

Female Mongolian gerbils (50-70 g) were obtained from Tumblebrook Farm (West Brookefield, MA, U.S.A.). Bilateral common carotid artery occlusion of 10 min duration in the gerbil was produced under anesthesia with 2 % halothane in 30 % $O_2$ / 70 % nitrous oxide by application of Heifetz aneurysm clips (Saito et al., 1993). Samples of brain regions were collected as previously described (Saito et al 1993). All samples were frozen for storage at -70°C until analysis. A dose of 100mg/kg 4-chloro-3-hydroxyanthranilate or 5mg/kg dexamethasone was admimistrated by intraperitoneal injection every 12hours for 7 days (the first injection of 4-chloro-3-hydroxyanthranilate or dexamethasone began at 12 hour post injury, Saito et al., 1994).

## Statistical Analyses

Results were analyzed by computerized statistical packages. Group mean values were compared by one-way analysis of variance and Dunnett's 't'-test for multiple comparisons as the *post hoc* test. The number of animals and samples is indicated by 'n'. A p value of <0.01 was considered statistically significant.

# RESULTS

The IDO activity and QUIN levels were increased in the hippocampus, striatum, cerebral cortex and thalamus following cerebral ischemia by 2 days. Notably, no significant changes in either IDO activity or QUIN concentration was found in the cerebellum, a region that does not become ischemic in the gerbil model (Figure 1).

The concentrations of QUIN in regional brain in gerbils treated with buffer alone, 4-chloro-3-hydroxyanthranilate and dexamethasone are presented in Figure 2. In the injured gerbils treated with buffer alone, QUIN levels on the 7 day post ischemia were elevated in hippocampus (20-fold), striatum (13-fold), cerebral cortex (3-fold) and thalamus (3-fold). 4-chloro-3-hydroxyanthranilate significantly reduced the accumulation of QUIN in all brain regions in response to cerebral ischemia (Figure 2).

In control gerbils, 4-chloro-3-hydroxyanthranilate slightly decreased brain QUIN levels, but had no effects on plasma QUIN concentrations. Dexamethasone also attenuated the increases in all regional brain QUIN concentrations response to ischemia (Figure 2).

# CONCLUSIONS

1. Delayed increases in the level of the NMDA receptor agonist, quinolinic acid (QUIN), occur in several brain regions following transient ischemia in the gerbils (Saito et al., 1993). These changes are particulary marked in regions which showed the most extensive brain damage (hippocampus and striatum).
2. Increased QUIN concentrations in *very specific* brain regions are linked with increases in the activities of indoleamine-2,3-dioxygenase (IDO) and other kynurenine pathway enzymes, which may have an important role during ischemia.

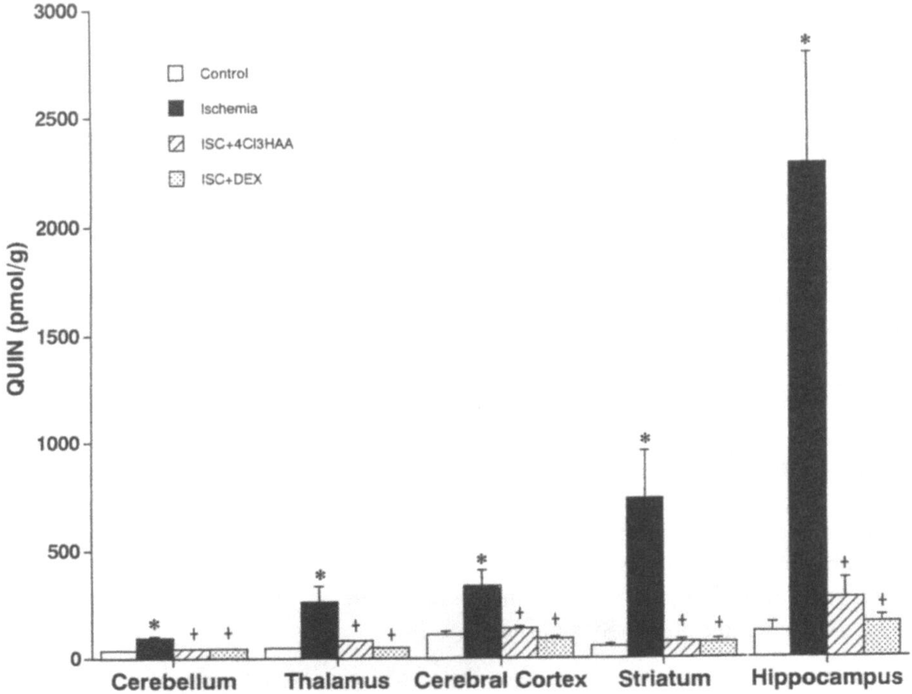


**Figure 2.** Effects of 4-chloro-3-hydroxyanthranilate or dexamethasone on quinolinic acid accumulations in brain following transient cerebral ischemia in the gerbil. Gerbils received an intraperitoneal injection of 4-chloro-3-hydroxyanthranilate (100mg/kg every 12 h beginning 6h post operation; n=8) or dexamethasone (5mg/kg every 12 h beginning 6h post operation; n=7) Samples of brain, CSF and serum were collected 7days later. *p<0.01 versus control; +p<0.01 versus ischemia alone.

3. We hypothesize that accelerated kynurenine pathway metabolism after ischemia reflects the presence of macrophages and other reactive cell populations at sites of injured brain.
4. The administration of 4-chloro-3-hydroxyanthranilate or dexamethasone *significantly reduced* the magnitude of the increases in QUIN levels in injured brain regions. Further studies on the consequences of increases in QUIN concentrations whether contribute neurologic dysfunction and neuropathology following cerebral ischemia are warranted.

## REFERENCES


Heyes M.P., Saito K., Crowley J., Davis L.E., Demitrak M.A., Der M., Dilling L., Kruesi M.J.P., Lackner A., Larsen S.A., Lee K., Leonard H., Markey S.P., Martin A., Milstien S., Mouradian M.M., Pranzatelli M.R., Quearry B.J., Salazar A., Smith M., Straus S.E., Sunderland T., Swedo S. and Tourtellotte W.W., 1992c, Quinolinic acid and kynurenine pathway metabolism in inflammatory and non-inflammatory neurologic disease *Brain* 115: 1249.

Saito K., Chen C.Y., Masana M., Crowley, J., Markey, S.P. and Heyes, M.P., 1993, 4-Chloro-3-hydroxyanthranilic acid, 6-chlorotryptophan and norharmane attenuate quinolinic acid formation by interferon-γ stimulated monocytes (THP-1) cells. *Biochem J* In Press.

Saito K., Lackner A., Markey S.P. and Heyes M.P., 1991a, Cerebral cortex and lung indoleamine-2,3-dioxygenase activity is increased in type-D retrovirus infected macaques *Brain Research* 540: 353-356.

Saito K., Markey S.P. and Heyes M.P., 1992, Effects of immune activation on quinolinic acid and kynurenine pathway metabolism in the mouse *Neuroscience* 51: 25.

Saito K., Nowak T.S., Jr., Markey S.P. and Heyes M.P., 1993, Mechanism of delayed increases in kynurenine pathway metabolism in damaged brain regions following transient cerebral ischemia *J Neurochem* 60: 180.

Saito K, Nowak T.S, Suyama K, Quearry B.J, Saito M, Crowley J.S *et al.*, 1993d, Kynurenine pathway enzymes in brain: responses to ischemic brain injury versus systemic immune activation. *J. Neurochem.,* 61: 2061

Stone T.W. and Perkins M.N., 1984, Actions of excitatory amino acids and kynurenic acid in the primate hippocampus: a preliminary study *Neurosci Lett* 52: 335.

Takikawa O., Yoshida R., Kido R. and Hayaishi O., 1986, Tryptophan degradation in mice initiated by indoleamine-2,3-dioxygenase *J Biol Chem* 261: 3648.

Taylor M.W. and Feng G., 1991, Relationship between interferon-γ, indoleamine-2,3-dioxygenase and tryptophan *FASEB J* 5: 2516.

Yasui H, Takai K, Yoshida R, Hayaishi O., 1986, Interferon enhances tryptophan metabolism by inducing pulmonary indoleamine-2,3-dioxygenase; Its possible occurrence in cancer patients. *Proc. Natl Acad. Sci. USA,* 83: 6622.

# THE EFFECT OF L-TRYPTOPHAN LOAD ON ENDOCRINE SYSTEM OF PANCREAS IN DIABETES MELLITUS

D. Martinsone, R. Ligere, and A. Martinsons

Latvia Institute of Experimental and Clinical Medicine
Latvia

## INTRODUCTION

Tryptophan inhibits hepatic glyconeogenesis in vivo. There are two main explanations: 1/ tryptophan administration stimulates the secretion of gastric inhibitory polypeptide (GIP), whose possible role in mediating enteric insulin release is known (Thomas et al., 1976); 2/ the inhibition of gluconeogenesis can occur secondarily to the formation of quinolinate, which blocks the reaction catalysed by phosphoenolpyruvate carboxykinase (Ray and Foster, 1966). It is known that liver cells from diabetic rats synthesize less quinolinate from tryptophan because of increased flux through the glutarate pathway (Smith and Pogson, 1980). Therefore, gluconeogenesis in such cells are less sensitive to tryptophan.

The aim of this study is to investigate the effect of L-tryptophan loading on endocrine system of pancreas in diabetes mellitus of human.

## PATIENTS AND METHODS

We have investigated 55 non-insulin dependent diabetes mellitus (NIDDM) patients and 10 healthy individuals. We have determined the serum levels of glucose (method of ortotoluidine) (Komarov and Korovin, 1981), insulin,glucagon and gastrin by radioimmunoassay (Tkachova and Balabolkin, 1983) on an empty stomach and 30; 60; 120 and 180 minutes after a peroral tryptophan loading in dose 30 mg/kg of the body weight. Total tryptophan in serum was determined basal and 60;120 and 180 minutes after the load (Spies and Chambers, 1948).

The results are expressed as mean ±SEM and their statistical significance calculated by Students t-test for unpaired data.

*Recent Advances in Tryptophan Research*, edited by Graziella Allegri Filippini et al.
Plenum Press, New York, 1996

**Table 1.** The levels of serum tryptophan (μmol/l) before and 60, 120, and 180 minutes after L-tryptophan load (30 mg/kg body weight peroral) in healthy individuals and in NIDDM patients, mean ±SEM

| | Minutes | | | |
|---|---|---|---|---|
| | 0 | 60 | 120 | 180 |
| Control group | 85 ± 15 | 211 ± 21 | 264 ± 28 | 201 ± 20 |
| NIDDM | 78 ± 12 | 204 ± 18 | 251 ± 22 | 188 ± 15 |

## RESULTS AND DISCUSSION

The levels of tryptophan in serum did not differ in both groups before and after the load (Table 1). We have found a decreased levels of blood glucose 60 and 120 minutes after L-tryptophan load in healthy individuals. These data agree with previous results (Thomas et al., 1976; Pogson et al., 1975) and confirm the thesis that tryptophan inhibits hepatic gluconeogenesis in mammals. In NIDDM patients L-tryptophan load resulted in consistent and significant rise in blood glucose concentration from 30 minutes onwards (Table 2).

Orally given tryptophan produced a rise in plasma glucagon levels in both groups, but in NIDDM patients this rise lasted longer - till 120 minutes after L-tryptophan loading (Table 3). The stimulatory effect of amino acids on pancreatic A-cells is well known although the mechanism and teleology of this effect remain contraversial. Prevention of postmeal aminogenic insulin induced hypoglicemia (Unger et al., 1969) and regulation of exogenous gluconeogenesis (Rocha et al., 1972) are two main explanations proposed so far. Why in NIDDM the stimulatory effect of L-tryptophan on glucagon delivery from A-cells is more expressed, is an unanswered question.

The insulinotropic effect of glucagon (Samols et al., 1965) we have seen in both groups. In diabetes mellitus insulin levels reached far higher mean values and lasted longer till 180 minutes as in control group (Table 4).

**Table 2.** Blood glucose levels (mmol/l) in response to L-tryptophan peroral load (30 mg/kg body weight) in control group and in NIDDM patients, mean±SEM

| | Minutes | | | | |
|---|---|---|---|---|---|
| | 0 | 30 | 60 | 120 | 180 |
| Control group | 4.5 ± 0.8 | 4.2 ± 0.7 | 3.1 ± 0.7* | 3.2 ± 0.6* | 4.2 ± 0.9 |
| NIDDM | 8.1 ± 1.9 | 13.8 ± 2.9** | 12.2 ± 2.3* | 11.4 ± 2.2* | 11.2 ± 2.1* |

*$p < 0.05$, **$p < 0.02$.

**Table 3.** Plasma glucagon levels ( pg/ml) in response to L-tryptophan peroral load (30 mg/kg body weight ) in control group and in NIDDM patients, mean ±SEM

| | Minutes | | | | |
|---|---|---|---|---|---|
| | 0 | 30 | 60 | 120 | 180 |
| Control group | 75 ± 14 | 120 ± 22* | 125 ± 24* | 84 ± 16 | 80 ± 12 |
| NIDDM | 92 ± 18 | 134 ± 24* | 144 ± 28** | 136 ± 24* | 94 ± 17 |

*$p < 0.05$, **$p < 0.02$.

**Table 4.** Plasma insulin levels (mekV/ml) in response to L-tryptophan per oral load (30 mg/kg body weight ) in control group and in NIDDM patients, mean ±SEM

| | Minutes | | | | |
|---|---|---|---|---|---|
| | 0 | 30 | 60 | 120 | 180 |
| Control group | 14 ± 2.5 | 24 ± 3.1* | 28 ± 4.0* | 26 ± 4.2* | 18 ± 2.7 |
| NIDDM | 26 ± 4.2 | 39 ± 4.4* | 44 ± 4.7** | 49 ± 4.9*** | 48 ± 4.7*** |

*p < 0.05, **p < 0.02, ***p < 0.01.

**Table 5.** Plasma gastrin levels μED/ml in response to L-tryptophan per oral load (30 mg/kg body weight) in control group and in NIDDM patients, mean ±SEM

| | Minutes | | | | |
|---|---|---|---|---|---|
| | 0 | 30 | 60 | 120 | 180 |
| Control group | 80 ± 12 | 122 ± 21* | 132 ± 28* | 91 ± 14 | 85 ± 15 |
| NIDDM | 92 ± 14 | 131 ± 22* | 151 ± 31** | 130 ± 21* | 102 ± 16 |

*p < 0.05, **p < 0.02.

These data correspond with literature (Unger et al., 1972) , that described the disturbances in the regulation of insulin-glucagon balance in diabetes mellitus.

L-Tryptophan load leads to rise in plasma gastrin levels in both groups. In NIDDM patients the values were higher and continued till 120 minutes after L-tryptophan administration (Table 5). These data can confirm that glucagon is the most important factor in the regulation of gastrin secretion (Becker et al., 1973).The elevated gastrin can lead to the stimulation of endocrine system of pancreas (Lefebvre, 1983).

The possible cause of the rise in insulin, glucagon and gastrin levels after L-tryptophan load can be less active degradation of these hormones in liver in diabetes mellitus.

Hyperinsulinemia is an often symptom in NIDDM. L-Tryptophan load can discover asymptomatic insulin resistance in such patiets, on the level of liver biochemistry.

The data obtained in the present investigation established a differential response of pancreatic endocrine cells to peroral L-tryptophan load in healthy individuals and in NIDDM patients.

## REFERENCES


Becker H.D., Reeder D.D., Thompson J.C., 1973, Effect of glucagon on circulating gastrin. *Gastroenterology*, vol.65, N.1, 28.

Komarov F., Korovin B., 1981, Biochemistry in the clinic. Leningrad, Medicina, 403 (Russ.).

Lefebvre F.J., 1983, Glucagon II. Handbook of Experimental Pharmacology, vol.66, N.11: 99.

Pogson C.I., Crisp D.M., Smith S.A., 1975, Effect of tryptophan and its metabolites on gluconeogenesis in mammalian tissues. *Acta Vitaminol. Enzimol.*, 232-236

Ray P.D., Foster H.A., 1966, Paths of Carbon in gluconeogenesis and lipogenesis. *J.Biol.Chem.* 241 : 3904.

Rocha D., Faloona G., Unger R., 1972, Glucagon - stimulating activity of 20 amino acids in dogs. *J.Clin.Invest.* 51 : 2346.

Samols E., Marri V., Marks V., 1965, Promotion of insulin secretion by glucagon. *Lancet ii* , 415.

Smith S., Pogson C., 1980, Tryptophan metabolism in diabetic rats . *Biochem. J.* 186 : 977.

Spies J.R., Chambers D.C., 1948, Chemical determination of tryptophan. Study of color formation of tryptophan, p-dimethylamino-benzaldehide and sodium nitrate in sulfuric acid solutions. *Analytical Chem.* 20. N1 : 30

Thomas F.B., Mazzaferri S.E., Crockett H.S., ,1976, Stimulation of secretion of gastric inhibitory polypeptide and insulin by intraduodenal amino acid perfusion. *Gastroenterology* 70 :523.

Tkachova G., Balabolkin M., ,1983, The methods of radioimmunoassay. *Moskva, Medicina*, 193 ( Russ.).

Unger R.H., Maddison L., Müller W., 1972, Abnormal alpha-cell funktion in diabetics. Response to insulin. *Diabetes*. vol.21, N.4, 301.

Unger R.H.,Ohneda E., Aguilar, 1969, A.The role of aminogenic glucagon secretion in blood glucose homeostasis. *J.Clin.Invest.* 48 : 810.

# THE RELATIONSHIP BETWEEN KYNURENINE, CATECHOLAMINES, AND ARTERIAL HYPERTENSION IN MESANGIOPROLIFERATIVE GLOMERULONEPHRITIS

A. Martinsons,[1] V. Rudzite,[2] E. Jurika,[2] and A. Silava[2]

[1] Latvia Medicine Academy
[2] Latvia Institute of Cardiology
Latvia

## INTRODUCTION

In our previous work we proved that hyperkynureninemia can participate in the development of mesangioproliferative glomerulonephritis (MPG) by impairing renal endotheliocytes and epitheliocytes (Rudzite et al.,1991). We have also found close association between serum kynurenine (K) and intensive hematuria and arterial hypertension (AH) in the patients with MPG (Martinsons et al.,1992 ). Although K is not a vasoconstrictor , it stimulates the effect of serotonin, norepinephrine and acetylcholine in blood vessels (Lapin and Umanskaja, 1980). K activates serum dopamine-beta-hydroxylase in vitro: the concentration of dopamine decreases and that of noradrenaline increases (Rudzite et al., 1987). Such a mechanism in the development of AH has been demonstrated in patients with essential AH (Rudzite et al., 1987).

The aim of this study is to investigate the relationship between K, catecholamines an blood presure in MPG patients.

## PATIENTS AND METHODS

36 Patients with MPG and 10 healthy individuals have been investigated. The diagnosis of MPG was proved by analyzing kidney bioptates.

Serum concentration of K (Joseph and Risby,1975), dopamine, norepinephrine and epinephrine (Bohuon , 1974) have been determined as well as systolic and diastolic arterial blood pressure examined at whole fasting and 3 hours after an L-tryptophan oral loading 30 mg/kg body weight. The statistical difference between the values obtained in the control

*Recent Advances in Tryptophan Research*, edited by Graziella Allegri Filippini et al.
Plenum Press, New York, 1996

**Table 1.** The serum levels of kynurenine (K),dopamine (DA), norepinephrine (NE), and epinephrine (E) as well as arterial blood pressure: systolic (S) and diastolic-before and 3 hours after peroral L-tryptophan load (30 mg/kg body weight) in healthy individuals (n=10) - group A, in patients with mesangiopriliferative glomerulonephritis with normal levels of K with an arterial hypertension (n=6) group B, and with normal blood pressure (n=10) group C (mean ± SEM)

| | Group A | | Group B | | Group C | |
|---|---|---|---|---|---|---|
| | Before load | 3 hr after load | Before load | 3 hr after load | Before load | 3 hr after load |
| K μmol/l | 3.2 ± 0.8 | 12.4 ± 2.5 | 3.6 ± 0.9 | 12.7 ± 2.8 | 3.1 ± 0.9 | 12.6 ± 2.4 |
| DA μg/l | 0.04 ± 0.002 | 0.07 ± 0.002[a] | 0.05 ± 0.003 | 0.08 ± 0.004[a] | 0.04 ± 0.004 | 0.07 ± 0.003[a] |
| E μg/l | 0.10 ± 0.009 | 0.22 ± 0.01[b] | 0.11 ± 0.01 | 0.24 ± 0.02[b] | 0.11 ± 0.01 | 0.25 ± 0.02[b] |
| NE μg/l | 0.14 ± 0.02 | 0.28 ± 0.04[c] | 0.15 ± 0.02 | 0.24 ± 0.04[b] | 0.14 ± 0.02 | 0.27 ± 0.04[b] |
| S mmHg | 122 ± 21 | 109 ± 18[a] | 171 ± 12 | 168 ± 10 | 121 ± 24 | 105 ± 21[a] |
| D mmHg | 72 ± 14 | 61 ± 11[a] | 99 ± 11 | 94 ± 10 | 79 ± 13 | 60 ± 9[a] |

[a]$p < 0.05$, [b]$p < 0.02$, [c]$p < 0.01$.

group and MPG patients were calculated. In cases of $p < 0.05$, the difference is statistically significant (Weber, 1957).

## RESULTS AND DISCUSSION

L-Tryptophan loading (30 mg/kg body weight) increased the serum concentration of K, with a peak 3 hours after the administration of the load. The peak of K in control group was accompanied by statistically significant increase of serum dopamine ($p< 0.05$), norepinephrine ($p<0.01$) and epinephrine ($p< 0.02$). We observed a decrease in systolic and diastolic blood pressure ($p<0.05$ ) (Table 1). These results agree with the data of literature (Feltkamp et al., 1984).

In 16 patients (44.4%) we have found a normal level of K basal and no significant differences were noted after L-tryptophan loading compared with the control group. The peak of K was accompanied also by statistically significant increase of serum dopamine ($p<0.05$), norepinephrine ($p<0.02$) and epinephrine ($p<0.02$). These laboratory changes were observed in both groups of patients with normal levels of K: with and without AH. In the group without basal AH we have observed a slight decrease in systolic and diastolic blood pressure ($p<0.05$) (Table 1).

In 20 patients (55.6%) we have found elevated levels of K: basal as well as after L-tryptophan loading. We have found an AH (blood pressure > 160/90 mm/Hg) in 70% of these patients. The peak of serum K in patients with AH was accompanied by a statistically significant decrease in dopamine ($p< 0.05$) and increase in norepinephrine ($p<0.02$): the changes in epinephrine after the tryptophan loading were not significant. In this group of patients we have observed a significant increase ($p<0.02$) in systolic and diastolic blood pressure after L-tryptophan loading (Table 2). These data correspond to the results obtained in patients with essential AH with hyperkynureninemia (Rudzite et al.,1987).

In the group of patients with hyperkynurinenimia, but without AH we have found no significant changes in serum levels of dopamine, epinephrine and norepinephrine. The blood pressure did not change either (Table 2). We suggest, that this may be a potential risk group for AH in future. Our findings suggest that K, a key metabolite of the tryptophan-nicotinic acid pathway, forms one of the pathogenetic basis for AH in the part of MPG patients by

**Table 2.** The serum levels of kynurenine (K), dopamine (DA), norepinephrine (NE) and epinephrine (E) as well as arterial blood pressure: systolic (S) and diastolic-before and 3 hours after peroral L-tryptophan load (30 mg/kg body weight) in healthy individuals (n= 10) -group A, in patients with mesangiopriliferative glomerulonephritis with basal hyperkynureninemia with an arterial hypertension (n=14) -group D, and with normal basal arterial blood pressure (n=6) -group E, (mean ± SEM)

| | Group A | | Group D | | Group E | |
|---|---|---|---|---|---|---|
| | Before load | 3 hr after load | Before load | 3 hr after load | Before load | 3 hr after load |
| K μmol/l | 3.2 ± 0.8 | 12.4 ± 2.5 | 5.84 ± 2.2 | 22.4 ± 4.8 | 5.91 ± 2.6 | 24.2 ± 4.4 |
| DA μg/l | 0.04 ± 0.002 | 0.07 ± 0.002[a] | 0.06 ± 0.002 | 0.02 ± 0.002[a] | 0.04 ± 0.002 | 0.03 ± 0.002 |
| E μg/l | 0.10 ± 0.009 | 0.22 ± 0.01[b] | 0.13 ± 0.01 | 0.17 ± 0.02 | 0.14 ± 0.02 | 0.13 ± 0.02 |
| NE μg/l | 0.14 ± 0.02 | 0.28 ± 0.04[c] | 0.18 ± 0.04 | 0.31 ± 0.08[b] | 0.19 ± 0.04 | 0.23 ± 0.05 |
| S mmHg | 122 ± 21 | 109 ± 18[a] | 174 ± 16 | 195 ± 17[b] | 145 ± 20 | 151 ± 22 |
| D mmHg | 72 ± 14 | 61 ± 11[a] | 98 ± 14 | 109 ± 16[b] | 80 ± 16 | 89 ± 18 |

[a]$p < 0.05$, [b]$p < 0.02$, [c]$p < 0.01$.

promoting the vasoconstruction through norepinephrine release as well as by insufficient vasodilatation by decreased formation of dopamine.

## REFERENCES


Bohuon C., 1974, Biologically active amines. Clin.Biochemistry (Curtins H.Ch. and Roth M. eds). Principles and Methods. Walter de Gruyter & Co, Berlin, 855 - 870.

Feldkamp H., Meurer K.A., Gedehart E., 1984, Tryptophan induced lowering of blood pressure and changes at serotonin uptake by platelets in patients with essential hypertension, *Klin. Wochenschr.*, 62: 1115.

Joseph M.H., Risby D., 1975 ,The determination of kynurenine in plasma. *Clin. Chem.Acta*, 63: 197.

Lapin I., Umanskaja L., 1980, The influence of kynurenine and its metabolites to blood vessels: effect of serotonin, noradrenalin and acethylholine, *Bul.Exp.Biol.Med.*, 83, 6: 704.

Martinsons A.,Rudzite V., Grom V et al. 1992, Kynurenine and neopterin in mesangioproliferative glomerulonephritis. Advances in Tryptophan Research, 389-390.

Rudzite V., Jurika E., Arajs J. et al., 1987, The relationship between kynurenine and arterial blood pressure. Progress in Tryptophan and Serotonin Research, 129-132.

Rudzite V., Grom V., Martinsons A.,1991, Kynurenine and mesangiopro-lipherative glomerulonephritis. Kynurenine and serotonin pathways. *Adv. Exper. Med. Biol.*, 294: 659.

Weber E. 1957 ,Grundriss der biologischen Statistik für Naturwissenshaftler, Landwirte und Mediziner, VEB Gustav Fischer Verlag, Jena.

65

# KYNURENINE PATHWAY METABOLISM IN THE GALACTOSAMINE MODEL OF HEPATIC INJURY

K. Saito,[1] M. Seishima,[1] A. Noma,[1] Y. Nagamura,[2] B. J. Quearry,[3] S. P. Markey,[3] and M. P. Heyes[3]

[1] Department of Laboratory Medicine, Gifu University School of Medicine Gifu, Gifu 500, Japan
[2] Department of Clinical Chemistry, School of Health Science Fujita Health University, Toyoake, Aichi 470-11, Japan
[3] Section on Analytical Biochemistry, Laboratory of Clinical Science National Institute of Mental Health, Bethesda, Maryland 20892

## INTRODUCTION

The mechanisms involved in producing hepatic encephalopathy and hepatic coma are unclear, but include an accumulation of neurotoxic metabolites. Elevated plasma concentrations of aromatic amino acids, including L-tryptophan, are characteristic of hepatic injury (James et al., 1976). Key metabolites of the L-tryptophan-kynurenine pathway are neuroactive within the central nervous system. Quinolinic acid (QUIN) is an agonist of N-methyl-D-aspartate receptors and an excitotoxin (Lapin, 1982; Schwarcz et al., 1983; Whetsell et al.,1989) while kynurenic acid (KYNA) is an antagonist of excitatory amino acid receptors and attenuates the excitotoxic effects of QUIN (Foster et al., 1984). L-Kynurenine (L-KYN) is a convulsant (Lapin, 1982), and 3-hydroxykynurenine may be neurotoxic (Eastman et al, 1989). Therefore, accumulation of neuroactive kynurenine pathway metabolites within the brain may be of functional or clinical significance. In addition, while the concentrations of QUIN, KYNA and L-KYN within the CNS are low in neurologically normal subjects and patients with chronic non-inflammatory neurological diseases, substantial increases occur in the CNS of humans and non-human primates with inflammatory neurological conditions, particularly those with inflammatory lesions and monocyte infiltration into the CNS (Heyes et al., 1992). In the present study, we have used systemic injections of galactosamine as a model of acute hepatic injury, with which to evaluate systemic and CNS kynurenine pathway responses. Galactosamine treatment has been used as a means to potentiate endotoxin-induced shock and multiple organ system failure. Since we have found that endotoxin is a potent stimulus to QUIN accumulations in brain and systemic tissues, we also evaluated the combined and separate effects of galactosamine and endotoxin.

*Recent Advances in Tryptophan Research,* edited by Graziella Allegri Filippini et al.
Plenum Press, New York, 1996 

# MATERIALS AND METHODS

## Materials

L-Tryptophan, L-KYN, QUIN, galactosamine, LPS from *Salmonella abortus equii* methylene blue, ascorbic acid and trichloroacetic acid were obtained from Sigma Chemical Company (St. Louis, MO). Catalase was obtained from Boehringer Mannheim. Other chemicals were of analytical grade and were purchased from Fisher Inc. (Fair Lawn, NJ). Female Mongolian gerbils (50-70 g) were obtained from Tumblebrook Farm (West Brookefield, MA) and housed in plexiglass cages for at least 7 days prior to study on a 12 h : 12 h dark-light cycle.

## Biochemical Techniques

For the QUIN assays, plasma (20 μL), CSF (30 μL), acidified tissue extracts (700 μL from tissue homogenized in 1 mL of 1 mol/L hydrochloric acid and centrifuged) and QUIN standards (in corresponding volumes of deionized water) were mixed with 300 μL of deionized water containing 30 pmol [$^{18}$O]-QUIN and freeze dried overnight. QUIN and [$^{18}$O]-QUIN were derivatized to their dihexafluoroisopropanol esters with complete retention of the isotope, washed with 250 μL of water, extracted into 300-1000 μL of heptane and quantified using a DB5 analytical column (J & W Scientific, Folsom, CA) and a Hewlett-Packard Model 5988 quadrupole mass spectrometer (electron capture negative chemical ionization mode). The molecular anions of QUIN (m/z 467), [$^{13}C_6$]-QUIN (m/z 473) and the [$^{18}$O]-QUIN internal standard (m/z 471) were monitored, and each peak area at the appropriate retention time was measured for quantification. L-KYN was quantified by high performance liquid chromatography and ultraviolet light absorption spectrometry in acidified extracts of plasma (1:1 with 0.1 mmol/L trichloroacetic acid) (Saito et al., 1992). The activity of IDO and TDO were quantified by the conversion of L-tryptophan to L-KYN (Saito et al., 1991, 1993). Enzyme activity was determined at saturating substrate concentrations, and expressed as the product formed/h/g wet tissue weight.

## Experimental Studies

Gerbils were given an intraperitineal injection of either sterile, pyrogen-free saline carrier alone as control, LPS alone (150 ng/Kg), galactosamine alone (50 mg/Kg) or a combination of both LPS and galactosamine. Samples of plasma, lung, liver, brain regions and cerebrospinal fluid (CSF) were collected 24 h later (Saito et al.,1993). To collect CSF, gerbils were anesthetized by an intraperitoneal injection of sodium pentobarbital (40 mg/Kg) and placed in a Kopf stereotaxic frame with the head tilted down at approximately 45 degrees. A 30-gauge needle attached to a 10 cm length of PE10 tubing and a 1 mL syringe was inserted into the cisterna magna, and 10-35 μL of CSF was withdrawn and transferred to a tared polypropylene tube. Lung was studied because it receives all of the cardiac output and because it is one of the most responsive organs to immune stimulation with respect to increased kynurenine pathway metabolism (Saito et. al., 1992, 1993). Liver was chosen because it has been viewed as an important regulatory organ of kynurenine pathway metabolism via tryptophan-2,3-dioxygenase activity (Bender, 1982). All samples were frozen for storage at -70°C until analysis.

**Table 1.** Combined and separate effects of GalN (650 mg/Kg) and LPS (150 ng/Kg) on systemic kynurenine pathway metabolism. Samples were collected 24 h after drug administration

| | Control | Galactosamine | LPS | Galactosamine + LPS |
|---|---|---|---|---|
| Plasma L-KYN (μM) | 1.0 ± 0.1 | 2.7 ± 0.2* | 0.9 ± 0.1 | 8.4 ± 1.1* |
| Plasma QUIN (nM) | 638 ± 70 | 764 ± 80 | 768 ± 36 | 3060 ± 1088* |
| Liver QUIN (nmol/Kg) | 6228 ± 375 | 9990 ± 4108 | 14776 ± 2662* | 46723 ± 5898* |
| Liver TDO (μmol/g/h) | 752 ± 66 | 204 ± 36* | 534 ± 29 | 375 ± 76* |
| Lung QUIN (nmol/Kg) | 3340 ± 408 | 7609 ± 821* | 8028 ± 637* | 21010 ± 3400* |
| Lung IDO (nmol/g/h) | 72 ± 9 | 72 ± 6 | 164 ± 26* | 176 ± 30* |

*p<0.05 vs control.

## RESULTS

Galactosamine treatment alone was associated with increased concentrations of QUIN in lung and liver (Table 1). QUIN concentrations in the five brain regions examined were also incresed (Figure 1). In liver, the activity of TDO was lower in galactosamine-treated gerbils, while lung IDO activity was unchanged (Table 1). Plasma L-KYN levels, but not plasma QUIN concentrations, were increased (Table 1). Treatment with LPS alone was associated with increases in brain, CSF, lung and liver QUIN concentrations. Liver TDO activity was reduced; lung IDO activity was increased, while plasma QUIN levels were unchanged. The co-administration of galactosamine and LPS was associated with significantly enhanced responses in brain, plasma, lung and liver QUIN concentrations, such that

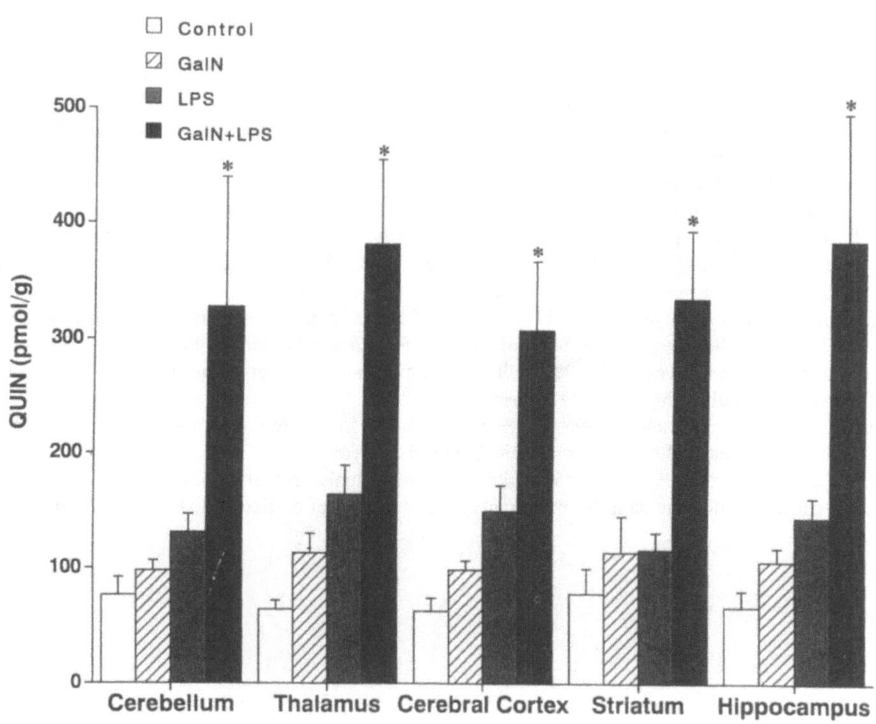


**Figure 1.** Combined and separate effects of GalN (650 mg/Kg) and LPS (150 ng/Kg) on regional brain quinolinic acid concentrations. Samples were collected 24 h after drug administration.

the magnitude of the increases was greater than simple addition of the stimulatory effects of galactosamine or LPS alone. Liver TDO activity was reduced to values intermediate between galactosamine and LPS treated groups alone. The increases in lung IDO activity was not significantly different from the effects of LPS alone. There were no statistically significant correlations between plasma QUIN concentrations and either CSF or regional brain QUIN levels in control, GalN alone or LPS alone treated gerbils. There were, however, significant correlations between plasma QUIN concentrations and both CSF and regional brain QUIN levels in gerbils treated with both GalN and LPS.

## CONCLUSIONS

The present study has used galactosamine-treated gerbils as a model of acute hepatic injury, and evaluated the effects of immune activation by endotoxin treatment on blood and brain QUIN levels. The results showed increased accumulation of QUIN in both systemic tissues and brain in response to either galactosamine or endotoxin. We conclude that increases in brain QUIN levels in conditions of hepatic injury may reflect increases in systemic kynurenine pathway metabolism and elevated blood QUIN levels. Immune activation in liver, lung and perhaps other systemic organs increases the activation of indoleamine-2,3-dioxygenase resulting in increases in systemic QUIN production.

## REFERENCES

Bender, D.A. and McCreanor, 1985, Kynurenine hydroxylase: a potential rate-limiting enzyme in tryptophan metabolism *Biochem Soc Trans* 13: 441.

Eastman C.L. and Guilarte T.R., 1989, Cytotoxicity of 3-hydroxykynurenine in a neuronal hybrid cell line. *Brain Research* 495: 225.

Foster A.C., Vezzani A., French E.D. and Schwarcz R., 1984, Kynurenic acid blocks neurotoxicity and seizures induced in rats by the related brain metabolite quinolinic acid. *Neurosci. Lett.* 48: 273.

Heyes, M.P., Saito, K., Crowley, J., Davis, L.E., Demitrak, M.A. et al.. 1992, Quinolinic acid and kynurenine pathway metabolism in inflammatory and non-inflammatory neurologic disease *Brain* 115: 1249.

James J.H., Hodgman J.M. Funovics J.M. Yoshimura N., and Fischer J.E., 1976, Brain tryptophan, plasma free tryptophan and distribution of plasma neutral amino acids. *Metabolism* 25: 471.

Lapin I.P.,1982, Convulsant action of intracerebroventricularly administered L-kynurenine sulphate, quinolinic acid and other derivatives of succinic acid, and effects of amino acids: structure-activity relationships. *Neuropharmacology* 21: 1227.

Saito, K., Markey, S.P. and Heyes, M.P., 1991, Chronic effects of gamma-interferon on quinolinic acid and indoleamine-2,3-dioxygenase in brain of C57BL6 mice *Brain Research* 546: 151.

Saito, K., Markey, S.P. and Heyes, M.P., 1992, Effects of immune activation on quinolinic acid and kynurenine pathway metabolism in the mouse *Neuroscience* 51: 25.

Schwarcz, R., Whetsell, W.O. and Mangano, R.E.M., 1983, Quinolinic acid: an endogenous metabolite can produce axon sparing lesions in rat brain *Science* 219: 316.

Whetsell W.O. and Schwarcz R., 1989, Prolonged exposure to submicromolar concentrations of quinolinic acid causes excitotoxic damage in organotypic cultures of rat corticostriatal system. *Neurosci. Lett. 97: 271.*

# IS AN ALTERED TRYPTOPHAN METABOLISM RESPONSIBLE FOR AGE-RELATED CATARACT?

A. J. Elderfield and R. J. W. Truscott

Biochemistry Department
Illawarra Regional Hospital
Wollongong, NSW, Australia and
The Australian Cataract Research Foundation
University of Wollongong
Northfields Avenue, Wollongong, NSW, Australia

## 1. INTRODUCTION

Senile cataract remains one of the major health concerns of the world with approximately 17 million people blind as a result of this disease (Ohrloff, 1989).

Although the cause of senile cataract remains largely unknown, physical and chemical changes have been identified as being associated with the formation of opacities within the lens nucleus. These opacities are accompanied by aggregation of the proteins into high molecular weight compounds with an increase in the urea-insoluble coloured component (Harding and Crabbe, 1984). Further evidence in the support of these changes is the loss of glutathione, a decrease in the concentration of protein thiols and an increase in the amount of methionine methoxide, all pointing to an oxidising environment within the nucleus of the lens (Truscott and Augustyn, 1977). The coloured insoluble proteins from the nucleus of the cataractous lens have also been associated with products from the oxidation of tryptophan (trp), namely kynurenine (kyn) and derivatives (Dilley and Pirie, 1974), anthranilic acid (Truscott et al., 1977) and possibly beta-carbolines (Dillon and Spector, 1976). Ogino and Ichihara (1957) reported the isolation of a trp metabolites, 5-hydroxyanthranilic acid (5OHA) from the urine of cataract patients but not from the urine of controls. Others have also found increased trp metabolites in the urine of cataract patients (Costa et al., 1982). Controversy arose when a 2-fold increases of trp itself were reported in the serum of cataract sufferers (Cotlier and Sharma, 1980), and other investigators (Allegri and Angi, 1981) found no such increase above that of normal controls. It has been postulated that some of the reactive metabolites of trp, which are known to be capable of modifying proteins could be implicated in the aetiology of this disease (Truscott et al., 1991).

We sought to help clarify the situation by examining if persons suffering from cataract actually metabolise trp differently from those who have no presence of cataract.

*Recent Advances in Tryptophan Research*, edited by Graziella Allegri Filippini et al.
Plenum Press, New York, 1996

## 2. MATERIAL AND METHODS

Volunteers were chosen from patients who because of impaired vision had presented themselves for surgical removal of cataract. Volunteers who had clear bilateral lenses were chosen for the control group. In both groups the degree of opacity of the lens was confirmed by slit-lamp examination. Subjects were fasted overnight, and blood was collected the following morning for serum biochemical and haematological analysis and for high-performance liquid chromatography analysis for trp and its kyn pathway metabolites. A dose of L-tryptophan was administered orally at a rate of 50 mg/kg body weight. Further blood samples were taken for trp metabolite analysis at 1 hr, 2 hr, 4 hr and 6 hr after the loading dose. Blood specimens for HPLC analysis were immediately precipitated with cold 10% trichloroacetic acid. The supernatant was frozen until analysis.

HPLC analysis was performed using an isocratic system using an C8 silica column and a 20 mM acetate buffer mobile phase, pH 4.5 (Elderfield et al., 1989). Quantification was achieved using a dual electrochemical and programmable fluorescence detection. Not all trp metabolites are electroactive, but some show fluorescent activity. In this way, trp and eight of its metabolites could be measured with considerable accuracy and sensitivity. The metabolites quantified, in addition to trp, were kyn, 3-hydroxyanthranilic acid, xanthurenic acid, kynurenic acid, 5-hydroxyanthranilic acid, anthranilic acid, 5-hydroxytryptophan and anthranilic acid. Free trp was also estimated by HPLC after separation of the albumin-bound amino acid.

## 3. RESULTS AND DISCUSSION

A total of 79 volunteers submitted themselves for the trp loading protocol. Of these, 42 were cataract patients comprising 24 males and 18 females with a median age of 71 years. The control group consisted of 37 volunteers, 26 males and 11 females.

Of the ten compounds estimated by HPLC only four displayed significantly different serum responses to the trp load between the cataract and control groups. These were total trp, kyn, kynurenic acid (KA) and 5-hydroxyanthranilic acid (5OHA).

Figure 1 shows the profiles of total trp for the cataract and control group. The total trp had increased 4-fold in the cataract group and 3 fold in the cataract group after 1 hr, with both groups peaking at 2 hr. The peak mean responses were 340 ± 87 μmol/L in the cataract

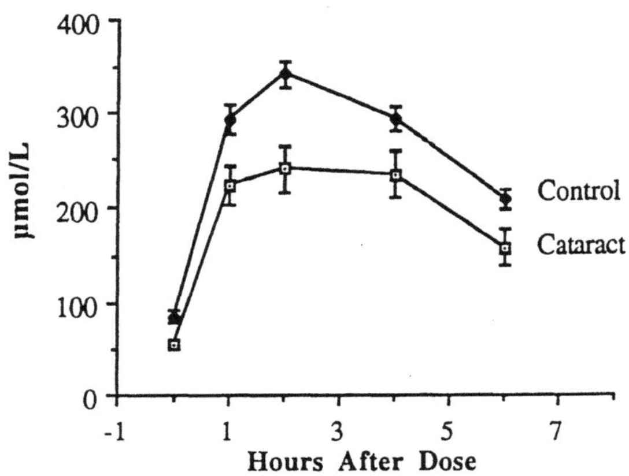


**Figure 1.** Total tryptophan response to the tryptophan load.

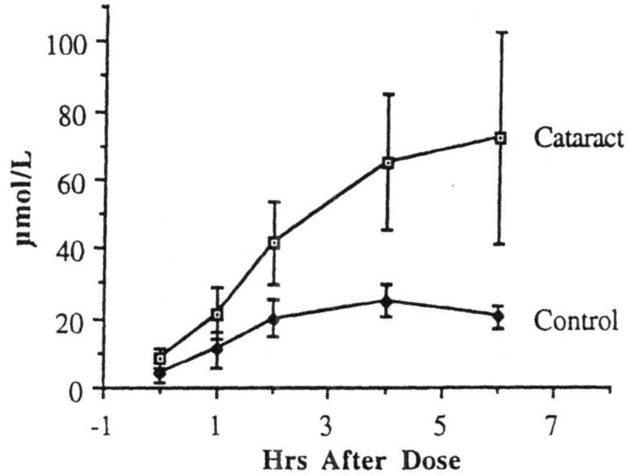


**Figure 2.** Kynurenine response to the tryptophan load.

group and 240 ± 24 µmol/L in the control group. In contrast, there was no significant difference in the mean fasting level or the post-load levels of the free trp in both groups, although a peak a 2 hr was observed in controls and in cataracts.

The mean serum kyn response is shown in Figure 2. The cataract group shows an increasing mean value at 6 hr, with each value higher than the same time point in the control group. The controls peaked at 4 hr with a narrower spread of data. There was no significant effect of sex in either group.

Showing a pattern similar to the kyn response, 5OHA (Figure 3) continued to increase over the 6 hr period with mean concentrations higher at each time point. These concentrations were significantly different from the control group which, unlike the kyn control group, showed an overall flat response.

Figure 4 illustrated the difference between cataract and control groups for the mean response of KA to the trp load. Both response peaked at 4 hr. In contrast to the kyn response the cataract group was lower at each time point. There was also a significantly different response between the males and females within the cataract group.

Of the plasma constituents (30 biochemical and 15 haematological) only serum magnesium, chloride, albumin and the total leucocyte count showed a significant difference

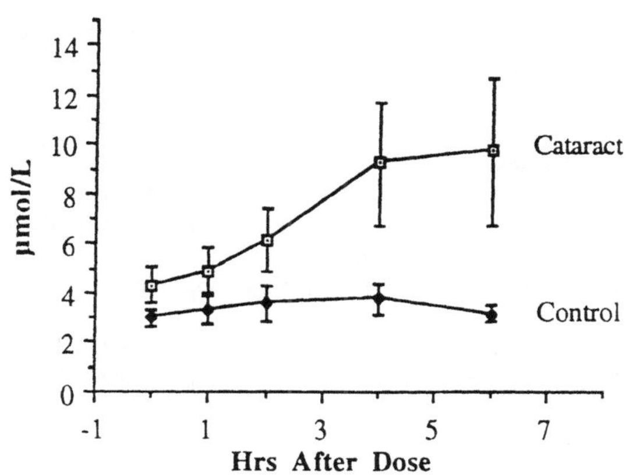


**Figure 3.** 5-Hydroxyanthranilic acid response to the tryptophan load.

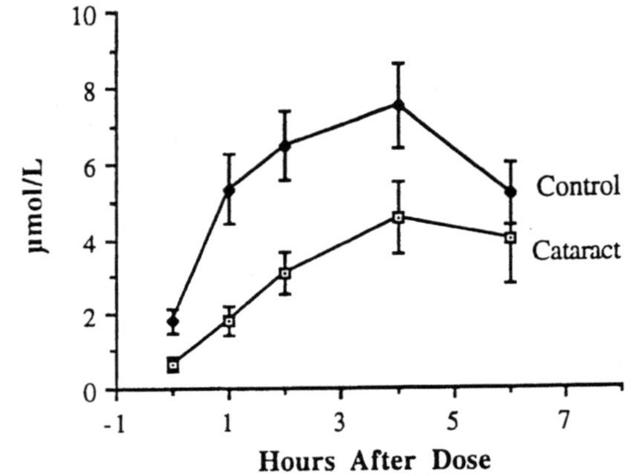


**Figure 4.** Kynurenic acid response to the tryptophan load.

between the two groups. The mean values in both cataracts and controls, however, were not outside normal reference intervals.

The discovery that such a clear difference in somatic trp metabolism in cataracts was surprising. It is highly unlikely that the cataract itself is responsible. It is possible that it may be related to the lenticular UV-filter pathway in primates which is based on the catabolism of trp. (Van Heyningen, 1971; Wood and Truscott, 1993) It is also possible that the two factors, cataract formation and a trp metabolic disturbance occur together as a result of a quite separate endogenous or acquired defect.

## 4. REFERENCES


Allegri G. and Angi M., 1981, Free and total tryptophan in senile cataract. *Lancet* 2:157.

Costa C., Angi M., De Cari M., Vanzan S. and Allegri G., 1982, Metabolism and serum levels of tryptophan in senile cataract patients. *Boll.Soc.It.Biol.Speri.* LVIII:333.

Cotlier E. and Sharma Y., 1980, Plasma tryptophan in senile cataract. *Lancet* 15:607.

Dilley K. and Pirie A., 1974, Changes to the proteins of the human lens nucleus in cataract. *Exp.Eye Res.* 19:59.

Dillon J. and Spector A., 1976, Identification of beta-carbolines isolated from fluorescent human lens proteins. *Nature* 259:422.

Elderfield A.J., Truscott R.J.W., Gan I., Schier G., 1989, A method for the separation of tryptophan metabolites by rp-HPLC with amperometric and fluorescence detection. *J.Chromatog.* 495:71.

Harding J., and Crabbe M., 1984, The Lens: development, proteins, metabolism and cataract *in*: The Eye, H.Davson, ed. Academic Press, Inc. London.

Ogino S. and Ichihara T., 1957, Biochemical studies on cataract: biochemical genesis of senile cataract. *Am.J.Ophthalmol.* 43:754.

Ohrloff C., 1989, Epidemiology of senile cataract. *Dev.Ophthalmol.* 17:1.

Truscott R.J.W. and Augustyn R.C., 1977, Changes in human lens nucleus in cataract. *Exp.Eye Res.* 24:159.

Truscott R.J.W., Faull K. and Augustyn R.C., 1977, The identification of anthranilic acid in proteolytic digests of cataractous lens proteins. *Ophthalmol.Res.* 9:263.

Truscott R.J.W., Pyne S. and Manthey M., 1991, Reactive metabolite hypothesis for human senile cataract. *Lens Eye Tox.Res.* 8:251.

Van Heyningen R., 1971, Fluorescent glucoside in the human lens. *Nature* 230:393.

Wood A. and Truscott R.J.W., 1993, UV filters in the human lens: tryptophan catabolism. *Exp. Eye Res.* 56:317.

# 67

# TRYPTOPHAN AND ATHEROSCLEROSIS

G. Baldo-Enzi,[1] M. R. Baiocchi,[1] A. Bertazzo,[2] C. V. L. Costa,[2] and G. Allegri[2]

[1] Institute of Internal Medicine
[2] Department of Pharmaceutical Sciences
University of Padova
I-35131 Padova, Italy

## INTRODUCTION

The association between hypercholesterolemia and atherosclerosis in humans is widely accepted (Goldstein and Brown, 1977; Steinberg, 1983; Nordoy and Goodnight, 1990). Some studies (Muldoon et al., 1990) have shown that a decrease in serum cholesterol in middle-aged subjects by diet, drugs or both, leads to a significant reduction of the incidence of coronary heart disease, although total mortality is not affected by treatment, while higher mortality due to suicide or violence is observed among people undergoing treatment to lower serum cholesterol concentrations. Therefore, Engelberg (1992) suggested association between serum cholesterol levels, neuronal function and behavioural predisposition. The same author found that anomalies in brain serotonin were associated to reduced control of instinctive impulses.

Heron et al. (1980) demonstrated that, by increasing the content of cholesterol in the synaptic membranes of mouse brain, serotonin receptors increased greatly. Instead, by lowering the cholesterol concentration, the number of serotonin receptors and binding of serotonin to the brain membranes was found.

These observations are compatible with a correlation between cholesterol and serotonin or with the concept that serum cholesterol may influence the function of serotonin in the central nervous system (Engelberg, 1992).

As tryptophan is the precursor of serotonin and is transported in blood mainly bound to albumin (McMenamy and Oncley, 1958), the content of this amino acid and its possible binding to lipoproteins which represent the transport system of cholesterol in blood, were investigated in the sera and lipoprotein fractions ($HDL_2$ and LDL) of healthy women. Total cholesterol, cholesterol bound to LDL and HDL lipoproteins and tryptophan concentrations were also determined in the sera of 16 patients (12 males and 4 females) subjected to surgical endoarteriectomy. Total cholesterol and tryptophan levels were assayed in atheromatous plaques excised from the carotid artery in the same patients.

*Recent Advances in Tryptophan Research,* edited by Graziella Allegri Filippini et al.
Plenum Press, New York, 1996

# MATERIALS AND METHODS

Two groups of subjects were studied:

a. Thirteen women with different levels of total and HDL cholesterol. From this group serum lipoprotein fractions (LDL and $HDL_2$) were obtained after double centrifugation in a preparative Beckman ultracentrifuge to remove albumin and after dialysis to remove salts from each isolated lipoprotein.
b. Sixteen patients (12 males and 4 females) subjected to surgical endoarteriectomy. Atheromatous plaques were excised from carotid arteries and homogenized in Tris-HCl buffer (pH 8.0) in the presence of EDTA and Tween 20. The homogenate was then centrifuged at 37,000 g for 75 min at 8°C in a preparative Beckman ultracentrifuge. The supernatant was dialysed and concentrated by Amicon Centricon 10. HDL lipoproteins were obtained from the sera of these patients after polyanion precipitation (Burstein et al., 1970), and LDL by the Friedewald formula (Friedewald et al., 1972). Total and lipoprotein cholesterol was assayed by the enzymatic method of Allain (Allain et al., 1974) and tryptophan by HPLC according to the method of Costa et al. (1987).

# RESULTS

Table 1 reports the mean values ± S.D. of total serum cholesterol and cholesterol bound to LDL and $HDL_2$ lipoproteins in 13 healthy women. All these subjects showed slight hypercholesterolemia and slightly high total HDL cholesterol levels.

Table 1 also shows tryptophan concentrations in serum and tryptophan bound to lipoprotein fractions. Tryptophan was found in both lipoprotein fractions examined (LDL and $HDL_2$). Comparing the ratio of LDL and $HDL_2$ tryptophan to total serum tryptophan, it appears that the amount bound to both lipoproteins is similar, with a slight prevalence of that bound to LDL. If we consider the ratio between tryptophan bound to each lipoprotein and cholesterol present in the same lipoprotein, it emerges that $HDL_2$ transports about four times more amino acid than LDL.

Table 2 reports the mean levels ± S.D. of total serum cholesterol, LDL and HDL cholesterol, serum tryptophan and cholesterol and tryptophan present in atheromatous plaques excised from 16 carotid arteries. Endoarteriectomized patients did not have hypercholesterolemia, while HDL cholesterol was lower when compared with our reference values from a group of healthy subjects matched for sex and age (54.2±10.1mg/dl).

The serum tryptophan concentrations are in the normal range (1.350±0.488 mg/dl). In the atheromatous plaques, cholesterol values were 12.0±4.1 mg/g of tissue and tryptophan 6.29±2.74 μg/g of tissue.

**Table 1.** Levels (mean values ± S.D.) of cholesterol and tryptophan in serum and in LDL and $HDL_2$ lipoproteins of 13 healthy females

| | C (mg/dl) | Trp (mg/dl) | Lipoprotein-Trp/$Trp_t$ (%) | Lipoprotein-Trp/ lipoprotein C (%) |
|---|---|---|---|---|
| Serum | 233 ± 42 | 1.53 ± 0.27 | | |
| $HDL_T$ | 61.3 ± 10.5 | / | | |
| LDL | 140 ± 33 | 0.025 ± 0.011 | 1.75 ± 0.49 | 0.022 ± 0.010 |
| $HDL_2$ | 28.3 ± 11.0 | 0.023 ± 0.005 | 1.48 ± 0.33 | 0.091 ± 0.039 |

C=cholesterol; $Trp_t$=total tryptophan; $HDL_T$=total HDL lipoproteins.

**Table 2.** Mean levels ± S.D. of total cholesterol, LDL and HDL cholesterol and tryptophan in serum; cholesterol and tryptophan in atheromatous plaques excised from 16 carotid arteries

| | Serum (mg/dl) | Atheromatous plaque (mg/g tissue) |
|---|---|---|
| TC | 191 ± 37 | 12.0 ± 4.1 |
| LDL-C | 129 ± 39 | / |
| HDL-C | 37 ± 9 | / |
| Trp | 1.263 ± 0.4 | 6.29 ± 2.74 μg/g<br>$Trp_p$/Cp 0.057+0.026 % |

TC = total cholesterol; LDL-C = cholesterol bound to LDL lipoprotein; HDL-C = cholesterol bound to HDL lipoprotein; Trp = tryptophan, $Trp_p$=plaque tryptophan, Cp=plaque cholesterol

# DISCUSSION

Many factors contribute to the development of atherosclerosis and to the formation of atheromatous plaques. Serotonin released from platelets present in the endothelial wall may also contribute to the development of plaques.

In this paper we investigated whether tryptophan, the precursor of serotonin, plays a role in the process of atherosclerosis.

One of the most interesting findings that emerged from these preliminary results was that tryptophan is not only present in atheromatous plaques in carotid arteries, but also appears to be bound to LDL and $HDL_2$ lipoprotein fractions. As is known, tryptophan is transported in the blood mostly bound to albumin (80-90%) only a small part being in the free form (10-20%) (McMenamy and Oncley, 1958).

The presence of tryptophan in serum lipoproteins may be interpreted either as an amino acid necessary to the rebuilding of damaged connectival tissue, or, in the case of that transported by the LDL fraction, as a source for serotonin biosynthesis in platelets.

Tryptophan appears to be prevalently transported by $HDL_2$ lipoproteins, as shown by the ratio between tryptophan and cholesterol bound to lipoproteins. This observation suggests the possible role of $HDL_2$ lipoproteins as a system for tryptophan removal, decreasing the availability of this precursor for serotonin biosynthesis.

# REFERENCES


Allain C.C., Poom L.S., Chan C.S., 1974, Enzymatic determination of total serum cholesterol, *Clin.Chem.*, 20: 470.

Burstein M., Scholnick H.R., Morgin R., 1970, Rapid method for the isolation of lipoproteins from human serum by precipitation with polyanions, *J.Lipid Res.*, 11: 583.

Costa C., Bettero A., Allegri G., 1987, Rapid measurements of tryptophan, 5-hydroxytrytophan, serotonin and 5-hydroxyindoleacetic acid in human CSF by selective fluorescence and HPLC, *Giorn. It.Chim.Clin.*, 12: 307.

Engelberg H., 1992, Low serum cholesterol and suicide, *Lancet*, 339: 727.

Friedewald W.T., Levy R.I., Fredrickson D.S., 1972, Estimation of plasma low density lipoproteins cholesterol concentration without use of the preparative ultracentrifuge, *Clin.Chem.* 18: 499.

Goldstein J.L., Brown M.S., 1977, Low density lipoprotein pathway and its relations to atherosclerosis, *Annu. Rev. Biochem.*, 46: 897.

Heron D.S., Shinitzky M., Herrshkowitz M., and Samuel D., 1980, Lipid fluidity markedly modulates the binding of serotonin to mouse brain membranes, *Proc.Natl.Acad.Sci.* USA, 77: 7463.

McMenamy R.H., Oncley J.L., 1958, The specific binding of tryptophan to serum albumin, *J.Biol.Chem.*, 233:1436.

Muldoon M.F., Manuck. S.B, Mathews K.M., 1990, Lowering cholesterol concentrations and mortality: a quantitative review of primary prevention trial, *Br. Med. J.*, 301: 309.

Nordoy A, Goodnight S.H., 1990, Dietary lipids and thrombosis, Relationship to atherosclerosis, *Arteriosclerosis*,10: 149.

Steinberg D., 1983, Lipoproteins and atherosclerosis. A look back and a look ahead, *Arteriosclerosis, 3: 283.*

# VARIATIONS OF FREE AND TOTAL TRYPTOPHAN SERUM LEVELS AFTER GN-RH ADMINISTRATION IN PATIENTS WITH AMENHORREA

## Preliminary Results

N. Carretti,[1] C. V. L. Costa,[2] and A. Bertazzo[2]

[1] Istituto di Clinica Ostetrica e Ginecologica dell'Università di Siena
Siena, Italy
[2] Dipartimento di Scienze Farmaceutiche dell'Università di Padova
Padova, Italy

## INTRODUCTION

Tryptophan is present in the serum in two forms: free (FT) and linked to the proteins, total tryptophan (TT) comprises both fractions. The fraction FT which is detached continuously from the proteinic link is able to pass the hemato-encephalic barrier and is utilized as a precursor for the formation of cerebral serotonin ( McMenamy et al., 1957; Knott and Curzon, 1972).

In the experimental animals the levels of FT in the serum reflect the quantity of serotonin present in the cerebral tissue (Tagliamonte et al., 1973)

In women the concentration of the receptors of the serotonin in the brain undergoes regular modifications during the menstrual cycle (Biegon et al., 1980).

In 15 normal women in reproductive age it was seen that FT and TT vary during the menstrual cycle and present a marked diminution at midcycle preceded and followed by two peaks which are of the same level for FT while, as far as TT is concerned, post-ovulatory peak is about 30% higher than pre-ovulatory one (Carretti et al., 1983 a).

Due to this behaviour there is an inverse elevated correlation between TT and FT with FSH-LH (respectively :-0.655-0.636/ -0.663-0.574). This correlation is complitely lost after menopause (Carretti et al., 1983 b).

In this study it was our aim to see if any modification of the levels of TT and FT might be present in amenhorroic girls under stimulation of Gn-RH (gonadotropin releasing hormone).

In two amenorrhoic women pulsatile modifications of FT and TT levels are described.

*Recent Advances in Tryptophan Research*, edited by Graziella Allegri Filippini et al.
Plenum Press, New York, 1996

# METHODS

Serum levels of TT, FT, gonadotropin LH and FSH were assayed every 20 minutes (0-20-40-60-80-100) until i.v. administration of 50 μg of Gn-RH (Gonadotropin releasing hormone) after 100 minutes from time 0.

Afterwards determinations were done every 15 minutes for four times (at minutes 115-130-145-160) and every 30 minutes the last two times (minutes 190 and 220). Plasma levels of hormones were measured by RIA in the Laboratory of Gynecologiyc Clinic of Siena University. Tryptophan levels were measured in the serum after ultrafiltration in the Department of Pharmaceutical Sciences of Padua University according to methods reported elsewhere (De Antoni et al., 1980; Carretti et al., 1983a).

Graphs and statistic were done with PC program.

Two cases of women with amenorrhea are reported in this study.

## Cinical History

*B.M.* 19 y, 1.68 m tall and weighs 49 Kg. Her menarche began at age 15 and her cycles have been irregular since that time, occurring approximately every 2-3 months. She describes problems with acne and has been treated by a dermatologist. Treated with estro-progestins birth control low disage pill for 1 year for "hormonal imbalance". She is used to voluntary dietary restrictions and has lost about 5 kg of weight during the assumption of the pill. At the suspension of this treatment secondary amenorrhea for about one year. At ecography diagnosis of PCO (Poli-Cystic-Ovary) and enlarged right ovary. Basal HPrl=19ng/ml.

*S.E.* 17 y, 1.64 tall, and weights 51 Kg. Menarche at 14 with cycles delayed (every 40-50 days) and scanty. Voluntary dietary restrictions with weight loss of about 10Kg at 16 years when she started to assume estroprogestins low dosage for 5-6 months. At the suspension of this treatment amenorrhea followed by a new assumption of pill for 1 month. Second episode of amenorrhea after the end of this treatment. At echography PCO like aspect of both ovaries. Basal HPrl = 5.9 ng/ml.

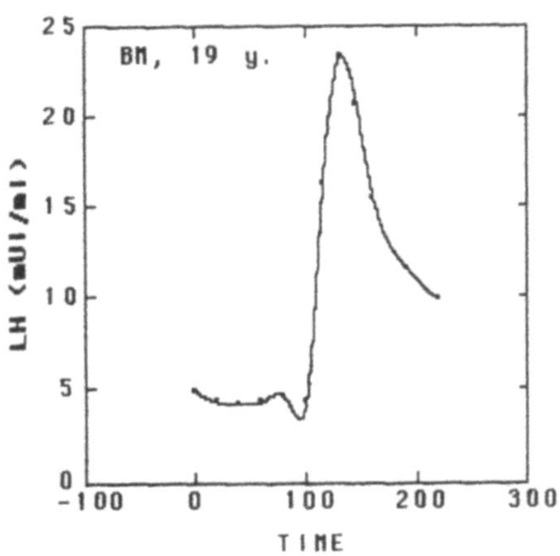


**Figure 1.** Level of LH values before and after GnRH stimulus drowed with interpolation spline curve in patient BM.

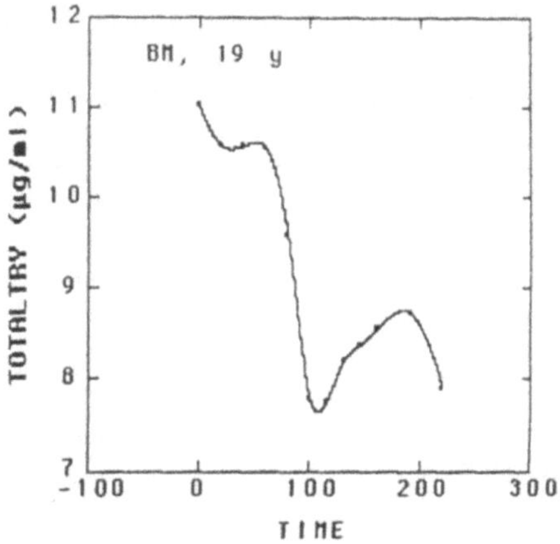


**Figure 2.** Level of Total Tryptophan values before and after GnRH stimulus drowed with interpolation spline curve in patient BM.

## RESULTS

1. *Interpolation Curves of the Mean Values:* The behaviour of LH, TT and FT before and after the administration of Gn-RH is reported in Fig. 1, 2, 3 in patients BM. The elevation of LH follows immediately the administration of Gn-RH, as it was expected. The shapes of TT and FT are described for the first time.

a. The descend of TT after Gn-RH is simultaneous to the elevation of LH and is about 30% of its initial value in 15 minutes. This lower TT level persists for 30 minutes and then shows a slight elevation but, in the remaining period of observation doesn't reach again its initial value.
b. Free tryptophan (FT) levels show a series of picks and decrements resembling a pulsatile behaviour. This behaviour was not evident for TT. If one makes equal to 10 the maximum values of TT and FT it can be seen that more FT is liberated after

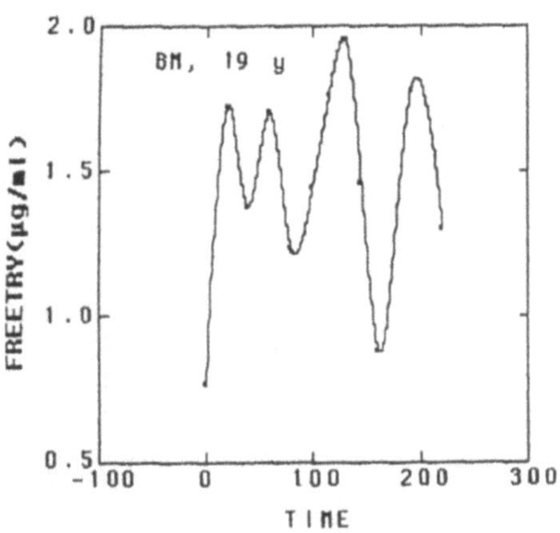


**Figure 3.** Level of Free Tryptophan values before and after GnRH stimulus drowed with interpolation spline curve in patient BM.

**Table 1.** Pearson's correlations matrix and probability matrix of the mean values of total tryptophan (TT) and LH of two amenorrhoic patients at time (in minutes: 0-20-40-60-80-100) before and after Gn-RH stimulation, against time: 115-130-145-160-190- 220 minutes

Number of observed: 12

| | Time | TT | LH |
|---|---|---|---|
| Time | 1.000 | | |
| TT | –0.812 | 1.000 | |
| LH | 0.580 | –0.641 | 1.000 |

Bartlett chi-square statistic: 14.896 DF= 3PROB=0.002
Matrix of probabilities

| | Time | TT | LH |
|---|---|---|---|
| Time | 0.000 | | |
| TT | 0.001 | 0.000 | |
| LH | 0.048 | 0.025 | 0.000 |

Gn-RH stimulus then before. The curve of FT shape overcome that of TT after Gn-RH. Picks after Gn-RH administration are higher then picks before Gn-RH stimulus.

The behaviour illustrated for patients BM is similar in patient S.E.

2. *Analysis of Correlations:* After 24 determinations (12 for each patient) of the two patients in exam, TT and LH levels shown an inverse (-0.641) significant (p 0.05) correlation.

The correlation between time, positive for LH and negative for TT are expression of the inverse link between these two variables both in dependence to the lapse of time. Correlation between time and TT is very elevated and highly significant. (Table 1).

3. By scatterplots matrix of the mean of TT, FT in the two patients and of the two LH determinations it is possible to have an idea of the entire phenomenon. (Fig. 4).

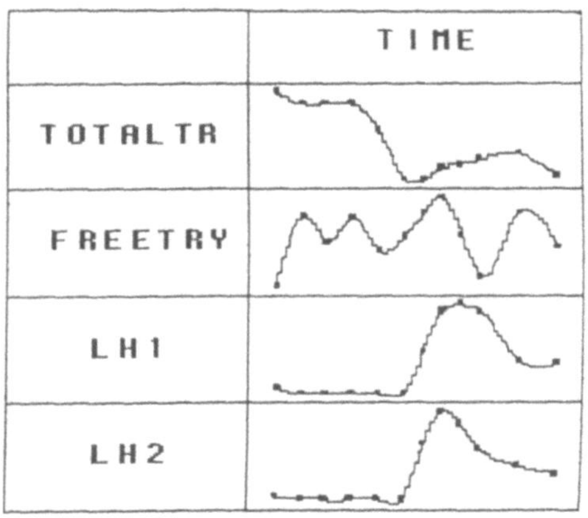


**Figure 4.** SPLOM (Scatterplot matrix) of the mean of tryptophan serum levels (TT/FT) and the LH behaviour before and after Gn-RH administration.

# DISCUSSION

This preliminary observation shows that tryptophan levels present short period modifications.

It is already known that TT and FT levels have an inverse significant correlation with gonadotrophin hormones during menstrual cycle. In particular during ovulatory phase there is an augmentation of LH/FSH levels and a decrease of TT and FT levels (Carretti et al., 1983a).

This study gives evidence that such a relationship is present following LH/FSH release after Gn-RH administration.

Correlation between TT and LH levels seem strictly time dependent. Another point is that FT presents a pulsatile aspects before and after Gn-RH administration. This aspect resembles that of LH.

In addition if we make equal to 10 the mx levels of TT and FT it can be seen that in both patients, after Gn-RH administration, FT curves and picks overcome those of TT: it looks like as more FT is liberated.

Since FT in the serum is the mirror of serotonin in the brain these data are in favour of the idea that there are hypothalamic sites of 5-HT action on gonadotrophin release (Ulrich et al., 1994).

Following this release a consumption of FT might take place derivated from TT which shows a diminution of its serum levels.

# REFERENCES

Biegon, A., Bercovitz, H., Samuel, D., 1980, Serotonin receptor concentration during the estrous cycle of the rat, *Brain Res.*, 187: 221.

Carretti, N., Bertolin, A., Costa, C., Allegri, G., Zilli, G., 1983a, Modifications of levels of total and free tryptophan during the menstrual cycle and their correlations with the hormonal levels of FSH, LH and $E_2$: a mathematical study with multivariate analysis, cluster and interpolation curves, in: Neuropeptides, neurotransmitters and regulation of endocrine processess, Endroczi et al.eds., Akademiai Kiado, Budapest, pp.527-534.

Carretti, N., Bertolin, A., Fiore, R., Allegri, G., Costa, C., 1983b, Variazioni ematiche dei livelli di triptofano totale e libero in menopausa, Atti LXII Congr.SIGO, Bologna 21-24 Settembre. pp.829-836.

De Antoni, A., Allegri, G., Costa, C., Vanzan, S., Bertolin, A., Carretti, N., Zanardo, V., 1980, Total and free tryptophan levels in serum of newborn infants. Relationships with the serotonin and nicotinic acid pathways,*Acta Vitamin. Enzymol.* 2: 17.

Knott, P.J. and Curzon, G., 1972, Free tryptophan in plasma and brain tryptophan metabolism, *Nature*, 239: 452.

McMenamy, R.H., Lund, C.C., Oncley, J.L., 1957, Unbound aminoacid concentrations in human blood plasma, *J.Clin.Invest.*, 36: 1672.

Tagliamonte, A., Biggio, G., Vargiu, L., Gessa, G.L., 1973, Free tryptophan in serum controls brain tryptophan level and serotonin synthesis, *Life Sci.*, 12: 277.

Ulrich, U., Nowara, I., Rossmanith, W.G., 1994, Serotoninergic control of gonadotrophin and prolactin secretion in women, *Clin.Endocrinol.*, 41: 779.

# *ENZYMES IN TRYPTOPHAN DEGRADATION*

# A TRYPTOPHAN PYRROLE-RING CLEAVAGE ENZYME IN THE MOST PRIMITIVE EUKARYOTE

Y. Iwamoto,[1] I. S. Matsui Lee,[1] M. Tsubaki,[2] and R. Kido[1]

[1] Department of Biochemistry
Wakayama Medical College
27 Kyubancho, Wakayama 640, Japan
[2] Department of Life Science, Faculty of Science
Himeji Institute of Technology
Kanaji 1479-1, Kamigoori-cho, Akou-gun, Hyogo 678-12, Japan

## INTRODUCTION

Two types of dioxygenases that catalyze the oxidative cleavage of the pyrrole ring of tryptophan by insertion of molecular oxygen to yield *N*-formylkynurenine have been reported (Feigelson and Brady, 1974). One is tryptophan 2,3-dioxygenase (TDO), and the other is indoleamine 2,3-dioxygenase (IDO). Although protoheme IX is a sole prosthetic group for both dioxygenases (Feigelson and Brady, 1974; Hirata and Hayaishi, 1975; Ishimura et al., 1980), these have proved to be distinct enzymes. TDO is a tetrameric protein (M.W. 120,000-167,000) metabolizing L-tryptophan specifically (Feigelson and Brady, 1974). This dioxygenase requires nonspecific reductants such as L-ascorbic acid for its activation *in vitro* (Feigelson and Brady, 1974). IDO is a monomeric protein (M.W. 41,000) (Shimizu et al., 1978) exhibiting a wide substrate specificity for various indoleamine derivatives including L-and D-tryptophan and serotonin (Ishimura et al., 1970). This enzyme can be also activated by a wide variety of reductants. In addition, methylene blue (or toluidine blue) is absolutely required as an electron mediator from a reductant to the ferric enzyme for its activation (Yamamoto and Hayaishi, 1967).

It is interesting to note the difference in distribution between the two dioxygenases. In mammals, TDO seems to exist exclusively in liver (Knox, 1955). On the other hand, IDO is not present in liver, but ubiquitous in extrahepatic tissues (Yamazaki et al., 1985). TDO has been also found in *Psudomonas acidovorans* (Feigelson and Brady, 1974), a prokaryotic cell. IDO has not been found in organisms other than mammals. Such a difference in distribution suggests a functional distinction between TDO and IDO. We have been interested in whether the tryptophan pyrrole-ring cleavage enzyme (TPCE) in yeast, a eukaryotic uicellular organism, is TDO or IDO.

*Recent Advances in Tryptophan Research*, edited by Graziella Allegri Filippini et al.
Plenum Press, New York, 1996

Cleavage of the tryptophan pyrrole ring is the initial step in the pathway for metabolism from tryptophan to NAD. In addition to its function as a cofactor in oxidation-reduction reactions, NAD is a well-known substrate in the ADP-ribosylation reactions (Ueda and Hayaishi, 1985) and in the synthesis of cADP-ribose (Galione, 1992), which mobilizes calcium from intracellular calcium pools (Dargie et al., 1990; Koshiyama et al., 1991). ADP-ribosylation (Shall, 1985) and intracellular calcium mobilization by cADP-ribose (Takasawa et al., 1993) have proved to be related to various cellular events in eukaryotic cells. Therefore, it is very important to take note of tryptophan as a metabolic source of NAD. In the present study, we considered *Saccharomyces cerevisiae* as a model for eukaryotic cells and investigated its TPCE. We purified the enzyme partially and identified it as TDO by its enzymatic characteristics.

## MATERIALS AND METHODS

### Yeast Strain and Culture

*Saccharomyces cerevisiae* (strain A553; **a** *leu* 2 *ura* 3) was cultured in the YPD medium [glucose (20 g/l) / polypeptone (20 g/l) / yeast extract (10 g/l)], with or without 0.1 % (w/v) L-tryptophan, with continuous agitation at 28°C. The cells, grown until the stationary phase, were used for the present study.

### Enzyme Assay

TPCE activity was assayed according to the method of Yamazaki et al. (1985) with some modifications.

### Induction of TPCE by L-Tryptophan

Five ml preculture of *Saccharomyces cerevisiae* in the YPD medium was added to 2 l of the YPD medium containing 0.1 % L-tryptophan, and incubated with continuous agitation at 28°C. An L-tryptophan concentration of 0.1 % was chosen because it afforded maximal induction of the enzyme activity in *Pseudomonas acidovorans* (Feigelson et al., 1965). Control cells were cultured in the YPD medium but without added L-tryptophan under the same condition. The cells, grown until the stationary phase, were harvested by centrifugation, and homogenized. The homogenizing and column-chromatographic procedures were carried out as will be described in *Partial purification of TPCE*. The eluate from a mini-column of phospho-cellulose (gel volume: 1 ml) was immediately assayed for enzyme activity, and subjected to two-dimensional polyacrylamide gel electrophoresis.

### Two-Dimensional Polyacrylamide Gel Electrophoresis

Two-dimensional polyacrylamide gel electrophoresis was performed using a Shimadzu, Model TEP-2, instrument as described previously (Nokihara et al., 1992) with some modifications. The isoelectric focusing gel contained 4 % Pharmalyte 3-10 and 2 % Pharmalyte 5-8. The SDS polyacrylamide gel electrphoresis was carried out on 10 % polyacrylamide gel.

## Partial Purification of TPCE

The TPCE was partially purified from *Saccharomyces cerevisiae* cells grown in the YPD medium containing 0.1 % L-tryptophan. All manipulations were performed at 0-4°C. Cells [22 g (wet weight)] were suspended in 1.5 volume of 50 mM potassium phosphate buffer (pH 6.5) containing 0.2 mM EDTA, 50 mM 2-mercaptoethanol, and 100 μg/ml PMSF. The cell suspension, containing 2 g of cells, was poured into the glass tube containing 5 g of glass beads. Homogenization was carried out by vortexing the glass tube with 7 cycles of 10-sec beating and 150-sec sitting. The homogenate was centrifuged at 10,000 x g for 30 min. The supernatant (34 ml) was diluted 2.5-fold with 0.2 mM EDTA containing 10 mM 2-mercaptoethanol and 10 μg/ml PMSF to give a potassium phosphate concentration of 20 mM, and loaded onto a preparative column of phospho-cellulose (gel volume: 17 ml, 1.3 x 13 cm) previously equilibrated with 20 mM potassium phosphate buffer (pH 6.0) containing 0.2 mM EDTA, 10 mM 2-mercaptoethanol, and 10 μg/ml PMSF. The column was washed with five column-volumes of 20 mM potassium phosphate buffer (pH 6.0) containing 0.2 mM EDTA, 1 mM L-tryptophan, 20 % (v/v) glycerol, 0.01 % (v/v) Triton X-100, 10 mM 2-mercaptoethanol, and 10 μg/ml PMSF. The TPCE was then eluted with the washing buffer containing 0.3 M NaCl. The TPCE-rich fractions were collected, and immediately used for the enzymatic characterizations.

## RESULTS AND DISCUSSION

Culturing of *Saccharomyces cerevisiae* cells in the YPD medium containing L-tryptophan resulted in a drastic induction of the TPCE activity. The TPCE specific activities (pmol kynurenine/min/mg protein) of the elutes from mini-columns of phospho-cellulose were compared. The eluate from the cells grown in the presence of 0.1 % L-tryptophan exhibited the specific activity three times as high as that from the control cells (Fig. 1). The same samples were also analyzed by two-dimensional polyacrylamide gel electrophoresis. We found five major protein spots represented as a, b, c, d, and e in the gel. As shown in Fig. 2, the intensity of the protein spot e increased strikingly by addition of L-tryptophan.

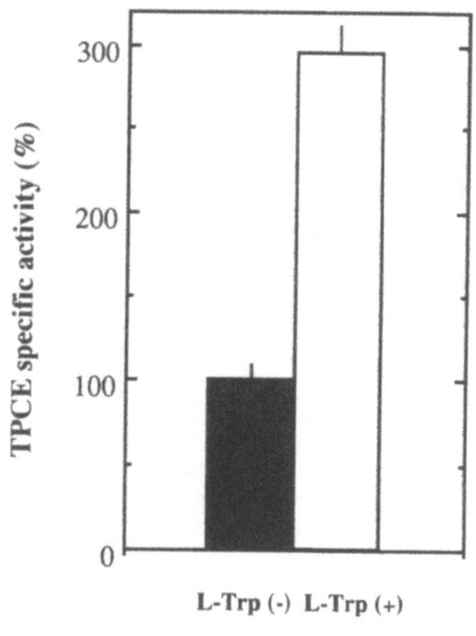


**Figure 1.** Induction of the yeast TPCE specific activity by L-tryptophan. The TPCE activity of the phospho-cellulose mini-column eluate from the control (solid bar) or the L-tryptophan-induced (open bar) cells was assayed with L-tryptophan as a substrate in the presence of 20 mM L-ascorbic acid and 25 μM methylene blue. The induction of the TPCE specific activity was expressed as a percentage of the value obtained from the control cells.

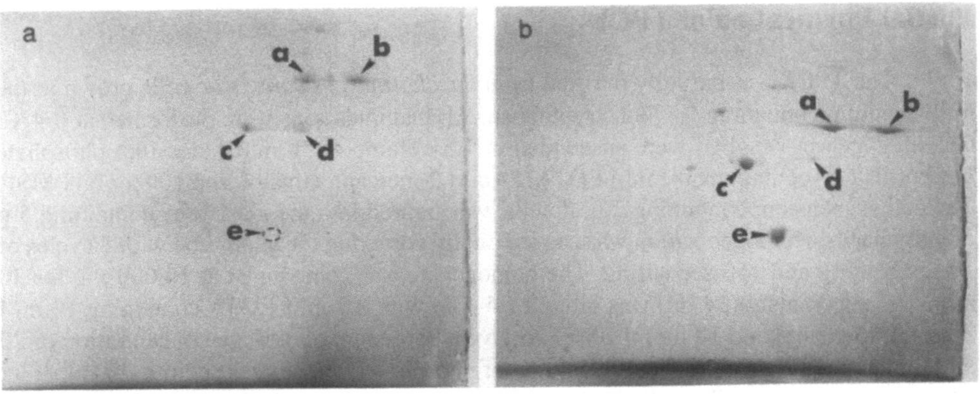


**Figure 2.** Two-dimensional polyacrylamide gel electrophoretic analyses of the eluates from the phospho-cellulose mini-column. The eluate (80 μg of protein) prepared from the control (a) or the L-tryptophan-induced cells (b) was applied. The proteins were separated first by isoelectric focusing in the horizontal dimension (the acidic end is on the left) and then by SDS-polyacrylamide gel electrophoresis in the vertical dimension.

This protein spot e corresponds to a molecular weight of approx. 38,000 and an isoelectric point of approx. pH 8.0. Thus, we used *Saccharomyces cerevisiae* cells cultured in the presence of 0.1 % L-tryptophan and started to purify the TPCE, taking aim at the polypeptide with a molecular weight of approx. 38,000.

The TPCE was partially purified by a preparative phospho-cellulose column chromatography. A major peak of the TPCE activity (peak I) was eluted from the column, followed by another minor peak (peak II) (Fig. 3). The TPCE activity of the peak II fraction was quite labile. Although the TPCE activity of the peak I fraction was much stabler than that of the peak II fraction, a 65 % decrease in its activity was observed after a cycle of freeze-thawing. We, therefore, used the peak I fraction for the enzymatic characterizations immediately after the column chromatography. The SDS-polyacrylamide gel electrophoretic pattern of the peak I fraction is presented in Fig. 3. The intensity of the protein-staining band with a molecular weight of approx. 38,000 (indicated by an arrow) correlated well with the TPCE activities of the peak I fractions (data not shown). The TPCE was purified 12-fold with an activity yield of 18 %. The final sample exhibited a specific activity of 14.2 pmol of kynurenine/min/mg protein.

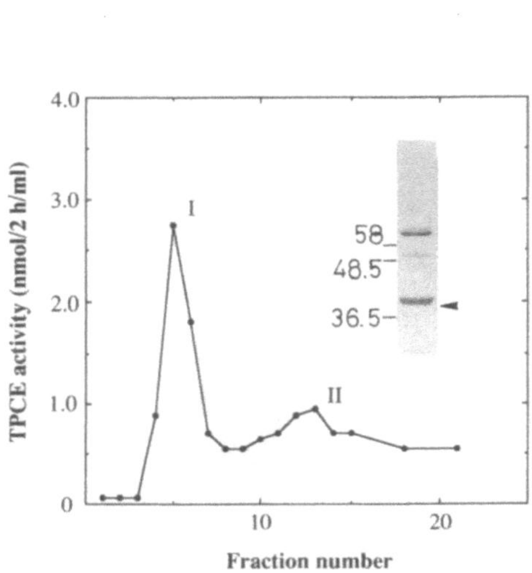


**Figure 3.** Elution profile of the preparative phospho-cellulose column chromatography. Fractions (5 ml each) were collected, and the TPCE activity was assayed with L-tryptophan as a substrate in the presence of 20 mM L-ascorbic acid and 25 μM methylene blue. I and II indicate the peaks of the enzymatic activity. The SDS-polyacrylamide gel electrophoretic pattern of the peak I fraction is also presented. 30 μg of protein was applied to the SDS-polyacrylamide gel.

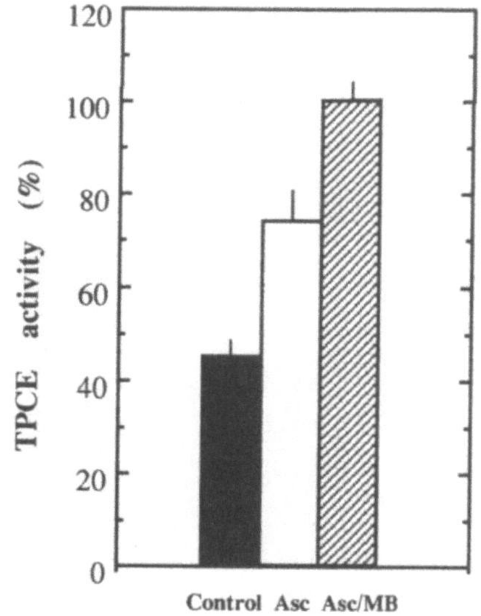


**Figure 4.** Cofactors-dependency of the yeast TPCE activity. The TPCE activity was assayed with L-tryptophan as a substrate in the absence of L-ascorbic acid and methylene blue (solid bar), or in the presence of 20 mM L-ascorbic acid solely (open bar) or together with 25 μM methylene blue (striped bar).

The partially purified enzyme activity was stimulated solely by L-ascorbic acid (Fig. 4), suggesting that the yeast TPCE is a TDO-type dioxygenase. Hereafter we call the yeast TPCE the yeast TDO. Addition of methylene blue in combination with L-ascorbic acid led to the further activation, indicating that methylene blue works as an electron mediator from L-ascorbic acid.

As shown in Fig. 5, the yeast enzyme cleaved L-tryptophan preferentially and D-tryptophan slightly, while TDO in mammalian liver or *Pseudomonas acidovorans* has been reported to metabolize L-tryptophan, but not D-tryptophan at all (Feigelson and Brady, 1974). The tryptophan-cleaving activity for D-tryptophan was only 14 % of that for L-tryptophan.

The heme exogenous ligands such as KCN and $NaN_3$ inhibited the yeast enzyme activity remarkably (Fig. 6). Addition of KCN inactivated the enzyme in a concentration-

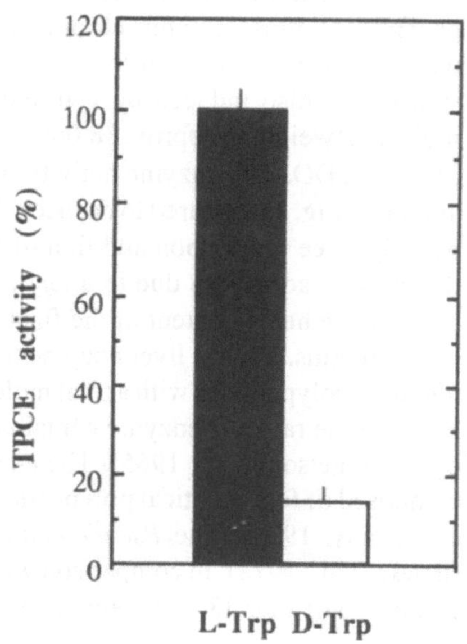


**Figure 5.** Substrate specificity of the yeast TPCE. The TPCE was eluted from the phospho-cellulose column with buffer without L-tryptophan, and assayed with L-tryptophan (solid bar) or D-tryptophan (open bar) as substrate in the presence of 20 mM L-ascorbic acid and 25 μM methylene blue.

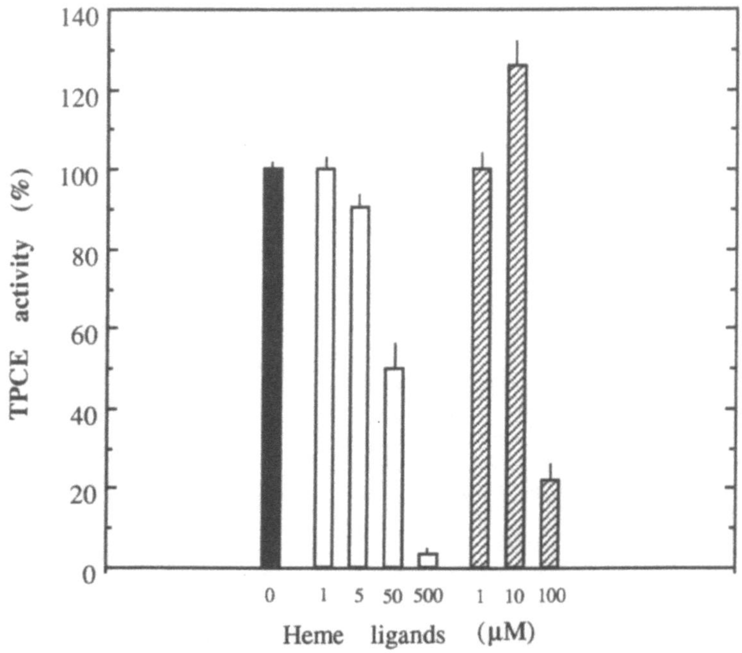


**Figure 6.** Inhibition of the yeast TPCE activity by heme exogenous ligands. The TPCE activity was assayed with L-tryptophan as substrate in the presence of 20 mM L-ascorbic acid and 25 μM methylene blue. The enzyme was incubated without heme ligand (solid bar), or with KCN (open bar) or $NaN_3$ (hatched bar).

dependent manner, and 500 μM KCN caused an almost complete inactivation. 100 μM $NaN_3$ inhibited 78 % of the original activity. These suggest that yeast TDO containes a heme(s) in its active site like other TDOs reported previously (Feigelson and Brady, 1974).

The optimum pH of yeast TDO was pH 6.5 (data not shown). This value is lower than the value for the TDOs from rat liver or *Pseudomonas acidovorans* (7.2 - 7.4).

Two-dimensional polyacrylamide gel electrophoresis of the peak I fractions after the phospho-cellulose column chromatography demonstrated the existence of a protein spot whose intensity changes in correlation with the TDO activities (data not shown). This protein spot E was also induced by L-tryptophan (Fig. 2). We consider that this protein, with a molecular weight of approx. 38,000 and an isoelectric point of approx. 8.0, might be a subunit of yeast TDO. The enzyme activity was so labile that the induction of this protein in high amounts (Fig. 2) appeared to be inconsistent with the finding that the TPCE activity increased by only three times upon addition of L-tryptophan (Fig. 1). We think that the instability of the enzyme activity is due to a tendency toward dissociation of heme(s) from the enzyme because the heme content of the final sample was too low to obtain the spectrum typical of hemoproteins. The rat liver enzyme has proved to be a tetrameric protein consisting of four identical polypeptides with a total molecular weight of 167,000 (Feigelson and Brady, 1974). Because the rat liver enzyme contains two protoheme IX per molecule, it is represented as $\alpha_2\beta_2$ (Feigelson et al., 1965). The enzyme of *Pseudomonas acidovorans* is also a tetramer composed of four identical polypeptides with a total molecular weight of 122,000 (Feigelson and Brady, 1974). The *Pseudomonas* enzyme contains two protoheme IX per molecule (Brady et al., 1972). In comparison with these TDOs, yeast TDO may be a tetrameric protein composed of four 38,000 identical polypeptides.

We consider *Saccharomyces cerevisiae* to be a good model for eukaryotic cells to investigate the role of TDO in tryptophan metabolism. Our present study represents the first step in elucidating the relationship between tryptophan metabolism and regulation of cellular functions.

## REFERENCES


Brady, F.O., Monaco, M.E., Forman, H.J., Schutz, G., and Feigelson, P., 1972, On the role of copper in activation of and catalysis by tryptophan-2,3-dioxygenase. 4. Indoleamine 2,3-dioxygenase, purification and some properties. *J. Biol. Chem.* 247: 7915.

Dargie, P.J., Agre, M.C., and Lee, H.C., 1990, Comparison of calcium mobilizing activities of cyclic ADP-ribose and inositol trisphosphate. *Cell Regul.* 1: 279.

Feigelson, P., Ishimura, Y., and Hayaishi, O., 1965, Studies on the role of hematin in the catalytic mechanism of tryptophan pyrrolase. *Biochim. Biophys. Acta* 96: 283.

Feigelson, P., and Brady, F.O., 1974, in *Molecular mechanisms of oxygen activation* (Hayaishi, O., ed.) pp.84-133, Academic Press, New York

Galione, A., 1992, Calcium-induced $Ca^{2+}$ release and its modulation by cyclic ADP-ribose. *Trends Pharmacol. Sci.* 13: 304.

Hirata, F., and Hayaishi, O., 1975, Studies on indoleamine 2,3-dioxygenase. 1 Superoxide anion as substrate. *J. Biol. Chem.* 250: 5960.

Ishimura, Y., Nozaki, M., and Hayaishi, O., 1970, The oxygenated form of L-tryptophan 2,3-dioxygenase as reaction intermediate., *J. Biol. Chem.* 245: 3593.

Ishimura, Y., Makino, R., Ueno, R., Sakaguchi, K., Brady, F.O., Feigelson, P., Aisen, P., and Hayaishi, O., 1980, Copper is not essential for the catalytic activity of L-tryptophan 2,3-dioxygenase, *J. Biol. Chem.* 255: 3835.

Knox, W.E., 1955, Tryptophan oxidation. *Methods Enzymol.* 2: 242.

Koshiyama, H., Lee, H.C., and Tashjian, A.H., 1991, Novel mechanism of intracellular calcium release in pituitary cells. *J. Biol. Chem.* 266: 16985.

Nokihara, K., Morita, N., and Kuriki, T., 1992, Applications of an automated apparatus for two-dimensional electrophoresis, Model TEP-1, for microsequence analyses of proteins. *Electrophoresis* 13: 701.

Shall, S., 1985, ADP-ribosylation as a cellular control mechanism, in: ADP-ribosylation of proteins (Althaus, F.R., Hilz, H., and Shall, S., ed.) pp.9-29, Springer-Verlag, Berlin

Shimizu, T., Nomiyama, S., Hirata, F., and Hayaishi, O., 1978, Indoleamine 2,3-dioxygenase, purification and some properties. *J. Biol. Chem.* 253: 4700.

Takasawa, S., Nata, K., Yonekura, H., and Okamoto, H., 1993, Cyclic ADP-ribose in insulin secretion from pancreatic β cells. *Science* (Washington, D.C.) 259: 370.

Ueda, K., and Hayaishi, O., 1985, ADP-ribosylation. *Annu. Rev. Biochem.* 54, 73.

Yamamoto, S., and Hayaishi, O., 1967, Tryptophan pyrrolase of rabbit intestine. *J. Biol. Chem.* 242: 5260.

Yamazaki, F., Kuroiwa, T., Takikawa, O., and Kido, R., 1985, Human indolylamine 2,3-doxygenase, its tissue distribution, and characterization of the placental enzyme. *Biochem. J.* 230: 635.

# CLONING OF RAT AND HUMAN KYNURENINE AMINOTRANSFERASE

M. Mosca,[1] L. Cozzi,[1] J. Breton,[1] N. Avanzi,[1] S. Toma,[1] E. Okuno,[3] R. Schwarcz,[4] C. Speciale,[2] S. Magagnin,[2] M. Mostardini,[2] and L. Benatti[2]

[1] BioScience Center
[2] R&D CNS
Pharmacia
Nerviano, Italy
[3] Wakayama Medical College, Department of Biochemistry
Wakayama, Japan
[4] Maryland Psychiatric Research Center, University of Maryland School of Medicine
Baltimore, Maryland

## 1. INTRODUCTION

Kynurenic acid (KYNA), a normal component of the mammalian brain, has recently gained attention as a broad spectrum antagonist of ionotropic excitatory amino acid receptors (Stone et al., 1993). While in mammalian peripheral organs, KYNA is biosynthesized from L-kynurenine by several rather unspecific aminotransferases (Noguchi et al., 1975, Okuno et al., 1980), it appears that a single enzyme, kynurenine aminotransferase (KAT), is responsible for KYNA synthesis in the rat brain (Okuno et al., 1991a). In the human brain, on the other hand, two enzymes (KAT I and KAT II) have been characterized (Okuno et al., 1991b). These two KATs have distinct catalytic characteristics and can be physically separated by conventional techniques.

We previously reported (Mosca et al., 1994) the cloning of a cDNA encoding cytosolic rat kidney KAT. In this study we describe the properties of recombinant rat cytosolic KAT and the cloning of a putative mitochondrial isoform of rat KAT. Moreover, we report the cloning and full length sequence of a cDNA coding for human brain KAT.

## 2. METHODS

### 2.1 Recombinant Activities in Cos-1 Cells

The expression plasmid encoding rat cytosolic KAT was constructed as described (Mosca et al., 1994). Cos-1 cells were transfected with 10 μg of plasmid by the calcium

*Recent Advances in Tryptophan Research*, edited by Graziella Allegri Filippini et al.
Plenum Press, New York, 1996

phosphate method (Sambrook et al. 1989). 72 hours after transfection, cells were disrupted by freezing and thawing. After centrifugation (15,000 g for 30 min), the supernatant was used for the determination of the enzymatic activities. KAT and glutamine transaminase K activities were determined as described (Mosca et al., 1994).

The β-lyase assay was a coupled assay as described by Abraham and Cooper (1991). The product of the β-lyase reaction (pyruvate) was assayed by measuring the oxidation of NADH during the transformation of pyruvate to lactate catalyzed by alanine dehydrogenase. The reaction mixture (200 µl in a microtiter plate) contained 2 mM S-(1,2-dichlorovinyl)-L-cysteine (DCVC), 0.5 mM MTB, 0.1 mM PLP and the enzyme in 100 mM Tris buffer pH 8.8, and was incubated at 37°C for 5, 10, 15 min prior the addition of 0.3 mM NADH, 7.3 U/ml alanine dehydrogenase, and 0.1 M ammonium acetate. Absorbance at 340 nm was measured using a microplate reader (Cerves uv900) and the NADH concentration was calculated using an $\varepsilon = 4200\ M^{-1}\ cm^{-1}$ .

## 2.2 in Vitro Transcription and Translation

The expression plasmids containing the full length sequence of the two cDNAs encoding cKAT and +14KAT were constructed as follows. PCR amplification was performed using three specific oligonucleotides with XhoI linkers. The two sense orientation oligonucleotides were design in order to amplify the entire 5′-untranslated region of both cDNAs, starting from nucleotide -618 (+14KAT: GATCCTCGAGGGGCGACTCTAGATTTTT) and -243 (cKAT: GATCCTCGAGG GGAAACTCCACCCAGAGGT). The antisense-orientation oligonucleotide was common to the two forms and complemented the sequence starting from the end of the 3′-untranslated region (GCTACTCGAGAGGAACAGTAGCAA-CATCC). The full length cDNAs obtained by PCR were verified by sequencing and cloned into pBluescript SK (Stratagene) in an orientation which yielded synthetic mRNA by transcription from the T7 promoter. Linearized plasmids were transcribed with T7 RNA polymerase according to the protocol of the manufacturer for cotranscriptional capping with 5′-7MeGpppG (Stratagene). In vitro translation with synthetic mRNAs in the presence of L-[3,4,5-$^3$H(N)]-leucine (DuPont NEN) was performed in nuclease-treated reticulocyte lysates as specified by the manufacturer (Promega).

## 2.3 5′-RACE

0.5 µg of polyA+ RNA from human brain was reverse transcribed (as described) with a primer (5′-CAGGGCCTGGAAGGTGTGA-3′), located at the N-terminal part of the longest cDNA clone isolated from the human brain cDNA library. 20 µl of the reverse transcribed reaction mixture were precipitated and resuspended in a mixture containing 0.2 mM dATP, buffer tailing (0.1 M potassium cacodylate pH 6.8, 1 mM $CoCl_2$, 100 mM DTT, 100 µg/ml BSA) and 15 U TdT enzyme (Gibco BRL). After incubation at 37°C for 10 min water was added to a final reaction volume of 250 µl. cDNA was mixed with 25 pmol of oligo (5′-ATAGCCACCAACAGTCACCA-3′), 10 pmol of oligo (5′-GACTCGAGTCGA-CATCGATTTTTTTTTTTTTTTTT-3′) and 25 pmol of oligo (5′-GACTCGAGTCGA-CATCGA-3′), 10 µl of 10x Taq polymerase buffer (500 mM Kcl/ 100mM Tris.HCl, pH 8.3), 8 µl of 25 mM $MgCl_2$, 8 µl of a dNTP solution (2.5 mM dNTP). The volume was brought to 100 µl with $H_2O$. The tube was heated at 95°C for 7 min and 0.5 µl (2.5 U) of Taq polymerase (Perkin Elmer Cetus) were added. Annealing was carried out for 2 min at 58°C and polymerization for 2.5 min at 72°C: The cycle was repeated 40 times. PCR products were blotted on a nitrocellulose filter and hybridized with a oligonucleotide probe (5′-AC-CACTGACGAAGATCCTGGCAAGTTTCTTTGGGG AGC-3′), based on the known sequence of the partial cDNA human clone. Probe was labelled with ($\gamma^{32}$P) dATP by T4 polynucleotide kinase (Boehringer) and purified on Nap5 column (Pharmacia). A positive band (330 bp) was re-amplified using oligo with SalI linkers and cloned in pUC18. DNA sequencing was performed on both strands confirming correspondence between the PCR fragment and the lacking 5′-part of the human KAT clone.

## 3. RESULTS AND DISCUSSION

We previously reported (Mosca et al., 1994) the cloning of a cDNA encoding a soluble form of rat kidney KAT and its identity with glutamine transaminase K (GTK). Comparison of the nucleotide sequence with the GenBank database indicated that it was also virtually identical with rat kidney cysteine conjugate β-lyase (Perry et al. 1993). To confirm that the isolated clone indeed encoded three different enzymatic activities, COS-1 cells were transfected with the cDNA as described (Mosca et al.,1994) and examined for the presence of KAT, GTK and β-lyase activities. As shown in Table 1, a significant increase in KAT, GTK and β-lyase activities was seen in cells transfected with the cDNA in the sense orientation (cKAT).

### 3.1 A Novel cDNA Encoding Rat KAT

Several hybridization positives encoding for KAT were isolated by the screening of a rat kidney cDNA library (Mosca et al., 1994). Restriction endonuclease mapping of the cDNA clones suggested heterogeneity at the 5′ termini of the cDNAs. Particularly a novel cDNA was identified which has an N-terminal extension of 14 amino acids that, together with the first 10 amino acids of cKAT, resembles a mitochondrial targeting peptide (Fig. 1) (von Heijne et al. 1989). This is consistent with previous biochemical observations showing KAT activity in different subcellular fractions such as mitochondria and cytosol (Stevens et al. 1988). Differences between the cytosolic (cKAT) and putative mitochondrial KAT (+14KAT) are present only upstream the ATG start codon of cKAT. N-terminal leader peptides are generally cleaved after or during import of the protein into the mitochondria. It is therefore likely that the two cDNAs share the same coding region for mature KAT.

Of interest, +14KAT differs from cKAT not only in having a putative mitochondrial targeting signal but also in lacking a strong secondary mRNA stem-loop structure present in the cKAT form which is likely involved in the control of cKAT mRNA translation. To verify this hypothesis, the two cDNAs, including their entire 5′-untranslated regions, were subcloned into the expression vector pBluescript. Synthetic RNAs were produced by in vitro transcription and then tested in the reticulocyte lysate translation system supplemented with [$^3$H]leucine (Fig. 2). The polypeptide synthesized by +14KAT transcript had an apparent Mr around 48000, which is in good agreement with the predicted

**Table 1.** Enzymatic activity in cytosolic extracts of COS-1 cells transfected with the rat cytosolic KAT (cKAT)

| | GTK nmole $min^{-1}mg^{-1}$ | KAT nmole $min^{-1}mg^{-1}$ | β-lyase nmole $min^{-1}mg^{-1}$ |
|---|---|---|---|
| cKAT | 27.0 | 2.4 | 7.4 |
| Antisense | 0.9 | N.D. | N.D. |
| Untransfected | N.D. | N.D. | N.D. |

N.D: not detected

**+14KAT**

-254 AAAATGGAAACATTCTTCAAGTGACCAGAGCACTCTGCCGGGGACAAAAGACTTTG

-198 CCTTGAACGCGTAGTGGAGAAGCTACAAACCCCAGGTCCCAGTGGCCTGATTGACTT

-141 AGGGTCTCAGCTGGCCCAAAACTCAGTGTGTAGATCAGACTGATCTCAAACTCACAG

- 84 AGATCTCCCTGCCTTTGCCTGCTGAGTCCTGGGATTAAAGGC**ATG**AATCACAGTACC
**Met** Asn His Ser Thr

- 27 TGGTGCCTTTTCTTTAAAAAGCTCACC**ATG**ACCAAACGGCTGCAGGCTCGGAGGCTG
Trp Cys Leu Phe Phe Lys Lys Leu Thr **Met** Thr Lys Arg Leu Gln Ala Arg Arg Leu

+ 31 GACGGGATTGATCAAAACCTCTGGGTGGAGTTTGGCAAACTGACCAAGGAGTAT-----
Asp Gly Ile Asp Gln Asn Leu Trp Val Glu Phe Gly Lys Leu Thr Lys Glu Tyr

**cKAT**

-243 AACTGACCAAGGAGTATGATCAATCCCGTCCAGCCTCCGAGCCTGCAGCCGTTTGGT

-186 CATGGTGAGCTGCTTCAGCTAACAATTGCACTGACAGTGCTCTTGAGCCAAGTTGCT

-129 TCTGGGACGGAAGTAGTCCATCTAGGGCTCGGCCTCTTTAAAGAAACAGACTTCTGC

- 72 AACCTTGGGACTACGTTTGGGGTCGCCGGCTATTGGACGGAGCAGCGCAATTGTTAG

- 15 CTGAAGCAGCTCACC**ATG**ACCAAACGGCTGCAGGCTCGGAGGCTGGACGGGATTGAT
**Met** Thr Lys Arg Leu Gln Ala Arg Arg Leu Asp Gly Ile Asp

+43 CAAAACCTCTGGGTGGAG TTTGGCAAACTGACCAAGGAGTAT------------------------------
Gln Asn Leu Trp Val Glu Phe Gly Lys Leu Thr Lys Glu Tyr

**Figure 1.** Nucleotide sequences of +14KAT 5′UTR and cKAT 5′UTR with partial deduced aminoacidic sequences. The starting point at +1 is referred to the first nucleotide of the cKAT coding region. The identity of the two sequences starts at nucleotide (nt) -8 and extends to nt +1577, in the 3′UTR region (not shown). Underlined nucleotide sequences indicate the two 70-nucleotides inverted repeats.

size of KAT. No bands of the expected size were observed when the cKAT RNA was translated.

Therefore, it is tempting to hypothesize that different subcellular localization or control of translation can play a role in modulating KAT activity.

## 3.2 Cloning of Human KAT I

Two KAT enzymes exist and have been characterized in the human brain. To gain insights into human KAT enzymes, the rat kidney cDNA clone was used to screen a λ ZapII human brain cDNA library (Stratagene). About 1,350,000 recombinant phages were plated on a lawn of *E. coli* XL1 blue cells and screening was performed as described (Benatti et al. 1993). Unfortunately, none of the positive clones contained a full length sequence. Therefore, in order to isolate the 5′ lacking sequence, a RACE protocol was applied (see Methods). A full length cDNA encoding human KAT was then assembled from the partial cDNA obtained by library screening, and the 5′-RACE fragment.

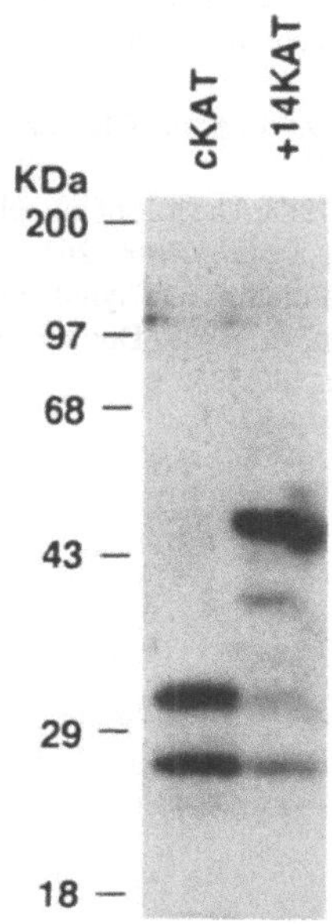


**Figure 2.** In vitro translation of cKAT and +14KAT transcripts with [$^3$H]leucine. Analysis of the polypeptides was by SDS-PAGE and fluorography. Positions of molecular mass markers (in kDa) are indicated on the right.

The deduced amino acid sequence (not shown) is 422 amino acids long and exhibits high similarity (82%) to the rat, with 90% similarity around the pyridoxal phosphate binding site.

## ACKNOWLEDGMENTS

We thank Jan Malyszko for the oligonucleotide synthesis and purification.

## 4. REFERENCES

Abraham, D.G. and Cooper, A. J. L., 1991, Glutamine transaminase K and cysteine S-conjugate-b-lyase activity stains, *Anal. Biochem.*, 197:421.

Benatti, L., Bonecchi, L., Cozzi, L., and Sarmientos, P., 1993, Two preproendothelin 1 mRNAs transcribed by alternative promoters, *J. Clin. Invest.*, 91:1149.

Mosca, M., Cozzi, L., Breton, J., Speciale, C., Okuno, E., Schwarcz, R. and Benatti, L., 1994, Molecular cloning of rat kynurenine aminotransferase: identity with glutamine transaminase K, *FEBS Lett.*, 353:21.

Noguchi, T., Minatogawa, Y., Okuno, E., Nakatani, M., Morimoto, M. and Kido, R., 1975, Purification and characterization of kynurenine-2-oxoglutarate aminotransferase from the liver, brain and small intestine of rats, *Biochem. J.*, 151:399.

Okuno, E., Minatogawa, Y., Nakamura, M., Kamoda, M., Nakanishi, J., Makino, M. and Kido, R., 1980, Crystallization and characterization of human liver kynurenine-glyoxylate aminotransferase, *Biochem. J.* 189:581.

Okuno, E., Schmidt, W., Parks, D.A., Nakamura, M. and Schwarcz, R., 1991a, Measurement of rat brain kynurenine aminotransferase at physiological kynurenine concentrations, *J. Neurochem.*, 57:553.

Okuno, E., Nakamura, M. and Schwarcz, R., 1991b, Two kynurenine aminotransferases in human brain, *Brain Res.*, 542:307.

Perry, S.J., Schofield, M.A., MacFarlane, M., Lock, E.A., King, L.J., Gibson, G.G., Goldfarb, P.S., 1993, Isolation and expression of a cDNA coding for rat kidney cytosolic cysteine conjugate β-lyase, *Mol. Pharmacol.*, 43:660.
Sambrook, J., Fritsch, E.F., Maniatis, T., 1989, Molecular Cloning: A Laboratory Manual, Cold Spring Harbor Laboratory, Cold Spring Harbor, NY.
Stevens, J.L., Ayoubi, N. and Robbins, J.D., 1988, The role of mitochondrial matrix enzymes in the metabolism and toxicity of cysteine conjugates, *J. Biol. Chem.*, 263:3395.
Stone, T.W. 1993, Neuropharmacology of quinolinic and kynurenic acids, *Pharmacol. Rev.*, 45:309.
von Heijne, G., Steppuhn, J. and Herrmann, G., 1989, Domain structure of mitochondrial and chloroplast targeting peptides, Eur. J. Biochem. 180:535.

# KYNURENINE AMINOTRANSFERASES IN THE RAT

## Localization and Characterization

E. Okuno, T. Nishikawa, and M. Nakamura

Wakayayama Medical College
Department of Biochemistry
27 Kyuban-cho Wakayama 640, Japan

## ABSTRACT

Kynurenine-2-oxoglutarate aminotransferase (kynurenine specific, designated here KAT-II) and kynurenine pyruvate aminotranssferase (designated here KAT-I) activities were detected in the kidney using 2 μM kynurenine. The majour activities were performed by KAT-II and the contribution of KAT-I is about 1/15. KAT-I activity was detected in all organs tested. The liver showed the highest KAT-I activity, however , the highest activity of glutamine transaminase-K (GTK) was detected in the kidney. KAT-I activity was well corresponded with GTK activity in all organs except liver. KAT-I or GTK activity of crude extract didn't inhibited by addition of glutamine either kynurenine. KAT-I or GTK activity of purified preparation, however, inhibited strongly by addition of glutamine either kynurenine. KAT-I and GTK showed different pH optimum profile, but purified and crude preparation of those were similar. Phenylpyruvate or 2-oxo-4-methiolbutyrate reduced the inhibition of purified KAT-I activity by glutamine using 2 μM kynurenine. Phenylpyruvate changed the Km value for kynurenine and Vmax, suggesting conformational change of the enzyme.

## INTRODUCTION

Kynurenine aminotransferase (L-kynurenine: 2-oxoglutarate aminotransferase (cyclizing), EC 2.6.1.7) catalyzes the irreversible transamination between L-kynurenine and 2-oxoglutarate to form kynurenic acid and L-glutamate. The kynurenine: 2-oxoglutarate aminotrnsferase in brain or small intestine is the same enzyme with mitochondrial aspartate aminotranaferase and also using phenylalanine and tyrosine for substrate (Noguchi et al., 1975). Liver, However, cotains another kynurenine specific aminotransferase Noguchi et al., 1975). Many reports were published about aromatic amino acids, including kynurenine, aminotransferase with different 2-oxo acids. Finally, It has been reported that four enzymes are capable of transamination of kynurenine, named m-aspartate aminotransferase, kynurenine-2-oxoglutarate aminotrans-ferase, glutamine transaminase-K and alanine-glyoxylate amino-transferase I (Kido, 1984) using mM order of kynurenine concentration for enzyme assay. Kynurenine-2-oxoglutarate aminotransferase has been reported that the enzyme is identical with aminoadipate aminotransferase in kidney by Tobes and Mason (1975; 1977), and in liver by Takeuchi et al. (1983). In 1991, Mawal and Deshmukh showed those activities were catalyzed by separate protein in kidney (Mawal and Deshmukh, 1991).

Kynurenic acid, the product of kynurenine aminotransferase, recognized physiolosically important substance in central nervous system for its ability to block excitatory amino acid receptor function in the mammalian brain. It has been reported that kynurenic acid can antagonize competitively N-methyl-D-aspartate induced neuronal excitation (Perkins and Stone, 1982), and also show the neuroprotective (Foster et al., 1984) and anti convulsant effect (Germano et al., 1987) to react with N-methyl-D-aspartate receptor for antagonist. Kynurenine-pyruvate aminotransferase(KAT-I) was reported in rat kidney and brain by Okuno et al. (1990). The chatacters of purified enzyme, higher activity with pyruvate than 2-oxoglutarate and inhibition by glutamine, were similar with human KAT-I (Baran et al., 1994).

At physiological kynurenine concentration, only one enzyme activity(KAT-I) could be measurable in rat brain using 1 mM 2-oxoglutarate or pyruvate (Okuno et al., 1991a). In human brain, however, two kynurenine aminotransferases were reported (Okuno et al., 1991b) using 3 mM or either 2 μM kynurenine (Schmidt et al., 1993). There were some similarities in charactors between rat KAT-I and human KAT-I, also rat KAT-II and human KAT-II. c-DNA clonig study showed rat KAT-I is the same enzyme with GTK (Mosca et al., 1994) and cysteine conjugate β-lyase (Alberati-Giani et al., 1994).

In this paper, we show the enzyme react independently from glutamine or methionine and also characterize the native and purified KAT-I.

# MATERIALS AND METHODS

## Animals

Male Sprague-Dawley rats (200-300 g), kept on a 12-h light/dark cycle with free access to food and water, were used in all experiments.

## Materials

Kynurenine(sulfate salt), 2-oxoglutarate, pyridoxal 5'-phosphate, and other amino acids were purchased from Nacalai Tesque Inc. (Kyoto, Japan). All other chemicals were obtained from different commercial suppliers were of the highest available purity. Custom-synthesized [$^3$H] kynurenine, prepared by an exchange procedure using tritiated water, were obtained from Amersham (Arlington Heights, IL, U.S.A.; specific activity was 335 mCi/mmol)

## Enzyme Assay

*KAT I Assay by Spectrophotometer.* Routinely, KAT-I activity was assayed by the method of Tobes and Mason (1977). The reaction mixture contained unless specified otherwise, 3 mM L-kynurenine, 5 mM pyruvate, 70 μM pyridoxal 5'-phosphate(PALP), 0.17 M potassium phosphate buffer, pH 8.0 and the enzyme preparation in a total volume 0.1 ml.

Newly synthesized kynurenic acid was determined at wave length 333 nm. The details were described in the cited reference (Okuno et al., 1990).

*KAT Assay by Using [$^3$H]Kynurenine.* The reaction mixture contained unless specified otherwise, [$^3$H]kynurenine(30 nCi), diluted with unlabelled kynurenine to yield a final concentration of 2 μM, 1 mM 2-oxoglutarate, 70 μM PALP, 150 mM Tris-acetate buffer, pH 8.0 and enzyme preparation in a total volume of 0.1 ml. Formed [$^3$H]kynurenic acid was separated by Dowex 50W column (0.5 x 1.0 cm: $H^+$form). The details were described in the cited reference (Okuno et al., 1991a).

*GTK Assay.* GTK assay was based on the method described by Cooper (1978). The reaction mixture contained 5 mM phenylalanine, 5mM 2-oxo-4-methiolbutyrate, 70 μM PALP, 170 mM potassium phosphate buffer pH 8.0 and enzyme preparation in a total volume 0.4 ml. Formed phenylpyruvate was measured for aromatic enol-2-oxo acid-borate complex (Lin et al., 1958). Details of the experimental procedure have been described previously (Noguchi et al., 1976).

*Purification of KAT I.* Purified KAT I used in this study, was prepared from rat kidney according the method by Okuno et al. (1991a). The purified enzyme showed single protein band by poly acrylamide gel electrophoresis and stored at -20 °C in 50% glycerol.

*Protein measurement.* Protein was assayed by the method of Lowry et al. (1951) using bovine serum albumin as a standard.

# RESULTS

## Organ Distribution of KAT-I and GTK Activities

KAT-I and GTK activities were measured in various organs showing Table 1. The highest KAT-I activity was detected in the liver. Kidney also contained higher KAT-I activity than others. However the highest GTK activity measured in kidney a half of kidney's activity

**Table 1.** Organ distribution of kynurenine-pyruvate aminotransferase (KAT-I) and glutamine aminotranferase-K (GTK) activities

| Organ | KAT-I activity nmol/h/mg | GTK activity nmol/h/mg |
|---|---|---|
| Liver | 17.6 | 420 |
| Kidney | 11.8 | 773 |
| Pancreas | 2.89 | 134 |
| Testis | 2.00 | 118 |
| Brain | 1.17 | 69 |
| Small intestine | 1.15 | 71 |
| Spleen | 1.13 | 67 |
| Stomach | 0.80 | 98 |
| Lung | 0.79 | 42 |
| Heart | 0.76 | 47 |
| Muscle | 0.57 | 38 |

KAT-I activity was detected with 3 mM of kynurenine and 5 mM of pyruvate. The details were described in the text. The values were average of three separate experiments.

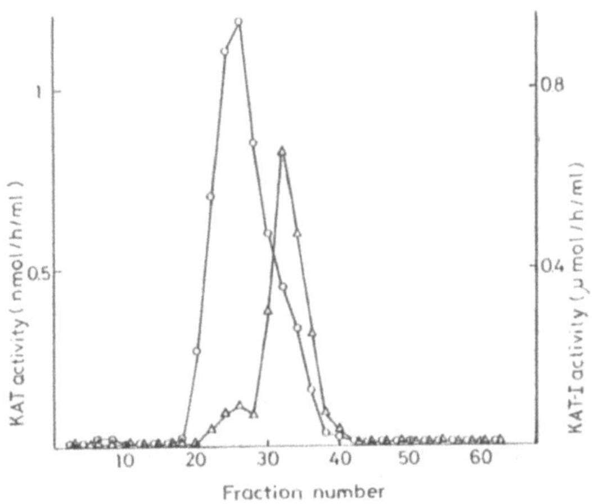


**Figure 1.** DEAE-Sepharose column chromatography of KAT. Rat kidney (2.9g) was homogenized with 5 vol of 5 mM potassium phosphate buffer, pH 7.5, containing 10 mM mercaptoethanol and 50 μM PALP. After centrifugation (15,000 rpm, 20 min), the resulting supernatant was applied to a DEAE-sepharose column (1.5 x 7 cm) equiribrated by homogenate buffer. The enzymes were eluted by linear gradient of 5 - 250 mM potassium phosphate buffer, pH 7.5, containig 10 mM mercaptoethanol. KAT activity (O) using 2 μM kynurenine and 1 mM 2-oxoglutarate for substrate and KAT-I activity (Δ) using 3 mM kynurenine and 5 mM pyruvate for substrate were shown.

was detected in liver. The lowest of both KAT-I and GTK activities were observed in muscle. The activity ratio of GTK/KAT-I was about 50-70 except liver.

## DEAE-Sepharose Profile of KAT Activities

Rat kidney contains kynurenine specific aminotransferase(KAT-II) and KAT-I. DEAE-sepharose column chromatography didn't show the clear separation by measuring KAT activity using 2 μM of kynurenine (Fig. 1). KAT-I activity showed the peak at the fraction 32 using 3mM of kynurenine and 5 mM of pyruvate because of KAT-II show the high affinity for 2-oxoglutarate and low affinity for pyruvate. The contribution of KAT-I to KAT activity was estimated about 1/10-1/20.

## Effect of 3-Hydroxykynurenine on KAT-I Activity

3-Hydroxykynurenine, which is reportedly converted to xanthurenic acid by KAT-I (Noguchi et al., 1973), was tested at three concentrations (Table 2). Ten times high concentration (20 μM) was virtually ineffective and 2 mM concentration of 3-hydroxykynurenine inhibit the KAT activity only 50 % with 2-oxoglutarate and 67 % with pyruvate. These results show that 3-hydroxykynurenine is not good substrate for KAT-I.

**Table 2.** Effect of 3-hydroxykynurenine on KAT-I activity

| | KAT-I activity (% of control) | |
|---|---|---|
| 3-Hydroxykynurenine | 2-Oxoglutarate | Pyruvate |
| 20 μM | 96 ± 1 | 84 ± 1 |
| 200 μM | 85 ± 1 | 54 ± 1 |
| 2 mM | 50 ± 1 | 33 ± 1 |

The data obtained from Okuno et al. (1991a).

**Table 3.** Effect of amino acids on KAT-I and GTK activities

| | KAT-I activity | | GTK activity | |
|---|---|---|---|---|
| Addition (5 mM) | Homogenate extract | Purified enzyme | Homogenete extract | Purified enzyme |
| — | 100 | 100 | 100 | 100 |
| DCVC | ND | 8 | 91 | 84 |
| Glutamate | 130 | 82 | 95 | 93 |
| Cysteine | 48 | 46 | 62 | 59 |
| Methionine | 72 | 77 | 100 | 85 |
| Phenylalanine | 79 | 17 | — | — |
| Glutamine | 127 | 20 | 87 | 72 |
| Kynurenine | — | — | 105 | 13 |

DCVC : S-dichloro-vinyl-L-cysteine, ND : notdetected. KAT-I and GTK activities were assayed with 3 mM kynurenine and 5 mM pyruvate, and 5 mM phenylalanine and 5 mM 2-oxo-4-methiol-butyrate, respectively. The details were described in the text.

# Effect of Various Amino Acids on KAT-I and GTK Activities

Effect of various amino acids was tested on KAT-I and GTK activities with crude and purified enzyme preparations (Table 3). Glutamine and phenylalanine strongly inhibited the KAT-I activity of purified enzyme, but not of crude enzyme. Kynurenine also inhibited the GTK activity of purified enzyme, but not crude enzyme. S-Dichlorovinyl-L-cysteine(DCVC) which is the substrate of cysteine conjugate β-lyase, inhibited the KAT-I activity of both crude and purified enzyme. GTK activity, however, did not inhibited more than 20%. These results showed that crude KAT-I was independent from the substrate of GTK. The inhibition by glutamine to purified enzyme confarmed further in Fig. 2. According to the increasing of glutamine concentration, purified KAT-I inhibited strongly. Crude enzyme activity, however, was stimulated by glutamine but not more than 35%.

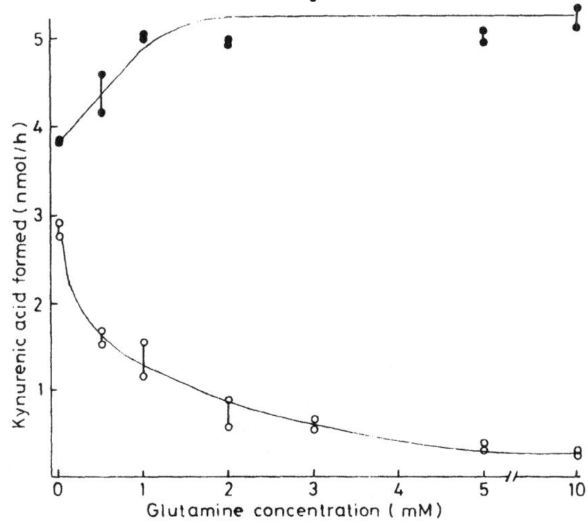


**Figure 2.** Effect of different concentration of glutamine on purified and crude KAT-I activity. Rat kidney was homogenized with 5 vol. of potassium phosphate buffer, pH 7.5, containing of 10 mM mercaptoethanol and 50 μM PALP, and centrifuged at 10,000 g, 30 min. The resulting supernatant was used for crude extract (●). Purified KAT-I from kidney, shown (O), was obtained according to the method by Okuno et al. (1990).

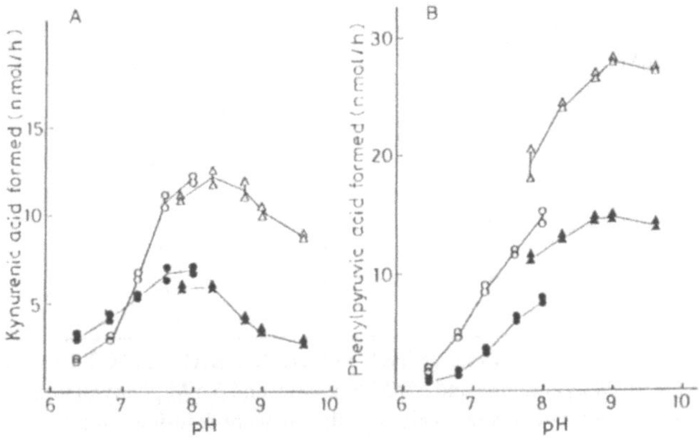


**Figure 3.** pH dependency of KAT(A) and GTK(B) activities. The enzyme used as the same of used in Fig. 2. Open simbol showed activity of purified enzyme and crosed simbol was that of crude enzyme. 0.15 M Potassium phosphate buffer (○, ●) and Tris-acetate buffer (Δ,▲) were shown. KAT activity was measured by spectrophotometrically using 3 mM kynurenine and 5 mM pyruvate for substrate.

## pH Optimum

The influence of pH on the activities of KAT-I and GTK was investigated over therange pH 6.3-9.7 with crude and purified enzyme preparations. The buffers employed 0.15 M potassium phosphate buffer and Tris-acetate buffer, showed pH optima of 8.3 for KAT-I and 9.0 for GTK activities (Fig. 3). No significant difference in pH optimum was observed between crude and purified preparations. However the difference was found between KAT-I and GTK activity. Tris-acetate buffer showed lower activity for KAT-I than phosphate buffer. On the contraly, GTK activity with Tris buffer was higher than that of phosphate buffer.

## Addition of Purified KAT-I

The effect of glutamine to crude or purified or both preparations of KAT-I activity were investigated. Glutamine stimulated the KAT-I activity of crude preparation slightly but

**Table 4.** Effect of glutamine on crude and purified KAT-I activity

| Enzyme | 3 mM glutamine | KAT-I activity (nmol/h) |
|---|---|---|
| Crude | – | 60 ± 4 |
| Purified | – | 27 ± 3 |
| Crude | + | 70 ± 2 |
| Purified | + | 5.6 ± 0.2 |
| Crude + purified | – | 118 ± 7 |
| Crude + purified | + | 114 ± 7 |

KAT-I activity was measured by spectrophotometrically using 3 mM kynurenine and 5 mM pyruvate for substrate.

inhibited 80% of purified enzyme activity. Glutamine didn't inhibit the purified enzyme activity when it was added to the crude preparation. It is suggesting that the addition of purified enzyme to the crude preparation changed the charactor of purified enzyme to crude one using 3 mM of kynurenine (Table 4). The radio enzymatic assay, using 2 μM of kynurenine, showed the added purified enzyme to crude preparation was inhibited completely by glutamine (data not shown).

## Effect of Various 2-Oxo Acids on Glutamine Inhibition

The effect of 0.1 mM of various 2-oxo acids to glutamine inhibition was investigated to KAT-I activity using 2 μM kynurenine and 0.2 mM glutamine with purified KAT-I (Table 5). It has been reported that 2-oxo-4-metholbutyrate and phenylpyruvate stimulate the activity of cysteine conjugate β-lyase activity. Phenylpyruvate or 2-oxo-4-methiolbutyrate itself didn't stimulate the KAT activity, however, these compounds showed the decreasing of glutamine effect. 2-Oxo-iso-caproate also showed the decreasing of glutamine effct about 1/2 of 2-oxo-4-methiolbutyrate.

## Effect of Phenylpyruvate on Km Value for Kynurenine

Km value of KAT-I for kynurenine was reported to 17 μM with 1 mM 2-oxoglutarate (Okuno et al., 1991a). Here we had the identical value, however, addition of 0.1 mM of phenylpyruvate brought high Km value (about 400 μM) and high Vmax (Fig. 4). Phenylpyruvate itself is not good substrate to KAT-I, because 10 μM or lower concentration of kynurenine, the enzyme activity with phenylpyruvate was lower than that without phenylpyruvate.

# DISCUSSION

It has been reported that four enzymes were capable of transamination of kynurenine by spectrophotometric assay (Kido, 1984). KAT-I activity, using 3 mM kynurenine and 5 mM pyruvate for substrate, well correspond with GTK activity except liver. Liver showed the highest KAT activity, but GTK activity was about 1/2 of that of kidney. KAT activity in liver may be refrect of the activity from alanine-glyoxylate aminotransferase I (Kido, 1984).

**Table 5.** Effect of 2-oxo acid on KAT-I activity

| Addition (0.1 mM) | Glutamine (0.2 mM) | Relative KAT activity |
|---|---|---|
| — | – | 100 |
| — | + | 2.8 ± 1.2 |
| Phenylpyruvate | + | 54 ± 0.9 |
| 2-Oxo-4-methiolbutyrate | + | 69 ± 8.2 |
| 2-Oxo-iso-caproate | + | 32 ± 0.9 |
| Pyruvate | + | 7.4 ± 1.1 |
| Indolpyruvate | + | 6.7 ± 0.2 |
| Glyoxylate | + | 6.2 ± 2.6 |
| 2-Oxoadipate | + | 2.7 ± 1.5 |
| Oxaloacetate | + | 2.5 ± 1.4 |

Purified KAT-I from kidney was used. KAT activity was measured with 2 μM kynurenine and 1 mM 2-oxoglutarate for substrate.

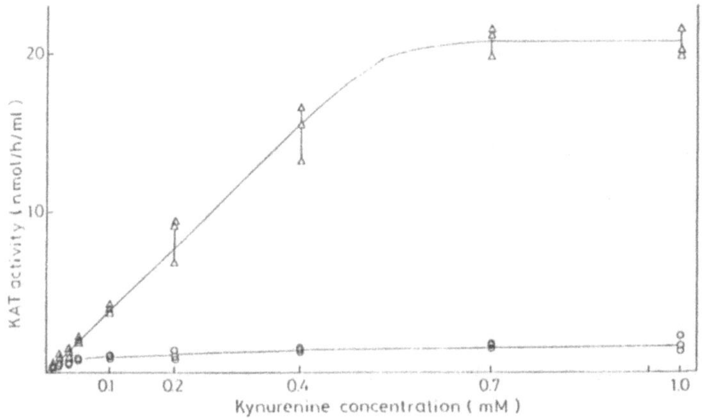


**Figure 4.** KAT-I activity with different concentration of kynurenine. Purified KAT-I from rat kidney was incubated with 1 mM of 2-oxoglutarate with (Δ) and without (O) of 0.1 mM phenylpyruvate. The activity was measured by radio enzymatic assay using 2 μM kynurenine and 1 mM 2-oxoglutarate for substrate.

Okuno et al. (1991a) showed, in rat brain the physiological concentration of kynurenine, m-aspartate aminotransferase did not show the detectable activity for kynurenine. Futher, under the same assay conditions purified alanine-glyoxylate aminotransferase I, present in the liver, did not give the effective activity with 2 μM kynurenine(unpublished data). These data suggest that two enzymes will be active in intact rat cells. The major activity of kynurenine aminotransferase may be catalyzed by KAT-II in kidney and liver under physiological conditions (Fig. 2). The products of kynurenine aminotransferase, kynurenic acid and xanthurenic acid, are excreted in the urine. When the animals got vitamin $B_6$ deficiency, the excretion of xanthurenic acid was increased. The convertion of 3-hydroxykynurenine to xanthurenic acid will be performed by KAT-II, because high activities for 3-hydroxykynurenine were reported by Takeuchi et al. (1983). Further, the enzyme in rat brain KAT-I activity did not inhibited more than 50% by 1000 times high concentration of 3-hydroxykynurenine, meaning the affinity of KAT-I for 3-hydroxykynurenine is low.

Kynurenine pyruvate aminotransferase (KAT-I) which is reportedly from charactor, identical with histidine-pyruvate and glutamine oxo-acid aminotransferase (Noguchi and Kido, 1976; Noguchi et al., 1977). Further study of cloning and sequencing of KAT-I confirmed the enzyme was the same protein with GTK and cysteine conjugate β-lyase. However, the KAT activity of native enzyme didn't inhibited by glutamine(Table 3). GTK activity also was not affect by kynurenine. The possible explanation is the exsistence of two active sites, one for kynurenine and the other for glutamine. The purification of enzyme may brought the conformational change and both active sites to be close. DCVC, which is the substrate of cystein conjugate β-lyase inhibited the KAT-I activity both crude and purified enzyme, but not GTK activity. The active site of KAT and cystein conjugate β-lyase may be the same.

The pH optimum of KAT-I activity slightly higher with phosphate buffer than Tris buffer, but GTK activity with Tris buffer was higher than phosphate buffer. Usually Tris buffer give the lower activity than phosphate buffer in transaminase reaction, because of amino group in Tris buffer compete to pyridoxal group. These data also support the existence of two active sites. The change from purified form to intact form is need the certain level of protein showing addition of purified enzyme to crude enzyme showed the change from purified to intact form (Table 4). Using 2 μM kynurenine for KAT assay, the added purified enzyme to crude preparation was inhibited by glutamine (data not shown). The required amount of enzyme by radio enzymatic assay is much less conpaired with spectrophotometrically assay. Phenylpyruvate or 2-oxo-4-methiolbutyrate was reported to stimulate the activity of cystein conjugate β-lyase (Stevens et al., 1986). In our experiment, these 2-oxo acids didn't stimulate the KAT activity, but reduced the inhibition by glutamine(Table 5). The effect of these 2-oxo acids might be influence the conformational change of the enzyme, showing change of Km and Vmax value (Fig. 4). The amino acid sequence of active site have been reported to -$S^{244}$-A-G-K-S-$F^{249}$-(Mosca et al., 1994). The possible second active site is coming up in down stream of sequence -$S^{345}$-D-F-K-S-$K^{350}$.

The clear proof of two active sites in one enzyme need more work. This paper show the majour activity of KAT and transamination of 3-hydroxykynurenine might be performed by KAT-II in liver and kidney. KAt-I exist all organs, and the activity is independent from another amino acids.

## REFERENCES


Alberati-Giani, D., Malherbe, P., Kohler, C., Lang, C., Kiefer, V., Lahm, H.-W. and Cesura, A.M., 1995, Cloning and characterisation of a soluble kynurenine aminotransferase from rat brain: identity with kidney cysteine conjugate b-lyase. *J. Neurochem.* 64, 1448.

Baran, H., Okuno, E., Kido, R. and Schwarcz, R., 1994, Purification and characterization of kynurenine aminotransferase I from human brain. *J. Neurochem.* 62: 730.

Cooper, A.J.L., 1978, Purification of soluble and mitochondrial glutamine transaminase K from rat kidney. Use of a sensitive assay involving transamination between L-phenylalanine and α-keto-γ-methiolbutyrate. *Anal. Biochem.* 89: 451.

Foster, A.C., Vezzani, A., French, E.D. and Schwarcz, R., 1984, Kynurenic acid blochs neurotoxicity and seizures induced in rats by the related brain metabolite quinolinic acid. *Neurosci. Lett.* 48: 273.

Germano, I.m., Pitts, L.H., Meldrum, B.S., Bartkowski, H.M. and Simon, R.P., 1987, Kynurenato inhibition of cell excitation decreases stroke size and deficits. *Ann. Neurol.* 22: 730

Kido, R., 1984, Kynurenine aminotransferase activity in human and other mammalian tissues. In:Proc. 4th Meeting on Progress in Tryptophan and Serotonin Research, H.G. Schlossberger, W. Kochen, B. Linzen, H. Steihart, Eds., W. De Gruyter Berlin, pp. 651-656

Lin, E.C.C., Pitt, B.M., Civen, M. and Knox, W.E., 1958, The assay of aromatic amino acid transaminations and keto acid oxidation by the enol borate-tautomerase method. *J. Biol. Chem.*, 233: 668.

Lowry, O.H., Rosebrough, N.J., Farr, A.L. and Randall, R.J., 1951, Protein measurement with the Folin phenol reagent. *J. Biol. Chem.*, 193: 265.

Mawal, M.R. and Deshmukh, D.R., 1991, α-Aminoadipate aminotransferase and kynurenine aminotransferase activities from rat kidney. *J. Biol. Chem.* 266: 2573.

Mosca, M., Cozzi, L., Breton, J., Speciale, C., Okuno, E. Schwarcz, R. and Benatti, L., 1994, Molecular cloning of rat kynurenine aminotransferase: identity with glutamine transaminase. *FEBS lett.* 353: 21.

Noguchi, T., Nakamura, J. and Kido, R., 1973, Kynurenine pyruvate transaminase and its inhibitor in rat intestine. *Life Sci.*, 13: 1001.

Noguchi, T., Minatogawa, Y., Okuno, E., Nakatani, M. and Kido,R., 1975, Purification and characterization of kynurenine-2-oxoglutarate aminotransferase from the liver, brain and small intestine of rats. *Biochem. J.* 151: 399.

Noguchi, T. and Kido, R., 1976, Identity of kynurenine pyruvate aminotransferase with histidine: pyruvate aminotransferase. *Hoppe-Seyler's Z. Physiol. Chem.* 357, 649.

Noguchi, T., Okuno, E., Minatogawa, Y. and Kido, R., 1976, Purification, characterization and identification of rat liver histidine-pyruvate aminotransferase isoenzymes. *Biochem. J.*, 155: 107.

Noguchi, T., Okuno, E. and Kido, R., 1977, Identity of rat kidney histidine-pyruvate aminotransferase with glutamine-oxo-acid amino transferase. *Biochem. J.* 161: 177.

Okuno, E., Du, F., Ishikawa, T.,Tsujimoto, M., Nakamura, M., Schwarcz, R., and Kido, R., 1990, Purification and characterization of kynurenine-pyruvate aminotransferase from rat kidney and brain. *Brain Res.* 534: 37.

Okuno, E., Schmidt, W., Parks, D.A., Nakamura, M. and Schwarcz, R., 1991a, Measurement of rat brain kynurenine aminotransferase at physiological kynurenine concentrations. *J. Neurochem.* 57: 533.

Okuno, E., Nakamura, M. and Schwarcz, R., 1991b, Two kynurenine aminotransferases in human brain. *Brain Res.* 542: 307.

Perkins, M.N. and Stone, T.W., 1982, An iontophoretic investigation of the actions of convulsant kynurenines and their interaction with the endogenous exitant quinolinic acid. *Brain Res.* 247: 184.

Schmidt, W., Guidetti, P., Okuno, E. and Schwarcz, R.., 1993, Characterization of human brain kynurenine aminotransferase using [$^3$H] kynurenine as substrate. *Neuroscience* 55: 177.

Stevens, J.L., Robbins, J.D. and Byrd, R.A., 1986 A purified cysteine cnjugate β-lyase from rat kidney cytosol. *J. Biol. Chem.* 261: 15529.

Takeuchi, F., Otsuka, H. and Shibata, Y., 1983, Purification, characterization and identification of rat liver mitochondrial kynurenine aminotransferase with α-aminoadipate aminotransferase. *Biochim. Biophys. Acta* 743: 323.

Tobes, M.C. and Mason, M., 1975, L-Kynurenine aminotransferase and L-α-aminoadipate aminotransferase. *Biochem. Biophys. Res.Commun.* 62: 390.

Tobes, M.C. and Mason, M., 1977, α-Aminoadipate aminotransferase: purification, characterisation, and further evidence for identity. *J. Biol. Chem.* 252: 4591.

# KYNURENINE AMINOTRANSFERASE IN HYPERCHOLESTEROLEMIC RATS

A. Bertazzo,[1] M. Biasiolo,[1] C. V. L. Costa,[1] G. Allegri,[1] E. Ragazzi,[2] and G. Fassina[2]

[1] Department of Pharmaceutical Sciences
[2] Department of Pharmacology
University of Padova
I-35131 Padova, Italy

## INTRODUCTION

Nicotinic acid is used in the treatment of atherosclerosis for its hypolipidemic effects (Kritchevsky, 1971; Levy, 1980; Grundy et al., 1981; Altschul et al., 1995), which reduce triglycerides and also cholesterol. In mammals, nicotinic acid derives from the metabolism of tryptophan along the kynurenine pathway. But kynurenine can also be metabolized to kynurenic acid by means of kynurenine aminotransferase (KAT). An abnormal activity of this enzyme should reduce kynurenine and consequently the biosynthesis in nicotinic acid.

The purpose of this study was to investigate whether kynurenine aminotransferase activity was increased in the liver and kidney of Pittsburg-Yoshida hyperlipidemic rats affected by hypercholesterolemia without vascular atherosclerotic damage. The Pittsburg-Yoshida rat is an inbred hypercholololesterolemic strain which has recently become available (Fantappié et al., 1992) and represents an interesting new animal model of endogenous dyslipidemia which may provide further information on the properties of drugs affecting lipid metabolism.

We therefore prepared kynurenine aminotransferase, which catalyses the formation of kynurenic acid from L-kynurenine, from the liver, kidney and brain of Pittsburg-Yoshida (YOS) rats, comparing enzyme activity with that obtained from age-matched, male, Brown-Norway normolipidemic rats, considered as controls.

## MATERIALS AND METHODS

### Animals

Male Pittsburg-Yoshida hypercholesterolemic rats (9 months of age), were purchased from Charles River Italia (Calco, Italy). Age-matched male Brown-Norway normolipidemic

*Recent Advances in Tryptophan Research*, edited by Graziella Allegri Filippini et al.
Plenum Press, New York, 1996 

rats, obtained from the same supplier were used as controls (Chinellato et. al., 1994). The animals were caged in groups of four. YOS and BN rats received a standard GLP diet (Mucedola, Settimo Milanese, Italy); water was supplied *ad libitum*. The rats were maintained at 21°C with a light-dark cycle of 12 h. Whole blood was obtained from rats fasted overnight before sacrifice. Serum was separated.

## Enzyme Preparation and Assay

YOS and BN rats were killed by decapitation, and their kidney, liver and brain were removed immediately and kept at 0° C for all subsequent manipulations. Tissues were cut into small pieces and homogenized in 4 volumes (wt/vol) of 5 mM potassium phosphate buffer, pH 7.3. The homogenate was centrifuged at 10,000 g for 10 min and the pellet was discarded. The supernatant was frozen and thawed and then centrifuged at 15,000 g for 60 min. The supernatant fluid was used for the assay and for determination of protein.

Kynurenine aminotransferase (L-kynurenine-2-oxoglutarate aminotransferase, EC 2.6.1.7.) activity was assayed by a modification of the method of Mason (1954). Briefly, KAT was assayed at 37°C in the total volume of 1.5 ml containing 50.0 mM potassium phosphate buffer, pH 7.3, 10.0 mM 2-oxoglutarate, 40.0 μM pyridoxal phosphate and 1.0 mM L-kynurenine. The reaction was started by the addition of kynurenine. After incubation for 1 h at 37°C, 7.5 ml of 1% (wt/vol) boric acid in ethanol were added, and the resulting precipitate was removed by centrifugation. The supernatant was used and the resulting kynurenic acid was determined spectrophotometrically at 333 nm. Blank samples were obtained by adding L-kynurenine immediately before 1% boric acid in ethanol. Specific activity was expressed as μmol of kynurenic acid/h per mg of protein.

## Protein Determination

Protein concentration was measured according to the method of Lowry et al. (1951) using bovine serum albumin as standard.

## Tryptophan Determination

Tryptophan was measured in serum by HPLC, according to the method of Costa et al. (1987). As in humans tryptophan circulates bound (80-90%) to albumin (McMenamy and Oncley, 1958), to determine the free fraction of tryptophan, ultrafiltration of serum was carried out in an Amicon model 12 ultrafiltration cell with a XM-50 membrane, using the method of Eccleston (1983).

# RESULTS

Tables 1 and 2 show kynurenine aminotransferase activity in the kidney, liver and brain of YOS hypercholesterolemic rats (Table 1) and BN rats (Table 2). Specific activity was expressed as μmol of kynurenic acid formed/h per mg of protein. Tables 1 and 2 also report activity expressed in grams of fresh tissue per hour.

Table 1 shows that KAT activity is four times higher, both as specific activity and activity/g fresh tissue in kidney, than in the liver of YOS rats, whereas in brain it is very low in both strains.

Table 2 shows that, in BN rats, KAT activity is from three (activity/g fresh tissue) to almost six times (specific activity) higher in kidney than in liver, whereas activity in brain is very low.

**Table 1.** Activity of kynurenine aminotransferase in kidney, liver, and brain of male Pittsburg-Yoshida hypercholesterolemic rats

| Tissue homogenate (Supernatant) | Volume (ml) | Protein (mg/ml) | Activity (μmol/h/ml) | Specific activity* (μmol/h/mg protein) | Activity/g fresh tissue (μmol/h) |
|---|---|---|---|---|---|
| Kidney (14.89g) | 70 | 20.60 | 1.665 | 0.080 | 7.747 |
| Liver (89.31g) | 310 | 28.20 | 0.492 | 0.017 | 1.663 |
| Brain (7.15g) | 27 | 6.98 | 0.029 | 0.004 | 0.102 |

*Specific activity expressed as μmol of kynurenic acid/h per mg of protein

**Table 2.** Activity of kynurenine aminotransferase in kidney, liver, and brain of male Brown-Norway rats

| Tissue homogenate (Supernatant) | Volume (ml) | Protein (mg/ml) | Activity (μmol/h/ml) | Specific activity* (μmol/h/mg protein) | Activity/g fresh tissue (μmol/h) |
|---|---|---|---|---|---|
| Kidney (8.29g) | 28 | 21.50 | 2.016 | 0.094 | 6.825 |
| Liver (43.85g) | 145 | 40.00 | 0.660 | 0.016 | 2.180 |
| Brain (8.06g) | 22 | 11.50 | 0.048 | 0.004 | 0.125 |

*Specific activity expressed as μmol of kynurenic acid/h per mg of protein

Comparing KAT activity in the two strains of rats, we observe that specific activity is similar in the kidney, liver and brain of YOS and BN rats, but considered as activity per gram of fresh tissue, activity is slightly higher in YOS (7.747 μmol/h) than in BN rat kidney (6.825 μmol/h). Activity in liver is higher in BN (2.180 μmol/h) than in YOS rats (1.663 μmol/h).

Table 3 reports the concentrations of total (albumin-bound + free) and free tryptophan (Trp) in the sera of YOS and BN rats. The levels of total and free serum tryptophan are clearly significantly higher (P 0.002) in BN (37.36±0.84 μg/ml for total and 4.1± 0.71 μg/ml for free Trp) than in YOS rats (29.53±1.24 μg/ml for total and 3.32± 0.40 μg/ml for free Trp).

## DISCUSSION

The specific activity of KAT in YOS hypercholesterolemic rats did not differ substantially from the values in BN normolipidemic rats. However, liver KAT activity in

**Table 3.** Serum concentrations of tryptophan (Trp) in YOS and BN rats. Mean values expressed as μg/ml ± S.E.

| Rat strain | Total Trp* (μg/ml) | Free Trp (μg/ml) |
|---|---|---|
| YOS | 29.53 ± 1.24 | 3.32 ± 0.40 |
| BN | 37.36 ± 0.84 | 4.10 ± 0.71 |
| | P < 0.002 | N.S. |

*Total Trp = albumin-bound + free

both strains appeared to be markedly lower when compared with that of kidney. Comparing KAT activity per gram of fresh tissue, some variations were observed. In the kidney of BN rats, KAT activity is slightly lower than in YOS rats, but this is compensated by higher activity in liver. Instead the activities of this enzyme differ among other strains of rats previously studied by us (Costa et al., 1982; Allegri et al., 1982).

Brain KAT showed very low activity in both groups of rats. It has been reported that rat brain contains two kinds of kynurenine aminotransferase: kynurenine-pyruvate aminotransferase (Minatogawa et al., 1974) and kynurenine-2-oxoglutarate aminotransferase (Noguchi et al.,1975). Minatogawa et al. (1974) demonstrated that KAT in rat brain is more active for pyruvate as an amino-acceptor than for 2-oxoglutarate, whereas in mammalian liver and kidney KAT showed a higher affinity for 2-oxoglutarate than for pyruvate. We here report KAT activity in brain using 2-oxoglutarate.

On the basis of the present findings, it may be concluded that specific KAT activity does not differ between the groups of rats examined.

Regarding concentrations of tryptophan in serum, BN rats have greater availability of tryptophan to be metabolized to nicotinic acid than YOS rats. This condition may play a role in the development of dyslipidemia in YOS rats, or may represent a consequence of an alteration in tryptophan metabolism. All these data suggest that further investigation on tryptophan function and also on the kynureninase activity in YOS rats should be pursued.

## REFERENCES

Allegri G., Costa C., De Antoni A., Baccichetti F., Vanzan S., and Rubaltelli F.F., 1982, Effect of exposure to light on enzyme activities and tryptophan metabolites of the kynurenine pathway in Wistar, in heterozygous and homozygous adult and newborn Gunn rats, *Photochem.Photobiol.*, 35: 691.

Altschul R., Hoffer A., and Stephen J.D., 1995, Influence of nicotinic acid on serum cholesterol in man, *Arch. Biochem.,* 54: 558.

Chinellato A., Ragazzi E., Pandolfo L., Palermo A., Alvano A., Froldi G., De Biasi M., Caparrotta L., Alievi G, and Fassina G., 1994, Functional and morphologic characterization of thoracic aorta in heritable hyperlipidemic Yoshida rats of different ages, *J.Cardiovasc. Pharmacol.*, 24: 216.

Costa C., De Antoni A., Baccichetti F., Vanzan S., Appodia M., Allegri G., 1982, Strain differences in the tryptophan metabolite excretion and enzyme activities along the kynurenine pathway in rats, *Ital.J.Biochemistry*, 31: 412.

Costa C., Bettero A., Allegri G., 1987, Rapid measurements of tryptophan, 5-hydroxytrytophan, serotonin and 5-hydroxyindoleacetic acid in human CSF by selective fluorescence and HPLC, *Giorn.It.Chim.Clin.*, 12: 307.

Eccleston E.G., 1983, A method for the estimation of free and total acid- soluble plsma tryptophan using an ultrafiltration technique, *Clin.Chim.Acta*, 48: 269.

Fantappié S., Catapano A.L., Cancellieri M., Fasoli L., Fabiani E.De, Bertolini M., and Bosisio E., 1992, Plasma lipoproteins and cholesterol metabolism in Yoshida rats: an animal model of spontaneous hyperlipidemia, *Life Sci.* 50: 1913.

Grundy S.M, Mok H.Y.I., Zech L. and Berman M., 1981, Influence of nicotinic acid on metabolism of cholesterol and tryglicerides in man, *J. Lipid Res.*, 22: 24.

Kritchevsky D., 1971, in Metabolic Effects of Nicotinic Acid and its Derivatives, K.F. Gey and L.A. Carlson, Eds., Bern, Hans Huber, p. 541.

Levy R.I., 1980, Drugs used in the treatment of hyperlipoproteinemias, in: The Pharmacological Basis of Therapeutics, 6th Ed., A.G. Gilman, L.S. Goodman and A.Gilman, Eds., New York, MacMillan Co., pp. 834-847.

Lowry O.H., Rosebrough N.J., Farr A.L., and Randall R.J., 1951, Protein measurement with the Folin phenol reagent, *J.Biol.Chem.*, 193: 265.

Mason M., 1954, The kynurenine transaminase of rat kidney, *J.Biol.Chem.,* 211: 839..

McMenamy R.H., Oncley J.L., 1958, The specific binding of tryptophan to serum albumin, *J.Biol.Chem.*, 233: 1436.

Minatogawa Y., Noguchi. T, Kido R., 1974, Kynurenine pyruvate transaminase in rat brain, *J.Neurochem*, 23: 271.

Noguchi T., Minatogawa Y., Okuno E., Nakatani M., Morimoto M. and Kido R., 1975, Purification and characterization of kynurenine-2-oxoglutarate aminotransferase from the liver, brain and small intestine of rats, *Biochem.J.*, 151: 399.

73

# LIVER SPECIFIC KYNURENINE (ALANINE)

## Glyoxylate Aminotransferase was Expressed in Kidney Cell Line

Y. Minatogawa, C. Kawai, S. Hatada, and M. Sato

Department of Biochemistry
Kawasaki Medical School
577 Matsushima, Kurashiki City, Okayama, 701-01 Japan

## SUMMARY

Kynurenine (Alanine): glyoxylate aminotransferase (AGT) expression plasmid in the COS-7 was constructed. pGV-C was used as a expression vector which contains SV-40 promoter and enhancer. pGV-C and human AGT clone, H1-2, were digested by Hind III/ Sma I separately. The AGT fragment was inserted into the digested pGV-C large fragment. The constructed plasmid was named as pGV-AGT. The constructed plasmid was transfected to COS-7 cultured cell by electroporation. The best electroporation condition was checked.

## INTRODUCTION

Kynurenine (Alanine): glyoxylate aminotransferase (AGT) has broad substrate specificity and it has a catalytic activity to kynurenine (Okuno et al., 1980). The other enzymes have also transamination activity to kynurenine such as α-aminoadipate: α-ketoglutarate aminotransferase and mitochondrial aspartate: α-ketoglutarate aminotransferase (Noguchi et al., 1975; Tobes and Mason, 1977). In these enzymes, AGT is the most effective enzyme for kynurenine, especially in human. The AGT is localized in liver peroxisomes, but not in the other organs in human (Nakatani et al., 1985). We have interest in the mechanisms of liver specific expression and of peroxisomal translocation of this enzyme. For the translocation analysis, this enzyme should be expressed in extra liver cells which has originally no AGT activity. In a preliminary examination, COS-7 which is a cell line from monkey kidney had no intrinsic AGT activity. Mammalian cell expression plasmid of AGT was constructed and was transfected to COS-7. The transformed COS-7 expressed AGT effectively in the cell.

*Recent Advances in Tryptophan Research,* edited by Graziella Allegri Filippini et al.
Plenum Press, New York, 1996

# MATERIALS AND METHODS

## Plasmid Construction

Vector plasmid, pGV-C, was purchased from Toyo Ink Co. Ltd., Tokyo, Japan. This vector contains SV-40 promoter and enhancer with a replication origin. pGV-C and human AGT clone, H1-2 (Takada et al., 1990), were digested by Hind III/ Sma I, separately. The AGT fragment was inserted into the digested pGV-C large fragment. The constructed plasmid was named as pGV-AGT (Fig. 1).

## Transformation of COS-7

Plasmid was introduced to COS-7 by the electroporation method by Gene pulser (Bio-Rad, Cal. USA). COS-7 was cultured with Dulbecco's modified Eagle's medium (DMEM) with 10 % fetal bovine serum. Just before the transformation, cells were harvested by trypsin, and the medium was changed to PBS. The cell concentration was adjusted to 5 x 106 / 0.8 ml PBS. Ten μg pGV-AGT and 300 μg of salmon sperm DNA as a carrier were dissolved in 50 μl PBS. DNAs were mixed with the cell suspension. Transformation by electroporation was carried out by 960 μF capacitance, 0.24-0.32 kV. After the electric shock, cells were immediately transferred to DMEM with 10 % fetal bovine serum, and cultured for two to three days. The AGT activity was transiently expressed.

## Enzyme Assay and Protein Determination

Kynurenine: glyoxylate aminotransferase activity of the transformed cell was not enough to assay. Alanine: glyoxylate aminotransferase activity, therefore, was assayed (Okuno et al., 1980) instead of kynurenine: glyoxylate aminotransferase activity. The assay was based on the determination of NADH consumption (Okuno et al., 1980).

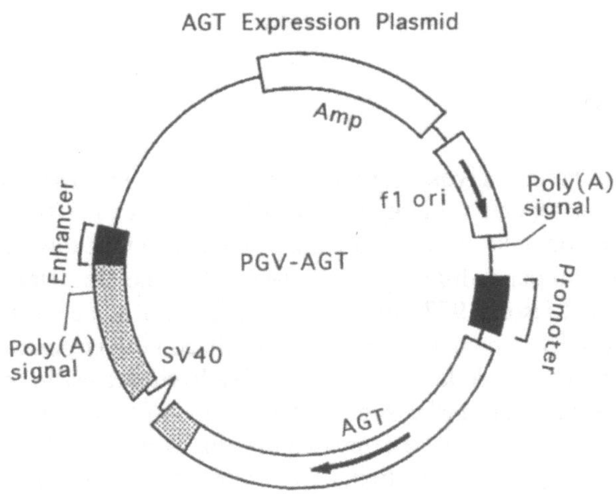


**Figure 1.** Construction of Kynurenine (Alanine): Glyoxylate Aminotransferase expression plasmid. Vector pGV-C contains promoter and enhancer with fi ori. Kynurenine (Alanine): glyoxylate aminotransferase (AGT) DNA fragment (1424 bp) was inserted in Hind III / Sma I site of the vector. The constructed expression plasmid (6046 bp) is called as pGV-AGT.

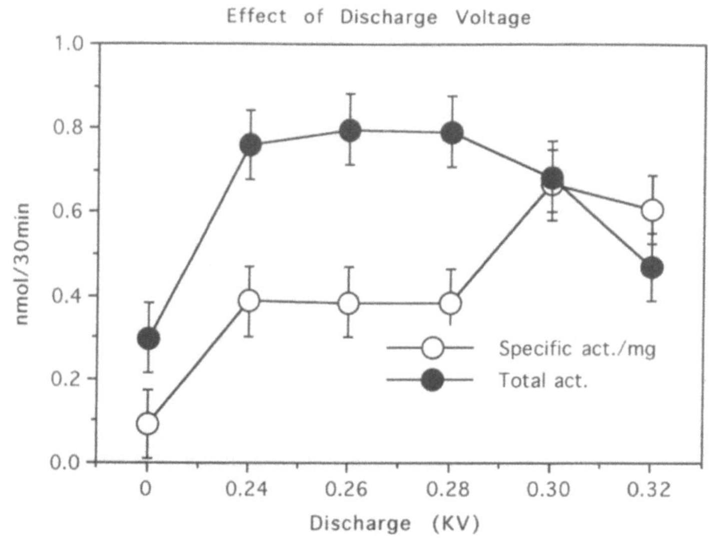


**Figure 2.** Transformation efficiency of discharged voltage for electroporation COS-7 cells ( 5 x 106 ) were suspended in 0.8 ml of PBS. The capacitance was fixed at 960 μ farad. The discharge voltages were varied from 0.24 to 0.32 kV. Two days after the electroporation, the transformed cells were homogenized and assayed the AGT activity. Enzyme assay was described in the text. Total activity (●) and specific activity (O) were determined.

Protein was determined by the method of Lowry et al.(1951). Bovine serum albumin was used as a standard.

## RESULTS AND DISCUSSION

Electroporation is a drastic method for the cultured cell. Higher voltage may result in better transfection, but it may result in lower viability of the cell. Appropriate voltage is required to get the best transformation efficiency. The discharged voltage was varied from 0.24 to 0.32 kV under 960 μF capacitance (Fig. 2). The highest expression efficiency per mg protein was obtained by 0.30 kV, but the total activity was decreased. It means that application of 0.30 kV introduced the expression plasmid most effectively to the COS-7 cell, but the lived cells were decreased dramatically by the electric shock (data not shown). By 0.24 to 0.28 kV, the specific activity of the expressed AGT was not so high, but the total activity was reached to the maximum. For the targeting analysis, high yield of AGT activity is needed, because the subcellular fractionation by density gradient centrifugation of the COS-7 homogenate should be done. We selected the application of 0.26 kV.

Twenty μg / 0.8 ml or more pGV-AGT showed the highest specific activity, but about 10 μg of pGV-AGT resulted in the effective expression efficiency as a total activity (Fig. 3). Ten μg/0.8 ml of pGV-AGT is the best choice for this experiment.

AGT was expressed with a small lag time in the transformed cell, and reached plateau at 2 days and maintained the same activity to 7 days (Fig. 4). The specific activity, however, was linearly increased to 2 days, but then it decreased rapidly, and reached to near basal level at 7 days. This result showed that in a 2 to 3 days after the electroporation, the viability of the cells was recovered and they were divided effectively with increasing synthesis of proteins, but the copy number of the transfected plasmid did not increased.

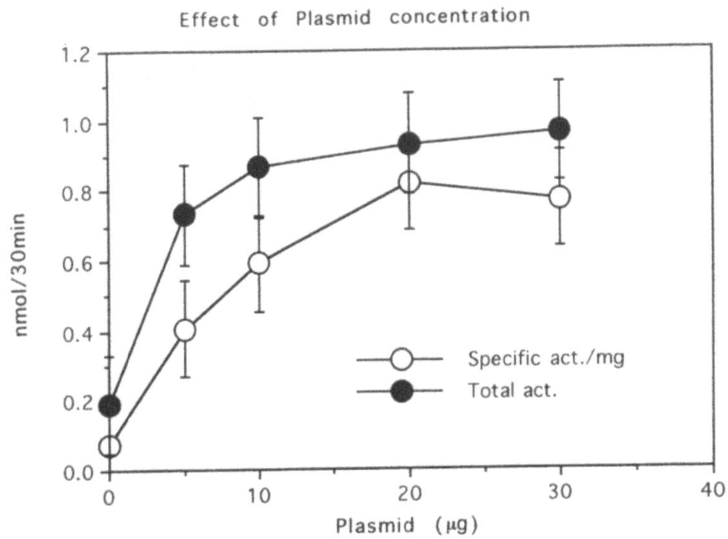


**Figure 3.** Transformation efficiency of expression plasmid concentration for electroporation. Variable expression plasmid, pGV-AGT, from 0 to 30 μg was used. Thirty μg salmon sperm DNA was concomitantly used as a carrier. Treatments after the electroporation were as same as described in Fig. 2. Total activity (●) and specific activity (O) were determined.

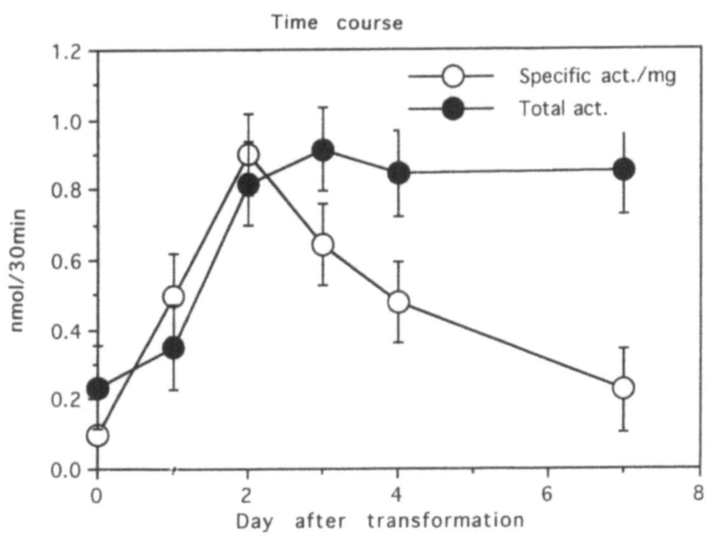


**Figure 4.** Effect the culture time after the electroporation. COS-7 ( 5 x 106 ) transformation was carried out in 10 mg pGV-AGT / 0.8 μl PBS with 0.26 kV and 960 μF. The time course of the transformants were examined. Treatments for the enzyme assay were as same as described in Fig. 2. Total activity (●) and specific activity (O) were determined.

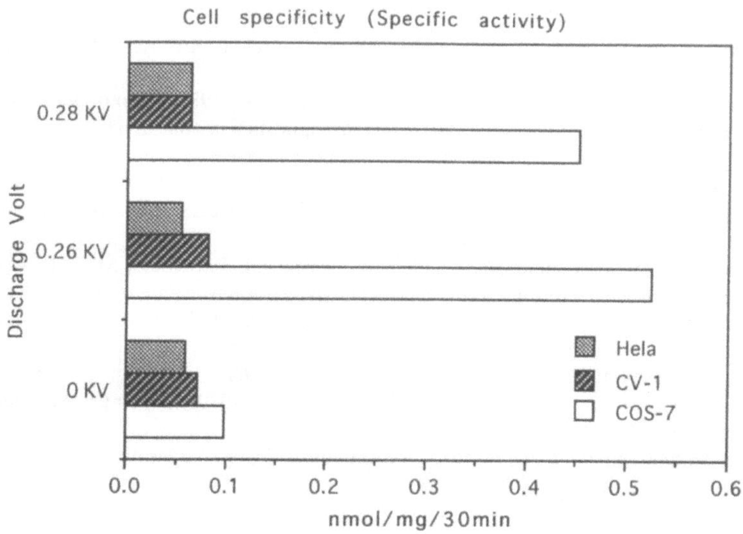


**Figure 5.** AGT expression efficiency by several cell lines by the use of electroporation method. Hela cell, cancer cell of human cervix uteri, CV-1 cell, monkey kidney cell, and COS-7 cell, transformed CV-1 cell, were transformed by pGV-AGT. The transformation conditions were described in the text. Only COS-7 showed detectable AGT activity.

From above results, transformation was carried out by 10 μg pGV-AGT with 30 μg salmon sperm DNA, with 0.26-0.28 kV / 960 μF, for 5 x 106 cells per 0.8 ml PBS, and resulted cells were cultured for further 2 days in DMEM with 10 % fetal bovine serum.

Transformation of the other cultured cells, Hella cell and CV-1 cell, were carried out (Fig. 5). But they did not express detectable amount of AGT. Only COS-7 expressed AGT effectively. CV-1 is a parent cell of COS-7 (Riken Cell Bank, 1992), but the expression of AGT is not enough. COS-7 is a transformed cell of CV-1 to express large T antigen of SV-40. The large T antigen results in increase the copy number of a plasmid which has the replication-origin sequence of SV-40 DNA. The pGV-AGT may increased the copy number in the COS-7 by the large T antigen.

This article is a report of a method to construct an expression plasmid of kynurenine (alanine): glyoxylate aminotransferase, and of several checks for transfection conditions by electroporation. The enzyme activity is not sensitive to assay under low expression condition. COS-7 is a proper cell line to get enough copy number of the expression plasmid and it results in high expression of the transformed enzyme to assay the enzyme activity under usual assay method. We are preparing antibody to the kynurenine (alanine): glyoxylate aminotransferase. This antibody and the transformation method to a heterogeneous cell line may help us to clarify the way and machineries of peroxisomal translocation of this enzyme.

## ACKNOWLEDGMENT

The excellent technical support for electroporation of Prof. Ayako Ueki and Assistant prof. Akio Harano are gratefully acknowledged. This study was supported in part by a grant (No. 6-101) of Kawasaki Medical School Research Project.

# REFERENCES

Lowry, O.H., Rosebrough, N.J., Farr, A.L. and Randall, R.J., 1951, Protein measurement with the Folin phenol reagent. *J. Biol. Chem.* 193: 265

Nakatani, T., Kawasaki, Y., Minatogawa, Y., Okuno, E. and Kido, R., 1985, Peroxisome localized human hepatic alanine-glyoxylate aminotransferase and its application to clinical diagnosis. *Clin. Biochem.* 18: 311.

Noguchi, T., Minatogawa, Y., Okuno, E., Nakatani, M., Morimoto, M. and Kido, R., 1975, Purification and characterization of kynurenine-2-oxoglutarate aminotransferase from the liver, brain and small intestine in rats. *Biochem. J.* 151: 399.

Okuno, E., Minatogawa, Y., Nakamura, M., Kamoda, N., Nakanishi, J., Makino, M. and Kido, R., 1980, Crystallization and characterization of human liver kynurenine-glyoxylate amino transferase: Identity with alanine-glyoxylate amino transferase and seine-pyruvate aminotransferase. *Biochem. J.* 189: 581.

Riken Cell Bank, Cell Line Catalog No5, 1992, Cell No. RCB539, p 98.

Takada, Y., Kaneko, N., Esumi, H., Purdue, P.E. and Danpure, C.J., 1990, Human peroxisomal L-alanine: glyoxylate aminotransferase. *Biochem. J.* 268: 517.

Tobes, M.C. and Mason, M., 1977, α-Aminoadipate aminotransferase: purification, characterisation, and further evidence for identity. *J. Biol. Chem.* 252: 4591.

# MOLECULAR CHARACTERISATION OF KYNURENINE PATHWAY ENZYMES

## 3-Hydroxyanthranilic-Acid Dioxygenase and Kynurenine Aminotransferase

A. M. Cesura,[1] D. Alberati-Giani,[1] R. Buchli,[1] C. Broger,[2] C. Köhler,[1] F. Vilbois,[3] H.-W. Lahm,[3] M.-P. Heitz,[1] and P. Malherbe[1]

[1] Pharma Division, Preclinical Research, Nervous System Diseases
[2] Computational Chemistry Department
[3] Gene Technology
F. Hoffmann-La Roche Ltd.
CH-4002 Basel, Switzerland

## 1. INTRODUCTION

The findings that the L-tryptophan metabolites kynurenic acid (KYNA) and quinolinic acid (QUIN) interfere with glutamate excitatory receptors and that these kynurenines are present in the brain of various animal species including man, has greatly stimulated research on their role in neuropathophysiology. These two compounds influence N-methyl-D-aspartate (NMDA) receptors in opposite way, with KYNA acting as an antagonist at the glycine co-agonist site, whereas QUIN is an agonist at the receptor. It has been postulated that KYNA and QUIN may exert a neuromodulatory function and that alteration of the kynurenine pathway may be among the causative factors leading to several neurological disorders (for review see Stone, 1993), in particular those associated with immune activation (Heyes, 1993). It has therefore been suggested that drugs interfering with this metabolic route may prove beneficial in treating such diseases (Pellicciari et al., 1994).

Knowledge on the amino acid sequences and on the genes encoding the enzymes of the kynurenine pathway is instrumental for the study of several aspects of the function of this metabolic route, including the regulation of the expression of these enzymes, their localisation within the brain, and role played in pathological conditions.

In this paper we report on the molecular characterisation of 3-hydroxyanthranilic-acid dioxygenase (3-HAO) and kynurenine aminotransferase (KAT), the enzymes directly responsible for the synthesis of QUIN and KYNA, respectively.

*Recent Advances in Tryptophan Research*, edited by Graziella Allegri Filippini et al.
Plenum Press, New York, 1996

## Abbreviations Used

3-HAO, 3-hydroxyanthranilic-acid dioxygenase; KAT, kynurenine aminotraneferase; GTK, glutamine transaminase K; AAT, aspartate aminotransferase; KYNA, kynurenic acid; QUIN, quinolinic acid; 3-HANA, 3-hydroxyanthranilic acid; PLP, pyridoxal-5′-phosphate; RT, reverse transcription; PCR, polymerase chain reaction; MS, mass spectrometry.

# 2. HUMAN 3-HYDROXYANTHRANILIC-ACID DIOXYGENASE: CLONING AND FUNCTIONAL EXPRESSION

3-HAO (EC 1.13.11.6) is a monomeric cytosolic protein belonging to the family of intramolecular dioxygenases containing non-heme ferrous iron ($Fe^{2+}$). The enzyme catalyses the cleavage of the benzene ring of 3-hydroxyanthranilic acid (3-HANA). The reaction product is α-amino-β-carboxymuconic acid ε-semialdehyde which can spontaneously rearrange to QUIN or be converted to picolinic acid after decarboxylation. 3-HAO is widely ditributed in peripheral organs and is also present in low amounts in the brain where it appears to be localised in astroglial cells (Köhler et al., 1988).

To obtain information on the amino acid sequence of 3-HAO, the protein was purified from rat liver following the procedure shown in Table 1 (Malherbe et al., 1994). As observed for other dioxygenases of similar function, 3-HAO was relatively unstable upon manipulation, possibly due to air oxidation and/or iron dissociation, but the enzyme activity could be restored after partial acid denaturation and renaturation in the presence of $Fe^{2+}$ and reducing agents. The purified protein showed a pI value of ≈5.2 and SDS-PAGE analysis indicated a molecular mass of ≈33 kDa. More precise determination of the molecular mass of rat liver 3-HAO could be achieved by ion-spray mass-spectometrical (MS) analysis of the purified protein which revealed a monoisotopic mass of 32,627±3 Da (Fig. 1).

To isolate a human cDNA clone encoding human 3-HAO, degenerate oligonucleotides were designed from the amino acid sequences of tryptic peptides of rat liver 3-HAO, and they were used as primers for reverse-transcription (RT)-polymerase chain reaction (PCR) of rat liver RNA. The resulting rat cDNA product was used to screen a human hepatoma cell line (Hep $G_2$) cDNA library. The isolated human clone was found to have an insert of 1276 nucleotides. The deduced primary structure of human 3-HAO is composed of 286 amino acid residues (Fig. 2) with the predicted molecular mass of 32,589 Da, a value very close to the mass obtained by MS analysis of the purified rat enzyme. The human sequence exhibits high similarity (94%) to the rat partial amino acid sequence deduced from the rat RT-PCR fragment (Fig. 2). Analysis of the hydrophilicity plot of the amino acid

**Table 1.** Summary of the purification steps for 3-HAO from rat liver

| | Total protein (mg) | Total activity (μmol/min) | Specific activity (μmol/min/mg protein) | |
|---|---|---|---|---|
| | | | | Activated[a] |
| Homogenate supernatant | 750 | 105 | 0.14 | 0.19 |
| DEAE-Sepharose | 12.7 | 19.8 | 1.56 | 3.09 |
| Sephadex G-75 | 4.6 | 17.6 | 3.79 | 5.30 |
| Mono Q | 0.4 | 3.6 | 8.95 | 53.80 |

Activity was measured with 0.1 mM 3-HANA for 15 sec at 30°C. [a]The activation procedure was performed as described by Malherbe et al., (1994).

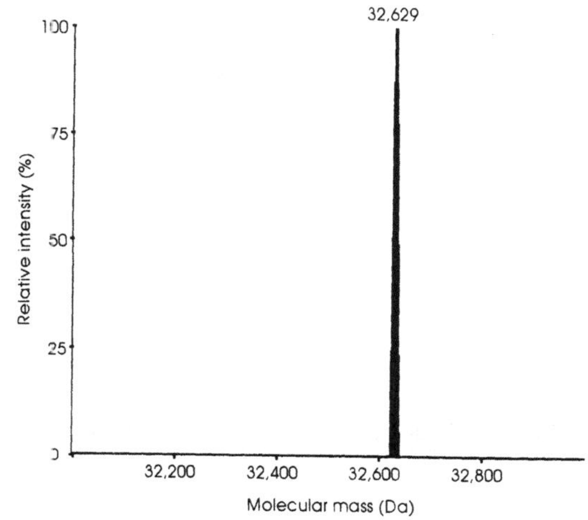


**Figure 1.** Reconstructed mass spectrum of native 3-HAO from rat liver. Purified 3-HAO was desalted by perfusion chromatography before MS analysis. Mass spectra were obtained in the positive ion mode using a triple-quadrupole mass spectrometer equipped with an ion-spray (pneumatically assisted electrospray) interface and an atmospheric pressure ion source.

sequence predicted for human 3-HAO did not reveal the existence of membrane-anchoring or spanning regions, consistent with the soluble nature of the protein. In addition, no significant similarity was found between the nucleotide and amino acid sequences of 3-HAO with any sequence in the GenBank™/EMBL and Swiss-Prot Data Banks. RNA blot analysis of human liver and Hep $G_2$ cells revealed one major species of human 3-HAO mRNA of ≈1.3 kb.

Insertion of the human 3-HAO-coding sequence into an eukaryotic expression vector yielded relatively high amounts of the active enzyme in human embryonic kidney 293 (HEK-293) cells. The $K_m$ value of 3-HANA for recombinant human 3-HAO (≈2 μM) was in good agreement with that reported for the native enzyme. The activity of the recombinant protein was potently inhibited by the substrate analogue 4-Br-3-HANA (Fig. 3), a specific inhibitor of 3-HAO (Walsh et al., 1991). In addition, QUIN could be identified as reaction product of the recombinant enzyme by capillary electrophoresis analysis.

```
h3-HAO: MERRLGVRAWVKENRGSFQPPVCNKLMHQEQLKVMFVGGPNTRKDYHIEEGEEVFYQLEG   60
r3-HAO:                                                       FYQLEG    6
                                                              ******

h3-HAO: DMVLRVLEQGKHRDVVIRQGEIFLLPARVPHSPQRFANTVGLVVERRRLETELDGLRYYV  120
r3-HAO: DMVLRVLEQGEHRDVVIRQGEIFLLPARVPHSPQRFANTMGLVIERRRMETELDGLRYYV   66
        **********.****************************.***.****.***********

h3-HAO: GDTMDVLFEKWFYCKDLGTQLAPIIQEFFSSEQYRTGKPIPDQLLKEPPFPLSTRSIMEP  180
r3-HAO: GDTEDVLFEKWFHCKDLGTQLAPIIQEFFHSEQYRTGKPNPDQLLKEPPFPLSTRSVMEP  126
        *** ********.**************** ********* ****************.***

h3-HAO: MSLDAWLDSHHRELQAGTPLSLFGDTYETQVIAYGQGSSEGLRQNVDVWLWQLEGSSVVT  240
r3-HAO: MSLKAWLESHSRELQAGTFLSLFCDSYETQ                                156
        ***.***.** ******* **** *.****

h3-HAO: MGGRRLSLAPDDSLLVLAGTSYAWERTQGSVALSVTQDPACKKPLG                286
```

**Figure 2.** Predicted amino acid sequence of human 3-HAO (h3-HAO) and comparison with the partial amino acid sequence of rat 3-HAO (r3-HAO) deduced from the PCR fragment. * Indicates amino acid identities, · , conservative amino acid substitutions. The tryptic peptides of r3-HAO sequenced are underlined.

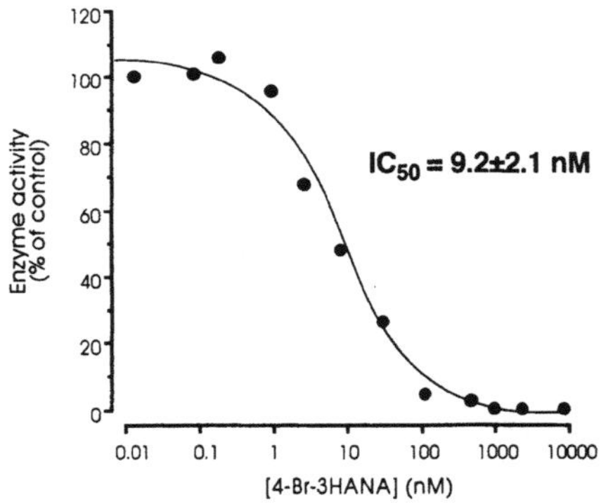


**Figure 3.** Inhibition of human recombinant 3-HAO by 4-Br-3HANA. The substrate concentration was 50 μM.

# 3. KYNURENINE AMINOTRANSFERASE

## 3. 1. KAT Isoforms in the Rat Brain

KAT (EC 2.6.1.7) catalyses the irreversible transamination of L-kynurenine to form KYNA. Since several pyridoxal-5′-phosphate (PLP)-dependent aminotransferases with broad substrate specificities are apparently able to catalyse this reaction, the identification of the isoenzymes responsible for the biosynthesis of cerebral kynurenic acid is of importance for clarifying the role of this metabolic pathway in pathophysiology. Due to the postulated role of KYNA as a putative endogenous modulator of glutamatergic neurotransmission (see above), particular attention has been devoted to the presence of KAT isoenzymes in cerebral tissues (Okuno et al., 1990; 1991). A cDNA clone from rat brain encoding a soluble enzyme with KAT activity has been isolated (Alberati-Giani et al., 1995) and it has been found that this protein corresponds to a cytosolic form of rat kidney glutamine transaminase K (GTK, EC 2.6.1.14) (Perry et al., 1993; Mosca 1994). This aminotransferase form displays substantial cysteine-S-conjugate β-lyasic activity (Perry et al., 1993) and has also been referred to under this nomenclature. Due to its β-lyasic activity, this enzyme may be involved in the renal and cerebral toxicity of several halogenated xenobiotics (Cooper, 1994). The human form of kidney cytosolic cysteine-S-conjugate β-lyase/GTK has been recently cloned and shown to have 82% amino acid similarity to the rat form (Perry et al., 1995).

In addition to the soluble cytosolic enzyme (S-KAT/GTK), the existence of a mitochondrial KAT/GTK (M-KAT/GTK) located in the matrix of mitochondria (Stevens et al., 1988) has also been reported. Interestingly, mitochondrial GTK activity appears to be predominant in rat brain, whereas the opposite is true for peripheral organs such as kidney (Cooper et al., 1993). We recently isolated a rat brain cDNA clone encoding a KAT/GTK enzyme identical to the cytosolic form but carrying an additional oligonucleotide stretch coding for 32 amino acids at the $NH_2$-terminus of the protein (Fig. 4) (Malherbe et al., 1995).

Several structural features of this amino acid sequence resemble those of leader peptides for mitochondrial import (von Heijne et al., 1989): i) positively charged amino acids are distributed along this sequence (predicted isoelectric point, 12.2); ii) the peptide is enriched in Ala (15.6%), Leu (15.6%), hydroxylated amino acids (Ser + Thr, 15.6%) and Arg (9.3%); iii) it lacks acidic amino acid residues and iv) contains an hydrophobic core. Its

```
        PCRI
-149  CTATTGGACGGAGCAGCGCAATTGTTAGCTGAAGCAGAACTGTGTGTGGACTCAGGCCCT
 -89  GGCTTGGAGCCATTTTCTGGGCTAGGCTGTCTGCCCTTCTGTCCCTCTGGAGGGGAAGCC
                                      +1
 -29  TGCAGTGCCTGTGGACCTACCTCAGAGGCATGTTCAGGAGTGCAGCAGCCCTCTCGGTGC
                                   M  F  R  S  A  A  A  L  S  V  H     11
  32  ACCTGATGTGGCCACTCTGGGGAAGGAAAGCTGGAGCCTCACTCACCCGGTGCTTGCACC
       L  M  W  P  L  W  G  R  K  A  G  A  S  L  T  R  C  L  H  Q     31
  92  AGTCTCTCACCATGACCAAACGGCTGCAGGCTCGGAGGCTGGACGGGATTGATCAAAACC
       S  L  T  M  T  K  R  L  Q  A  R  R  L  D  G  I  D  Q  N  L     51

1292  TCCGATTCTGTTTTGTCAAGGACAAGGCCACACTCCAGGCCATGGATGAGAGACTGCGCA
       R  F  C  F  V  K  D  K  A  T  L  Q  A  M  D  E  R  L  R  K     451
1352  AGTGGAAAGAGCTCCAACCCTGAGGAGGCTGCCCTCAGCCCCACCTCGAACACAGGCCTC
       W  K  E  L  Q  P                                               457
                                                       PCRII
1412  AGCTAT
```

**Figure 4.** Nucleotide and deduced amino acid sequence of the rat brain M-KAT/GTK cDNA clone. The 208 bp/32 amino acid insertion sequence is boxed. The two in-frame initiation codons (ATG) for M- and S-KAT/GTK are shown in bold characters. The nucleotide sequence is numbered from the codon for initiator methionine (+1) for M-KAT/GTK and indicated at the left of each line. The amino acid numbers are indicated at the right of each line. The two oligonucleotide probes (PCRI and PCRII) which were used as primers in RT-PCR of rat brain Poly(A)$^+$RNA are indicated by arrows. The central part of the sequence is not shown in the figure.

$NH_2$-terminal domain has the potential to fold into a moderately amphiphilic α-helix similar to that of the leader peptide of mitochondrial aspartate aminotransferase from both human and rat. These characteristics strongly suggested that the newly isolated cDNA encodes the mitochondrial-associated form of KAT/GTK. Evidence for this was obtained after transfection of HEK-293 cells with the cDNA coding for this longer KAT/GTK isoenzyme. In fact, a significant enrichment of both KAT and GTK enzymatic activities was found in the crude mitochondrial fraction of the transfected cells (Malherbe et al., 1995). Table 2 shows the kinetic properties of the M-and S-form of KAT/GTK, transiently expressed in HEK-293 cells, assayed for KAT activity in the presence of pyruvate and 2-oxoglutarate as amino

**Table 2.** Comparison of KAT activity kinetic properties of the M- and S-forms of KAT/GTK recombinantly expressed in HEK-293 cells

| | M-KAT/GTK | | S-KAT/GTK | |
|---|---|---|---|---|
| Co-substrate (5 mM) | $K_m$ | $V_{max}$ | $K_m$ | $V_{max}$ |
| Pyruvate | 1630 ± 210 | 10.8 ± 0.58 | 1200 ± 300 | 11.4 ± 0.9 |
| 2-oxoglutarate | 96.2 ± 26.1 | 0.39 ± 0.05 | 86.3 ± 3.4 | 0.147 ± 0.3 |

$K_m$, μM; $V_{max}$, nmoles/min/mg protein. KAT activity of M-KAT/GTK and S-KAT/GTK was measured in the crude mitochondrial and soluble fractions, respectively, of the transfected cells.

acceptors. Similar to what has been previosly reported for the cytosolic enzyme (Alberati-Giani et al., 1995), the $K_m$ value of L-kynurenine for M-KAT/GTK was lower in the presence of 2-oxoglutarate than with pyruvate, but the $V_{max}$ value was higher in the presence of the latter amino acceptor. The ≈2.5-fold higher $V_{max}$ value observed for M-KAT/GTK in comparison to the S-form in the presence of 2-oxoglutarate may suggest the existence of some subtle differences in the catalytic properties of the two enzyme forms.

It has been claimed that, in rats, a single pyruvate-preferring enzyme with KAT activity, probably corresponding to the identified KAT/GTK form (Mosca et al., 1994; Alberati-Giani, 1995), is of physiological relevance in the cerebral synthesis of KYNA (Okuno et al., 1991). However, whether in rat brain other aminotransferase forms are involved in the biosynthesis of kynurenic acid remains to be established. Noteworthy, we recently isolated from peripheral tissue a rat cDNA clone encoding another aminotransferase enzyme with KAT activity (Buchli et al., submitted). Binary alignment of this protein with KAT/GTK showed only 18.4% amino acid identity between the two KAT isoforms, with 14 gaps maily located in the $NH_2$- and COOH-termini. Northern-blot analysis of mRNA extracted from rat brain revealed that this enzyme is also present in rat cerebral tissues. This suggest that, in analogy to the situation in the human brain, which contains at least two distinct KAT isoforms (Schmidt et al., 1993), this newly identified aminotransferase may represent a second synthetic enzyme for KYNA in the brain of rats. It should be emphasised that several factors may determine the role of the various KAT forms in KYNA biosynthesis, such as the different affinity constants of these enzymes for L-kynurenine, and their regional and cellular distribution in the different organs.

## 3.2. Molecular Modeling of S-KAT/GTK Structure

A multiple sequence alignment of aspartate aminotransferases (AAT) from various organisms (Swiss-Prot database) with S-KAT/GTK showed that several residues in the central region of the sequences are totally conserved. This suggests that the corresponding three-dimensional structures might also be conserved. We therefore attempted to model the structure of S-KAT/GTK based on that of *E. coli* aspartate aminotransferase (AAT), whose three-dimensional structure has been solved (Jaeger et al., 1994). *E. coli* aspartate aminotranferase (AAT) consist of three parts, an $NH_2$-terminal loose segment and two compact domains. The alignment of the larger domain of *E. coli* AAT (from $Thr^{47}$ to $Ser^{299}$, Swiss-Prot numbering) shows only 13.3% sequence identity with S-KAT/GTK. A preliminary model for this domain of S-KAT/GTK was built by exchanging the amino acids of *E. coli* AAT for those of rat S-KAT/GTK according to the sequence alignment and optimised using the Moloc molecular modeling software. The scores calculated by the 3D-profile method (3d profile, Lüthy et al., 1992) indicated that the model exhibits features of a correctly folded protein. $Lys^{281}$ of S-KAT/GTK correspond to the lysine residue (position 246) of the *E. coli* enzyme, which binds the PLP cofactor. Sequence similarities in the $NH_2$- and COOH-terminal regions were too low for molecular modeling. This suggest that these regions may participate in determining the different substrate specificities of aminotransferase isoforms.

# 4. CONCLUSIONS

The availability of the cDNA clones for 3-HAO and KAT isoforms will certainly contribute to the further clarification of QUIN and KYNA disposition in peripheral organs and in the central nervous system, and of their possible involvement in diseases. In addition, heterologous overexpression of the kynurenine enzymes in sufficient quantities needed for structural and functional analysis will allow the rational design of potent and selective inhibitors which could be used as research tools as well as therapeutics agents.

## 5. REFERENCES

Alberati-Giani, D., Malherbe, P., Köhler, C., Lang, G., Kiefer, V., and Cesura, A. M., 1995, Cloning and characterisation of a soluble kynurenine aminotransferase from rat brain: identity with kidney cysteine conjugate β-lyase, *J. Neurochem.* 64:1448

Cooper A.J., Abraham D.G., Gelbard A.S., Lai J.C.K., and Petito C.K., 1993, High activities of glutamine transaminase K (dichlorovinylcysteine β-lyase) and ω-amidase in the choroid plexus of rat brain, *J. Neurochem.* 61:1731.

Cooper A.J.L., 1994, Enzymology of cysteine S-conjugate β-lyases, *Adv. Pharmacol.* 27:71.

Heyes, M.P., 1993, Metabolism and neuropathologic significance of quinolinic acid and kynurenic acid, *Biochem. Soc. Transac.* 21:83.

Jaeger, J., Moser, M., sauder, U., and Jansonius, J.N., 1994, Crystal structure of Escherichia coli aspartate aminotransferase in two conformations. Comparison of an unliganded and two liganded closed forms, *J. Mol. Biol.* 239:285.

Köhler, C., Eriksson, L.G., Okuno, E., and Schwarcz, R., 1988, Localisation of quinolinic acid metabolising enzymes in the rat brain. Immunohistochemical studies using antibodies to 3-hydroxyanthranilic acid oxygenase and quinolinic acid phosphoribosyltransferase, *Neuroscience* 27:49.

Lüthy, R., Bowie, J.U. and Eisenberg, D., 1992, Assessment of protein models with three-dimensional profiles, *Nature* 356:83.

Malherbe P., Köhler C., Da Prada M., Lang, G., Kiefer V., Schwarcz R., Lahm H.-W and Cesura A.M., 1994, Molecular cloning and functional expression of human 3-hydroxyanthranilic-acid dioxygenase, *J. Biol. Chem.* 269:13792.

Malherbe, P., Alberati-Giani, D., Köhler, C, and Cesura A.M., 1995, Identification of a mitochondrial form of kynurenine aminotransferase/glutamine transaminase K from rat brain, *FEBS Lett.*, in press.

Mosca, M., Cozzi, L., Breton, J., Speciale, C., Okuno, E., Schwarcz, R., and Benatti, L., 1994, Molecular cloning of rat kynurenine aminotransferase: identity with glutamine transaminase K, *FEBS Letters* 353:21.

Okuno E., Du F., Ishikawa T., Tsujimoto M., Nakamura M., Schwarcz R., and Kydo R., 1990, Purification and characterization of kynurenine-pyruvate aminotransferase from rat kidney and brain, *Brain Res.* 534:37.

Okuno E., Schmidt W., Parks D.A., Nakamura M., and Schwarcz R., 1991, Measurement of rat brain kynurenine aminotransferase at physiological kynurenine concentrations, *J. Neurochem.* 57:533.

Pellicciari R., Natalini B., Costantino G., Mahmoud M.R., Mattoli L., Sadeghpour B.M., Moroni F., Chiarugi A., and Carpenedo R., 1994, Modulation of the kynurenine pathway in search for new neuroprotective agents. Synthesis and preliminary evaluation of (m-Nitrobenzoyl)alanine, a potent inhibitor of kynurenine-3-hydroxylase, *J. Med. Chem.* 37:647.

Perry S.J., Schofield M.A., MacFarlane M., Lock E.A., King L.J., Gibson G., and Goldfarb P.S., 1993, Isolation and expression of a cDNA coding for rat kidney cytosolic cysteine conjugate β-lyase, *Mol. Pharmacol.* 43:660.

Perry, S., Harries, H., Scholfield, C., Lock, T., Ling, L., Gibson, G., and Goldfarb, P., 1995, Molecular cloning and expression of a cDNA for human kidney cysteine conjugate β-lyase, *FEBS Lett.* 360:277.

Schmidt W., Giudetti P., Okuno E., and Schwarcz R., 1993, Characterization of human brain kynurenine aminotransferase using [$^3$H]kynurenine as substrate, *Neuroscience* 55,:177.

Stevens J.L., Ayoubi N., and Robbins J.D., 1988, The role of mitochondrial matrix enzymes in the metabolism and toxicity of cysteine conjugates, *J Biol. Chem.* 263:3395.

Stone T.W., 1993, Neuropharmacology of quinolinic and kynurenic acids, *Pharmacol.Rev.* 45:309.

von Heijne, G., 1989, Domain structure of mitochondrial and chloroplast targeting peptide, *Eur. J. Biochem.* 180:535.

Walsh, J.L., Todd, W.P., Carpenter, B.K., and Schwarcz, R., 1991, 4-Halo-3-hydroxyanthranilic acids: potent competitive inhibitors of 3-hydroxyanthralic acid oxygenase in vitro, *Biochem. Pharmacol.* 42:985.

# 75

# KYNURENINE PATHWAY ENZYMES IN BRAIN

## Properties of Enzymes and Regulation of Quinolinic Acid Synthesis

K. Saito[1] and M. P. Heyes[2]

[1] Department of Laboratory Medicine
Gifu University School of Medicine
Gifu 500, Japan
[2] Section on Analytical Biochemistry, Laboratory of Clinical Science
National Institute of Mental Health
Bethesda, Maryland 20892

The kynurenine pathway is the metabolic route by which the essential amino acid, L-tryptophan is catabolized to a number of metabolites for excretion and convertion to nicotinamide-containing nucleotides (Bender, 1982). The liver is viewed as the major source of kynurenine pathway metabolites in blood and urine. Certain kynurenine pathway metabolites are also neuroactive within the central nervous system (CNS), including L-kynurenine (L-KYN) (Lapin, 1982), kynurenic acid (KYNA), 3-hydroxykynurenine (Eastman, 1989) and quinolinic acid (QUIN) (Lapin, 1982; Schwarcz et al., 1983; Rios et al, 1991). Accumulations of these neuroactive metabolites have been implicated in a number of neurodegenerative, convulsant and neurologic diseases, including complex partial seizures, HIV-associated neurologic disease, as well as a broad spectrum of inflammatory conditions that result from microbiological infections, autoimmune processes and brain injury (Lapin, 1982; Heyes et al, 1991; Heyes et al., 1992; Stone, 1993). Normally, the concentrations of kynurenine pathway metabolites within the CNS are low, and only a few metabolic steps have been demonstrated to occur within the normal brain. While direct conversion of L-tryptophan to QUIN has not been demonstrated in brain from normal animals, direct conversion does occur in brain tissue from 4 days following transient cerebral ischemia or macaques from poliovirus infected spinal cord.

In this review, the role of kynurenine pathway enzymes [indoleamine-2,3-dioxygenase (IDO), kynurenine aminotransferase, kynurenine-3-hydroxylase, kynureninase, 3-hydroxyanthranilate-3,4-dioxygenase] in the synthesis of QUIN or kynurenic acid by the CNS were described.

*Recent Advances in Tryptophan Research,* edited by Graziella Allegri Filippini et al.
Plenum Press, New York, 1996

## QUANTITATION OF KYNURENINE PATHWAY ENZYMES

Tissues were disrupted by a Polytron in 3-10 volumes of ice-cold 0.14 M KCl/20 mM potassium phosphate buffer (pH 7.0) unless otherwise stated. For the enzyme assays, the samples were incubated at 37° C while shaking at 100 strokes/min., using the reagents and reaction conditions indicated for individual enzymes. After incubation, samples were acidified, centrifuged at 12,000 g for 10 min., and the enzymatic products were measured. Enzyme activity was calculated from the difference between product content before and after incubation, and expressed as the amount formed/h/g wet tissue weight.

### Indoleamine-2,3-Dioxygenase

The activity of IDO was quantified by the conversion of L-tryptophan to L-KYN (Saito, 1991; Saito, 1993). Tissue homogenates were centrifuged at 25,000 g for 30 min at 4° C; a sample of supernatant was collected. The reaction mixture consisted of 50 μL of tissue homogenate supernatant, and 50 μL of substrate solution containing 100 mM potassium phosphate buffer (pH 6.5), 50 μM methylene blue, 10 μg catalase, 50 mM ascorbate and 0.4 mM L-tryptophan. The enzymatic reaction was terminated after 60 min. by the addition of 0.1 ml of 20% (w/v) trichloroacetic acid, and the mixture was incubated at 50° C for an additional 30 min. to hydrolyze N-formylkynurenine to L-KYN, and the concentration of L-KYN was quantified.

### Kynureninase

The activity of kynureninase was measured by the conversion of L-KYN to anthranilic acid with minor modifications (Takikawa et al., 1986; Saito et al., 1993). Tissue homogenates were centrifuged at 25,000 g for 30 min at 4° C; a sample of supernatant was collected. The reaction mixture consisted of 50 μL of tissue homogenate supernatant, and 50 μL of substrate solution containing 200 mM Tris-HCl buffer (pH 8.0), 100 μM pyridoxal phosphate and 2.0 mM L-KYN. The reaction was terminated after 30 min by the addition of 0.1 ml of 20% (w/v) trichloroacetic acid, and the concentration of anthranilic acid was quantified.

### Kynurenine Aminotransferase

The activity of kynurenine aminotransferase was measured by the conversion of L-KYN to kynurenic acid (Takikawa et al., 1986; Saito et al., 1993) with minor modifications. The reaction mixture consisted of 50 μL of tissue homogenate supernatant, and 50 μL of substrate solution containing 200 mM potassium phosphate buffer (pH 8.0), 200 μM pyridoxal phosphate, 20 mM α-ketoglutarate and 2mM L-KYN. The reaction was terminated after 60 min by the addition of 0.1 ml of 20% (w/v) trichloroacetic acid, and the concentration of kynurenic acid was quantified.

### Kynurenine 3-Hydroxylase

The activity of kynurenine 3-hydroxylase activity was quantified by the conversion of L-KYN to 3-hydroxykynurenine (Saito et al., 1993). Tissues were homogenized in 10 volumes of ice-cold 0.32 M sucrose. Homogenates were centrifuged at 12,000 g for 30 min. at 4° C and the pellet was washed 3 times with 0.32 M sucrose by centrifugation. The pellet was suspended and sonicated in ice-cold 0.14 M KCl / 20 mM potassium phosphate buffer (pH 7.0). The reaction mixture consisted of 50 μL of suspended tissue homogenate, and 50 μL of a substrate solution containing 100 mM potassium phosphate buffer (pH 7.5), 4 mM $MgCl_2$, 3 mM glucose-6-phosphate, 0.4 units of glucose-6-phosphate dehydrogenase, 0.8 mM NADP, and 2.0 mM L-KYN. After 5 min. incubation, the reaction was terminated by the addition of 0.1 ml of 20 % (w/v) trichloroacetic acid, and the concentration of 3-hydroxykynurenine was quantified.

## 3-Hydroxyanthranilate-3,4-Dioxygenase

The activity of 3-hydroxyanthranilate-3,4-dioxygenase activity was quantified by the conversion of 3-hydroxyanthranilic acid to QUIN (Okuno et al., 1987; Saito et al., 1993) with the following modifications. Tissue homogenates were centrifuged at 25,000 g for 30 min at 4° C; a sample of supernatant was collected. The reaction mixture consisted of 50 μL of tissue homogenate supernatant and 50 μL of substrate solution containing 100 mM MES buffer (pH 6.5), 10 μM ascorbate, 6 mM $FeSO_4$ and 10 μM 3-hydroxyanthranilic acid. After 30 min. incubation, the reaction was terminated by addition of 0.1 ml of 3.0 M HCl, and the concentration of QUIN was quantified.

# PROPERTIES OF KYNURENINE PATHWAY ENZYMES

## Indoleamine-2,3-Dioxygenase

Indoleamine-2,3-dioxygenase cleaves the indole ring of tryptophan and other indoles, uses superoxide as oxygen donor and is distinct from hepatic tryptophan-2,3-dioxygenase (Tone, et al.,1990). Indoleamine-2,3-dioxygenase has a widespread organ distribution, and is induced by interferon-γ and other immune stimuli , whereas hepatic tryptophan-2,3-dioxygenase activity is unaffected or reduced following immune activation (Takikawa, et al., 1986; Saito,et al., 1992; Saito, et al., 1991; Taylor, at al., 1991; Saito, et al., 1991; Yasui, et al., 1986). Such induction occurs in patients with cancer, infectious disease and following therapeutic interferon administration, and has been implicated in the anti-proliferative and anti-microbial effects of interferon-γ (Taylor, et al., 1991). Previously, we have demonstrated that IDO activity increases within the CNS during local inflammation, and that in poliovirus-infected macaques, IDO is predominantly localized in glia and inflammatory cell infiltrates (Heyes, et al., 1992; Saito, et al., 1991). Marked increases in L-KYN synthesis from L-tryptophan occurred in cultures consisting of human astrocytes and neurons that were stimulated by interferon-γ. Similar observations have also been noted in rat microglia, oligodendrocytes and astrocytes *in vitro* (Saito, et al., 1993). Collectively, these observations suggest that IDO induction and L-KYN synthesis occurs in several different CNS cell types. Indoleamine-2,3-dioxygenase from the intestine has a $K_m$ for L-tryptophan of 20 μM, which is close to the normal L-tryptophan concentration in CNS tissue (10 μM in spinal cord). Indoleamine-2,3-dioxygenase may be an important regulatory enzyme in the synthesis of L-KYN by CNS tissue in some circumstance.

Several studies have led to the hypothesis that a component of the antimicrobial and antiproliferative effects of interferon-γ in some circumstances is dependent on depletion of L-tryptophan secondary to IDO induction (Carlin et al., 1989; Taylor et al., 1991). Further studies on the effects of IDO in the context of the antimicrobial and antiproliferative effects of interferon-γ are warranted. Another potentially beneficial consequence of IDO induction is the consumption of free radicals. It is also possible that QUIN and other kynurenine pathway metabolites serve as an immunologic 'signaling' system (Varesio et al., 1990) in an analogous way to cytokines, or have antimicrobial effects.

## Kynurenine Aminotransferase

The levels of KYNA in the CSF of poliovirus-infected macaques increased in proportion to both IDO induction and L-KYN accumulation. There was, however, no significant increase in spinal cord kynurenine aminotransferase activity, or correlation between CSF KYNA levels and kynurenine aminotransferase activity. These observations demonstrate that in non-human primates, increases in IDO activity influence KYNA synthesis, and that KYNA formation can be substrate-driven by the availability of L-KYN. The $K_m$ values for brain kynurenine aminotransferase are reported to be 17 μM for 2-oxoglutarate as co-factor and 910 μM for pyruvate as co-factor (Okuno et al., 1991), as compared to L-KYN levels which are in the low nanomolar range in CNS (Saito et al., 1992). Kynurenine aminotransferase, therefore, does not appear to be rate-limiting in KYNA synthesis from L-KYN in poliovirus-infected macaques. This interpretation is also consistent with observations in normal rats, where direct application of L-KYN to brain tissue is associated with *de novo* synthesis of KYNA (Heyes et al., 1990; Turski et al., 1989; Swartz et al., 1990). The absence of changes in kynurenine aminotransferase activity is consistent with the relative lack of astrogliosis in the spinal cord of poliovirus-infected macaques, and evidence that kynurenine aminotransferase activity is localized in astrocytes (Turski et al., 1989; Okuno et al., 1990).

# KYNURENINE-3-HYDROXYLASE AND KYNURENINASE

## Kynurenine-3-Hydroxylase

Kynurenine 3-hydroxylase is a L-tryptophan to kynurenine pathway enzyme that catalyzes the conversion of L-kynurenine to 3-hydroxykynurenine (3-HKYN). The enzyme is FAD dependent and uses NADPH as an electron donor. Although the rate of L-tryptophan flux through the kynurenine pathway to L-kynurenine and nicotinamide-containing nucleotides in the liver is regulated by the activity of tryptophan-2,3-dioxygenase, the activity of kynurenine 3-hydroxylase may also be rate-limiting in some circumstances.

## Kynureninase

Kynureninase catalyzes the hydrolysis of L-kynurenine to anthranilic acid, and the hydrolysis of 3-hydroxykynurenine to 3-hydroxyanthranilic acid. The enzyme is found predominantly in the cytosol fraction. Liver has the highest activity, whereas brain, lung and skeletal muscle have the lowest kynureninase activity. Kynureninase requires pyridoxal 5'-phosphate (vitamin $B_6$) as a cofactor, and its activity is very sensitive to pyridoxal 5'-phosphate availability. In mammals, the $V_{max}$ is higher and the Km is lower for 3-hydroxykynurenine than L-kynurenine, which favors the conversion of L-tryptophan to 3-hydroxyanthranilic acid via 3-hydroxykynurenine rather than anthranilic acid.

Kynurenine-3-hydroxylase and kynureninase activities were increased in the poliovirus-infected macaques, in proportion to the accumulations of QUIN, lesion scores. While L-tryptophan and L-KYN are reported not to be converted to QUIN by normal CNS tissue (Speciale et al., 1989; Saito et al., 1993; Saito et al, 1992) lumbar spinal cord obtained from poliovirus infected macaques can convert [$^{13}C_6$]-L-tryptophan to [$^{13}C_6$]-QUIN (Heyes et al., 1992).

In addition, direct injection of [$^{13}C_6$]-L-tryptophan into the cisterna magna of poliovirus-infected macaques and 4 days post-ischemic gerbil in brain were also associated with the appearance of [$^{13}C_6$]-QUIN in the CSF in proportion to already existing QUIN levels in the CSF. Human and macaque macrophages stimulated by cytokines (i.e. interferon-γ) converted [$^{13}C_6$]-L-tryptophan to [$^{13}C_6$]-QUIN (Heyes et al, 1992). In contrast, although cultures of human astrocytes and neurons responded to interferon-γ with IDO induction and the synthesis of L-KYN, no conversion to QUIN was found (Saito et al, 1993). Clearly, monocyte-derived cells contain kynurenine-3-hydroxylase, kynureninase and 3-HAD, as well as IDO. The increased activities of kynurenine-3-hydroxylase and kynureninase (Saito et al, 1993) may therefore be secondary to infiltration of enzyme-containing monocytes into the CNS, as well as induction in resident CNS cells, and explain the ability of injured CNS tissue to synthesize [$^{13}C_6$]-QUIN from [$^{13}C_6$]-L-tryptophan. It is important to recognize, however, that exchange of substrates between cell types, as well as the production of cytokines, may render a cell that is unable to synthesize QUIN directly from L-tryptophan, able to affect QUIN synthesis by other cell types. An important but unresolved issue is whether human microglia can convert L-tryptophan to QUIN. No synthesis of [$^{13}C_6$]-QUIN was detected by rat microglia (Saito et al., 1993), although it is of note that this species is very resistant to increased kynurenine pathway metabolism in response to immune stimuli (Saito et al., 1992).

The magnitude of the increases in kynurenine-3-hydroxylase and kynureninase were substantially smaller than that of IDO, although the activities of IDO, kynurenine-3-hydroxylase and kynureninase were highly correlated. These parallel changes suggest a co-localization of these enzymes to responsive cells, or response to a common stimulus, such as cytokines. Perhaps the larger increases in IDO activity in poliovirus-infected macaques represents IDO induction in many CNS cell types, whereas increased kynurenine-3-hydroxylase and kynureninase activities occur predominantly in the monocyte infiltrates. The presence of activated macrophage infiltrates can account for the ability of CNS tissue to convert [$^{13}C_6$]-L-tryptophan to [$^{13}C_6$]-QUIN.

Measurable activities of IDO, kynurenine-3-hydroxylase, kynureninase and 3-HAD were present in uninfected control macaque samples (Heyes et al., 1993). Further, the $K_m$ values for kynurenine-3-hydroxylase (20 μM) and kynureninase (300 μM) are substantially higher than endogenous L-KYN and 3-hydroxykynurenine levels (nanomolar) (Saito et al., 1993). Theoretically then, L-KYN catabolism through kynurenine-3-hydroxylase and kynureninase should be substrate driven. The apparent inability of normal CNS tissue to convert L-tryptophan and L-KYN to QUIN, however, may indicate differential cellular and/or sub-cellular localization of these enzymes and substrates, which compartmentalizes substrate in such a way as to render the kynurenine pathway effectively 'incomplete' within any given cell of the CNS. In this context it is of note that IDO, kynureninase and 3-HAD are found predominantly in cytosol extracts of brain (Foster et al., 1986; Saito et al., 1993), whereas kynurenine-3-hydroxylase is predominantly localized in mitochondria (Battie et al., 1981; Okamoto et al., 1967; Saito et al., 1993).

It is not known whether the activities of either kynurenine-3-hydroxylase or kynureninase or both enzymes are rate-limiting in the conversion of L-KYN to QUIN. No accumulation of 3-hydroxykynurenine was found, however, in the spinal cord of poliovirus-infected macaques. This suggests that 3-hydroxykynurenine is quickly catabolized to 3-hydroxyanthranilic acid, or was transported out of the CNS. Alternatively, 3-hydroxykynurenine may not have been formed. Anthranilic acid is a product on interferon-γ stimulated macrophages (Werner-Felmayer et al., 1989) and there are reports in liver that anthranilic acid may be hydroxylated to 3-hydroxyanthranilic acid and 5-hydroxyanthranilic acid (Kashiwamata et al., 1966; Ueda et al., 1978). Small quantities of [$^{13}C_6$]-QUIN are formed from [$^{13}C_6$]-anthranilic acid by slices of gerbil liver and lung *in vitro,* although the amounts of [$^{13}C_6$]-QUIN formed were not enhanced by prior immune stimulation (Saito et al., 1993). One study has suggested that anthranilic acid also serves as a substrate for the synthesis of 3-hydroxyanthranilic acid and thereby QUIN in the rat brain (Baran et al., 1990). If this finding is correct, then the kynurenine-3-hydroxylase step would be bypassed and 3-hydroxykynurenine would not have been synthesized. However, this report in normal rat CNS tissue has not been confirmed. Further, conversion of [$^{13}C_6$]-anthranilic acid to [$^{13}C_6$]-QUIN was not detected in unstimulated or interferon-γ stimulated THP-1 cells (Saito et al., 1993), or following an intracisternal administration of [$^{13}C_6$]-anthranilic acid into gerbils following ischemic brain injury (Saito et al., 1993). In both conditions, [$^{13}C_6$]-L-tryptophan was shown to be converted to [$^{13}C_6$]-QUIN. At the present time it is not possible to definitively conclude that the anthranilic acid is the preferred route of L-KYN conversion to 3-hydroxyanthranilic acid in brain.

## 3-Hydroxyanthranilate-3,4-Dioxygenase

3-Hydroxyanthranilate-3,4-dioxygenase converts 3-hydroxyanthranilic acid to 2-amino-3-carboxymuconate semialdehyde, an unstable intermediate which spontaneously cyclizes in a non-enzymatic reaction to form QUIN. Several studies have proposed that the activity of 3-HAD controls QUIN formation (Okuno et al., 1987; Schwarcz et al., 1988; Speciale et al., 1993). In poliovirus-infected macaques, 3-HAD activity was not significantly increased, even though local spinal cord QUIN levels were markedly elevated. The activity of 3-HAD did not correlate with the concentrations of QUIN in either the CSF or spinal cord, or with the activities of IDO, kynurenine-3-hydroxylase, kynureninase or kynurenine aminotransferase. Immunocytochemical studies have shown that 3-HAD is predominantly localized in astrocytes, and that brain lesions are associated with increased 3-HAD activity and increased 3-HAD staining (Okuno et al., 1987). The absence of significant increases in 3-HAD activity in the spinal cord of the poliovirus infected macaques is consistent with the relative lack of astrogliosis in the poliovirus model of spinal cord inflammation. Studies in experimental animals have shown that 3-hydroxyanthranilic acid is readily converted to QUIN by uninfected brain tissue and that 3-HAD is not saturated with substrate. The Km for 3-HAD in rat brain is 3 μM (Foster et al., 1986; Okuno et al., 1987), whereas the endogenous concentration of 3-hydroxyanthranilic acid is substantially lower, and ranges from 8 to 22 nM (Baran et al., 1990). Conversely, increases in 3-HAD activity in patients with Huntington's disease (Schwarcz et al., 1988) are not associated with QUIN accumulation in either brain tissue or CSF (Schwarcz et al., 1988; Reynolds et al., 1988; Heyes et al., 1991; Heyes et al., 1992). Therefore, increased 3-HAD activity is not sufficient to increase QUIN formation, and induction of 3-HAD is not essential for accelerating the conversion of 3-hydroxyanthranilic acid to QUIN. Increases in brain 3-HAD activity may facilitate QUIN synthesis if activated macrophages are present, as macrophages synthesize 3-hydroxyanthranilic acid when stimulated by interferon-γ (Werner et al., 1987). The available evidence does not, however, support the hypothesis that 3-HAD is a rate-limiting enzyme in the synthesis of QUIN by brain tissue. Nevertheless, 3-HAD activity is still required to convert 3-hydroxyanthranilic acid to QUIN, and is a target for inhibition to reduce QUIN synthesis.

# CONCLUSION

Clinical studies of patients with immune activation that is present in either CNS and/or systemic tissues have shown substantial accumulations of neuroactive kynurenine pathway metabolites in the CNS and blood. Conceivably, these metabolites may be common mediators of neuronal dysfunction and neurodegeneration in a broad spectrum of inflammatory neurologic disease. The mechanisms responsible for their accumulations have always been viewed as complex, and include *de novo* synthesis in the brain, metabolism in infiltrating cells, entry of systematically synthesized precursors into the CNS, as well as breakdown of the blood-brain barrier. These mechanisms may be disease-selective, and be manifested to different degrees, according to the prevailing pathophysiologic responses, and the stage of disease. In the previous results also support roles for macrophage infiltration, and increased activities and co-localization of IDO, kynurenine-3-hydroxylase and kynureninase in conveying the ability of brain tissue to convert L-tryptophan to QUIN in conditions of brains inflammation. Importantly, kynurenine pathway metabolism withing the CNS is modified following immune activation in a manner that is dependent on whether inflammation is localized in the central compartment or systemic compartment, or both.

## REFERENCES

Bender D.A., 1982, Biochemistry of tryptophan in health and disease. *Molec. Aspects Med.*, 6: 101.

Baran H., Schwarcz R., 1990, Presence of 3-hydroxyanthranilic acid in rat tissues and evidence for its production from anthranilic acid in the brain. *J. Neurochem.*, 55: 738.

Battie C., Verity M.A., 1981, Presence of kynurenine hydroxylase in developing ret brain. *J. Neurochem.* 36: 1308.

Carlin J.M., Ozaki Y., Byrne G.I., Brown R.R., Borden E.C., 1989, Interferons and indoleamine 2,3-dioxygenase: role in antimicrobial and antitumor effects. *Experientia*, 45: 535.

Foster A.C., White R.J., Schwarcz R., 1986, Synthesis of quinolinic acid by 3-hydroxyanthranilic acid oxygenase in rat brain tissue in vitro. *J. Neurochem.* 47: 23.

Heyes M.P., Swartz K.J., Markey S.P., Beal M.F., 1991, Regional brain and cerebrospinal fluid quinolinic acid concentrations in Huntington's disease. *Neurosci. Letters*, 122: 265.

Heyes M.P., Brew B.J., Saito K., Quearry B.J., Price R.W., Bhalla R.B. *et al.*, 1992, Interlationship between neuroactive kynurenines, neopterin $\beta_2$-microglobulin in cerebrospinal fluid and serum of HIV-1 infected patients. *J. Neuroimmunol.*, 40: 71

Heyes M.P., Jordan E.K., Lee K., Saito K., Frank J.A., Snoy P.J. *et al.*, 1992, Relationship of neurologic status in macaques infected with the simian immunodeficiency virus to cerebrospinal fluif and serum quinolinic acid and kynurenic acid. *Brain Res., Amsterdam*, 570: 237.

Heyes M.P., Saito K., Crowley J., Davis L.E., Demitrak M.A., Der M *et al.*, 1992, Quinolinic acid and kynurenine pathway metabolism in inflammatory and non-inflammatory neurologic disease. *Brain*, 115: 1249.

Heyes M.P., Saito K., Jacobowitz D., Takikawa O., Markey S.P., Vickers J., 1992, Poliovirus induces indoleamine-2,3-dioxygenase and quinolinic acid synthesis in macaque brain. *FASEB J.* 6: 2977.

Kashiwamata S., Nakashima K., Kotake Y., 1966, Anthranilic acid hydroxylation by rabbit liver microsomes. *Biochim. Biopfys. Acta*, 113: 244.

Lapin I.P., 1982, Convulsant action of intracerebroventricularly administered L-kynurenine sulphate, quinolinic acid and other derivatives of succinic acid, and effects of amino acids: structure-activity relationship. *Neuropharmacology*, 21: 1227.

Okamoto H., Yamamoto S., Nozake M., Hayaishi O., 1967, On the sub-mitochondrial localization of L-kynurenine-3-hydroxylase. *Biochem. Biophys. Res. Comm.*, 26: 309.

Okuno E., Kohler C., Schwarcz R., 1987, Rat 3-hydroxyanthranilic acid oxygenase: Purification from the liver and immunocytochemical localization in the brain. *J. Neurochem.* 49: 771.

Okuno E., Du F., Ishikawa T., Tsujimoto M., Nakamura M., Schwarcz R. et al., 1990, Purification and characterization of kynurenine-pyruvate aminotransferase from rat kidney and brain. *Brain Res., Amsterdam* 542: 37.

Okuno E., Nakamura M., Schwarcz R., 1991, Two kynurenine aminotransferases in human brain. *Brain Res., Amsterdam* 542: 307.

Okuno E., Schmidt W., Parks D.A., Nakamura M., Schwarcz R., 1991, Measurement of rat brain kynurenine aminotransferase at physiological kynurenine concentrations. *J. Neueochem.* 57: 533.

Reynolds G.P., Pearson S.J., Halket J., Sandler M., 1988, Brain quinolinic acid in Huntington's disease. *J. Neurochem.* 50: 1959.

Rios C., Santamaria A., 1991, Quinolinic acid is a potent lipid peroxidant in rat brain homogenates. *Neurochem. Res.*, 16: 1139.

Saito K., Lackner A., Markey S.P., Heyes M.P., 1991, Cerebral cortex and lung indoleamine-2,3-dioxygenase activity is increased in type-D retrovirus infected macaques. *Brain Res. Amsterdam*, 540: 353.

Saito K., Markey S.P., Heyes M.P., 1991, Chronic effects of gamma-interferon on quinolinic acid and indoleamine-2,3-dioxygenase in brain of C57BL6 mice. *Brain Res., Amsterdam*, 546: 151.

Saito K., Markey S.P., Heyes M.P., 1992, Effects of immune activation on quinolinic acid and kynurenine pathway metabolism in the mouse. *Neurosci.*, 51: 25.

Saito K., Chen C.Y., Masana M., Crowley J., Markey S.P., Heyes M.P., 1993, 4-Chloro-3-hydroxyanthranilic acid, 6-chlorotryptophan and norharmane attenuate quinolinic acid formation by interferon-γ stimulated monocytes (THP-1) cells. *Biochem.J.*, 291: 11.

Saito K., Crowley J.S., Markey S.P., Heyes M.P., 1993, A mechanism for increased quinolinic acid formation following acute systemic immune stimulation. *J. Biol. Chem.*, 268(21): 15496.

Saito K., Nowak T.S. Jr., Markey S.P., Heyes M.P., 1993, Delayed increases in kynurenine pathway metabolism in damaged brain regions following transient cerebral ischemia. *J. Neurochem.*, 60: 180.

Saito K., Nowak T.S., Suyama K., Quearry B.J., Saito M., Crowley J.S. *et al.*, 1993, Kynurenine pathway enzymes in brain: responses to ischemic brain injury versus systemic immune activation. *J. Neurochem.*, 61: 2061.

Sakai N., Saito K., Kaufman S., Heyes M.P. Milstien S., 1993, Induction of pterin synthesis is not required for cytokine-stimulated tryptophan metabolism. *Biochem.* J., 295(2): 543.

Schwarcz R., Whetsell W.O., Mangano R.E.M., 1983, Quinolinic acid: an endogenous metabolite can produce axon sparing lesions in rat brain. *Science,* 219: 316.

Schwarcz R., Okuno E., White R.J., Bird E.D., Whetsell W.O., 1988, 3-Hydroxyanthranilate oxygenase activity is increased in the brains of Huntington's disease victims. *Proc. Nat. Acad. Sci. USA*, 85: 4079.

Speciale C., Ungerstedt U., Schwarcz R., 1989, Production of extracellular quinolinic acid in the striatum studied by microdialysis in unanesthetized rats. *Neuroscience Letters*, 104: 345.

Speciale C., Schwarcz R., 1993, On the production and disposition of quinolinic acid in rat brain and liver. *J. Neurochem.*, 60: 212.

Stone T.W., Perkins M.N., 1984, Action of excitatory amino acids and kynurenic acid in the primate hippocampus: a preliminary study. *Neuroscience Letters*, 52: 335.

Stone T.W., 1993, Neuropharmacology of quinolinic and quinolinic and kynurenic acids. *Pharmacol. Rev.*, 45: 309.

Takikawa O., Yoshida R., Kido R., Hayaishi O., 1986, Tryptophan degradation in mice initiated by indoleamine-2,3-dioxygenase. *J. Biol. Chem.*, 261: 3648.

Taylor M.W., Feng G., 1991, Relationship between interferon-γ, indoleamine-2,3-dioxygenase and tryptophan. *FASEB J.*, 5: 2516.

Tone S., Takikawa O., Habara-Ohkubo A., Kadoya A., Yoshida R., Kido R., 1990, Primary structure of human indoleamine-2,3-dioxygenase deduced from the nucleotide sequence of its cDNA. *Nucleic Acid Res.*, 18: 367.

Turski W.A., Gramsbergen J.B., Traitler H., Schwarcz R., 1989, Rat brain slices produce and liberate kynurenic acid upon exposure to L-kynurenine. *J. Neurochem.*, 52: 1629.

Ueda T., Otsuka H., Goda K., Ishiguro I., Naito J., Kotake Y., 1978, The metabolism of [carboxyl-$^{14}$C]-anthranilic acid. I The incorporation of radioactivity into $NAD^+$ and $NADP^+$. *J. Biochem.*, 84: 687.

Varesio L., Clayton M., Blasi E., Ruffman R., Radzioch D., 1990, Picolinic acid, a catabolite of L-tryptophan as the second signal in the activation of IFN-γ-primed macrophages. *J. Immunol.* 145: 4265.

Werner E.R., Bitterlich G., Fuchs D., Hausen A., Reibnegger G., Szabo G. et al., 1987, Human macrophages degrade tryptophan upon induction by interferon-gamma. *Life Sci.*, 41: 273.

Werner-Felmayer G., Werner E.R., Fuchs D., Hausen A., Reibnegger G., Wachter H., 1989, Tumor necrosis factor-alpha and lipopolisaccharide enhance interferon-induced tryptophan degradation and pteridine synthesis in human cells. *Biol. Chem. Hoppe Seyler*, 370: 1063

Yasui H., Takai K., Yoshida R., Hayaishi O., 1986, Interferon enhanches tryptophan metabolism by inducing pulmonary indoleamine-2,3-dioxygenase. Its possible occurrence in cancer patients. *Proc. Nat. Acad. Sci. USA.*, 83: 6622.

# KYNURENINE-3-HYDROXYLASE AND ITS SELECTIVE INHIBITORS

## Molecular Modeling Studies

G. Costantino,[1] L. Mattoli,[1] F. Moroni,[2] B. Natalini,[1] and R. Pellicciari[1]

[1] Istituto di Chimica e Tecnologia del Farmaco
Università degli Studi di Perugia
Via del Liceo, 1. 06123 Perugia, Italy
[2] Dipartimento di Farmacologia Preclinica e Clinica "Mario Aiazzi Mancini"
Università di Firenze
Via Morgagni 65. 50134 Firenze, Italy

## 1. INTRODUCTION

The modulation of the kynurenine (KYN) catabolic pathway of L-Tryptophan (L-Trp) is gaining a considerable interest since the discovery that two metabolites along the route, kynurenic acid (KYNA) and quinolinic acid (QUIN), act in an opposite manner at excitatory amino acid (EAA) receptors (Schwarcz, 1993). KYNA is a non selective antagonist at the strychnine-insensitive glycine site of the NMDA receptor complex and is endowed with neuroprotective properties (Foster et al., 1984). QUIN, on the other hand, is a selective agonist at the NMDA site of the NMDA receptor complex and is a relatively potent neurotoxin implicated in the pathogenesis of a number of neurodegenerative diseases.

The modulation of the KYNA/QUIN balance is therefore a considerable goal toward the pharmacological control of central nervous system excitatory pathways. Among the strategies aimed at favourably altering the KYNA/QUIN balance, the one focused on the increase of KYNA production or on the decrease of QUIN production by inhibiting key enzymes (Scheme I) of the KYN pathway is particularly promising also due to recent advances in the elucidation of the physiology and biochemistry of the KYN pathway. In particular, when kynurenine is available the enzyme kynurenine-aminotransferase (KAT) may significantly increase the cerebral concentration of KYNA, thus yielding pharmacological effects compatible with an antagonism of the NMDA receptor complex (Nozaki and Beal, 1992). The aim of our work, indeed, has been the design and synthesis of selective inhibitors of enzymes of the KYN pathway and the evaluation of the physiological and pharmacological consequences of selectively inhibiting enzymes of the KYN pathway of the L-Trp metabolism.

*Recent Advances in Tryptophan Research,* edited by Graziella Allegri Filippini et al.
Plenum Press, New York, 1996

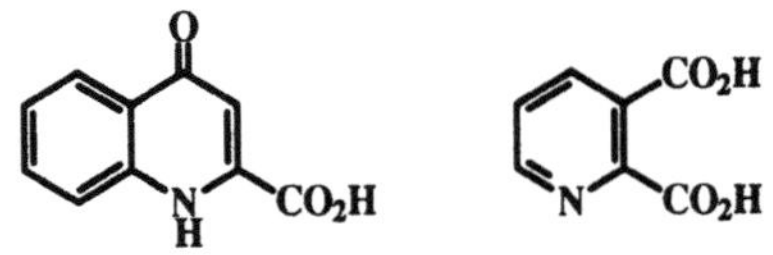


It has been previously reported that nicotinoylalanine (NAL, **1** ), a synthetic KYN analog, is able to increase significantly the cerebral concentration of KYNA and that this effect is likely to be due to an inhibition of the both kynureninase and kynurenine-3-hydroxylase (Pellicciari et al., 1992; Russi et al., 1992). Having proved that by inhibiting these enzymes it is possible to increase the KYN availability and, as a consequence the cerebral concentration of KYNA, we have started a program aimed at developing KYN analogs more potent and selective than **1**. Among the large number of compounds synthesized, a bioisoster of **1**, (±)-(m-nitrobenzoyl)alanine ((±)mNBA, **2**) revealed to be a potent and selective inhibitor of kynurenine-3-hydroxylase, the first so far reported, able to significantly increase the KYNA concentration in the blood, in the liver as well as in the brain (Pellicciari et al, 1994).

## 2. MOLECULAR MODELING STUDY

Although the KYN metabolic pathway has been the object of several studies which have allowed to gain knowledge about its physiology and biochemistry, information are not yet available about the tridimensional (3D) structure of key enzymes such as kynurenine-3-hydroxylase and kynureninase. The availability of a potent and selective inhibitor of kynurenine-3-hydroxylase such as **2**, on the other hand, makes it interesting to investigate

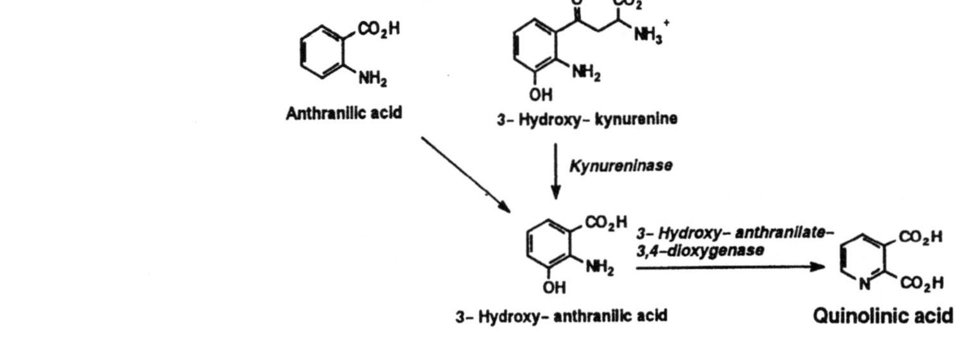


Scheme 1

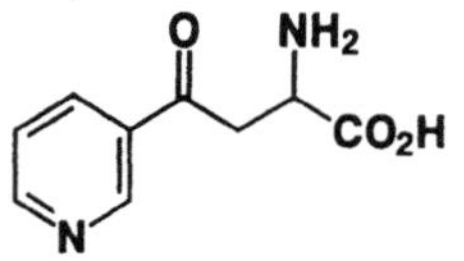
1

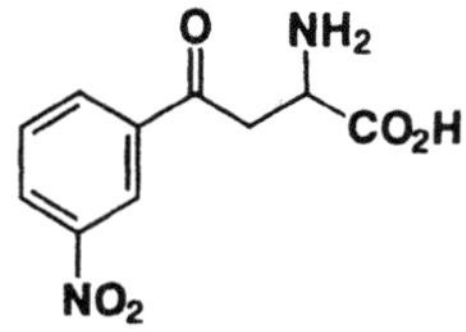
2

the nature of the inhibitor-enzyme interaction in order to understand the mechanism of action of this enzyme and to use the information thus acquired for the rational design of more potent and selective inhibitor.

In the absence of direct information on the 3D structure of the substrate-enzyme complex, some clues about the mode of interaction can be obtained by analysing enzymes, functionally or structurally related, whose tridimensional structure is known and by relating the data thereby obtained with the structure of known ligands. This approach can be considered as a link between the two main ligand design strategies, namely structure based and ligand based molecular design.

## 2.1 Mechanism of Action of Kynurenine-3-Hydroxylase

Kynurenine-3-hydroxylase is a NAD(P)H-dependent flavoprotein which belongs to the class of external monooxygenases. Kynurenine-3-hydroxylase catalyzes the insertion of molecular oxygen into the aromatic ring of kynurenine to give 3-hydroxy-kynurenine.

A number of crystallographic and spectroscopic studies carried out on different aromatic monooxygenases allow to postulate a mechanistic pathway for the hydroxylation reaction. In particular, NADPH reduces the FAD portion of the enzyme which then reacts with molecular oxygen to form the C(4a)-hydroperoxyflavin moiety as a key intermediate. This intermediate is the electrophilic source for the subsequent hydroxylation reaction. The existence of such intermediate has been demonstrated by spectroscopic, kinetic and molecular modeling studies (Entsch et al., 1976). Although no similar informations are directly available for kynurenine-3-hydroxylase, it is conceivable that an analogue mechanism is operative for this enzyme.

## 2.2 Pseudo-Active Site as a Model for Enzyme Inhibition

Among the several known monooxygenases, p-hydroxybenzoate hydroxylase (PHBH) is one of the best characterized. In particular, the structures of the free enzyme, the complex substrate-enzyme and product-enzyme have been crystallographically solved with good resolution (Schreuder et al., 1988). Despite the lack of any known structural homology between PHBH and kynurenine-3-hydroxylase these two enzymes are functionally very related. Both are NAD(P)H-dependent flavoproteins and both catalyze the same reaction on aromatic substrates. Assuming that these two monooxygenases share a common mechanistic pathway, it can be assumed that in the case of kynurenine-3-hydroxylase a network of non-covalent interactions will cooperate to keep the substrate in the proper orientation in order to undergo electrophilic attack. It is reasonable that this orientation will mimic to a large extent that of p-hydroxybenzoic acid into the active site of PHBH. Accordingly, we have selected from the X-ray structure of PHBH the FAD moiety (in its reduced form) and the substrate. The aromatic ring of kynurenine was then forced to match the corresponding

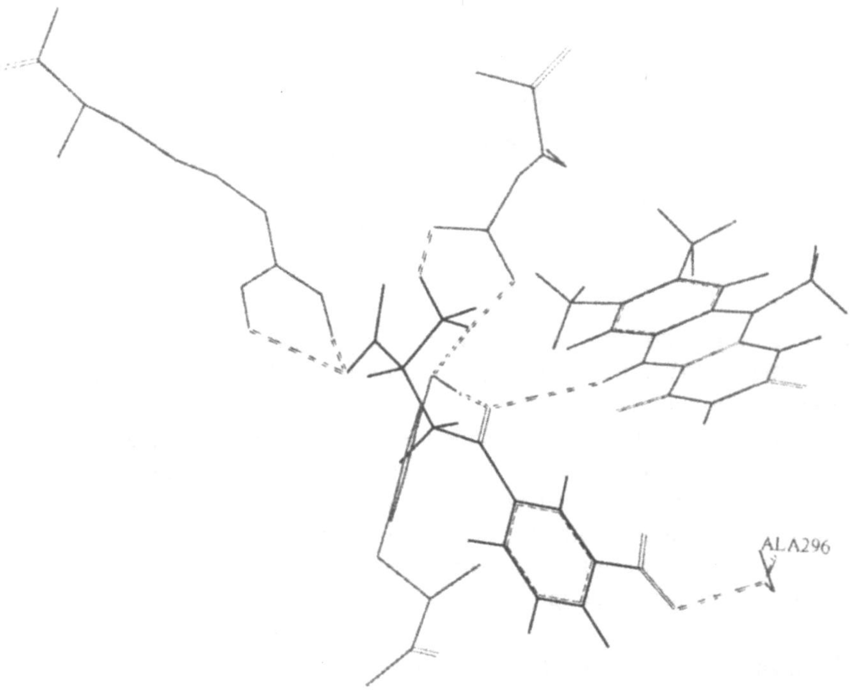


**Figure 1.** (±)-mNBA docked into the pseudo-active site model.

aromatic ring of p-hydroxy benzoic acid. The triad Gly-46, Val-47-Ala-296, which has been shown to have mechanistic importance has been selected as well. As a next step, some prototypic aminoacids were manually added to the complex in order to accommodate the major interactions between kynurenine and the active site of kynurenine-3-hydroxylase. In particular, glutamine, asparagine and tyrosine residues were placed close to the carboxy group, the amino group and the carbonyl group of kynurenine, respectively. The resulting molecular framework can be considered as a mimic of the active site of kynurenine-3-hydroxylase, able to account at least for the major interaction pattern (Snyder, et al., 1993;), and therefore can be considered as a "pseudo-active site of kynurenine-3-hydroxylase. The complex pseudo-active site/kynurenine was submitted to energy minimization and the

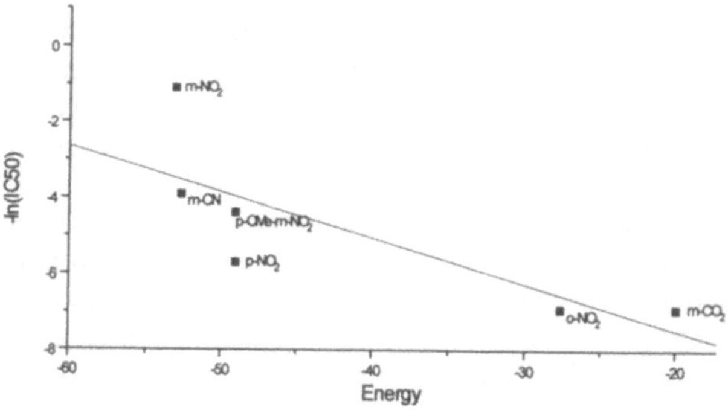


**Figure 2.** Prediction of inhibitory potency from calculated interaction energies.

resulting structure was shown to be consistent with the proposed monoxygenase mechanistic pattern. (±)-mNBA (**2**) was then docked into the pseudo-active site model in the place of kynurenine and the resulting complex was energy minimized. As shown in Figure 1, **2** adopts a stable disposition into the active site cavity. In particular one of the oxygen of the nitro group is involved in a hydrogen bonding with the NH group of Ala-296 residue (residue numbering refers to PHBH).

In order to further validate the reliability of our model, a number of other KYN analogs having potency as kynurenine-3-hydroxylase inhibitors ranging from low micromolar to no activity were inserted into the model. The interaction energy between ligands and the pseudo-active site were evaluated and correlated with the experimental inhibitory potency. The semiquantitative agreement (Figure 2) is a confirmation of the validity, at least as a working model, of the pseudo-active site model. It is conceivable that the goodness of the relationship might be improved by adding to the pseudo-active site model additional features that will take into account some other interactions (e.g. dipole-dipole interaction with positive ends of $\alpha$-helices) which are not explicitly considered in the present model.

## 3. CONCLUSION

By taking advantage of the functional similarity between kynurenine-3-hydroxylase and other monooxygenases whose tridimensional structure is known, a working model of the active site of the former enzyme has been constructed. This model is able to explain the observed kynurenine-3-hydroxylase inhibitory potency of a series of KYN analogs in a semiquantitative way. Although a major help to the design of more potent inhibitors will be given by the determination of the primary sequence and by the knowledge of the tridimensional structure of kynurenine-3-hydroxylase, our pseudo-active site construct can be considered as an useful model to rationalize existing data and for envisaging rational chemical modification of the KYN core toward new selective and potent inhibitors.

## 4. REFERENCES

Entsch, B.; Ballou, D. P.; Massey, V., 1976, Flavin oxygen derivatives involved in hydroxylation by p-hydroxybenzoate hydroxylase. *J. Biol. Chem.* 251: 2550.

Foster, A. C.; Vezzani, A.; French, E. D.; Schwarcz, R., 1984, Kynurenic acid blocks neurotoxicity and seizures induced in rat by the related brain metabolite quinolinic acid. *Neurosci. Lett.* 48: 273

Nozaki, K.; Beal, M., F., 1992, Neuroprotective effects of L-Kynurenine on hypoxia-ischemia and NMDA lesions in neonatal rats. *J. Cerebr. Blood Flow Metab.* 12: 400

Pellicciari, R.; Gallo-Mezo, M. A.; Natalini, B.; Amer, A. M., 1992, Stereospecific synthesis of the enantiomers of nicotinoylalanine, a neuroprotective agent. *Tetrahedron Lett.* 33: 3003

Pellicciari, R.; Natalini, B.; Costantino, G.; Mahmoud, R. M.; Mattoli, L.; Sadeghpour, B. M.; Moroni, F.; Chiarugi, A.; Carpenedo, R., 1994, Modulation of the kynurenine pathway in search for new neuroprotective agents. Synthesis and preliminary evaluation of (m-nitrobenzoyl)alanine, a potent inhibitor of kynurenine-3-hydroxylase. *J. Med. Chem.* 37: 647

Russi, P.; Alesiani, M.; Lombardi G.; Davolio, P.; Pellicciari, R.; Moroni, F., 1992, Nicotinoylalanine increases the formation of the kynurenic acid in the brain and antagonizes convulsions. *J. Neurochem.* 59: 2076

Schreuder, H. A.; van der Laan, J. M.; Hol, W. G. J.; Drenth, J., 1988, Crystal structure of p-hydroxybenzoate hydroxylase complexed with its reaction product 3,4-dihydroxybenzoate. *J. Mol. Biol.* 199: 637

Schwarcz, R., 1993, Metabolism and function of brain kynurenines. *Biochem. Soc. Trans.* 21:77

Snyder, J. P., Rao, S. N.; Koehler, K. F.; Vedani, A., 1993, Minireceptors and Pseudoreceptors in "3D-QSAR in Drug Design", Kubinyi H., Ed. Escom, Leiden.

# BENZOYLALANINE ANALOGUES AS INHIBITORS OF RAT BRAIN KYNURENINASE AND KYNURENINE 3-HYDROXYLASE

A. Giordani,[1] L. Corti,[1] M. Cini,[1] M. Marconi,[1] A. Pillan,[1] R. Ferrario,[1] R. Schwarcz,[2] P. Guidetti,[2] C. Speciale,[1] and M. Varasi[1]

[1] Pharmacia, R&D-CNS
20014 Nerviano, Milano, Italy
[2] Maryland Psychiatric Research Center
University of Maryland School of Medicine
Baltimore, Maryland 21228

## 1. INTRODUCTION

Alterations in glutamatergic neurotransmission mediated by excitatory amino acid (EAA) receptors are involved in the pathogenesis of several neurological disorders caused by either chronic or acute events (Meldrum, 1990). Accordingly, new neuroprotective strategies aimed at reducing glutamatergic receptor overstimulation have been proposed.

The major pathway of tryptophan catabolism in mammals is the kynurenine (KYN) pathway. Several lines of evidence have shown that two of the metabolites formed along the pathway may play a significant role in the CNS through interaction with the glutamatergic system (Stone, 1993). The neuroprotective properties of kynurenic acid (KYNA) (Kessler et al., 1989; Foster et al., 1984), as well as the excitotoxic effects elicited by quinolinic acid (QUIN) (Schwarcz et al., 1983) via an interaction with EAA receptors are well established, and the two compounds are synthesised in the CNS (Schwarcz and Du, 1991). In particular, whereas these tryptophan metabolites are normally present in mammalian cerebrospinal fluid in nanomolar concentrations with relatively little variation between species, substantial changes in their concentrations and balance occur in several human neurodegenerative disorders (Schwarcz et al., 1992). This, in addition to the immunohistochemical localisation of QUIN- and KYNA-synthesising enzymes in close proximity of glutamatergic synapses, suggested that at the synaptic cleft QUIN and KYNA may modulate glutamatergic neurotransmission by affecting neuronal excitability and vulnerability.

The possibility to interfere with the KYN pathway by pharmacological alteration of the balance between the two functionally opposite metabolites (i.e. to effect increases in KYNA and/or decreases in QUIN) offers a novel approach for developing neuroprotective

*Recent Advances in Tryptophan Research,* edited by Graziella Allegri Filippini et al.
Plenum Press, New York, 1996

agents (Speciale et al., this meeting). Thus, since QUIN and KYNA share common metabolic precursors yet are produced along separate branches of the KYN pathway, selective inhibition of appropriate enzymes may influence the conversion of KYN into QUIN and KYNA, respectively.

In the KYN pathway, kynurenine aminotransferase (KAT) is the enzyme responsible for the synthesis of KYNA from KYN. Kynurenine 3-hydroxylase (KYN 3-OHase) provides 3-OH-KYN, and kynureninase (KYNase) converts 3-hydroxy-kynurenine (3-OH-KYN) to 3-hydroxyanthranilic acid, which is further metabolised to QUIN. We reasoned that potent inhibitors of KYNase and/or KYN 3-OHase, if devoid of KAT inhibitory activity, could be useful tools to increase the levels of KYNA in the brain.

## 2. RESULTS

Structural modification of the natural substrate is the classical approach to obtain enzyme inhibitors. The most obvious modifications of KYN are aromatic ring replacement or substitutions, modifications of the carbonyl group and aminoacidic moiety manipulations. At the time our studies were initiated, nicotinylalanine (NAL) (Decker et al., 1963) was the only compound known to interfere with the KYN metabolic pathway (Table 1).

NAL could be considered as an example of aromatic ring replacement (pyridin-3-yl instead of 2-aminophenyl). However, NAL is endowed with very low potency and selectivity. As far as selective KYNase inhibitors are concerned, a group of transition state analogues (i.e. dihydrokynurenine and S-(2-aminophenyl)-L-cysteine S,S-dioxide), representing examples of carbonyl group modifications, was reported by Phillips and Dua (1991) and Dua et al. (1993). However, although these compounds proved quite effective in inhibiting bacterial kynureninase ($K_i$ values in the nM range), less significant inhibition was found against mammalian (rat) KYNase ($K_i$ in the high μM range). Furthermore, during the course of our work, a potent and selective inhibitor of KYN 3-OHase, 3′-nitro-benzoylalanine (mNBA), was discovered by Pellicciari et al. (1994).

Our previous work on the influence exerted by modifications of the KYN structure on KAT activity (Varasi et al., in press) suggested that the presence of the aromatic amino group is essential for interference with KAT. Consequently, benzoylalanine (FCE 27377, Table 2), a KYN analogue lacking the aromatic amino group, was devoid of KAT inhibitory activity and was found to be a relatively potent and very selective KYN 3-OHase inhibitor. After enantioselective synthesis (Giordani et al., this meeting), the S-enantiomer (FCE 28468) turned out to be responsible for the activity, whereas the R-enantiomer (FCE 28469) was inactive (Table 2).

**Table 1.** Nicotinylalanine (NAL)

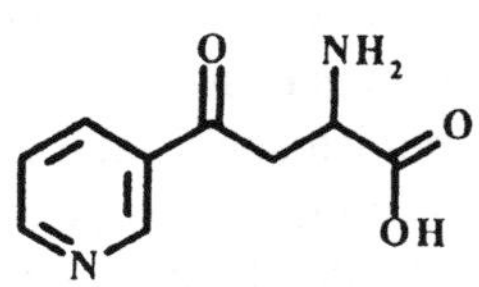


| | KAT | | KYNase | | KYN 3-OHase | |
|---|---|---|---|---|---|---|
| Compound | [c] μM | % inhibition | [c] μM | % inhibition | [c] μM | % inhibition |
| NAL | 1000 | 0 | 1000 | 80 | 1000 | 74 |
| | | | 100 | 50 | 100 | 29 |

**Table 2.** 2-substituted benzoylalanines

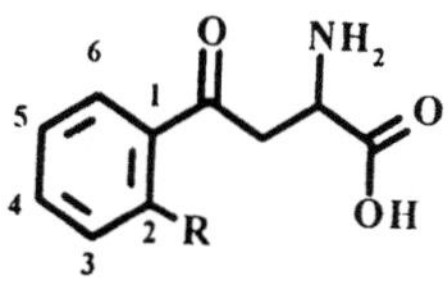

| R | FCE | KAT $IC_{50}$ μM | KYNase $IC_{50}$ μM (% inhib. at 100 μM) | KYN 3-OHase $IC_{50}$ μM (% inhib. at 100 μM) |
|---|---|---|---|---|
| H | 27377 | >1000 | >100 | 42 |
| H | (S)-28468 | >100 | >100 | 16 |
| H | (R)-28469 | >100 | >100 | >100 |
| F | 28631 | >1000 | (52%) | >100 |
| Cl | 28753 | >1000 | (80%) | >100 |
| Br | 29069 | >1000 | (85%) 29 | >100 |
| $CH_3$ | 28630 | >1000 | >100 | (81%) 23 |
| OH | 27378 | >1000 | >100 | >100 |
| $OCH_3$ | 27384 | >1000 | (95%) 4.65 | >100 |
| $OCH_2CH_3$ | 28866 | nt | (75%) | >100 |
| $CF_3$ | 28655 | >1000 | >100 | >100 |

nt: not tested.

This finding prompted us to analyse extensively the effect of various substituents of the aromatic ring of benzoylalanine on the activities of KAT, KYNase and KYN 3-OHase. In vitro inhibition of the compounds was evaluated in rat brain homogenate. KAT activity was assayed by the method of Okuno et al. (1991) using $^3$H-KYN as substrate and measuring $^3$H-KYNA formation. KYN 3-OHase activity was quantified by the conversion of KYN to 3-OH-KYN according to Saito et al. (1993). KYNase activity was measured by the conversion of KYN to anthranilic acid (Takikawa et al., 1986).

Substitution of the aromatic amino group of the natural substrate (KYN) with alkoxy groups or halogen atoms was analysed first (Table 2). The most striking feature of this new class of compounds was their remarkable KYNase inhibitory property and their lack of effect on KAT activity. Concerning KYNase/KYN 3-OHase selectivity, alkoxy substituted inhibitors revealed better characteristics than the corresponding halogen derivatives. The 2′-methoxy-derivative (FCE 27384; $IC_{50}$= 4.65μM) was the most potent and selective KYNase inhibitor (see also Moroni et al., this meeting). The inhibitory activity was stereoselective with the S-enantiomer the eutomer, even though this enantioselectivity was characterised by a low eudismic ratio (Giordani et al., this meeting). Conversely, aromatic amino group replacement either with a trifluoromethyl group or with a methyl group did not give rise to active compounds.

Introduction of a fluorine atom in position 3′ (Table 3) led to a potent and selective KYN 3-OHase inhibitor, while substitution with chlorine gave rise to a more potent but slightly less selective agent. A further increase of the size of the substituent (bromine) decreased both potency and selectivity. Less interesting results were obtained with other substituents in position 3′.

Substitution in position 4′ (Table 4) yielded selective KYN 3-OHase inhibitors with both fluorine and chlorine, though with lower potency in comparison with the corresponding 3′-substituted derivatives. With the exception of the 4′-methoxy-derivative, other substituents were unable to produce any inhibitory effect. The good selectivity in inhibiting KYN 3-OHase displayed by the 4′-chloro-derivative prompted us to synthesise the corresponding

**Table 3.** 3-substituted benzoylalanines

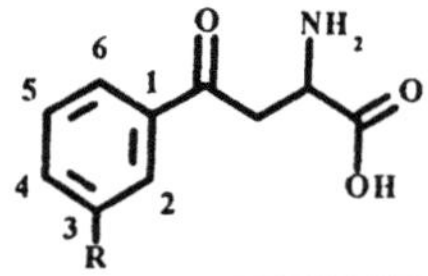

| R | FCE | KAT $IC_{50}$ μM | KYNase $IC_{50}$ μM (% inhib. at 100 μM) | KYN 3-OHase $IC_{50}$ μM (% inhib. at 100 μM) |
|---|---|---|---|---|
| F | 28764 | >1000 | >100 | 0.9 |
| Cl | 28752 | >1000 | (58%) | 0.4 |
| Br | 29068 | >1000 | (67%) | 1.3 |
| $CH_3$ | 29070 | >1000 | (60%) | >100 |
| $OCH_3$ | 28629 | >1000 | >100 | >100 |
| $CF_3$ | 28680 | >1000 | (55%) | >100 |

3′,4′-dichloro-derivative. This compound (FCE 28833, $IC_{50}$=0.2 μM) turned out to be the most potent and selective KYN 3-OHase inhibitor described so far. FCE 28833 was found to inhibit KYN 3-OHase in an enantioselective fashion, with the S-form of 3′,4′-dichloro-benzoylalanine being more effective (Giordani et al., this meeting).

The in vivo effects of FCE 28833 were investigated in the gerbil (Cini et al., this meeting). A dose-related increase in the concentrations of KYN and KYNA was observed in both brain and periphery following oral administration of the inhibitor. In particular, brain KYNA content was elevated up to 44-fold following a single oral dose of the compound. The effect of the inhibitor was long lasting, i.e. KYNA levels were still elevated after 14 hours in all tissues.

**Table 4.** 4-substituted benzoylalanines

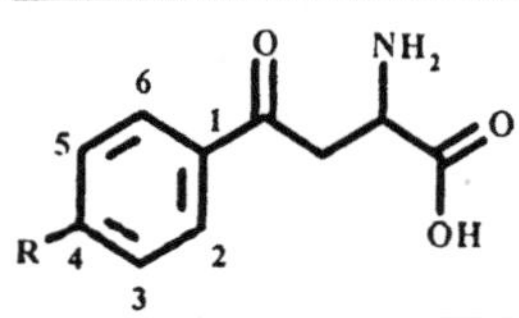

| R | FCE | KAT $IC_{50}$ μM | KYNase $IC_{50}$ μM | KYN 3-OHase $IC_{50}$ μM |
|---|---|---|---|---|
| F | 28835 | >1000 | >100 | 4.5 |
| Cl | 28751 | >1000 | >100 | 7 |
| Br | 29525 | >1000 | >100 | 8.9 |
| $CH_3$ | 29071 | >1000 | >100 | >100 |
| $OCH_3$ | 28628 | >1000 | >100 | 30 |
| $CF_3$ | 28765 | >1000 | >100 | >100 |
| 3,4-Cl | 28833 | >1000 | >100 | 0.2 |

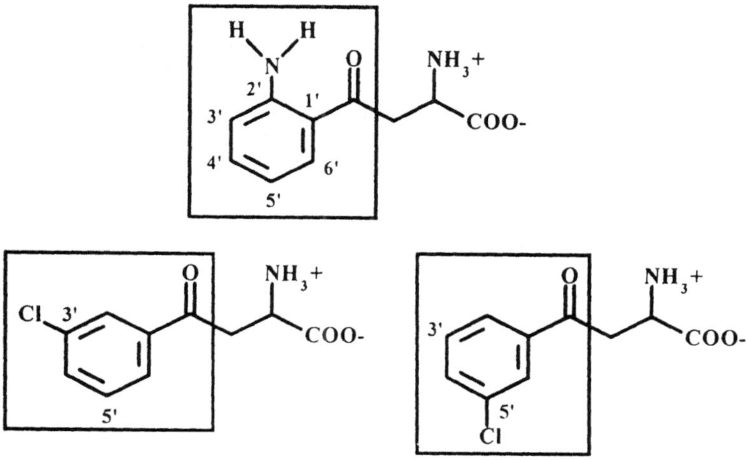


**Figure 1**

# 3. MOLECULAR MODELING

The bioactive conformation of the above reported KYN 3-OHase inhibitors was analysed using the molecular electrostatic potential (MEP) approach, adopting the MEP of the substrate as the reference point. The MEP of a molecule is a dominant feature in molecular recognition, since the initial long-range interactions between the ligand and the target protein are entirely electrostatic. This approach was satisfactory for a qualitative evaluation of our competitive inhibitors' binding mode within the enzyme's active site, as compared to the binding of the natural substrate. An important feature of the KYN 3-OHase inhibitors (e.g. meta-chloro-benzoylalanine) is that the aromatic ring is endowed with a symmetric substitution pattern, i.e. the substituent could be either in position 3′ or in the formal position 5′. This fact becomes relevant when the molecule is modeled in three dimensions. Thus, we were faced with the question which of the two conformations was preferred at the enzyme's recognition site (Figure 1).

The conformational analysis of KYN indicated that, among the several possible conformations, the one leading to an intramolecular hydrogen bond between the carbonyl moiety and the aromatic amino group was the preferred one. This conclusion was supported by molecular mechanics calculations (Discover), and confirmed by a semiempirical quantum mechanics AM1 method. In addition, the same conformation was found for 2-N-acetyl-kynurenine as elucidated by an X-ray crystallographic study (Cambridge database). A preliminary analysis comparing the MEP of KYN's aromatic ring (in its preferred conformation) with the MEPs of the two possible conformations of the inhibitor (meta-chloro-benzoylalanine) suggested that interaction with the active site is more likely with the substituent in the formal position 5′ than in position 3′.

# 4. CONCLUSIONS

Structural modifications (substitution) carried out on the aromatic moiety of KYN led us to obtain potent and selective inhibitors of KYN-ase (FCE 27384) and KYN 3-OHase (FCE 28833) which did not affect KAT activity. These two compounds are the most potent

**Table 5.** Selected FCE and related compounds

| Structure | FCE code | KAT $IC_{50}$ μM (%in./100μM) | KYNase $IC_{50}$ μM (%in./100μM) | KYN 3-OHase $IC_{50}$ μM (%in./100μM) |
|---|---|---|---|---|
| [Graphic here] | 27984 (NAL) | >1000 | (50%) | >100 |
| [Graphic here] | 27994 (Dua et al.) | >1000 | 28.7 | >100 |
| [Graphic here] | 28865 (Pellicciari et al.) | >1000 | (72%) | 3.1 |
| [Graphic here] | 27384 | >1000 | 4.5 | >100 |
| [Graphic here] | 28833 | >1000 | >100 | 0.2 |

and selective inhibitors described so far (Table 5). The selective blockade of KYN 3-OHase considerably enhanced the endogenous production of the neuroprotectant KYNA.

## 5. REFERENCES

Decker R.H., Brown R.R., and Pierce J.M., 1963, Studies on the biological activity of nicotinylalanine, an analogue of kynurenine, *J. Biol. Chem.* 238: 1049.

Dua R.K., Taylor E.W., and Phillips R., 1993, S-aryl-Lcysteine-S,S-dioxides: design and evaluation of a new class of inhibitors of kynureninase, *J. Am. Chem. Soc.* 115: 1264.

Foster A.C., Vezzani A., French E.D., and Schwarcz R., 1984, Kynurenic acid blocks neurotoxicity and seizures induced in rats by the related brain metabolite quinolinic acid, *Neurosci. Lett.* 48: 273.

Kessler K., Terramani T., Lynch G., and Baudry M., 1989, A glycine site associated with N-methyl-D-aspartic acid receptors: characterisation and identification of a new class of antagonists, *J. Neurochem.* 52: 1319.

Meldrum B., 1990, Excitatory amino acid neurotoxicity and neurodegenerative diseases, *Trends Neurosci* 11: 379.

Okuno E., Schmidt W., Parks D.A., Nakamura M., and Schwarcz R., 1991, Measurement of rat brain kynurenine aminotransferase at physiological concentrations, *J. Neurochem.* 57:533.

Pellicciari R., Natalini B., Costantino G., Mahmoud M.R., Mattoli L., Sadeghpour B., Moroni F., Chiarugi A., and Carpenedo R., 1994, Modulation of kynurenine pathway in search for new neuroprotective agents. Synthesis and preliminary evaluation of (m-nitrobenzoyl)alanine, a potent inhibitor of kynurenine hydroxylase, *J. Med. Chem.* 37:647.

Phillips R. and Dua R.K., 1991, Stereochemistry and mechanism of aldol reactions catalysed by kynureninase, *J. Am. Chem. Soc.* 113: 735.

Saito K., Nowak T.S., Markey S.P., and Heyes M.P., 1993, Mechanism of delayed increases in kynurenine pathway metabolism in damaged brain regions following transient cerebral ischemia. *J. Neurochem.* 60: 180.

Schwarcz R., Whetsell W.O., and Mangano R.M., 1983, Quinolinic Acid: an endogenous metabolite that produces axon-sparing lesions in rat brain, *Science* 219: 316.

Schwarcz R. and Du F., 1991, Quinolinic acid and kynurenic acid in the mammalian brain, in *Kynurenine and Serotonin Pathways,* R. Schwarcz et al. eds,. Plenum Press, New York, p.185.

Schwarcz R., Du F., Schmidt W., Turski W.A., Gramsbergen J.B., Okuno E., and Roberts R.C., 1992, Kynurenic acid: a potential pathogen in brain disorders, in "Neurotoxins and Neurodegenerative Disease", J.W. Langston and A. Young eds., *Ann. N. Y. Acad. Sci.* 648:140.

Stone T.W., 1993, Neuropharmacology of quinolinic and kynurenic acids, *Pharmacol. Rev.* 45: 309.

Takikawa O., Yoshida R., Kido R., and Hayashi O., 1986, Tryptophan degradation in mice initiated by indoleamine 2,3-dioxygenase, *J. Biol. Chem.* 261:3648.

Varasi M., Della Torre A., Heidempergher F., Pevarello P., Speciale C., Guidetti P., Wells D.R., and Schwarcz R., Derivatives of kynurenine as inhibitors of rat brain kynurenine aminotransferase, *Eur. J. Med. Chem.,* in press.

# NICOTINAMIDE METHYLTRANSFERASE ACTIVITY AND CELL-DEATH IN THE LIVER OF THE MOUSE BEARING EHRLICH ASCITES TUMOR

S. Fujimura,[1] A. Okamura,[1] Y. Ohmura,[1] Y. Moriyama,[1] H. Ohwada,[2] K. Horitsu,[3] and M. Ohkubo[4]

[1] Department of Biochemistry
Chiba University School of Medicine
[2] Department of Pathology, Institute for Lung Cancer Research
Chiba University School of Medicine
1-8-1 Inohana, Chuo-ku, Chiba 260 Japan
[3] Faculty of Home Economics, Tokyo Kasei University
1-18-1 Kaga, Itabashi-ku, Tokyo 173 Japan
[4] Kagoshima Immaculate Heart College
4-22-1 Toso, Kagoshima 890 Japan

## 1. INTRODUCTION

In order to develop new ways of tumor diagnosis and therapy, it is important to know the metabolic changes occurring while bearing the tumor, since these ways should be applied to the cancer patient, who is, after all, a tumor bearer. Various metabolic changes in the tumor host tissues, especially in the liver, have been observed to take place together with tumor-growth (Fujimura and Shimizu, 1977; Shimizu and Fujimura, 1978; 1984; Asanagi et al., 1988; Tanaka et al., 1991; Nakagawa et al., 1991; Fujimura et al., 1992a; 1992b; Hanazawa et al., 1994; Fujimura et al., 1995).

Nicotinamide is converted to 1-methylnicotinamide via a reaction catalyzed by nicotinamide methyltransferase (NMT; Nicotinamide: S-adenosylmethionine methyltransferase; EC 2.1.1.1.), which transfers the methyl group from S-adenosylmethionine to nicotinamide coupling to produce S-adenosylhomocysteine, and localizes predominantly in the liver of the mammals. The 1-methylnicotinamide produced was further metabolized to two pyridones ($N^1$-methyl-2-pyridone-5-carboxamide and $N^1$-methyl-4-pyridone-5-carboxamide) by aldehyde oxidase. Even though these metabolites are excreted in the urine of the normal rat, it was reported that urinary excretion of 1-methylnicotinamide increased in the rat bearing Walker 256 carcinosarcoma (Clark et al., 1975).

*Recent Advances in Tryptophan Research,* edited by Graziella Allegri Filippini et al.
Plenum Press, New York, 1996

After *i.p.* transplantation of Ehrlich ascites tumor into the mouse, we revealed that the activity of NMT in the liver increased gradually with tumor development, then maintained a plateau up to death, as opposed to the steady decrease of catalase activity which has been well known to be a marker of tumor cachexia (Fujimura et al., 1992a). NMT activity in the liver of various tumor host animals was 2 to 7-fold higher than that of each respective control one. This increase was observed in different strain of the same animal species (*e.g.*, a mouse) bearing different tumor and even in different species of the animal (*e.g.*, a rat) (Hanazawa et al., 1994). Subsequently, from the cell-free ascitic fluid of the mouse bearing Ehrlich ascites tumor, we purified a factor partially which stimulated NMT activity, and at the same time, decreased catalase activity in the liver of the normal mouse 24 h after *i.p.* injection (Fujimura et al., 1992b).

In this study, we proposed the occurrence of the apoptosis-like morphological changes, correlating with NMT activity, in the liver of the tumor-bearing animal.

## 2. MATERIALS AND METHODS

### 2. 1. Animals and Tumors

Ehrlich ascites tumor was maintained *i.p.* in a male, ddy strain mouse (5 weeks old) by weekly serial transplantation of the ascites (1 x $10^7$ tumor cells). Ascitic type sarcoma S-180 was maintained by weekly serial *i.p.* transplantation of the ascites (5 x $10^6$ ascites sarcoma cells) into a male, ICR / J strain mouse (6 weeks old). Solid sarcoma S-180 was produced by *s.c.* transplantation of the ascites (5 x $10^6$ ascites sarcoma cells) into the axilla of a male, ddy strain mouse (4 weeks old). Walker 256 carcinosarcoma was maintained by fortnightly serial *i.m.* transplantation of the tumor (0.25 ml of 20 % suspension of the tumor crush in the saline) into the thigh of a male, Wistar (Imamichi) strain rat (6 weeks old).

### 2. 2. DNA Fragmentation

DNA fragments in the apoptotic cells were detected either by SDS-PAGE as described before with modifications (Fukuda et al., 1993), or quantitatively by a kit purchased from Boehringer Mannheim GmbH, Mannheim, Germany as histone-associated DNA fragments. Briefly, mouse monoclonal anti-histone-antibody was fixed adsorptively on the bottom of the micro-titer plate modules, and DNA fragment was made sandwich with anti-DNA antibody, which was conjugated with peroxidase. The amount of peroxidase retained in the immunocomplex was determined photometrically at 405 nm with ABTS (2,2′-azino-di-[3-ethylbenzthiazoline sulfonate]) as a substrate.

### 2. 3. Purification of the NMT Activity Stimulating Factor

The factor was purified partially from the cell-free ascitic fluid of the mouse bearing Ehrlich ascites tumor as described previously (Fujimura et al., 1992b). After dialysis against 10 mM sodium-phosphate buffer (pH 7.0), the ascitic fluid was applied to a DEAE-Sephacel column, which was then washed sufficiently with the same buffer. The whole eluate thus obtained, was applied to a CM-Sepharose column, followed by serial washing with the plain phosphate buffer as above and the buffer containing 0.5 M NaCl. The last eluate was used as the partially purified factor.

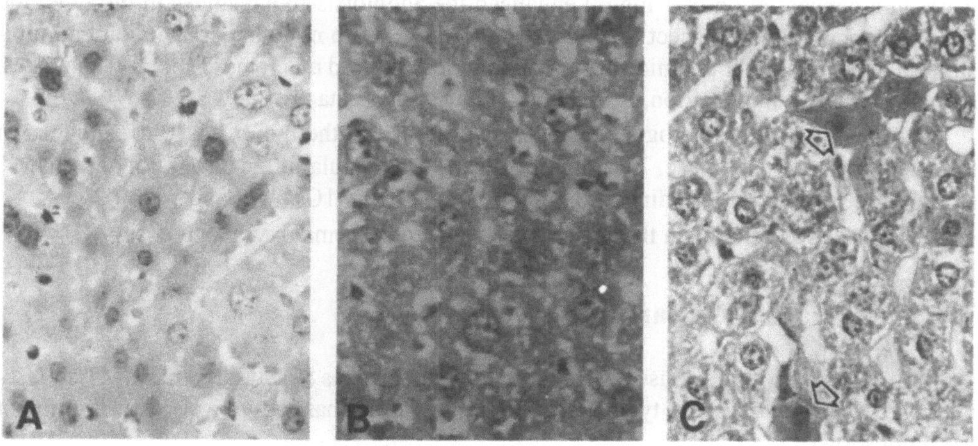


**Figure 1.** A. Representative example of the apoptotic morphological changes in a liver section from the mouse 1 week after *i.p.* transplantation of Ehrlich ascites tumor (400 x). B. A liver section from the mouse 2 weeks after transplantation of Ehrlich ascites tumor (400 x). C. A liver section from the mouse 78 h after *i.p.* inoculation of the NMT stimulating factor (400 x). The similar morphological changes as shown in Figure A were observed (arrow).

# 3. RESULTS AND DISCUSSION

## 3. 1. Ehrlich Ascites Tumor-Bearing Mouse

One week after transplantation of Ehrlich ascites tumor into the mouse, contour of the hepatocytes was unclear and cytoplasm was more acidophilic. Nuclei showed well defined pyknosis; *i.e.*, nuclei were intensely stained and rather shrunk, leaving the nuclear matrix and the nucleoli indistinctly demarcated. Other areas of the liver supposed to be normal; but variation in size of the nuclei in the hepatocytes were remarkable (Fig. 1A).

These changes in figure seemed to be apoptotic. In fact, fragmentation of DNA in the liver was detected by SDS-PAGE (data not shown). Apoptotic bodies were observed in acinar zone near the terminal hepatic veines of normal human and rat liver, possibly as an expression of physiological cell renewal (Benedetti et al., 1988). Fas antigen mRNA is expressed in the tissues of the mouse, including the liver. Since *i.p.* administration of the anti-Fas antibody rapidly killed the mouse with severe damage and apoptosis solely in the liver (Ogasawara et al., 1993), it is feasible that tumor cachexia can be caused by apoptosis in the liver. Stimulation of the expression of Fas antigen in the liver of the tumor bearer is under investigation.

Two weeks after transplantation of the tumor, such morphological changes in the liver were advanced parallely (Fig. 1B), and NMT activity was then increasing (Hanazawa et al., 1994). Histone-associated DNA fragments (mono- and oligonucleosome) in the cytoplasm of the liver were determined quantitatively by ELISA method as described in Materials and Methods, which was more prominent after the second week of transplantation than earlier (data not shown).

Every other day after transplantation of Ehrlich ascites tumor, NMT activity in the liver of the mouse was measured 4 h after *s.c.* injection of nicotinamide at a dose of 50 mg/Kg body weight. Then, the activity was enhanced in parallel with that of control tumor-bearing mouse (without nicotinamide loading), which increased continuously until death (Nkagawa et al., 1991). Administration of nicotinamide to the mouse bearing this tumor with drinking water at a concentration of 1 mg/ml enhanced the apoptotsis-like changes in the liver and DNA fragmentation in respect to control. It is interesting to note that concomitant administration of 1-methylnicotinamide and nicotinamide enhanced much more apoptotic changes, as well as DNA fragmentation, than nicotinamide alone (data not shown).

The similar morphological changes as observed in the tumor host were induced in the liver of the normal mouse, 78 h after the second *i.p.* inoculation of an aliquot of the NMT stimulating factor corresponding to 0.5 mg of protein (Fig. 1C).

All sections shown in the figures were stained by hematoxylin and eosin.

## 3. 2. Sarcoma S-180-Bearing Mouse

In the liver of the mouse bearing a solid tumor such as S-180 sarcoma, NMT activity was lower than that of ascites type tumor and the apoptotic changes were also less prominent (data not shown).

## 3. 3. Walker 256 Carcinosarcoma-Bearing Rat

Two weeks after transplantation of solid Walker 256 carcinosarcoma into the rat, the similar apoptotic changes as seen in the mouse bearing Ehrlich ascites tumor were observed in the liver. When nicotinamide was injected *s.c.* into the rat at a dose of 100 mg / Kg body weight, 1, 4, and 6 days after transplantation of the tumor, such apoptotic morphological changes were augmented in the two weeks period (data not shown).

Increase of NMT activity could be considered to cause the deficiency of S-adenosylmethionine and nicotinamide in the liver (Hanazawa et al., 1994), and consequently induce the cachexia possibly resulted from the liver damage because of the deficiency of these two components.

During ethanol metabolism in the rat, it was reported that aldehyde oxidase might be responsible for superoxide anion radical generation by oxidation of not only acetaldehyde but also $NADH^+$ which were produced by alcohol dehydrogenase (Maria et al., 1995). Since 1-methylncotinamide plausibly accumulated by the increased NMT activity in the liver of the tumor host can be metabolized into pyridones via aldehyde oxidase, the more enhanced activity of the latter enzyme might induce the more liver damage via the more superoxide anion radicals produced. However, in the liver of the mouse bearing Ehrlich ascites tumor, this enzyme activity rather decreased compared with the normal one (unpublished data). In fact, in the liver slice of this tumor host mouse, conversion of nicotinamide to the pyridones was not stimulated (1.2-fold compared with the normal) so much as 1-methyl nicotinamide (3.4-fold) (Hanazawa et al., 1994).

# 4. CONCLUSIONS

In this study, we have highlighted the occurrence of the morphological changes resembling apoptosis in the liver of the tumor-bearing animal, and its correlation with NMT activity which has been observed to increase as tumor-growth proceeded.

Finally, it is suggested that such apoptotic morphological changes might be caused by a factor which stimulated NMT activity in the liver.

Increase of NMT activity is still remained to be elucidated in its significance in the tumor cachexia, however, could be a good marker for the tumor cachexia.

# 5. REFERENCES


Asanagi M., Moriyama Y., and Fujimura S., 1988, Purification and characterization of thymidylate synthetase in the liver of the mouse bearing Ehrlich ascites tumor, *Arch. Biochem. Biophys*. 267:749.

Benedetti A., Jezequel A.M., and Orlandi F., 1988, Preferential distribution of apoptotic bodies in acinar zone 3 of normal human and rat liver, *J. Hepatol*. 7:319.

Clark B.R., Murai J.T., Pomeranz A., Mills P.A., and Smith R.A., 1975, Altered distribution and excretion of $N^1$-methylnicotinamide in rats with Walker 256 carcinosarcoma, *Cancer Res*. 35:1727.

Fujimura S., and Shimizu M., 1977, Enhanced activity of tRNA-pseudouridine synthetase in Yoshida sarcoma, *Biochem. Biophys. Res. Commun*. 79:763.. ,,

Fujimura S., Kato N., Han J.-Y., Hanazawa Y., and Nakagawa K., 1992a, Preferential increase of nicotinamide methyltransferase activity in the liver of the tumor host, in: "Frontiers and new horizons in amino acid research," K. Takai , ed., Elsevier Science Publishers B.V., Amsterdam.

Fujimura S., Kato N., Han J.-Y., Hanazawa Y., Nakagawa K., and Horitsu K., 1992b, Increase of nicotinamide methyltransferase activity in the liver of the mouse after transplantation of Ehrlich ascites tumor, in: "Advances in Tryptophan Research," I. Ishiguro et al. eds., Fujita Health Univ. Press, Toyoake.

Fujimura S., Kato N., Tanaka N., and Moriyama Y., 1995, Acidic $a_1$-proteinase inhibitor species in the ascitic fluid of the mouse bearing Ehrlich ascites tumor and the endometrial cancer patient, in: "Proteases involved in cancer," M. Suzuki and T. Hiwasa, eds., Monduzzi Editore, Bologna.

Fukuda K., Kojima M., and Chiu J.-F., 1993, Demonstration of extensive chromatin cleavage in transplanted Morris Hepatoma 777 tissue: apoptosis or not? *Am. J. Pathol*. 142:935.

Hanazawa Y., Sato K., Kuroiwa N., Ogawa M., Kuriyama A., Asanagi M., Kato N., Moriyama Y., Horitsu K., and Fujimura S., 1994, Characterization of nicotinamide methyltransferase in livers of mice bearing Ehrlich ascites tumors; Preferential increase of activity, *Tumor Biol*. 15:7.

Maria L., Maria L., Barreira L., and Manso C.F., 1995, Evidence for free radical generation due to NADH oxidation by aldehyde oxidase during ethanol metabolism, *Arch. Biochem. Biophys*. 318:53.

Nakagawa K., Miyazaki M., Okui K., Kato N., Moriyama Y., and Fujimura S., 1991, $N^1$-Methylnicotinamide level in the blood after nicotinamide loading as further evidence for malignant tumor burden, *Jpn. J. Cancer Res*. 82:1277.

Ogasawara J., Watanabe-Fukunaga R., Adachi M., Matsuzawa A., Kasugai T., Kitamura Y., Ito N., Suda T. and Nagata S., 1993, Lethal effect of the anti-Fas antibody in mice, *Nature* 364:808.

Shimizu M., and Fujimura S., 1978, Studies on the abnormal excretion of pyrimidine nucleosides in the urine of Yoshida ascites sarcoma-bearing rats, *Biochim. Biophys. Acta*. 517:277.

Shimizu M., and Fujimura S., 1984, Origin of increased deoxycytidine excretion into urine of rats bearing Yoshida ascites sarcoma, *Cancer Res*. 44:2387.

Tanaka N., Sekiya S., Takamizawa H., Kato N., Moriyama Y., and Fujimura S., 1991, Characterization of a 54kDa, $a_1$-antitrypsin- like protein isolated from ascitic fluid of an endometrial cancer patient, *Jpn. J. Cancer Res*. 82:693.

# TRYPTOPHAN HYDROXYLASE ACTIVITY IN SEROTONIN PRODUCING MAST CELLS

## Dependence on Intracellular Iron Concentration Manipulated by Permeable Chelators

H. Hasegawa,[1] Y. Iida,[1] K. Oguro,[1] M. Kojima,[1] and A. Ichiyama[2]

[1] Department of Bioscience
The Nishi-Tokyo University
Yamanashi 409-01, Japan
[2] Department of Biochemistry
Hamamatsu University School of Medicine
Hamamatsu 431-31, Japan

## 1. INTRODUCTION

The monoamine, serotonin, has long been known to function as a hormone, neurotransmitter, and local chemical mediator. Enterochromaffin cells of gastrointestinal mucosa are believed to be the predominant source in the periphery and, in rodents, mast cells are also a source. Tryptophan hydroxylase (TPH) is the rate limiting enzyme in the biosynthesis of this monoamine. TPH of peripheral origin and brain TPH are known to differ with respect to their various properties (Nakata and Fujisawa, 1982; Hasegawa et al., 1987; Kim et al., 1991). The phosphorylated brain enzyme is known to be activated in the presence of 14-3-3 protein (Yamauchi et al., 1981), however, little is known concerning regulation of the enzyme of peripheral origin such as from mast cells and gastrointestinal enterochromaffin cells. In vitro activation of peripheral TPH by anaerobic incubation with DTT and $Fe^{2+}$ included in routine measurement of this enzyme since the enzyme is very labile and must be re-activated prior to the assay (Ichiyama et al., 1974). With regard to the mechanism of activation, accumulated data from in vitro studies suggest that this activation requires supplementation with ferrous iron. Hence "DTT-activation" of the enzyme could be termed "ferrous iron activation". Some degree of uncertainty has remained regarding whether the iron concentration in the living cells is inadequate for TPH activation. The question arose as to whether the intracellular enzyme is actually supplied with sufficient ferrous iron or even whether "ferrous iron activation" really occurs. The aim of this work was to answer this question. Intracellular enzyme active was assessed under conditions in which the iron concentration was manipulated using permeable chelators. Mouse bone marrow derived mast cells (BMMC) in culture were generally used in this work as a source of natural mast cells.

*Recent Advances in Tryptophan Research,* edited by Graziella Allegri Filippini et al.
Plenum Press, New York, 1996

# 2. MATERIALS AND METHODS

Salicylaldehyde isonicotinoyl hydrazone (SIH) was a gift of Dr. Premsyl Ponka of the Lady Davis Institute, Jewish General Hospital, Montreal, Canada and iron loaded SIH was prepared according to his method (personal communication). Hemin was dissolved at 10 mM in 0.2 M $Na_2CO_3$-$NaHCO_3$ buffer, then neutralized and diluted with culture medium prior to use. Bone marrow derived mast cells were obtained according to T. Tsujimura and Y. Kitamura of Osaka University (personal communication) as follows. Femoral bones were obtained from C57BL/6crSlc mice of either sex (10±2 weeks old). Bone marrow cells were isolated and cultured in the Alpha-Modification of EAGLE'S Medium(AMEM, ICN Biomedicals,Inc.) containing 100 U/ml penicillin, 100 µg/ml of streptomycin, and 10% each of fetal calf serum and a conditioned medium in which mouse spleen cells had been cultured for 5 days in the presence of pokeweed mitogen (PWM , Gibco BRL). In order to elimimate fibroblast-type cells from the culture, adherent cells were removed every week. By about one month, no cells of this type were distinguishable. Cells were used after 1 to 3 months of culture. On the day of analysis, 18 hours after the last medium change, cells were plated on multiple well plates (V-shaped 96 wells, No.3897, Costar) at a density of 1 x $10^5$ cells per well in 200 µl of AMEM (serum free).

## Metabolic Assay of TPH

Tryptophan hydroxylase activity in cells was measured by arresting metabolism of 5HTP (Nagatsu et al., 1983). 5HTP was allowed to accumulate for 30 min in the presenc NSD1015, an inhibitor of aromatic L-amino acid decarboxylase. Culturing was terminated by adding 0.2 vol of 1 N HCl containing 10% ascorbic acid. Tryptophan hydroxylase activity in an optimized cell-free system was assayed essentially as described previously (Hasegawa et al., 1987).

# 3. RESULTS AND DISCUSSION

## 3.1 Bone Marrow Derived Mouse Mast Cells and Metabolic Assay of TPH

The serotonin content of the bone marrow cells increased for about 1 month in the activated splenocyte-conditioned medium, suggesting differentiation of the cells into mast cells. The serotonin content reached to 8.2 ± 1.2 nmol/$10^6$ cells by 30 days in culture. These bone marrow derived mast cells are rich in serotonin when compared with 0.4 nmol/$10^6$ cells in RBL2H3 cells, a rat basophilic leukemia derived mast cell line. TPH activity, measured using the cell-free assay was 0.5 pmol 5HTP/min/$10^6$ cells at 5th day and reached a plateu of roughly 20 pmol/min/$10^6$ cells at 30th day. The TPH activity of the differentiated cells varied from culture to culture for unknown reasons. The enzyme activity was maintained for about 3 months with a slight decrease (20-30%).

When NSD1015 was added to the culture medium at various concentrations, 5HTP accumulation was observed, as shown in Fig.1a. The accumulation of 5HTP continued for at least 40 min in the presence of NSD1015 (Fig. 1b). The amount of 5HTP detected in the cells (pellets obtained by centrifuging the suspension culture) before and after incubation with NSD1015 was almost constant at 15 ±2 pmol/$10^6$ cells. Most of the accumulated 5HTP was found in the medium (outside the cells) even in the absence of NSD1015 (Fig. 1b). These results suggest that considerable amount of 5HTP had leaked out of the cells before beeing

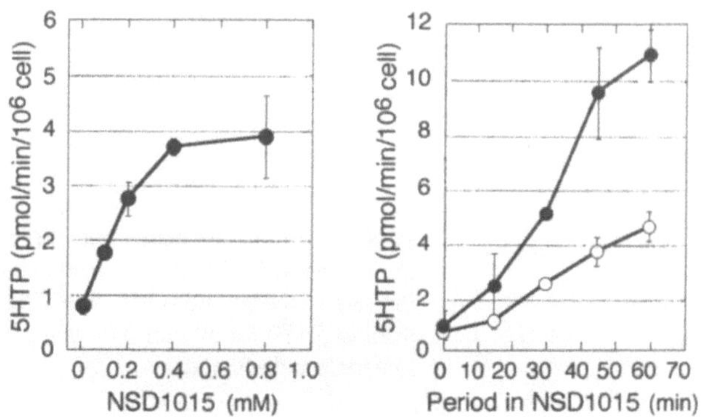


**Figure 1.** Accumulation of 5HTP in the presence of NSD1015, an inhibitor of aromatic L-amino acid decarboxylase, as a measure of tryptophan hydroxylase activity in mouse bone marrow derived mast cells (BMMC) in culture. a) Dependence of 5HTP accumulation on NSD1015 concentration. BMMC were kept in the presence of the indicated concentrations of NSD1015 for 30 min. Cells and culture medium were collected together and subjected to 5HTP determination. b) Time course of 5HTP accumulation. Culture medium was replaced with fresh medium with (closed circles) or without 400 μM NSD1015 (open circles). Culture was terminated at the indicated times and the cells and the medium were collected together and subjected to 5HTP determination. Values are means ± S.D. (n=6).

decarboxylated leaving it unconverted. To assess the intracellular conversion of tryptophan to 5HTP, we measured 5HTP accumulation in the cell and medium mixture for 30 min in the presence of 400 μM NSD1015.

The measured conversion rate of tryptophan to 5HTP was around 30-40 pmol per 30 min. It is known that the intracellular concentration of the enzyme substrates, tryptophan and tetrahydrobiopterin, are not always saturated. When the tryptophan concentration increased to 270 μM in the medium (70 μM using the usual method), the conversion rate increased to 2.13 ±0.85 fold. In addition, the tetrahydrobiopterin concentration was increased by adding sepiapterin which is taken up by cells and converted to tetrahydrobiopterin and 1.68 ±0.09 fold higher activity was observed. This value represented about 36% of the TPH activity obtained under optimized cell-free conditions.

## 3.2 Manipulation of TPH Activity with Permeable Chelators

Hemin is known to elevate intracellular iron concentration (Rouault et al., 1985). Stimulation of the intracellular conversion of tryptophan to 5HTP was observed at a hemin concentration of around 10 μM (2 fold, Fig. 2). However, stimulation was antagonized using hemin itself at higher concentrations. In the case of RBL2H3 cells, the conversion was saturable at around 30 μM and no significant decrease was observed up to 100 μM. The reason for this difference between the two types of cells was not investigated. In the work using RBL2H3 cells, stimulation with hemin was effectively antagonized with simultaneous addition of permeable iron chelators such as desferrioxamine and salicylaldehyde isonicotinoyl hydrazone (SIH). These compounds may compete with thecommon iron pool within the cells since the above cherators are unable to remove iron from hemin in extracellular solutions such as culture medium.

SIH, a hydrophobic $Fe^{3+}$-chelator, has no inhibitory effect on TPH activity under cell-free assay conditions, at concentrations up to 1 mM, consistent with the specific reqirement of the enzyme for $Fe^{2+}$. When SIH was administered to BMMC, the intracellular activity of TPH was effectively suppressed at concentrations less than 30 μM (Fig. 3). Using

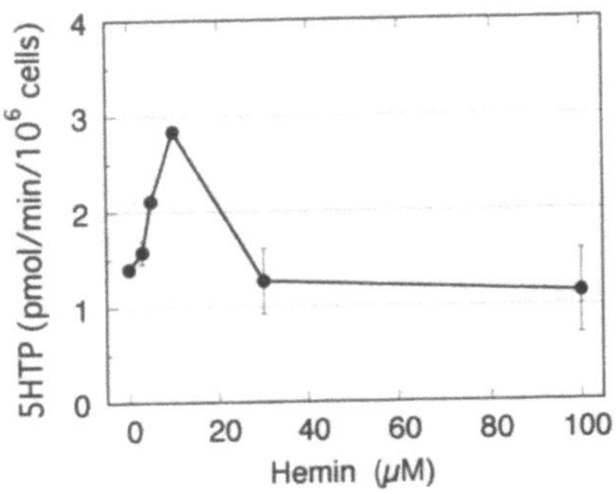


**Figure 2.** Effect of hemin on intracellular conversion of tryptophan to 5HTP. BMMC were exposed for 2 hours to the indicated concentrations of hemin. The conversion of tryptophan to 5HTP for 30 min was then mesured as described in Materials and Methods.

200 μM SIH, the conversion rate was lowered to its minimum within 1 hour of exposure to the cherator; accumulation of 5HTP after 1 hour was below significant detection. This result suggested that the chelator suppressed $Fe^{2+}$ concentration indirectly via masking $Fe^{3+}$. $Fe^{2+}/Fe^{3+}$ ratio may be determined by dominating intracellular homeostatic mechanisms and the decrease in $Fe^{3+}$ may pull the $Fe^{2+}$ concentration down to a level which TPH can not usitize.

Iron loaded SIH (Fe-SIH) was also effective, but only in stimulating cellular conversion of tryptophan to 5HTP (Fig. 2, closed circle). The conversion rate was around 2.5 fold of the initial level. Higher concentrations of Fe-SIH up to 200 μM, or longer incubations of up to 5 hours did not cause further stimulation. Assuming a subsaturating supply to the cell of tryptophan and tetrahydrobiopterin, as mentioned earlier, this value was calculated to be about 90% of the potential activity of the enzyme.

## 4. CONCLUSION

The intracellular conversion rate of tryptophan to 5HTP under ordinary culture conditions is lower than the potential activity of TPH. Since the conversion rate increases after supplying cells with iron, the intracellular concentration of $Fe^{2+}$ might have been much lower than optimal for TPH. The intracellular $Fe^{2+}$ concentration could be raised or lowered. The $Fe^{2+}/Fe^{3+}$ balance in the cell may be conserved by a dominating homeostatic equilibrium which allows for conversion from $Fe^{3+}$ to $Fe^{2+}$ when $Fe^{3+}$ is supplied or removed. SIH is a useful agent for manipulation of the intracellular concentration since it can either raise or lower it as the situation requires.

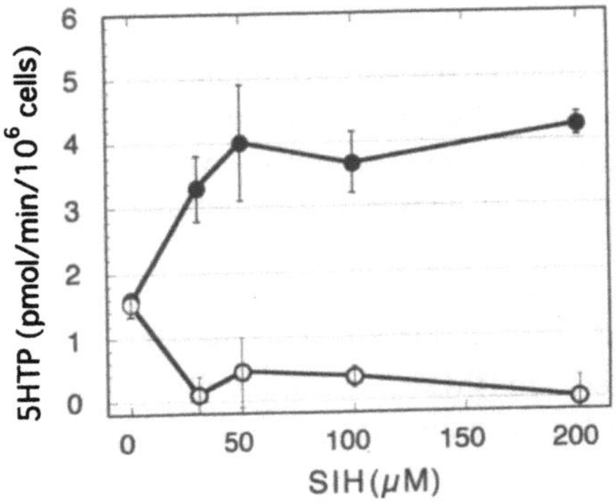


**Figure 3.** Tryptophan conversion to 5HTP in BMMC in the presence of various concentrations of SIH. Cells were fed with fresh medium (AMEM) containing the indicated concentrations of SIH (open circles) or Fe-SIH (closed circles) and cultured for 2 hours. Accumulation of 5HTP for 30 min in the presence of NSD1015 was then measured.

# REFERENCES

Hasegawa H., Yanagisawa M., Inoue F., Yanaihara N., and Ichiyama A., 1987, Demonstration of non-neural tryptophan 5-mono-oxygenase in mouse intestinal mucosa, *Biochem. J.* 248: 501.

Ichiyama A., Hori S., Mashimo Y., Nukiwa T., and Makuuchi H., 1974, The activation of bovine pineal tryptophan 5-monooxygenase, *FEBS Lett* 40: 88.

Kim K.S., Wessel T.C., Stone D.M., Carver C.H., Joh T.H., and Park D.H., 1991, Molecular cloning and characterization of cDNA encoding tryptophan hydroxylase from rat central serotonergic neurons, *Brain Res Mol Brain Res* 9: 277.

Nagatsu T., Sawada M., and Yamaguchi T., 1983, Tryptophan hydroxylase system in brain tissue slices assayed by hight-performance liquid chromatography, *Neurochem. Int.* 5: 603.

Nakata H., and Fujisawa H., 1982, Tryptophan 5-monooxygenase from mouse mastocytoma P815. A simple purification and general properties, *Eur J Biochem* 124: 595.

Rouault T., Rao K., Harford J., Mattia E., and Klausner R.D., 1985, Hemin, chelatable iron, and the regulation of transferrin receptor biosynthesis, *J Biol Chem* 260: 14862.

Yamauchi T., Nakata H., and Fujisawa H., 1981, A new activator protein that activates tryptophan 5-monooxygenase and tyrosine 3-monooxygenase in the presence of Ca2+-, calmodulin-dependent protein kinase, *J. Biol. Chem.* 256: 5405.

# OCCURRENCE OF INDOLEAMINE 2,3-DIOXYGENASE IN HUMAN CUTANEOUS MALIGNANT MELANOMA

A. Bertazzo,[1] C. V. L. Costa,[1] M. Lise,[2] D. Nitti,[2]
and G. Allegri[1]

[1] Department of Pharmaceutical Sciences
[2] Institute of Surgical Clinic II
University of Padova
I-35131 Padova, Italy

## INTRODUCTION

Yamamoto and Hayaishi (1967) first reported the presence of indoleamine 2,3-dioxygenase [IDO; indoleamine: oxygen 2,3-oxidoreductase (decyclizing)] in rabbit intestine. This enzyme was also later found in mice (Yoshida et al., 1980), rats (Cook et al., 1980) and humans (Yamazaki et al., 1985), in which the highest activity was observed in the placenta, followed by lung, small intestine, and various other organs. IDO was markedly induced by various pathogens such as viruses and bacterial endotoxins (Yoshida and Hayaishi, 1978; Yoshida et al., 1979). Induction appears to be mediated by interferon-γ and causes a marked increase in tryptophan catabolism (Takikawa et al., 1986). Enzyme induction was also observed by intraperitoneal transplantation of tumor cells to allogeneic strains of mice (Yoshida et al., 1984). A marked increase in IDO activity was also found in human lungs bearing cancer (Yasui et al., 1986).

In this context, it was of interest to examine whether IDO activity was increased in the skin of patients with cutaneous malignant melanoma. In the present paper, both melanoma and surrounding skin were analysed. Samples were obtained from patients who had undergone surgical resection. Peripheral portions of resected skin tissues, macroscopically normal, were excised, frozen on dry ice, and stored at -75° until assayed for enzyme activity.Cancer tissues were processed in the same way.

IDO activity was assayed using the method of Shimizu et al. (1978). Aliquots from 50 to 200 μl of crude homogenate supernatant fractions were subjected to protein determination according to Lowry et al. (1951), with bovine serum albumin as standard.

*Recent Advances in Tryptophan Research*, edited by Graziella Allegri Filippini et al.
Plenum Press, New York, 1996

**Table 1.** IDO activity in tissues of patients bearing cutaneous malignant melanoma*

| Case n° | Skin | Melanoma | Lymphonode | Serum | Trp in serum µg/ml |
|---|---|---|---|---|---|
| 1 | 0.6143 | 0.0602 | | 0.0280 | 11.32 |
| 2 | 0.6770 | 0.1810 | 0.6666 | 0.1346 | 5.55 |
| 3 | 0.3301 | 0.1237 | | | |
| 4 | 0.500 | 0.1797 | | | |

*Specific actvity expressed as µmol formylkynurenine/ mg of protein/min.

## RESULTS

IDO activity was determined in supernatant fractions of human cutaneous malignant melanoma tissue homogenate and in portions of macroscopically normal skin at a distance from the melanoma, with L-tryptophan as substrate.

As shown in Table 1, the skin exhibited IDO activity from three to ten times higher than activity in melanoma tissue.

Lymphonode tissue also showed IDO activity at the same level as in skin.

Although individual variations in the activity of the various analysed tissues were found, enzyme activity was higher in the skin surrounding melanoma.

In cases 1 and 2, IDO activity was also assayed in serum, in which the tryptophan concentration was also determined. Figure 1 shows that, when IDO activity in serum is high, it reduces the availability of tryptophan in serum.

## DISCUSSION

Naito et al. (1991) demonstrated the presence of indoleamine 2,3-dioxygenase in rat skin. Our results report the presence of IDO in human skin, in melanoma and lymphonode

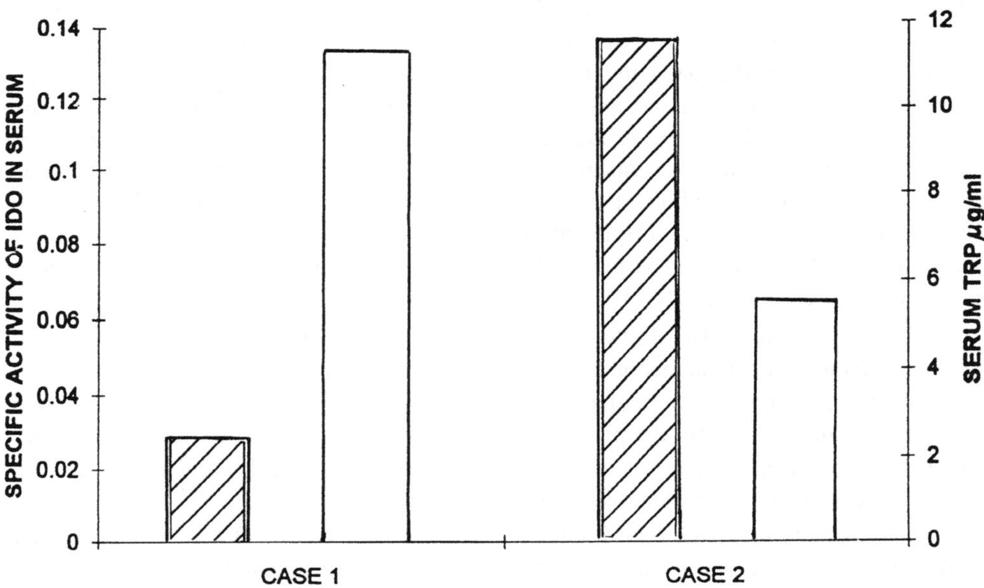


**Figure 1.** Specific activity of IDO expressed as µmol formylkynurenine/min per mg of protein: (diagonal bar). Tryptophan concentration in µg/ml: (open bar).

tissues, and in the serum of patients bearing cutaneous malignant melanoma. Induction of IDO activity takes place "in vivo" in cutaneous malignant melanoma as a response to the tumour, leading to marked catabolism of tryptophan along the kynurenine pathway. In a previous paper (Nitti et al., 1990), we demonstrated the abnormal urinary excretion of tryptophan metabolites in patients with cutaneous malignant melanoma. However, abnormal tryptophan catabolism has been observed in many patients bearing tumours (Quagliarello et al., 1961; Benassi et al., 1963; Musajo and Benassi, 1964; DeGeorge and Brown, 1970), probably due to increased IDO activity, as the activity of liver tryptophan 2,3-dioxygenase determined in biopsies did not show a relationship with urinary kynurenine (Gailani et al., 1974).

We postulate that increased IDO activity in serum leads to a decrease in serum tryptophan concentration, as shown in Fig. 1, accounting for the substantial increase in urinary metabolites of the kynurenine pathway observed in patients with cutaneous malignant melanoma (Nitti et al., 1990).

## REFERENCES

Benassi C.A., Perissinotto B., and Allegri G., 1963, The metabolism of tryptophan in patients with bladder cancer and other urological diseases, *Clin.Chim.Acta*, 8: 822.

Cook J.S., Pogson C.I., Smith S.A., 1980, Indoleamine 2,3-dioxygenase: a new, rapid, sensitive radiometric assay and its application to the study of the enzyme in rat tissues, *Biochem. J.*, 189: 461.

DeGeorge F.V. and Brown R.R., 1970, Difference in tryptophan metabolism between breast cancer patients with and without cancer at other sites, *Cancer* 26: 767.

Gailani S., Ezdinli E., Nussbaum A., Silvernail P., and Elias E.G., 1974, Studies on tryptophan metabolism in patients with lymphoma, *Cancer Res.*, 34: 1664.

Lowry O.H., Rosebrough N.J., Farr A.L. and Randall R.J., 1951, Protein measurement with the Folin phenol reagent, *J.Biol.Chem.*, 193:265.

Musajo L., Benassi C.A., 1964, Aspects of disorders of the kynurenine pathway of tryptophan metabolism in man, *Adv. Clin. Chem.*, 7: 63.

Naito J., Ishiguro I., Nagamura Y., and Ogawa H., 1991, Rat skin tryptophan 2,3-dioxygenase, in Kinurenine and Serotonin Pathways, R. Schwarcz et al., eds., Plenum Press, New York, p. 527.

Nitti D., Lise M., Costa C., Cardin de Stefani E.L., Allegri G., 1990, L-Tryptophan metabolism in human cutaneous malignant melanoma, *Giorn.It.Chim.Clin.*, 15: 211.

Quagliarello E., Tancredi F., Fedele L., and Saccone C., 1961, Tryptophan-nicotinic acid metabolism in patients with tumors of the bladder. Changes in the excretory products after treatment with nicotinamide and vitamin $B_6$, *Brit. J.Cancer*, 15: 367.

Shimizu T., Nomiyama S., Hirata F., and Hayaishi O., 1978, Idoleamine 2,3-dioxygenase, purification and some properties, *J.Biol.Chem.* 253: 4700.

Takikawa O., Yoshida R., Kido R., and Hayaishi O., 1986, Tryptophan degradation in mice initiated by indoleamine 2,3-dioxygenase, *J.Biol.Chem.*, 261: 3648.

Yamamoto S. and Hayaishi O., 1967, Tryptophan pyrrolase of rabbit intestine, *J.Biol.Chem.*, 242: 5260.

Yamazaki F., Kuroiwa T., Takikawa O., and Kido R, 1985, Human indolyl-amine 2,3-dioxygenase, its tissue distribution, and characterization of the placental enzyme, *Biochem. J.*, 230: 635.

Yasui H., Takai K., Yoshida R., and Hayaishi O., 1986, Interferon enhances tryptophan metabolism by inducing pulmonary indoleamine 2,3-dioxygenase: its possible occurrence in cancer patients, *Proc. Natl. Acad. Sci.*, USA, 83: 6622.

Yoshida R. and Hayaishi O., 1978, Induction of pulmonary indoleamine 2,3-dioxygenase by intraperitoneal injection of bacterial lipopolysaccharide, *Proc. Natl. Acad. Sci.*, USA, 75: 3998.

Yoshida R., Urade Y., Tokuda. M, and Hayaishi O., 1979, Induction of indoleamine 2,3-dioxygenase in mouse lung during infection, *Proc. Natl.Acad. Sci.*USA, 76: 4084.

Yoshida R., Nukiwa T., Watanabe Y., Fujiwara M., Hirata F. and Hayaishi O., 1980, Regulation of indoleamine 2,3-dioxygenase activity in the small intestine and epididymis of mice, *Arch. Biochem.Biophys.* 203: 343.

Yoshida R., Takikawa O., Yasui H., Hayaishi O., and Yasuhira K., 1984, Role of indoleamine 2,3-dioxygenase in the defense mechanism against tumor growth, in Progress in Tryptophan and Serotonin Research, Walter de Gruyter ed., Berlin, p. 513.

# THE INFLUENCE OF ANTAGONISTS OF POLY(ADP-RIBOSE)POLYMERASE ON THE ACTIVITY OF ANTIRHEUMATIC DRUGS ON THE DEVELOPMENT OF ADJUVANT ARTHRITIS IN RATS AND ON THE INDUCTION OF TYROSINE AMINOTRANSFERASE IN THE LIVER OF RATS

H. Kröger, A. Dietrich, R. Grätz, M. Klewer, H. Wohlert, and W. Ehrlich

Deutsches Rheumaforschungszentrum Berlin and
Robert Koch-Institut
Nordufer 20, D-13353-Berlin, Germany

## INTRODUCTION

Poly-adenoribosylation is a posttranslational modification of chromatin-bound proteins. Recently, there has been increasing evidence that poly-adenoribosylation is involved in the genesis of cancer but also in the genesis of autoimmune diseases such as rheumatoid arthritis. The synthesis of poly(ADP-ribose) by the enzyme poly(ADP-ribose)polymerase depends on the intracellular levels of NAD and this depends on the supply with the precursers tryptophan, nicotinamide, and nicotinic acid. Poly-adenoribosylation is a protective cellular response to DNA damage which leads to a decrease in the NAD content of cells (Kröger et al., 1960). The consequence is a depletion of NAD with a reduced poly(ADP-ribose) accumulation and altered biological responses such as the expression of disease specific genes (Jacobson et al., 1992, Ehrlich, et. al., 1995). The question has been raised whether agents or conditions that influence the NAD-poly(ADP-ribose)metabolism do have an effect on the pathogenesis of diseases such as rheumatoid arthritis. Previously, we reported that a combination of benzamide or a diet without tryptophan and nicotinamide in combination with D-penicillamine leads to an improvement of adjuvant arthritis in rats (Kröger and Ehrlich, 1994). We continued these studies and investigated the effects of the poly(ADP-ribose)polymerase antagonists nicotinamide, benzamide, caffeine-derivatives and a diet lacking tryptophan and nicotinamide alone and in combination with the antirheumatic drugs

*Recent Advances in Tryptophan Research*, edited by Graziella Allegri Filippini et al.
Plenum Press, New York, 1996 

endoxan (cyclophosphamide), D-penicillamine, auranofin, voltaren, and methotrexate on the development of adjuvant arthritis of rats. In another set of experiments we studied the influence of D-penicillamine, benzamide, and cofpropamine (a caffeine-derivative) on the induction of tyrosine aminotransferase in the liver of rats with adjuvant arthritis set on a diet without tryptophan and nicotinamide.

# MATERIALS AND METHODS

## Adjuvant Arthritis

For the experiments male rats of the strain Wistar-AF/Han-EMD-SPF from A. Wild, Merck, Darmstadt, Germany, were used. Arthritis was induced by injecion of adjuvant (a suspension of heat inactivated mycobacterium tuberculosis) into the joint of the paw of the right hind leg (Mohr et al., 1975; Mohr and Wild, 1976). Subsequently, the rats showed arthritic reactions in the left paw, the ears, and at the tail. The bodyweight and the increase of volume of the right and the left hind paws as a measure of arthritic reactions was followed for 24 days. The volume of the paws was determined by displacement of water. For the experiments three animals each were used.

## Treatment Protocolls

NaCl 0.9 % 20 x 10 ml (disease control). Benzamide 20 x 100 mg/kg i.p. Nicotinamide 20 x 700 mg/kg i.p. Cofpropamine (8-dimethylaminopropoxy-caffeine, Knoll AG, Germany) 10 x 150 mg/kg i.p. Diet without L-tryptophan and nicotinamide. Auranofin 10 x 10 mg/kg p.o., auranofin 7 x 10 mg/kg p.o. + 7 x nicotinamide 500 mg/kg i.p.; auranofin 9 x 10 mg/kg p.o. + 9 x benzamide 25 mg/kg i.p. Endoxan (cyclophosphamide) 1x 50 mg/kg i.p., endoxan 1x 50 mg/kg i.p. + 3x benzamide 25 mg/kg i.p., endoxan 1x50 mg/kg i.p. + 3x nicotinamide 100 mg/kg i.p., endocofamine II (endoxan/cofpropamine 1:1) 100 mg/kg i.p. 2x , endocofamine IV (endoxan/cofpropamine 1:3) 3 x 200 mg/kg i.p. Methotrexate 7 x 2.5 mg/kg i.p.;methotrexate 7 x 2.5 mg/kg+cofpropamine 10 x 150 mg/kg i.p. D-Penicillamine 20 x 100 mg/kg i.p. ; D-penicillamine 20 x 100 mg/kg i.p. + 20 x nicotinamide 500 mg/kg i.p. ; D-penicillamine 20 x 100 mg/kg i.p. + benzamide 20 x 25 mg/kg i.p. Diet without L-tryptophan and nicotinicamide and D-penicillamine 20 x 100 mg/kg i.p. Voltaren 20 x 10 mg/kg i.p.; voltaren 20 x 10 mg/kg i.p. + benzamide 20 x 25 mg/kg daily i.p.

## Induction of Tyrosine Aminotransferase (TAT) in the Rat Liver

On day 25 after injection of adjuvant rats were sacrificed and TAT was determined as described by Kröger and Greuer, 1966. *Treatment protocoll*: NaCl 0.9 % 20 x 10 ml/kg (disease control), Diet without L-tryptophan and nicotinamide, diet without L-tryptophan and nicotinamide + D-penicillamine 20 x 100 mg/kg i.p., diet without L-tryptophan and nicotinamide + cofpropamine · HCl 20 x 25 mg/kg i.p., diet without L-tryptophan and nicotinamide + benzamide 20 x 25 mg/kg i.p, diet without L-tryptophan and nicotinamide + D-penicillamine 20 x 100 mg/kg i.p. + cofpropamine 20 x 25 mg/kg i.p., diet without L-tryptophan and nicotinamide + D-penicillamine 20 x 100 mg/kg i.p. + benzamide 20 x 25 mg/kg i.p.

**Table 1.** The influence of antagonists of poly(ADP-ribose)polymerase (PARP) on the therapeutic effect of antirheumatic drugs

| | No PARP antagonist | Nicotinamide | Benzamide | Caffeine-derivatives | Diet without L-tryptophan and nicotinamide |
|---|---|---|---|---|---|
| No drug | | ++ | + | – [1] | ++ |
| Endoxan | – | ++ | ++ | +++ [2,3] | |
| D-penicillamine | – | ++ | ++ | | +++ |
| Auranofin | ++ | ++ | +++ | | |
| Voltaren | +++ | | +++ | | |
| Methotrexate | ++ | | | +++ [1] | |

[1]cofpropamine, [2]endocofamine II, [3]endocofamine IV - no antiarthritic effect; + weak antiarthritic effect; ++ medium antiarthritic effect; +++ strong antiarthritic effect.

# RESULTS AND DICUSSION

Table I gives a survey of the results. Treatment with nicotinamide, benzamide, or a diet without tryptophan and nicotinamide clearly reduces the development of adjuvant arthritis in rats. This effect is enhanced when adjuvant arthritis is treated with poly(ADP-ribose)polymerase antagonists in combination with the antirheumatic drugs endoxan, D-penicillamine, auranofin, or methotrexate (Table 1). It can be concluded that poly-adenoribosylation is involved in the genesis of arthritis induced with adjuvant in rats. Thus, substances that modulate the NAD-poly-adenoribosylation metabolism can be useful for the treatment of rheumatoid arthritis by their potential to increase the therapeutic effect of drugs used in rheumatoid arthritis therapy.

Previously, we have reported that poly-adenoribosylation is involved in the induction of tyrosine aminotransferase (TAT) in liver (Kröger, and Grätz; 1980; Cihak. and Kröger, 1981; Kröger and Grätz, 1984). The results shown in Figure 1 underline this. TAT is induced in the liver of rats with adjuvant arthritis. It is reduced when rats with adjuvant arthritis are

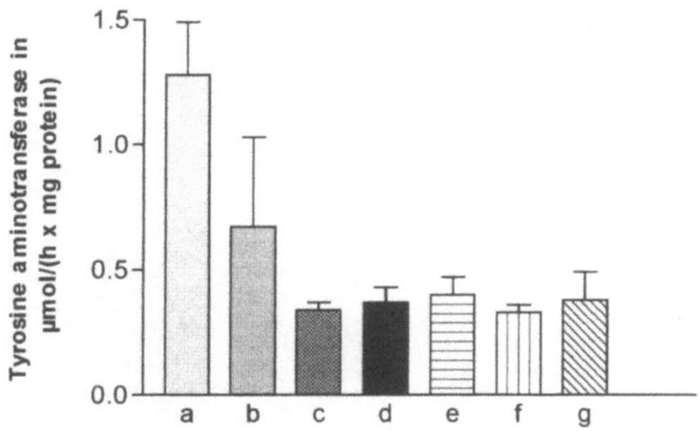


**Figure 1.** Induction of tyrosine amino transferase in rats with adjuvant athritis treated with NaCl ( a ), and in rats with adjuvant arthritis set on a diet without L-tryptophan and nicotinamide and without additional treatment ( b ), and additional treatment with D-penicillamine ( c ), benzamide ( d ),cofpropamine ( e ) cofpropamine and D-penicillamine ( f ), and benzamide and D-penicillamine ( g ). For treatment protocol see Materials and Methods.

set on a diet without tryptophan and nicotinamide. It is further reduced by additional treatment with benzamide, cofpropamine, or D-penicillamine.

## ACKOWLEDGMENTS

We thank Prof. Dr. A. Wild, Darmstadt for the supply with rats with adjuvant arthritis. Cofpropamine was originally synthesized by Dr. J. Klosa, Berlin. This work was supported by a grant from the Völkl Stiftung.

## REFERENCES

Cihak, A. and Kröger, H., 1981, Tryptophan-mediated induction of hepatic enzymes stimulated by 5-azacytidine pretreatment, *Biomedicine*, 34:18.

Ehrlich, W. Huser, H. Kröger, H., 1995, Inhibition of the induction of collagenase by interleukin-1b in cultured rabbit synovial fibroblasts after treatment with the poly(ADP-ribose)polymerase inhibitor 3-aminobenzamide, *Rheumatol. Int.*, in press

Jacobson, E.L., Nhumbhakdi - Craig, V., Smith, D.G., Chen, H.Y., Wasson, B.L., Jacobson, M.K.1992, ADP-ribose polymer metabolism: Implications for human nutrition ,in: "ADP-Ribosylation Reactions", G.G. Poirier, P. Moreau, ed., Springer Verlag, New York

Kröger, H., Rotthauwe, H.W. Ulrich, B. Holzer, H., 1960, Zum Einfluß von Carcinostatika auf den DPN-Stoffwechsel der Tumoren, *Biochem. Z.* 333: 148.

Kröger, H. and Greuer, B., 1966, Influence of cortisone on the substrate-induction of enzymes in rat liver, *Nature* 210: 200.

Kröger, H. and Grätz, R., 1980, On the induction of tyrosine aminotransferase (TAT) by tryptophan and methionine. *Int. J. Biochem.* 12: 575.

Kröger, H. and Grätz, R., 1984, Induction of tyrosine aminotransferase under the influence of D-galactosamine. *Int. J. Biochem.*16:103.

Kröger, H. and Ehrlich, W., 1994, The influence of tryptophan and nicotinamide on acetaminophenon hepatotoxicity in mice potentiated by ethanol and on adjuvant arthritis in rats, in: "L-Tryptophan: Current Prospects in Medicine and Drug Safety", W. Kochen, H. Steinhart, ed.,W. de Gruyter Verlag, Berlin

Mohr, W., Hersener, J., Wild, A., Paulini, K., 1975, Untersuchungen zum Knorpelabbau bei der Adjuvanskrankheit der Ratte, *Z. Rheumatol.* 34: 234.

Mohr, W. and Wild, A., 1976, Der Ablauf der Adjuvanskrankheit bei Ratten des Stammes AF/Han-EMD-SPF, *Z. Rheumatol.* 35:77.

# PURIFICATION OF L-KYNURENINE 3-MONOOXYGENASE FROM PIG LIVER MITOCHONDRIAL OUTER MEMBRANE

T. Uemura and K. Hirai

Department of Neurochemistry
Tokyo Institute of Psychiatry
2-1-8, Kamikitazawa, Setagaya-ku, Tokyo, 156, Japan

## INTRODUCTION*

L-Kynurenine 3-monooxygenase is an FAD-dependent mixed function oxidase located in the mitochondrial outer membrane (Okamoto and Hayaishi, 1969). The purification of the mammalian enzyme has been done only from rat liver mitochondria (Horn et al., 1971; Nishimoto et al., 1977; 1979) and never from other mammalian livers. In this paper we describe the method for the purification of this enzyme from pig liver mitochondrial outer membrane. It consists of (1) the extraction of the enzyme from the Chaps*-insoluble outer membrane material by the nonionic detergent, β-Octylglucoside, (2) ammonium sulfate fractionation, (3) DEAE-Sepharose CL-6B chromatography, and (4) affinity chromatography (mercurial-Sepharose).

## METHODS

Crude mitochondrial outer membranes were isolated from pig liver by hypotonic swelling method (Parsons et al., 1966), suspended in 0.2 M Tris-HCl buffer pH 8.0, containing 10% glycerol, 0.2 mM EDTA/DTT, 50 μM FAD and treated with the zwitterionic detergent, Chaps, at a final concentration of 0.5% at 4° C for 30 min followed by the addition of protamine sulfate at a final concentration of 0.07% for 20 min. After centrifugation at 170,000xg for 60 min, the Chaps-insoluble membrane pellets were treated with the nonionic detergent, OGS, at a final concentration of 2% for 30 min. After centrifugation as described above, the supernatant was subjected to the 40-80% satuation of $(NH_4)_2SO_4$ fractionation.

* **Abbreviations used**: Chaps: 3-(3-cholamidopropyl)-dimethylammonio-1-propane sulfonate; OGS:β -Octylglucoside; DTT: Dithiothreitol; MAO: Monoamine oxidase; 3-OHKYN: 3-Hydroxykynurenine; HPLC-ECD: High Performance Liquid Chromatography with Electrochemical Detector.

*Recent Advances in Tryptophan Research*, edited by Graziella Allegri Filippini et al.
Plenum Press, New York, 1996

The floating pellet having about 40% of the outer membrane activity was dissolved in 10 mM Tris-acetate buffer, pH 8.0, containing 10% glycerol, 0.1% Chaps, 0.5 mM EDTA, 0.2 mM DTT, 50 μM FAD and dialyzed against the same buffer without Chaps for 16 hr. After centrifugation at 25,000xg for 30 min, the supernatant was applied to a DEAE-Sepharose CL-6B column equilibrated with the above buffer containing 0.05% OGS. After washing the column with the equilibration buffer, the enzyme was eluted with linear gradient of 0 to 1.0 M KCl and then of 0.05 to 0.6% OGS in equilibration buffer. The enzyme found in OGS fractions having higher activity than that in KCl fractions was pooled, dialyzed for 6 hr against 10 mM phosphate buffer, pH 8.0 containing 10% glycerol, 0.1% OGS, 0.1 mM EDTA, 50 μM FAD and applied to an Affi-Gel 501 (organomercurial Agarose). After washing the column with 0.1M phosphate buffer, pH 8.0, containing the same components as above except FAD, the enzyme was eluted with linear gradient of 0 to 20 mM DTT in washing buffer. The enzyme found in DTT fractions having higher activity than that in pass-through fractions was pooled, concentrated with an Amicon Centriprep-30. L-Kynurene 3-monooxygenase activity was measured spectrophotometrically at room temperature by the disappearance of NADPH at 340 nm (Hayaishi, 1962) with an extinction coefficient of 6.22 $mM^{-1}$ $cm^{-1}$ or at 37° C by HPLC-ECD (Uemura et al., 1991) except that a known standard was composed of enzyme protein, 557 fmol of 3-OHKYN with potential of a Glassy Carbon electrode at 0.6 V. The MAO activity was measured fluorometrically using kynuramine as substrate (Krajl, 1965). Native polyacrylamide-gel disc electrophoresis was performed according to the method of Davis 1964). Gels were stained with either Coomassie blue R-250 or MAO activity staining (Youdim and Collins, 1971). Protein determinations were carried out by the dye-binding method of Bradford (1976) or by the method of Lowry et al (1951) using Bio-Rad protein assay kits with bovine serum albumin as standard.

## RESULTS

Table 1 shows a summary of the purification procedure showing the specific activity, purification factor and recoveries at upto the step of anion exchange chromatography. After DEAE-Sepharose CL-6B column chromatography, the OGS-eluate fraction showed a specific activity of about 60 nmoles/min/mg protein. The increase in specific activity was about 20 times that of the outer mitochondrial membrane.

Further purification was performed by an Affi-Gel 501 column equilibrated with 0.1 M phosphate buffer, pH 8.0, containing 10% glycerol, 0.1 mM EDTA, 0.1 M NaCl, and 0.1% OGS. Despite no further increase in the specific activity by a mercurial-Sepharose column, this step served the removal of the bulk of the MAO activity from the DTT fractions. Unfortunately, however, the best DTT fractions (IV') showing the highest specific activity

**Table 1.** Purification of L-kynurenine 3-monooxygenase from pig liver mitochondrial outer membrane

| Step | Total volume (ml) | Total protein (mg) | S.A. | Purification factor | Yield (%) |
|---|---|---|---|---|---|
| Outer membrane | 117.3 | 1173 | 2.8 | 1 | 100 |
| Chaps-protamine sulfate ppt. | 75.1 | 751 | 3.6 | 1.3 | 82 |
| OGS-supernatant | 84 | 185 | 8.7 | 3.1 | 49 |
| Ammonium sulfate (40-80%) | 17.2 | 126 | 10.6 | 3.8 | 41 |
| DEAE-sepharose CI-6B OGS-eluate | 26.5 | 12 | 61 | 22 | 22 |

S.A. = nmoles of NADPH oxidized/mean/mg protein

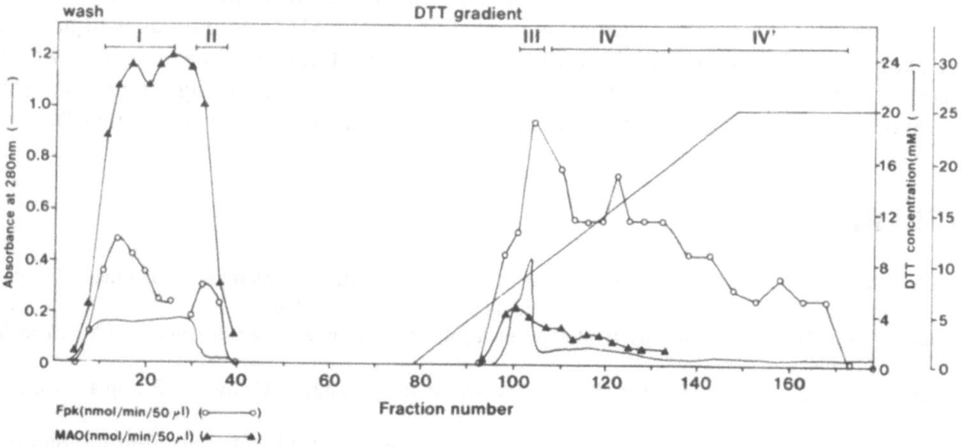


**Figure 1.** Elution profile of L-kynurenine 3-monooxygenase and MAO from an Affi-Gel 501 column. Column, Affi-Gel 501 (10 mmx8.3 cm); OGS eluate from DEAE-Sepharose CL-6B (ca. 13 mg of protein) was applied to the column. Flow rate 0.3 ml/min; Specific activity of the monooxygenase (nmoles of NADPH oxidized/min/mg protein): I=3, II=6, III=33, IV=68, IV'=65 and that of MAO (μmoles of 4-hydroxyquinoline formed/min/mg protein): I=2.7, II=2.8,III=0.86, IV=0.71, IV'=0.85.

of the monooxygenase (65nmoles/min/mg protein) contained still some MAO activity (Fig. 1).

Fig. 2 shows native polyacrylamide-gel disc electrophoretic pattern. The concentrated DTT fractions obtained by another experiment showing somewhat lower specific activity (46 nmoles/min/mg protein) than the preparation IV' were electrophoresed. Ten protein bands migrated from cathode to anode (band 1-10) on 6.6% polyacrylamide gels. None of them, however, showed the monooxygenase activity and one band remaining at the origin of the separation gel (band 11) showed almost the same specific activity as the charged enzyme preparation. Concomitantly, however, the band 11 was stained by purple-pink formazan (data not shown), indicating the presence of MAO.

## DISCUSSION

As reported for pig brain mitochondrial MAO B purified by mercurial-Sepharose (Minamiura and Yasunobu, 1978), the monooxygenase did not penetrate the gels used in ordinary disc electrophoresis which suggests that the enzyme is isolated as polymer having a very high molecular weight. The DTT fractions having higher monooxygenase activity

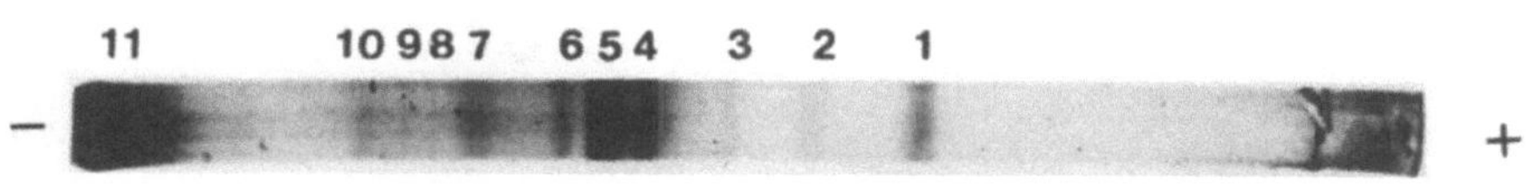


**Figure 2.** Polyacrylamide gel electrophoresis of preparation of pig liver L-kynurenine 3-monooxygenase. Protein from the DTT fractions (76 μg) was placed on 2.5% spacer gels. Disc gel electrophoresis was carried out at a constant current of 2.5 mA per tube for about 2.5 hr at 4°C. After electrophoresis, one gel was stained by Coomassie blue and protein bands from other 7gels were cut with a sharp razor and the gels were homogenzied in 0.2 M Tris-acetate buffer, pH 8.0 to elute the protein. After centrifugation at 40,000xg for 45 min at 4 °C, the supernatant was used to determine the monooxygenase activity measured by HPLC-ECD.

accompanied simultaneously MAO activity which prevented a spectral analysis of the monooxygenase. Further purification is needed to remove MAO activity as well as outer membrane cytochrome b from the DTT fractions for spectral as well as molecular weight analysis of the enzyme. We are trying an affinity chromatography using a potent enzyme inhibitor such as m-nitrobenzoylalanine recently synthesized by Pellicciari et al (1994).

## REFERENCES

Bradford MM. 1976, A rapid and sensitive method for the quantitation of microgram quantities of protein utilizing the principle of protein-dye binding, *Anal Biochem* 72: 248

Davis BJ., 1964, Disc electrophoresis. II. Method and application to human serum proteins, *Ann N Y Acad Sci* 121: 404

Hayaishi O., 1962, Kynurenine hydroxylase. In: Methods in Enzymology V, Colowick SP, Kaplan NO, eds., Academic Press, 807-809

Horn U, Ullrich V, Staudinger Hj., 1971, Purification and characterization of L-kynurenine 3-hydroxylase (EC 1.14.1.2.) from rat liver, *Hoppe-Seyler's Z Physiol Chem* 352: 837.

Krajl M., 1965, A rapid microfluorimetric determination of monoamine oxidase, *Biochem Pharmacol* 14: 1684.

Lowry OH, Rosebrough NJ, Farr AL, Randall RJ., 1951, Protein measurement with Folin phenol reagent. *J Biol Chem* 193: 265.

Minamiura N, Yasunobu KT., 1978, Purification and some properties of porcine brain mitochondrial monoamine oxidase B, *Biochem Pharmacol* 27: 2737

Nishimoto Y, Takeuchi F, Shibata Y., 1977, Molecular properties of L-kynurenine 3-hydroxylase from rat liver mitochondria, *J Biochem* 1977; 81: 1413

Nishimoto Y, Takeuchi F, Shibata Y., 1979, Purification of L-kynurenine 3-hydroxylase by affinity chromatography, *J Chromatogr* 169: 357.

Okamoto H, Hayaishi, O., 1969, Solubilization and partial purification of kynurenine hydroxylase from mitochondrial outer membrane and its electron donors. *Arch Biochem Biophys* 131: 603

Parsons DF, Williams GR, Chance B., 1966, Characteristics of isolated and purified preparations of the outer and inner membranes of mitochondria, *Ann N Y Acad Sci* 1966; 137: 643

Pellicciari R, Natalini B, Costantino G, Mahmoud MR, Mattoli L, Sadeghpour BM., 1994, Modulation of the kynurenine pathway in search for new neuroprotective agents. Synthesis and preliminary evaluation of (m-nitrobenzoyl) alanine, a potent inhibitor of kynurenine-3-hydroxylase. *J Med Chem* 37: 647.

Uemura T, Hirai K., 1991, Kynurenine 3-monooxygenase activity of rat brain mitochondria determined by high performance liquid chromatography with electrochemical detection. In: Kynurenine and Serotonin Pathways, Schwarcz R., Young S.N. and Brown R.R., eds., Plenum Press, 531-534

Youdim MBH, Collins GGS., 1971, The dissociation and reassociation of rat liver mitochondrial monoamine oxidase, *Eur J Biochem* 18: 73

# ENANTIOSPECIFIC SYNTHESIS AND *IN VITRO* ACTIVITY OF SELECTIVE INHIBITORS OF RAT BRAIN KYNURENINASE AND KYNURENINE-3-HYDROXYLASE

A. Giordani, L. Corti, M. Cini, R. Bormetti, M. Marconi, O. Veneroni, C. Speciale, and M. Varasi

Pharmacia, R&D-CNS
20014 Nerviano Milano, Italy

## 1. INTRODUCTION

Chirality, a phenomenon of dissimetry, is a fundamental characteristic of biological systems. A large amount of endogenous compounds including drug-interacting macromulecules are chiral, thus giving rise to differences in the intermolecular energy involved in the interaction of the chiral drug and the biological target. As a result of such chiral discrimination, drug enantiomers may differ in their pharmacodynamic and pharmacokinetic effects. Due to such differences, particularly *in vivo* the use of 1:1 mixture of enantiomers (racemates), often fails to display, and occasionally masks the properties of each enantiomer. In searching of selective inhibitors of tryptophane catabolism enzymes, we identified two classes of potent and selective inhibitors represented by 2′-methoxy-benzoylalanine (R,S)-**1** and 3′,4′-dichloro-benzoylalanine (R,S)-**2**, acting respectively as kynureninase and kynurenine 3-hydroxylase selective inhibitors with no activity on the kynurenic acid synthesising enzyme (Kynurenine aminotransferase). Moreover, it is noteworthy that the unsubstituted benzoyl-alanine (R,S)-**3**, even though endowed with lower potency and selectivity, is by itself an inhibitor of kynurenine-3-hydroxylase. The need to analyse the relationship between the activities of the two enantiomers (Eudismic ratio), for each of the above reported compounds, along with the identification of the corresponding absolute configuration, prompted exploration of enantiospecific synthetic routes for these selected molecules:

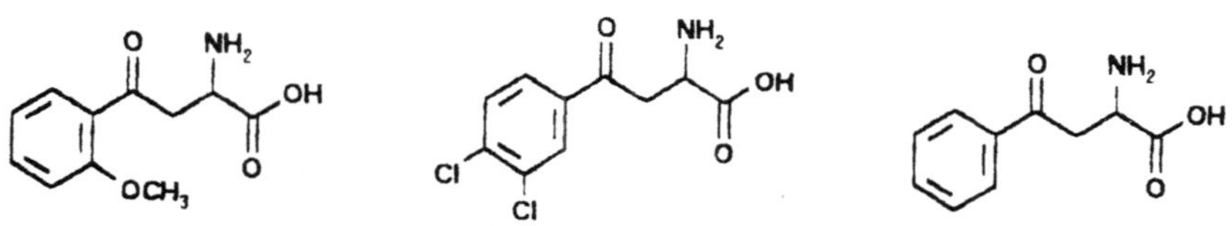


(R,S) - **1** (S) - **1a** (R) - **1b**

(R,S) - **2** (S) - **2a** (R) - **2b**

(R,S) - **3** (S) - **3a** (R) - **3b**

**Scheme 1**

*Recent Advances in Tryptophan Research,* edited by Graziella Allegri Filippini et al.
Plenum Press, New York, 1996

# 2. METHODS

## 2.1 Chemistry

Racemic compounds **1**, **2** and **3** were prepared from the corresponding bromoacetophenones and acetamidomalonate, followed by hydrolysis and decarboxylation. Two different approaches, both using aspartic acid as chirality source, were applied to the enantiospecific synthesis of the enantiomers of 2-methoxy-benzoylalanine (**1**) (Scheme 2a) and 3,4-dichloro-benzoylalanine (**2**) (Scheme 2b).

In a recently reported synthesis of nicotinyl-alanine's enantiomers, Pellicciari and co-workers (1992) showed an elegant application of Stille's ketone synthesis using the oxazolidinone (**5**) derived from aspartic acid as chiral building block. That work prompted us to adapt such a methodology for the synthesis of **1a** and **1b**. Thus metallation of

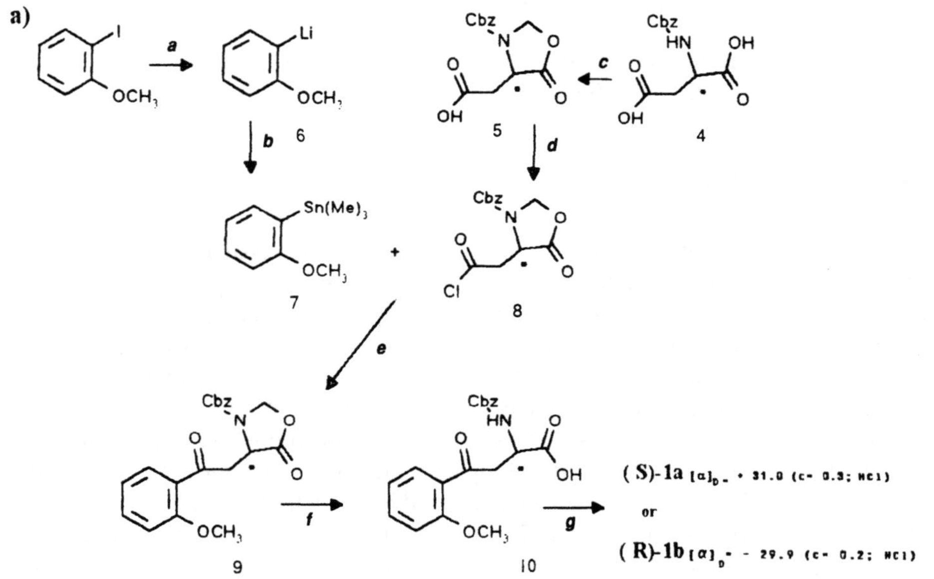


*a) n-BuLi, $Et_2O$, 0°C, 2hrs.*
*b) $Me_3Sn$-Cl, THF, 0°, 2 hrs., 89%.*
*c) Paraformaldehyde, toluene, PTSA, 110°C, 4 hrs, 80%.*
*d) SOCl2, Toluene, r.t., 4 hrs.*
*e) $((Ph)_3P)_2PdCl_2$, Toluene, 80°C, 8 hrs, 40%.*
*f) NaOH 1N, EtOH, 0°C, 30 min., then $H^+$, 70%.*
*g) HCl 6N, 80°C, 36 hrs, 70%.*

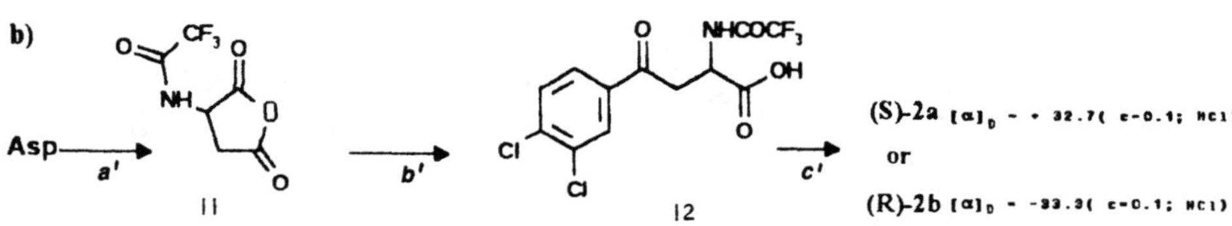


*a') $(CF_3CO)_2O$, AcOEt, 60°C, 4 hrs, 88%.*
*b') 1,2-Dichlorobenzene, $AlCl_3$, 70°C, 8 hrs., 68%.*
*c') HCl 6N, 50°C, 4 hrs, 82%.*

**Scheme 2**

2-iodoanisole afforded the derivative (**6**), that by reaction with the trimethyl stannyl chloride provided the organostannane (**7**). The acylchloride (**8**), was obtained from homochiral N-carbobenzyloxy-aspartic acid (**4**) by cyclization to oxazolidinone (**5**), and underwent palladium catalysed coupling with **7**, affording the protected derivative (**9**). Basic hydrolysis of the oxazolidinone ring, followed by acidic urethane hydrolysis of **10** gave (S)- **1a** or (R)-**1b**.

Although the above reported route could be extended to the synthesis of (S)-**2a** and (R)-**2b** the starting appropriate iodo-dichlorobenzene is not easily available. Recently, Friedel-Crafts acylation of 1,2-dichlorobenzene with succinic anhydride was reported (Quallich et al., 1990) to proceed completely with the desired regiochemistry. Moreover aspartic anhydride was shown to undergo Friedel-Crafts acylation selectively at the β-carbonyl position, independently from the nitrogen protective group (Reifenrath et al., 1976; Nordlander et al., 1985) and without affecting the chiral centre (Nordlander et al., 1985; Melillo et al., 1987). These reports induced us to carry out the Friedel-Crafts acylation of 1,2-dichlorobenzene with N-trifluoroacetyl aspartic anhydride. Thus, reaction of homochiral aspartic acid with trifluoroacetic anhydride gave the protected anhydride **11** (Nordlander et al., 1985). Aluminium trichloride catalysed reaction of **11** with 1,2-dichlorobenzene afforded the protected benzoyl-alanine **12**, which after acidic hydrolysis gave (S)- **2a** or (R)- **2b**.

An analogue approach was followed for the synthesis of **3a** and **3b**, using ethoxycarbonyl (Melillo et al., 1987) instead of trifluoroacetyl as protecting group; no difference in the optical purity of the obtained enantiomers was observed using these protections, being the optical purity of both **2a/2b** and **3a/3c** greater than 98% as confirmed by chiral HPLC analysis.

## 2.2 Assays of Enzyme Activities in Brain Homogenates

Male Wistar rats (Charles River ITALY) weighing 250-300 g were used. Animals were sacrified by decapitation, brains were rapidly excised, frozen on dry ice and stored at -80°C until used. Kynurenine 3-hydroxylase activity was quantified according to Saito et al. (1993), with minor modifications. Kynureninase activity was quantified by the conversion of L-kynurenine to anthranilic acid (Saito et al., 1993). Anthranilic acid was quantified by an HPLC equiped with fluorimetric detection (ex. 313 nm; em. 420nm). The column was a 30 cm C18 reverse phase. The mobile phase consisted of 45 % acetonitrile in 80 mM potassium phosphate buffer at pH 2.5; flow rate was 1 ml/min.

# 3. RESULTS

The inhibitory effect of the different molecules synthetised are reported in Table 1. All the S-isomers showed a greater inhibitory activity than the corresponding R-isomers, while mantaining the same order of magnitude of activity and selectivity of their respective racemates.

# 4. CONCLUSIONS

The enantiomers of benzoylalanine, 2-methoxy-benzoylalanine and 3,4-dichlorobenzoylalanine have been prepared using two different synthetic approaches in good overall yield.

In vivo experiments (cf. Cini et al,. and Speciale et al., this meeting) have confirmed that 3,4-dichloro-benzoylalanine, the most potent inhibitor of kynurenine 3-hydroxylase

**Table 1.** Effect of the selected compounds (racemates and the respective S and R enantiomers) on the brain kynureninase and kynurenine 3-hydroxylase activities

| | | Kynureninase ($IC_{50}$ μM) | Kynurenine 3-OH ase ($IC_{50}$ μM) |
|---|---|---|---|
| 1 | R,S | 4.2 | >100 |
| 1a | S | 4.6 | >100 |
| 1b | R | 31 | >100 |
| 2 | R,S | >100 | 0.1 |
| 2a | S | >100 | 0.3 |
| 2b | R | >100 | 2.1 |
| 3 | R,S | >100 | 42 |
| 3a | S | >100 | 16 |
| 3b | R | >100 | >100 |

Each value was calculated from at least two separate experiments. Six or seven different concentrations were used for the calculation of the $IC_{50}$.

described so far, provides a useful tool to evaluate the neuroprotective effects possibly deriving from the pharmacological modulation of neuroactive kynurenines in the brain. Notably, in recent experiments (cf. Speciale et al., this meeting) we have shown that this compound significantly prevents the nerve cell loss occurring in the model of global ischaemia in the gerbil.

## 5. REFERENCES

Melillo D.G., Larsen R.D., Mathre D.J., Shukis W.F., Wood A.W., and Colleluori J.R., 1987, Pratical enantioselective synthesis of a homotyrosine derivative and (R,R)-4-propyl-9-hydroxynaphtoxazine, a potent dopamine agonist, *J.Org.Chem.* 52 : 5143.

Nordlander D.S., Payne M.J., Njorge F.G., Vishwanath V.M., Han G.R., Laikos G.D., and Balk M.A., 1985, A short enantiospecific synthesis of 2-amino-6,7-dihydroxy-1,2,3,4-tetrahydronaphtalene, *J.Org.Chem.*, 50 : 3619..

Pellicciari R., Gallo-Mezo M.A., and Amer A.M., 1992, Stereospecific synthesis of the enantiomers of nicotinylalanine, a neuroprotecting agent, *Tetrahedron Letters*, 33: 3003.

Quallich G.J, Williams M.T., and Friedman R.C., 1990, Friedel-Crafts synthesis of 4-(3,4-dichlorophenyl)-3,4-dihydro-1(2H)-naphtalenone, a key intermediate in the preparation of the antidepressant Sertaline, *J.Org.Chem.* 55:4971.

Reifenrath W.D., Bertelli D.J., and Fries D.S., 1976, Stereochemistry of Friedel-Crafts addition of phtaloylaspartic anhydride to benzene, *Tetrahedron Letters*, 23: 1959.

Saito K., Nowak T.S., Markey S.P., and Heyes M.P., 1993, Mechanism of delayed increases in kynurenine pathway metabolism in damaged brain regions following transient cerebral ischemia. *J Neurochem.* 60: 180.

# 28833, A SELECTIVE KYNURENINE 3-HYDROXYLASE INHIBITOR

## Effects on the Kynurenine Pathway Metabolites in the Gerbil

M. Cini,[1] R. Bormetti,[1] M. Marconi,[1] A. Molinari,[1] O. Veneroni,[1] M. Varasi,[1] R. Schwarcz,[2] and C. Speciale[1]

[1] Pharmacia, R&D CNS
Nerviano Milano, Italy
[2] Maryland Psychiatric Research Center
University of Maryland School of Medicine
Baltimore Maryland 21228

## 1. INTRODUCTION

An imbalance of the physiological equilibrium between the neuroactive kynurenines kynurenic acid (KYNA) and quinolinic acid (QUIN) in the brain has been tentatively associated with the development of neurodegenerative processes (Schwarcz et al., 1992). Thus, it has been suggested that a decrease in KYNA formation may favor the occurrence of the excitotoxic related events (Schwarcz et al., 1992). On the other hand, an increase in QUIN brain content in the absence of any significant change in KYNA levels occurs in the gerbil model of experimental ischemia, supporting a role for QUIN as an endogenous neurotoxin (Saito et al., 1993). Thus, pharmacological interventions aimed at increasing KYNA formation and/or blocking QUIN synthesis hold promise for the therapy of excitotoxic brain disease.

In the kynurenine pathway, kynurenine 3-hydroxylase (KYN 3-OHase) is the first enzyme involved in the formation of QUIN from kynurenine (KYN). Inhibitors of this enzyme are therefore expected to block KYN metabolism to QUIN, and may at the same time, by causing an accumulation of KYN, increase the formation of KYNA. Indeed, the KYN 3-OHase inhibitor m-nitrobenzoylalanine ($IC_{50}$: 3 μM) was shown to be able to increase the content of KYNA in the rat brain approximately 5-fold (Carpenedo et al., 1994).

In our search for more active inhibitors, we have identified 28833 [(R,S)-2-amino-4-oxo-4-(3',4'-dichlorophenyl) butanoic acid], as a selective KYN 3-OHase inhibitor with an $IC_{50}$ of 0.1 μM in brain homogenate. In the present communication, we describe the effects of this compound in the gerbil on the plasma and brain content of KYN, KYNA, and anthranilic acid (AA).

*Recent Advances in Tryptophan Research,* edited by Graziella Allegri Filippini et al.
Plenum Press, New York, 1996

# 2. METHODS

## 2.1 Animals

Female Mongolian gerbils (Morini, Italy; 45-60 g) were used. For the determination of kynurenine aminotransferase (KAT) activity, male Sprague Dawley rats (Charles River, USA; 200-250 g) were used.

## 2.2 Assay of Enzyme Activities and Metabolites

Brain KYN 3-OHase activity was measured according to Saito et al.(1993) by monitoring the production of 3-hydroxy-kynurenine (3-OH-KYN). The activity of kynureninase (KYNase) was tested in brain homogenates by assessing the conversion of L-KYN to anthranilic acid (AA) (Saito et al., 1993). Brain KAT activity was measured according to Okuno et al. (1991).

KYN was determined by HPLC with UV detection (340 nm). The mobile phase contained 3.6 mM heptanesulphonic acid in 50 mM ammonium acetate buffer:methanol (97:3), pH 6, and was used at a flow-rate of 1 ml/min with a 10 cm $C_{18}$ reversed phase (3 μm) column. 3-OH-KYN was determined by HPLC coupled to a coulometric detector (+0.2 V) (Heyes and Quearry, 1988). AA was determined by HPLC with fluorescence detection (excitation: 313 nm, emission: 420 nm) using a 30 cm $C_{18}$ reversed phase (5 μm) column, with a mobile phase containing 80 mM potassium phosphate buffer:acetonitrile (83:17) at pH 2.5. The flow rate was 1 ml/min. KYNA was determined by HPLC with a zinc acetate post-column derivatization and fluorescence detection (excitation. 254 nm, emission. 404 nm) (Swartz et al., 1990; Heyes and Quearry, 1990).

## 2.2 In Vivo Studies

All experiments were performed after a single oral treatment with 28833 (suspended in methocel). Control animals were treated with the vehicle alone. A time curve was established using the dose of 400 mg/kg. Dose dependency was tested by using 50 to 400 mg/kg. Animals were sacrificed by decapitation and plasma recovered. Brains were rapidly excised, frozen on dry ice and stored at -80°C. On the day of the assay, tissues were thawed and homogenised in 2.5-3 volumes of 0.5 M $HClO_4$ with a Polytron homogenizer (setting 8, twice for 15 sec). The samples were then centrifuged (12,500 x g for 15 min), and the supernatants were subjected to HPLC for metabolite analysis. Plasma samples were mixed with 2 volumes of $HClO_4$ (final concentration: 0.5 M) and then processed like the tissues. KYN 3-OHase activity was measured ex vivo in the brain homogenates (1:10, w/v) of treated (400 mg/kg) and control gerbils.

# 3. RESULTS

In vitro, 28833 had the following $IC_{50}$ values: 0.1μM for KYN 3-OHase, and >200 μM for KYNase and KAT.

A single oral administration of 28833 (400 mg/kg), caused a several-fold, long-lasting increase of KYN and KYNA both in the plasma and in the brain (Table 1). Additionally, a dose-related increase of KYN and KYNA in the plasma and in the brain was observed (Table 1). In both series of experiments, AA was also found to be increased, although more moderately (data not shown).

**Table 1.** Time and dose related effects of 28833 (400 mg/kg, p.o.) on KYN and KYNA content in plasma and brain tissue from Mongolian gerbils

| | Kynurenine | | Kynurenic acid | |
|---|---|---|---|---|
| | Plasma | Brain | Plasma | Brain |
| Time (h) | | | | |
| Control | 907 ± 77 | 307 ± 40 | 31 ± 3 | 4 ± 1 |
| 2 | 24912 ± 2530** | 14771 ± 997** | 16178 ± 2494** | 137 ± 16** |
| 4 | 22424 ± 1704** | 16605 ± 1185** | 20816 ± 3648** | 225 ± 46** |
| 8 | 25148 ± 2764** | 16795 ± 2120** | 24207 ± 6435** | 176 ± 27** |
| 14 | 17966 ± 1783** | 13262 ± 1606** | 18953 ± 2608** | 156 ± 32** |
| Dose (mg/kg) | | | | |
| Control | 1052 ± 82 | 447 ± 30 | 18 ± 2 | 2 ± 1 |
| 50 | 3567 ± 661** | 2051 ± 551** | 241 ± 83** | 4 ± 2 |
| 100 | 7826 ± 1066** | 4548 ± 948** | 1889 ± 622** | 15 ± 4** |
| 200 | 15558 ± 2190** | 13693 ± 2649** | 9207 ± 3791** | 93 ± 23** |
| 400 | 21450 ± 1837** | 21471 ± 2103** | 28580 ± 4699** | 192 ± 42** |

The time related effects were studied by treating the animals orally with 400 mg/kg. For the dose dependency experiments, gerbils were sacrificed 8 hours after the administration of the compound. Values related to plasma and brain tissue are expressed as pmoles/ml and pmoles/g, respectively. Data are the means ± S.E.M. of at least five individual animals per group. **$p < 0.01$ as compared to untreated control gerbils (Dunnett's test).

In order to examine the activity of 28833 in the brain, KYN 3-OHase activity was tested ex vivo in brain homogenates of 28833-treated animals. In animals killed 2 hours after the administration of the compound, KYN 3-OHase activity was inhibited by approximately 70% when the substrate (KYN) concentration used in the assay was 50 μM (Table 2). No such inhibitory effect was observed when 500 μM KYN were used in the assay.

## 4. DISCUSSION

28833, a molecule developed in our laboratories, is the most potent and selective inhibitor of KYN 3-OHase discovered so far. In vivo, a p.o. dose of 400 mg/kg increased the brain tissue content of KYNA by approximately 40-60 and 12-14-fold in the gerbil and in the rat (Speciale et al., this volume), respectively. By measuring KYN 3-OHase activity in gerbils treated orally with the compound, the enzyme was found to be significantly

**Table 2.** Activity of kynurenine 3-hydroxylase in brain homogenates of gerbils treated with 28833 (400 mg/kg, p.o.)

| | Enzyme activity (pmoles/h/mg protein) | |
|---|---|---|
| Time (h) | Kynurenine (50 μM) | Kynurenine (500 μM) |
| 0 | 172 ± 12 | 276 ± 24 |
| 2 | 49 ± 5 | 234 ± 8 |

Brain homogenates were incubated for 30 min at 37°C in the presence of the substrate L-kynurenine at the indicated concentrations. Values are the means ± S.E.M. of at least five animals per group.

inhibited after 2 hours, indicating that the molecule {or its metabolite(s)} can cross the blood-brain barrier and affect the functioning of the pathway in the brain. In addition, the lack of inhibition observed when the substrate concentration used in the enzyme activity assay was raised to 500 μM suggests that 28833 may compete with KYN for the catalytic site of the enzyme.

In conclusion, 28833 is a novel Kyn 3-OHase inhibitor which causes long lasting incresases in brain KYNA content. Additional studies are currently in progress to explore the effect of the compound on the synthesis of brain QUIN. Notably, we have recently shown that 28833 can prevent hippocampal cell loss in the gerbil model of global ischemia (cf. Speciale et al., this volume), thus supporting the hypothesis that the modulation of the kynurenine pathway constitutes a viable strategy for the development of novel neuroprotective molecules. Taken together, these data suggest that 28833 provides a useful tool to probe the role of kynurenines in brain physiology and pathology, and that the compound also appears to have neuroprotective properties which could be exploited for the treatment of excitotoxic brain diseases.

## 5. REFERENCES

Carpenedo R., Chiarugi., A., Russi P., Lombardi G., Carlà V., Pellicciari R., Mattioli L.,and Moroni F., 1994, Inhibitors of kynurenine hydroxylase and kynureninase increase cerebral formation of kynurenate and have sedative and anticonvulsant activities. *Neuroscience*, 61: 237.

Heyes M.P. and Quearry B.J., 1988, Quantification of 3-hydroxykynurenine in brain by high-performance liquid chromatography and electrochemical detection. *J. Chrom.* 428: 340.

Heyes M.P. and Quearry B.J., 1990, Quantification of kynurenic acid in cerebrospinal fluid: effects of systemic and central L-kynurenine administration. *J. Chrom.* 350: 108.

Okuno E., Schmidt W., Parks D.A., Nakamura M., and Schwarcz R., 1991, Measurement of rat brain kynurenine aminotransferase at physiological kynurenine concentrations. *J. Neurochem.* 57: 533.

Saito K., Nowak T.S., Markey S.P., and Heyes M.P., 1993, Mechanism of delayed increases in kynurenine pathway metabolism in damaged brain regions following transient cerebral ischemia. *J Neurochem.* 60: 180.

Schwarcz R., Du F., Schmidt W., Turski W.A., Gramsbergen J.B.P., Okuno E., and Roberts R.C., 1992, Kynurenic acid: a potential pathogen in brain disorders, *in*: "Neurotoxins and Neurodegenerative Disease", J.W. Langston and A. Young eds., Academic Press, New York, p. 140.

Swartz K.J., Matson W.R., MacGarvey U., Ryan E.A., and Beal M.F., 1990, Measurement of kynurenic acid in mammalian brain extracts and cerebrospinal fluid by high-performance liquid chromatography with fluorimetric detection and coulometric electrode array detection. *Anal. Biochem. 185: 363.*

# 85

# ASPECTS OF HEPATIC TRYPTOPHAN DIOXYGENASE ACTIVITY AND LEVELS OF CEREBRAL TRYPTOPHAN METABOLITES IN TWO IMBRED LINES OF MICE

D. Ghosh

Biology Department
Texas Southern University
Houston, Texas

Hepatic tryptophan 2,3-dioxygenase [tryptophan pyrrolase; tryptophan oxygenase (TO)] catalyzes the oxidative breakdown of most (~70%) of dietary L-tryptophan. TO activity is subject to genic (i.e., increased apoenzyme induction through corticosteroid-mediated m-RNA synthesis) and several cytoplasmic regulations. One cytoplasmic regulation seems to be related to catalase activity. In our assays for cytosolic TO and catalase in BALB/cAn (albino) and C57BL/6 (dark coat color) male mice, we found a highly significant inverse relationship as follows:

| Mouse strain | Catalase activity (millimoles $H_2O_2$ decomposed per gram liver per minute) | Tryptophan dioxygenase activity (micromoles kynurenine formed per gram liver per hour) |
|---|---|---|
| I. BALB/cAn Tex. | 14.4 (=100%; *n*=5)↑ | 1.37 (=54%; *n*=5)↓ |
| II. C57BL/6 Tex. | 5.01 (=35%; *n*=5)↓ | 2.53 (=100%: *n*=5)↑ |

HPLC/electrochemical detection revealed that the whole brain concentrations of both 5-hydroxytryptamine (5-HT) and 5-hydroxyindole-3-acetic acid (5-HIAA) to be ~15% lower in C57BL/6 than BALB/cAn mice. Badawy et al. [Biochem.J., 264:597 (1989)] also found that alcohol-preferring C57BL/10 ScSn (with higher serum corticosterone levels) had lower 5-HT by ~24%) and 5-HIAA (by ~10%) than non-alcohol preferring CBA/Ca (agouti) mice, although they did not report any catalase activity - which seems to be another factor in regulating TO activity. Several enzymatic reactions produce $H_2O_2$ which is normally decomposed by fast-acting catalase into $H_2O$ and $O_2$, but in the absence of (or, with low) catalase activity, $H_2O_2$ slowly breaks down into $2H^+ + O_2 + 2e^-$; electrons, so generated, keep the iron in the active ferrous state in the heme coenzyme of TO, thus increasing its activity. Thus, increased hepatic TO activity due to low catalase activity decreases the level of circulating tryptophan, which in turn lowers the concentration of 5-HT in the brain, and

*Recent Advances in Tryptophan Research*, edited by Graziella Allegri Filippini et al.
Plenum Press, New York, 1996

possibly that of melatonin in the pineal gland as well; this could results in eliciting some specific physiological and behavioural effects.

# TRYPTOPHAN-DERIVED COFACTORS FUNCTIONING IN OXIDOREDUCTASES

J. A. Duine

Department of Microbiology and Enzymology
Delft University of Technology
Julianalaan 67, 2628 BC Delft, The Netherlands

During the past few years, two novel types of redox cofactors have emerged, quinones and the free radical form of a specific amino acid residue in the enzyme protein chain. Quinone cofactors (Duine, 1991) (see Figure 1) can be either derived from tyrosine (PQQ, TPQ) or from tryptophan (TTQ). PQQ is non-covalently attached to the enzyme while TPQ and TTQ are protein-chain-integrated. TTQ consists of two tryptophans far apart in the protein chain and attached to each other via a covalent bond between their indole rings. It has sofar been discovered only in bacterial amine dehydrogenases (e.g. methylamine dehydrogenase, EC 1.4.99.3), enzymes involved in the oxidation of primary amines to the corresponding aldehydes and ammonia. Several amino acids have now been found which act as cofactor when oxidized to the free radical form in the protein chain. Most of them are derived from the aromatic amino acids tyr and trp and the cofactor TRP$^{\bullet}$ (a tryptophyl residue in free radical form) has been discovered in several enzymes (Prince and George, 1990). The present status of this cofactor will be discussed.

All available evidence indicates that formation of TTQ and TRP$^{\bullet}$ occurs by selective oxidation of the tryptophan(s) concerned in a precursor protein. The genes involved in TTQ

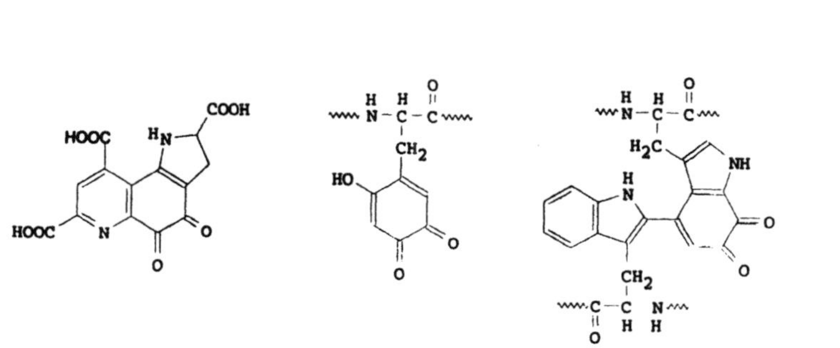


**Figure 1**

*Recent Advances in Tryptophan Research,* edited by Graziella Allegri Filippini et al.
Plenum Press, New York, 1996

biosynthesis by *Paracoccus denitrificans* have been cloned and sequenced (Spanning et al., 1994). A mutant protein has been isolated which lacks activity, does not show absorbance above 300 nm, and gives a negative response in the quinone staining assay. One of the enzymes involved in biosynthesis of TTQ, having a peroxidase-like nature, has been overexpressed. The present status of the biosynthesis of TTQ as well as its physiological role will be highlighted.

## REFERENCES

Duine J.A., 1991, *Eur. J. Biochem.*, 200: 271

Prince R.C. and George G.N., 1990, *Trends Biochem. Sci.*, 15: 170.

van Spanning R.J.M., van der Paalen C.J.N.M., Slotboom D-J., Reijnders W.N.M., Stouthamer A.H., and Duine J.A.,1994, *Eur. J. Biochem.*, 226: 202.

# *TRYPTOPHAN AND EATING BEHAVIOUR AND NUTRITIONAL ASPECTS*

# 87

# TRYPTOPHAN AND SECONDARY ANOREXIA

F. Rossi Fanelli, A. Laviano, I. Preziosa, A. Cascino, M. Muscaritoli, and C. Cangiano

Department of Clinical Medicine
University of Rome "La Sapienza"
Rome, Italy

## 1. INTRODUCTION

Anorexia, defined as the loss of the desire to eat, represents one of the most typical symptoms of a modified central regulation of eating behavior. Anorexia and the consequent reduction of food intake may represent per se a pathological status like in anorexia nervosa (i.e., primary anorexia), or it may be observed in numerous pathological conditions like cancer, chronic liver and renal failure, sepsis, etc. In this case anorexia may be defined as secondary anorexia.

The high prevalence of secondary anorexia in different diseases deserves carefully consideration by physicians. In fact, the reduction of food intake leading to a negative nitrogen balance, and reduction of body weight heavily contributes to the severity of the prognosis. Moreover, the presence of such malnutrition limits the use of therapeutic tools and may even lead to a delay in adequate treatment. Thus, a better understanding of the pathophysiology of secondary anorexia and its possible improvement would definitely enhance the efficacy of specific treatments.

## 2. CENTRAL REGULATION OF EATING BEHAVIOR

Experimental findings suggest that under normal circumstances, food intake is centrally regulated by two hypothalamic centers, a lateral "feeding center" and a ventromedial "satiety center"; the latter inhibits the feeding center after a meal leading to the sensation of satiety (Lytle, 1977). Nevertheless, it is still not clear which specific event provides the initial stimulus that leads to food ingestion, from which site this event originates, and how it is communicated to the brain.

Among the different hypothesized mechanisms proposed to answer these questions, the role of amino acids in the regulation of food intake has been confirmed by experimental data supporting the hypothesis that changes in plasma amino acid levels affect brain availability of neurotransmitter amino acid precursor and may thus control food intake (Li and Anderson, 1983). The two major neurotransmitter systems, i.e., catecholaminergic and

*Recent Advances in Tryptophan Research*, edited by Graziella Allegri Filippini et al.
Plenum Press, New York, 1996

serotoninergic, are both involved in the regulation of eating behavior, but with rather specific and opposite functional roles (Li and Anderson, 1983; Leibowitz, 1990). Particularly, serotonin's action is believed to influence both the energy balance and the circadian patterns of eating by activating satiety neurons localized in the medial hypothalamus (Leibowitz et al., 1989). In this process, serotonin seems to interact antagonistically with norepinephrine and its alpha2-noradrenergic receptors (Leibowitz, 1988).

Under physiological conditions, the rate of serotonin synthesis is primarily dependent on the cerebral concentration of its precursor. Brain trytpophan levels are, in turn, determined by at least three factors: 1) plasma free tryptophan (i.e., the quota not bound to albumin) (Tagliamonte et al., 1973); 2) the molar ratio in plasma between free tryptophan and the other large neutral amino acids competing with tryptophan for brain entry (Perez-Cruet et al., 1974); and 3) the activity of the common transport system across the blood brain barrier (Cangiano et al., 1983). Consequently, any clinical or experimental situation that influences one or more of these factors, by affecting tryptophan availability within the brain, may result in altered eating behavior.

## 3. TRYPTOPHAN AND SECONDARY ANOREXIA IN DIFFERENT DISEASES

A possible association between increased brain tryptophan and serotonin and the presence of anorexia was first reported by Krause et al. (1981). They observed that rats inoculated with an experimental tumor progressively reduce their food intake when compared to controls, and that this change in eating behavior is associated with a significant increase in brain concentrations of tryptophan, serotonin and 5-hydroxy-indoleactic acid, the final metabolite of serotonin. More recently, we were able to confirm Krause's data, using another experimental tumor model (Meguid et al., 1992). In collaboration with the group directed by Dr. Meguid at the State University of New York Health Science Center at Syracuse, we showed that the reduction of food intake in anorectic tumor bearing rats results from the selective reduction of meal size, which is the animal equivalent for early satiety. And more important, this reduction is paralleled by an increase in the plasma ratio between free tryptophan and competing amino acids, and by an increase in brain trytpophan and serotonin content. Thus, experimental evidence consistently indicates that increased brain trytpophan availability and serotonin activity are strongly related to cancer anorexia. But is this the case also in anorectic cancer patients?

To answer this question we designed a series of clinical studies aiming at investigating whether in anorectic cancer patients, reduced food intake is associated with increased plasma and cerebro-spinal fluid concentrations of tryptophan. We studied 45 cancer patients with tumor of different origin (Rossi Fanelli et al., 1986), mainly of the lung and breast. These patients were stratified according to the presence of anorexia. To this end, we developed a specific questionnaire, which assesses the presence of two different groups of symptoms. As mentioned before, anorexia is a complex eating disorder, and different symptoms contribute to its definition. In the first group of symptoms or major symptoms, we investigate the presence of disturbances of eating behavior which are related to a derangement of the central regulatory system. These symptoms are meat aversion, altered smell or taste, nausea and/or vomiting, and early satiety. In the second group of symptoms, or minor symptoms, we investigate the presence of several factors affecting food intake, but not related to the central regulatory system. By definition, patients reporting one or more of the major symptoms are defined as anorectic. Using this questionnaire, we stratified our cancer patients according to the presence of anorexia, and we observed that in anorectic cancer patients, brain tryptophan availability, as expressed by the plasma ratio between free tryptophan and the other competing amino acids, was significantly greater than in non-anorectic cancer patients and controls. By inference, we postulated that in anorectic cancer patients brain tryptophan concentrations were significantly incresaed. This supposition was subsequently confirmed in a study looking at the relationship between tryptophan cerebro-spinal concentrations and cancer anorexia (Cangiano et al., 1990). In this study, we found that anorectic cancer patients have increased cerebro-spinal fluid levels of tryptophan than non-anorectic cancer patients and controls. In addition, we were able to find a significant positive relationship between the plasma ratio free tryptophan over large neutral amino acids and cerebro-spinal fluid tryptophan concentrations. Thus, it seems that in both animals and humans, cancer anorexia might be mediated by increased brain tryptophan availability and serotonin activity.

However this mechanism appears to be effective in at least two other chronic diseases. Both in chronic liver and renal failure, changes in plasma amino acid profile have been reported, basically consisting of an increase of aromatic amino acids, including tryptophan, and a reduction of branched-chain amino acids, which compete with tryptophan for brain entry (Kopple et al., 1978; Rossi Fanelli et al., 1987). It is therefore conceivable that the anorexia often reported by cirrhotic and uremic patients might be mediated by increased brain tryptophan availability and serotonin activity. Indeed, in a recently completed study we showed that during liver cirrhosis, plasma free tryptophan concentrations as well as the ratio between free tryptophan and competing amino acids are significantly increased in patients with anorexia with respect to non-anorectic patients (Laviano et al., 1995 in press). Similar findings were observed in uremic patients (Cangiano et al., 1987), indicating that anorexia and reduced food intake are associated with increased plasma free tryptophan levels and brain tryptophan availability.

There are many other diseases where anorexia and changes of plasma amino acid profile, leading to increased brain tryptophan availability, can be simultaneously present. Among them, sepsis, trauma, chronic inflammatory diseases and chronic hypercapnia, and studies investigating the possible relationship between trytpophan and anorexia are currently on their way.

Considering all these experimental evidence, we hypothesized that a common pathogenic mechanism may be operating in inducing anorexia in different diseases (Rossi Fanelli et al., 1991). It is likely set out by complex metabolic alterations which eventually result in a greater availability of tryptophan to the brain, achieved via either increased plasma tryptophan concentrations or decreased concentrations of competing amino acids. Increased brain trytpophan concentrations result in a marked increase in cerebral serotonin synthesis, leading to anorexia.

## 4. TREATMENT OF SECONDARY ANOREXIA

If this hypothesis holds true, it is conceivable that by reducing the entry of tryptophan into the brain as well as by decreasing the cerebral synthesis of serotonin, an improvement of anorexia may be induced and patients' food intake may be restored to normal, or at least improved.This may be achieved via different strategies:

1. acting directly on the specific mechanisms responsible for the increase in free tryptophan or the reduction of the competing amino acid levels. Unfortunately, this task is difficult to accomplish, since the mechanism responsible for the derangement of plasma amino acid profile are still poorly understood;
2. it is also possible to modify brain serotonin turnover, by blocking serotonin receptors in the brain. We recently demonstrated that this approach is effective in

counteracting anorexia in an experimental tumor model, since anorectic tumor bearing rats injected in the hypothalamus with a serotonin antagonist improves their food intake, while the intrahypothalamic injection of the vehicle does not affect the progressive reduction of caloric intake (Laviano et al., 1995 submitted).However, when we move to the clinical setting we must consider that most of the patients suffering form anorexia, has a severe impairment of brain activity, as is the case of hepatic and renal encephalopathy, and the possible sedative effects of drugs interfering with serotonin metabolism should be carefully considered;

3. finally, one may reduce brain trytpophan availability. This could be achieved by oral or parenteral administration of the large neutral amino acids competing with tryptophan for transport across the blood-brain barrier.

At present, the administration of the competing amino acids seems to be the safest approach in the treatment of anorexia in many pathological conditions. And recent data show that it is also highly effective. In a pilot study, we tested the effects on anorexia and caloric intake of the oral administration of branched-chain amino acids, which compete with tryptophan for brain entry, in anorectic cancer patients (Cangiano et al., 1995). Twenty-five well-nourished anorectic cancer patients were randomly assigned to receive an oral supplementation with branched-chain amino acids or with an isonitrogenous, isocaloric amount of glycine, for seven consecutive days. Caloric intake was daily monitored, while the presence of anorexia was assessed at the outset and at the end of the study period. The results obtained show that branched-chain amino acid supplementation significantly reduced the incidence of anorexia, but more important significantly improved caloric intake when compared to controls. These data further support the role of increased brain tryptophan availability in the pathogenesis of secondary anorexia. They also suggest that plasma amino acid manipulation may represent a new therapeutic approach to anorexia associated with different diseases, probably leading to an improvement in the quality of life, and a reduction of morbidity and mortality associated with a number of pathological conditions.

## REFERENCES

Cangiano C, Cardelli-Cangiano P, Cascino A, et al., 1983, On the stimulation by insulin of tryptophan transport across the blood-brain barrier. *Biochem. Intern.* 7:617.

Cangiano C, Alegiani F, Bartoli R, et al., 1987, Effetti clinici e nutrizionali della somministrazione di una nuova miscela di amino acidi essenziali in pazienti in trattamento dialitico continuato. *Riv. It. Nutr. Clin.* 5:27.

Cangiano C, Cascino A, Ceci F, et al., 1990, Plasma and CSF tryptophan in cancer anorexia. *J. Neural Transm.* 81:225.

Cangiano C, Rossi Fanelli F, Laviano A, et al., 1995, Effects of the oral administration of branched-chain amino acids on anorexia and food intake in cancer patients. *J. Natl. Cancer Inst.*, (submitted)

Kopple JD, Jones M, Fukuda S, et al., 1978, Amino acid and protein metabolism in renal failure. *Am. J. Clin. Nutr.* 31:1532.

Krause R, Humphrey C, von Meyenfeldt M, et al., 1981, A central mechanism for anorexia in cancer: a hypothesis. *Cancer Treat. Rep.* 65:15.

Laviano A, Cangiano C, Preziosa I, et al., 1995, Plasma tryptophan and anorexia in liver cirrhosis. *Int. J. Eating Disord.*, (in press)

Laviano A, Gleason JR, Meguid MM, et al., 1995, Effects of intra-VMH mianserin, dopamine and IL-1ra on cancer anorexia in the tumor bearing rat. *J. Clin. Invest.* (submitted)

Leibowitz SF., 1988, Hypothalamic paraventricular nucleus: interaction between alpha2-noradrenergic system and circulating hormones and nutrients in relation to energy balance. *Neurosci. Biobehav. Rev.* 12:101.

Leibowitz SF, Weiss GF, Walsh UA, et al., 1989, Medial hypothalamic serotonin: role of circadian patterns of feeding and macronutrient selection. *Brain Res.* 503:132.

Leibowitz SF., 1990, The role of serotonin in eating disorders. *Drugs* 39 (suppl. 3):33.

Li ETS, Anderson GH., 1983, Amino acids in the regulation of food intake. *Nutr. Abs. Rev.* 53:169

Lytle LD, 1977, Control of eating behavior. In: Wurtman RJ, Wurtman JJ, eds. "Nutrition and the Brain". Raven Press, New York.

Meguid MM, Muscaritoli M, Beverly JL, et al., 1992, The early cancer anorexia paradigm: changes in plasma free tryptophan and feeding indexes. JPEN 16 (suppl. 6): 56S.

Perez-Cruet J, Chase TN, Murphy DL., 1974, Dietary regulation of brain tryptophan metabolism by plasma ratio of free trytptophan and neutral amino acids in humans. *Nature* (Lond.) 248:693.

Rossi Fanelli F, Cangiano C, Ceci F, et al., 1986, Plasma tryptophan and anorexia in human cancer. *Eur. J. Cancer Clin. Oncol.* 22:89.

Rossi Fanelli F, Cascino A, Strom R, et al., 1987, Amino acids and hepatic encephalopathy. *Prog. Neurobiol.* 28:277.

Rossi Fanelli F, Cangiano C., 1991, Increased availability of trytpophan in brain as common pathogenic mechanism for anorexia associated with different diseases. *Nutrition* 7:364.

Tagliamonte A, Biggio G, Vargiu L, et al., 1973, Free tryptophan in serum controls brain tryptophan levels and serotonin synthesis. *Life Sci.* 12:277.

# SEROTONINERGIC BLOCK IN THE VENTROMEDIAL NUCLEUS OF HYPOTHALAMUS IMPROVES FOOD INTAKE IN ANORECTIC TUMOR BEARING RATS

A. Laviano,[1] C. Cangiano,[1] I. Preziosa,[1] M. M. Meguid,[2] M. Muscaritoli,[1] L. Conversano,[1] A. Cascino,[1] G. F. Torelli,[1] S. Cherubini,[1] and F. Rossi Fanelli[1]

[1] Department of Clinical Medicine
University of Rome "La Sapienza", Rome, Italy
[2] Surgical Metabolism and Nutrition Laboratory
SUNY Health Science Center at Syracuse, New York

## 1. INTRODUCTION

In animal models, cancer associated anorexia has been extensively investigated. We recently showed that cancer anorexia is primarily a centrally mediated phenomenon, in which the ventromedial nucleus of hypothalamus (VMH) plays a significant role (Laviano et al., 1995). Under normal conditions, VMH is known to significantly participates in mediating satiety, primarily via the neurotransmitter serotonin (Leibowitz, 1990). Serotonin has also been postulated to mediate cancer anorexia (Rossi Fanelli and Cangiano, 1991), and the previously reported involvement of VMH, a predominantly serotoninergic nucleus, in the development of reduced food intake during tumor growth is in agreement with this supposition (Laviano et al., 1995).

To further support the role of serotonin in inducing cancer anorexia, we designed the present study, which aims at characterizing the effects on food intake of the intra-VMH microinjection of the serotonin antagonist mianserin (MIA) were investigated using the Fischer rat-MCA sarcoma model (Landel et al., 1987). Mianserin is a serotonin antagonist acting on the $5\text{-HT}_{1C/2}$ receptors (Glennon, 1990), which inhibits serotonin-mediated reduction of food intake (Grignaschi and Samanin, 1992). Food intake was recorded daily via an Automated Computerized Rat Eater Meter (ACREM), developed in our laboratory (Meguid et al., 1990).

## 2. MATERIALS AND METHODS

Acclimated male Fischer-344 rats, with a purchase weight of 280-300g, were randomly assigned to receive either a subcutaneous injection of 1ml of $10^6$ trypan blue vital

*Recent Advances in Tryptophan Research,* edited by Graziella Allegri Filippini et al.
Plenum Press, New York, 1996

MCA sarcoma cells (TB groups) into the right flank or an equal amount of saline (NTB groups). Then rats were placed in individual cages equipped with the ACREM, previously described in detail (Meguid et al., 1990). Rat chow and tap water were available ad libitum. Tumor bearing rats were defined as anorectic when their daily food intake per 100g body weight was at least 1g lower than the mean daily intake of NTB rats for three consecutive days (Chance et al., 1991).

On day 0, when TB rats were defined as being anorectic, TB rats and their NTB controls were randomly assigned to the study or control groups. Briefly, under ether anesthesia the rat's head was mounted in a Kopf stereotaxic apparatus; through a burr hole in the skull, 200mMol of MIA (TB-MIA, n=6; NTB-MIA, n=7) dissolved in 200nl of saline were bilaterally injected into the VMH. Control rats in groups TB-NS (n=6) and NTB-NS (n=6) were injected with an isovolumic and isotonic amount of saline. Then rats were placed into their individual cages and food intake measured via the ACREM for the next 7 days. At the end of the study, rats were terminally anesthetized, brains removed and fixed in formaldheyde for histologic verification of correct injection site.

Data were analyzed using ANOVA, Student's t-test and Welch T-test. Data are expressed as mean±standard error (SE).

## 3. RESULTS AND DISCUSSION

Brain histology confirmed correct VMH injection sites in all rats studied. In NTB-NS and NTB-MIA rats, food intake did not change significantly throughout the study period, averaging approximately at 14.2 g/day. In TB rats, food intake progressively declined, and on the day before the VMH injection it was significantly lower than in NTB-NS and NTB-MIA rats (TB-NS: 9.0±1.5 g/day; TB-MIA: 10.6±0.7 g/day). In TB-NS group, the intra-VMH injection of normal saline did not reverse the decline of food intake. In contrast, food intake in TB-MIA rats significantly improved: on day 0, it raised to 12.1+1.6 g/day ($p<0.01$ vs TB-NS) and on day 1 it was 8.6+1.0 g/day ($p<0.05$ vs TB-NS). On day 2, food intake in TB-MIA was not any more significantly different from that TB-NS, despite being almost the double.

This study shows that in normal rats, the intra-VMH block of serotoninergic activity does not affect food intake, while in anorectic tumor bearing rats, this improves food intake. Although the role of the central serotoninergic system in the regulation of spontaneous food intake and in the pathogenesis of reduced food intake remains to be more clearly understood, our study suggests that VMH serotonin is not critical in the control of food intake under normal conditions. However, data obtained in the present investigation are in agreement with previously reported evidence from our laboratory (Rossi Fanelli et al., 1986; Cangiano et al., 1990; Meguid et al., 1992) and indicate that VMH serotonin critically mediates the onset of anorexia in tumor bearing rats.

## REFERENCES


Cangiano C, Cascino A, Ceci F, et al., 1990, Plasma and CSF tryptophan in cancer anorexia. *J. Neural Transm.-Gen. Sect.* 81:225.

Chance WT, Zhang F-S, Foley-Nelson T, et al., 1991, Hyperammoniemia and anorexia in Morris hepatoma-bearing rats. *Physiol. Behav.* 50:397.

Glennon RA., 1990, Serotonin receptors: clinical implications. *Neurosci. Biobehav. Rev.* 14:35.

Grignaschi G, Samanin R., 1992, Role of serotonin and catecholamines in brain in the feeding suppressant effect of fluoxetine. *Neuropharmacol.* 31:445.

Landel AM, Lo CC, Meguid MM, et al., 1987, Effect of methylcholanthrene-induced sarcoma and its removal on rat plasma and intracellular free amino acid content. *Surg. Res. Commun.* 1:273.

Laviano A, Meguid MM, Gleason JR, et al., 1995, Hypothalamic sites of cancer anorexia in the tumor bearing rat. *Cancer Res.*, (submitted)

Leibowitz SF., 1990, The role of serotonin in eating disorders. *Drugs* 39 (suppl. 3):33.

Meguid MM, Kawashima Y, Campos ACL, et al., 1990, Automated Computerized Rat Eater Meter: description and application. *Physiol. Behav.* 48:759.

Meguid MM, Muscaritoli M, Beverly JL, et al., 1992, The early cancer anorexia paradigm: changes in plasma free tryptophan and feeding indexes. *JPEN* 16 (suppl. 6):56S.

Rossi Fanelli F, Cangiano C, Ceci F, et al., 1986, Plasma tryptophan and anorexia in human cancer. *Eur. J. Cancer Clin. Oncol.* 22:89.

Rossi Fanelli F, Cangiano C., 1991, Increased availability of tryptophan in brain as a common pathogenic mechanism for anorexia associated with different diseases. *Nutrition* 7:364.

# THE EFFECTS OF PERIPHERAL SEROTONIN$_2$ RECEPTOR AGONIST ON FOOD INTAKE OF RATS

J. Yamada, Y. Sugimoto, T. Yoshikawa, T. Noma, and K. Horisaka

Department of Pharmacology
Kobe Pharmaceutical University
Motoyamakita–machi, Higashinada–ku, Kobe 658 Japan

## 1. INTRODUCTION

It has been reported that the serotonin$_2$ (5–HT$_2$) receptor is involved in several physiological effects of 5–HT such as platelet aggregation and hypertension (Fozard, 1989). It has been suggested that 5–HT participates in the regulation of food intake and that activation of the serotonergic system decreases food intake in rats (Blundell, 1984; Curzon, 1990). It has been reported that the central 5–HT$_{2A/2C}$ receptor agonist DOI or 5–HT$_{2C}$ receptor agonist mCPP induces a decrease in food intake in rats and that these receptors may be mediated by the central effects of 5–HT on food intake (Aulakh et al., 1992; Schechter and Simansky, 1988; Kennet and Curzon, 1991). However, the involvement of the peripheral 5–HT$_2$ receptor in food intake is not well established and the effects of the peripheral 5–HT$_2$ receptor agonist is not yet clear. Therefore, in this study, we investigated the effects of the peripheral 5–HT$_2$ receptor agonist α–methyl–5–HT on the food intake of rats.

2–Deoxy–D–glucose (2–DG) is a glucose analog, which prevents utilization of intracellular glucose, and is known to elicit eating in rats and 2–DG–induced feeding is an animal model of hyperphagia (Smith and Epstein, 1969). To clarify the involvement of the peripheral 5–HT$_2$ receptor in food intake, we investigated the effects of the peripheral 5–HT$_2$ receptor agonist α–methyl–5–HT on 2–DG–induced hyperphagia in rats.

## 2. MATERIALS AND METHODS

Male Sprague–Dawley rats ( 200–240 g, SLC Japan) were housed in individual stainless wire cages. They were maintained under a controlled 12 hr/ 12 hr light dark cycle (light from 7:00 a.m. to 7:00 p.m.), with room temperature at 24 ±1°C and humidity at 55 ± 5 % for at least 7 days prior to the experiment. Rats had free access to food and water. All drugs were administered i.p. Preweighed food was placed in the cage and the remaining amount was weighed 1 hr after the injection of α–methyl–5–HT.

*Recent Advances in Tryptophan Research*, edited by Graziella Allegri Filippini et al.
Plenum Press, New York, 1996

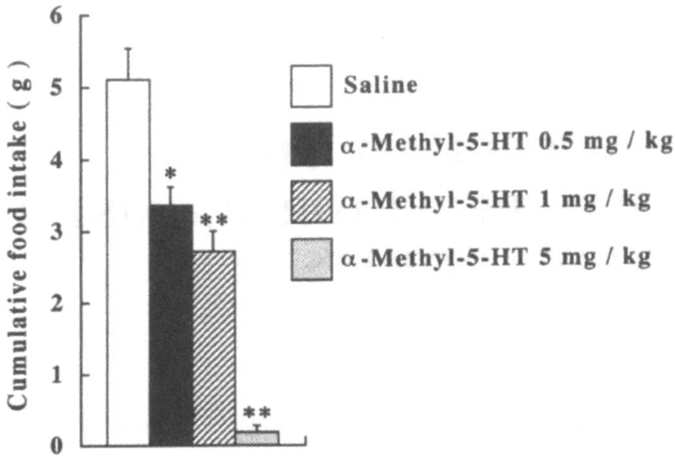


**Figure 1.** Effects of α–methyl–5–HT on food intake of rats. Results are shown as mean ±S.E. (N = 7–9). Rats were starved for 20 hr. α–Methyl–5–HT was given i.p. * $P < 0.05$, ** $P < 0.01$.

Statistical significance was evaluated by Student's t–test between two groups and by Dunnett's test among three or more groups.

## 3. RESULTS AND DISCUSSION

The effects of a peripheral 5–$HT_2$ receptor agonist α–methyl–5–HT are shown in Fig. 1. Rats were deprived of food, but not water for 20 hr before the experiments. α–Methyl–5–HT induced significant dose–dependent anorexia in starved rats (Fig. 1). α–Methyl–5–HT at a dose of 1 mg/kg induced a 46.8 % decrease in food intake for 1 hr. α–Methyl–5–HT–induced anorexia was attenuated by pretreatment with the 5–$HT_{2A}$ antagonist ketanserin. Ketanserin alone did not affect the food intake of rats. These results suggest that α–methyl–5–HT induced decrease in food intake of rats are mediated by the peripheral 5–$HT_{2A}$ receptor.

To clarify the involvement of the peripheral 5–$HT_2$ receptor in food intake, we further examined the effects of α–methyl–5–HT on 2–DG–induced hyperphagia. 2–DG i.p. at a dose of 750 mg/kg induced an apparent hyperphagia in free–feeding rats. The administration of α-methyl–5–HT significantly inhibited 2–DG–induced hyperphagia (Fig. 2). The inhibitory effects of α–methyl–5–HT on 2–DG–induced hyperphagia were antagonized by the 5–$HT_{2A}$ receptor antagonist ketanserin.

It was reported that peripherally administered 5–HT induces anorexia in rats and these effects were antagonized by the peripheral 5–$HT_2$ receptor antagonist xylamidine (Edwards and Stevens, 1989). Our results demonstrated that the peripheral 5–$HT_2$ receptor agonist α-methyl–5–HT induces hypophagia by activation of the peripheral 5–$HT_{2A}$ receptor. These findings indicate that peripheral 5–HT may be involved in the regulation of food intake and that the peripheral 5–$HT_{2A}$ receptor is related to the effects of 5–HT.

It is well recognized that the 5–HT–releasing drug fenfluramine induces anorexia. It has been proposed that some peripheral mechanisms are involved in the action of fenfluramine (Baker et al., 1988; Rowland and Carlton, 1982). Therefore, the peripheral 5–$HT_{2A}$ receptor may be associated with the effects of fenfluramine.

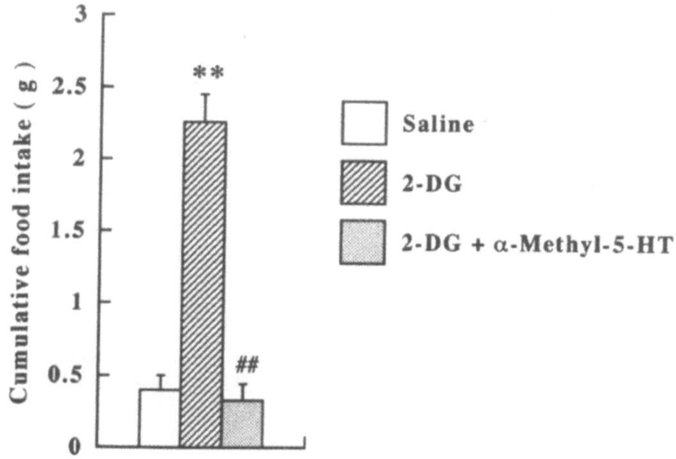


**Figure 2.** Effects of α–methyl–5–HT on 2–DG–induced hyperphagia in rats. Results are shown as mean ±S.E. 2–DG at 750 mg/kg was given i.p. (N =6–7 ). ** $P < 0.01$ vs. saline-treated group. ## $P < 0.01$ vs. 2–DG–treated group.

## 4. REFERENCES


Aulakh, C.S., Hill, J.L., Yoney, H.T., and Murphy, D.L., 1992, Evidence for involvement of 5–HT1C and 5–HT2 receptors in the food intake suppressant effects of 1–(2, 5–dimethoxy 4–iodophenyl)–2–aminopropane (DOI ), *Psychopharmacology* 109: 444.

Baker, B.C., Duggan, J.P., and Booth, D.A., 1988, Effects of DL–fenfluramine and Üxylamidine on gastric emptying of maintenance diet in freely feeding rats, *Eur. J. Pharmacol.* 150: 137.

Blundell, J.E., 1984, Serotonin and appetite, *Neuropharmacology* 23: 1537.

Curzon, G., 1990, Serotonin and appetite, Ann. N.Y. Acad. Sci. 600: 521. Fozard, J.R., 1989, " The Peripheral Actions of 5–Hydroxytryptamine," Oxford University Press, Oxford.

Kennet, G.A., and Curzon, G., 1991, Potencies of antagonists indicate that 5–HT1 receptors mediate 1–3(chlorophenyl)piperazine–induced hypophagia, *Br. J. Pharmacol.* 103:8.

Rowlland, N., and Carlton, J., 1986, Inhibition of gastric emptying by peripheral and central fenfluramine in rats: Correlation with anorexia, *Life Sci.* 34: 2495.

Schechter,L.E., and Simansky, K.J., 1988, 1–(2,5–Dimethoxy–4–iodophenyl)–2-aminopropane (DOI) exerts an anorexic action that is blocked by 5–HT2 antagonists in rats, *Psychopharmacology* 94: 342.

Smith, G.P., and Epstein, A.N., 1969, Increased feeding response to decreased glucose Üutilization in the rat and monkey, *Am. J. Physiol.* 217: 1083.

90

# PREDICTING VALUE OF TRYPTOPHAN/LARGE NEUTRAL AMINO ACIDS RATIO ON SELECTIVE SEROTONIN REUPTAKE INHIBITORS

V. Lucini, A. Lucca, M. Catalano, and E. Smeraldi

IRCCS H. San Raffaele
Dipartimento di Scienze Neuropsichiche
Università degli Studi di Milano
Via Prinetti 29, Milano, Italia

The molar ratio of total plasma Tryptophan (TRP) to the sum of other Large Neutral Amino Acids (LNAAs), thought to reflect brain serotonin (5-HT) formation, was estimated in 45 patients with major depression before and after 1 week of treatment with different serotonin reuptake inhibition, fluvoxamine (29), fluoxetine (10) and citalopram (7). We found significant difference among good, middle and poor responders for the percentual variation of TRP/LNAAs ratio after 1 week of therapy.

*Recent Advances in Tryptophan Research,* edited by Graziella Allegri Filippini et al.
Plenum Press, New York, 1996 

# THE INVOLVEMENT OF SEROTONIN IN THE CATECHOLAMINE RELEASE FROM THE ADRENAL MEDULLA

Y. Sugimoto, J. Yamada, T. Yoshikawa, F. Nishikawa, T. Noma, and K. Horisaka

Department of Pharmacology
Kobe Pharmaceutical University
Motoyamakita-Machi, Higashinada–Ku, Kobe 658, Japan

## 1. INTRODUCTION

Previous studies have suggested that central serotonin (5–HT) receptors are involved in adrenaline release from the adrenal gland. The central 5–HT receptor agonists including the 5-$HT_{1A}$ agonist 8–OH–DPAT, the 5–$HT_2$ receptor agonist DOI and the 5–HT2C receptor agonist mCPP can elevate plasma adrenaline levels in rats (Bagdy et al., 1989; Chaouloff et al., 1990). These effects are thought to be mediated by facilitation of centrally located receptors. We previously found that peripherally administered 5–HT induces hyperglycemia in rats and that these hyperglycemic effects were abolished by adrenodemedullation(Yamada et al., 1995). It suggests that 5–HT–induced hyperglycemia is related to adrenaline release from the adrenal gland and the peripheral 5–HT receptor may also be connected to adrenaline release. However, the peripheral effects of 5–HT on catecholamine release are not yet clear. In this study, we investigated the effects of peripherally administered 5–HT on the plasma adrenaline levels in rats. Furthermore, we examined the effects of 5–HT on catecholamine release in cultured bovine adrenal chromaffin cells.

## 2. MATERIALS AND METHODS

Male Sprague–Dawley rats (180 –230 g) were obtained from SLC Japan Inc. They were housed under a controlled 12 hr/ 12 hr light dark cycle ( light from 7:00 a.m. to 7:00 p.m.), with room temperature at 24 ±1 °C and humidity at 55 ±5 %. Rats had free access to food and water. 5–HT was administered i.p. Ketanserin was injected i.p. 30 min before the injection of 5–HT. Plasma adrenaline levels were determined by HPLC with electrochemical detection following adsorption on aluminum oxide.

Chromaffin cells were prepared from bovine adrenal medulla by collagenase digestion and purified on Percoll gradients (Kilpatrick et al., 1980). Chromaffin cells were cultured in Dulbecoo's modified Eagle's medium supplemented with 10 % fetal bovine serum and maintained at 37 in 5% $CO_2$/ 95% air for 4–6 days prior to experiments.

Cells washed with Krebs–Ringer bicarbonate buffer (pH 7.4) and incubated with drug containing buffer at 37° C for 10 min. The amount of catecholamines released in medium and that left in cells were determined by fluorometry. Catecholamine secretion was expressed as the percentage of total cellular catecholamine contents.

Determination of $[Ca^{2+}]$ i was performed by loading fura–2/AM at 37° C for 30 min. Changes in fluorescence ratio were calibrated to changes in $[Ca^{2+}]$i using the previously described method (Grynkieewicz et al., 1985).

Statistical significance was evaluated by Student's t–test between two groups and by Dunnett's test among three or more groups.

## 3. RESULTS AND DISCUSSION

5–HT induced an increase in plasma adrenaline levels in rats after injection of 5–HT (Fig. 1).Since 5–HT can not cross the blood brain barrier (Oldendorf, 1971), our results suggest that peripheral 5–HT is also involved in adrenaline release. We found that 5–HT–induced hyperglycemic effects were inhibited by both the 5–$HT_{2A}$ receptor antagonist ketanserin and adrenodemedullation (Yamada et al., 1995). Therefore, we studied the effects of ketanserin on 5–HT–induced adrenaline release. 5–HT–induced elevation of adrenaline levels was antagonized by ketanserin. These results suggest that the increase in adrenaline levels induced by 5–HT is mediated by peripheral 5–$HT_{2A}$ receptors and this receptor is related to adrenaline release from the adrenal medulla.

We further studied the effects of 5–HT on catecholamine release in cultured bovine adrenal chromaffin cells to elucidate the effects of 5–HT on catecholamine release from the adrenal medulla. In cultured bovine chromaffin cells, 5–HT did not change basal catecholamine release. It is well known that catecholamine release is regulated by the cholinergic nicotinic receptor. Thus, we examined the effects of 5–HT on nicotinic receptor–stimulated catecholamine release. As shown in Fig. 2, 5–HT inhibited nicotine–induced catecholamine release. 5–HT also suppressed the increase in intracellular $Ca^{2+}$ elicited by nicotine. It indicates that in cultured bovine chromaffin cells, 5–HT inhibits nicotine receptor–mediated catecholamine release by blocking $Ca^{2+}$ influx into the cells.

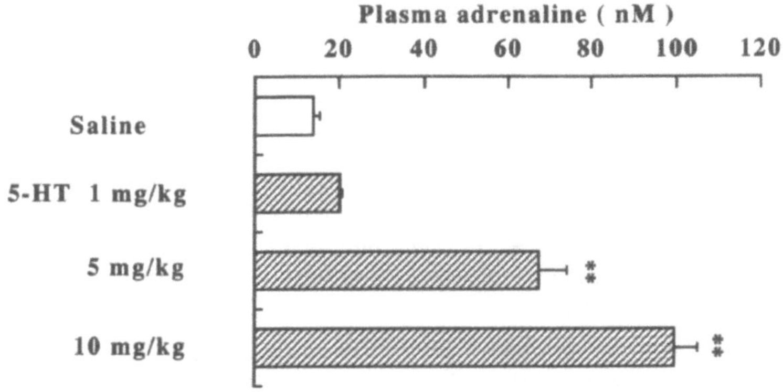


**Figure 1.** Effects of 5–HT on the plasma adrenaline levels in rats. Results are shown as mean ±S.E. (N = 6). Plasma adrenaline levels were determined 15 min after the injection of 5–HT.** $P < 0.01$.

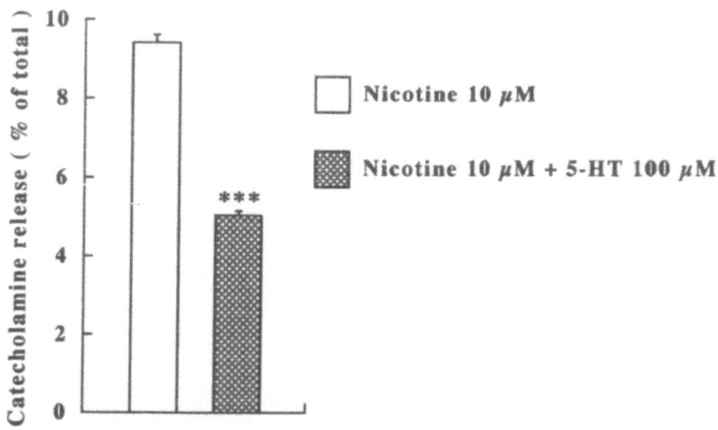


**Figure 2.** Effects of 5–HT on catecholamine release induced by nicotine from cultured bovine adrenal chromaffin cells. Results are shown as mean ±S.E. ( N = 5 – 7 ). Chromaffin cells were incubated at 37° C for 10 min. Spontaneously released CA was subtracted in each group.** $P < 0.01$.

In summary, our results indicate that in rats, peripherally administered 5–HT increases adrenaline release mediated by the peripheral 5–$HT_{2A}$ receptors. It suggests that the peripheral 5–$HT_{2A}$ receptor is involved in the control of adrenaline release from the adrenal medulla. However, in cultured bovine adrenal chromaffin cells, 5–HT did not increase catecholamine release. 5–HT inhibited nicotinic receptor mediated catecholamine release by decreasing influx of $Ca^{2+}$ into the cells. Therefore, there seems to be the difference between the effects of 5–HT on catecholamine release in vivo and in vitro. The reason remains unclear but species difference may have affected the release.

## 4. REFERENCES


Bagdy, G., Calogero, A.E., Murphy, D.L., and Szemeredi, K., 1989, Serotonin agonists Îcause parallel activation of the sympathoadrenomedullary system and the hypothalamo pituitary–adrenocortical axis in conscious rats, *Endocrinology* 125: 2664.

Chaouloff, F., Baudrie, V., and Laude, D., 1990, Evidence that 5–$HT_{1A}$ receptors are involved in the adrenaline–releasing effects of 8–OH–DPAT in the conscious rats, *Naunyn-Schmiedeberg's Arch. Pharmacol* 341: 381.

Grynkiewicz, G., Poenie, M., and Tsien, R.Y., 1985, A new generation of $Ca^{2+}$ indicators with greatly improved fluorescence properties, *J. Biol. Chem.*260 : 3440.

Kilpatrick, D.L., Ledbetter, F.H., Carson, K.A., Kirshner, A.G., Slepetis, R., and Kirshner, N., 1980, Stability of bovine adrenal medulla cells in culture, *J. Neurochem.* 35:679

Yamada, J., Sugimoto, Y., Yoshikawa, T., Kimura, I., and Horisaka, K., 1995, The involvement of the peripheral 5–$HT_{2A}$ receptor in peripherally administered serotonin–induced hyperglycemia in rats, *Life Sci.* in press.

# TRYPTOPHAN AND NIACIN NUTRITION—IS THERE A PROBLEM?

D. A. Bender

Department of Biochemistry and Molecular Biology
University College London
Gower Street, London WC1E 6BT, United Kingdom

There is general agreement amongst national and international authorities that an appropriate intake of niacin is 6.6 mg niacin equivalents (ne) /1000 kcal energy expenditure, where total dietary niacin is the sum of preformed nicotinic acid and nicotinamide plus 1/60 of the tryptophan intake (Bender & Bender, 1986). This concordance is unsurprising, since all expert committees have depended largely on the depletion/repletion study of Horwitt *et al* (1956). It is noteworthy that the original authors stated that there was no scientific basis for relating niacin requirements to energy expenditure; nevertheless, we continue to set reference intakes for niacin, and to assess diets, on the basis of energy expenditure.

A weighed dietary intake study of adults in Britain (Gregory *et al*, 1990) showed that the mean intake was 39.9mg ne /day, equivalent to 16mg ne /1000kcal, with a 95% range of 21.6 - 62.2mg /day. There would thus seem to be no cause for concern over niacin intakes in developed countries, and indeed pellagra due to dietary deficiency of tryptophan and niacin is now extremely rare, except for outbreaks associated with acute famine as a result of war. If the biologically unavailable niacin of cereals (27% of average British intake) is excluded, then 40% of the total comes from preformed niacin, and 60% from tryptophan.

A number of factors that affect the oxidative metabolism of tryptophan or increase the turnover of NAD(P) may impair niacin nutritional status. Such factors include inadequate intakes of vitamins $B_2$, $B_6$ and C, iron and zinc, as well as diabetes mellitus, endogenous or exogenous oestrogens, oxidative stress, alcohol abuse and lead exposure.

## TRYPTOPHAN-NIACIN EQUIVALENCE

Although Horwitt *et al* (1956) concluded that 60mg dietary tryptophan was equivalent to 1mg preformed niacin, and this is the value that is used to calculate niacin intakes, it was a deliberate over-estimate of the mean equivalence, so as to cover individual variation and provide a margin of safety.

The relationship between tryptophan intake, tryptophan metabolism and NAD(P) synthesis is not a simple one. The branch point in the tryptophan oxidative pathway which determines the fate of aminocarboxymuconic semialdehyde is controlled by an enzyme-cat-

*Recent Advances in Tryptophan Research*, edited by Graziella Allegri Filippini et al.
Plenum Press, New York, 1996 

alysed reaction (picolinate carboxylase), leading to complete oxidation via acetyl CoA, in competition with non-enzymic cyclisation to yield quinolinic acid as a precursor for NAD(P) synthesis. Ikeda *et al* (1965) showed that the ability to synthesise NAD(P) from tryptophan in various species was inversely related to the activity of picolinic carboxylase, and Salih *et al* (1985) have reported a child with a hereditary pellagra-like condition and a complete inability to synthesise NAD(P) from tryptophan, apparently due to elevated picolinic carboxylase activity.

Patterson *et al* (1980) demonstrated a 3-fold greater increase in urinary excretion of $N^{1}$-methylnicotinamide (NMN) /100mg tryptophan intake, and a 2.5-fold greater increase in methyl pyridone carboxamide (MPCX), when tryptophan intake was increased from 548 - 845 mg /day than when it was increased from 245 to 548 mg /d.

Mehler *et al* (1958) first demonstrated the inverse relationship between NMN excretion and liver picolinic carboxylase activity in studies with rats rendered diabetic by alloxan. Salter and Pogson (1985) demonstrated that while there was increased entry of tryptophan into the oxidative pathway in hepatocytes prepared from streptozotocin-diabetic rats, this was countered by a proportionally greater increase in flux through acetyl CoA - diversion of aminocarboxymuconic semialdehyde away from NAD(P) synthesis. It is thus likely that diabetes may impair endogenous synthesis of NAD(P) from tryptophan, and so compromise niacin nutritional status.

## OXIDATIVE STRESS AND CANCER

McCreanor & Bender (1983) demonstrated an increase in liver NAD(P) in rats maintained under hypoxic conditions, and an increase in urinary excretion of NMN, but no change in liver NAD(P), suggesting increase turnover of the coenzymes, in animals maintained under conditions of hyperoxia. At the time this was interpreted solely in terms of the specificity of NADase for the oxidised coenzymes; NAD(P)H is not a substrate. Janero *et al* (1993) have demonstrated a considerable depletion of NAD(P) in cardiomyocytes in culture, in response to oxidative stress induced by hydrogen peroxide. They showed that this was related to DNA strand breakage, and the subsequent activation or induction of ADP-ribosyltransferase.

It is unlikely that a significant proportion of the nicotinamide released by NADase and ADP-ribosyltransferases can be re-utilised for NAD(P) synthesis; nicotinamide and nicotinic acid phosphoribosyltransferases and nicotinamide deamidase have values of $K_m$ suggesting that they normally operate at or near $V_{max}$, while quinolinate phosphoribosyltransferase has a very high $K_m$ compared with the normal intracellular concentration of its substrate (Bender *et al*, 1982). Feeding rats with diets providing very high levels of nicotinic acid or nicotinamide does not increase the liver content of NAD(P), while there is a considerable increase in liver NAD(P) when they are fed a high tryptophan diet (McCreanor & Bender, 1986). Similarly, incubation of isolated hepatocytes with nicotinic acid or nicotinamide does not permit maintenance of intracellular NAD(P), while incubation with tryptophan does (Bender & Olufunwa, 1988). This suggests that oxidative stress or chronic exposure to DNA-damaging mycotoxins may adversely affect niacin nutritional status.

Jacob *et al* (1991) have demonstrated depletion of erythrocyte NAD(P) in response to short-term (35d) vitamin C depletion, with restoration on repletion with vitamin C. This NAD(P) depletion was accompanied by a doubling of the content of 8-hydroxyguanine in sperm DNA, suggesting impairment of the ADP-ribosyltransferase-dependent DNA repair mechanism. This may be important; Jacobson (1993) has reported that 15% of women with cancer in the Malmö study had a red cell NAD: total nicotinamide nucleotide ratio below the 2.5 centile of the reference range, suggesting niacin depletion associated with cancer. It is, however, not possible to interpret this observation as suggesting a relationship between niacin status and the initiation of cancer.

## IMPAIRMENT OF KYNURENINE HYDROXYLASE AND KYNURENINASE ACTIVITIES

The activities of kynurenine hydroxylase and kynureninase are close to the basal activity of tryptophan dioxygenase, and it is therefore likely that impairment of either enzyme will reduce the rate of tryptophan oxidative metabolism, and so increase the proportion of aminocarboxymuconic semialdehyde that undergoes oxidation rather than being used for NAD(P) synthesis. It has been known since the pioneering studies of Perrini in Padua in the 1940s that vitamin $B_6$ deficiency impairs kynureninase action and leads to increased excretion of xanthurenic and kynurenic acids. The increased excretion of xanthurenic and kynurenic acids in vitamin $B_6$ deficient animals is indeed accompanied by markedly reduced excretion of NMN and MPCX (Bender *et al*, 1990). Since some 10 - 30% of the population of developed countries show biochemical signs of inadequate vitamin $B_6$ nutritional status (Bender 1989), this suggests that there is some cause for concern over apparently adequate intakes of tryptophan and niacin.

Similarly, riboflavin deficiency, which is widespread in less developed countries, will impair the activity of kynurenine hydroxylase. There is considerable evidence from a number of studies that reduced kynurenine hydroxylase activity is associated with reduced endogenous synthesis of NAD(P) (Bender & McCreanor, 1985).

There is a 2:1 excess of women affected by pellagra (Miller, 1978); part of this can be accounted for by the inhibition of kynureninase by oestrogen metabolites (Bender & Wynick, 1981). In addition, administration of oestrogens to experimental animals leads to severe impairment of kynurenine hydroxylase (although there is no effect of oestrogens or their metabolites on the enzyme *in vitro*. This is accompanied by a significant fall in liver NAD(P) (Bender & Totoe, 1984). It is thus likely that both administered and endogenous oestrogens may compromise niacin nutritional status.

## TRACE MINERALS: ZINC AND IRON DEFICIENCY, LEAD EXPOSURE

Vannucchi *et al* (1986) showed that in rats fed a zinc, niacin and vitamin $B_6$ deficient diet, and given alcohol to drink, there was partial restoration of the low excretion of NMN and MPCX in response to repletion with zinc alone. In alcoholic pellagrins, Vannucchi and Moreno (1989) showed that addition of zinc to a defined enteral feeding mixture that was devoid of tryptophan and niacin led to a significant increase in urinary NMN and MPCX, associated with a fall in plasma tryptophan. Their results suggest that zinc deficiency (a well known complication of alcohol abuse) may be a factor in the aetiology of pellagra among alcoholics. It is likely that this reflects the rôle of zinc in glucocorticoid (and other steroid hormone) receptors. The increased metabolism of tryptophan in response to zinc supplementation is presumably a result of increased synthesis of receptor, thus enhancing the induction of tryptophan dioxygenase by cortisol, increasing flux through the pathway and therefore increasing the proportion of aminocarboxymuconic semialdehyde that undergoes non-enzymic cyclisation to quinolinic acid.

In studies of the effects of moderate iron deficiency in chicks, Oduho *et al* (1994) have shown that there is increased growth in response to a high intake of preformed niacin at all levels of iron intake, but there is only a response to increased tryptophan at high levels of iron intake. This suggests that even moderate iron deficiency may impair the synthesis of haem, and thus reduce the entry of tryptophan into the oxidative pathway.

Zerez *et al* (1990) have shown that purified NAD synthetase is inhibited by low levels of lead acetate *in vitro*. They also showed that erythrocyte NAD synthetase was significantly lower than normal in three subjects who had been occupationally exposed to lead. They proposed that measurement of erythrocyte NAD synthetase might be a useful index of lead exposure; it is equally likely that normal environmental and occupational exposure to lead may adversely affect niacin status.

## IMPAIRMENT OF COGNITIVE FUNCTION WITH NIACIN DEPLETION

It is apparent from the discussion above that a wide variety of common metabolic stresses and environmental hazards may compromise niacin nutritional status. Preliminary data from Ribaya-Mercado and coworkers at Tufts University (personal communication) indicate that marginal niacin depletion may be functionally important. They have fed a small group of healthy elderly people a diet providing 4mg ne /day (compared with a reference intake for this group of 14mg ne /day). Although there was the expected fall in NMN excretion, all subjects continued to excrete an amount considerably greater than levels associated with niacin deficiency. Nevertheless, after 20 days they showed a 40% impairment of delayed recall of a list of 10 words and a 35% decrease in performance in the backwards digit span test (but not the forward digit span test). Both were normalised by repletion with an adequate intake of niacin.

Despite the apparent adequacy of intakes of tryptophan and niacin in developed countries, it is possible that functional deficiency occurs, and may be a cause for concern.

## ACKNOWLEDGMENT

I am grateful to Drs Ribaya-Mercado, Riggs and Gershoff of the Jean Mayer USDA Human Nutrition Research Center on Aging at Tufts University for permission to cite their unpublished data.

## REFERENCES

Bender, D.A.,1989, Vitamin $B_6$ requirements and recommendations, *Eur. J. Clin. Nutr.* 43: 289.

Bender, D.A. and Bender, A.E., 1986, Niacin and trypotophan metabolism: the biochemical basis of niacin requirements and recommendations, *Nutr. Abs. Rev.* (Ser. A) 56: 695.

Bender, D.A. and McCreanor, G.M.,1985, Kynurenine hydroxylase: a potential rate-limiting enzyme in tryptophan metabolism, *Biochem. Soc. Transact.* 13: 441.

Bender, D.A. and Olufunwa, R., 1988, Utilization of tryptophan, nicotinamide and nicotinic acid as precursors for nicotinamide nucleotide synthesis in isolated rat liver cells, *Br. J. Nutr.* 59: 279.

Bender, D.A. and Totoe, L., 1984, Inhibition of tryptophan metabolism by oestrogens in the rat - a factor in the aetiology of pellagra, *Br. J. Nutr.* 51: 219.

Bender, D.A. and Wynick, D., 1981, Inhibition of kynureninase by oestrone sulphate: an alternative explanation for abnormal results of tryptophan load tests in women receiving oestrogenic steroids, *Br. J. Nutr.* 45: 269.

Bender, D.A., Magboul, B.I. and Wynick, D., 1982, Probable mechanisms of regulation of dietary tryptophan, nicotinamide and nicotinic acid as precursors of nicotinamide nucleotides in the rat, *Br. J. Nutr.* 48: 119.

Bender, D.A., Njagi, E.N.M. and Danielian, P.S., 1990, Tryptophan metabolism in vitamin $B_6$ deficient mice, *Br. J. Nutr.* 63: 27.

Gregory, J., Foster, K., Tyler, H. and Wiseman, M., 1990, "The Dietary and Nutritional Survey of British Adults", Social Survey Division, HMSO, Lonson.

Horwitt, M.K., Harvey, C.C., Rothwell, W.S., Cutler, J.L. and Haffron, D., 1956, Tryptophan-niacin relationships in man, *J. Nutr.* 60: suppl 1: 1.

Ikeda, M., Tsuji, H., Nakamura, S., Ichiyama, A., Nishizuka, Y and Hayaishi, O., 1965, Studies on the biosynthesis of nicotinamide adenine dinucleotide: (ii) a role of picolinate carboxylase in the biosynthesis of nicotinamide adenine dinucleotide from tryptophan in mammals, *J. Biol. Chem.* 240: 1395.

Jacob, R.A., Kelley, D.S., Pianalto, F.S., Swendseid, M.E., Henning, S.M., Zhang, J.Z., Ames, B.N., Fraga, C.G. and Peters, J.H., 1991, Immunocompetence and oxidant defense during ascorbate depletion of healthy men, *Am. J. Clin. Nutr.* 54: 1302s.

Jacobson, E.L., 1993, Niacin deficiency and cancer in women, *J. Am. Coll. Nutr.* 12: 412.

Janero, D.A., Hreniuk, D., Sharif, H.M. and Prout, K.C. 1993, Hydroperoxide-induced oxidative stress alters pyridine nucleotide metabolism in neonatal heart muscle cells, *Am. J. Physiol.* 264: C1401.

McCreanor, G.M. and Bender, D.A., 1983, The role of catabolism in controlling tissue concentrations of nicotinamide nucleotide coenzymes, *Biochim. Biophys. Acta* 759: 222.

McCreanor, G.M. and Bender, D.A., 1986, The metabolism of high intakes of tryptophan, nicotinamide and nicotinic acid in the rat, *Br. J. Nutr.* 56: 577.

Mehler, A.H., McDaniel, E.G. and Hundley, J.M., Changes in the enzymatic composition of liver: (i) increase of picolinic carboxylase in diabetes, *J. Biol. Chem.* 232: 331.

Miller, D.F., 1978, Pellagra deaths in the United States, *Am. J. Clin. Nutr.* 31: 558.

Oduho, G.W., Han, Y. and Baker, D.H., 1994, Iron deficiency reduces efficacy of tryptophan as a niacin precursor, *J. Nutr.* 124: 444.

Patterson, J.I., Brown, R.R., Linkswiler, H. and Harper, A.E., 1980, Excretion of tryptophan-niacin metabolites by young men: effects of tryptophan, leucine and vitamin $B_6$ intakes, *Am. J. Clin. Nutr.* 33: 2157.

Salih, M.A.M., Bender, D.A. and McCreanor, G.M., 1985, A lethal familial pellagra-like skin lesion associated with neurological and developmental impairment and the development of cataracts, *Pediatrics* 76: 787.

Salter, M. and Pogson, C.I., 1985, The role of tryptophan 2,3-dioxygenase in the hormonal control of tryptophan metabolism in isolated rat liver cells, *Biochem. J.* 229: 499.

Vannucchi, H. and Moreno, F.S., 1989, Interaction of niacin and zinc metabolism in patients with alcoholic pellagra, *Am. J. Clin. Nutr.* 50: 364.

Vannucchi, H., Kutnink, M. and Sauberlich, H.E., 1986, Interaction among niacin, vitamin $B_6$ and zinc in rats receiving ethanol, *Int. J. Vit. Nutr. Res.* 56: 355.

Zerez, C.R., Wong, M.D. and Tanaka, K.R., 1990, Synthetase from human erythrocytes: evidence that enzyme activity is a sensitive indicator of lead exposure, *Blood* 75: 1576.

# NON-PROTEIN TRYPTOPHAN IN FOODS

M. Biasiolo, A. Bertazzo, C. V. L. Costa, and G. Allegri

Department of Pharmaceutical Sciences
University of Padova
I-35131 Padova, Italy

## INTRODUCTION

Our previous papers (Allegri et al., 1993; Biasiolo et al., 1995) have shown that milk and yoghurt contain non-protein tryptophan. It was known that these foods contain free amino acids, although free tryptophan was not determined for methodological reasons (George and Lebenthal, 1981; Miller and Kandler, 1964; 1966; 1967a; 1967b).

In this paper we report the concentrations of total non-protein (protein-bound and free) and free tryptophan in yoghurts prepared from cow's milk, with or without fruit, cereals or coffee, and in other foods.

Both fractions of non-protein tryptophan were determined by selective fluorescence and HPLC according to the method of Bettero et al. (1984) and Costa et al. (1987). To obtain the free form of tryptophan, aliquots of samples dissolved in water (Biasiolo et al., 1995) were ultrafiltered by an Amicon cell with a XM-50 Diaflo membrane, according to the method of Eccleston (1973).

## RESULTS

Figure 1 reports the contents of total non-protein (protein-bound and free) and free tryptophan (mg/Kg) in plain and skimmed yoghurts derived from cow's milk. The concentrations of both fractions of tryptophan in cow's milk are also given for comparison. As the figure shows, the values of non-protein tryptophan in yoghurt are significantly different when compared with those of milk ($P<0.001$). No difference was found between the values of plain and skimmed yoghurt, for either form of non-protein tryptophan.

In yoghurts with fruit, the concentrations of non-protein tryptophan were quite similar to those of plain yoghurt, although they varied within each type of yoghurt. However, the highest levels of both total non-protein and free tryptophan (Table 1) were present in yoghurts with coffee or cereals, due to their contribution. These values are clearly much higher than those of plain yoghurt.

Table 2 reports the contents of total non-protein and free tryptophan (mg/Kg) in other kinds of foods. Here too the values of tryptophan in these foods are much higher, particularly

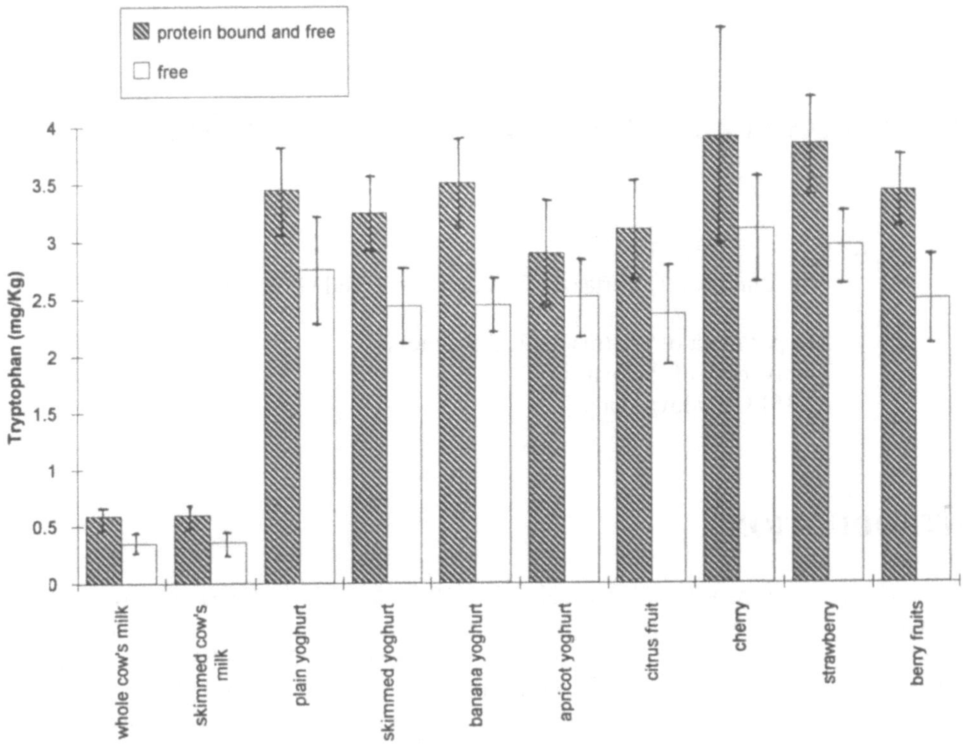


**Figure 1.** Levels of total non protein (protein-bound and free) and free tryptophan (mg/Kg) in cow's milk and in yoghurts with or without fruits.

in barley, with differences depending on the various samples of the stimulating beverages, cocoa shows a higher content of tryptophan with respect to coffee and tea.

## DISCUSSION

Miller and Kandler (1967a) observed that the amount of free amino acids in yoghurt was influenced by the proteolytic activity of the microorganisms with which the milk is processed to obtain yoghurt. The higher increase in the contents of free amino acids occurred during the lag phase of growth of the microorganisms (Miller and Kandler, 1967b).

The results obtained in the present study indicate that non-protein tryptophan too, which was not previously determined in foods, increases in processed products of milk such as yoghurt, the values of both fractions (protein-bound and free) being markedly higher than in the original milk.

**Table 1.** Concentrations of total non protein (protein-bound, and free) and free tryptophan (mg/Kg) in yoghurts with cereals or coffee (mean values ±S.D.)

| | Tryptophan (mg/Kg) | |
|---|---|---|
| | Total | Free |
| Yoghurt with cereals | 7.80 ± 1.64 | 5.32 ± 1.26 |
| Yoghurt with coffee | 8.92 ± 2.11 | 6.94 ± 1.17 |
| Yoghurt with malt | 8.91 ± 1.68 | 6.00 ± 1.60 |

**Table 2.** Contents (mean values±S.D.) of total non protein (protein-bound and free) and free tryptophan (mg/Kg) in some foods

| | Tryptophan (mg/Kg) | |
|---|---|---|
| | Total | Free |
| Coffee | 16.08 ± 3.28<br>(range 10.40–20.80) | 14.67 ± 3.01<br>(range 10.00–19.60) |
| Barley | 230.40 ± 49.36<br>(range 157–279) | 138.60 ± 18.13<br>(range 120–167) |
| Cocoa | 69.57 ± 17.64<br>(range 46–97) | 56.57 ± 14.19<br>(range 36–80) |
| Horlicks | 115.00 ± 30.11<br>(range 91–165) | 73.50 ± 7.76<br>(range 65–85) |
| Tea | 9.56 + 1.84<br>(range 7.00–13.00) | 9.56 ± 1.84<br>(range 7.00–13.00) |

In the fruit-flavoured yoghurts, we did not find great differences in the levels of free and total non-protein tryptophan with respect to yoghurt without fruit, while a marked increase appeared when other foods, such as coffee or cereals, are added to yoghurt, due to their contribution. In fact, determining non-protein tryptophan in coffee, malt and horlicks, we found high concentrations of both fractions of this amino acid.

The increase in free tryptophan in processed milk products makes more tryptophan available to brain (Tagliamonte et al., 1973). Tryptophan is the only amino acid bound 80-90% to plasma albumin (McMenamy and Oncley, 1958). The other 10-20%, the free fraction, is the only part capable of crossing the blood-brain barrier and of being utilized in the cerebral synthesis of the neurotransmitter serotonin (Tagliamonte et al., 1973; Fernstrom and Wurtman, 1971; Knott and Curzon, 1972; Gessa and Tagliamonte, 1974). The contribution of free tryprophan with foods is therefore of great interest.

## REFERENCES

Allegri G., Biasiolo M., Costa C., Bettero A., Bertazzo A., 1993, Content of non-protein tryptophan in human milk, bovine milk and milk- and soy-based formulas. *Food Chem.* 47: 23..

Bettero A., Costa C., Allegri G., 1984, Monitoring of tryptophan levels in biological fluids by HPLC: an improved method. In Progress in Tryptophan and Serotonin Research, eds H. Schlossberger, W. Kochen, B. Linzen & H. Steinhart. W. de Gruyter, Berlin, pp. 61-66.

Biasiolo M., Bertazzo A., Costa C., Beghetto A., Allegri G., 1995, Determination of non-protein tryptophan in yoghurts by selective fluorescence and HPLC. *Food Chem.* 52: 87.

Costa C., Bettero A., Allegri G., 1987, Rapid measurement of tryptophan, 5-hydroxytryptophan, serotonin and 5-hydroxyindoleacetic acid in human CSF by selective fluorescence and HPLC. *Gior. It. Chim. Clin.* 12: 307.

Eccleston E.G. A, 1973, method for the extimation of free and total acid-soluble plasma tryptophan using an ultrafiltration technique. *Clin. Chim. Acta*, 48: 269.

Fernstrom J.D. & Wurtman R.J., 1971, Brain serotonin content: physiological dependence on plasma tryptophan levels. *Science* 173: 149.

George D.E. & Lebenthal E., ??? Human breast milk in comparison to cow's milk. In Textbook of Gastroenterology and Nutrition in Infancy, ed. E. Lebenthal. Raven Press, New York, pp. 295-320.

Gessa G.L. & Tagliamonte A., 1974, Possible role of free serum tryptophan in the control of brain tryptophan level and serotonin synthesis. *Adv. Biochem. Psychopharmacol.* 11: 119.

Knott P.J. & Curzon G., 1972, Free tryptophan in plasma and brain tryptophan metabolism. *Nature* 239: 452.

McMenamy R.H. & Oncley J.L., 1958, The specific binding of L-tryptophan to serum albumin. *J. Biol. Chem.* 233: 1436.
Miller I. & Kandler O., 1964, Breakdown of protein in sour milks. I. Free amino acids in yoghurt, bioyoghurt, and acidophilus milk. *Med. Ernaehrung* 5: 100.
Miller I. & Kandler O., 1966, The free amino acids in fermented milks. In Proc. 17th Int: Dairy Congress, Munich, Vol.5, pp. 625-635.
Miller I. & Kandler O., 1967a, Proteolysis and liberation of free amino acids by lactic acid bacteria in milk. *Milchwissenschaft*, 22: 150.
Miller I. & Kandler O., 1967b, Influence of the raw milk flora on the content of free amino acids and other soluble N fractions. *Milchwissenschaft*, 22: 435.
Tagliamonte A., Biggio G., Vargiu L., Gessa G.L., 1973, Free tryptophan in serum controls: brain tryptophan level and serotonin synthesis. *Life Sci.* 12: 277.

# BIOCHEMICAL AND MORPHOLOGICAL CHANGES OBSERVED IN RAT MUSCLES FOLLOWING CONSUMPTION OF EXCESSIVE L-TRYPTOPHAN AND ATHEROGENIC DIETS

B. Gross,[1] N. Ronen,[2] A. Reznick,[2] S. Mokady,[3] S. Honigman,[1] and E. Livne[2]

[1] Department of Neurology, Lady Davis Carmel Medical Center and
The Bruce Rappaport Faculty of Medicine
Technion-Israel Institute of Technology, Haifa, 31096 Israel
[2] Division of Morphological Sciences
The Bruce Rappaport Faculty of Medicine
Technion-Israel Institute of Technology, Haifa 31096, Israel
[3] Department of Food Engineering and Biotechnology
Technion-Israel Institute of Technology, Haifa 32000, Israel

## 1. INTRODUCTION

Recently there has been an explosion of information regarding nutrition - related diseases. Eosinophilia Myalgia Syndrome (EMS), which is characterized by scleroderma-like induration, neuromuscular manifestations, arthralgia, pneumonitis and thromboembolic events has been associated with a preceding history of oral intake of L-tryptophan (Trp) (Kafman and Seidman, 1991). Even though epidemiologic studies indicated that a contaminant, 1,1'ethylidenebis Trp was involved in EMS (Mayeno et al., 1990), abnormalities in the metabolism of Trp associated with tissue fibrosis were also documented in a number of syndromes such as carcinoid syndrome, scleroderma, or rheumatoid arthritis (Fries et al., 1973; Houpt, 1991; Blauvelt and Falanga, 1991). In addition it has been demonstrated that some cell and tissue functions depend on fatty acid composition and that specific types of dietary lipids can alter the function of the immune system (Robinson, 1991) and of high plasma cholesterol levels associated with atherosclerosis. Dietary tryptophan was previously found to increase platelet aggregation (Mokady et al., 1990), and there was a synergistic effect between atherogenic diet and Trp on induction of macrophage uptake of plasma cholesterol (Aviram et al., 1991). Rudzite and Jurika (1991) also reported on the interaction between lipid metabolism and kynurenine.

*Recent Advances in Tryptophan Research*, edited by Graziella Allegri Filippini et al.
Plenum Press, New York, 1996

The purpose of the present study was to investigate the effect of saturated fat diet enriched with Trp on biochemical and histopathological changes in skeletal muscles.

## 2. METHODS

24 female weanling rats (Charles River C.D.) were divided into 4 groups of 6 animals each and were fed on a diet containing 10% protein, derived from casein supplemented with either 8% soybean oil (control) or with 8% coconut oil, 1% cholesterol, and 0.5% cholic acid (atherogenic). The diets were further supplemented with 0.4% L-tryptophan (Sigma, St. Louis, MO.). Food and water were supplied ad libitum. After a feeding period of 3 weeks, animals were sacrificed. Gastrocnemius muscles were dissected and quickly frozen in isopentane bathed in liquid nitrogen (-196°C) for the histochemical and biochemical studies. Samples of heart muscle were fixed in 10% formalin and stained with hematoxylin and eosin for general histologica observation. Other gastrocnemius muscle samples were homogenized in Tris-HCl buffer (pH 7.4) and the activities of alkaline and acid phosphatase were measured using para-nitrophenol phosphate (PNPP) as substrate at pH 5.0 and 9.5 respectively (Kwan and Ito, 1987). Protein oxidation assay was determined in supernatants of muscle samples

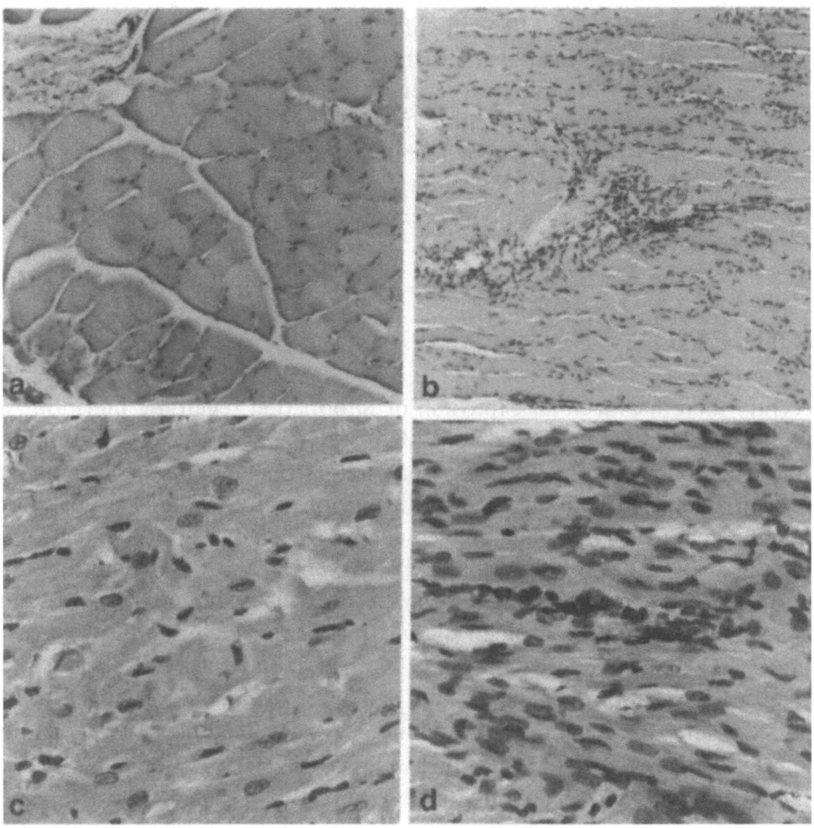


**Figure 1.** Gastrocnemius muscle from a rat control animal (a), and from an animal fed on atherogenic diet + 0.4% Trp (b) .H&E X90. **Note**: excessive infiltration of cells throughout the tissue and along the blood vessels in b. Heart muscle in control rat (c) and from an animal fed on atherogenic diet +0.4% Trp (d). H&E X142. **Note**: excessive cellular infiltration in d.

that were minced gently in 0.1 M phosphate buffer pH 7.4, containing 0.1% digitonin, using the spectrophotometeric assay for the reduction of dinitrophenylhydrazine (DNPH) to diphenylhydrazone (Levine et al., 1990). Data were expressed as nmole of carbonyl per mg protein in tissue. For all measurements statistical analysis was performed using unpaired Student's test and results are reported as mean ± SEM.

## 3. RESULTS AND DISCUSSION

Infiltration of inflammatory cells was observed in the fascia and perimysium of skeletal and heart muscle biopsies from rats that received tryptophan and atherogenic diet (Figure 1). These cells appeared to be clustered around perimysial small vessels. An overall increase of connective tissue and inflammatory cells, including infiltration into muscle spindle capsules and intramuscular nerves, was also seen. It is known that some cell functions and tissue phospholipid depend on fatty acid composition of dietary lipids and that specific type of dietary lipid can alter the function of the immune system (Robinson, 1991). Thus it may be speculated that saturated fatty acid and abnormalities in the metabolism of Trp could play a role in the induction of certain inflammatory reactions. L-Trp is the precursor of several neuroactive compounds (Schwartz et al., 1984). Elevated plasma levels of L-Trp were shown to augment the pathway of oxidation to formylkinurenine and synthesis of quinolinic acid which is known to be a potent endogenous neurotoxin. Excessive levels of Trp may also cause a shift to another metabolic pathway which is hydroxylated to serotonin, neurotransmitter and a potent vasoconstrictor. Mustant et al., (1977) reported that serotonin induce myopathy. Inflammatory infiltration into muscle might have been due to excessive serotonin production or due to activation of macrophages by the atherogenic diet (Aviram et al., 1991), or due to increased platelet aggregation (Mokady et al., 1990). Platelet activation on the other hand initiates several processes such as smooth muscle cell proliferation, endothelial injury, activation of macrophages and it may also accelerate atherosclerosis.

Table 1 shows the effect of tryptrophan and atherogenic diet on alkaline and acid phosphatase activities in skeletal muscle. The increased activity of alkaline phosphatase in muscle could reflect a relative increase of connective tissue elements,damage to the cell membrane, and as a result, a leakage of this membrane - bound enzyme. It may also indicate de novo formation of blood capillaries, which were consistently shown to contain large amounts of this enzyme.

The increased activities of acid phosphatase in muscle tissue may also indicate an increased number of connective tissue elements, among them macrophages and increased activity of lysosomes.

**Table 1.** Effect of atherogenic diet with excessive tryptophan on alkaline and acid phosphatase enzymes and protein oxidation in rat muscle

| | Control | Atherogenic | Control +0.4% Trp | Atherogenic +0.4% Trp |
|---|---|---|---|---|
| Alkaline phosphatase U/mg protein | 4.75 ± 1.2 | 5.5 ± 1.5 | 8.38 ± 2.9* | 7.48 ± 2.9* |
| Acid phosphatase U/mg protein | 4.05 ± 1.5 | 11.74 ± 1.07** | 13.34 ± 3.4** | 15.8 ± 5.9** |
| Protein oxidation nmol carbonyl/ mg protein | 2.2 ± 0.51 | 2.11 ± 0.27 | 4.3 ± 0.17** | 8.16 ± 1.04** |

*$p < 0.05$ compared to control; ** $p < 0.01$ compared to control.

A significant increase of protein oxidation was observed in gastrocnemius muscles from rats fed with excessive Trp when compared to controls (Table 1). This may be due to an increase in the number of fibroblasts which contain more oxidized proteins. Also increased protein peroxidation in hypercholesterolemic rats might result from relatively reduced levels of lipoprotein antioxidants, increased platelet aggregation and induction of microischemic changes.

In conclusion, this study demonstrates that excessive dietary tryptophan, when consumed with saturated fatty acids and cholesterol, might mediate atherogenic processes, inflammation, tissue damage and fibrosis.

## ACKNOWLEDGMENT

This study was supported by Israel Ministry of Health Grant No.181-681.

## 3. REFERENCES

Aviram M., Cogan U. and Mokady S., 1991,Excessive dietary tryptophan enhances plasma lipid peroxidation in rats. *Atherosclerosis* 88:29.

Blauvelt A.,and Falanga V., 1991, Idiopathic and L-tryptophan associated Eosinophilic fasciitis before and after L-tryptophan contamination. *Arch Dermatol* 127:1159.

Fries J.F., Lindgren J.A. and Bull J.M., 1973, Scleroderma-like lesions and the carcinoid syndrome. *Arch Intern Med* 131:550.

Houpt J.B., 1991,Tryptophan-New questions for an old amino acid. *J Rheumatol* 17:1431.

Kaufman L.D. and Seidman R.J., 1991, L-tryptophan associated eosinophilia myalgia syndrome: perspective of a new illness. *Rheum Dis Clin North Am* 17:427.

Kwan C.Y. and Ito H., 1987, Acid and alkaline phosphatase activities in vascular smooth muscle: species differences and subcellular distribution. *Comp Biochem Physiol* 87B:367.

Levine R.E., Garland D., Oliver C.N., Amici A., Climent I., Anke G.L., Shaltiel A. and Stadtman E.R, 1990, Determination of carbonyl content in oxidatively modified proteins. *Meth in Enzymol* 186:464.

Mayeno A.N., Lin F., Foote C.S., Loegering D.A., Hedberg C.W. and Gleich G.J., 1990, Characterization of "peak E" a novel amino acid associated with Eeosinophilia myalgia syndrome. *Science* 250:1707.

Mokady S., Cogan U. and Aviram M., 1990, Dietary tryptophan enhances platelet aggregation in rats. *J Nutr Sci Vitaminol* 36(suppl)S177.

Mustant T.L., Hudgson P. and Johnson M.A., 1977, Experimental serotonin myopathy. *Neurology* 27:772.

Peter J.C., 1991, Tryptophan nutrition and metabolism: an overreview. *Adv Exp Med Biol* 294 :345.

Robinson D.R., 1991, Alleviation of autoimmune disease by dietary lipids containing omega-3 fatty acid. *Rheum Dis Clin North Am* 17:213.

Rudzite V. and Jurika E., 1991, Kynurenine and lipid metabolism. *Adv Exp Med Biol* 294: 463

Schwartz R., Whestell W.O. and Mangano R.M., 1984, Quinolinic acid: an endogenous metabolite that produces axon sparing lesions in rat brain. *Science* 219:316.

# AGE DEPENDENCE OF LARGE NEUTRAL AMINO ACID LEVELS IN PLASMA

## Focus on Tryptophan

J. Demling,[1] K. Langer,[2] and M.Q. Mehr[1]

[1] Department of Psychiatry
University of Erlangen-Nürnberg
Schwabachanlage 6, D-91054 Erlangen, Germany
[2] Pharmacia GmbH
Hofmannstraße 26, D-91052 Erlangen, Germany

## INTRODUCTION

Low levels of plasma tryptophan (TRP) and low values of plasma TRP ratios (total or free TRP concentration divided by the sum of the concentrations of the other large neutral amino acids [LNAA] valine [VAL], leucine [LEU], isoleucine [ILEU], tyrosine [TYR], and phenylalanine [PHE]) have been attributed to depressive disorders (Møller et al. 1976, Maes et al. 1993), and it is well known that older people are especially prone to depression and suicide. Some studies have addressed the issue of physiological changes of plasma amino acid levels in connection with increasing age but there are only few data allowing for conclusions on TRP ratios (e.g., Rudman et al. 1989, Caballero et al. 1991). Age-dependent decrease of TRP and TRP ratio may contribute to the metabolic basis of affective disorders in elderly people. Therefore it seemed interesting to compare the fasting plasma TRP levels and TRP ratios of two different age groups with no psychiatric abnormalities.

## SUBJECTS AND METHODS

### Subjects

The younger population, medical students and members of our clinical staff, included 23 males, age range 22-37 (mean age 28) years, and 32 females, age range 18-38 (mean age 27) years.

The older subjects were patients from the university department of orthopedics: 18 males, age range 61-78 (mean age 70) years, and 18 females, age range 60-84 (mean age 68)

*Recent Advances in Tryptophan Research*, edited by Graziella Allegri Filippini et al.
Plenum Press, New York, 1996

years. Most patients were taking medications, mainly cardiovascular (digitalis and antihypertensive) and nonsteroidal anti-inflammatory. All were in good medical condition, none was on a special diet or had a history of psychiatric disorder. Laboratory parameters (electrolytes, glucose, liver and kidney parameters) were within normal ranges. Approximately half of the patients were in a preoperative state, the other half had been operated on (mainly hip replacement) not less than 10 days before this investigation; some of these were receiving heparin. Amino acid levels of these two groups did not differ significantly.

### Procedure

Using an EDTA treated syringe, venous blood samples were obtained between 6:30 and 8:30 a.m. after the subjects had fasted overnight.

### Chemical Assays

Total and free TRP was measured by high performance liquid chromatography, the other LNAA by cation exchange chromatography, according to the specifications described by Demling et al. (1993).

### Statistics

Differences of means were tested by the two-tailed Student's t-test for independent samples.

## RESULTS

We found the following significant ($p<0.05$) differences of means between age and sex groups, respectively (cf. tables 1 and 2):

- Total TRP and total TRP ratio were lower in the older than in the younger population from both sexes.
- All LNAA except total TRP were higher in older than in younger men, significance being reached by VAL and PHE, whereas only a slight tendency was found in older vs. younger women.
- Irrespective of the age group, total TRP, VAL, LEU, ILEU, and PHE were significantly higher in men than in women, TYR showed a tendency in the same direction.
- There was no difference in total TRP ratio between the sexes of the same age group, only a slight tendency to lower values in men.
- Older men showed a significantly lower free TRP ratio than the older women.

## DISCUSSION

Our results confirm the finding of Thomas et al. (1986) that total TRP is higher in young men than in young women. This sex difference is maintained with the increase in age, in accordance to Phipps and Powell (1985) but in contrast to Thomas et al. (1986) and Caballero et al. (1991). Most investigators found a marked decline of total TRP with increasing age either in both sexes (Partsch et al. 1977, Koh and Cha 1983, Phipps and Powell 1985, this study) or in males only (Thomas et al. 1986, Fukagawa et al. 1987 [who studied

**Table 1.** Mean and standard deviations ($\bar{x}\pm$SD) for the large neutral amino acids (except tryptophan) and for the sum of these compounds, specified for the sexes and age groups. For significances see results section

| Amino acids (μmol/l) | $\bar{x} \pm$ SD | | | |
|---|---|---|---|---|
| | Males | | Females | |
| | Younger | Older | Younger | Older |
| Valine | 235 ± 30 | 262 ± 51 | 197 ± 60 | 207 ± 29 |
| Leucine | 140 ± 25 | 152 ± 28 | 106 ± 29 | 114 ± 19 |
| Isoleucine | 72 ± 14 | 81 ± 18 | 59 ± 18 | 63 ± 11 |
| Phenylalanine | 59 ± 11 | 68 ± 12 | 50 ± 8 | 54 ± 9 |
| Tyrosine | 63 ± 12 | 69 ± 10 | 57 ± 16 | 64 ± 14 |
| Total | 569 ± 76 | 631 ± 105 | 470 ± 118 | 502 ± 61 |

**Table 2.** Mean and standard deviations ($\bar{x}\pm$SD) of the tryptophan related parameters, specified for the sexes and age groups. For significances see results section

| Amino acids (μmol/l) and ratios | $\bar{x} \pm$ SD | | | |
|---|---|---|---|---|
| | Males | | Females | |
| | Younger | Older | Younger | Older |
| Total tryptophan | 60.0 ± 15.6 | 41.4 ± 7.4 | 54.8 ± 11.7 | 34.4 ± 6.1 |
| Free tryptophan | 5.6 ± 2.9 | 5.5 ± 1.5 | 5.9 ± 2.0 | 5.9 ± 1.7 |
| Total tryptophan ratio | 0.106 | 0.066 | 0.121 | 0.069 |
| Free tryptophan ratio | 0.009 | 0.008 | 0.013 | 0.011 |

only males], Caballero et al. 1991) whereas Rudman et al. (1989) were unable to show an age connected change in their male sample. As for the other LNAA, Cynober et al. (1983) and Koh and Cha (1983) discovered, in accordance with our results, higher values in men than in women of all ages, whereas with Caballero et al. (1991), the sex difference of VAL, LEU, and ILEU found in young persons had disappeared in persons with higher age. This was due to increased concentrations of these amino acids in elderly females only, a finding described already by Galante et al. (1978). We found a significant increase for VAL and PHE (with a tendency for the other LNAA except TRP) in elderly men but only a tendency in women. In contrast, Rudman et al. showed a significant decrease of VAL, LEU, ILEU, and PHE in older compared to younger men.

Our results are in fairly good accordance with the data described in the literature. Similar to the findings of Caballero et al (1991), although caused by somewhat different mechanisms, the changes of total TRP and the other LNAA described in this study result in a significant net decrease of total TRP ratio with increasing age in both sexes. It seems interesting that this finding could be replicated in our study which involved institutionalized elderly subjects. Thus, lower plasma total TRP and total TRP ratios in older subjects appear to be fairly robust parameters in blood biochemistry. A sex difference of free TRP ratio in elderly subjects has not been reported hitherto. If confirmed, it might account for a lower TRP supply to the brain especially in older men, contributing to the known higher suicide rates of the male compared to the female elderly.

## Conclusion

The data support the hypothesis that lower TRP levels and TRP ratios in plasma, with the consequence of a decreased influx of TRP into the brain and diminished synthesis of serotonin, may act as a vulnerability factor for depressed states in older age.

## REFERENCES

Caballero, B., Gleason, R.E., and Wurtman, R.J., 1991, Plasma amino acid concentrations in healthy elderly men and women, *Am J Clin Nutr* 53:1249

Cynober, L., Blonde, F., Nguyen Dinh, F., Gerbet, D., and Giboudeau, J., 1983, Mesure des acides amines plasmatiques et urinaires par chromatographie en phase gazeuse chez des sujets sains. Variations en fonction de l'age et du sexe, *Ann Biol Clin (Paris)* 41:33

Demling, J., Langer, K., Wörthmüller, M., and Yusufu, V., 1993, Individual day-to-day variations in plasma amino acid levels in healthy persons, *Amino Acids* 5:253

Fukagawa, N.K., Minaker, K.L., Rowe, J.W., and Young, V.R., 1987, Plasma tryptophan and total neutral amino acid levels in men: influence of hyperinsulinemia and age, *Metabolism* 36:683

Galante, A., Angelico, F., Crocchioni, G., and Pennetti, V., 1978, Intersexual differences in the serum free amino acid pattern of young adults, normal and obese aged subjects, *Nutr Metab* 22:119

Koh, E.T., and Cha, C.J., 1983, Comparison of plasma amino acids by race, sex and age, *Nutr Rep Int* 28:8

Maes, M., Meltzer, H.Y., Scharpe, S., Bosmans, E., Suy, E., De Meester, I., Calabrese, J., and Cosyns, P., 1993, Relationships between lower plasma L-tryptophan levels and immune-inflammatory variables in depression, *Psychiatry Res* 49:151

Møller, S.E., Kirk, L., and Fremming, K.H., 1976, Plasma amino acids as an index for subgroups in manic depressive psychosis: correlation to effect of tryptophan, *Psychopharmacology* 49:205

Partsch, G., Schwarzer, C., Tausch, G., and Eberl, R., 1977, Untersuchungen über die Plasmaaminosäuren bei alten Patienten mit chronischer Polyarthritis, *Akt Gerontol* 7:621

Phipps, D.A., and Powell, C., 1985, Plasma tryptophan, age and depression, *Age Ageing* 14:71

Rudman, D., Mattson, D.E., Feller, A.G., Cotter, R., and Johnson, R.C., 1989, Fasting plasma amino acids in elderly men, *Am J Clin Nutr* 49:559

Thomas, D.R., Shaw, D.M., and Tidmarsh, S.F., 1986, Effect of aging on plasma tryptophan, *J Clin Psychopharmacol 6:252*

# ROLE OF SERUM FREE FATTY ACIDS IN L-TRYPTOPHAN TRANSPORT INTO THE LIVER

E. Sasaki,[1] Y. Ohta,[1] R. Shinohara,[2] and I. Ishiguro[1]

[1] Department of Biochemistry
School of Medicine
[2] School of Health Sciences
Fujita Health University
Toyoake, Aichi 470-11, Japan

## INTRODUCTION

A large part of L-tryptophan (Trp) present in the plasma or serum is bound to albumin (McMenamy et al., 1957). It is known (Curzon et al., 1974; Kragh-Hansen, 1983) that this binding of Trp to albumin is affected by various substances such as free fatty acids (FFA). It has been shown (Madras et al., 1974; Curzon et al., 1972) that in the plasma of rats under feeding of a high fat diet, starvation, and immobilization, free Trp levels are enhnaced by an increase in FFA levels. Madras et al. (1974) have demonstrated that in rats fed a protein-carbonhydrate diet with various amounts of fat, liver Trp content dose not correlate with an increase in free serum Trp levels consequent to increased serum FFA levels. Recently, we have reported (Sasaki et al., 1993) that in rats fed a 40% lard-containing diet, there are little changes in the disappearance of blood Trp and the content of liver Trp under a marked increase in free serum Trp levels, and have suggested that under physiological levels of serum Trp and albumin, increased serum FFA levels do not have a large influence on Trp transport into the liver. However, it is still unclear whether or not serum FFA plays a facilitating role in Trp transport into the liver of rats, although increased serum FFA levels seem to have little effect on Trp transport into the liver under physiological levels of serum Trp and albumin. We, therefore, attempted to further clarify the role of serum FFA in Trp transport into the liver using analbuminemic rats, which are the mutant strain established from Sprague-Dawley rats and are genetically lacking in plasma albumin (Nagase et al., 1979).

## MATERIALS AND METHODS

Analbuminemic rats bred in our laboratory were used throughout. Under ether anesthesia, analbuminemic rats weighing about 200g were injected with a Trp solution (20 μmol/kg)

*Recent Advances in Tryptophan Research*, edited by Graziella Allegri Filippini et al.
Plenum Press, New York, 1996

containing fatty acid-free bovine serum albumin (BSA) (1.5 g/kg) (Trp-BSA solution) with and without oleic acid (200 µmol/kg) at a dose volume of 6.0 ml/kg into the right jugular vein. Trp, BSA, and oleic acid were dissolved in phosphate buffered saline (pH 7.4). Blood was withdrawn from the left jugular vein of the animals immediately before and 0.5, 1, 3, 5, 10, and 20 min after injection with the Trp-BSA solution with and without oleic acid. Livers were collected 20 min after injection with the Trp-BSA solution with and without oleic acid and used for measurements of Trp, protein, and tryptophan 2,3-dioxygenase (TDO) activity. Serum was separated from the collected blood by centifugation and used for measurements of total and free Trp, FFA, and albumin. Free serum Trp was separated by centrifugation at 4°C through an ultrafiltration membrane, Ultrafree C3TK (Millipore Co., Tokyo) as described in our previous report (Sasaki et al., 1993). Trp in serum and liver was determined by high-performance liquid chromatography with electrochemical detection as described in our previous report (Sasaki et al., 1993). Albumin and FFA were measured using commercial clinical test kits of abumin assay based on the bromcresol green method and FFA assay based on the enzymatic method, respectively. Liver protein was measured by the method of Lowry et al. (1951). Liver TDO was assayed in the presence of 16 µmM haematin by the method of Metzler et al. (1982). This enzyme activity is expressed as the amount of kynurenine (Kyn.) produced per g liver tissue for one hour at 37°C.

Values are expressed as means ±S.D. Results obtained were statistically analyzed by Student's *t*-test if necessary.

## RESULTS AND DISCUSSION

We have shown (Sasaki et al., 1993) that in rats fed a 40% lard-containing diet, there are little changes in the disappearance of blood Trp and the content of liver Trp under a

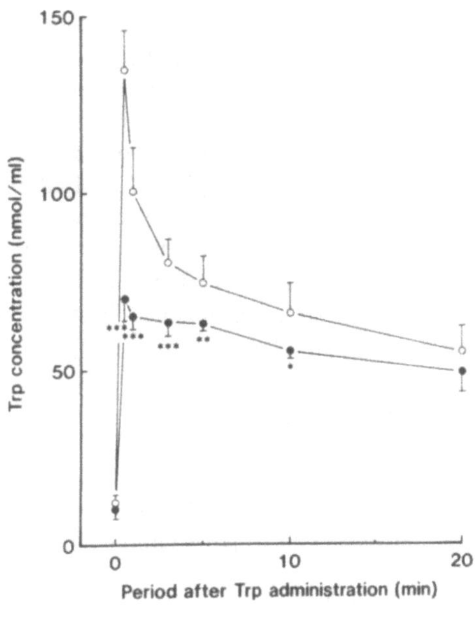


**Figure 1.** Changes in total serum Trp levels in analbuminemic rats following injection with a Trp-BSA solution with and without oleic acid. Analbuminemic rats were injected i.v. with a Trp-BSA solution with (-•-) and without (-o-) and total serum Trp was measured as described in Materials and Methods. Values are means ± S.D. (n = 4-7). Vs. without oleic acid: *, p < 0.05; **, p < 0.01; ***, p < 0.001.

marked increase in free serum Trp levels, and have suggested that under physiological levels of serum Trp and albumin, increased serum FFA levels do not have a large influence on Trp transport into the liver. However, it is still unclear whether or not serum FFA plays a stimulating role in Trp transport into the liver of rats.

We, therefore, attempted to further clarify the role of serum FFA in Trp transport into the liver using analbuminemic rats, which are known to be genetically lacking in plasma albumin (Nagase et al., 1979). When the disappearance of blood Trp was examined in the animals injected i.v. with a Trp (20 μmol/kg)-BSA (1.5 g/kg) solution with and without oleic acid (200 μmol/kg) over a 20 min period, total serum Trp levels in the animals injected with the Trp-BSA solution with oleic acid were significantly lower than those in the animals injected with the Trp-BSA solution without oleic acid during the period except for the time point of 20 min (Figure 1). The concentration of total serum Trp in the animals injected with the Trp-BSA solution without oleic acid was approximately twice in the animals injected with the Trp-BSA solution with oleic acid at 0.5 min after the injection (Figure 1). Thus, the addition of oleic acid to the injected Trp-BSA solution was found to enhance the disappearance of blood Trp. The concentration of free serum Trp in the animals injected with the Trp-BSA solution with oleic acid was 63.8 ± 7.8 nmol/ml (n = 4) at 0.5 min after the injection, while that concentration in the animals injected with the Trp-BSA solution without oleic acid was 24.0 ± 3.0 nmol/ml (n=4) at 0.5 min after the injection.

Thus, free serum Trp BSA solution without oleic acid at 0.5 min after the injection. In the animals injected with the Trp-BSA solution with oleic acid, a marked increase in serum FFA levels was found 0.5 min after the injection and the increased serum FFA levels decreased thereafter, reaching near the original level at 20 min (Figure 2). There was no difference in the levels of disappeared serum albumin between the animals injected with the Trp-BSA solution with and without oleic acid (data not shown). These results suggest that the enhancement of the disappearance of blood Trp by the addtion of oleic acid to the injected Trp-BSA solution should be due to an increase in free serum Trp levels consequent to increased serum FFA levels.

It is known (Bender, 1982) that in rats, more than 90% of total Trp present in the plasma is degraded to Kyn. and subsequently CoA and NAD in the liver through the kynurenine pathway in which TDO takes part under physiological conditions. Trp itself has been shown to stimulate liver protein synthesis in rats (Sidransky et al., 1971). Hence, we determined liver Trp and protein contents and TDO activity in analbuminemic rats injected with the Trp-BSA solution with and without oleic acid at 20 min after the injection. As shown in Table 1, not onlly Trp content but also protein content and TDO activity in livers of the animals injected with the Trp-BSA solution with oleic acid were not different from those of the animals injected with the Trp-BSA solution without oleic acid. Thus, the addition of oleic acid to the injected Trp-BSA solution was found to fail to increase liver Trp content in spite

**Table 1.** Liver Trp and protein contents and TDO activity in analbuminemic rats at 20 min after inejction with a Trp-BSA solution with and without oleic acid

| | Without oleic acid | With oleic acid | P-value |
|---|---|---|---|
| Trp content (nmol/g tissue) | 44.1 ± 5.6[a] | 45.6 ± 3.3 | N.S.[b] |
| Protein content (mg/g tissue) | 207.3 ± 7.4 | 203.1 ± 9.4 | N.S. |
| TDO activity (μmol Kyn./h/g tissue) | 3.36 ± 0.30 | 3.82 ± 0.25 | N.S. |

[a]Values are means ± S.D. (n = 3-4). [b]N.S., not significant.

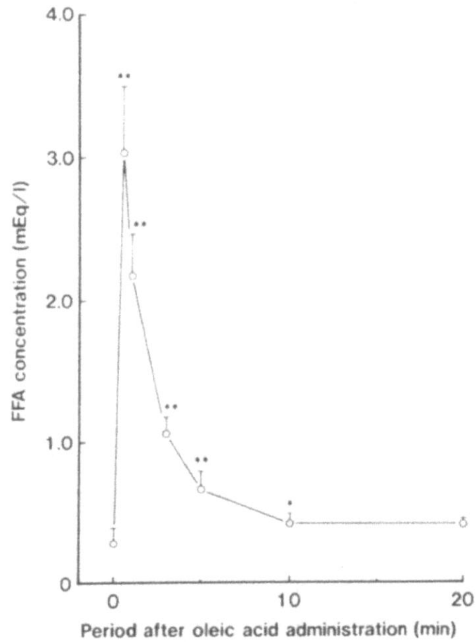


**Figure 2.** Change in serum FFA levels in analbuminemic rats following injection with a Trp-BSA solution. Analbuminemic rats were injected i.v. with a Trp-BSA solution with oleic acid and serum FFA was measured as described in Materials and Methods. Values are means ±S.D. (n = 4-7). Vs. values at 0 min: *, $p < 0.01$; **, $p < 0.001$.

of no change in liver protein content and TDO activity. These results suggest that an increase in free serum Trp consequent to increased serum FFA levels should be unable to facilitate Trp transport into the liver of analbuminemic rats.

In conclusion, the present results indicate that serum FFA should be unable to play a facilitating role in Trp transport into the liver of rats. However, the mechanism by which an increase in serum FFA levels cannot facilitate the transport of a large amount of free Trp liberated from Trp bound to albumin into the liver is required for further investigation.

## REFERENCES

Bender D.A., 1982, Biochemistry of tryptophan in health and disease, *Mol. Aspects Med.*, 6: 101.

Curzon G., Joseph M.H., and Knott P.J., 1972, Effects of immobilization and food deprivation on rat brain tryptophan metabolism, *J. Neucrochem.*, 19: 1967.

Curzon G., Friedel J., Katamaneni M.C., Greenwood H., and Lader M.H., 1974, Unesterified fatty acids and the binding of tryptophan in human plasma, *Clin. Sci. Mol. Med.*, 47: 415.

Kragh-Hansen U., 1983, Relation betwen high-affinity binding site for L-tryptophan, diazepam, salicylate and phenol red on human serum albumin, *Biochem. J.*, 209: 135.

Lowry O.H., Rosebrough N.J., Farr A.L., and Randall R.J., 1951, Protein measurement with the Folin phenol reagent, *J. Biol. Chem.*, 183: 265.

Madras B.K., Cohen E.L., Messing R., Munro H.N., and Wurthman R.J., 1974, Relevance of free tryptophan in serum to tissue tryptophan concentrations, *Metabolism*, 23: 1107.

McMenamy R.H., Lund C.C., and Oncley J.L., 1957, Unbound amino acid concentrations in human blood plasmas, *J. Clin. Invest.*, 36: 1672.

Metzler H., Gehardt R., and Mecke Y., 1982, A convenient and highly sensitive spectrophotometeric assay for tryptophan 2,3-dioxygenase, *Anal. Biochem.*, 121: 10.

Nagase S., Shimamune K., and Shumiya S., 1979, Albumin-deficient rat mutant, *Science*, 205: 590.

Sasaki E., Ohta Y., Shinohara R., and Ishiguro I., 1993, Effect of serum fatty acid levels on disappearance of blood L-tryptophan in rats fed lard, *J. Jpn. Sco. Nutr. Food Sci.*, 46: 487.

Sidransky H., Verney E., and Sarma D.S.R., 1971, Effect of tryptophan on polyribosomes and protein synthesis in liver, *Am. J. Clin. Nutr.*, 24: 779.

# L-TRYPTOPHAN TRANSPORT INTO THE LIVER OF VITAMIN $B_6$-DEFICIENT RATS

Y. Ohta, E. Sasaki, K. Nishida, and I. Ishiguro

Department of Biochemistry
School of Medicine
Fujita Health University
Toyoake, Aichi 470-11, Japan

## INTRODUCTION

We have demonstrated (Saito et al., 1986) that in isolated rat hepatocytes, L-tryptophan (Trp) uptake occurs via at least two saturable transport systems [one high-affinity (Km = 14 μM) and one low-affinity (Km = 2.1 mM)] and one non-saturable one. It has also been shown (Saito et al., 1990; Ohta et al., 1992) that the activity of Trp uptake via the high-affinity transport system into hepatocytes increases in rats with induction of liver tryptophan 2,3-dioxygenase (TDO) by Trp or cortisone treatment, while that uptake activity decreases in rats with depression of liver TDO activity by chronic nicotinamide administration. These findings have suggested that liver Trp uptake via the high-affinity transport system is closely related to liver Trp metabolism via TDO in rats.

It is known (Otsuka et al., 1981; 1982; Takeuchi and Shibata, 1984; Kondo and Okada, 1985) that vitamin $B_6$ ($B_6$) deficiency has a variety of effects on liver Trp metabolism via TDO, kynureninase, and kynurenine aminotransferase in rats. It is also known (Okada and Suzuki, 1974) that $B_6$ deficiency affects plasma levels of branched-hyphone chain amino acids such as leucine (Leu), isoleucine (Ile), and valine (Val) which are transported into liver cells via the L-system common to Trp. We, therefore, examined the effect of $B_6$ deficiency on the transport of Trp into the liver of rats.

## MATERIALS AND METHODS

Male Wistar rats aged four weeks were maintained on a $B_6$-deficient diet (Oriental Yeast Co.) *ad libitum* for 4 and 9 weeks as described by Otsuka et al. (1981). The control animals were fed a normal diet ($B_6$, 80 mg/100 g) during the same period. Under ether anesthesia, livers of the animals were perfused well with ice-cold 0.15 M KCl to remove blood and then cut off. Immediately before liver perfusion, blood was collected from vena cava caudalis and separated into serum. Livers were used for Trp measurement. Serum was

*Recent Advances in Tryptophan Research*, edited by Graziella Allegri Filippini et al.
Plenum Press, New York, 1996

used for measurements of free fatty acids (FFA), albumin, and amino acids such as Trp (total and free), Leu, Ile, Val, tyrosine (Tyr), and phenylalanine (Phe). Trp in serum and liver was measured by high-performance liquid chromatography with electrochemical detection as described in our previous reports (Saito et al., 1986; 1990; Ohta et al., 1992). Free serum Trp was separated by centrifugation through an ultrafiltration membrane, Ultrafree C3TK (Millipore Co.) as described in our previous reports (Saito et al., 1986; 1990; Ohta et al., 1992). Serum albumin and FFA were measured using commercial clinical test kits. Leu, Ile, Val, Tyr, and Phe in serum were measured in an amino acid analyzer, Hitachi L-8500 (Hitachi Co.). The uptake rates of Trp, Leu, Try, and Phe into isolated rat hepactocytes were determined as described in our previous reports (Saito et al., 1986; 1990; Ohta et al., 1992). Isolated hepatocytes (1 x $10^6$ cells/ml) were incubated with various concentrations of unlabeled each amino acid and about 1 µCi $^3$H-labeled each corresponding amino aicd at 37°C for the initial first minute in 0.5 ml of Hanks' balanced salt solution saturated with 95% $O_2$-5% $CO_2$ gas with shaking at 100 strokes/min. The uptake of each amino acid was estimated by the amount of the $^3$H-labeled amino acid taken up per $10^6$ cells per min. Rat hepatocytes were prepared by the method of Seglen (1976). Cell viability, estimated by trypan blue exclusion test, was above 90%.

Values are means ± S.D. Results obtained were statistically analyzed by Student's *t*-test if necessary.

## RESULTS AND DISCUSSION

The body and liver weights of rats fed a $B_6$-deficient diet for 4 and 9 weeks were significantly lighter than those of rats fed a normal diet for the corresponding period as having been shown by Otsuka et al (1981) (data not shown). However, there was little difference in liver weight based on gram per 100 g body weight between rats fed the $B_6$-deficient and normal diets for 4 or 9 weeks (data not shown). In order to clarify the effect of $B_6$ deficiency on Trp transport into the liver, we first examined total and free serum Trp levels in rats fed $B_6$-deficient and normal diets for 4 and 9 weeks. As shown in Table 1, there was no difference in total serum Trp levels between rats fed the $B_6$-deficient and normal diets for 4 or 9 weeks. The concentration of serum free Trp levels in rats fed the $B_6$-deficient diet for 9 weeks was significantly lower than that in rats fed the normal diet for the same period, although there was no difference in the level between rats fed the $B_6$-deficient and normal diets for 4 weeks (Table 1). Since serum albumin and FFA are known to affect the binding of Trp to albumin (Curzon et al., 1974), serum albumin and FFA levels were checked in rats fed $B_6$-deficient and normal diets for 4 and 9 weeks. There was no diffirence in serum

**Table 1.** Levels of total and free Trp, albumin, and FFA in serum of rats fed normal and $B_6$-deficient diets for 4 and 9 weeks

| Period & groups | Total Trp (nmol/ml) | Free Trp (nmol/ml) | Albumin (g/100 ml) | FFA (µEq/l) |
|---|---|---|---|---|
| 4 weeks | | | | |
| Control | 124.3 ± 8.3 | 3.5 ± 0.63. | 3.2 ± 0.1 | 281.3 ± 34.9 |
| $B_6$-deficient | 127.0 ± 16.4 | 3.1 ± 0.6 | 3.2 ± 0.2 | 277.0 ± 47.0 |
| 9 weeks | | | | |
| Control | 137.0 ± 13.9 | 4.0 ± 0.3 | 3.2 ± 0.1 | 381.8 ± 73.0 |
| $B_6$-deficient | 136.9 ± 6.3 | 2.7 ± 0.3** | 3.2 ± 0.1 | 284.2 ± 55.3* |

Values are means ± S.D. (n = 5-7).
Vs. control (normal diet-fed): *, $p < 0.05$; **, $p < 0.001$.

albumin levels between rats fed the $B_6$-deficient and normal diets for 4 or 9 weeks (Table 1). The concentration of serum FFA in rats fed the $B_6$-deficient diet for 9 weeks was significantly lower than that in rats fed the normal diet for the same period, although there was no difference in the level between rats fed the $B_6$-deficient and normal diets for 4 weeks (Table 1). From these results, it is suggested that an decrease in free serum Trp levels in rats fed a $B_6$-deficient diet for 9 weeks could be due to a decrease in serum FFA levels.

It has been shown (Saito et al., 1986; Salter et al., 1986) that liver Trp uptake occurs via not only the L-system common to branched-chain amino acids such as Leu, Ile, and Val, but also the T-system common to aromatic amino acids such as Try and Phe. In addition, it is known (Okada and Suzuki, 1974) that $B_6$ deficiency increases plasma levels of branched-hyphone chain amino acids such as Leu, Ile, and Val. Hence, we examined the levels of Leu, Ile, Val, Try, and Phe in the serum of rats fed $B_6$-deficient and normal diets for 4 and 9 weeks. Serum Val and Tyr levels in rats fed the $B_6$-deficient diet for 4 weeks were significantly higher than those in rats fed the normal diet for the same period, while serum Phe levels in the former were significantly lower than thsoe in the latter (data not shown). Serum Val, Ile, Leu, and Tyr levels in rats fed the $B_6$-deficient diet for 9 weeks were significantly higher than those in rats fed the normal diet for the same period (data not shown). These results suggest that increases in serum Leu, Ile, and Val, and Tyr levels following prolonged $B_6$ deficiency could affect Trp uptake into the liver.

We next checked liver Trp content in rats fed $B_6$-deficient and normal diets for 4 and 9 weeks. Liver Trp content in rats fed the $B_6$-deficient diet for 4 weeks was not significantly different from that in rats fed the normal diet for the same period, while that content in rats fed the $B_6$-deficient diet for 9 weeks was significantly lower than that in rats fed the normal diet for the same period (Figure 1). Similar results have been reported by Takeuchi and Shibata (1984). This decrease in liver Trp content following prolonged $B_6$ deficiency seems to be partly due to its degradation via TDO because $B_6$ deficiency is known to enhnace liver TDO activity under decreased kynureninase and kynurenine aminotransferase activities (Otsuka et al., 1981; 1982). However, it is likely that depression of Trp transport into the liver partly contributes to this decrease in liver Trp content because prolonged $B_6$ deficiency caused not only a decrease in free serum Trp levels, but also increases in the levels of serum branchaed-chain and aromatic amino acids as described above.

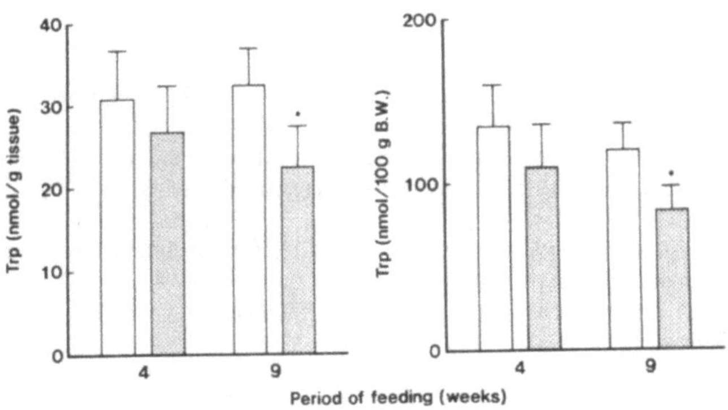


**Figure 1.** Liver Trp levels in rats fed normal and $B_6$-deficient diets for4 and 9 weeks. Liver Trp was determined in rats fed normal (open bar) and $B_6$-deficient (shaded bar) diets for 4 and 9 weeks as described in Materials and Methods. Values are means ±S.D. (n = 5-8). Vs. control (normal diet-fed): *, $p < 0.01$.

**Table 2.** Trp uptake into hepatocytes isolated from rats fed normal and $B_6$-deficient diets for 4 and 9 weeks

| Trp (μM) | Uptake rate (pmol/min/$10^6$ cells) | | | |
|---|---|---|---|---|
| | 4 weeks | | 9 weeks | |
| | Control | $B_6$-deficient | Control | $B_6$-deficient |
| 10 | 32.6 ± 3.0 | 27.5 ± 2.1 | 31.7 ± 5.6 | 21.7 ± 0.3* |
| 50 | 140.7 ± 24.4 | 116.2 ± 21.3 | 127.4 ± 38.2 | 103.4 ± 4.6 |
| 100 | 282.8 ± 29.8 | 230.5 ± 30.6 | 250.8 ± 51.3 | 225.8 ± 41.0 |
| 1000 | 2503.5 ± 111.6 | 2885.6 ± 271.4 | 2587.6 ± 612.5 | 2485.2 ± 543.1 |

Values are means ± S.D. with three different preparations.
Vs. control (normal diet-fed): *, $p < 0.05$.

We further examined the effect of $B_6$ deficiency on Trp uptake into the liver of rats using isolated hepatocytes. The results are shown in Table 2. Trp uptake activity was measured in hepatocytes isolated from rats fed $B_6$-deficient and normal diets for 4 and 9 week using 10 to 1000 μM Trp as the substrate. A significant decrease in Trp uptake rate was found only when hepatocytes from rats fed the $B_6$-deficient diet for 9 weeks were incubated with 10 μM Trp. It has been shown (Saito et al., 1986) that in isolated rat hepatocytes, Trp uptake occurs via at least two saturable transport systems [one high-affinity (Km = 14 μM) and one low-affinity (Km = 2.1 mM)] and one non-saturable one. In addition, the uptake rates of Leu, Tyr, and Phe into hepatocytes from rats fed the $B_6$-deficient diet for 4 or 9 weeks did not differ from those from rats fed the normal diet for the corresponding period (data not shown). These amino acids are known to be taken up into rat liver cells via the transport systems common to Trp (Saito et al., 1986; Salter et al., 1986). Accordingly, it is suggested that prolonged $B_6$ deficiency should impair liver Trp uptake via the high-affinity transport system.

In conclusion, the present results indicate that prolonged $B_6$ deficiency impairs Trp transport into the liver of rats both by decreasing the activity of Trp uptake via the high-affinity transport system and free serum Trp levels and by increasing the serum levels of Leu, Ile, Val, and Tyr which are taken up into the liver via the transport systems common to Trp.

## REFERENCES


Curzon G., Friedel J., Katamaneni M.C., Greenwood H., and Lader M.H., 1974, Unesterified fatty acids and the binding of tryptophan in human plasma, *Clin. Sci. Mol. Med.*, 47: 415.

Kondo T. and Okada M., 1985, Effect of pyridoxine administration on the induction of cytosolic aspartate aminotransferase in the liver of rats treated with hydrocortisone, *J. Nutr. Sci. Vitaminol.*, 31: 509 (1985).

Ohta Y., Nishida K., Otsuji K., Sasaki E., and Ishiguro I., 1992, Effects of chronic nicotinamide administration and its withdrawal on L-tryptophan transport into the liver of rats, in: "Advances in Tryptophan Research 1992", I. Ishiguro, R. Kido, T. Nagatsu, Y. Nagamura, Y. Ohta, eds., Fujita Health University Press, Toyoake.

Okada M. and Suzuki K., 1974, Amino acid metabolism in rats fed a high protein diet without pyridoxine, *J. Nutr.*, 104: 287.

Otsuka H., Yokomine R., Nomura S., Takeuchi F., and Shibata Y., 1981, Tryptophan metabolism in the perfused liver isolated from normal and vitamin B6-deficient rats, *J. Aichi Med. Univ. Assoc.*, 9: 226.

Otsuka H., Nomura S., and Shibata Y., 1982, Relation between substrate and enzyme on L-tryptophan metabolism in liver, *J. Aichi Med. Univ. Assoc.*, 10: 270.

Saito K., Sasaki E., Ohta Y., Nagamura Y., and Ishiguro I., 1986, Mode of L-tryptophan uptake into rat hepatocytes via trypsin-sensitive high-affinity transport system, *Biochem. Inter.*, 13: 873.

Saito K., Ohata Y., Nagamura Y., Sasaki E., and Ishiguro I., 1990, Relationship between L-tryptophan uptake and L-tryptophan 2,3-dioxygenase activity in rat hepatocytes, *Biochem. Inter.*, 20: 71.

Salter M., Knowles R.G., and Pogson C.I., 1986, Transport of the aromatic amino acids into isolated rat liver cells, *Biochem. J.*, 233: 499.

Seglen P.O., 1976, Preparation of isolated rat liver cells, *Methods Cell Biol.*, 13: 29.

Takeuchi F. and Shibata Y., 1984, Kynurenine metabolism in vitamin B-6-deficient rat liver after tryptophan injection, *Biochem. J.*, 220: 693.

# 98

# TRANSPORT OF VARIOUS FORMS OF L-TRYPTOPHAN IN THE SMALL INTESTINE OF CHICKS

## Mechanism and Regulation

N. Basova

Institute of Biology
Latvian Academy of Sciences
Salaspils LV-2169, Latvia

## INTRODUCTION

Tryptophan belongs under the irreplacable amino acids, the presence of which is needed in the ration for keeping up the life of organism. The ways of their absorption in mammal intestines were studied by Munck and Rasmussen (1975). They had stated that the entrance of L-tryptophan across the apical membrane of enterocytes was performed by the carriers of neutral amino acids. At its exit across the basolateral membrane a mechanism is used common with L-lysine. The study of mechanism for L-tryptophan absorption has been carried out on chicks in our laboratory as well. At this investigation the active character of L-tryptophan transport process was found out, its stereospecificity was revealed as well as dependence of the presence of sodium and hydrogen ions in the incubation medium and the presence of oxygen in it. A stimulating impact of a number of L-amino acids and carbohydrates was observed on the absorption of this irreplacable amino acid (Kushak et al., 1984; Kushak and Basova, 1987; Basova and Kushak, 1987).

In the recent years it was found out that the absorption of amino acids in the small intestine was performed by sodium-dependent and sodium-independent transport mechanisms. The role of sodium in peptide uptake is rather contradictory.

The present work gives data characterizing the absorption of free L-tryptophan in the small intestine of 1 to 65-day old chicks, as well as of amino acids released at the hydrolysis of glycyl-L-tryptophan (peptide L-tryptophan). To find out the ways of their transport the dynamics of free and peptide amino acid transport was studied at the presence of sodium ions in the incubation medium and at their replacement by potassium ions. Kinetics of L-tryptophan transport under these conditions and the effect of inhibitors on transport processes have also been studied.

*Recent Advances in Tryptophan Research,* edited by Graziella Allegri Filippini et al.
Plenum Press, New York, 1996 

# METHODS

Experiments were performed on the 175 cocks of the Highsex White cross strain, 1-65 days old, fed on a standard mash or diets with increasing amounts of sodium chloride (0%, 0,3%, 1,5%, 3%, 6%). The transport of free amino acid and those released at the hydrolysis of peptide was investigated by the accumulation technique (AMP) (Ugolev et al., 1970). AMP were incubated in 5 ml of substrate (0,24 mM of L-tryptophan or glycyl-L-tryptophan) for 3-60 min at 41° C and constant oxygenation. After incubation the concentration of tryptophan was determi-ned in the intestinal tissue (Messineo and Musarra, 1972).

Studying the role of sodium ion in the absorption of free and peptide L-tryptophan, the incubation was carried out in a medium (Ringer solution) containing sodium (140 mM) or in a solution where sodium was replaced by an equivalent amount of potassium. The concentration (mM) of amino acids in the medium containing the sodium characterizes amount of the total transport flux ($I_{Na}$) and the concentration of amino acid in a potassium-containing medium shows the amount of sodium-independent transport flux ($I_K$). The differences between them ($I_{Na}$ - $I_K$) reflect the amount of sodium-dependent flux. It is supposed that sodium-independent flux also included a diffusion component (Stevens et al., 1982).

Studying the kinetics of transport the AMP were incubated in L-tryptophan solutions of different concentrations (0,16-2,86 mM) under conditions similar to those described above. $K_t$ and V of the transport were estimated according to the least square method.

Some L-amino acids (threonine, valine, α-alanine) in a 100 mM concentration prepared on Ringer solution (pH 7,4) or similar solution where sodium was replaced by potassium were used as modifiers of the transport of free and peptide tryptophan. The statistical significance was calculated by the Student's test. At least 6 chickens were used for each experiment.

# RESULTS

According to the investigation results the absorption of various forms of L-tryptophan in the small intestine of chicks was performed by sodium-dependent and sodium-independent transport mechanisms already the first days after hatching. Intensity of each of them depended on the age of chicken and it differed for the compounds. The sodium-dependent transport is decisive for the transport of free L-tryptophan. The transport of peptide L-tryptophan during the first 15 min of incubation occurred mainly by sodium-independent mechanism, and further on the sodium-dependent mechanism was dominating. So, in case of peptide tryptophan the power of both ways is about the same. Similar data proving the decisive role of sodium-dependent transport of L-tryptophan were obtained in experiments on rabbits (Cooke et al., 1980).

The study of kinetics of L-tryptophan accumulation showed that the replacement of sodium ions in the incubation medium is accompanied by a decrease of $K_t$ from 2,23 to 0,84 mM and that of V from 4,54 to 0,84mM/ min/ cm. Due to the differentiation of sodium-dependent and sodium-independent ways of transport it was possible to state that L-valine, L-threonine and L-α-alanine in 100 mM concentrations caused highest inhibition (78-96%) of the sodium-dependent transport of free and peptide L-tryptophan. The amount of the observed effect was considerably higher than in the case of the total transport flux. Sodium-independent flux of peptide L-tryptophan at the presence of modifiers undergoes no changes, but that of free tryptophan is inhibited only at the presence of L-valine.

Chicks feeding on diets containing increasing amounts of sodium chloride (0%, 0,3%, 1,5%, 3%, 6%) revealed that the transport of various L-tryptophan forms was done mainly in a sodium-dependent way, and it was not related with the level of sodium chloride in the diet.

Thus, sodium-dependent and sodium-independent pathways of transport should be distinguished not only for free amino acids but also for amino acids released at the hydrolysis of peptides. Their analysis is important both for understanding the mechanism of transport of these compounds and the high rates of their absorption.

## REFERENCES

Basova N.A., Kushak R.I., 1987, Absorption of L-tryptophan in the small intestine of chicks, *in*: " Progress in Tryptophan and Serotonin Research, 1986 ", Bender-Joseph-Walter-Steinhart, ed., Walter de Gruyter & Co., Berlin-New York, 175.

Cooke H.J., Phankuche L., Cooke A.R., 1980, Tryptophan transport by isolated newborn rabbit jejunum, *Amer.J.Physiol.*, 239: 6306.

Kushak R., Basova N., Beker V., Feldmane A., 1984, Stimulation of L-tryptophan by di- and polysaccharides in the small intestine of chicks. *Comp.Biochem.Physiol.*, 79A: 185.

Kushak R., Basova N., 1987, Role of exchange transport in the absorption of L-tryptophan in the small intestine of chicks, *Comp.Biochem.Physiol.*, 87A: 951.

Messineo L., Musarra E., 1972, A sensitive spectrophotometric method for the determination of free or bound tryptophan, *Intern.J.Biochem.*, 3: 700.

Munck B.G., Rasmussen S.N., 1975, Characteristics of rat jejunal transport of tryptophan. *Biochim.Biophys.Acta*, 389: 261.

Stevens B.R., Ross H.J., Wright E.M., 1982, Multiple transport pathways for neutral amino acids in rabbit jejunal brush border vesicles, *J.Membr.Biol.*,66: 213.

Ugolev A.M., Zhigure D.R., Nurks J.J., 1970, Accumulating mucosal preparation - a new method of investigation of the initial stages of substance transport across the intestinal wall, *Physiol.Journal*, 56: 1638.

# URINARY EXCRETION OF N[1]-METHYL-2-PYRIDONE-5-CARBOXYLIC ACID AND THE FATE OF REMAINING OF TRIGONELLINE

S. Yuyama and Y. Kawano

Department of Nutrition and Biochemisty
The Institute of Public Health
Tokyo, 108 Japan

## INTRODUCTION

During the development of a procedure for the determination of trigonelline (Tg) in human urine, an organic acid with a similar ultraviolet spectrum was isolated. The compound was found to be N[1]-methyl-2-pyridone-5-carboxylic acid (Tg-2Py) (Lindenblad et al. 1956), which Holman et al. (1950) were unable to detected in normal human urine. Tg-2Py was determined by using high performance thin-layer chromatography (HPTLC), and the level of Tg-2Py excretion was about 1.5 times of Tg or N[1]-methylnicotinamide(MNA) in 9 normal human urine (Yuyama et al, 1991). When Tg was orally administered in 3 human subjects, only 20% of the dose recovered in 24hr urine as unchanged Tg, and about 10% of the dose was converted into Tg-2Py (Yuyama et al. 1991). To explain the fate of missing Tg (60-70%),the possible mechanism for metabolism of Tg are as follows; 1) Tg may be metabolized further, expect Tg-2Py, 2) the ring fission could occur to pyridinium compounds, 3) Tg may be retained in the organ(s).

In the paper, to recomfirm whether Tg is converted into Tg-2Py, the changes of Tg-2Py, Tg and MNA excretions were examined after oral administration of Tg. Also it is indicated that Tg-derivatives, expect Tg-2Py were isolated by using ion-exchanges, aluminium oxide column chromatography, and verified by using ultraviolet, HPTLC, HPLC and infrared absorption spectra, etc.

## METHODS

### I. Subjects and Diet

Nine female volunteers, aged 20-21 years, who had no history of sevious renal, gastrointestinal or metabolic disease, were used in the studies. They were kept at an ordinary

*Recent Advances in Tryptophan Research*, edited by Graziella Allegri Filippini et al.
Plenum Press, New York, 1996 

diet for 3 days before and 2 days during the experimental period. After collecting on 24hr urine (before oral administration of Tg), 50 mg (288 μmole) of Tg was given to each of the 9 subjects and 24hr urine was collected for 2 days. All samples were stored at -20°C untile analysis.

Extraction and Determination of Tg-2Py and Tg from Urine. One-two per cent of 24hr urine was diluted to 50 ml with water. The extraction and determination of Tg-2Py, and Tg and MNA from human urine has been discribed previously (Yuyama et al., 1991, 1989).

## II. Isolation and Characterization of Tg-Derivative in Human Urine

Aliquots of urine corresponding to 3-4% of 24hr urine was diluted to 80 ml with water. The diluted urine was absorbed to column of Dowex 1-X8 (Cl-) (2.5 x 10 cm) and the column was washed with 150 ml of water and then eluted with 300 ml of 0.1N hydrochloric acid. The effluent and washing were discarded. The eluate was evaporated to dryness under pressure in a water bath at 50°C. The residue was dissolved in 20 ml of water. The sample was applied to column of Dowex 50W-X12 (H+) (1.5 x 10 cm) and then washed with 150 ml of water. The effluent and washing was evaporated to dryness, and then dissolved in 20 ml of water. This sample was purified moreover with aluminium oxide column (1 x 10 cm, 10 g) chromatography.

# RESULTS AND DISCUSSION

## I. Excretion of Tg-2Py, Tg and MNA by Nine Healthy Subjects before and after Oral Administration of 50 mg (288 μmole) of Tg

The data describing the excretion of Tg-2Py, Tg and MNA by 9 healthy subjects before and after the andministration of Tg are summarized in Table 1. The daily excretion of Tg-2Py in this group (before the administration) ranged between 28.8 and 41.0 μmole with an average of 35.7 μmole. The excretion of Tg and MNA in urine ranged between 20.7 and 28.3, and 19.8 and 26.5 μmole, respectively. The range of values for Tg-2Py in the group of subjects is of the same order of magunited as that found by Lindenblad et al. (1956) in 3 male adults (3-6 mg, 19.6-39.1 μmole). The excretion of Tg in 9 subjects given the ordinary diet was 26.8 μmole and the level was almost identical to that of MNA excretion. The daily excretion of Tg-2Py in the administered Tg group for 1 day increased on the first day and decrease on the second day. The excretion of Tg-2Py in this group returned almost to the value of the ordinary diet after 2 days. About 10% of the dose of Tg was excreted in the urine as Tg-2Py within 24hr. The excretion of Tg in this group was 86.4 μmole per day, and about

**Table 1.** Urinary excretion of various metabolites of nicotinic acid before and after oral administration of 50 mg (288 μmole) of trigonelline

| Derivatives of nicotinic acid | Subjects | Urinary excretion (μmole/day) Before | After | % of dose excreted in the urine |
|---|---|---|---|---|
| Tg-2Py | ( 9 ) | 35.7 ± 4.2 | 64.5 ± 6.1* | 10.0 ± 0.8 |
| Tg | ( 9 ) | 26.8 ± 1.9 | 86.4 ± 3.4* | 20.7 ± 1.2 |
| MNA | ( 9 ) | 24.9 ± 2.7 | 24.6 ± 3.1 | — |

Percent of the dose was calculated as follows; (After value – Before value/dose) × 100.
*Significantly different from corresponding ordinary diet (Student's t test):P<0.01.

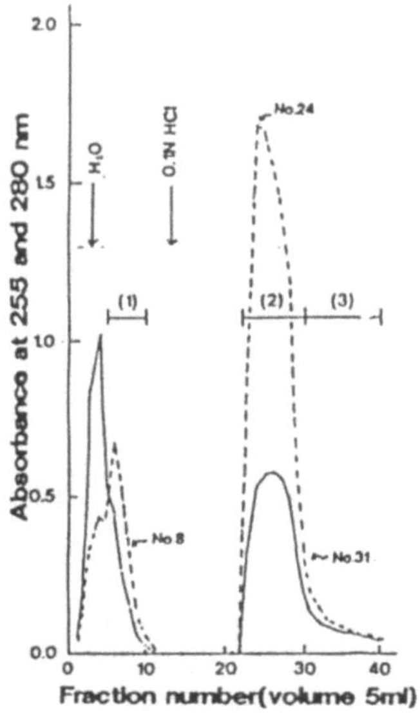


**Figure 1.** Elution pattern of aluminium oxide column 255 nm (—), 280 nm (- - -).

21% of the dose was recovered in the 24hr urine. No increase of MNA excretion in the urine was observed.

Therefore it was suggested that Tg is neither directly converted into MNA (amidation) nor cicuitously converted into MNA (demethylation). Since the actual conversion of Tg into Tg-2Py was about 10% of the dose and the recovery of unchanged Tg was very low, we investigated the fate of missing Tg (60-70%).

## II. Isolation and Characterization from Human Urine

*1. Isolation from Human Urine.* After aliquots of urine was treated with Dowex 1 and Dowex 50W columns, the samle was purified moreover with a Aluminium oxide column (1 x 10 cm). Fig. 1 shows the elution- ans ultraviolt-pattern. The effluent and washing was collected and then the absorbed compounds were eluted with 0.1N hydrochloric acid. Fraction I ( No. 8) and Fraction II (No. 24) exhibited a maximum at 230 and 290 nm, and a maximum at 235 and 282.5 nm, respectively.

*2. Characterization of Acid and Alkaline of the Compound Isolated from Urine.* The absorption spectra of the fraction II in $H_2O$ exhibited a maximum at 235 and 290 nm; in 1N HCI, a maximum at 235 and 282.5 nm and in 1N NaOH, a maximum at 295 nm and at the 235 nm disapperared (Fig 2). The compound is quite stable to neutral and acid conditions, and is very soluble in $H_2O$ and slightly in methanol and ethanol.

*3. HPTLC Behavior.* The isolated compound disappeared for alkaline solvent system and not deveroped for acidic solvent system. For the neutral solvent system (methanol: benzen: butanol: water, v/v/v/v, 4:1:1:1), the Rf value was 0.25. The Rf value (0.25) area was scraped off and then extracted with $H_2O$. The absorption spectra of the compound exhibited a maximum at 235 and 290 nm (Fig. 3).

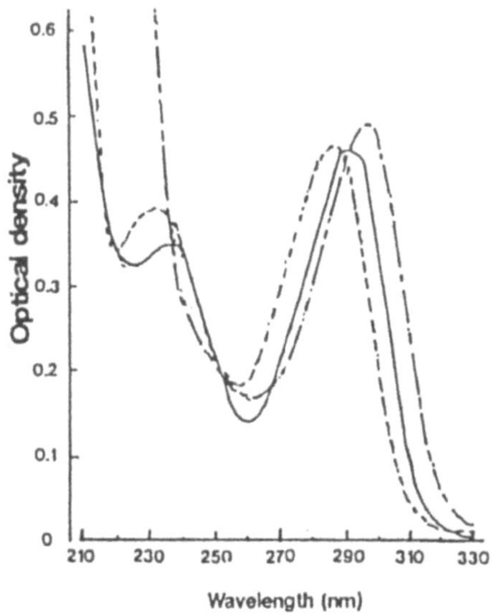


**Figure 2.** Characterization of acid and alkaline of fraction II. Water (—), 1 N HCl (- - -), 1 N NaOH (– - –).

*4. HPLC Behavior.* The fraction I (A) and fraction II (B) shows the elution profiles obtained when the material isolated from urine was applied to 7-ODS-L column (150 x4.6 mm). The fraction II exhibited one peak (Fig. 4).

*5. Infrared Absorption Spectrum.* The fraction II (No. 22-30) was evaporated to dryness and then the residue was treated with a IR. The infrared absorption spectra of this compound showed a medium absorption band at $1337^{-1}$ cm and a strong absorption band at $1665^{-1}$ cm. These two band at $1337^{-1}$ cm and $1665^{-1}$ cm are characterized of $=N\text{-}CH_3$ functional group and the carboxyl or ketone functional group, respectively (Fig. 5).

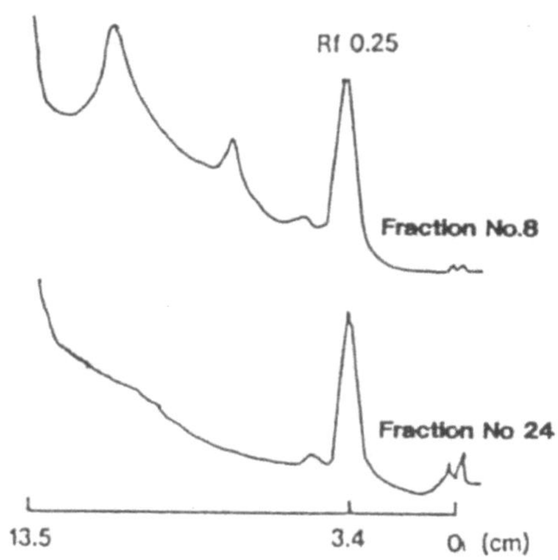


**Figure 3.** HPTLC scanning pattern.

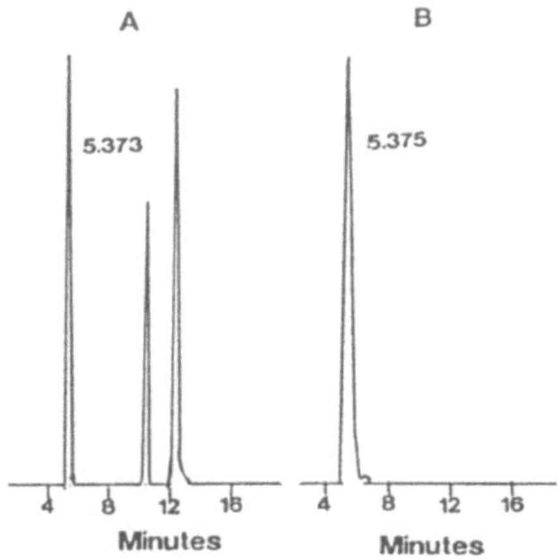


**Figure 4.** Chromatograms of fractions I and II.

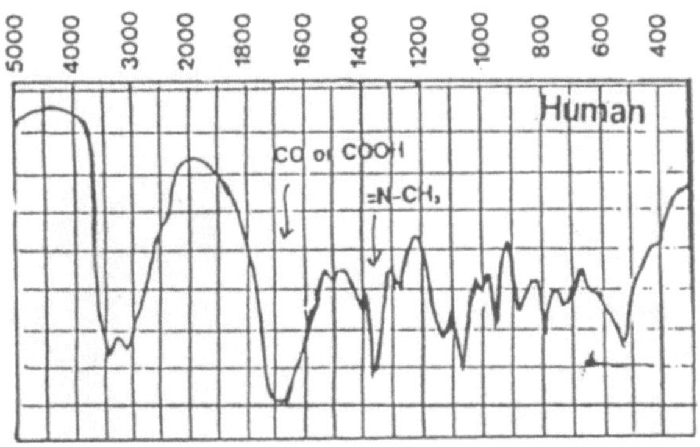


**Figure 5.** Infrared spectrum in KBr of fraction II.

From these results, nicotinic acid can be converted into Tg and Tg-2Py (Yuyama et al. 1991) and Tg into Tg-2Py. These changes do not appear to be reversible, and it is interesting to note that the methylated compound (Tg) can be oxidized to a pyridone. Since the recovery of unchanged Tg was very low, we investigated the fate of missing Tg. As results, it is indicated that Tg-derivative was excreted in humen urine.

## REFERENCES

Lindenblad G.E., Kaihara M. and Price J.M., 1956, The occurrence of N-methyl-2-pyridone-5-carboxylic acid and its glycine conjugate in normal human urine. *J. Biol. Chem.*, 219: 893.

Holman W.I.M, and Lange D.J., 1950, Metabolism of nicotinic acid and related compounds by humans. *Nature*. 165: 604.

Yuyama, S. and Suzuki, T. ,1989, *Bull. Inst. Public Health*, 38: 11.

Yuyama S, and Suzuki T., 1991, The excretion of N[1]-methyl-2-pyridone-5-carboxylic acid and related compounds in human subjects after oral administration of nicotinic acid, trigonelline and N[1]-methyl-2-pyridone-5-carboxylic acid. *Adv. Exp. Med. Biol.*, 294: 475.

# *TRYPTOPHAN IN PEPTIDES AND PROTEINS*

# THE ROLES OF TRYPTOPHAN RESIDUES IN THE STRUCTURE, FUNCTION, AND FOLDING OF THE GRAMICIDIN TRANSMEMBRANE ION CHANNEL

B. A. Wallace

Department of Crystallography
Birkbeck College
University of London
London WC1E 7HX, United Kingdom

Gramicidin is a polypeptide antibiotic which forms ion channels in phospholipid membranes. Four out if its 15 residues are tryptophans, and hence it is not surprising that they play major role in its structure, function and folding. Gramicidin adopts two major types of ion-conducting structures that have very different conformational and functional properties; these two structures are known as the "pore" and the "channel" forms. The pore consists of two intertwined polypeptide strands which form an antiparallel double helix that spans the membrane, while the channel is formed by two single helices that abut each other in the centre of the membrane forming a helical dimer. However, in addition to having entirely different backbone folds, the pore and the channel differ distinctly in the conformation and orientation of their tryptophan side chains: in the pore, all tryptophans lie essentially parallel to the helix axis, creating a smooth, uniform cylindrical shape to the molecule, while in the channel the tryptophan side chains lie nearly perpendicular to the helical axis. Because the tryptophans are located near the C-termini of the two monomers, this means the channel has a head-to-head double conical shape.

Chemical modification studies have shown that the tryptophan side chains are crucial to gramicidin's channel activity and replacement, even with other aromatic (phenylalanine) or hydrophobic (leucine) side chains, vastly alters its conductance activity, even though none of the side chains are directly involved in ion binding. Hence the tryptophans must be exerting a long-distance effect on its functional properties.

Finally, the tryptophan residues appear to be crucial in the process of insertion and folding of the channels in membranes. The conversion of the pore to the channel conformation can be dissected into several distinct steps by choice of physical conditions. The initial step in the process appears to be dissociation into monomers and realignment of the tryptophan residues from the parallel to the perpendicular orientation upon binding lipid molecules; this is then followed by the refolding of the polypeptide backbone into the channel

*Recent Advances in Tryptophan Research*, edited by Graziella Allegri Filippini et al.
Plenum Press, New York, 1996

secondary structure in bilayers. It thus appears that the tryptophan/lipid interaction may be an important feature in the folding process.

In summary, the tryptophan residues in gramicidin appear to be crucial for both its structural and functional characteristics as a transmembrane ion channel.

## 1. INTRODUCTION

Gramicidin A is a polypeptide antibiotic that functions as a cation-specific transmembrane channel in phospholipid bilayers (Wallace, 1990; Woolley and Wallace, 1992a). Active conducting forms of the molecule are homodimeric in nature. Each monomer consists of 15 amino acids with alternating L- and D- configurations. The N-terminus is blocked with a formyl group and the C-terminus with an ethanolamine group. Gramicidin A has the following sequence (Sarges and Witkop, 1965a):

Formyl-L-Val-Gly-L-Ala-D-Leu-L-Ala-D-Val-L-Val-D-Val-L-Trp-D-Leu-L-Trp-D-Leu-L-Trp-
D-Leu-L-Trp-ethanolamine

It is obvious that gramicidin is exceedingly rich in hydrophobic amino acids, with tryptophans comprising more than 25% of the molecule, all clustered in the C-terminal half of the polypeptide. It is not surprising, therefore, that the tryptophans play major roles in the structure and functioning of the molecule, as well as being involved in the process of polypeptide folding and lipid interactions.

## 2. THE ROLE OF TRYPTOPHAN IN GRAMICIDIN STRUCTURES

Gramicidin exhibits considerable conformational polymorphism, with different conformations having different functional properties (Wallace, 1991). In phospholipid membranes composed of lipids with saturated fatty acid chains, the principal form is a helical dimer (Figure 1a) known as the "channel" form; in membranes made of lipids with unsaturated fatty acid chains and in organic solvents, the principal forms are double helices; in the presence of ions, a conducting form of the double helix exists which has been designated the "pore" form (Figure 1b); in the absence of ions, a different type of double helix is found, which is a "non-conducting" species (Figure 1c) because its central pore does not have a sufficiently large-sized lumen to allow ions to pass through (Smart et al., 1993). The pore and channel forms have different conductance properties: the pore produces long-lived channels because its monomers are held together by 28 hydrogen bonds, whereas the channel produces shorter-lived channels because the interface between its monomers consists of only six hydrogen bonds. The channel form is found in dynamic equilibrium with an inactive monomeric species, and its opening and closing occurs as the two monomers come together and then separate.

The pore and non-conducting double helical forms have been characterized at high resolution by X-ray crystallography, and their backbone conformations and the precise location of their tryptophan residues are well defined.

The pore is an antiparallel beta double helix with 6.4 residues per turn (Wallace and Ravikumar, 1988) (Figure 1b). It is a 26 A long tube with a central pore that has a maximum diameter of approximately 4.4 A (Smart et al., 1993) and a helical pitch of 10.4 A/turn. The lumen of the pore is formed by the polypeptide backbone atoms, with the carbonyl oxygens forming the binding sites for the cations. Because gramicidin consists of alternating L- and D- amino acids, all of its side chains are located on the external surface of the pore. The

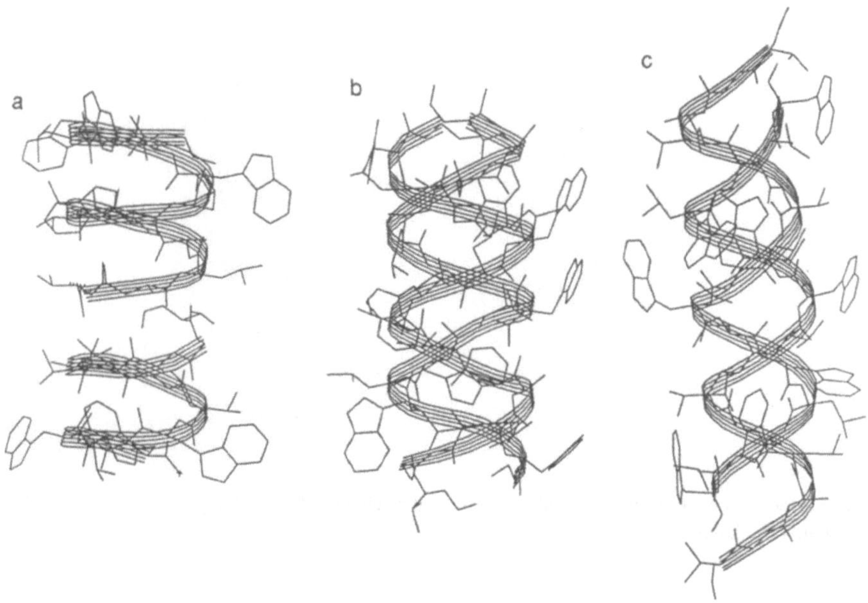


**Figure 1.** Molecular structures of gramicidin: a) an NMR structure of the helical dimer (channel) form (from Ketchem et al., 1993), b) the X-ray structure of the ion-containing double helix (pore) form (from Wallace and Ravikumar, 1988), and c) the X-ray structure of the ion-free double helix (non-conducting) form (from Langs, 1988).

planes of the tryptophan side chains lie more or less along the direction of the helix axis. Because the two polypeptide chains in the double helix run antiparallel to each other with their C-termini being located at the ends of the pore, each chain spans the length of the tube. As a result the tryptophans are spread along the entire cylinder, and they must therefore contact both the headgroups and fatty acid chains of the lipids.

Earlier model structures for the pore had suggested that the tryptophan residues would have an orientation perpendicular to the helix axis, and parallel to each other, thus producing aromatic ring stacking interactions. It can clearly be seen from the crystal structure that is not the case, and in retrospect, a parallel orientation for the tryptophans relative to the helix axis is more sensible in terms of packing into a membrane. As a result of this sort of tryptophan orientation, the pore structure has a relatively smooth cylindrical surface (Figure 2a) which means it could pack closely against the nearly cylindrical lipid fatty acid chains in a membrane. A perpendicular tryptophan orientation would have produced a "lumpy" surface that would leave gaps when packed against lipids, thereby creating membranes which are leaky, a phenomenon not observed for membranes containing this molecule. The question can then be asked if the tryptophan positions in the crystal structure, though precisely defined, are representative of their orientation in solution. Circular dichroism (CD) spectroscopic studies of the pore form (Woolley and Wallace, 1992b) do not provide any evidence for exciton interactions (Figure 3), hence mitigating against the possibility of stacked aromatic rings, a result that is at least consistent with the X-ray results.

The non-conducting form is also an antiparallel beta helix, but in this case, the number of residues per turn is 5.6 instead of the 6.4 found in the pore. This produces a longer (32 A) but narrower cylinder (Langs, 1988) (Figure 1c). Like the pore form, its lumen is formed from the polypeptide backbone, but in this case, the central hole is much more irregular in diameter (Smart et al., 1993), and at some places it is too narrow to permit an ion to be

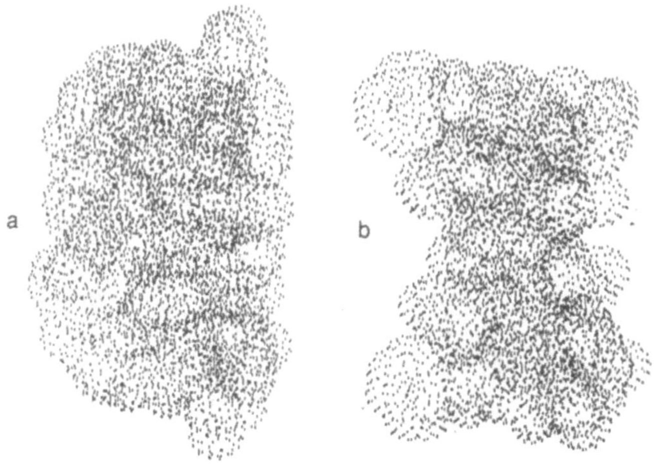


**Figure 2.** Surface features of the gramicidin a) pore and b) channel forms. The former is a relatively smooth cylinder, whilst the latter has a more irregular surface.

accommodated. Nevertheless, its tryptophan side chains have a similar orientation to those in the pore, lying nearly parallel to the helix axis. Another crystal form (from a different solvent), also of the non-conducting form, has been reported (Langs et al., 1991). It differs somewhat from the structure shown in Figure 1c in that the diameter of the central pore is more regular, and the tryptophans are oriented at greater angles relative to the helix axis. This difference could be attributed to different packing forces in the two different crystal forms which influence the tryptophan side chains that lie at the interface between asymmetric units in the crystal. The tryptophans, in turn, could influence the folding of the polypeptide backbone, making it (and the lumen diameter) more or less irregular depending on their orientations.

The channel form has been characterized by NMR spectroscopy both in detergent micelles (Arseniev et al., 1985) and in the solid state (Ketchem et al., 1993). The former studies have produced a model for the folding of the polypeptide backbone, whilst the latter have provided further details on the polypeptide backbone as well as information on the orientation and flexibility of the tryptophan side chains. The channel is composed of two single right-handed beta helices joined by hydrogen-bonding between their N-termini which

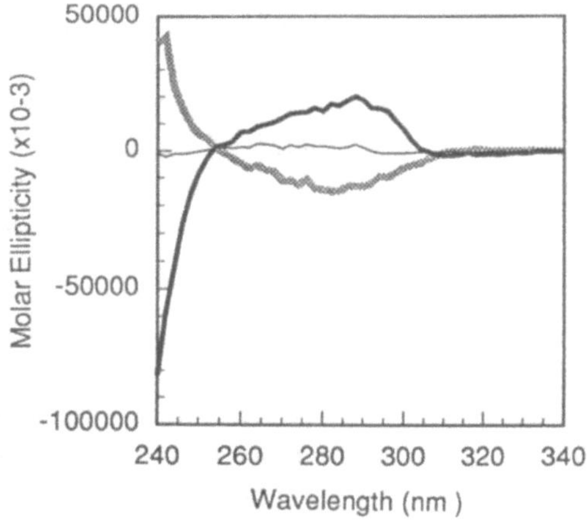


**Figure 3.** Near UV circular dichroism spectra of gramicidin showing the aromatic regions for the double helical (thick line) and helical dimer forms (grey line) of gramicidin (from Woolley and Wallace, 1992), and the monomeric form found in trifluoroethanol (thin line) (from Woolley et al., 1992).

are located in the centre of the membrane; hence this form is described as a helical dimer. (Figure 1a). The polypeptide backbones have 6.3 residues per turn, but the pitch of this type of helix is roughly half that of the double helix. The two monomers together make a channel of length approximate 27 A, with a internal maximum diameter of about 4.3 A. Although topologically different from the double helical forms, the helical dimer also has its tryptophans on the outer surface of the channel, but because of the backbone fold, the tryptophans tend to be concentrated at the two ends of the channel, mostly in contact with the lipid headgroup interfacial region of the membrane. In the channel form, there is considerable flexibility of the tryptophan side chains, but on average they all appear to lie at approximately the same angle relative to the helix axis, (closer to the helix normal than in the double helical forms). However, there is no evidence for stacking interactions between tryptophans. Earlier models for the channel (Venkatachalam and Urry, 1983) had suggested substantial stacking between Trp-9 and Trp-15, but this is not consistent with the NMR data, nor with fluorescence data for the channel in fluid-state lipids (Scarlata, 1988).

The outer diameter of the channel form is considerably less regular (Figure 2b) and significantly larger than the pore, as a consequence of the tryptophan orientations. However, in this case, the substantial flexibility of the side chains may mean that efficient packing against the lipids could be accommodated by the side chain and membrane dynamics (Crouzy et al., 1992).

That the arrangements of the tryptophans in the helical dimer and double helical structures are indeed different can be seen from CD spectra (Woolley and Wallace, 1992b) (Figure 3) of their aromatic regions which are not only different in shape and magnitude but also in sign (positive for the double helix, negative for the helical dimer), indicative of the presence of different conformational states. For the monomeric species in trifluoroethanol, the aromatic region is featureless with a magnitude near zero (Figure 3).

In summary, gramicidin adopts several distinctly different conformations, depending on its environment (Wallace, 1983). The tryptophan side chain orientations also differ considerably in these different forms. The orientations of the tryptophans appear to have important implications for polypeptide/lipid packing interactions and surface features.

## 3. THE ROLE OF TRYPTOPHAN IN GRAMICIDIN FUNCTION

Although the tryptophans, being located at the periphery of both the pore and channel, are not directly involved in the ion transport process, they appear to have considerable influence on the conductance properties of gramicidin. There are three naturally occurring variants of gramicidin which differ in the nature of the amino acid at position 11. In gramicidin A, it is tryptophan, in gramicidin B, it is phenylalanine and in gramicidin C it is tyrosine (Sarges and Witkop, 1965b, 1965c). These occur in the naturally isolated mixture in a ratio of approximately 80:5:15 (Weinstein et al., 1980) and can be separated by chromatographic means, thus affording the opportunity to examine the influence of the tryptophan at position 11 on the conductance properties of the channel. Conductance measurements have shown clear effects of replacement of this tryptophan, even by the other aromatic residues, on both mean channel lifetime and single channel conductance (Sawyer et al., 1990). Hence, even though it is at a substantial distance from the ion binding sites, this tryptophan is clearly influential in the transport process. This is likely due to its influence on the polypeptide chain flexibility and the conformational constraints it imposes which affect the geometry of the carbonyl binding sites and the diameter of the channel lumen, as described in the previous section.

Because gramicidin A is a relatively small polypeptide, it has also been possible to systematically vary its sequence by peptide synthesis techniques, in order to examine the role of each type of amino acid at various positions along the chain. Most dramatic effects have been observed when all four tryptophans are changed, even to the "conservative replacement" of phenylalanines (Becker et al., 1991). This reduces conductivity by a factor of twenty. Replacement of individual tryptophans one-by-one has an incremental effect on the conductance, and it has been suggested that is may indicate all tryptophans are in roughly the same orientation. When the tryptophans are replaced by tyrosines (which have similar dipole moments), however, the conductance effects are much less (Heitz et al., 1988). It has thus been suggested that the tryptophan side chains must have orientations which produce dipoles whose primary component is axial, with their negative end towards the channel centre (middle of the bilayer) and that the conductance effect is one produced by the lowering of the energy of central barrier of the channel (Hu et al., 1993). Other alterations, such as formylation of the tryptophan indole rings (Killian et al., 1985) or tryptophan photolysis (Jones et al., 1986), have been shown to produce inactive channels, and to interfere with the association of the polypeptide with lipids.

Thus it can be seen that the tryptophans in gramicidin have substantial, and additive, functional influences at all the naturally-occurring positions along the entire length of the molecule. Their conformational states, mobility, and indole ring dipoles play important roles in the electrical characteristics of the channel.

## 4. THE ROLE OF TRYPTOPHAN IN GRAMICIDIN FOLDING

As discussed above, gramicidin adopts a number of different conformations and can convert between these conformers, depending on its environment. In membranes formed from saturated-chain phospholipids, there is a dynamic equilibrium which strongly favours the channel form. The double helical forms can be converted to the helical dimer form upon exposure to lipids containing saturated fatty acid chains, but the rate and manner in which the conversion occurs depend on the solvent, ionic strength, temperature, and other conditions present (Wallace, 1984). Since the tryptophan side chains are in direct contact with lipid, they can respond to the bilayer properties (tension, thickness, fluidity), thereby modulating not only function but also the folding equilibrium. The tryptophan residues have been shown to be essential for the membrane-modifying properties of gramicidin, including effects on lipid order, phase properties and membrane fusion (Tournois et al., 1990).

The process of refolding from the pore form to the channel form is a dramatic one, involving the breakage of 28 hydrogen bonds and the formation of 28 new hydrogen bonds, along with the topological refolding of the polypeptide backbones and the reorientation of the tryptophan side chains. The tryptophan residues appear to play a major role in this transformation. This process has been modeled computationally (Smart, 1994) and dissected experimentally in a way which permits examination of the various parts of the process separately (Woolley et al., 1992).

It was shown by CD spectroscopy (Woolley et al., 1992) that it is possible to first re-orient the tryptophans merely by association with lipids, prior to refolding the polypeptide backbone, and then by organizing the lipids into membranes, to independently refold the polypeptide chains. The tryptophan reorientations occur upon initial association with lipids prior to organization into bilayers. This process is shown schematically in Figure 4.

Studies on synthetic derivatives of gramicidin also suggest the importance of the tryptophans in the initial association with lipids. If the tryptophans are modified or replaced with other aromatic side chains, this association is perturbed, and the ultimate end-product helical dimer form is not obtained (Killian, 1992).

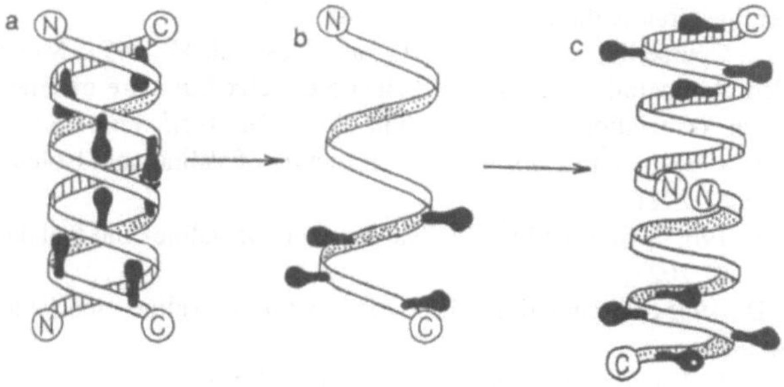


**Figure 4.** Schematic diagram depicting conversion of the pore to the channel form. In the first step, the tryptophans realign from an orientation parallel to the helix axis in the double helix (a) to one perpendicular to the helix axis as the monomers separate upon association with lipid (b); in the second step, the polypeptide backbones reassociate and fold into the helical dimer when the lipids reorganize to form bilayers (c). The lollipop-like shapes represent the tryptophan residues.

In summary, the high content of tryptophan residues in gramicidin means this amino acid plays a particularly important role in all aspects of the structure, function, and dynamic equilibria of this ion-conducting antibiotic molecule.

## ACKNOWLEDGMENTS

I thank past and present members of my group, especially Drew Woolley and Declan Doyle, for helpful discussions, and Bob Janes for drawing Figure 4. This work was supported by a project and Biomolecular Sciences program grant from the BBSRC.

## REFERENCES

Arseniev, A.S., Barsukov, I.L., Bystrov, V.F., Lomize, A.L., and Ovchinnikov, Y.A., 1985, $^{1}$H-NMR Study of Gramicidin A Transmembrane Ion Channel: Head-to-Head Right-Handed, Single Stranded Helices, *FEBS Lett.* 186:168.

Becker, M.D., Greathouse, D.V., Koeppe, R.E. II, and Andersen, O.S., 1991, Amino Acid Sequence Modulation of Gramicidin Channel Function: Effects of Tryptophan-to-Phenylalanine Substitutions on the Single-Channel Conductance and Duration, *Biochemistry* 30:8830.

Crouzy, S., Janes, R.W., Roux, B., and Wallace, B.A., 1992, Modeling Gramicidin/Lipid Interactions in a Crystalline Complex. Report of the CECAM Workshop on Molecular Modeling of Membrane Proteins and Lipids, p. 1-4.

Heitz, F., Daumas, P., van Mau, N., Lazaro, R., Trudelle, Y., Etchebest, C., and Pullman, A., 1988, Linear Gramicidins: Influence of the Nature of the Aromatic Side Chains on the Channel Conductance, in: "Transport Through Membranes: Carriers, Channels, and Pumps," Pullman, A., Jortner, J., and Pullman, B., eds., p. 147-175.

Hu, W., Lee, K.C., and Cross, T.A., 1993, Tryptophans in Membrane Proteins - Indole Ring Orientations and Functional Implications in the Gramicidin Channel, *Biochemistry* 32:7035.

Jones, D., Hayon, E., and Busath, D., 1986, Tryptophan Photolysis is Responsible for Gramicidin Channel Inactivation by Ultraviolet Light, *Biochim. Biophys. Acta* 861:62.

Ketchem, R.R., Hu, W., and Cross, T.A., 1993, High-Resolution Conformation of Gramicidin A in a Lipid Bilayer by Solid-State NMR, *Science* 261:1457.

Killian, J.A., 1992, Gramicidin and Gramicidin-Lipid Interactions, *Biochim. Biophys. Acta* 1113:391..

Killian, J.A., Timmermans, J.W., Keur, S., and de Kruijff, B., 1985, The Tryptophans of Gramicidin are Essential for the Lipid Structure Modulating Effect of the Peptide, *Biochim. Biophys. Acta* 820:154..

Langs, D.A., 1988, Three-Dimensional Structure at 0.86 A of the Uncomplexed Form of the Transmembrane Ion Channel Peptide Gramicidin A, *Science* 241:188..

Langs, D.A., Smith, G.D., Courseille, C., Precigoux, G., and Hospital, M., 1991, Monoclinic Uncomplexed Double-Stranded, Antiparallel, Left-Handed Beta 5.6 Helix Structure of Gramicidin A - Alternate Patterns of Helical Association and Deformation, *Proc. Nat. Acad. Sci. (USA)* 88:5345.

Sarges, R. and Witkop, B., 1965a, Gramicidin A. V. The Structure of Valine- and Isoleucine-Gramicidin A, *J. Amer. Chem. Soc.* 87:2011..

Sarges, R. and Witkop, B., 1965b, Gramicidin. VII. The Structure of Valine- and Isoleucine-Gramicidin B, *J. Amer. Chem. Soc.* 87:2027..

Sarges, R. and Witkop, B., 1965c, Gramicidin. VIII. The Structure of Valine- and Isoleucine-Gramicidin C, *Biochemistry* 4:2491.

Sawyer, D.B., Williams, L.P., Whaley, W.L., Koeppe, R.E. II, and Andersen, O.S., 1990, Gramicidins A, B, and C Form Structurally Equivalent Ion Channels, *Biophysical J.* 58:1207.

Scarlata, S.F., 1988, The Effects of Viscosity on Gramicidin Tryptophan Rotational motion, *Biophysical J.* 54:1149..

Smart, O.S., 1994, A New Method to Calculate Reaction Paths for Conformational Transitions of Large Molecules, *Chem. Phys. Lett.* 222:503.

Smart, O.S., Goodfellow, J.M., and Wallace, B.A., 1993, The Pore Dimensions of Gramicidin A, *Biophysical J.* 65:2455.

Tournois, H., Fabrie, C.H.J.P., Burger, K.N.J., Mandersloot, J., Hilgers, P., van Dalen, H., de Gier, J., and De Kruijff, B., 1990, Gramicidin A Induced Fusion of Large Unilamellar Dioleoylphosphatidylcholine Vesicles and its Relation to the Induction of Type II Nonbilayer Structures, *Biochemistry* 19:8297.

Venkatachalam, C.M. and Urry, D.W., 1983, Theoretical Conformational Analysis of the Gramicidin A Transmembrane Channel. 1. Helical Sense and Energetics of Head-to-Head Dimerization, *J. Computational Chemistry* 4:461.

Wallace, B.A., 1983, Gramicidin Adopts Distinctly Different Conformations in Organic Solvents and in Membranes, *Biopolymers* 22:397.

Wallace, B.A., 1984, Ion-Bound Forms of the Gramicidin A Transmembrane Channel, *Biophysical J.* 45:114.

Wallace, B.A., 1990, Gramicidin Channels and Pores, *Ann. Rev. Biophysics and Biophysical Chemistry* 19:127.

Wallace, B.A., 1991, Alternate Folding Motifs for Gramicidin: Crystallographic and Spectroscopic Analysis of Polymorphism, in: "Conformations and Forces in Protein Folding," Nall, B.T. and Dill, K.A., eds., p. 188-195.

Wallace, B.A. and Ravikumar, K., 1988, The Gramicidin Pore: Crystal Structure of a Cesium Complex, *Science* 241:182.

Weinstein, S., Wallace, B.A., Morrow, J., and Veatch, W.R., 1980, Conformation of Gramicidin A Transmembrane Channel: A 13C NMR Study of 13C-enriched Gramicidin in Phosphatidylcholine Vesicles, *J. Mol. Biol.* 143:1.

Woolley, G.A. and Wallace, B.A., 1992a, Model Ion Channels: Gramicidin and Alamethicin, *J. Memb. Biol.* 129:109.

Woolley, G.A. and Wallace, B.A., 1992b, Circular Dichroism Studies of Tryptophan Residues in Gramicidin, in: "Peptides Chemistry and Biology," Smith, J.A. and Rivier, J.E., eds., p. 247-249.

Woolley, G.A., Dunn, A., and Wallace, B.A., 1992, Gramicidin-Lipid Interactions Induce Specific Tryptophan Side-Chain Conformations, *Biochemical Society Transactions 20:864.*

# SYNTHETIC APPROACH TO NON-ISOSTERIC PEPTIDOMIMETICS RELATED TO HUMAN ENDOTHELIN

F. D'Angeli, P. Marchetti, R. Rondanin, S. Salvadori, and G. Balboni

Department of Pharmaceutical Sciences
University of Ferrara
44100-Ferrara, Italy

In the last decades, a number of biochemical transmitters or diversely active molecules proved to consist of peptides, fragments from those peptides proved to be endowed with authentic biological activity(ies), and single aminoacids were identified as active sites interacting with pertinent receptors. Structure-activity studies stand on the synthesis and screening of "pseudopeptides" where i) one (or more) peptide (CONH) group is substituted by an isosteric group such as methyleneamine ($CH_2NH$), thioamide (CSNH), ketomethylene ($COCH_2$), retroinverso (NHCO), imidazoline, or ii) an aminodicarboxyl unit is present as in ACE inhibitors (Wu et al., 1985) or ANF potentiating factors (Haslanger et al., 1989), or iii) the structure and/or number of side chains is varied. Eventually altered peptide structures (peptidomimetics) display agonist or antagonist activity, and/or resist enzymatic attack (Gilbert et al., 1995).

As part of our research on the stereochemistry of substitution reactions in 2-haloamides, we revisited Emil Fisher's pioneer work on peptide synthesis, consisting of three consecutive reactions: i) diazotization-bromination of natural aminoacids to 2-bromoacids, ii) formation of 2-bromoacylaminoacids, iii) reaction with ammonia. However, we extended that rationale to reactions of bromoacid derivatives with amines, sugars, or aminoacids, and found conditions suitable to the stereoselective synthesis of aminoamides (D'Angeli et al., 1991), ethers or glycosides (Moftakhari et al., 1994), and depsipeptides (Cavicchioni et al., 1988). Our present efforts deal with the synthesis of 2-bromoacylaminoacid derivatives and their reaction with the free aminogroup of a suitably protected aminoacid. The alkylation step provides a secondary amino group common to two *alfa* carbons, in a built-in aminodicarboxylicacid amide structure (3-azaglutaramide; 3-AGA) and, eventually, in a diamide: the aminoacid side chains are linked to C2 and C4 of 3-AGA. Whereas ACE inhibitors are obtained as diastereomeric mixtures in aminative-reduction reactions and the ANF potentiating factors in substitution reaction with inversion of configuration, the NH function in 3-AGA arises with overall major retention of configuration in the two substitutions, the latter reaction being promoted by $Ag_2O$ (see above and Experimental). The -NH- is *non-isosteric* with a -CONH- group, it is shorter and more basic. It introduces profound conformational implications, as well as chemical changes, *e.g.* it is not prone to hydrolytic attack.

*Recent Advances in Tryptophan Research*, edited by Graziella Allegri Filippini et al.
Plenum Press, New York, 1996

More or less extensively modified peptidomimetics may be obtained by combining the alkylation step with peptide bond formation and other reactions. Non-isosteric Ψ(NH) modified dermorphine and deltorphine peptides, where N-terminal tyrosine was turned into carboxy terminal tyrosine, showed affinity to opioid receptors and antagonistic activity (D'Angeli et al., 1993; Balboni et al., 1995).

James et al. (1994) reported the final x-ray structure of human endothelin and stressed that carboxy terminal hexapeptide ET-1 (16-21), crucial for receptor binding and vasoactivity, is helical in crystal structure, but not in any of the NMR structures. The (16-21) sequence (Doherty, 1992) has been altered in a variety of ways, and antagonist analogues were obtained (Cody et al., 1992; 1995). We considered ET-1 (16-21) (His-Leu-Asp-Ile-Ile-Trp) expedient for testing our methodology and for providing new tryptophan peptidomimetics, possibly with a variety of alterations. Tryptophan, however, challenged our synthetic protocols that require, for maximum performance, that an aminoacid enters the modification either with its amino group or a Cα-Br function originated from the C-$NH_2$. Infact, the presence of the indole moiety hampers the diazotization-bromination (Gortatowski and Armstrong, 1957). However, the free amino-group of tryptophan ester (**1**) may behave as a nucleophile in the following two ways: i) in reactions with a 2-bromoamide (**2**) (**Chart 1**), to yield the 3-azaglutaroyl isoleucine block (**3**), where the amino group of tryptophan is now common to tryptophan itself and isoleucine; ii) to make a peptide link between **1** and an activated 2-bromoacid, resulting in a tryptophan bromoamide (**4**) (**Chart 2**). Compound **4**, in turn, reacts with the free amino group of isoleucine benzyl ester (**5**) to yield the trimeric sequence **6**, where the amino group of **5** is common to the two isoleucine units.

The trimeric sequence **3**, bearing the Ψ(NH) modification at position 20-21, was linked to the tripeptide unit H-Asp(OtBu)-Leu-Phe-OMe, through classic steps of peptide synthesis.

The hexameric modified sequence **7a** (Xaa = Phe) (**Fig. 1**) was obtained, having the expected $M^+$ (m/z) and yielding, by routine aminoacid analysis, only Ile, Asp, Leu, and Phe.

By the same rationale, we are carrying out the synthesis of modified peptides **7b,c**, **8** and **9** (**Fig. 2, 3**); the aminoacid at position 16 in compound **7-9** will be Phe, Trp, or His, the non-isosteric NH link varying from position 20-21 to position 19-20 or 18-19. All relevant products will be tested as ET agonists or antagonists, and for other activities (Ragazzi et al., unpuplished; Regoli et al., unpublished).

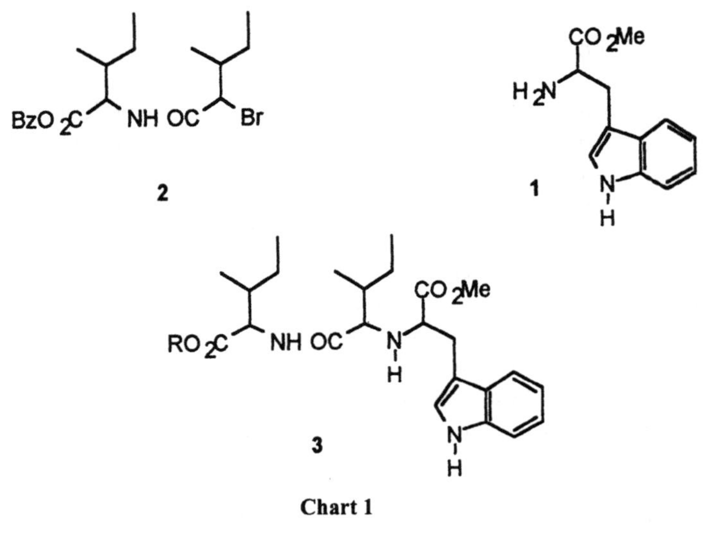


**Chart 1**

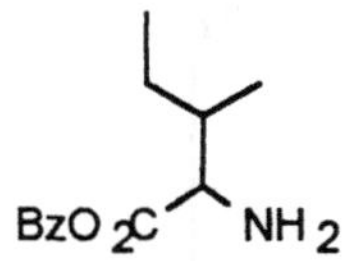

5

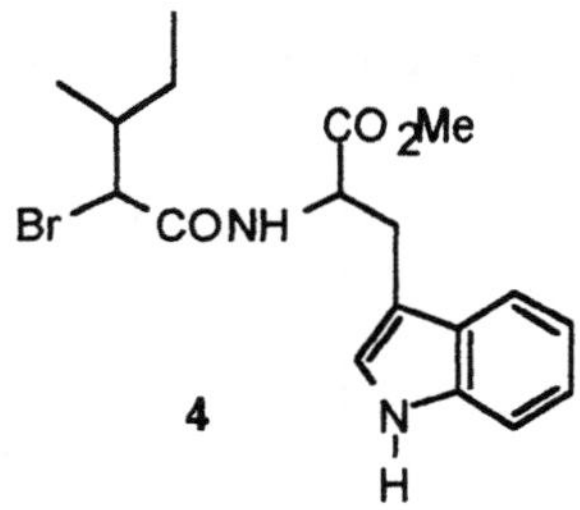

4

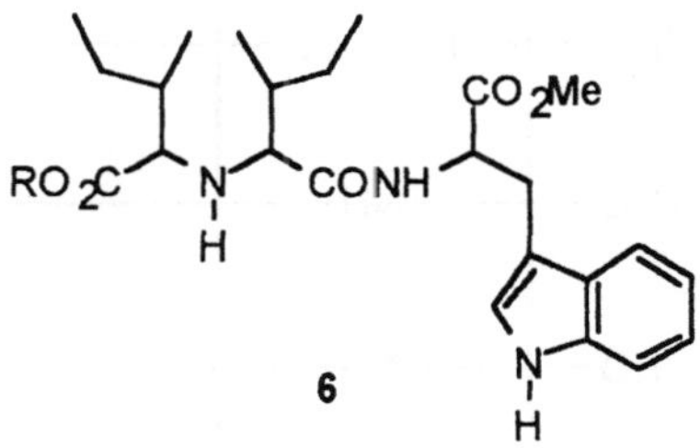

6

Chart 2

## EXPERIMENTAL (D'angeli et al., 1993)

All aminoacids and derivatives used were of *(L)* configuration. Analytical and preparative HPLC were performed with the mobil phases: A (10% v/v acetonitrile in 0.1 % TFA); B (60 % v/v acetonitrile in 0.1 % TFA) and the same solvent system gradient: 50-100 % B in 25 min at flow rates of 30 mL/min (preparative HPLC) and of 1 mL/min (analytical HPLC).

## *(S)*-2-Bromo-*(S)*-3-methylpentanoyl *iso*Leucine, Benzyl Ester (2)

To a solution of *(S)*-2-bromo-*(S)*-3-methylpentanoic acid (1.58 g, 8.1 mmol) in DMF (30 mL), N-methylmorpholine (1.85 mL, 17 mmol) and isobutylchloroformate (1.05 mL,

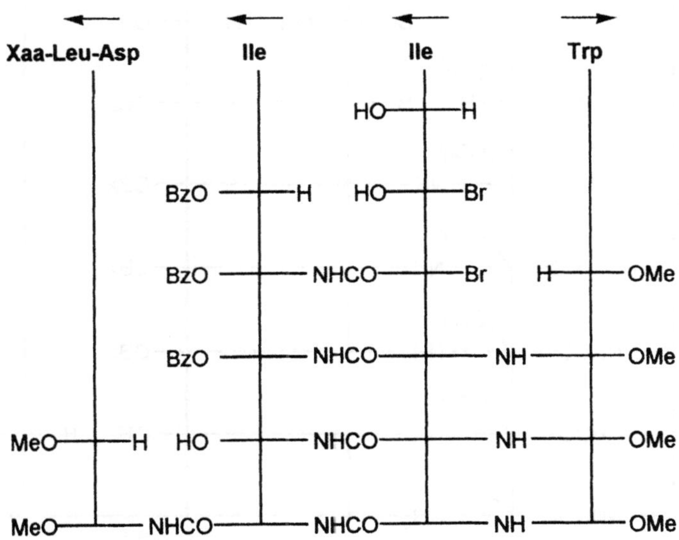


**Figure 1.** Synthetic approach to a 20-21-Ψ(NH) terminal ET-1 hexapeptide (**7a**).

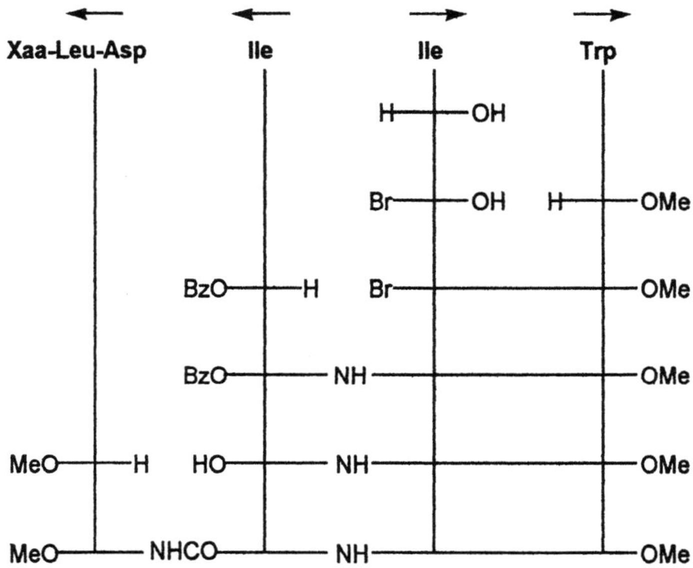


**Figure 2.** Synthetic approach to a 19-20-Ψ(NH) terminal ET-1 hexapeptide (**8**).

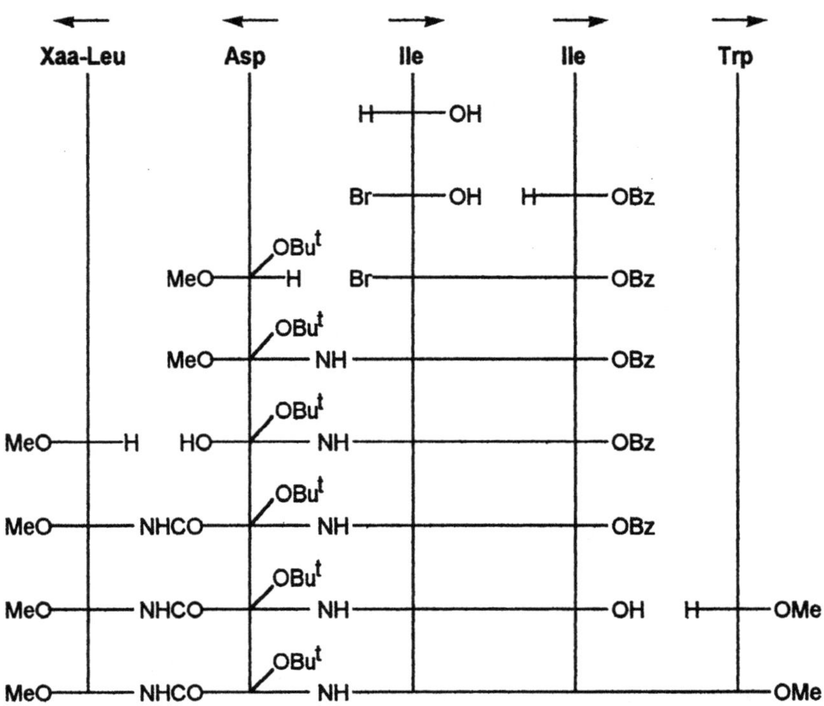


**Figure 3.** Synthetic approach to an 18-19-Ψ(NH) terminal ET-1 hexapeptide (**9**).

8.1 mmol) were added at -20°C. After 10 min. stirring, isoleucine benzyl ester *p*-tosylate (3.5 g, 8.9 mmol) was added and stirring was continued at -20°C (1 h) and at r. t. (4 h). The mixture was evaporated and the residue taken up with ethyl acetate (250 mL). The solution was washed with 0.5 N $KHSO_4$, brine, 5% $NaHCO_3$, brine (10 mL twice each). After drying ($Na_2SO_4$), the solvent was evaporated and the residue was purified on column ($SiO_2$, petroleum ether 40°-60°C/ethyl acetate (P/A) 2:1). Oil 2.2 g, 69%); Rf (P/A 2:1) = 0.80.

$^1$H-NMR: 0.85-1.05: (12H, 2t + 2d), 1.1-1.65: (4H, 2m); 1.85-2.25: (2H, 2m); 4.32: (1H, d); 4.63: (1H, dd); 5.18: (2H, AB); 7.0: (1H, d); 7.36:(5H, s).

## *(S)*-2-*sec*-Butyl-*(S)*-4-(3′-indolylmethyl)-3-azaglutaroyl *iso*Leucine Benzyl Ester, 5-Methyl Ester (3a)

A suspension of tryptophan methyl ester hydrochloride (0.51 g, 2 mmol) and Amberlyst A-21 (Fluka) (2 g) in ethyl acetate (10 mL) was stirred 30 min. The resin was filtered off and the solution was concentrated. The oil (0.45 g) and bromoamide (**2**) (400 mg, 1 mmol) were dissolved in toluene (5 mL). $Ag_2O$ (232 mg, 1 mmol) was added and the mixture was sonicated 45 min, allowing the temperature to rise (ab. 70°C). Centrifugation and evaporation of the solution gave an oil (800 mg) that was purified by column chromatography (silica gel, P/A 2:1) . Oil (320 mg, 60%) Rf = 0.35.

$^1$H-NMR: 0.81-0.93: (12H, 2t + 2d); 1.1-1.85: (6H, m); 3.0: (1H, d); 3.15: (2H, d); 3.58: (4H, s); 4.58: (1H, dd); 5.16: (2H, AB); 6.98-7.62: (11H, m); 8.05: (1H, s). $^{13}$C-NMR: 11.43, 11.82, 15.64, 15.85, 24.80, 25.40, 29.78, 37.17, 38.65, 51.70, 56.12, 61.14, 66.03, 66.89, 110.68, 111.13, 118.97, 119.26, 121.67, 122.97, 127.71, 128.43, 128.58, 135.52, 136.11, 171.83, 173.15, 175.23.

## *(S)*-2-*sec*-Butyl-4-(3′-indolylmethyl)-3-azaglutaroyl *iso*Leucine, 5-Methyl Ester (3b)

A sample of diester (**3a**) (230 mg, 0.43 mmol) in methanol (20mL) with Pd/C (10%, 30 mg) was hydrogenated (2.8 atm, Parr) 30 min. The catalyst was filtered off and the solution was evaporated. Solid(184 mg, 96%), mp 68-73°C, Rf (P/A 1:1) = 0.32. $^1$H-NMR: 0.79-0.97: (12H, 2t + 2d); 1.2 e 1.5: (4H, 2m); 1.72: (2H, m); 3.0: (1H, d); 3.11: (2H, AB); 3.55: (4H, s); 4.34: (1H, dd); 4.8: (2H, s, broad); 7.0-7.6: (6H, m); 8.48: (1H, s).

## *(S)*-2-Bromo-*(S)*-3-methylpentanoyl Tryptophan, Methyl Ester (4)

*(S)*-2-Bromo-*(S)*-3-methylpentanoic acid (1.685 g, 8.64 mmol), N-methylmorpholine (1.98 mL, 18.15 mmol), isobutylchloroformate (1.12 mL, 8.64 mmol) and tryptophan methyl ester hydrochloride (2.421 g, 9.51 mmol) were allowed to react as for **2**. Working up gave a solid (1.36 g, 40%). Prisms from toluene-*n*-hexane: mp 127°-129° C, Rf (P/A 1:1) = 0.67. $^1$H-NMR: 0.82: (3H, t); 0.93: (3H, d); 1.2 e 1.4: (2H, 2m); 2.1: (1H, m); 3.33: (2H, dd); 3.68: (3H, s); 4.19: (1H, d); 4.92: (1H, dd); 6.89: (1H, d); 7.04-7.59: (5H, m); 8.22: (1H, s).

## *(S)*-2-*(S)*-4-Di-*sec*-butyl-3-azaglutaroyl Tryptophan Methyl Ester, 5-Benzyl Ester (6)

Isoleucine benzyl ester *p*-tosylate (393 mg, 1 mmol), Amberlyst A-21 (Fluka) (1 g), bromoamide (**4**) (198 mg, 0.5 mmol) and $Ag_2O$ (116 mg, 0.5 mmol) were allowed to react as for **3a**. Working up gave an oil (128 mg, 48%). Rf (cyclohexane/ethyl acetate 2:1) = 0.28.

$^1$H-NMR: 0.64-0.87: (12H, 2t + 2d); 1.0-1.8: (6H, m); 2.86: (1H, d); 3.08: (1H, d); 3.31: (2H, AB); 3.64: (3H, s); 4.93: (1H, dd); 5.08: (2H, AB); 6.98-7.62: (11H, m); 8.35: (1H, s).

## Z-Asp(OtBu)-Leu-Phe-OMe

To a solution of TFA H-Leu-Phe-OMe (obtained by treating 30 min Boc-Leu-Phe-OMe (Beyerman et al., 1973) with TFA, at 20° C), (967 mg, 2.38 mmol) in DMF (20 mL) Z-Asp($OBu^t$)-OH $H_2O$ (812 mg, 2.38 mmol), N-methylmorpholine (24 mg, 2.38 mmol), N-hydroxybenzotriazole (401 mg, 2.62 mmol) and DCC (541 mg, 2.62 mmol) were added in the order. The mixture was magnetically stirred overnight. The solvent was removed in vacuo and the residue was taken up with ethyl acetate. The solution was washed (10% citric acid, brine, 5 mL twice each), dried ($Na_2SO_4$) and concentrated. The solid (1.1 g, 77%) was purified by crystallization with *n*-hexane/$Et_2O$ 2/1: prisms, mp 103-05° C; $[\alpha]_D^{20}$ (c = 1, MeOH) = - 30.8.

## *(S)*-2-*iso*Butyl-*(S)*-4-(3′-indolylmethyl)-3-azaglutaroyl-Ile-Asp$^{t.Bu}$-Leu-Phe, Methyl Ester (7a)

A solution of compound **3b** (174 mg, 0.39 mmol) in DMF (10 mL) was cooled at 0° C and H-Asp-Leu-Phe-OMe acetate (obtained by deblocking the Z-tripeptide with hydrogen in the presence of 5% Pd/C) (205 mg, 0.39 mmol), N-methylmorpholine (0.08 mg, 0.78 mmol) and a mixture of DCC (89 mg, 0.43 mmol) and N-hydroxybenzotriazole (66 mg 0.43 mmol) were added. The mixture was magnetically stirred overnight and treated as described for the above tripeptide. The solid (314 mg, 82%) was purified by reverse phase HPLC (D'Angeli et al., 1993) (K′ = 11.06). MS (MALDI-tof) m/z = 891 $[M + H^+]^+$, 835. Aminoacid analysis gave: Asp = 1, Leu = 1, Phe = 1, Ileu = 0.6 (D'Angeli et al., 1993).

## ACKNOWLEDGMENTS

This work was financially supported by the Consiglio Nazionale delle Ricerche (CNR) and Ministero dell'Università e della Ricerca Scientifica e Tecnologica (MURST). We are grateful to Dr. Roberta Seraglia (CNR, Padova) for the MALDI-tof mass spectrum of **7a**.

## REFERENCES

Balboni G., Salvadori S., D'Angeli F., Marchetti P., Lazarus L.H., Bryant S.D., and Bianchi C., 1995, Single diastereomeric desaminotyrosylalanyl tetra- and heptapeptides with opioid antagonistic activity, *Int. J. Peptide Protein Res.* 45:187.

Beyerman H.C., De Leer E.W.B., and Floor J., 1973, Repetitive excess mixed anhydride method for the synthesis of peptides. Synthesis of the sequence 1-10 of human growth hormon, *Rec. Trav. Chim. Pays-Bas* 92:481.

Cavicchioni G., D'Angeli F., Casolari A., and Orlandini P., 1988, Synthesis of O-(α-aminoacyl)glycolic and -lactic amides from 2-bromoacetamides or -propanamides with N-protected amino acids, *Synthesis*, 947.

Cody W.L., Doherty A.M., He J.X., De Pue P.L., Rapundalo S.T., Hingorani G.A., Major T.C., Panek R.L., Dudley D.T., Haleen S.J., La Douceur D., Hill K.E., Flynn M.A., and Reynolds E.E., 1992, Design of a functional hexapeptide antagonist of endotelin, *J. Med. Chem.* 35:3301.

Cody W.L, He. J.X., Doherty A.M., De Pue P.L., Kaltenbronn J.S., Reisdorph B.R., Walker D.M., Welch K.M., Haleen S.J., Reynolds E.E., Tse E., Reyner E.L., and Stewart B. H., 1995, The design of potent hexapeptide endothelin antagonist stable to proteolysis, *Peptides 1994*, H. L. S. Maia Eds., ESCOM Science Publishers, Leiden, The Nederlands.

D'Angeli F., Marchetti P., Cavicchioni G., Bertolasi V., and Maran F., 1991, Enantioselective reactions of 2-bromopropanamides with primary, secondary, or tertiary amines. Synthesis of some alaninamides, *Tetrahedron: Asymmetry* 2:1111.

D'Angeli F., Marchetti P., Salvadori S., and Balboni G., 1993, 2-Bromoamides as synthons for pseudopeptides containing aminodicarboxy units, *J. Chem. Soc. Chem. Commun.* 3:304.

Doherty A.M., 1992, Endothelin: A New Challenge, *J. Med. Chem.* 35:1493.

Gilbert I.H., Rees D.C., Crockett A.K., and Jones R.C.F., 1995, Imidazolines as amide bond replacements, *Tetrahedron* 51:6315, and references therein.

Gortatowski M.J., and Armstrong M.D., 1975, Convenient synthesis for β-(3-indolyl)-DL-lactic acids, *J. Org. Chem.* 22:1217.

Haslanger M. F., Sybertz E. J., Neustadt B. R., Smith E. M., Nechuta T. JL., and Berger J., 1989, Carboxyalkyl dipeptides with atrial natriuretic factor potentiating and antihypertensive activity, *J. Med. Chem.* 32:737.

James R.W., Peapus D.H., and Wallace B.A., 1994, The crystal structure of human endothelin, *Structural biology* 1:311 and references therein.

Moftakhari Kamrani Nejad F., Marchetti P., Rondanin R., Barili P.L., and Catelani G., 1994, Stereoselective formation of glycoconjugated lactamides from chiral 2-bromopropanamides, *Gazz. Chim. Ital.* 124:65.

Ragazzi E. et al., unpublished.

Regoli D. et al., , unpublished.

Wu M. T., Douglas A. W., Ondeyka D. L., Payne L. G., Ikeler T. J., Joshua H., and. Patchett A. A, 1985, Synthesis of $N^2$-[(S)-1-carboxy-3-phenylpropyl]-L-lysyl-L-proline (Lisinopril), *J. Pharm. Sc.* 74:352.

# TRYPTOPHAN-CONTAINING POLYPEPTIDE $3_{10}$-HELICES

## Side-Chain Interactions and Molecular Recognition Studies

M. Crisma,[1] A. Bianco,[1] F. Formaggio,[1] C. Toniolo,[1] C. George,[2] and J. L. Flippen-Anderson[2]

[1] Biopolymer Research Center, C.N.R.
Department of Organic Chemistry, University of Padova
35131 Padova, Italy
[2] Laboratory for the Structure of Matter
Naval Research Laboratory
Washington, DC 20375

## 1. INTRODUCTION

A proper understanding of the mechanisms of side-chain interactions and molecular recognition depends heavily upon the ability to design and build conformationally constrained structures whose intercomponent geometry is well defined. As appropriate templates, we chose to focus on structurally restricted $3_{10}$-helical oligopeptides, which fold to bring into close proximity two partners positioned one turn apart.

The $3_{10}$-helix, first predicted as a reasonably stable polypeptide secondary structure more than fifty years ago (Taylor, 1941), has only recently attracted the attention of structural biochemists and protein crystallographers. It represents the third principal structure occurring in globular proteins (in addition to the α-helix and the β-pleated sheet structure) and has been described at atomic resolution in model peptides and in peptaibol antibiotics (Toniolo and Benedetti, 1991a).

The $3_{10}$-helical peptide architecture is known to provide access to the investigation of through-space interactions between side chains of two amino acid residues incorporated at the *i* and *i*+3 relative positions (Toniolo et al., 1995; Augspurger et al., 1995). By exploiting the conformationally restricted Aib (α-aminoisobutyric acid) residue and other members of the $C^{\alpha,\alpha}$-disubstituted glycine family as the main building blocks, it is possible to achieve stable $3_{10}$-helicity and excellent structural rigidity at very short peptide main-chain lengths (Toniolo and Benedetti, 1991b; Toniolo et al., 1993).

In the work described in this communication we have extended this approach to the structural characterization and exploitation of Trp - containing polypeptide $3_{10}$-helices.

*Recent Advances in Tryptophan Research*, edited by Graziella Allegri Filippini et al.
Plenum Press, New York, 1996

## 2. RESULTS AND DISCUSSION

A host, terminally blocked, Aib peptide series ranging from three to five amino acid residues and containing one L-Trp residue and, in one case, the nitroxide spin labelled amino acid TOAC (2,2,6,6-tetramethylpiperidine-1-oxyl-4-amino-4-carboxylic acid) as a fluorescence quencher, was synthesized by solution methods and fully characterized. The TOAC helicogenic properties (Toniolo et al., 1995) are known to closely parallel those of its prototype Aib and its cycloaliphatic analogue $Ac_6c$ (1-aminocyclohexane-1-carboxylic acid) (Toniolo and Benedetti, 1991b).

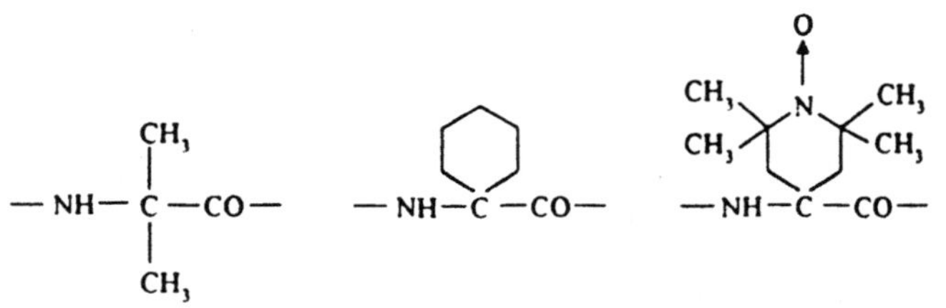

Aib $Ac_6c$ TOAC

A X-ray diffraction structural analysis of Z-Aib-L-Trp-Aib-OMe, Z-$(Aib)_2$-L-Trp-Aib-OMe (Fig. 1), and *t*-Boc-$(Aib)_3$-L-Trp-Aib-OMe (Z, benzyloxycarbonyl; *t*-Boc, *tert*-butyloxycarbonyl; OMe, methoxy) showed that these peptides are folded into right-handed $3_{10}$-helices stabilized by consecutive N-H...O=C intramolecular H-bonds of the β-turn type. The C-terminal β-turn (type-I β-turn) of the tetra- and pentapeptides is slightly distorted from the regular, helical type-III β-turn. The C- terminal Aib residue of all three peptides,

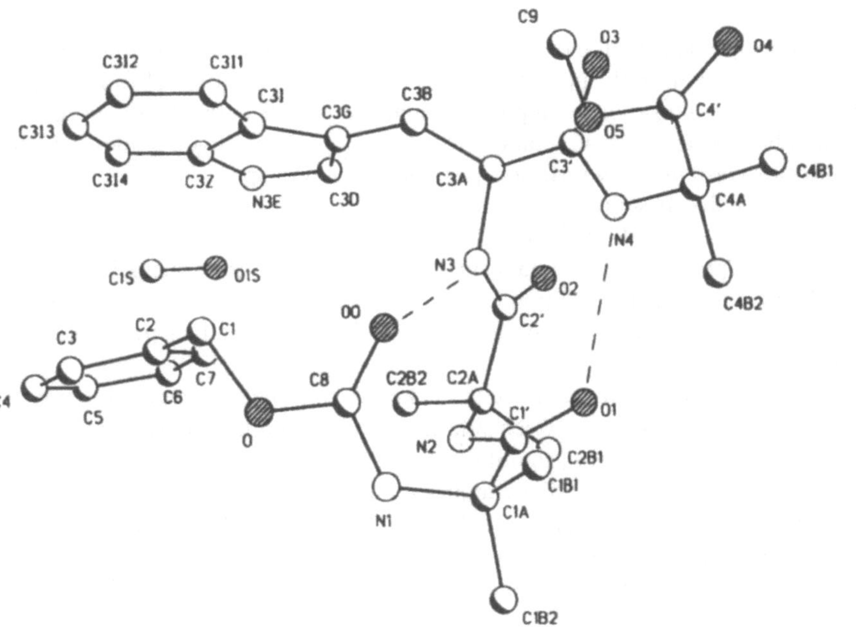

**Figure 1.** X-ray diffraction structure of Z-$(Aib)_2$ -L-Trp-Aib-OMe with numbering of the atoms. The two intramolecular H-bonds are indicated by dashed lines.

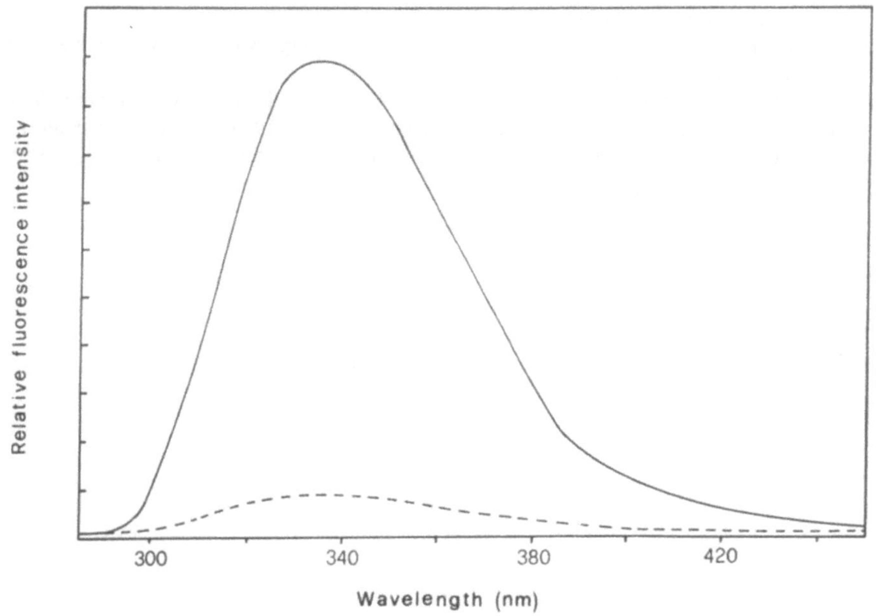


**Figure 2.** Fluorescence spectra of *t*-Boc-(Aib)$_3$-L-Trp-Aib-OMe (—) and Ac-TOAC-(Aib)$_2$-L-Trp-Aib-OMe (- - -) in ethanol (*c* 5.0 x $10^{-6}$M).

although lying outside the $3_{10}$-helix, is also helical. Interestingly, its helix screw sense is opposite to that of the preceding residues in the tri- and tetrapeptides, where it is the same in the pentapeptide. The L-Trp side-chain $\chi^1$ (N-C$^\alpha$-C$^\beta$-C$^\gamma$) values are $g^+$ for the tripeptide, while $g^-$ for the tetra- and pentapeptides. On the other hand, the L-Trp side-chain $\chi^{2,1}$ and $\chi^{2,2}$(C$^\alpha$-C$^\beta$-C$^\gamma$-C$^{\delta1}$) and (C$^\alpha$-C$^\beta$-C$^\gamma$-C$^{\delta2}$) values are close to ±90° in the tri- and pentapeptides, whereas they are close to 0°, 180° in the tetrapeptide. The N- terminal $3_{10}$-helical sequence Z-(Aib)$_2$-L-Trp-NH- of the tetrapeptide generates a cavity between the indole and the benzyloxycarbonyl moieties (Fig. 1). The two ring systems are approximately parallel to one another (the angle between the planes is 13.7°) and about 3.9 Å apart. The methanol (solvent) molecule is H-bonded to the Trp side-clain N$^{ind}$- atom.

In a related guest peptide with the sequence Ac-TOAC-(Aib)$_2$-L-Trp-Aib-OMe the nitroxide and indole side chains, located on the same side of the $3_{10}$-helix after one complete turn, strongly interact, producing a dramatic quenching of the Trp fluorescence compared to that of the host peptide *t*-Boc-(Aib)$_3$-L-Trp-Aib-OMe (Fig. 2).

It may be concluded that Trp-containing, conformationally constrained $3_{10}$-helical peptides are excellent tools for studies of fluorescence quenching and molecular recognition provided that suitable blocking groups and side-chain functionalities are incorporated in the sequence at appropriate relative positions.

## 3. REFERENCES

Augspurger J.D., Bindra V.A., Scheraga H.A., and Kuki A., 1995, Helical stability of *de novo* designed α-aminoisobutyric acid-rich peptides at high temperatures, *Biochemistry* 34:2566.

Taylor H.S., 1941, Polypeptide chain configuration, *Proc. Am. Phyl. Soc.* 85:1.

Toniolo C. and Benedetti E., 1991a, The polypeptide $3_{10}$-helix, *Trends Biochem. Sci.* (*TIBS*) 16:350.

Toniolo C. and Benedetti E., 1991b, Structures of polypeptides from α-amino acids disubstituted at the α-carbon, *Macromolecules* 24 : 4004.

Toniolo C., Crisma M., Formaggio F., Valle G., Cavicchioni G., Précigoux G., Aubry A., and Kamphuis J., 1993, *Biopolymers* 33: 1061.

Toniolo C., Valente E., Formaggio F., Crisma M., Pilloni G., Corvaja C., Toffoletti A., Martinez G.V., Hanson M.P., Millhauser G.L., George C., and Flippen-Anderson J.L., 1995, Synthesis and conformational studies of peptides containing TOAC, a spin-labelled $C^{\alpha,\alpha}$ disubstituted glycine, *J. Pept. Sci.* 1:45.

# THE CORE DOMAIN OF HIRUDIN FROM THE LEECH *Hirudinaria manillensis*

## Chemical Modification of a Tryptophan-Containing Synthetic Peptide Analog

F. De Antoni, V. De Filippis, L. Altichieri, A. Vindigni,
P. Polverino de Laureto, and A. Fontana

CRIBI Biotechnology Centre, University of Padua
via Trieste 75, 35121 Padua, Italy

## 1. INTRODUCTION

Hirudin is a small (~ 7 kDa) disulfide-crosslinked polypeptide known as the most potent and specific inhibitor of thrombin, a serine protease that plays a key role in the coagulation cascade and pathology of thrombotic diseases (Tapparelli et al., 1993). We have previously shown that the N-terminal proteolytic fragment 1-47 of hirudin HM2 from *Hirudinaria manillensis* (Fig. 1) maintains inhibitory action towards thrombin (Vindigni et al., 1994). An analog of fragment 1-47 with a $Tyr^3 \rightarrow Trp$ exchange (Y3W) was prepared by solid-phase chemical synthesis and shown to be ~5-fold more active than the $Tyr^3$ natural fragment 1-47 in inhibiting thrombin (De Filippis et al., 1995). In this study, we report the results of experiments of chemical modification and peptide bond cleavage at the level of $Trp^3$ of the synthetic Y3W peptide 1-47 with the aim to prepare modified and truncated species of the Y3W peptide analog. Modification reactions were carried out also on fragment 1-41, prepared by V8-protease digestion of the synthetic Y3W peptide 1-47 at $Glu^{41}$. All peptide derivatives prepared in the course of this work were isolated to homogeneity and assayed for biological activity. The results of this study revealed the critical role of the N-terminal portion of peptide 1-47 in the inhibition of thrombin, providing experimental support for the proposed mechanism of interaction between hirudin and thrombin advanced in previous studies.

## 2. RESULTS

### 2.1. Preparation of Fragment 1-41

The synthetic Y3W peptide was reacted with endoprotease V8 from *Staphylococcus aureus*, which selectively cleaves peptide bonds at the C-terminus of Glu-residues (Drapeau,

*Recent Advances in Tryptophan Research,* edited by Graziella Allegri Filippini et al.
Plenum Press, New York, 1996

```
1        5                             10                  15
Val-Ser-Trp-Thr-Asp-Cys-Thr-Glu-Ser-Gly-Gln-Asn-Tyr-Cys-Leu-

16          20                25                     30
Cys-Val-Gly-Ser-Asn-Val-Cys-Gly-Glu-Gly-Lys-Asn-Cys-Gln-Leu-

31           35                    40                    45
Ser-Ser-Ser-Gly-Asn-Gln-Cys-Val-His-Gly-Glu-Gly-Thr-Pro-Lys-

46  47
Pro-Lys
```

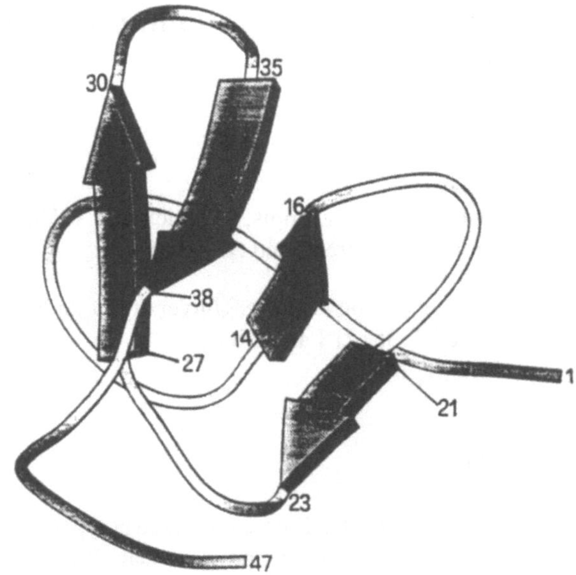


**Figure 1.** (*Top*) Amino acid sequence of the Y3W analog of the N-terminal fragment 1-47 of hirudin HM2 from *Hirudinaria manillensis.* (*Bottom*) Schematic representation of the 3D structure of fragment 1-47 of hirudin HM2.

1977). Even if the Y3W peptide contains three Glu-residues at chain positions 8, 24 and 41 (see Fig. 1), peptide bond fission occurs at Glu[41] only and the fragment 1-41 was obtained in high yield (~95 %) by micropreparative reverse-phase HPLC (Fig. 2). Minor amounts of a fragment species deriving from internal nicking by V8 protease at the Glu[8]-Ser[9] were also obtained (fragment 1-41Y3W*).

## 2.2. Chemical Modification with o-Nitrophenylsulfenyl Chloride (NPS-Cl)

The peptide species 1-47Y3W and 1-41Y3W were modified at Trp[3] with NPS-Cl. This reagent allows the selective modification of Trp-residues in peptides and proteins not containing free Cys-residues, leading to the covalent binding of the NPS moiety at the 2-position of the indole nucleus (Scoffone et al., 1968; Fontana and Scoffone, 1972). After dilution with water and extraction of excess reagent with ethyl acetate, the reaction mixture was applied to a reverse-phase HPLC column. As shown in Fig. 3, both peptide species were cleanly and in high yield converted to the more hydrophobic NPS-derivatives. The modified peptides were isolated and characterized by amino acid analysis after acid hydrolysis, N-terminal sequencing and MALDI-TOF mass spectrometry. The results of these analyses

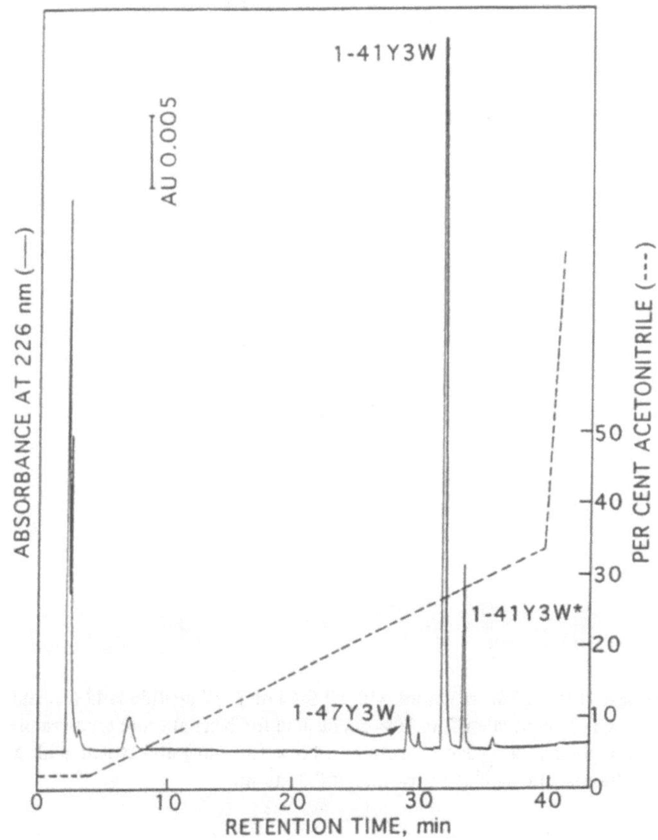


**Figure 2.** Reverse-phase HPLC of the proteolytic reaction of the synthetic peptide (1-47Y3W analog) with Glu-C protease from *Staphylococcus aureus* V8. The peptide was digested with V8-protease for 1-h at 37°C at an enzyme:substrate ratio of 1:10 (by weight). After digestion, an aliquot of the reaction mixture was applied to a Vydac $C_{18}$ column (4.6 x 150 mm) eluted with a linear gradient of acetonitrile-0.05% trifluoroacetic acid at a flow rate of 0.8 ml/min. Identities of the proteolytic fragments were determined by amino acid analysis after acid hydrolysis, N-terminal sequencing and MALDI-TOF mass spectrometry.

established the purity and identity of the two NPS-derivatives, which showed absorption spectra characterized by a maximum at 365 nm, typical of the NPS-Trp chromophore (Scoffone et al., 1968; Fontana and Scoffone, 1972).

## 2.3. Chemical Cleavage with BNPS-Skatole

Both peptides 1-47 and 1-41 were cleaved at $Trp^3$ with BNPS-skatole (Fontana, 1972; Fontana et al., 1973). Reactions were carried out in aqueous acetic acid with a peptide:reagent molar ratio of 1:100. Phenol (100-equivalents) was added to the reaction mixture as a scavenger for Tyr bromination. After dilution with water and extraction of reagent and by-products with ethyl acetate, the mixture was analyzed by reverse-phase HPLC. The chromatograms shown in Fig. 4 appear to be quite complex, but nevertheless major peaks corresponding to the truncated peptides 4-47 and 4-41 are seen, whereas other peaks contained non peptidic material, deriving from BNPS-skatole, phenol and other side products. Fragments 4-47 and 4-41 were isolated after HPLC in about 50 % yield. The homoge-

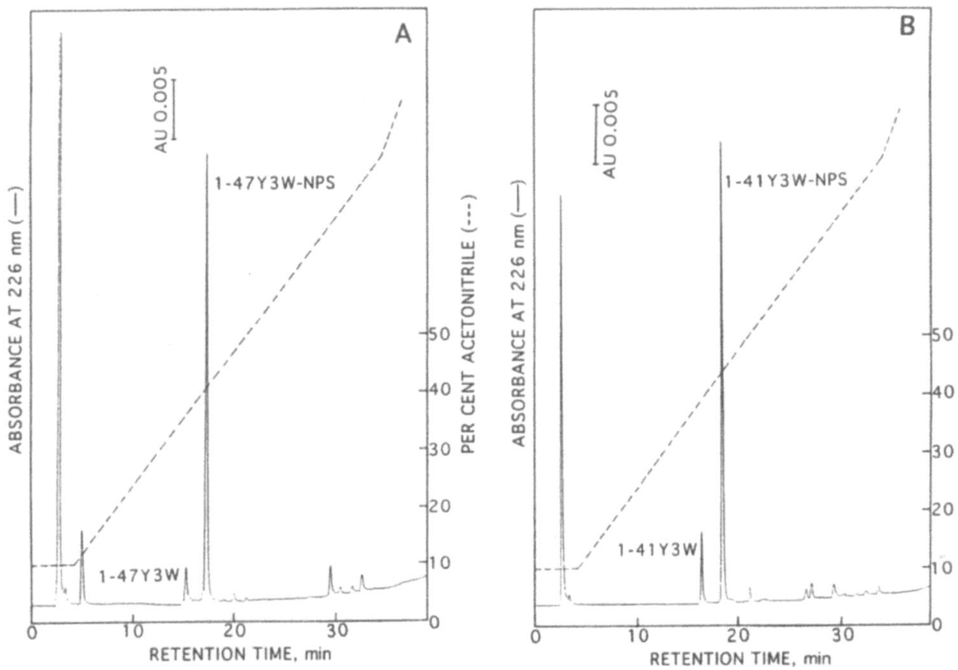


**Figure 3.** Reverse-phase HPLC of the reactions with NPS-Cl of Y3W peptide 1-47 (A) and 1-41 (B). Peptides were treated with NPS-Cl (25 equivalents) in 70% acetic acid for 2-hrs at room temperature. An aliquot of the reaction mixture was applied to a Vydac $C_4$ column (4.6 x 150 mm) and eluted with a linear gradient of acetonitrile-0.05% trifluoroacetic acid at a flow rate of 0.8 ml/min.

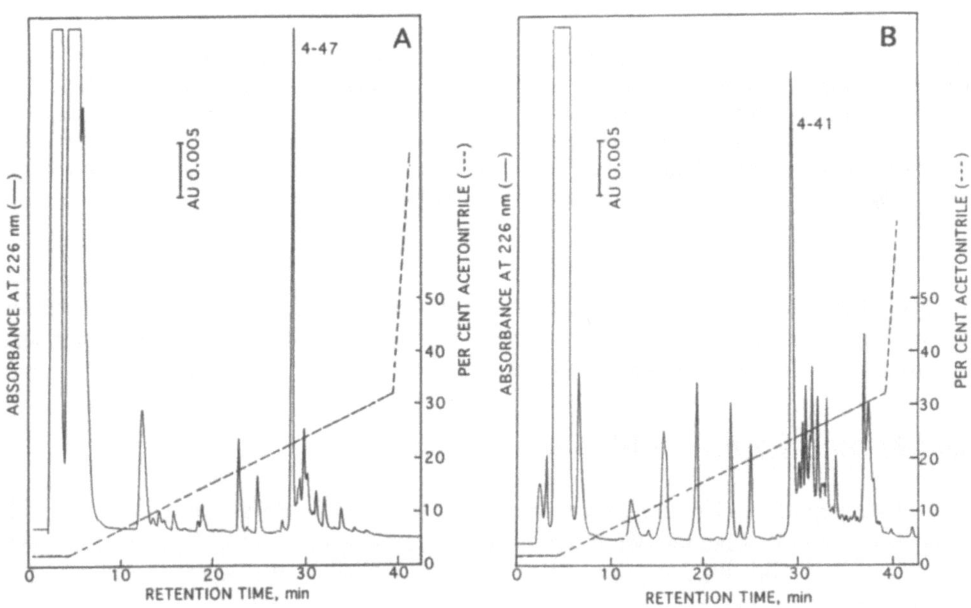


**Figure 4.** Reverse-phase HPLC of the fragmentation reaction at $Trp^3$ with BNPS-skatole of Y3W peptides 1-47 (A) and 1-41 (B). Peptides were cleaved with BNPS-skatole (100 equivalents) in 66% acetic acid for 2 days at 37°C in the dark. The reaction mixture was diluted with water and excess BNPS-skatole removed by three extractions with ethyl acetate. Aliquots of the aqueous solutions were applied to a Vydac $C_{18}$ column (4.6 x 150 mm) eluted with a linear gradient of acetonitrile-0.05% trifluoroacetic acid at a flow rate of 0.8 ml/min.

neity and identity of the truncated peptides were established by a number of protein chemistry analytical methods, including mass spectrometry.

## 2.4. Chemical Oxidation

The Trp-residue in peptides and proteins can be selectively oxidized using the dimethylsulfoxide(DMSO)-HCl mixture in acetic acid (Savige and Fontana, 1980), leading to the oxindolyalanine(Oia)-derivative. Since the Oia-residue contains a chiral $C^3$ at the level of the indole mojety, diasteroisomeric Oia-peptides are expected. Indeed, diasteroisomers can be obtained from short peptides oxidized with the DMSO-HCl mixture, which can be separated by reverse-phase HPLC (unpublished results). Two diastereoisomers were separated by HPLC utilizing a $C_{18}$ column after oxidation of peptide 1-41, while the diasteroisomers of peptide 1-47 were not resolved by HPLC. Mass spectrometry analyses of the peptide species separated by HPLC revealed that, besides the Oia-derivatives, also minor amounts of acetylated species were present in the mixture. The Oia-derivatives were isolated in low yields, thus not allowing further biological assays of the oxidized peptide species.

## 2.5. Antithrombin Activity

The competitive inhibition of thrombin by all peptide species herewith prepared were studied utilizing the synthetic substrate H-D-Phe-pipecolyl-Arg-p-nitroanilide (S-2238) (Stone and Hofsteenge, 1986). Fig. 5 shows the data of the inhibitory activity of the peptide species, given in terms of the reciprocal of the dissociation costant $K_i$ of the thrombin inhibitor complex and thus bars are proportional to the inhibitory activity. As previously reported (De Filippis et al., 1995), the Trp-analog is ~5-fold more active than the natural Tyr-derivative and the Trp-containing 1-41 peptide displays an inhibitory action similar to

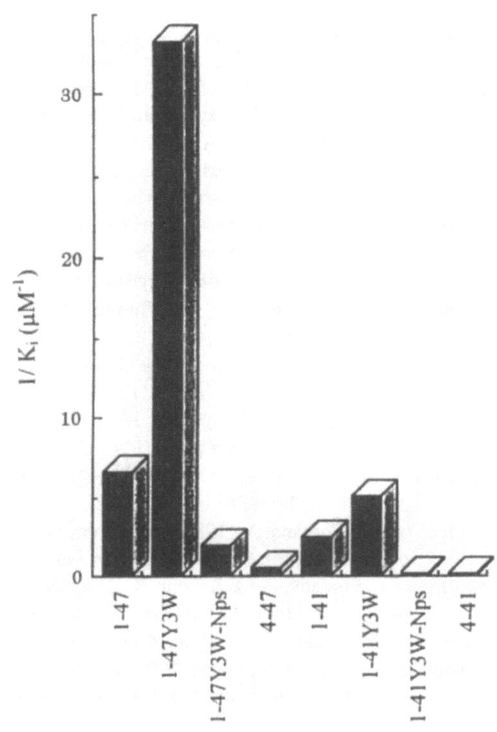


**Figure 5.** Inhibitory activity of hirudin HM2 fragments and related peptide species towards the amidolytic activity of thrombin. Assays were performed using H-D-Phe-pipecolyl-Arg-p-nitroanilide (S-2238) as a substrate for thrombin (Stone and Hofsteenge, 1986). Kinetic parameters were calculated by the Michaelis-Menten treatment of competitive inhibition. Bars show the reciprocal of the dissociation constant $K_i$ of the thrombin-inhibitor complex and are proportional to the inhibitory activity.

that of the longer 1-47 natural fragment. The NPS-derivatives show a significant reduction in inhibitory power and the truncated peptides 4-47 and 4-41 show negligible activity.

## 3. DISCUSSION

Chemical modification of the Trp-residue was used to study the functional role of the N-terminal portion of hirudin in inhibiting thrombin. The N-terminal tripeptide represents a key element for the binding of hirudin to thrombin (Braun et al., 1988; Lazar et al., 1991; Winant et al., 1991; Rydel et al., 1991; Betz et al., 1992). The X-ray structure of the hirudin-thrombin complex (Rydel et al., 1991) reveals that, upon binding, the N-terminal tripeptide of hirudin penetrates into the active-site region forming a parallel b-strand with $Ser^{214}$ to $Gly^{219}$ of thrombin, with $Tyr^3$ becoming deeply buried in the hydrophobic pocket of the thrombin active site. Replacement of $Tyr^3$ in the natural fragment by a Trp-residue leads to an enhanced inhibitory activity, due to the higher hydrophobicity of Trp (De Filippis et al., 1995). Sulfenylaton of $Trp^3$ reduces the inhibitory action of fragments 1-47 and 1-41, which indicates that, despite the increase in hydrophobicity due the introduction of the NPS group, the bulky Trp-NPS mojety inhibits a productive interaction of the modified peptide species within the hydrophobic pocket of the thrombin active site. Even more drastic effects are caused by removal of the N-terminal tripeptide by cleaving at $Trp^3$ with BNPS-skatole, thus implying the key functional role of the N-terminal portion of hirudin. Overall, the results of this study are in agreement with the proposed mechanism of interaction of hirudin with thrombin derived from X-ray (Rydel et al., 1991) and NMR (Szyperski et al., 1992) measurements.

## 4. REFERENCES


Betz, A., Hofsteenge, J. & Stone, S.R., 1992, Interaction of the N-terminal region of hirudin with the active-site cleft of thrombin, *Biochemistry* 31:4557.

Braun, P.J., Dennis, S., Hofsteenge, J. & Stone, S.R., 1988, Use of site-directed mutagenesis to investigate the basis for the specificity of hirudin, *Biochemistry* 27:6517.

De Filippis, V., Vindigni, A., Altichieri, L. & Fontana, A., 1995, The core domain of hirudin from the leech *Hirudinaria manillensis*: chemical synthesis, purification, and characterization of a $Trp^3$ analog of fragment 1-47, *Biochemistry* in press.

Drapeau, G.R., 1977, Cleavage at glutamic acid with staphylococcal protease, *Methods Enzymol.* 47:189.

Fontana, A., 1972, Modification of tryptophan with BNPS-skatole [2-(2-nitrophenylsulfenyl)-3-methyl-3-bromoindolenine], *Methods Enzymol.* 25:482.

Fontana, A. & Scoffone, E., 1972, Sulfenyl halides as modifyng reagents for polypeptides and proteins, *Methods Enzymol.* 25:482.

Fontana, A., Vita, C. & Toniolo, C., 1973, Selective cleavage of the single tryptophanyl peptide bond in horse cytochrome c, *FEBS Lett.* 32:133.

Lazar, J.B., Winant, R.C. & Johnson, P.H., 1991, Hirudin: amino-terminal residues play a major rule in the interaction with thrombin, *J. Biol. Chem.* 266:685.

Rydel, T.J., Tulinsky, A., Bode, W. & Huber, R., 1991, Refined structure of the hirudin-thrombin complex, *J. Mol. Biol.* 221:583.

Savige, W.E. & Fontana, A., 1980, Oxidation of tryptophan to oxindolylalanine by dimethyl sulfoxide-hydrochloric acid, *Int. J. Peptide Protein Res.* 15:285.

Scoffone, E., Fontana, A. & Rocchi, R., 1968, Sulfenyl halides as modifying reagents for polypeptides and proteins. I. Modification of tryptophan residues, *Biochemistry* 7:971.

Stone, S.R. & Hofsteenge, J., 1986, Kinetics of the inhibition of thrombin by hirudin, *Biochemistry* 25:4622.

Szyperski, T., Güntert, P., Stone, S.R., Tulinski, A., Bode, W., Huber, R. & Wütrich, K., 1992, Impact of protein-protein contacts on the conformation of thrombin-bound hirudin studied by comparison with the nuclear magnetic resonance solution structure of hirudin, *J. Mol. Biol.* 228:1206.

Tapparelli, C., Metternich, R., Ehrhardt, C.& Cook, N.S., 1993, Synthetic low molecular mass thrombin inhibitors: Molecular design and pharmacological profile, *Trends Pharmacol. Sci.* 14:366.

Vindigni, A., De Filippis, V., Zanotti, G., Visco, C. & Fontana, A., 1994, Probing the structure of hirudin from *Hirudinaria manillensis* by limited proteolysis, *Eur. J. Biochem.* 226:323.

Winant, R.C., Lazar, J.B. & Johnson, P.H., 1991, Chemical modification and amino acid substitutions in recombinant hirudin that increase hirudin-thrombin affinity, *Biochemistry* 30:1271.

# ADVANCES IN TRYPTOPHAN ANALYSIS IN PEPTIDES AND PROTEINS

I. Molnár-Perl

Institute of Inorganic and Analytical Chemistry
L. Eötvös University
H-1518, Budapest 112, P.O. Box 32, Hungary

## INTRODUCTION

The accurate and reliable analysis of tryptophan was, it is and it will be one of the most challenging task for analytical chemist. For two main reasons:

1. because of its importance in numerous aspects (Curzon, 1992), (being nutritionally essential for many animals and for man, and, taking part in a number of biological functions), simultaneously,
2. because of its 'hard approachability' in its intact form in protein hydrolyzates. (being tremendously sensitive to redox interactions, particularly in the presence of acids, needed classically to protein hydrolysis).

This last phenomena proved to be the object of the overwhelming part of all those paramount efforts (Friedman and Cuq, 1988; Molnár-Perl and Pintér-Szakács, 1989; Molnár-Perl et al., 1987; 1992; 1993; Michalczuk et al., 1992; Gjerde et al., 1987; Peterson et al., 1987; Kim and Angyal, 1993; Morita et al., 1985; 1990; Caruso et al., 1994; Berardino et al., 1990; Aymard et al., 1994; Servillo et al., 1982; Nozaki, 1990; Fletouris et al., 1993; Botsoglou et al., 1993; Spies and Chambers, 1949; 1950; Spies, 1950; 1967; Al-Tamrah, 1989; Iizuka and Yajima, 1985; 1993; Kell et al., 1989; Opienska-Blauth et al., 1963; Concon, 1975; Gorinstein et al., 1988; Gaitonde and Dovey, 1970; Zahnley and Davis; 1973; Sodek et al., 1975; Pintér-Szakács and Molnár-Perl, 1990; Dickman and Crockett, 1956; Barman and Koshland, 1967; Verma and Jain, 1988; Chrastil, 1986; Hugly and Moore, 1972; Sato et al., 1984; 1987; Knox et al., 1970; Werner, 1986; Bech-Andersen, 1991) Delhaye and Landry, 1986; 1992; 1993; Landry et al., 1988; 1992; Landry and Delhaye, 1992; 1993; 1994a; 1994b; Aiple und Menke, 1990; Lucas and Sotelo, 1980; Nielsen and Hurrell, 1985; Allred and MacDonald, 1988; Ranfft and Faure, 1993; Garcia and Baxter, 1992; Simpson et al., 1976; Chiou and Wang, 1988; 1989; Malmer and Schroeder, 1990; Liu and Chang, 1971; Pecavar et al., 1990; Penke et al., 1974; Gardner, 1984; Csapó et al., 1986; 1994; Felker, 1976; Wong et al., 1984; Matsubara and Sasaki, 1969; Ohta and Nakai, 1979; Ashworth, 1987; Yano et al., 1990; Sottrup-

*Recent Advances in Tryptophan Research*, edited by Graziella Allegri Filippini et al.
Plenum Press, New York, 1996

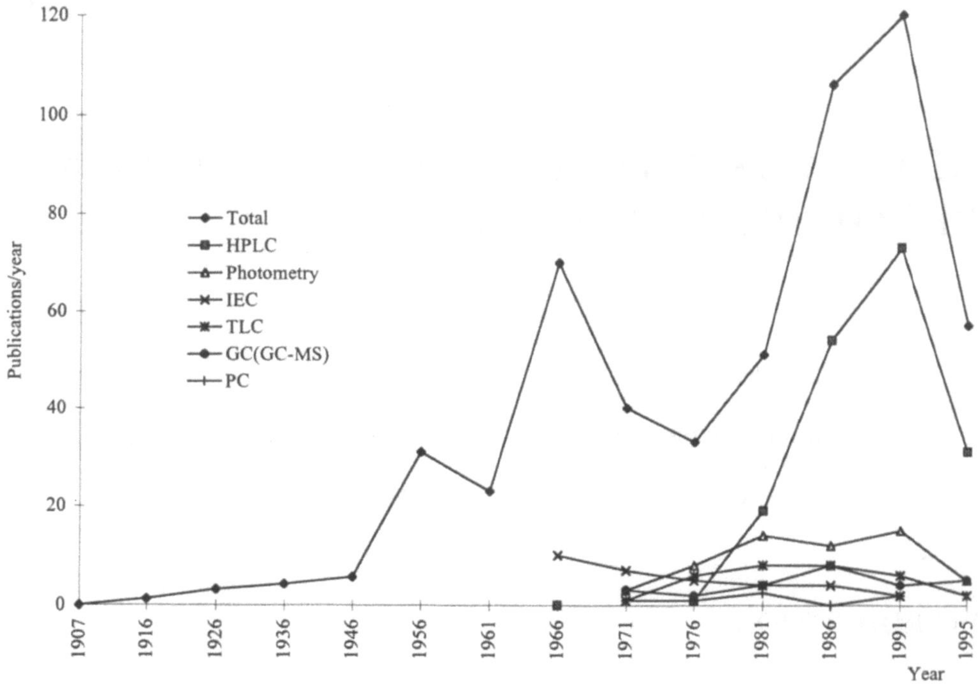


**Figure 1.** Development of methods in tryptophan analysis.

Jensen, 1993; Ng et al., 1987; Muramoto and Kamiya, 1990; Molnár-Perl and Fábián-Vonsik, 1988; Fábián et al., 1990; 1991; Morvai et al., 1992; Molnár-Perl and Morvai, 1992; Molnár-Perl and Khalifa, 1993; 1994; Tatár et al., 1994; Molnár-Perl, 1994) which have been put into the field of tryptophan quantitation in this century, performed in order to find the optimum solution for its maximum recovery and selective determination. After an exhaustive 'general' literature overview, based on the Chemical Abstracts, (Figure 1) it became clear

i. that scientists still are looking for satisfactory alternatives, and

ii. the increasing amounts of newly developed analytical procedures are mainly chromatographic ones, representing, in the 'peak period' of method development, in 1991, the 73 % of the total: in accordance with the 'boom' of high performance liquid chromatography (HPLC) (Figure 1.). This distribution proved to be characteristic to methods published not only for quantitation of tryptophan in free form, including its metabolites, but also for those of proteins and peptides, in thorough connection with the corresponding hydrolysis procedures they are originating from.

Thus, based on the above detailed tendency, this paper aims to be an update of the earlier compilation (Friedman and Cuq, 1988) with particular attention to the development of chromatographic methods.

# 1. FREE TRYPTOPHAN AND ITS METABOLITES

## 1.1. Gas Chromatography (GC)

It is very well known that amino acids are to be present in impurity and water free condition in order of their acylation and esterification prior to GC analysis (Molnár-Perl et al., 1992. In addition, the quantitative acylation of amino- and indolyl groups of tryptophan arose special attention (composition of acylation solution, temperature and time, proved to be critical). Due to latter difficulties only few researcher utilize the advantage of the very high sensitivity and selectivity of GC. The free tryptophan content of 20-100 mg crude extracts obtained from various plant samples, subsequently to solid phase extraction and HPLC separation have been unambiguously identified and determined as their N-acetyl (m/z = $M^+$ = 260) and N-diacetyl (m/z = $M^+$ = 302) tryptophan methyl ester derivatives (Michalczuk et al., 1992).

## 1.2. High Performance Liquid Chromatography (HPLC): *Tryptophan Only*

High performance liquid chromatography without derivatization (Gjerde et al., 1987; Peterson et al., 1987; Kim and Angyal, 1993; Morita et al., 1985; 1990; Caruso et al., 1994; Berardino et al., 1990; Aymard et al., 1994; Servillo et al., 1982; Nozaki, 1990) offers in quantitation of tryptophan (Gjerde et al., 1987; Peterson et al., 1987; Kim and Angyal, 1993) and also in those of their metabolites (Morita et al., 1985; 1990; Caruso et al., 1994; Berardino et al., 1990; Aymard et al., 1994; Servillo et al., 1982). (i) Particularly comfortable (previously only deproteinization is to be carried out), and, (ii) relatively selective conditions (both UV and fluorescens detection can be applied).

Tryptophan content of blood samples (Gjerde et al., 1987; Peterson et al., 1987) were measured using isocratic elution, (column: 25cm x 4.6mm), by UV detection, within 6 min, in the range of 0.1-100 μmole (Gjerde et al., 1987), while miniaturized conditions (column: 8.3cm x 4.6mm, 3 μm) provided considerably lower detection limit: 5 pmoles using UV detection and 1 pmoles applying fluorescence one (Peterson et al., 1987). In order to determine the L-tryptophan content of tablets and capsules by HPLC, in comparison to the official AOAC microbiological method, its phenylisothiocianate derivative was measured (Kim and Angyal, 1993). Results revealed that out of the 'warm' and cold' microbiological procedure (if any) the cold is to be followed (the 'warm' leads to 4-8 % tryptophan losses), and HPLC having the advantages of using a smaller test portion and having a shorter analysis time.

## Tryptophan Together with Its Metabolites

Being tryptophan an essential amino acid whose metabolism involves several pathways, its quantitation together with its numerous metabolites (Morita et al., 1985; 1990; Caruso et al., 1994), or together with tyrosine (Berardino et al., 1990) only, and/or with 5-hydroxytryptamine (Aymard et al., 1994) only, proved to be of primary importance by HPLC (Morita et al., 1985; 1990; Caruso et al., 1994; Berardino et al., 1990; Aymard et al., 1994).

Major tryptophan metabolites of kynurenine and that of indole pathways simultaneously (Morita et al., 1985, 1990) have been determined by UV and fluorescence detections: first (Morita et al., 1985) from 50-500 μl, later (Morita et al., 1990) from 100 μl direct plasma injections. Performing a special, automated precolumn deproteinization, within 40 min, 17 metabolites, present in human serum and/or plasma, have been measured in pmol ranges. In order of their elution: **1** 5-hydroxytryptophan, **2** hydroxytryptamine, **3** 5-hydroxytryptophol, **4** tryptophan **5** 5-hydroxyindole-3-acetic acid, **6** indole-3-lactic acid, **7** indole-3-acetic acid, **8** tryptophol, **9** indole-3-propionic acid, **10** indole-3-butyric acid, **11** 3-hydroxykinurenine, and **12** kinurenine by fluorescence detection, while **13** 3-hydroxyanthranilic acid, **14** xanthurenic acid, **15** kynurenic acid, **16** quinaldic acid and **17** anthranilic acid by UV detection). Multiple ion detection system (Caruso et al., 1994) (UV diode array detector connected on line with fluorescence one) provided simultanaeously UV spectra in the 190 - 450 nm range and fluorescence spectra setting two different excitation (280-, 340 nm) and emission (340-, 440 nm) wavelengths: 13 metabolites have been evaluated in 23 min. Practical utilization was not given. Tryptophan and tyrosine content of 250 μl blood (Berardino et al., 1990), from perchloric acid deproteinized plasma, were quantitated by fluorescence detection.

Hydroxytryptamine and tryptophan (Aymard et al., 1994) were determined from 450 μl whole blood sample within 10 min, (column: 15cm x 4.6mm) applying fluorescence detection. The method was utilized in the differentiation of healthy control subjects and depressed patients of major mental disorders.

## 2. TRYPTOPHAN IN INTACT PROTEINS: SPECTROPHOTOMETRY

### 2.1 Derivative Spectrophotometry without Chemical Derivatization

The characteristic absorbances of aromatic amino acids have been quantitated in the low UV region: based on the distances between the single positive/negative maxima, applying second derivative spectrometry, in solutions of pH close to seven (Servillo et al., 1982; Nozaki, 1990). An early approach (Servillo et al., 1982) utilized the ratios of the vertical distances of the adjacent peaks and throughs of tyrosine/tryptophan, as a measure of their second derivative absorbancies: being these distances dependent of each other, consequently their amounts were not linear with their concentrations. Improvements (Nozaki, 1990) have been reported in the simultaneous evaluation by second derivative spectrometry of tryptophan (at ~290 nm), tyrosine (at~283 nm) and phenylalanine (~258 nm and ~264), applying a new principle, utilizing the vertical distance from the baseline to the second derivative spectral curves only. Tryptophan's (Fletouris et al., 1993) and tyrosine's (Botsoglou et al., 1993) quantitation on the base of their fourth derivative UV absorption spectra, can be regarded as a considerably new approach. Spectra evaluations obtained in 0.1 M sodium hydroxide solution or in supernatants of protein extracts provided excellent results. Tryptophan (Fletouris et al., 1993) and tyrosine (Botsoglou et al., 1993) contents both of standard proteins and those of protein containing matrices proved to be in thorough accordance with literature data.

### 2.2. Spectrophotometric Determinations Subsequently to Chemical Derivatizations

*2.2.1. Chemical Reactions with Aldehydes.* A major part of spectrophotometric methods of tryptophan quantitation are based on the interaction of the indole ring with aldehydes (Spies and Chambers, 1949; 1950; Spies, 1950; 1967; Al-Tamrah, 1989; Iizuka and Yajima, 1985; 1993; Kell et al., 1989; Opienska-Blauth et al., 1963; Concon, 1975; Gorinstein et al., 1988; Gaitonde and Dovey, 1970; Zahnley and Davis; 1973; Sodek et al., 1975; Molnár-Perl, and Pintér-Szakács, 1989; Pintér-Szakács and Molnár-Perl, 1990). The reaction of p-dimethylaminobenzaldehyde (DABA) in sulphuric acid media (Spies and

Chambers, 1949; 1950; Spies, 1950; 1967) has been exhaustively examined and widely utilized in its determination also in hydrolyzates (Spies, 1950; 1967). A simplified version of the DABA method using 10 M hydrochloric acid instead of sulphuric acid was reported for free tryptophan only (Al-Tamrah, 1989). Fluorescence derivatives, formed from tryptophan both with chloracetaldehyde (Iizuka and Yajima, 1985) and methoxyaldehyde (Iizuka and Yajima, 1993), remained on the level of model reactions (Iizuka and Yajima, 1985; 1993). Similarly, the kinetic study of the reaction of tryptophan with formaldehyde proved to be of theoretical importance (Kell et al., 1989) only. Practical utilizations have been published (Opienska-Blauth et al., 1963; Concon, 1975; Gorinstein et al., 1988) based on reaction between acetic acid/ferric chloride, (in concentrated sulphuric acid resulted in glyoxylic acid) and tryptophan residues of proteins. The tryptophan contents of γ-globulins and casein (Opienska-Blauth et al., 1963), as well as those of numerous cereal grains (Concon, 1975) and potatoes (Gorinstein et al., 1988) have been quantitated measuring the absorbances of the resulting violet solution at 545 nm in all cases (Opienska-Blauth et al., 1963; Concon, 1975; Gorinstein et al., 1988).

The most promising tryptophan/aldehyde reaction performed with the intact protein matrix proved to be the 'acid ninhydrin method': because of its selectivity, general applicability and outstanding reproducibility. Earlier works (Gaitonde and Dovey, 1970; Zahnley and Davis, 1973; Sodek et al., 1975) proposed long lasting (16-20 hrs) incubation of intact proteins and insufficient reaction time for the quantitative interaction of tryptophan(residue) with the ninhydrin reagent. Recently (Molnár-Perl, and Pintér-Szakács, 1989; Pintér-Szakács and Molnár-Perl, 1990), exhaustive kinetic study was performed (with tryptophan, tyrosine and lysozyme) in order to obtain optimum analytical conditions providing: (i) stoichiometric reaction both with free tryptophan and tryptophan residues of proteins (including also protein containing matrices, such as feed- and food stuffs), (ii) quantitative elimination of the disturbing effect of tyrosine. Applying optimum analytical conditions (35 °C, 90 min, 1% ninhydrin prepared with a solution containing HCl acid (37%) and formic acid (82-92%) in a volume ratio of 2/3). Optimum conditions allowed the direct interaction of samples, using the same molar absorptivity values for free and peptide linked tryptophan. The practical utilization of the improved procedure was shown by the tryptophan quantitation of numerous standard proteins (lysozyme, human and bovine serum albumin, casein, α-chymotrypsine, trypsine) and protein matrices (soy bean, wheat, maize, barley, rye, sun flower, fish and meat meals).

*2.2.2. Colour Reactions: by Means of Others Than Aldehydes.* The acid stable product of tryptophan and xanthydrol (Dickman and Crockett, 1956) proved to be proper to the quantitation of free tryptophan only. Xanthydryltryptophan could not be obtained from tryptophan residues in a reasonable reaction time without the interference of tyrosine. As a highly selective tryptophan reagent (Barman and Koshland, 1967) 2-hydroxy-5-nitrobenzyl bromide was reported. It could not achieve wide acceptance in analysis of tryptophan residues requiring a long lasting (16-20 hrs) incubation of proteins, in 10 M urea/HCl acid buffer of pH = 2.7. Nitrous acid oxidation of the heterocyclic nucleus in tryptophan (Verma and Jain, 1988), resulting in a nitro compound, in alkaline solution became golden yellow: proper for spectrophotometric determination, having a maximum at 400 nm. A new combination (Chrastil, 1986) was reported for the simultaneous tryptophan/tyrosine quantitation: (i) the coloured derivative originating from the formaldehyde/tryptophan reaction was measured at 450 nm, while (ii) the total of tryptophan and tyrosine, based on their reaction with hydroxyl amine and Ce(IV), in sulphuric acid media, at 355 nm. The reaction time for both reactions were 3-5 hrs; Both colours proved to be stable and the reactions stoichiometric, but applicable for dissolvable peptides and proteins only (myoglobuline, papain, ribonuclease, ovalbumin, and chymotrypsin).

# 3. TRYPTOPHAN IN PROTEIN HYDROLYZATES

Those efforts which aim to determine tryptophan in hydrolyzates should be divided into two basically different groups. Aiming the quantitation of tryptophan only, the use of various alkalines (NaOH, KOH, LiOH, $Ba(OH)_2$) have been investigated and proposed. In order to determine tryptophan together with all other amino acids, hydrolyses by enzymes, by organic acids, and also by hydrochloric acid with different additives have been suggested. In general, it can be stated that the cumbersome, time consuming and tedious hydrolyses procedures proved to be worth to undertaken in those cases exclusively when hydrolyses will be followed by one of the selective chromatographic techniques. Tryptophan monitoring in hydrolyzates by any of the colour reactions, followed by spectrophotometry can be regarded as an unnecessary luxury. Consequently, special problems in chromatography of tryptophan alone and together with all other amino acids, (subsequently to their various hydrolyses techniques), being the most relevant topic, will be given in details: applying ion exchange chromatography (IEC), (carried out with amino acid analyzators), GC and HPLC.

## 3.1. Chromatography of Tryptophan Only

This field of researches, including inter laboratory comparative studies, covers optimatization tests concerning simultaneously the quality and quantity of alkalines, the temperature and time of hydrolyses and the conditions of chromatographic elution and quantitation procedures as well.

*3.1.1. Ion Exchange Chromatography.* The selective identification and quantitation of tryptophan in alkaline hydrolyzates started in the early seventies by means of IEC (Hugly and Moore, 1972; Sato et al., 1984; 1987; Knox et al., 1970). At that time, taking into account the required conditions of IEC, sodium hydroxide (Hugly and Moore, 1972; Sato et al., 1984; 1987)(4.2 M, at 110 °C, 22 hrs (Hugly and Moore, 1972) and 20 hrs (Sato et al., 1984; 1987)), providing 85-95% tryptophan recovery was preferred to barium hydroxide (Knox et al., 1970) (4.2 M, 110 °C,16 hrs) that needed an elimination step in form of $BaSO_4$, or $BaCO_3$ resulting in additional 4-10 % losses of tryptophan, prior to its elution by IEC. The elution procedure only, by IEC, carried out at 52 °C, in all four cited cases, needed also on short columns (0.9 x 7 cm), as an optimum achievement, 30 min (Knox et al., 1970), while others, remaining on the safe side, in sake of better resolution, performed 90 min (Sato et al., 1984; 1987)-120 min (Hugly and Moore, 1972) elution times (regeneration not included).

*3.1.2. High Performance Liquid Chromatography.* In the light of the above mentioned (paragraph 3.1.1.), no doubt about it, a considerably faster chromatographic technique of higher sensitivity, opened a new chapter for tryptophan analysis alone, in alkaline protein hydrolyzates: in respect of practical and scientific aspects equally. The faster obtainable chromatographic information promoted a repeated review also in optimatization (Werner, 1986; Bech-Andersen, 1991) and in comparison (Lucas and Sotelo, 1980; Nielsen and Hurrell, 1985; Allred and MacDonald, 1988; Ranfft and Faure, 1993; Landry et al., 1992) of hydrolysis techniques in thorough connection with chromatography.

**Lithium hydroxide** (4 M, 110-120 °C, 22-26 hrs) (Werner, 1986) has been advised for the quantitation of trytophan in numerous food- and feed stuff hydrolyzates, obtained by direct injections, with linear responses, by fluorescence detection, in the 0.2-100 ng range. Tryptophan was eluted with 5 min retention time, recovery of the method was not reported.

**Sodium hydroxide** was proposed (4,2 M, 110 °C, 16 hrs, in autoclave) (Bech-Andersen, 1991), for hydrolysis of feed stuffs in the presence of lactose (applying as protecting agent against tryptophan losses). In calculation of tryptophan values, in order of correcting losses, α-methyl-tryptophan was applied as internal standard. Recovery data were not given.

The renascence of **barium hydroxide** digestions (Delhaye and Landry, 1986; 1993; Landry et al., 1988; Landry and Delhaye, 1992; 1993; Aiple und Menke, 1990) was associated with the development of HPLC, in particular by fluorescence detection. In the possesion of this very sensitive and selective technique, - due to the possibility of direct injections of diluted hydrolyzates, barium ion elimination became unnecessary. Consequently, the main disadvantage, the additional tryptophan loss contributed to its adsorption on barium precipitates counted no longer. The poor solubility of barium hydroxide still remained a problem. Exhaustive basic studies (Delhaye and Landry, 1986; 1993; Landry et al., 1988; Landry and Delhaye, 1992) performed by barium hydroxide hydrolyses clarified important characteristics. Investigating the peak purity parameters by diode array detection (Delhaye and Landry, 1986) of tryptophan, obtained from feedstuffs (rich in UV absorbing constituents), it could be stated, that in barytic hydrolyzates (~1.66 M $Ba(OH)_2$, 125°C, 16 hrs) injected after neutralization directly (without any previous pretreatment), no UV interfering substance was coeluted with tryptophan. Lower barytic (1.27 M) concentrations (Landry et al., 1988), in the presence of 5 % thiodiglycol, were performed at 105- and 125°C temperatures. The advantage of HPLC with fluorescence detection has been repeatedly shown: low detectability (20-2000 ng) and short analysis time (5 sample/hr). In the hydrolysis of 14 different feed stuffs 99.7 ±1 % tryptophan recoveries were reported, indicating neither losses nor incomplete liberation occurred at conditions advised (based on additive hydrolysis of lysozyme). The statement of quantitative tryptophan recovery has been later on modified by others (Lucas and Sotelo, 1980; Nielsen and Hurrell, 1985) and also by the same authors (Allred and MacDonald, 1988; Ranfft and Faure, 1993; Landry et al.,1992; Delhaye and Landry, 1992; Landry and Delhaye, 1994a; 1994b). In addition of several simplifications (Landry and Delhaye, 1992; 1993; Delhaye and Landry, 1993), such as, decreasing the injection's volume (Landry and Delhaye, 1992), the hydrolysis time simplifications (Landry and Delhaye, 1993; Delhaye and Landry, 1993), (16-, 8-, 4 hrs) by increasing the hydrolysis temperature (Landry and Delhaye, 1993; Delhaye and Landry, 1993), (110-. 125-, 140 °C) at higher barium hydroxide concentrations (2.7 M) (Delhaye and Landry, 1993; Landry and Delhaye, 1993) and 3.3 M (Aiple und Menke, 1990), simultaneously, the use of internal standards (5-methyl-tryptophan (Landry and Delhaye, 1992), and/or both α-methyl- and 5-methyl-tryptophan (Landry and Delhaye, 1993)) was ultimately introduced. Comparisons, based on intra- (Lucas aan Sotelo, 1980; Nielsen and Hurrell, 1985; Delhaye and Landry, 1992; Landry and Delhaye, 1994b) and inter (Allred and MacDonald, 1988; Ranfft and Faure, 1993; Landry et al., 1992; Landry and Delhaye, 1994a) laboratory studies revealed that: (i) 100 % tryptophan recovery can not be obtained by any of the alkaline hydrolysis procedures (Lucas and Sotelo, 1980; Nielsen and Hurrell, 1985; Allred and MacDonald, 1988; Ranfft and Faure, 1993; Landry et al., 1992; Delhaye and Landry, 1992; Landry and delhaye, 1994a; 1994b). (ii) Perfoming IEC (Lucas and Sotelo, 1980), subsequently to the hydrolysis (4 - 8 hrs), best results (5 % losses) have been obtained by 4 M LiOH, in the presence of 5 % $SnCl_2$. Applying for separation and quantitative evaluation HPLC, subsequently to sodium hydroxide hydrolyses, though the base of comparison (5-methyl-tryptophan) was the same, contradictory experiences have been reported: Accordingly earlier studies (Nielsen and Hurrell, 1985) NaOH and LiOH provided similar tryptophan recoveries, (between 88.1 % and 98.5 % ), while recently published data published also by the same authors (Landry et al., 1992; Landry and Delhaye, 1994b) revealed the outstanding necessity of oxygen elimination. Without applying autoclave purging, in comparison of 1.3 M $Ba(OH)_2$ (125 °C, 16 hrs) and 4 M LiOH (110 °C, 24 hrs) they proved to be as alkalines of identical value (Landry et al., 1992): ensuring 82.7-85.2 % recovery, without correction. Later on, performing $Ba(OH)_2$, allowing the quantitative oxygen elimination before hydrolysis, by autoclave purging at 100 °C, 'peak tryptophan recoveries' have been reported (96.1-102.4 %) (Delhaye and Landry, 1992).

**Inter laboratory studies** (Allred and MacDonald, 1988; Ranfft and Faure, 1993; Landry and Delhaye, 1994a) presented less promising results. In a comprehensive work (Allred and MacDonald, 1988) applying partly IEC, partly HPLC, performing hydrolyses uniformly by 4.2 M NaOH, within and between laboratories quite the same variabilities have been reported, in order of listing 3.87 -16.1 % and 3.87-16.5 %, respectively. Regarding the data of 14 laboratories (Ranfft and Faure, 1993), as a function of the matrix analyzed, they varied between 3.87 and 26.4 % (CV %). The overall means in the case of other 23 laboratories (Landry and Delhaye, 1994a) proved to be 86.5 ± 5.6 %.

## 3.2. Chromatography of Tryptophan Together with All Amino Acids Present in Protein Hydrolyzates Obtained by Different Ways

*3.2.1. By Enzymes.* Neglecting the earlier efforts (Curzon, 1992; Friedman and Cuq, 1988), relying on a recently presented work only (Garcia and Baxter, 1992), pronase digestion provided excellent characteristics. Enzymatic digestions have been accomplished in less than 6 hrs under mild conditions (pH 8.5, 50 °C). Spectacular chromatographic condition was shown: injection-to-injection times of 8 min, including tryptophan and 5-methyl-tryptophan (internal standard). Quantitation based both on UV (diode array proven) and fluorescence spectra could be carried out without disturbing interferences.

*3.2.2. By Sulfonic Acids + Additives.* **Methanesulfonic acid.** Accordingly the first proposal (Simpson et al., 1976) 4 M methanesulfonic acid containing 0.2 % 3-(2-aminoethyl)-indole (henceforth: tryptamine) have been used also in subsequent papers (Chiou and Wang, 1988; 1989; Malmer and Schroeder, 1990). Since a competition has been observed between the decomposition of serine, threonine, tryptophan and the release of valine, as a result of preliminary studies as optimum condition 115 °C, 22-72 hrs was suggested. The quantitation of 18 amino acids, including cysteine and tryptophan, has been reported by IEC, from lysozyme hydrolyzate: theoretical values were indicated without reproducibility data (Simpson et al., 1976). Performing microwave irradiation (Chiou and Wang, 1988; 1989) followed by IEC, the time of hydrolyses could be dramatically shortened: at 160 °C 45 min (Chiou and Wang, 1988), while at 178 °C and at 186 °C 4 min (Chiou and Wang, 1989) proved to be satisfactory. ~ 98 % tryptophan recoveries have been reported in both cases (Chiou and Wang, 1988; 1989). Methanesulfonic acid hydrolysis of cytochrome c (Malmer and Schroeder, 1990) was carried out at 115 °C for 22- and 70 hrs and at 150 °C for 22 hrs. Results proved that 115 °C for 22 hrs should be chosen as optimum condition. 16 amino acids, including cysteine and tryptophan have been quantitated by HPLC, subsequently to their direct derivatization in the hydrolyzate by 9-fluorenylmethylchloroformate.

**p-Toluenesulfonic acid.** The principle of the use of 3 M p-toluenesulfonic (Liu and Chang, 1971)- and 4 M methanesulfonic (Simpson et al., 1976) acids, both in the presence of 0.2 % tryptamine, have been introduced by the same researchers: they stated that p-toluenesulfonic acid is inferior to methanesulfonic one providing 'lower' tryptophan recovery which proved to be by IEC between 91 and 98 %. This early experience (Liu and Chang, 1971) has been repeatedly tested by microwave irradiation followed by HPLC after preliminary derivatization with o-phthalaldehyde (Pecavar et al., 1990). Tryptophan recovery of model solution, irradiated by 2450 MHz frequency for 15 min (corresponding temperature was not given), proved to be 98.6%.

**Mercaptoethanesulfonic acid**. The advantages of the use of 3 M mercaptoethane sulfonic acid without additives (Penke et al., 1974) has been presented in comparison with p-toluenesulfonic acid + 0.2 % tryptamine and with 6 M hydrochloric acid + 4 % thiolycolic acid: all three type of hydrolyses have been followed for 22- and 72 hrs, at 110 ± 2°C with lysozyme, chymotrypsine and cytochrome c. Data obtained proved that in all cases tested maximum tryptophan recovery were obtained at 110°C after 22 hrs by mercaptoethanesulfonic acid: in order of listing above, tryptophan contents proved to be 94.5 %, 96 % and 93 %, while hydrochloric acid with thioglycolic acid provided the lowest recovery, presented in the same order by 86.5 %, 81.5 % and 75 %. Later on difficulties have been reported (Gardner, 1984; Csapó et al., 1986) in quantitation by IEC. Due (i) to the formation of a mixed disulfide from mercaptoethanesulfonic acid and cysteine (Csapó et al., 1986), as well as (ii) to the coelution of cysteine with proline (Csapó et al., 1994). These shortages could be solved by modifying the elution program of IEC (Csapó et al., 1986; 1994).

*3.2.3. By Hydrochloric Acid + Additives.* In the classical 6 M hydrochloric acid hydrolyses the overwhelming part of tryptophan (70 -100 % of the total present in protein) get distroyed. The reason for that needs still further explanation. Being hydrochloric acid the most advantageous hydrolysing agent from numerous point of view researchers tried to prevent tryptophan destruction applying hydrochloric acid + different inhibitors.

**l1,2 ethane-dithiol** (Felker, 1976), in absence of oxygen, was proposed for preventing destruction of cyst(e)in, tryptophan and histidine during HCl acid hydrolyses of lysozyme as model compound. Resulting amino acids have been measured by GC as their N-acetylated methyl esters. Correction for losses during hydrolysis have been done on the base of isotope dilution assay: tryptophan's recovery proved to be 75.4 %.

**Pyridine borane** (Wong et al., 1984) reduces tryptophan in the presence of trifluoroacetic acid to dihydrotryptophan which appeared as a doublet between lysine and histidine measured by IEC. As to their HPLC quantitation difficulties have been reported: the phenylthiocarbamyl derivatives could not be detected, (likely covered by the reagent peaks), while as dabsyl derivatives the first eluting member of the doublet coeluted with tyrosine. Thus, hydrolysis studies based only on IEC confirmed quite quantitative recoveries compared to theoretical values in the case of twelve standard proteins.

**Thioglycolic acid** (Matsubara and Sasaki, 1969; Ohta and Nakai, 1979; Ashworth, 1987; Yano et al., 1990; Sottrup-Jensen, 1993) was tested at first (Matsubara et al., 1969) in 0.5 - 6 % (v/v) concentration added to 6 M HCl. Maximum tryptophan yield from lysozyme (84 %) and cytochrome c (86 %) were obtained with 2 % thioglycolic acid containing HCl performing hydrolyses in evacuated tubes for 24 Hs at 108-110°C, measured by IEC. Thioglycolic acid or hydroquinone have been applied in a theoretical study (Ohta and Nakai, 1979) aiming to clarify the mechanism of tryptophan decomposition. Results revealed that probably cystine is responsible for tryptophan destruction: due to its transformation during HCl acid hydrolyses to bis-(2-amino-2-carboxyethyl) trisulfide: its formation, in the presence of hydroquinone or thioglycolic acid proved to be completely inhibited. With shortened hydrolysis time (Ashworth, 1987) (4hrs, 145°C), applying the first introduced procedure (Matsubara and Sasaki, 1969), the principle has been extended for the analysis of various meet proteins by IEC and by HPLC, simultaneously. Hydrolyses in vapour phase (Yano et al., 1990) with 7 M hydrochloric acid containing 10 % trifluoroacetic- and 20 % thioglycolic (Bp: 123°C, under 26 mm Hg) acids were carried out at 166°C, for 15 min and 30 min. After 15 min hydrolysis time tryptophan recovery from model solutions proved to be 86 %, while in the case of myoglobine, papaine and lysozyme, in order of listing, 95-, 80- and 80 % were obtained. From reducing hydrolyzates (Sottrup-Jensen, 1993), with 6 M HCl, containing 5 % thioglycolic acid + 0.1 % phenol, at 118°C, after 18 hrs 77-, 86- and 82 % tryptophan recoveries have been reported from chymotrypsine, trypsine, and $\alpha_2$- macroglobulin subunit.

**β-Mercaptoethanol** (0.2 %) in 6 M HCl (Ng et al., 1987), performed for 24 hrs, at 110°C in order to determine all amino acids, including tryptophan in numerous protein hydrolyzates. Amino acid constituents of several proteins were quantitated as their phenylthiocarbamyl derivatives by HPLC: tryptophan recoveries varied between 93 and 102%.

**Phenol** (Ng et al., 1987) added in 3 v/v % to 6 M HCl 'largely' prevented the destruction of tryptophan during rapid hydrolyses at 166°C, for 25 min, or at 145°C, for 4 hrs. The largely prevented status indicates at least 80 % recovery.

**Hydrochloric acid + tryptamine.** (Molnár-Perl and Fábián-Vonsik, 1988; Molnár-Perl et al., 1987; 1992; 1993; Fábián et al., 1990; 1991; Morvai et al., 1992; Molnár-Perl and Morvai, 1992; Molnár-Perl and Khalifa, 1993; 1994; Tatár et al., 1994; Molnár-Perl, 1994); Exhaustive studies have been performed in order to optimize the preferred (i) chromatographic procedure, (GC (Molnár-Perl and Fábián-Vonsik, 1988; Molnár-Perl et al., 1987; 1992; 1993; Fábián et al., 1990; 1991) or HPLC (Morvai et al., 1992; Molnár-Perl and Morvai, 1992)), (ii) the ratios of tryptamine to proteins (Fábián et al., 1990; Molnár-Perl et al., 1993; Molnár-Perl and Khalifa, 1994),(1/1 = w/w ratios proved to be the optimum), (iii) the shortest hydrolysis time coupled with the lowest necessary temperature in liquid phase-(Fábián et al., 1990; 1991) and in vapour phase (Molnár-Perl et al., 1993; Molnár-Perl and Khalifa, 1993; 1994; Molnár-Perl, 1994) equally, (iv) both by simple heat and by microwave irradiation (Tatár et al., 1994) as well. Based on hundreds and hundreds of hydrolyses, followed by GC of the N,O,(S)trifluoroacetyl n- and isobutyl amino acid esters (Molnár-Perl and Fábián-Vonsik, 1988; Molnár-Perl et al., 1987; Fábián et al., 1990; 1991), and, by HPLC of the phenylthiocarbamyl amino acid derivatives (Morvai et al., 1992; Molnár-Perl and Morvai, 1992; Molnár-Perl et al., 1992; 1993; Molnár-Perl and Khalifa, 1993; 1994; Tatár et al., 1994; Molnár-Perl, 1994), (carried out with standard proteins and protein matrices, such as feed and food stuffs), it can be concluded as follows: (i) tryptamine protection can be achieved between 80 and 100 %. (ii) In liquid phase hydrolyses, followed by GC, the tryptophan contents of standard proteins and protein matrices have been recovered equally: because of the coelution of the lysine and tryptamine derivatives by GC, prior to derivatization, the cation exchange clean up of the hydrolyzates, in order of tryptamine's elimination could not be avoided. (iii) Applying vapour phase hydrolyses and HPLC by the quantitation of phenylthiocarbamyl derivatives no interferences of tryptamine could be detected (Molnár-Perl et al., 1993; Molnár-Perl and Khalifa, 1993): tryptophan recovery of pure proteins and peptides proved to be ≥80 %. (iv) Concerning microwave digestion (Tatár et al., 1994) its advantages, reported by others (Chiou and Wang, 1988, 1989), were not confirmed.

In summary it can be stated that

1. Selective chromatographic techniques, preferably HPLC should be followed for pure proteins and/or peptides, in all those tasks when either the amounts of
   - free tryptophan and its metabolites are to be measured, or,
   - the quality and quantity of all amino acids, including tryptophan, present in hydrolyzates are to be known.
2. The tryptophan quantitation of feed and foodstuffs, in particular of those of high saccharide containing ones, can be performed most advantageously with intact proteins: applying the acid-ninhydrin method (Michalczuk et al., 1992).

## ACKNOWLEDGMENT

This work was supported by the Hungarian Academy of Sciences (Project No.: OTKA I/4, T5053.

# 4. REFERENCES


Aiple K. und Menke K.H., 1990, Tryptophanbestimmung in futtermitteln mit hilfe der HPLC wiederfindung von tryptophan aus peptiden und lysozym, *Agribiol. Res.*, 43:200.

Allred M.C. and MacDonald J.L., 1988, Determination of sulfur amino acids and tryptophan in foods and food and feed ingredients: Collaborative study, *J. Assoc. Off. Anal. Chem.*, 71:603.

Al-Tamrah S.A., 1989, Kinetic studies on the determination of pyrroles and tryptophan by Ehrlich reaction, *Anal. Lett.*, 22:387.

Ashworth R.B., 1987, Amino acid analysis for meat protein evaluation, *J. Assoc. Off. Anal. Chem.*, 70:80.

Aymard N., Honore P. and Carbuccia I., 1994, Determination of 5-hydroxytryptamine and tryptophan by liquid chromatography in whole blood. Its interest for the exploration of mental disorders, *Prog. Neuro-Psychopharmacol. & Biol. Psychiat.*, 18:77.

Barman T.E. and Koshland D.E., 1967, A colorimetric procedure for the quantitative determination of tryptophan residues in proteins, *J. Biol. Chem.*, 242:5771.

Bech-Andersen S., 1991, Determination of tryptophan with HPLC after alkaline hydrolysis in autoclave using methyl-tryptophan as internal standard, *Acta Agric. Scand.*, 41:305.

Berardino M.B., Roingeard F.C. and Fukagawa N.K., 1990, Plasma tryptophan and tyrosine concentrations: determination using high performance liquid chromatography and fluorometric detection, *J.Nutr. Biochem.*, 1:220.

Botsoglou N.A., Fletouris D.J., Papageorgiou G.E. and Mantis A.J., 1993, Derivative spectrophotometric method for the analysis of tyrosine in unhydrolyzed protein, food, and feedstuff samples, *J. Agric. Food Chem.*, 41:1635.

Caruso U., Fowler B., Minniti G. and Romano C., 1994, Determination of tryptophan and ten of its metabolites in a single analysis by high-performance liquid chromatography with multiple detection, *J. Chromatogr.*, 661:101.

Chiou S.H. and Wang K.T., 1988, Simplified protein hydrolysis with methanesulphonic acid at elevated temperature for the complete amino acid analysis of proteins, *J. Chromatogr.*, 448:404.

Chiou S.H. and Wang K.T., 1989, Peptide and protein hydrolysis by microwave irradiation, *J. Chromatogr.*, 491:424.

Chrastil J., 1986, Spectrophotometric determination of tryptophan and tyrosine in peptides and proteins based on new color reactions, *Anal. Biochem.*, 158:443.

Concon J.M., 1975, Rapid and simple method for the determination of tryptophan in cereal grains, *Anal. Biochem.*, 67:206.

Csapó J., Penke B., Tóth-Pósfai I. and Csapó-Kiss Zs., 1986, Determination of the cystine content of foods and feeds by mercapto-ethane-sulfonic acid hydrolysis, *Acta Alimentaria*, 15:227.

Csapó J., Csapó-Kiss Zs., Folestad S. and Tivesten A., 1994, Mercaptoethanesulphonic acid as a protecting and hydrolysing agent for the determination of the amino acid composition of proteins using an elevated temperature for protein hydrolysis, *Anal. Chim. Acta*, 289:105.

Curzon G., 1992, Tryptophan as precursor of brain 5-hydroxytryptamine: An update, in: Advances in Tryptophan Research 1992, I. Ishiguro, R. Kido, T. Nagatsu, Y. Nagamura, Y. Ohta, eds., (Fujita Health University Press)

Delhaye S. and Landry J., 1986, High-performance liquid chromatography and ultraviolet spectrophotometry for quantitation of tryptophan in barytic hydrolysates, *Anal. Biochem.*, 159:175.

Delhaye S. and Landry J., 1992, Determination of tryptophan in pure proteins and plant material by three methods, *Analyst*, 117:1875.

Delhaye S. and Landry J., 1993, Quantitative determination of tryptophan in food and feedstuffs: Practical considerations on autoclaving samples for hydrolysis, *J. Agric. Food Chem.*, 41:1633.

Dickman S.R. and Crockett A.L., 1956, Reactions of Xanthydrol. IV. Determination of tryptophan in blood plasma and in proteins, *J. Biol. Chem.*, 220:957.

Fábián V., Pintér-Szakács M. and Molnár-Perl I., 1990, Gas chromatography of tryptophan together with other amino acids in hydrochloric acid hydrolysates, *J. Chromatogr.*, 520:193.

Fábián V., Morvai M., Pintér-Szakács M. and Molnár-Perl I., 1991, Criticisme and standardization of the cation-exchange clean-up procedure prior to GLC analysis of amino acids present in protein hydrolyzates, *J.Chromatogr.*, 553:87.

Felker P., 1976, A gas-liquid chromatographic-isotope dilution analysis of cysteine, histidine and tryptophan in acid-hydrolyzed protein, *Anal. Biochem.*, 76:192.

Fletouris D.J., Botsoglou N.A., Papageorgiou G.E. and Mantis A.J., 1993, Rapid determination of tryptophan in intact proteins by derivative spectrophotometry, *J. AOAC Intern.*, 76:1168.

Friedman M. and Cuq J.-L., 1988, Chemistry, analysis, nutritional value, and toxicology of trytophan in food. A review, *J. Agric. Food Chem.*, 36:1079.

Gaitonde M.K. and Dovey T., 1970, A rapid and direct method for the quantitative determination of tryptophan in the intact protein, *Biochem. J.*, 117:907.

Garcia S.E. and Baxter J.H., 1992, Determination of tryptophan content in infant formulas and medical nutritionals, *J. AOAC Intern.*, 75:1112.

Gardner M.L.G., 1984, Cysteine: A potential source of error in amino acid analysis of mercaptoethane sulfonic or hydrochloric acid hydrolysates of proteins and peptides, *Anal. Biochem.*, 141:429.

Gjerde H., Normann P.T. and Morland J., 1987, Determination of serum tryptophan by isocratic high-performance liquid chromatography and ultraviolet detection, *Biomed. Biochim. Acta*, 46:53.

Gorinstein S., Shin Y., Hadziyev D., 1988, Tryptophan, cystine & cysteine contents of raw and granulated potatoes: quantitative importance and nutritional value, *Nutr. Rep. Intern.*, 37:397.

Hugly T.E. and Moore S., 1972, Determination of the tryptophan content of proteins by ion exchange chromatography of alkaline hydrolysates, *J. Biol. Chem.*, 247:2828.

Iizuka H. and Yajima T., 1985, Fluorometric determination of L-tryptophan and tryptamine with chloroacetaldehyde, *Chem. Pharm. Bull.*, 33:2591.

Iizuka H. and Yajima T., 1993, Fluorometric determination of L-tryptophan with methoxyacetaldehyde, *Biol. Pharm. Bull.*, 16:103.

Kell G., Holscher W. and Steinhart H., 1989, Reaction mechanism of the kinetic method of tryptophan determination, *Landwirtsch. Forschung*, 42:237.

Kim H.S. and Angyal G., 1993, Comparison of liquid chromatographic method to AOAC microbiological method for determination of L-tryptophan in tablets and capsules, *J. AOAC Intern.*, 76:414.

Knox R., Kohler G.O., Palter R. and Walker H.G., 1970, Determination of tryptophan in feeds, *Anal. Biochem.*, 36:136.

Landry J., Delhaye S. and Viroben G., 1988, Tryptophan content of feedstuffs as determined from three procedures using chromatography of barytic hydrolysates, *J. Agric. Food Chem.*, 36:51.

Landry J. and Delhaye S., 1992, Simplified procedure for the determination of tryptophan of foods and feedstuffs from barytic hydrolysis, *J. Agric. Food Chem.*, 40:776.

Landry J., Delhaye S. and Jones D.G., 1992, Determination of tryptophan in feedstuffs: Comparison of two methods of hydrolysis prior to HPLC analysis, *J. Sci. Food Agric.*, 58:439.

Landry J. and Delhaye S., 1993, HPLC determination of tryptophan in foodstuffs using barium hydroxide hydrolysis at 140°C, *Amino Acids*, 5:317.

Landry J. and Delhaye S., 1994a, Tryptophan determination in feedstuffs: A critical examination of data from two collaborative studies through the evaluation of tryptophan recovery, *J. Agric. Food Chem.*, 42:1717.

Landry J. and Delhaye S., 1994b, Determination of tryptophan in feedstuffs: comparison of sodium hydroxide and barium hydroxide as hydrolysis agents, *Food Chemistry*, 49:95.

Liu T.-Y. and Chang Y.H., 1971, Hydrolysis of proteins with p-toluenesulfonic acid, *J. Biol. Chem.*, 246:2842.

Lucas B. and Sotelo A., 1980, Effect of different alkalies, temperature, and hydrolysis times on tryptophan determination of pure proteins and of foods, *Anal. Biochem.*, 109:192.

Malmer M.F. and Schroeder L.A., 1990, Amino acid analysis by high-performance liquid chromatography with methanesulfonic acid hydrolysis and 9-fluorenylmethylchloroformate derivatization, *J. Chromatogr.*, 514:227.

Matsubara H. and Sasaki R.M., 1969, High recovery of tryptophan from acid hydrolysates of proteins, *Biochem. and Biophys. Res. Comm.*, 35:175.

Michalczuk L., Bialek K. and Cohen J.D., 1992, Rapid determination of tree tryptophan in plant samples by gas chromatography-selected ion monitoring mass spectrometry, *J. Chromatogr.*, 596:294.

Molnár-Perl I., 1994, Advances in the HPLC quantitative of the phenyltiocarbamyl amino acids, *J. Chromatogr.*, 661:43.

Molnár-Perl I. and Fábián-Vonsik V., 1988, Gas chromatographic analysis of histidine, *J. Chromatogr.*, 446:231.

Molnár-Perl I. and Khalifa M., 1993, Tryptophan analysis simultaneously with other amino acids in gas phase hydrochloric acid hydrolyzates using the pico tag work station by HPLC, *Chromatographia*, 36:43.

Molnár-Perl I. and Khalifa M., 1994, Vapor phase hydrolysis of solid samples in the Pico Tag$^{TM}$ Work Station, *LC/GC International*, 7:395.

Molnár-Perl I. and Morvai M., 1992, Cystine and cysteine analysis as the same phenylthiocarbamyl derivatives by HPLC in protein hydrolyzates, *Chromatographia*, 34:132.

Molnár-Perl I., Pintér-Szakács M. and Fábián-Vonsik V., 1987, Esterification of amino acids with thionyl chloride acidified butanol for their gas chromatographic analysis, *J. Chromatogr.*, 93:434.

Molnár-Perl I. and Pintér-Szakács M., 1989, Spectrophotometric determination of tryptophan in intact proteins by the acid ninhydrin method, *Anal. Biochem.*, 177:16.

Molnár-Perl I., Pintér-Szakács M., Morvai M. and Fábián V., 1992, Advances in the tryptophan analysis: In intact proteins and protein matrices in: Advances in Tryptophan Research 1992, I. Ishiguro, R. Kido, T. Nagatsu, Y. Nagamura, Y. Ohta, eds., Fujita Health University Press

Molnár-Perl I., Pintér-Szakács M., Morvai M. and Fábián V., 1992, Advances in the tryptophan analysis: In intact proteins and protein matrices; In advances in tryptophan research 1992 (Eds. I. Ishiguro, R. Kido, T. Nagatsu, Y. Nagamura, Y. Ohta) Fujita Health University Press, Toyoake, p. 195-198.

Molnár-Perl I., Pintér-Szakács M. and Khalifa M., 1993, HPLC of tryptophan together with other amino acids in hydrochloric acid hydrolyzates, *J. Chromatogr.*, 632:57.

Morita I., Masujima T., Yoshida H. and Imai H., 1985, Direct plasma injection method for the analysis of tryptophan metabolites by high-performance liquid chromatography coupled with precolumn deproteinization, *Anal. Biochem.*, 151:358.

Morita I., Kawamoto M., Hattori M., Eguchi K., Sekiba K. and Yoshida H., 1990, Determination of tryptophan and its metabolites in human plasma and serum by high-performance liquid chromatography with automated sample clean-up system, *J. Chromatogr.*, 526:367.

Morvai M., Fábián V. and Molnár-Perl I., 1992, Buffer and pH dependence of phenylthiocarbamyl amino acids retention in RP-HPLC, *J. Chromatogr.*, 600:87.

Muramoto K. and Kamiya H., 1990, Recovery of tryptophan in peptides and proteins by high temperature and short-term acid hydrolysis in the presence of phenol, *Anal. Biochem.*, 189:223.

Ng L.T., Pascaud A. and Pascaud M., 1987, Hydrochloric acid hydrolysis of proteins and determination of tryptophan by reversed-phase high-performance liquid chromatography, *Anal. Biochem.*, 167:47.

Nielsen H.K. and Hurrell R.F., 1985, Tryptophan determination of food proteins by h.p.l.c. after alkaline hydrolysis, *J. Sci. Food Agric.*, 36:893.

Nozaki Y., 1990, Determination of tryptophan, tyrosine, and phenylalanine by second derivative spectrophotometry, *Arc. of Biochem. and Biophys.*, 277:324.

Ohta T. and Nakai T., 1979, Reaction of cystine with tryptophan under the conditions of acid hydrolysis of proteins: Mechanism of action of cystine, *Agric. Biol. Chem.*, 43:2419.

Opienska-Blauth J., Charezinski M. and Berbec H., 1963, A new, rapid method of determining tryptophan, *Anal. Biochem.*, 6:69.

Pecavar A., Prosek M., Temeljotov D.F. and Marsel J., 1990, Quantitative evaluation of amino acids using microwave accelerated hydrolysis, *Chromatographia*, 30:159.

Penke B., Ferenczi R. and Kovács K., 1974, A new acid hydrolysis method for determining tryptophan in peptides and proteins, *Anal. Biochem.*, 60:45.

Peterson L.C., Dwyer M.E. and Brown P.R., 1987, Miniaturized RPLC for the rapid determination of total tryptophan in pediatric plasma samples, *Chromatographia*, 24: 309.

Pintér-Szakács M. and Molnár-Perl I., 1990, Determination of tryptophan in unhydrolyzed food and feedstuffs by the acid ninhydrin method, *J. Agric. Food Chem.*, 38:720.

Ranfft K. and Faure U., 1993, An intercomparison of methods for the determination of tryptophan in feedstuffs, *Agribiol. Res.*, 46:65.

Sato H., Seino T., Kobayashi T., Murai A. and Yugari Y., 1984, Determination of the tryptophan content of feed and feedstuffs by ion exchange liquid chromatography, *Agric. Biol. Chem.*, 48:2961.

Sato H., Kobayashi T., Jones R.W. and Easter R.A., 1987, Tryptophan availability of some feedstuffs determined by pig growth assay, *J. Anim. Sci.*, 64:191.

Servillo L., Colonna G., Balestrieri C., Ragone R. and Irace G., 1982, Simultaneous determination of tyrosine and tryptophan residues in proteins by second-derivative spectroscopy, *Anal. Biochem.*, 126:251.

Simpson R.J., Neuberger M.R. and Liu T.-Y., 1976, Complete amino acids analysis of proteins from a single hydrolysate, *J. Biol. Chem.*, 251:1936.

Sodek L., Vecchia P.T.D. and Lima M.L.G.P., 1975, Rapid determination of tryptophan in beans (Phaseolus vulgaris) by the acid ninhydrin method, *J. Agic. Food Chem.*, 23:1147.

Sottrup-Jensen L., 1993, Determination of halfcystine in proteins as cysteine from reducing hydrolyzates, *Biochem. and Molecular Biol. Int.*, 30:789.

Spies J.R., 1950, Determination of tryptophan with p-dimethylaminobenzaldehyde. Elimination of interference by chloride and bisulfite ions, *Anal. Chem.*, 22:1447 (1950).

Spies J.R., 1967, Determination of tryptophan in proteins, *Anal. Chem.*, 39:1412.

Spies J.R. and Chambers D.C., 1949, Chemical determination of tryptophan in proteins, *Anal. Chem.*, 21:1249.

Spies J.R. and Chambers D.C., 1950, Determination of tryptophan with p-dimethylaminobenzaldehyde using photochemical development of color, *Anal. Chem.*, 22:1209.

Tatár E., Khalifa M., Záray Gy. and Molnár-Perl I., 1994, Comparison of the recovery of amino acids vapor phase hydrolyzates proteins performed in the Pico Tag$^{TM}$ Work Station and in the CEM microwave apparatus, *J. Chromatogr.*, 672:109.

Verma K.K. and Jain A., 1988, Spectrophotometric determination of tryptophan by reaction with nitrous acid, *Talanta*, 35:35.

Werner G., 1986, Eine HPLC-Methode zur Bestimmung von Tryptophan in Futtermitteln und Getreide, *Landwirtsch. Forschung*, 39:109.

Wong W.S.D., Osuga D.T., Burcham T.S. and Feeney R.E., 1984, Determination of tryptophan as the reduced derivative by acid hydrolysis and chromatography, *Anal. Biochem.*, 143:62.

Yano H., Aso K. and Tsugita A., 1990, Further study on gas phase acid hydrolysis of protein: Improvement of recoveries for tryptophan, tyrosine, and methionine, *J. Biochem.*, 108:579.

Zahnley J.C. and Davis J.G., 1973, Effect of high tyrosine content on the determination of tryptophan in protein by the acid ninhydrin method, *Biochem. J.*, 135:59.

# DETERMINATION OF $\gamma$-RADIATION INDUCED PRODUCTS IN FREE AND PEPTIDE BOUND TRYPTOPHAN

B. van Wickern, K. Kleeberg, D. Rogosch, and H. Steinhart

Institute of Biochemistry and Food Chemistry
University of Hamburg
Grindelallee 117D-20146 Hamburg

## 1. INTRODUCTION

Conservation of food by ionizing radiation has become legally accepted in 37 countries of the world. Although some methods for the identificaton of irradiated food have been developed, there is a deficit in the detection of the irradiation in food with a high protein content. Up to now, only a single method based on the detection of o- and m-tyrosine which develop in addition to the physiologically formed p-tyrosine during the radiation induced hydroxylation of phenylalanine, has been developed (Chuaqhi-Offermanns and McDougall, 1991). Corresponding to this method we wished to find some irradiation induced products of tryptophan which could serve as markers for irradiated food with a high protein content.

## 2. MATERIALS AND METHODS

### 2.1 Irradiation

The irradiation with doses of 10 and 20 kGy was carried out using an industrial Co-60 source at Beiersdorf GmbH Hamburg with a dose rate of 0.5 kGy/h.

The irradiation with the doses of 1, 3 and 5 kGy was performed with a Cs-137 source used for the irradiation of blood at Beiersdorf GmbH Hamburg and a dose rate of 0.7 kGy/h.

### 2.2 Sample Treatment

Tryptophan (TRP) (200, 400 mg/l) and the different peptides (500, 1000 mg/l), respectively, were dissolved in bidistilled water and filled into 20 ml polyethylene vials for irradiation. After the treatment the samples were deep frozen at -20 °C. The irradiated TRP solutions were used directly for the HPLC analysis. The irradiated peptide solutions were subjected to an enzymatic hydrolysis with Pronase E (Merck).

*Recent Advances in Tryptophan Research,* edited by Graziella Allegri Filippini et al.
Plenum Press, New York, 1996

All experiments were carried out at least in triplicate, and there was one unirradiated sample additional.

## 2.3 Enzymatic Hydrolysis

A mixture of 1ml peptide solution, 1 ml trizyma buffer (0.1 M, pH 8.5), 0.25 ml 5-methyltryptophan solution (IS) (0.5 mg/ml in buffer solution) and 0.25 ml of Pronase E solution (2 mg/ml) was incubated at 40 °C for 30 min. The hydrolysis was stopped by cooling in an ice bath and the solution was used directly for the HPLC measurement.

## 2.4 HPLC-System

| | |
|---|---|
| Pump: | Merck Hitachi L 6200 gradient pump |
| Detector: | Merck Hitachi L 4000 UV-detector (260 nm) |
| | Merck Hitachi F 1080 fluorescence-detector ($E_m$ 315 nm, $E_x$ 270 nm) |
| Column: | Nucleosil 120-3 RP 18, 250 x 4 mm |
| Eluent: | eluent A: 0.1 % trifluoroacetic acid in bidistilled water |
| | eluent B: methanol |
| | eluent C: acetonitrile |

Gradient:

| | | | | | | |
|---|---|---|---|---|---|---|
| | 0 | 10 | 25 | 26 | 30 | t [min] |
| A [%] | 95 | 86 | 66 | 46 | 46 | |
| B [%] | 5 | 14 | 14 | 14 | 14 | |
| C [%] | 0 | 0 | 20 | 40 | 40 | |

Flow: 1 ml/min

Temperature: 35° C

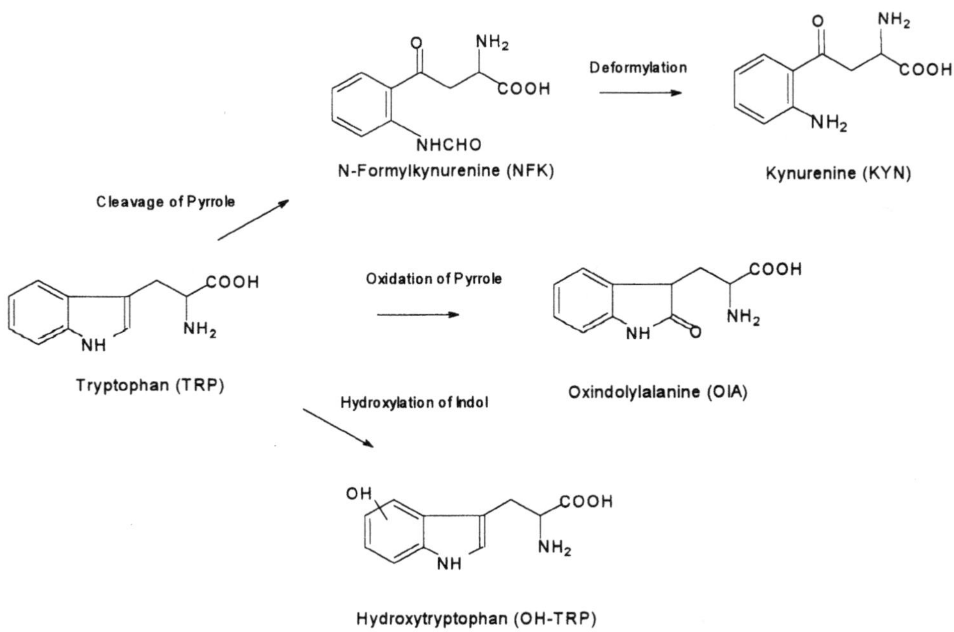


**Figure 1.** γ-Radiation induced products of Tryptophan.

# 3. RESULTS

Following the results of different authors (Simic, 1978; Troeder and Steinhart, 1988; Maskos et al., 1992), we expected to find N-formylkynurenine (NFK), kynurenine (KYN), oxindolylalanine (OIA) and 5-hydroxytryptophan (5-OH-TRP) as the main radiation induced products (see figure 1). In addition we looked for the other hydroxytryptophan isomers (4-, 6- and 7-OH-TRP).

Identification of these products took place through comparison of the retention times with those of reference substances and by standard addition. To confirm these results, we

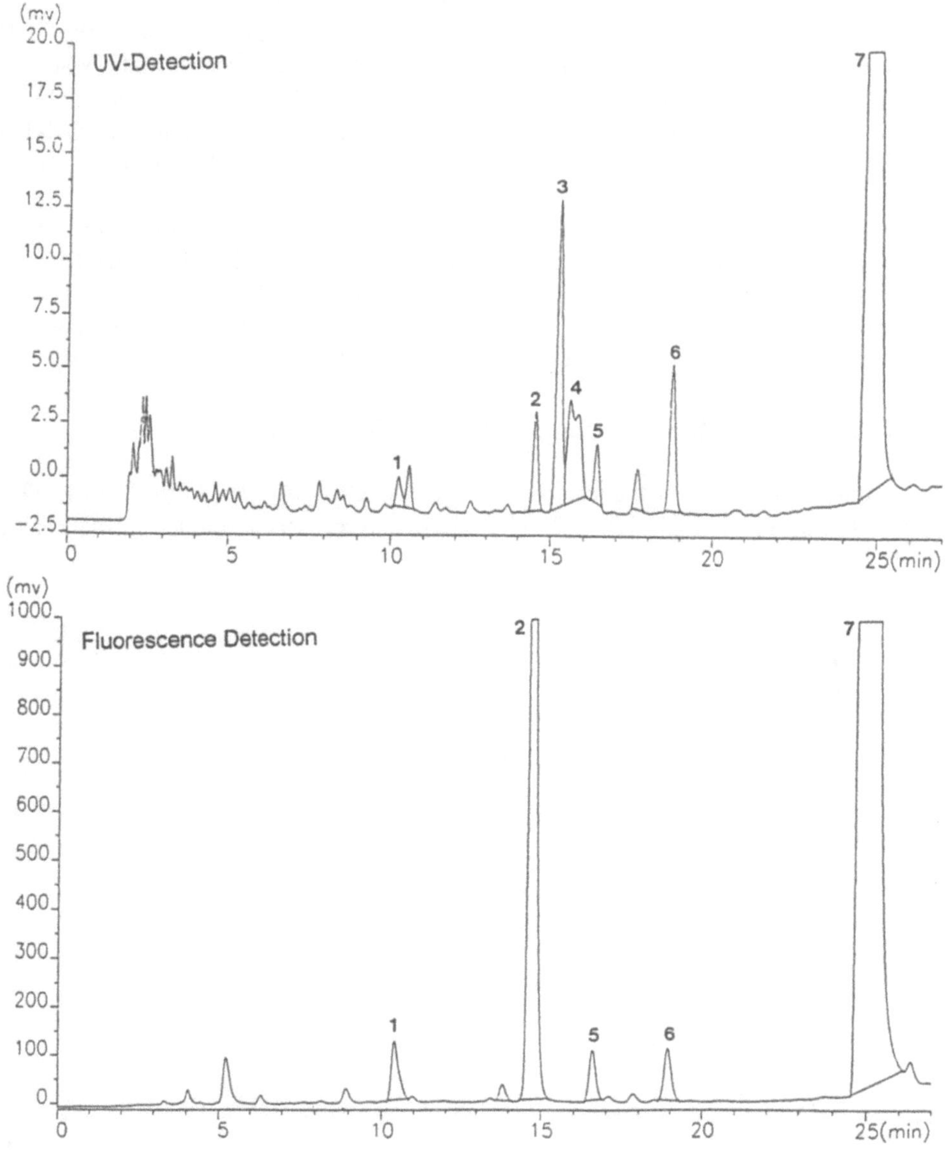


**Figure 2.** Chromatograms of an irradiated TRP solution (400 mg/l, 10 kGy). 1. 4-OH-TRP?, 2. 5-OH-TRP, 3. NFK, 4. OIA, 5. 6-OH-TRP?, 6. 7-OH-TRP, 7. TRP.

compared the fluorescence- and the UV-spectra of the products in the irradiated solutions with those of reference substances. NFK, KYN, OIA and 5-OH-TRP as reference substances were partly purchased and partly synthesized by our workgroup. 7-OH-Trp was synthesized by treating TRP with a Fenton system and was isolated by semipreparative separation on HPLC according to the method of Maskos et al., 1992. Unfortunately it was not possible to synthesize 4- and 6-OH-TRP in the same way.

## 3.1 Irradiation Products of Free Tryptophan

In figure 2 chromatograms of an irradiated TRP solution (400 mg/l, 10 kGy) are illustrated.

Apart from OIA and NFK, two hydroxylated TRP isomers (5-OH-TRP, 7-OH-TRP) were identified. In addition to these, two further hydroxylated substances, probably 4-OH-TRP and 6-OH-TRP, were detected and characterized by their fluorescence and UV-spectra.

The quantification of NFK, OIA, 5-OH-TRP and 7-OH-TRP was carried out by the method of external standard. NFK and OIA were quantified with UV-detection; the OH-TRP isomers, with fluorescence detection. Quantification of the other OH-TRP-isomers was not possible .

Figure 3 displays the amounts of the radiation induced products in solutions of free TRP in relation to the different radiation dose. The contents of OIA, 5-OH-TRP and 7-OH-TRP increase with increasing radiation doses up to 5 kGy. Between 5 and 10 kGy the increase flattens, and between 10 and 20 kGy the amounts of the products remain constant. The content of NFK increases up to a dose of 3 kGy and then decreases continuously.

## 3.2 Irradiation Products of Peptide Bound Tryptophan

Aqueous solutions of the tripeptides H-ALA-TRP-ALA-OH (ATA), H-LEU-TRP-LEU-OH (LTL) and H-LEU-TRP-MET-OH (LTM) were used as model substances for

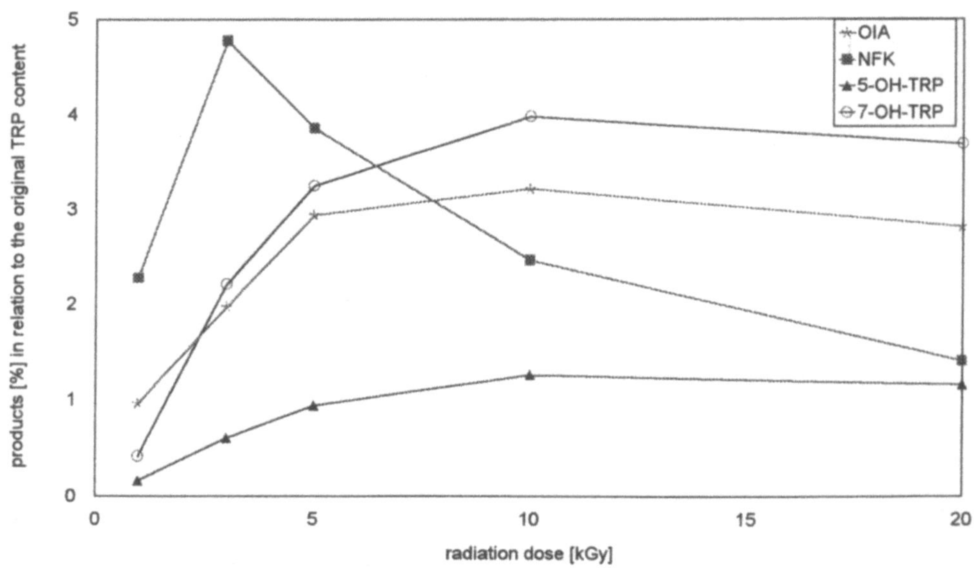


**Figure 3.** Amounts of radiation induced products of free TRP in relation to the radiation dose applied.

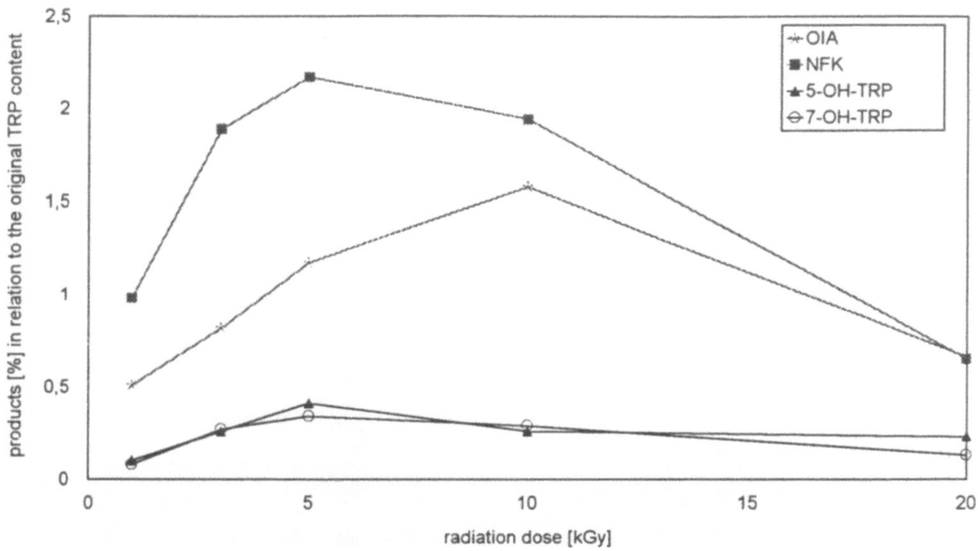


**Figure 4.** Amounts of radiation induced products of peptide bound TRP (i.e. LTM, 500 mg/l) in relation to the radiation dose applied.

peptide bound TRP. The same radiation products as described for free TRP were found, but in smaller amounts (see Fig. 4).

For all investigated peptides the amounts of these products increase with increasing radiation doses up to 5 kGy. Between 5 and 20 kGy the amounts of the OH-TRP isomers and NFK stay on the same level or even decrease. OIA increases with increasing radiation dose up to 10 kGy and then decreases. Only small differences in the amounts of the products were observed between the different peptides.

## 4. CONCLUSIONS

The same irradiation products are formed in solutions of free and of peptide bound TRP. As NFK and 5-OH-TRP develop during normal biological processes, and OIA is probably formed during food processing, the presence of the aphysiological 4-, 6- and 7-OH-TRP isomers could possibly be used as markers for irradiated, protein containing food. The next step in our investigation is to be the synthesis of 4-OH-TRP and 6-OH-TRP as reference substances. In addition to that, the behaviour of protein bound TRP is to be observed.

## REFERENCES

Chuaqui-Offermanns,N., McDougall, T., 1991, An HPLC method to determine o-tyrosine in chicken meat, *J. Agric. Food Chem.* 39:300.

Maskos,Z., Rush, J.D., Koppenol, W.H., 1992, The hydroxylation of tryptophan, *Arch. Biochem. Biophys.* 296:514.

Simic,M.G., 1978, Radiation chemistry of amino acids and peptides in aqueous solutions, *J. Agric. Food Chem.* 26:6.

Troeder, U., Steinhart, H., 1988, Verhalten von Tryptophan und tryptophanhaltigen Peptiden gegenüber ionisierender Bestrahlung, *Lebensm. Gerichtl. Chem.* 42:141.

# OXIDATION OF FREE AND PEPTIDE BOUND TRYPTOPHAN

T. Simat, K. Meyer, B. Stöver, and H. Steinhart

Institute of Biochemistry and Food Chemistry
University of Hamburg
Grindelallee 117, 20146 Hamburg, Germany

## 1. INTRODUCTION

Food is often processed or stored under conditions suitable for the oxidative damage of proteins, in particular of the amino acids methionine, lysine, histidine and tryptophan (Trp). The forming products can affect sensory quality, nutritional value and food safety. Focusing on Trp as an essential amino acid, most authors have described the decrease of free or protein bound Trp during oxidative treatment such as the addition of hydrogen peroxide ($H_2O_2$) (de Weck et al., 1987) or of oxidizing lipids (Matoba et al., 1984), or have determined the bioavailability of Trp in oxidized food proteins (Nielsen et al., 1985). Although many oxidation products of Trp are known (Savige, 1975), little is known about their formation in proteins. Krogull and Fennema (1987) determined N-formylkynurenine (NFK) and kynurenine (Kyn) after lipid oxidation; Holt et al. (1977) identified oxindolylalanine (Oia), aspartic acid (Asp) and Kyn after photo oxidation of Trp containing peptides. The aim of the present study was to identify and compare the degradation products of free, peptide and protein bound Trp formed in a simple model system with $H_2O_2$ as oxidative agent.

## 2. EXPERIMENTAL PROCEDURES

### 2.1. Materials

Dioxindolylalanine (DiOia), Oia and NFK were synthesized as has been published previously (Simat et al. 1994), 3a-hydroxy-1,2,3,3a,8,8a-hexahydropyrroloindole-2-carboxylic acid (diastereomere A) (PIC-A) was synthesized according to Savige (1975). Pronase E was immobilised to CNBr-Sepharose and had an aminopeptidase acitivity of 2,7 U/mg Pronase (1 U = conversion of 1 µmol leucine-4-nitroanilide / min).

## 2.2. Oxidation with Hydrogen Peroxide ($H_2O_2$)

Oxidation was performed for 1 - 6 h in 2 ml borate buffer pH 8.3 at 40°C. The buffer solution contained 0.002 M free, peptide or protein bound Trp. L-Alanyl-L-tryptophyl-L-alanine (H-Ala-Trp-Ala-OH) served as model substrate for peptide bound; lysozyme, as model substrate for protein bound Trp.

For oxidation 40 µl of $H_2O_2$ (30 %) was added to the sample every hour. The reaction was stopped by addition of catalase (EC 1.11.1.6) and the absence of $H_2O_2$ subsequently checked by potassium iodide test. In the case of lysozyme a few drops of octanol were added before catalase treatment to prevent foaming.

## 2.3. Hydrolysis of H-Ala-Trp-Ala-OH and Lysozyme

Both oxidized and untreated H-Ala-Trp-Ala-OH solutions were hydrolysed enzymatically in a cartridge containing immobilised Pronase E for 16 h at room temperature. To release the imino acid PIC, the hydrolysate was additionally incubated with Prolidase (EC 3.4.13.9).

Both oxidized and untreated lysozyme were submitted to 3 different hydrolysis procedures: Alkaline hydrolysis was performed according to Nielsen and Hurrell (1985) (polypropylene tubes, 4.2 N NaOH, 20 h, 110°C); acid hydrolysis was carried out with 6 N HCl, 24 h, at 110°C in vacuo, and enzymatic hydrolysis (Pepsin (EC 3.4.23.1), Pronase E, Aminopeptidase M (EC 3.4.11.2), Prolidase (EC 3.4.13.9), 40°C) was performed as described by Schmitz et al. (1976).

## 2.4. Analysis of Trp and Its Degradation Products

The RP-HPLC / UV method, as described by Simat et al. 1994, was used for monitoring the loss of Trp and the formation of its oxidation products. Confirmation of

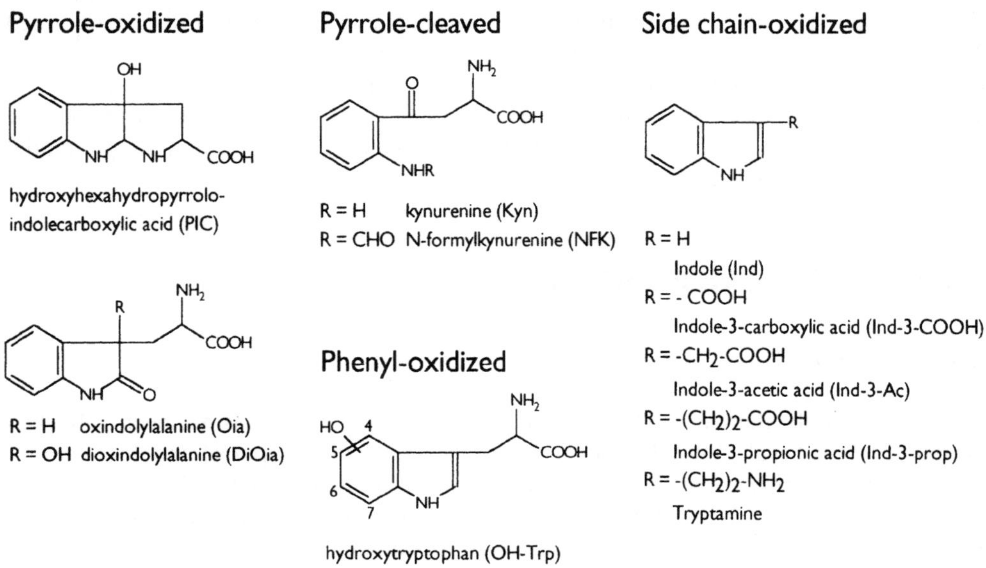


**Figure 1.** Classification of Trp oxidation products.

identity was done by comparison of retention times with standard substances, by standard addition, by DAD-detection and by comparison of UV- and fluorescence response.

## 3. RESULTS AND DISCUSSION

The oxidized samples were examined for the loss of Trp and for the formation of 15 known Trp oxidation products, which can be classified in 5 groups (Fig. 1).

All substances could be reproducibly separated by RP-HPLC (Fig. 2). Further known Trp degradation products such as alanine (Ala), serine (Ser), glycine (Gly) and Asp (Kell and Steinhart, 1990) could not be detected and are disregarded in this paper.

The oxidation rate of Trp, free and in H-Ala-Trp-Ala-OH, was much higher than in lysozyme (Fig. 3), which is consistent with the results of de Weck et al. (1987). This may be due to the steric hindrance of the protein, the absence of free amino or carboxyl groups and the antioxidative effect of the other amino acids.

The detected degradation products of free Trp oxidized under the described conditions belonged either to the groups with oxidative attack on pyrrole ring (PIC A, DiOia A/B, NFK, Kyn) or to the phenyl oxidation products (5-OH-Trp)(Fig. 4). No exclusively side chain oxidation products could be detected. All identified compounds, especially PIC A, are more susceptible to oxidation than is Trp itself.

Oxidation products of Trp in H-Ala-Trp-Ala-OH were identified after complete enzymatic hydrolysis with Pronase E and Prolidase. Oia, NFK and Kyn were the main oxidation compounds (Fig. 5). This result could be confirmed by RP-HPLC of the unhydrolysed oxidized peptide and assignment of the peaks to the standard substances by DAD spectra. Formation of the imino acid PIC A seems to be hindered because the amino group in peptide bound Trp is not free.

Identification of Trp degradation products in oxidized lysozyme proved to be difficult because of their instability towards hydrolysis conditions. Kyn and Oia are stable to 6 N HCl hydrolysis, however Trp is oxidized to Oia under these conditions. During alkaline hydroly-

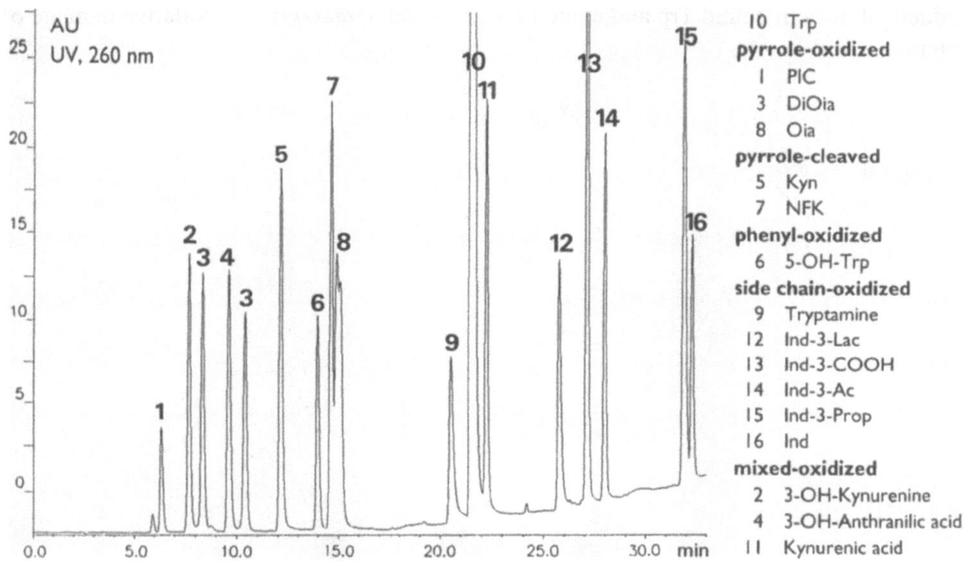


Figure 2: RP-HPLC separation of Trp and its oxidation products

**Figure 2.** RP-HPLC separation of Trp and its oxidation products.

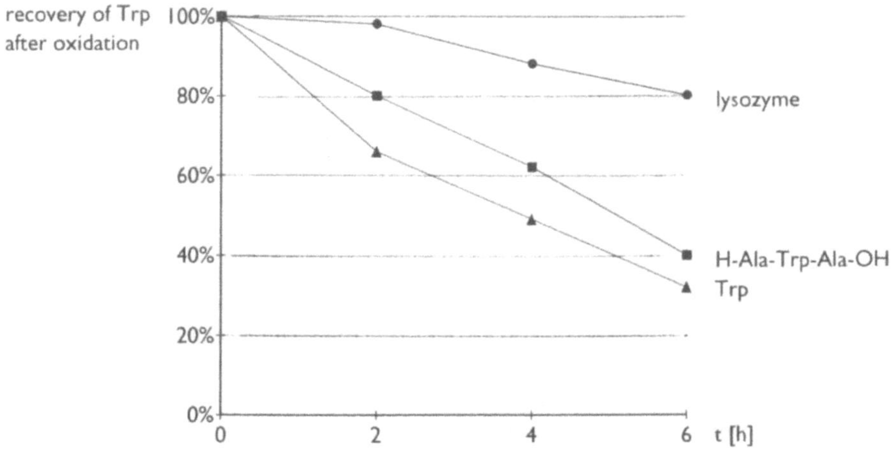


**Figure 3.** Oxidation rates of free and peptide bound Trp.

sis Oia is partially oxidized, yielding DiOia. NFK is hydrolysed forming Kyn during acid or enzymatic hydrolysis. The results for Kyn / NFK obtained by enzymatic or 6 N HCl hydrolysis were consistent; hydrolysis for determination of DiOia and Oia needs further optimisation (Tab. 1).

## 4. CONCLUSION

Six Trp oxidation products (PIC A, DiOia, Oia, Kyn, NFK, 5-OH-Trp), all formed by oxidative attack on the indole ring could be identified after oxidation of peptide bound Trp by hydrogen peroxide. Kyn / NFK and Oia seem to be the major primary degradation products of protein bound Trp and could serve as suitable markers for oxidative damage of proteins.

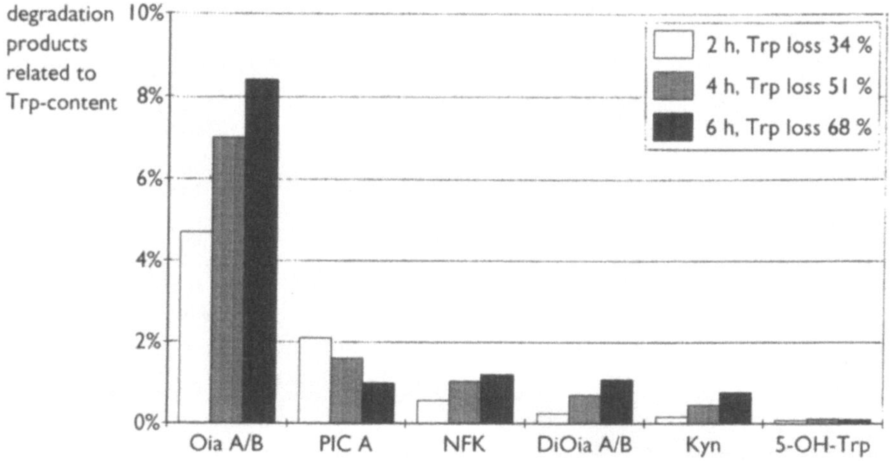


**Figure 4.** Formation of degradation compounds of free Trp.

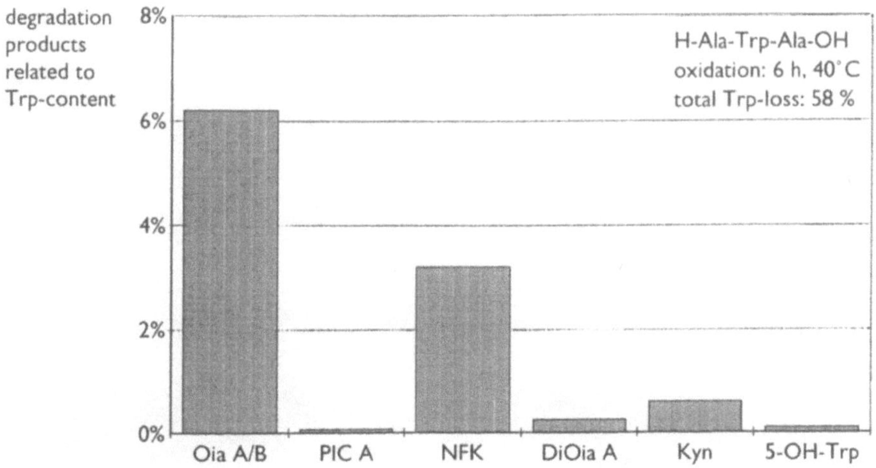


**Figure 5.** Formation of Trp degradation products in H-Ala-Trp-Ala-OH.

**Table 1.** Trp degradation products in oxidized lysozyme (6 h, 40°C)

| | Degradation products related to Trp-content | | | | | |
|---|---|---|---|---|---|---|
| Hydrolysis | PIC-A | DiOia A/B | Kyn/NFK | Oia A/B | 5-OH-Trp | Trp |
| Enzymatic | < | 0.2% | 2.4% | ∅ | < | 64% |
| Acid | ∅ | ∅ | 2.3% | (33%) | (0.04%) | ∅ |
| Alkaline | ∅ | (4.6%) | ∅ | (1.8%) | ∅ | 79% |

∅=not stable; < = below detection limit

# 5. REFERENCES


Holt, L.A., Milligan, B., Rivett, D.E., Stewart, F.H.C., 1977, The photodecomposition of tryptophan peptides, *Biochim. Biophys. Acta* 499:131

Kell, G., Steinhart, H., 1990, Oxidation of tryptophan by $H_2O_2$ in model systems, *J. Food Sci.* 55:1120

Krogull, M.K., Fennema, O., 1987, Oxidation of tryptophan in the presence of oxidizing methyl linoleate, *J. Agric. Food Chem.* 35:66

Matoba, T., Yonezawa, D., Nair, B., Kito, M., 1984, Damage of amino acid residues of proteins after reaction with oxidizing lipids: Estimation by proteolytic enzymes, *J. Food Sci.* 49:1082

Nielsen, H.K., Finot, P.-A., Hurrell, R.F., 1985, Reaction of proteins with oxidizing lipids. 2. Influence on protein quality and on the bioavailability of lysine, methionine, cyst(e)ine and tryptophan as measured in rat assays, *Br. J. Nutr.* 53:75

Nielsen, H.K., Hurrell, R.F., 1985, Tryptophan determination of food proteins by HPLC after alkaline hydrolysis, *J. Sci. Food Agric.* 36:893

Savige, W.E., 1975, New oxidation products of tryptophan, *Aust. J. Chem.* 28: 2275

Simat, T., Meyer, K., Steinhart, H., 1994, Synthesis and analysis of oxidation and carbonyl condensation compounds of tryptophan, *J. Chromatogr. A* 661:93

Schmitz, I., Zahn, H., Klostermeyer, H., Rabbel, K., Watanabe, K., 1976, Zum Vorkommen von Isopeptidbindungen in erhitztem Milcheiweiß, *Z. Lebensm. Unters.-Forsch.* 160:377

Weck, de D., Nielsen, H.K., Finot, P.-A., 1987, Oxidation rate of free and protein-bound tryptophan by hydrogen peroxide and the bioavailability of the oxidation products, *J. Sci. Food Agric.* 41:179

# THE TRYPTOPHAN CONTENT IN PROTEIN OF CEREAL GRAINS AND LEGUME SEEDS AS A FUNCTION OF NITROGEN CONTENT

## A Reappraisal of Tryptophan Score

S. Delhaye and J. Landry

Laboratoire de Chimie Biologique
INRA, INA-PG
F-78850 Thiverval Grignon, France

## ABSTRACT

The tryptophan content ([Trp]) in the dry matter (DM) of wheat, maize, barley, sorghum, rice, pearl millet grains, pea and broad bean seeds is evaluated from samples, with various nitrogen contents ($[N]_{DM}$) and genotypes, using a procedure that has been shown to be strictly quantitative.

The determination of linear correlations between $[Trp]_{DM}$ and $[N]_{DM}$, and between the tryptophan content in protein ($[Trp]_N$) and $[N]_{DM}$ or $1/[N]_{DM}$ for every species leads to the following observations: (1) $[Trp]_{DM}$ and $[N]_{DM}$ are linearly related. The data show that previous reports of similar relationships underestimate tryptophan by 10 ± 5 % owing to tryptophan degradation during alkaline hydrolysis preparatory to analysis; (2) the linear correlations between $[Trp]_N$ and $1/[N]_{DM}$, resulting from linear relationships $[Trp]_{DM}$ and $[N]_{DM}$ display coefficients of determination ($r^2$) far lower than 1 and similar to those found for linear correlations between $[Trp]_N$ and $[N]_{DM}$; (3) $[Trp]_N$ increases with $[N]_{DM}$ increasing for rice and pearl millet while it decreases for all other species.

In conclusion: (1) linear relationships between tryptophan and nitrogen have a low predictive value; (2) the nutritional score of tryptophan of foods and feeds, as calculated from the determination of tryptophan using a procedure involving alkaline hydrolysis, is generally underestimated by 10%.

## INTRODUCTION

The amino acid composition of seeds originating from diverse crop species has been the subject of many analyses. It has been shown that changes in the content of any amino acid in the dry matter of seed of a given species was linearly related to the variations of

*Recent Advances in Tryptophan Research*, edited by Graziella Allegri Filippini et al.
Plenum Press, New York, 1996

nitrogen content (N). Regarding tryptophan (Trp), the regression lines reported in literature are doubtful since most of specific procedures used for assaying this amino acid are not quantitative. The present work seeks these uncertainties to be assessed by determining regression lines from data obtained using a reliable assay for tryptophan.

## MATERIAL AND METHODS

Seed samples originated from commercial and experimental cultivars kindly supplied by ours colleagues of INRA, ITCF (Institut Français des Céréales et Fourrages), AGPM (Association Générale des Producteurs de Maïs) and CIRAD (Centre de coopération internationale en recherche agronomique pour le développement).

Nitrogen was assayed by a Kjeldahl method.

Tryptophan (Trp) was assayed according to Landry and Delhaye (1992) with slight modifications. Milled sample containing 50mg protein was mixed with 4.2g barium hydroxide octahydrate and 8ml water. The mixture was transferred in a benchtop autoclave (10 litre sterilizer) when water was boiling and hydrolyzed at 125°C for 10hr. Three μmoles of α-methyltryptophan (α-MT) as internal standard were added and mixed to hydrolysate. An aliquot of hydrolysate was diluted with a pH 3 solution. Trp and α-MT were separated by reverse phase high performance liquid chromatography and quantitated from their native fluorescence.

## RESULTS AND DISCUSSION

Table 1 reports the statistical parameters of regression lines between the levels of Trp and N in seed dry matter for eight cereal and legume species. The cofficients of determination ($r^2$) varied from 0.56 to 0.98. The value of 0.56 indicated a scatter diagramm for sorghum and that of 0.98 a perfect line for millet (probably related to the sampling).

The chemical score of Trp relative to a seed of a given species is proportional to its content in seed proteins since it is expressed as the quotient of this content to that in standard protein. Therefore $[\mathrm{Trp}]_N$ (g/16g N) was regressed against 1/N or N. The coefficients of determination given in Table 2 varied from 0.01 to 0.69. For all species but maize the level of Trp in protein appears to be virtually independent of N.

The data of present work (laboratory L1) were compared with those published by laboratories L2 (Mossé and Huet, 1990) and L3 (Heimbeck and Balschukat, 1990). Trp was assayed by laboratory L2 using a hydrolytic procedure similar to that applied by us, but the

**Table 1.** Statistical parameters of regression lines.
**Trp** (g/100g DM) = *a***N** (g/100g DM) + *b*

| | n[a] | *a* | *b* | $r^2$ |
|---|---|---|---|---|
| Barley | 20 | 0.069 | 0.034 | 0.73 |
| Maize | 40 | 0.025 | 0.037 | 0.85 |
| Millet | 17 | 0.131 | −0.045 | 0.98 |
| Rice | 13 | 0.071 | 0.017 | 0.70 |
| Sorghum | 26 | 0.053 | 0.029 | 0.56 |
| Wheat | 67 | 0.052 | 0.062 | 0.73 |
| Pea | 23 | 0.064 | −0.023 | 0.84 |
| Broad bean | 42 | 0.052 | 0.003 | 0.68 |

[a]n = number of samples

**Table 2.** Regression $[Trp]_N$ as a function of N or 1/N. Coefficient of determination ($r^2$)

| Regression | $[Trp]_N = aN + b$ | $[Trp]_N = a + b/N$ |
|---|---|---|
| Barley | 0.47 | 0.46 |
| Maize | 0.69 | 0.61 |
| Millet | 0.47 | 0.53 |
| Rice | 0.05 | 0.06 |
| Sorghum | 0.30 | 0.25 |
| Wheat | 0.51 | 0.46 |
| Pea | 0.03 | 0.01 |
| Broad bean | 0.01 | 0.01 |

mixtures to be hydrolysed were probably transferred in benchtop autoclave when water was cold. Trp was determined by laboratory L3 from a hydrolysate got with lithium hydroxide (Whiteacre et al, 1986). As shown in Table 3, the values from laboratories L2 and L3 averaged 90% from those determined by laboratory L1. The high values (5%) for standard deviation confirmed the dispersion of data.

## CONCLUSION

The present study led the following conclusions to be drawn:

- the linear dependence between Trp and N levels in the dry matter of cereal and legume seeds was far from being exact, making the prediction of Trp content from regression equations to be approximative only;
- the values of Trp given in literature were 10% underestimated, in the same order as stated by Slump et al (1991). The same held for Trp chemical score if the Trp content of standard protein was known with accuracy.

**Table 3.** Relative tryptophan contents calculated using the regression equations determined by our laboratory (L1) and two others (L2 and L3) for a given nitrogen content (percentages on the basis of 100% for the values of corresponding samples as determined in our laboratory)

| | N% | L1 | L2 | L3 |
|---|---|---|---|---|
| Barley | 2.00 | 100 | 94.6 | 85.7 |
| Maize | 1.65 | 100 | 84.6 | 94.9 |
| Millet | 1.80 | 100 | 100.0[a] | nd[b] |
| Rice | 1.60 | 100 | 91.6 | nd[b] |
| Sorghum | 1.60 | 100 | 91.2 | 94.5 |
| Wheat | 2.30 | 100 | 92.3 | 91.8 |
| Pea | 4.00 | 100 | 86.3 | 93.2 |
| Broad bean | 5.00 | 100 | 97.0[a] | 86.4 |
| Mean | | | 90.1 ±3.8 | 91.1 ±4.1 |

[a]The high percentages, resulting from the use of a procedure for preparing and hydrolyzing samples identical to that applied in the laboratory L1 are not considered in the determination of mean.

[b] not determined.

# REFERENCES

Heimbeck, W. and Balschukat, D., 1990, *in:* "The Amino Acid Composition of Feedstuffs", Degussa, Hanau, Germany.

Landry, J. and Delhaye, S., 1992, Simplified procedure for the determination of tryptophan of foods and feedstuffs from barytic hydrolysis, *J. Agric. Food Chem.*, 40:776

Mossé, J., and Huet, J. C., 1990, Amino acid composition and nutritional score for ten cereals and six legumes or oilseeds: causes and ranges of variations according to species and to seed nitrogen content, *Sci. Aliments*, 10:151

Slump, P., Flissebaalje, T. D. and Haaksman, I. K., 1991, Tryptophan in food proteins: a comparison of two hydrolytic procedures, *J. Sci. Food Agric.*, 55:493

# *CHEMICAL AND ANALYTICAL ASPECTS OF TRYPTOPHAN AND RELATED COMPOUNDS*

# SYNTHESIS AND ANALYSIS OF CONTAMINANTS IN EMS-RELATED TRYPTOPHAN

H. Steinhart, B. van Wickern, K. Meyer, and T. Simat

Institute of Biochemistry and Food Chemistry
University of Hamburg
Grindelallee 117, 20146 Hamburg, Germany

## 1. INTRODUCTION

Eosinophilia-Myalgia-Syndrome (EMS) has been linked to the ingestion of Tryptophan (Trp) manufactured by a single company containing more than 60 UV-detectable minor contaminants (Belongia et al., 1990). Hill et al. (1993) showed by statistical comparison of case and non-case associated lots that six of these contaminants can be associated with EMS: 1,1′-Ethylidenebis-L-tryptophan (1,1′-EBT) (Smith et al., 1991) and 3-(Phenylamino)alanine (PAA)(Toyo'oka et al., 1991, Goda et al., 1992) have already been structurally characterized and linked with the occurrence of EMS. 2-(3-indolylmethyl)-L-tryptophan (Peak 200) has been structurally determined by Müller et al. (1991). The structures of the remaining 3 EMS related contaminant peaks are still unknown (Hill et al., 1993). However there is still no animal model that establishes all features of EMS and there is still a lack of statistical significance for the association of EMS with 1,1′-EBT, Peak 200 (Philen et al., 1993) and PAA (Hill et al., 1993). That raises the possibility that the etiologic agent is still unknown because it could not be found by RP-HPLC and UV-detection. That is applicable, for example, for substances exhibiting little UV-response or coeluting with the large L-Trp peak (Hertzman et al., 1991).

For biochemical or toxicological investigations it is essential that the identified EMS related contaminants are available in sufficient quantity and purity. Their in-vitro formation conditions have to be elucidated in order to explain their presence in the implicated Trp lots and to prevent their formation by a prospective manufacturing practice. Since the use of Trp and 5-OH-Trp is obligatory for certain diets and parenteral nutrition, new requirements for the purity of Trp raw materials after the occurrence of EMS have to be established. The monograph for Trp in the German pharmacopoeia DAB 10 was changed in 1994 and now prescribes an RP-HPLC/UV (220nm) examination for the purity control of pharmaceutical grade Trp. Contaminant peaks with the retention time of 1,1′-EBT must not exceed 8 ppm (calculated as 1,1′-EBT); the sum of contaminant peaks eluting prior to Trp, 100 ppm (calculated with N-acetyltryptophan as internal standard), and those eluting after Trp, 300

*Recent Advances in Tryptophan Research*, edited by Graziella Allegri Filippini et al.
Plenum Press, New York, 1996

ppm (Feiden, 1994). Hence it seemed to be useful to compare the pattern and the amount of contaminants in various current manufactured pharmaceutical grade Trp raw materials. Additionally the question arises whether typical Trp-contaminants could be identified in Trp containing infusion solutions.

# 2. EXPERIMENTAL PROCEDURES

## 2.1. Materials

Dioxindolylalanine (DiOia), Oxindolylalanine (Oia), N-formylkynurenine (NFK), 1,2,3,4-tetrahydro-ß-carboline-3-carboxylic acid (THCC) and 1-methyl-1,2,3,4-tetrahydro-ß-carboline-3-carboxylic acid (MTHCC) were synthesized as had been published previously (Simat et al., 1994). 3a-hydroxy-1,2,3,3a,8,8a-hexahydropyrroloindole-2-carboxylic acid (PIC) was synthesized according to Savige (1975). 1,1′-EBT and PAA were synthesized as described briefly below; detailed synthesis will be published soon (Meyer et al., in prep.).

Tryptophanase (EC 4.1.99.1) was purchased from Sigma and was used both free and immobilized to CNBr-Sepharose.

Pharmaceutical grade Trp raw material was made available by two different manufacturers; EMS associated Trp (Showa Denko K.K., Lot 67236201), by Centers of Disease Control (CDC). 8 Trp containing infusion solutions of different composition were made available by a German company.

Other standard substances were obtained from commercial distributors and of analytical grade.

## 2.2. Methods

*2.2.1. HPLC Conditions.* The RP-HPLC/UV (260 nm) method, as described by Simat et al. (1994), was slightly modified for the determination of contaminants in Trp raw material and will be published in detail shortly (Simat et al., in prep.). Confirmation of identity was done by comparison of retention times with standard substances, by standard addition, by DAD-detection and by comparison of UV- and fluorescence response.

The sum parameters of contaminants prior to and after the Trp peak were calculated as Trp by addition of peak areas (UV 260 nm) and by assigning them to an external Trp standard.

*2.2.2. Selective Cleavage of Tryptophan by Tryptophanase (EC 4.1.99.1).* EMS related and unrelated Trp raw materials were subjected to an incubation with tryptophanase. The digestion was carried out for 76 h at 37 °C, using a volatile ammonium carbonate buffer (pH 8.3, 0.1 M) and pyridoxal-5-phosphate. The resulting solutions were lyophilized several times in order to remove the buffer substance, indole, and ammonia.

# 3. RESULTS AND DISCUSSION

## 3.1. Contaminants in an EMS Related Trp Lot (SD 67236201)

RP-HPLC analysis of an EMS related Trp sample manufactured by Showa Denko K.K. was performed to identify Trp metabolites and non-physiological Trp oxidation and carbonyl condensation compounds.

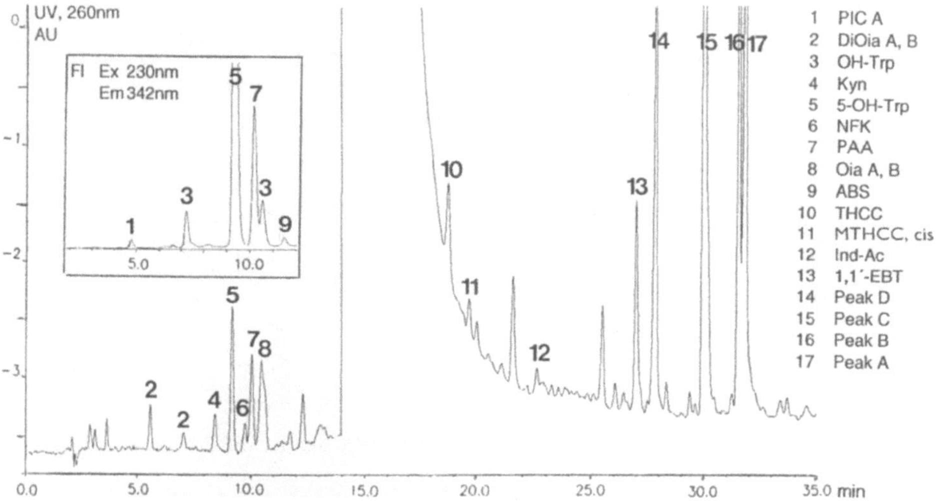


**Figure 1.** RP-HPLC / UV (260 nm) of SD-Trp, identification of contaminants.

Anthranilic acid was the starting material for the biosynthesis of Trp in the fermentation process using a genetically modified strain of Bacillus amyloliquefaciens (Fig. 7). Traces of the Trp metabolites kynurenine (Kyn), 5-hydroxytryptophan (5-OH-Trp) (Toyo'oka et al., 1991), NFK, anthranilic acid (ABS), indole-3-acetic acid (Ind-Ac) and indole (Toyo'oka et al., 1991) were identified in the examined Trp lot (Fig. 1). Non-physiological oxidation of Trp occurs during heat or acidic treatment of Trp in the presence of air. PIC-A, DiOia and Oia are suitable indicators for oxidative treatment (Simat et al., this issue) and were detected in the implicated Showa Denko K.K. Trp (SD-Trp) in small amounts (10 - 100 ppm)(Fig. 1). Their formation may be due to the conditions chosen in the purification process (Tab. 1)(Sakimoto and Torigoe, 1994).

Acetaldehyde is a by-product formed during fermentation with B. amyloliquefaciens (Sakimoto and Torigoe, 1994). Aldehydes react with Trp under slightly acidic conditions forming 1,2,3,4-tetrahydro-ß-carboline-3-carboxylic acids (Fig. 2)(Meyer et al., this issue). Formaldehyde and acetaldehyde condensation products (THCC and MTHCC) were identified in SD-Trp (Müller et al., 1991)(Fig. 1). Peaks A - D were structurally determined by Müller et al. (1991), Peak C is identical with 2-(3-indolylmethyl)-L-tryptophan mentioned above.

**Table 1.** Formation of 1,1'-EBT in Showa Denko K.K. manufacturing process (Sakimoto and Torigoe, 1991)

| Operating step | 1,1'-EBT-formation |
|---|---|
| Fermentation with B. amyloliquefaciens, Strains III (not EMS related) and V | not detectable |
| Anion IEC-process (pH 10.5, 2.5 h) | traces |
| Cation IEC-process (pH 1.7, 10 h) | 88 mg/kg Trp |
| Activated carbon (20 min, pH 4.5, 85°C) | slight decrease |

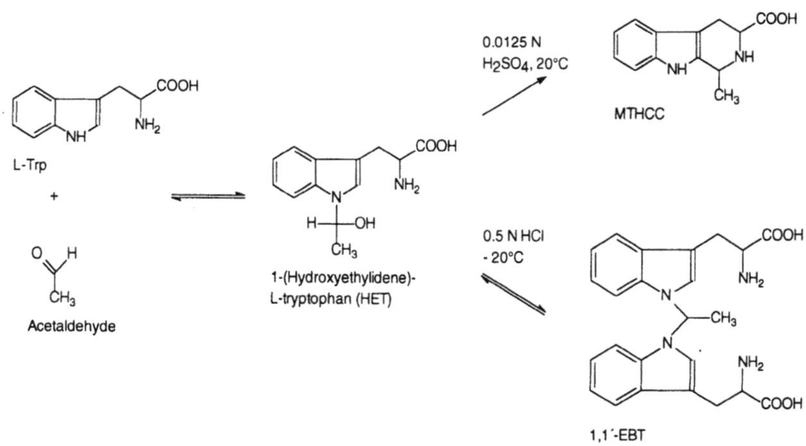


**Figure 2.** Formation conditions of MTHCC and 1,1′-EBT.

## 3.2. Formation of 1,1′-EBT and PAA

1,1′-EBT is formed by condensation reaction from acetaldehyde and Trp as is MTHCC (Fig. 2). Both reactions are catalysed by protonation of the aldehyde under acidic conditions (Fig. 2).

The yield of 1,1′-EBT is strongly influenced by temperature and state of aggregation (Fig. 3): 1,1′-EBT formation is preferred at low temperatures. An 1,1′-EBT yield of more than 60% and a purity of > 98% (RP-HPLC, 280nm) can be achieved by performing the reaction in a frozen aqueous solution at -20°C. This astonishing result may be explained by the assumption that Trp is over saturated in micelles formed by freezing the solution, which may promote dimerisation reactions.

These findings can be correlated with the results of Sakimoto and Torigoe (1994) who described the formation of 1,1′-EBT in the production process of Showa Denko K.K. (Table 1).

They found no 1,1′-EBT in the fermentation broth of the EMS-related B. amyloliquefaciens strain V, but were able to determine reasonable amounts of this impurity after cation ion exchange chromatographic (cation IEC) process at pH 1,7 regardless of the B. amyloliquefaciens strain used. The highest 1,1′-EBT amounts were found in SD-Trp lots that

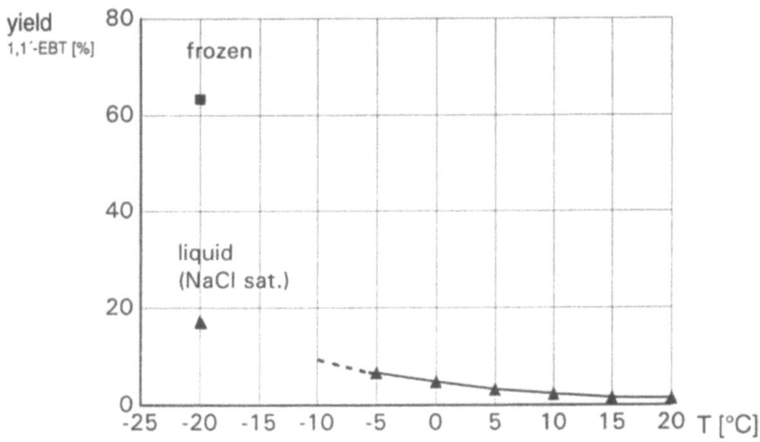


**Figure 3.** Synthesis of 1,1′-EBT, dependence on temperature.

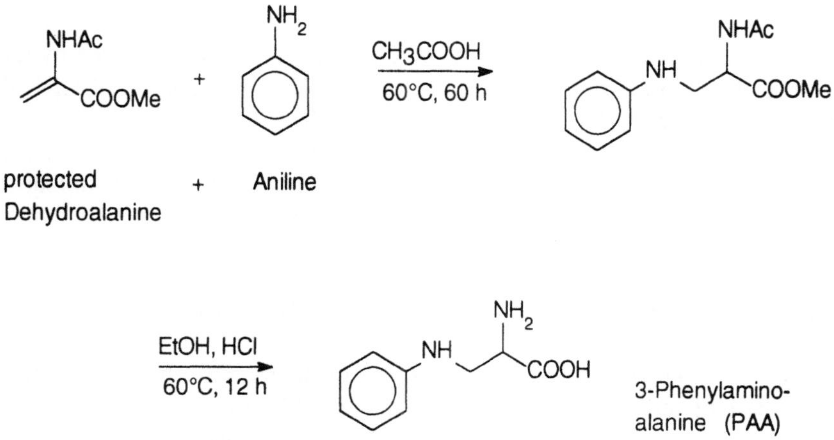


**Figure 4.** Synthesis of 3-Phenylaminoalanine (PAA).

were manufactured in winter. This might be due to the fact that the IEC columns of the commercial plant were not thermally insulated and followed the ambient temperature.

PAA can be synthesized in good yield (> 70 %) and purity (> 90 %, RP-HPLC, 240 nm) by reaction of protected dehydroalanine and aniline (Fig. 4).

The origin of PAA in SD-Trp may be revealed by the principle of this reaction and the conditions of the purification process. Dehydroalanine is formed by dehydration of L-serine, a precursor of Trp (Fig. 7), under basic conditions, while anthranilic acid can be decarboxylated yielding aniline under acidic conditions (Toyoda et al., 1994). This might correlate with the basic conditions during the anion IEC and the acidic conditions during the cation IEC process. The formation of PAA could take place under the acidic conditions of cation IEC (10 h) or during the treatment with activated carbon (Tab. 1).

According to the results described above, the formation of both EMS related contaminants, 1,1′-EBT and PAA, most probably occurs during the purification process and is not due to the genetically modified microorganism used.

## 3.3. Removal of L-tryptophan by Selective Cleavage by Tryptophanase (EC 4.1.99.1)

For the examination of the hypothesis that contaminants could coelute with the large Trp peak when RP-HPLC is used for purity control of Trp, our aim was to remove Trp most selectively by using the specificity of an enzyme. Tryptophanase was chosen because it is

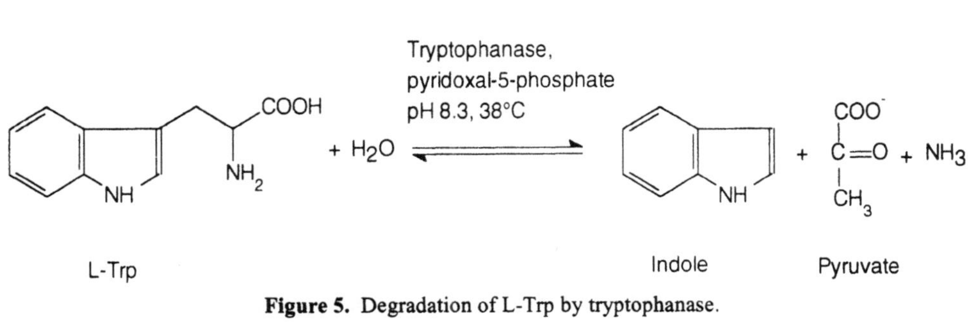


**Figure 5.** Degradation of L-Trp by tryptophanase.

commercially available and one reaction product, indole, can be removed by lyophylisation of the incubated Trp solution (Fig. 5).

However, activity of this enzyme is low and it is additionally inhibited by reaction products, especially indole. After 3 days incubation only 70 - 80 % removal of Trp had been achieved and several new by-products had been formed (Fig. 6).

Nevertheless, 2 peaks eluting prior to Trp could be detected in the incubated SD-Trp, but not after incubation of non-EMS associated Trp. This result might point out that further efforts should be made to reveal whether impurities coeluting with Trp have been overlooked hitherto.

## 3.4. Quality of Recent Trp Raw Materials and Preparations

SD-Trp and Trp raw material of 2 further manufacturers were chosen to compare the amount and pattern of impurities (Fig. 7). Manufacturer 1 (M1) uses glucose and ammonium salts as substrates for a de novo synthesis of Trp, manufacturer 3 (M 3) performs a biotransformation process with indole and D,L-serine as substrates. Even purification procedures, as far as they were specified by the manufacturer, are markedly different.

The total amount of contaminants in the Trp of M1 and M3 was very low compared to SD-Trp (Fig. 8). A unique pattern of impurities, distinctive for the product of M 1 or M 3, could not be discerned. In both recently manufactured Trp lots, PAA and 1,1′-EBT were not detectable (< 1 ppm). Oxidation (Oia) and carbonyl condensation compounds (THCC and MTHCC) were detected in much lower amounts than in SD-Trp.

It is obvious that the amount of impurities in Trp may be reduced drastically by employing suitable purification procedures. With respect to the results described above (3.2), the Showa Denko K.K. purification step seems to play a decisive role in the etiology of EMS.

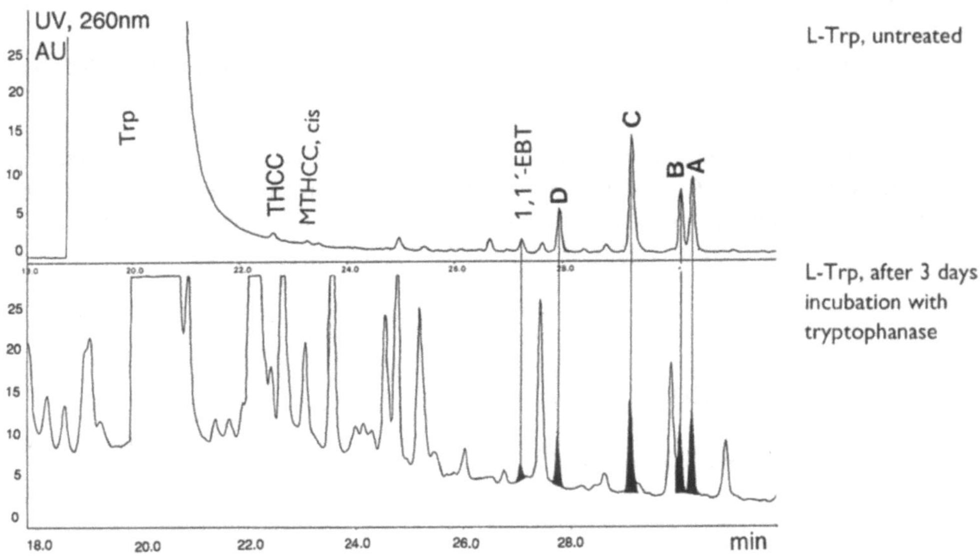


**Figure 6.** RP-HPLC of EMS related SD-Trp prior to and after tryptophanase treatment.

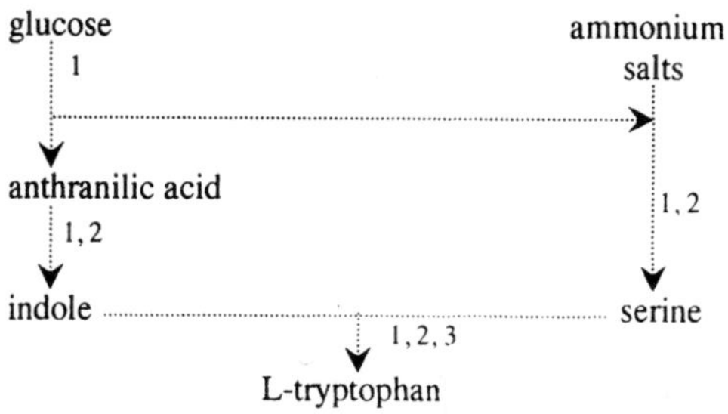


| manufacturer | microorganism | substrates | purification process |
|---|---|---|---|
| 1 | unknown | glucose, ammonium salts | unknown |
| Showa Denko K.K. (2) | B. amyloliquefaciens | anthranilic acid, ammonium salts | centrifugation, reverse osmosis, anion IEC, cation IEC, activated carbon, crystallization |
| 3 | E. Coli | indole, D,L-serine | centrifugation, evaporation crystallization, activated carbon |

**Figure 7.** Manufacturing processes of 3 different pharmaceutical grade L-Trp raw materials.

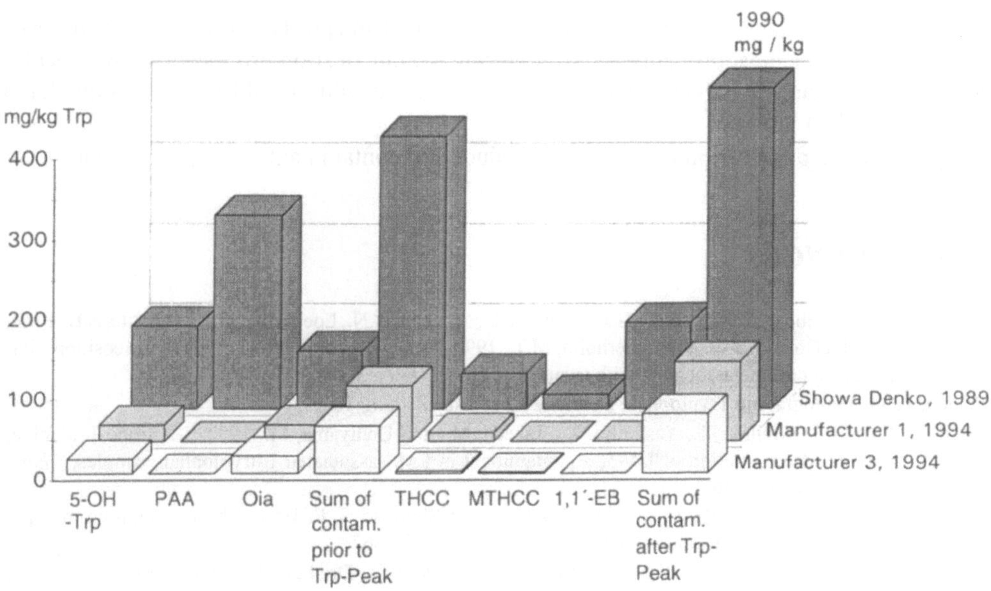


**Figure 8.** Contaminants in EMS related SD-Trp and 2 recently manufactured Trp lots.

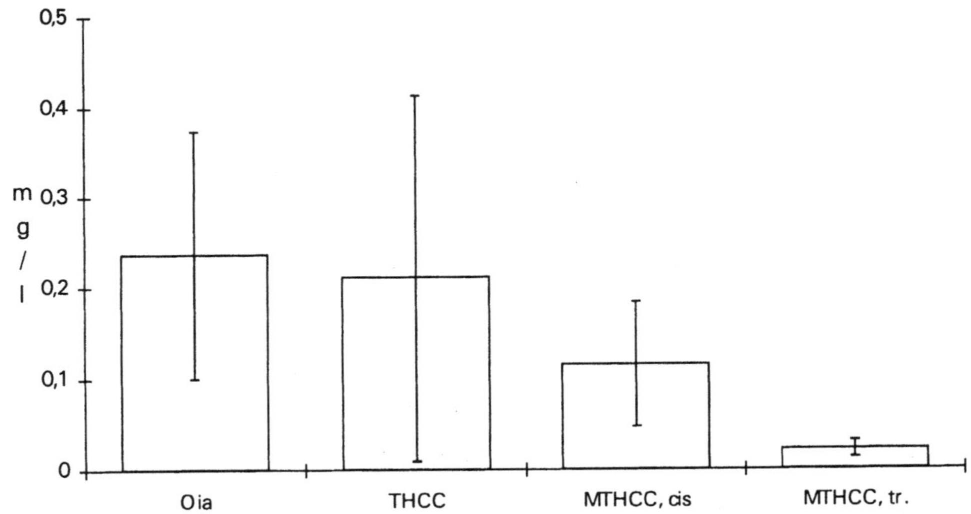


**Figure 9.** Contaminants in L-Trp containing infusions.

Infusion solutions of different composition and different Trp content (0,6 - 2,3 g/l) were analysed to identify Trp related impurities. Oxidation (Oia) and carbonyl condensation compounds (THCC and MTHCC cis, trans) were detected in varying amounts (Fig. 9).

It remains to be established whether they were formed during the preparation of the infusion and whether their amount is influenced by storage time and storage conditions and by the composition of the infusion (e.g. reducing carbohydrates).

## 4. CONCLUSIONS

1,1′-EBT and PAA were synthesized in high yield and purity. Formation of both EMS related substances most probably occurred during the purification process of Showa Denko K.K.. Contaminants in SD-Trp coeluting with Trp under conditions of RP-HPLC might have been overlooked hitherto.

Suitable purification processes can reduce Trp contaminants to a great extent.

## 5. REFERENCES

Belongia, E.A., Hedberg, C.W., Gleich, G.J., White, K.E., Mayeno, A.N., Loegering, D.A., Dunette, S.L., Pirie, P.L., MacDonald, K.L., and Osterholm, M.T., 1990, An investigation of the cause of the eosinophilia-myalgia syndrome associated with tryptophan use, *N. Engl. J. Med.* 323:357.

Feiden, 1994, Monographie Tryptophan, *Deutsche Apotheker Zeitung* 134:3361.

Goda, Y., Suzuki, J., Maitani, T., Yoshihira, K., Takeda, M., and Uchiyama, M., 1992, 3-Anilino-L-alanine, structural determination of UV-5, a contaminant in EMS-associated L-tryptophan samples, *Chem. Pharm. Bull.* 40:2236.

Hertzman, P.A., Falk, H., Kilbourne, E.M., Page, S., and Shulman, L.E.,1991, The eosinophilia-myalgia syndrome: The Los Alamos conference, *J. Rheumatol.* 18:867.

Hill, R.H. Jr., Caudill, S.P., Philen, R.M., Bailey, S.L., Flanders, W.D., Driskell, W.J., Kamb, M.L., Needham, L.L., and Sampson, E.J., 1993, Contaminants in L-tryptophan associated with eosinophilia myalgia syndrome, *Arch. Environ. Contam. Toxicol.* 25:134.

Müller, A., Busker, E., Günther, K., and Hoppe B., 1991. Characterization of byproducts in L-tryptophan, *Bioforum* 14:350.

Philen, R.M., Hill, R.F. Jr., Flanders, W.D., Caudill, S.P., Needham, L., Sewell, L., Sampson, E.J., Falk, H., Kilbourne, E.M., and the Eosinophilia-Myalgia Studies of Oregon, New York, and New Mexico, 1993, Tryptophan contaminants associated with eosinophilia-myalgia syndrome, *Am. J. of Epidemiol.* 138:154.

Savige, W.E., 1975, New oxidation products of tryptophan, *Aust. J. Chem.* 28:2275.

Sakimoto, K. and Torigoe, Y., 1994, Study of mechanism of peak E substance formation in process for manufacture of L-tryptophan, *in*: L-Tryptophan Current Prospects in Medicine and Drug Safety, W. Kochen and H. Steinhart, eds.,Walter de Gruyter Berlin, New York.

Simat, T., Meyer, K., and Steinhart, H., 1994, Synthesis and analysis of oxidation and carbonyl condensation compounds of tryptophan, *J. Chromatogr.* A 661:93.

Smith, M.J., Mazzola, E.P., Farrell, T.J., Sphon, J.A., Page, S.W., Ashley, D., Sirimanne, S.R., Hill, R.H. Jr., and Needham, L.L., 1991, 1,1'Ethylidenebis(L-tryptophan), structure determination of contaminant "97" - implicated in the eosinophilia.myalgia syndrome (EMS), *Tetrahedron Lett.* 32:991.

Toyoda, M., Saito, Y., Uchiyama, M., Troy, A.L., Trucksess, M.W., and Page, S.W., 1994, Formation of 3-(phenylamino)alanine contaminant in EMS-associated L-tryptophan, *Biosci. Biotech. Biochem.* 58:1318.

Toyo'oka, T., Yamazaki, T., Tanimoto, T., Sato, K., Sato, M., K., Toyoda, M., Ishibashi, M., Yoshihira, K., and Uchiyama, M., 1991, Characterization of contaminants in EMS-associated L-tryptophan samples by high-performance liquid chromatography, *Chem. Pharm. Bull.* 39:820.

# INDOLE DERIVATIVES AS PROBES TO STUDY THE BENZODIAZEPINE BINDING SITE IN GABA RECEPTOR COMPLEX

A. Da Settimo,[1] G. Primofiore,[1] F. Da Settimo,[1] E. Novellino,[2] G. Greco,[2] C. Martini,[3] G. Senatore,[3] and A. Lucacchini[3]

[1] Dipartimento di Scienze Farmaceutiche, Università di Pisa
Via Bonanno 6, 56126 Pisa, Italy
[2] Dipartimento di Chimica Farmaceutica e Tossicologica, Università di Napoli
Via Domenico Montesano, 49, 80131 Napoli, Italy
[3] Istituto Policattedra di Discipline Biologiche, Università di Pisa
Via Bonanno 6, 56126 Pisa, Italy

γ-Aminobutyric acid (GABA), one of the most ubiquitous inhibitory neurotrasmitter in the central nervous system (CNS), binds to its recognition sites, GABA receptor, and decreases neuronal excitability by increasing membrane chloride conductance. $GABA_A$ receptors are supermolecular receptor-chloride channel complexes of multiple subunits (α, β and γ) with distinct but interacting recognition sites for the neurotrasmitter, for convulsants (such as picrotoxin and bicyclophosphorothionates), and for depressants (such as benzodiazepines, barbiturates and neurosteroids) (MacDonald and Olsen, 1994).

1,4-Benzodiazepines (Bz) are a class of effective and widely-prescribed agents for the treatment of anxiety and sleep disorders. Several substances, with a chemical structure different from Bz, bind with a high affinity at BzR on GABA receptor complex and exhibit a wide variety of pharmacological actions, ranging in a continuum from a Bz like (full agonist) to a Bz opposite action (inverse agonist). In the last few years many attempts have been made to correlate the common features of the various BzR ligands in order to gain insight into the structure of the receptor, and to aid in the design of new therapeutic agents. Despite recent advances, the exact structural nature of the site which accomodates agonist, inverse agonist and antagonist ligands remains unresolved.

Moreover, scientific interest has been focused on the question of whether endogenous compounds exist which interact with the BzR and modulate GABA action on $Cl^-$ conductance. Until nowdays this search has not reached any definitive results, but several endogenous substances, called endozepines, have been isolated and characterized.

Several β-carboline derivatives have received the most attention because of their high affinity at BzR. n-Butyl-β-carboline (β-CCB), which behaves like an inverse agonist, has been isolated and characterized in the CNS and the existence of β-CCB transport into rat

*Recent Advances in Tryptophan Research*, edited by Graziella Allegri Filippini et al.
Plenum Press, New York, 1996 

cerebral cortex has been suggested. While the complete biosynthetic pathway of β-carbolines in animals or in man is not known, it is quite reasonable to think that these compounds are derivatives of tryptophan, tryptamine or, more generally, of a (2-aminoethyl)indole structure.

In the last few years our research has been directed to the search for new indole compounds that could be derived from biological pathways showing a high affinity at the BzR. These products were designed to clarify the structural requirements of the Bz. binding site and to give some indications about the structure of the still unknown endogenous ligand at the BzR. With this in mind, we have prepared and reported affinity data at the BzR and the activity profile of several derivatives **I** in which an α-amino acid, a dipeptide, a biogenic amine such as tryptamine, tyramine or dopamine, or a benzylamine was linked to the indole system by an oxalyl bridge (Primofiore et al., 1989; Bianucci et al., 1992; Da Settimo et al., 1993).

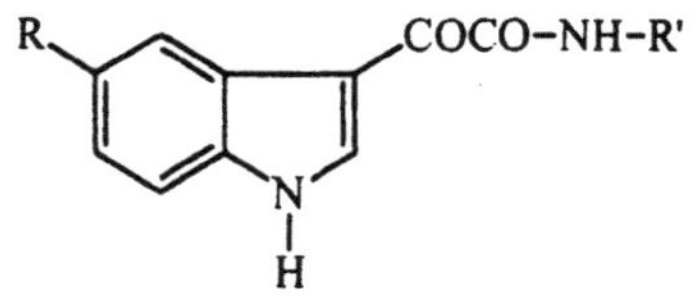


**I**

R = H, Cl, $NO_2$
NH-R' = α-aminoacid, dipeptide, phenylethylamine, tyramine, dopamine, tryptamine, benzylamine

The affinity of the latter products depended directly on the distance between the side phenyl ring and the amide bond. In general, the most active compounds were those containing a benzylamine residue in the side chain, with $K_i$ values in the nanomolar order (Table 1). The GABA ratio values lower, equal and higher than unity, showed for these compounds a pharmacological profile ranging from inverse agonist, to antagonist and to partial agonist depending on the nature of the substituents on the side phenyl ring.

Through the study of a great number of structurally different ligands (β-carbolines, pyridodiindoles, pyrazoloquinolinones, benzodiazepines, triazolophthalazines, etc.), Zhang et al. (1995) have proposed a model of the pharmacophore for benzodiazepine receptor binding. Thus, ligand interaction with a hydrogen bond acceptor group ($A_2$) and a donor site ($H_1$), as well as with a lipophilic area ($L_1$) is required to elicit potent inverse agonist activity. The interaction with two hydrogen bond donor sites, $H_1$ and $H_2$, at a distance of 5.0-7.5 Å, together with the occupation of two lipophilic regions designated $L_1$ and $L_2$ is required for partial agonist activity, while the further occupation of $L_3$ results in a full agonist activity.

On the basis of this model, it was supposed for our compounds showing an inverse agonist/antagonist profile, that the interaction with the receptor site should occur via the indole NH and the C=O (2) of the oxalyl bridge, forming hydrogen bonds with the acceptor ($A_2$) and the donor ($H_1$) sites on the receptor, respectively. Moreover, the phenyl ring of the side chain should interact with the lipophilic region $L_1$ (Figure 1a).

The good affinity and the interesting pharmacological profile shown by these derivatives prompted us to pursue this research. In this report, we describe the synthesis and the affinity at the BzR of a number of indolylglyoxylylphenylhydrazine derivatives of general formula **II**, in which the methylene spacer has been substituted by an isoster NH group. In our intention this group should have maintained the suitable distance between the phenyl ring and the amide bond for a high affinity and, moreover, could have also produced a further point of binding at the receptor site through the formation of a hydrogen bond.

**Table 1.** *In vitro* data of N-[(5-substituted indol-3-yl)glyoxylyl]benzylamine derivatives

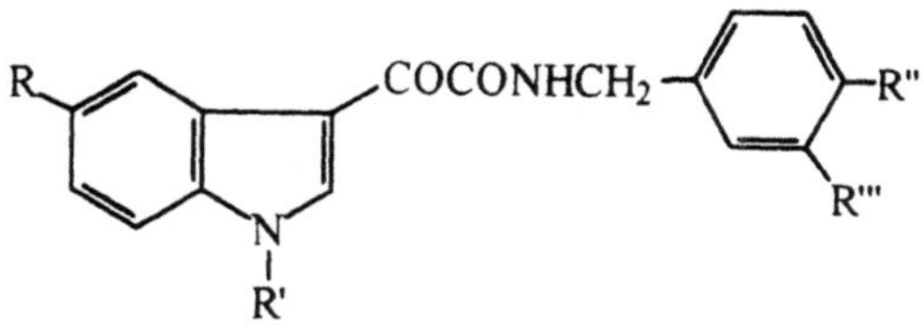


| n | R | R′ | R″ | R‴ | Inib.%[a] (10 μM) | $K_i$[b] (μM) | GABA ratio[c] |
|---|---|---|---|---|---|---|---|
| 1 | H | H | H | H | 98 | 0.117 | 1.00 |
| 2 | Cl | H | H | H | 34[d] | 0.490 | 1.00 |
| 3 | $NO_2$ | H | H | H | 55[d] | 0.117 | 0.95 |
| 4 | H | H | $OCH_3$ | H | 97 | 0.163 | 1.00 |
| 5 | Cl | H | $OCH_3$ | H | 81 | 0.107 | 1.00 |
| 6 | $NO_2$ | H | $OCH_3$ | H | 71 | 0.053 | 0.70 |
| 7 | H | H | H | $OCH_3$ | 97 | 0.290 | 0.74 |
| 8 | Cl | H | H | $OCH_3$ | 89 | 0.162 | 0.79 |
| 9 | $NO_2$ | H | H | $OCH_3$ | 84 | 0.038 | 0.91 |
| 10 | H | H | $OCH_3$ | $OCH_3$ | 99 | 0.094 | 0.80 |
| 11 | Cl | H | $OCH_3$ | $OCH_3$ | 94 | 0.030 | 0.60 |
| 12 | $NO_2$ | H | $OCH_3$ | $OCH_3$ | 97 | 0.011 | 0.92 |
| 13 | H | $CH_3$ | $OCH_3$ | $OCH_3$ | 70 | 2.70 | 0.88 |
| 14 | H | H | Cl | H | 99 | 0.067 | 0.85 |
| 15 | Cl | H | Cl | H | 38 | — | — |
| 16 | $NO_2$ | H | Cl | H | 2 | — | — |
| 17 | H | $CH_3$ | Cl | H | 86 | 1.00 | 1.00 |
| 18 | H | H | H | Cl | 96 | 0.16 | 0.95 |
| 19 | Cl | H | H | Cl | 14 | — | — |
| 20 | $NO_2$ | H | H | Cl | 0 | — | — |
| 21 | H | H | OH | H | 96 | 0.43 | 1.60 |
| 22 | Cl | H | OH | H | 98 | 0.15 | 1.30 |
| 23 | $NO_2$ | H | OH | H | 100 | 0.012 | 1.25 |
| 24 | H | H | H | OH | 91 | 0.82 | 1.35 |
| 25 | Cl | H | H | OH | 95 | 0.57 | 1.45 |
| 26 | $NO_2$ | H | H | OH | 95 | 0.065 | 1.38 |
| 27 | H | H | OH | OH | 95 | 0.349 | 1.56 |
| 28 | Cl | H | OH | OH | 96 | 0.18 | 1.33 |
| 29 | $NO_2$ | H | OH | OH | 98 | 0.048 | 1.34 |
| 30 | H | H | OH | $OCH_3$ | 96 | 0.14 | 1.07 |
| 31 | Cl | H | OH | $OCH_3$ | 99 | 0.027 | 1.00 |
| 32 | $NO_2$ | H | OH | $OCH_3$ | 100 | 0.013 | 1.00 |

[a]Percents of inhibition of specific [$^3$H]flunitrazepam binding at 10 μM compound concentration are means ± SEM of five determinations.

[b]Ki values are means ± SEM of three determinations.

[c]GABA ratio = Ki without GABA/Ki with GABA.

[d]Determined at 200 nM concentration of the compound due to its insolubility at 10 μM.

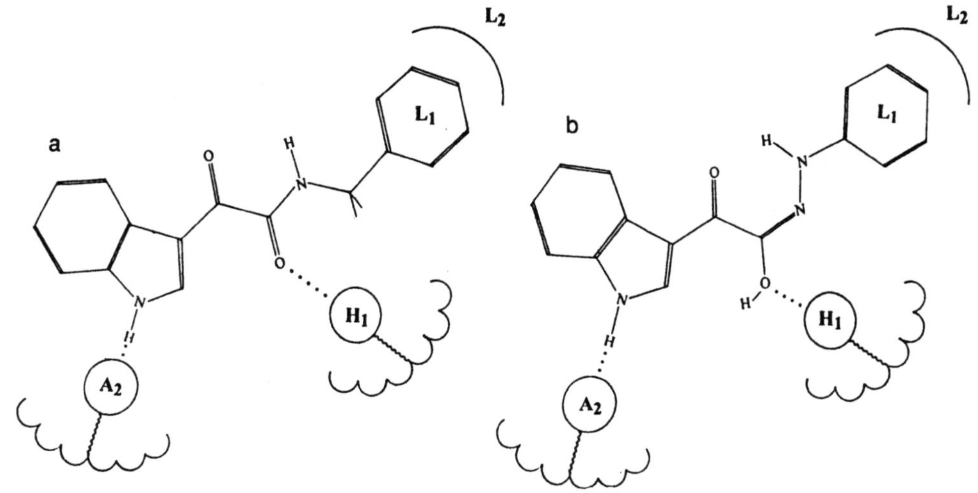


**Figure 1.** Interaction with the BzR in the inverse agonist/antagonist orientation of the benzylamine (**a**) and the hydrazine (**b**) derivatives.

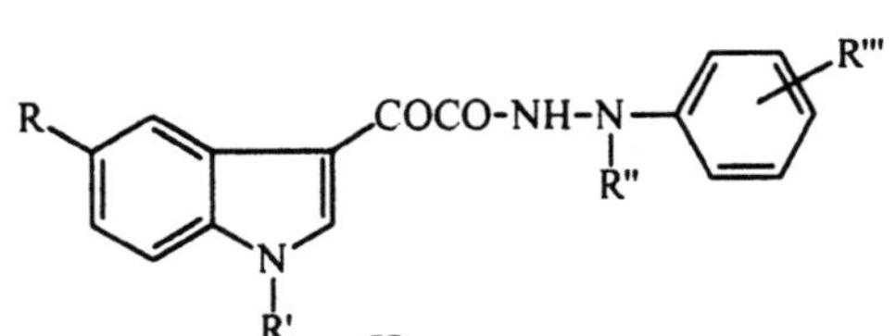

**II**

R = H, Cl, $NO_2$
R', R" = H, $CH_3$
R"' = H, Cl, $CH_3$, $OCH_3$, $NO_2$

The synthesis of compounds **II** was carried out by adding a suspension of the appropriate indol-3-ylglyoxylyl chloride in anhydrous benzene to a suspension of the opportunely substituted phenylhydrazine hydrochloride in the same solvent, in the presence of triethylamine.

All the products **II** were tested for their ability to displace specific [$^3$H]flunitrazepam binding from bovine brain membranes. The data reported in Table 2 showed for these compounds an affinity similar to that of the benzylamine derivatives. The GABA ratio values equal or lower than unity indicated a pharmacological profile of inverse agonist/antagonist. Generally, the most active indolylglyoxylylamine derivatives **I** with an inverse agonist/antagonist profile were those bearing an electron-withdrawing group at position 5 of the indole nucleus, as this group facilitates the indole NH in its interaction with the hydrogen bond acceptor site ($A_2$) on the receptor, via polarization of the indole NH bond. Instead, for the hydrazines **II**, the most active compounds were those unsubstituted at position 5 of the indole (**33**, **38**, **41**, **44**, **47**, **50**), and bearing an electron-withdrawing group in the para position of the side phenyl ring (**38** and **50**).

It is well known that, to show a high affinity at the BzR, ligands must have a planar or pseudoplanar conformation to penetrate sufficiently into the cleft of the receptor and interact with the binding sites. Our compounds **II** are not planar, due to the arylhydrazide

**Table 2.** Inhibition of [$^3$H]-Ro 15-1788 specific binding from bovine brain membranes of N-[(5-substituted indol-3-yl)glyoxylyl] phenylhydrazine derivatives

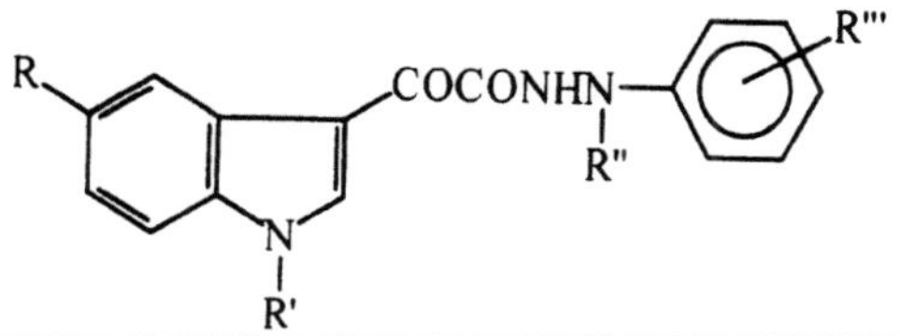


| n° | R | R′ | R″ | R‴ | Inib.% [a] (10 μM) | Ki [b] (μM) | GABA ratio[c] |
|---|---|---|---|---|---|---|---|
| 33 | H | H | H | H | 98 | 0.203 | 0.88 |
| 34 | Cl | H | H | H | 60 | — | — |
| 35 | $NO_2$ | H | H | H | 48 | — | — |
| 36 | H | $CH_3$ | H | H | 81 | 1.20 | 0.70 |
| 37 | $NO_2$ | $CH_3$ | H | H | 16 | — | — |
| 38 | H | H | H | p-Cl | 100 | 0.062 | 0.70 |
| 39 | Cl | H | H | p-Cl | 54 | — | — |
| 40 | $NO_2$ | H | H | p-Cl | 13 | — | — |
| 41 | H | H | H | p-$CH_3$ | 96 | 0.29 | 0.72 |
| 42 | Cl | H | H | p-$CH_3$ | 48 | — | — |
| 43 | $NO_2$ | H | H | p-$CH_3$ | 61 | — | — |
| 44 | H | H | H | m-$CH_3$ | 91 | 0.51 | 1.0 |
| 45 | Cl | H | H | m-$CH_3$ | 52 | — | — |
| 46 | $NO_2$ | H | H | m-$CH_3$ | 38 | — | — |
| 47 | H | H | H | p-$OCH_3$ | 94 | 0.43 | 0.84 |
| 48 | Cl | H | H | p-$OCH_3$ | 68 | — | — |
| 49 | $NO_2$ | H | H | p-$OCH_3$ | 31 | — | — |
| 50 | H | H | H | p-$NO_2$ | 96 | 0.056 | 0.75 |
| 51 | Cl | H | H | p-$NO_2$ | 30 | — | — |
| 52 | $NO_2$ | H | H | p-$NO_2$ | 3 | — | — |
| 53 | H | $CH_3$ | H | p-$NO_2$ | 85 | 0.455 | 0.87 |
| 54 | H | H | $CH_3$ | H | 51 | — | — |
| 55 | Cl | H | $CH_3$ | H | 56 | — | — |
| 56 | $NO_2$ | H | $CH_3$ | H | 73 | 2.40 | 1.30 |
| 57 | H | $CH_3$ | $CH_3$ | H | 7 | — | — |
| 58 | $NO_2$ | $CH_3$ | $CH_3$ | H | 5 | — | — |

[a]Percents of inhibition of specific [$^3$H]flunitrazepam binding at 10 μM compound concentration are means ± SEM of five determinations.
[b]Ki values are means ± SEM of three determinations.
[c]GABA ratio = Ki without GABA/Ki with GABA.

moiety that is arranged in a conformation in which the torsion angle about the N-N bond has values different from those necessary for a coplanar conformation. However, literature data and NMR studies in solution seem to suggest the hypothesis that the arylhydrazide residue can assume a planar tautomeric geometry such as **III**, stabilized by a hydrogen bond between the oxygen atom of the CO(1) of the oxalyl bridge and the aniline nitrogen atom.

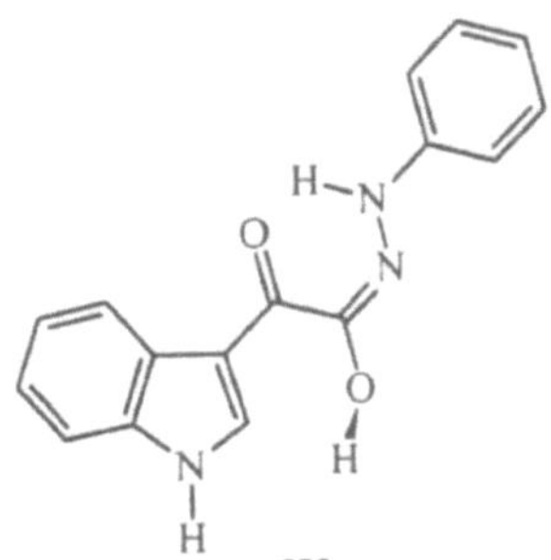

III

Therefore, it is plausible to suppose that our inverse agonist/antagonist compounds **II** may anchor at the receptor site in this planar tautomeric conformation by interacting at the acceptor site $A_2$ with the indole NH, at the donor site $H_1$ by forming a three-centered hydrogen bond involving the enolic oxygen and the imine nitrogen, and at the lipophilic area $L_1$ with the side phenyl ring (Figure 1b).

Figure 2 shows the superimposition of the hydrazide tautomer **III** on a benzylamine derivative **I** in the receptor site. As can be seen, the two structures share the principal points of interaction with the receptor. However, some differences may arise from the fact that the hydrazide derivative has a folded conformation, while the benzylamine one is completely unfolded. In particular, the substituents at position 5 of the indole nucleus and the indole NH

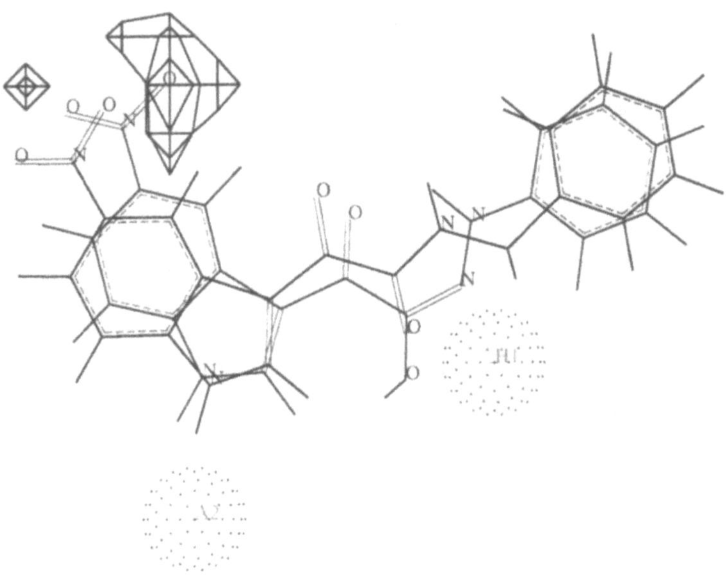

**Figure 2.** Superimposition of a benzylamine and a hydrazine derivative interacting at the BzR in the inverse agonist/antagonist orientation.

bond do not have the same orientation in the receptor site for the two series. These differences may explain the different trend of affinity shown by the two classes of ligands. With the increasing dimensions of the substituents at position 5 of the indole ring, affinity decreases progressively in the hydrazine series. This should be caused by a different orientation of these ligands in the receptor site that directs the 5-substituent to a region of negative steric repulsion, a thing that does not occur with the benzylamine analogues (Figure 2).

The methylation of the indole NH determines a lowering of affinity in both series. This effect is less dramatic in the hydrazine one because the alignment of the vector NH-$A_2$ does not have the optimal orientation.

The methylation of the aniline NH impedes the formation of the planar tautomer **III** and, therefore, causes a dramatic decrease in affinity. Clearly, compounds with both the NH methylated do not show any potency at the BzR.

These results are in agreement with the previously proposed inverse agonist/antagonist pharmacophore model.

Clinically, there is a need for partial (or selective) inverse agonists devoid of proconvulsant/convulsant activity, as they could be useful for the treatment of barbiturate alcohol-induced CNS depression, hepatic encephalopathy, and for cognition enhancement, among others. In general, recent research efforts have been directed towards the discovery of a new class of compounds able to discriminate among the various effects of the benzodiazepines, namely anxiolytic, sedative, anticonvulsant and myorelaxant, and that show a reduced propensity to cause one or more of their unwanted side effects including ataxia, interactions with alcohol, sedation, dizziness and, particularly, risk of dependence with long term use.

This result may be reached on the one hand by means of compounds selectively interacting with different Bz receptor sites such as $Bz_1$ and $Bz_2$, and on the other with products showing a pharmacological profile of partial inverse agonist or partial agonist. For these reasons, the most active hydrazine derivatives **II** were assayed with tests able to discriminate between the two receptor types, but they showed only a low selectivity.

Therefore our future goal is to rationally modify molecular portions or substituents of our molecules so that they may have the appropriate features to show selectivity in the spectrum of behavioural effects seen with the BzR ligands.

## REFERENCES

Bianucci A.M., Da Settimo A., Da Settimo F., Primofiore G., Martini C., Giannaccini G. and Lucacchini A., 1992, Benzodiazepine receptor affinity and interaction of some N-(indol-3-ylglyoxylyl)amine derivatives, *J. Med. Chem.* 35: 2214.

Da Settimo A., Primofiore G., Da Settimo F., Lucacchini A., Martini C. and Senatore G., 1993, Benzodiazepine receptor affinity of new indole derivatives, *Farmaco* 48: 285.

MacDonald R.L. and Olsen R.W., 1994, $GABA_A$ receptor channels, *Annu. Rev. Neurosci.* 17: 569.

Primofiore G., Marini A.M., Da Settimo F., Martini C., Bardellini A., Giannaccini G. and Lucacchini A., 1989, Specific inhibition of benzodiazepine receptor binding by some N-(indol-3-ylglyoxylyl)amino acid derivatives: stereoselective interactions, *J. Med. Chem.* 32: 2514.

Zhang W., Koehler K.F., Zhang P. and Cook J.M., 1995, Development of a comprehensive pharmacophore model for the benzodiazepine receptor, *Drug Design and Discovery* 12: 193.

# INDOLE ALKALOIDS AND OTHER CONSTITUENTS FROM THE PLANT *Securidaca longipedunculata fres*

J. T. Wróbel,[1] M. Matuszewska,[1] J. Szychowski,[1] A. Bertazzo,[2] P. Traldi,[3] C. V. L. Costa,[2] and G. Allegri[2]

[1] Department of Chemistry, University of Warsaw
Warsaw, Poland
[2] Dipartimento di Scienze Farmaceutiche, Univerisità di Padova
Padova, Italy
[3] C.N.R., Area della Ricerca
Padova, Italy

It is known since many centuries that the Polygalaceae family represents plants used by natives in Africa as a remedy for very wide spectrum of diseases starting with malaria and wound dressing finishing with psychotropic and narcotic activity.

*Securidaca Longipedunculata Fres.* as a species in this family was of medical interest by witch doctors and many local tribes mostly in West and South-West Africa. Also this species is listed as a remedy against almost twenty different illnesses starting with a trivial ones like cough and toothache finishing with remedy against snake-bite, psychotropic activity or drug against "evil spirits".

No wonder that for more than last seven decades this plant become also object of interest of chemists, pharmacologists and chemotaxonomists. Soon it was found that in the Polygalaceae family and in the species *Securidaca Longipedunculata Fres* about ten different types of constituents are present including compounds like saponines (Sezik and Yesilada, 1985), carbohydrates (De Tommasi et al., 1993), heterocyclic compounds and alkaloids. Most of the compounds found, were similar in Polygalaceae and Securidaca, however the alkaloids detected in the species (*Securidaca longipedunculata*) and in the family (Polygalaceae) were different and in the first one were classified as ergot group and in the second as beta-carboline alkaloids. β-carboline alkaloids were described by the Chinese and Korean researchers (Jin and Park, 1993; Park et al., 1993). Fig.1 gives two examples of these alkaloids.

Ergot alkaloids were detected by Costa et al.(1992a; 1992b) and Scandola et al. (1994),. Basic extracts ($NH_3$/MeOH) obtained from the roots of S.l. (Fres) collected in West Africa in the dry season, were analyzed by supercritical fluid chromatography/mass spectrometry. Analogy of fragmentations found between the spectra of model compounds like lysergol and elymoclavine and the spectra of the fractions from the examined plant (EI, FAB,

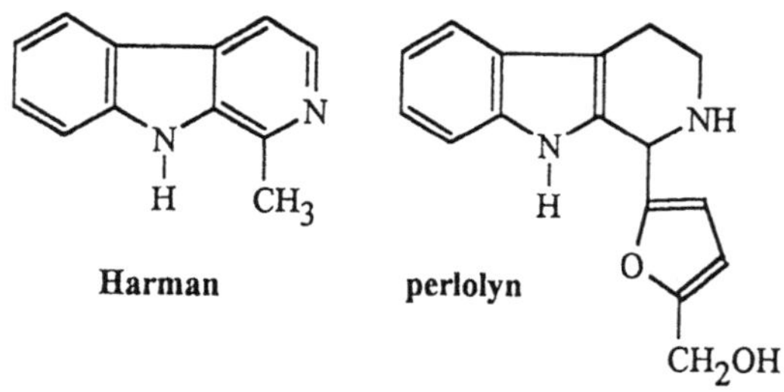


Figure 1

MIKE) made it possible to propose six structures for suggested ergot alkaloids. The ergot alkaloids detected in the S.l. (Fres.) are shown in Fig.2.

This observation together with known, wide use of the plant in the folks medicine, increased our interest not only to the isolation of further chemical constituents, but also to chemotaxonomy and even more to the allelochemical properties of alkaloids and other constituents in this plant. It is known from the natural products chemistry that there is a very strong biochemical and ecological motive for the alkaloids to be present in the plant at the right side and at the right time. This is strictly connected with the season of the year, with observed medical importance and with the necessity of self-difference of the plant against microorganisms, against competing plants and also against some herbivors.

Supporting strongly the general view that alkaloids are not waste and functionless products we made the next step in our research and examined the chemical constituents present in the roots of S.l. (Fres.) collected this time in the wet season. The result was surprising but not unexpected. The alkaloids were not present whatsoever. Instead some other compounds were isolated and identified. Extraction procedure in basic media (ammonia and methanol) applied to the roots, followed by TLC chromatography, resulted in eight compounds identified by MS, NMR, IR, UV and elemental analysis as cis- and trans- derivatives

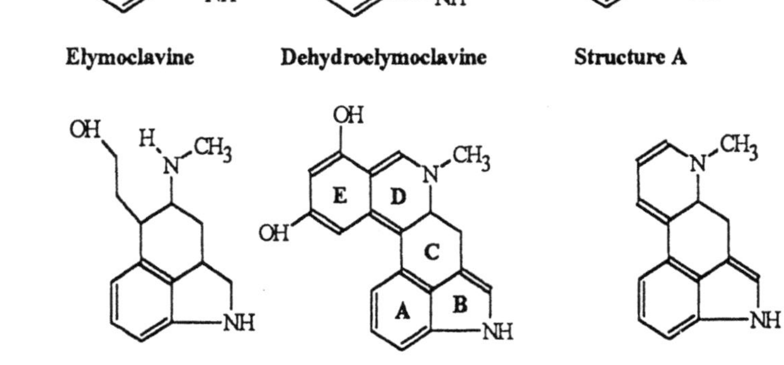


Figure 2

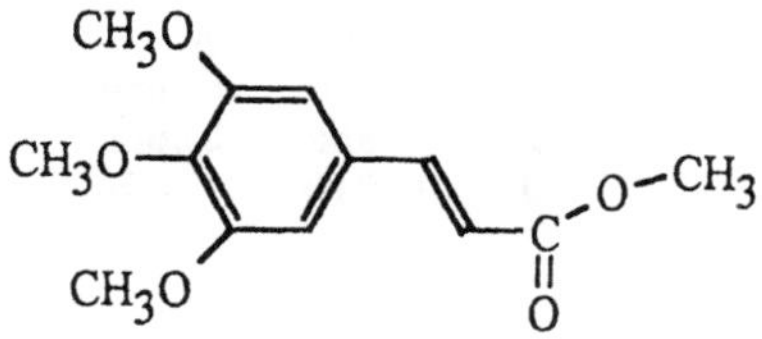

**trans-3,4,5-trimethoxycinnamic acid methyl ester (main comp)**

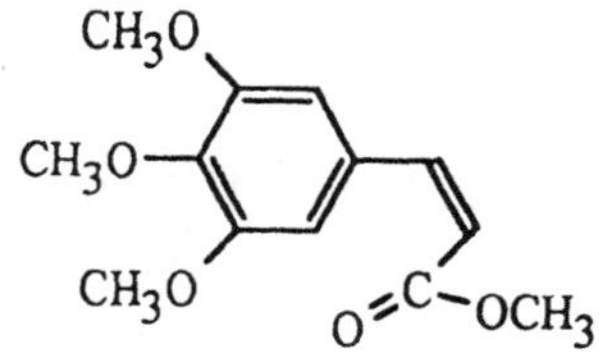

**cis-3,4,5-trimethoxycinnamic acis methyl ester**

**252**

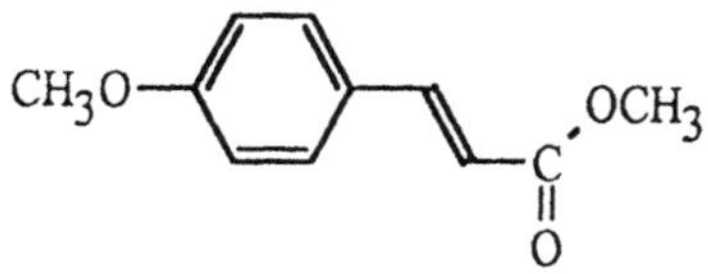

**trans-4-methoxycinnamic acid methyl ester**

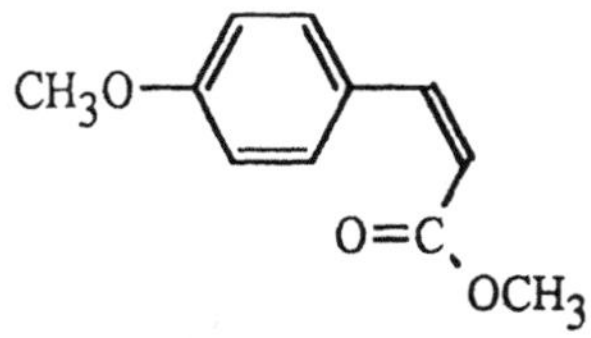

**cis-4-methoxycinnamic acid methyl ester**

**192**

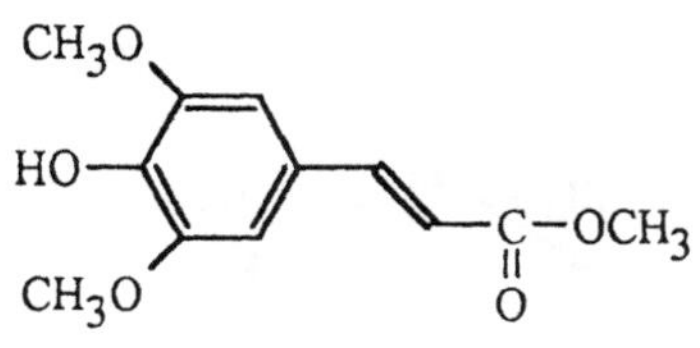

**trans-sinapoic acid methyl ester**

**cis acid present (not isolated)**

**238**

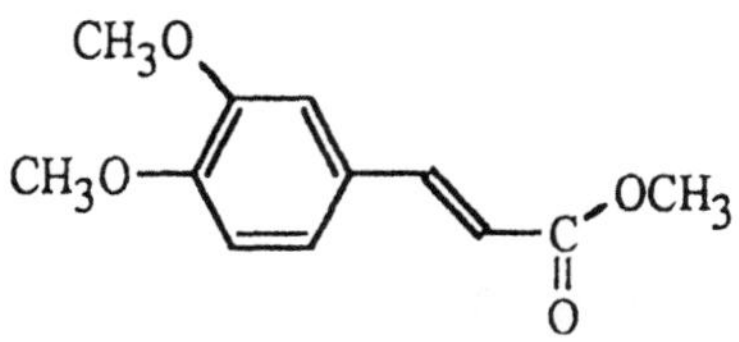

**trans-3,4-dimethoxycinnamic acid methyl ester**

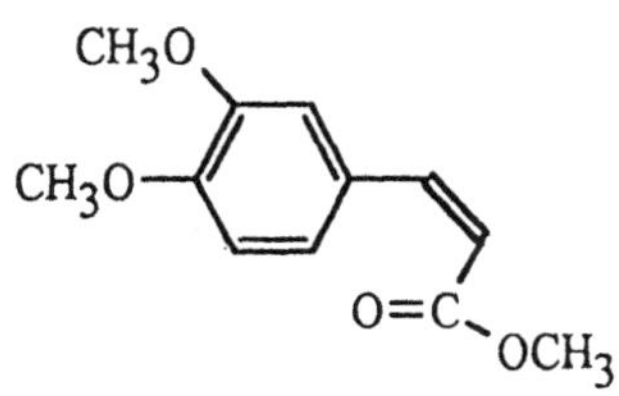

**cis-3,4-dimethoxycinnamic acid methyl ester**

**222**

**Figure 3**

of cinnamic acid ester with one, two or three of methoxy or hydroxy groups and various substitution pattern in the aromatic ring (Fig. 3).

Particularly usefull were $^1$H-NMR technique in homo-, hetero- and long range heteronuclear correlation experiments (COSY, HETCOR, LR HETCOR) which allowed to establish stereochemistry around the double bond and the positions of substitution in the aromatic ring (Fig. 4).

UV spectra showed the maxima at 227 nm and 309 nm with e value 1090 and 2016 respectively. Elemental analysis was fully in agreement with calculated figures and also with accurate mass measurements. EI mass fragmentation was characterized mostly by loss of $CH_3$, $OCH_3$ and $CO_2CH_3$ fragments for all the examined derivatives of cinnamic acid esters. This is exemplified on Fig. 5.

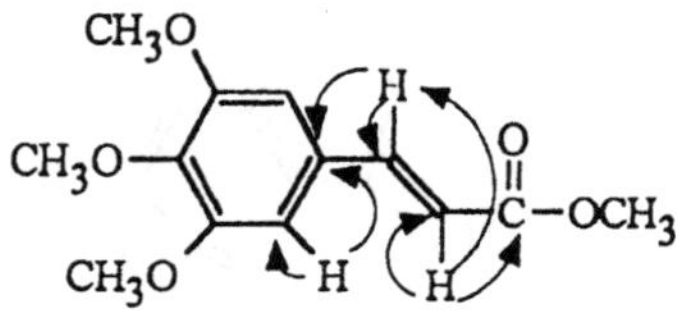

| | trans | cis |
|---|---|---|
| 1 H (200 MHz) H-C | 7.6ppm | 6.9ppm |
| H-C | 6.3ppm | 5.8ppm |
| J | 15.8 Hz | 12.4 Hz |

COSY correletion H-H

HECTOR correlation H-C

LR HECTOR correlation H-C

Figure 4

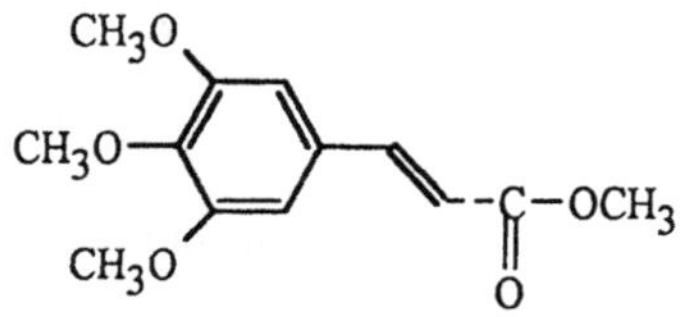

M = 252 $C_{13}H_{16}O_5$ (252.093214)

252: $-CH_3O$ → 221 $C_{12}H_{13}O_4$ (221.078692); $-CH_3$ → 237 $C_{12}H_{13}O_5$ (237.072791)

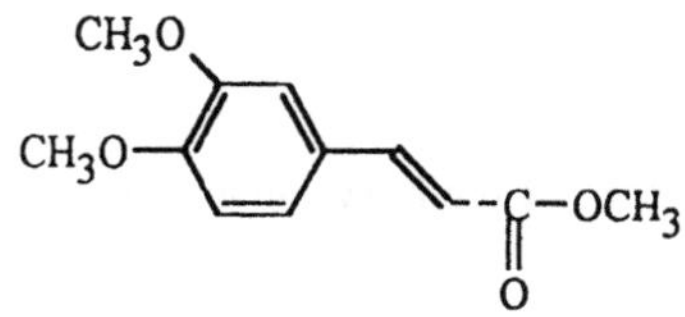

M =222 $C_{12}H_{14}O_4$ (222.0893)

222: $-CH_3O$ → 191 $C_{11}H_{11}O_3$ (191.07062); $-CH_3$ → 207 $C_{11}H_{11}O_4$ (207.06556)

Figure 5

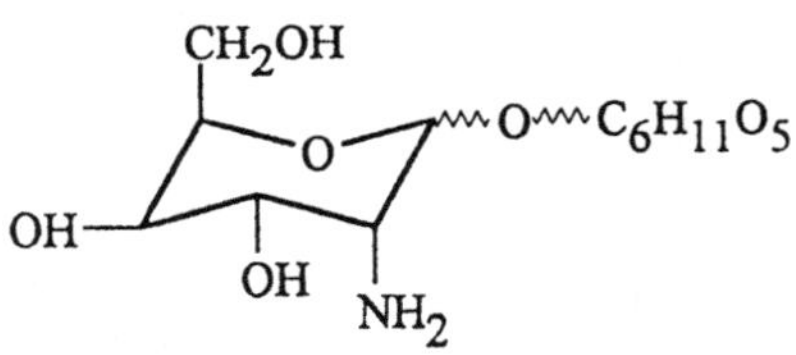

Figure 6

Independently, the powderal roots were extracted by modified procedure with sequence of organic solvents of different polarity, in neutral media. Fractions thus obtained were purified by preparative TLC resulting in saccharose, complex sugars (structure determination in progress) and amino sugars. $^1$H-, $^{13}$C-NMR, IR, UV, MS spectroscopy and elemental analysis were applied for structures elucidation. Morgan-Elson and Elson-Morgan tests in respect to aminosugars correlated with UV spectroscopy allowed one of the isolated aminosugar to describe as 2-amino-2-desoxy-glucosamine (Fig. 6 ) as working hypothesis.

## REFERENCES

Costa C., Bertazzo A., Curcuruto O., Traldi P., 1992a, Preliminary studies for identification of alkaloids from Securidaca Longipedunculata, *Il Farmaco* 47: 121.

Costa C., Bertazzo A., Allegri G., Curcuruto O., Traldi P., 1992b, Indole alkaloids from the roots of an African plant Securidaca Longipedunculata. I. Isolation by column chromatography and preliminary structural characterization by mass spectrometry, *J. Heterocyclic Chem*., 29: 1641

De Tommasi N., Piacente S., De Simone F., and Pizza C., 1993, New sucrose derivatives from the bark of Securidaca Longipedunculata, *J. Nat. Prod*. 56(1): 137.

Jin B., Park J., 1993, Alkaloidal components of Poligala tenuifolia Willd, *Zhongguo Zhongyao Zazhi*, 18: 675

Park M.K., Park J.H., Kim B.Y., Kim J.M., Liem K.J., Han B.H., 1993, Analysis of alkaloids in Polygala tenuifolia by HPLC, *Anal. Sci. Technol*., 6: 255.

Scandola M., Games D.E., Costa C., Allegri G., Bertazzo A., Curcuruto O., Traldi P., 1994, Structural study of alkaloids from Securidaca Longipedunculata roots. II. Isolation and characterization by supercritical fluid chromatography/mass spectrometry, *J. Heterocyclic Chem*., 31: 219.

Sezik E., Yesilada E., 1985, A new saponin from the roots of Polygala pruinosas, *Fitoterapia* 56: 159.

# 111

# MELATONIN MEASUREMENTS IN TISSUES

## Different Recovery of Endogenously Contained and Exogenously Added Melatonin

A. George, B. Poeggeler, and G. Huether

Department of Psychiatry
University of Göttingen
von-Siebold-Str. 5, D-37075 Göttingen

## 1. INTRODUCTION

The important role of melatonin in mediating photoperiodic information is now well established. The occurrence of melatonin and a broad spectrum of metabolic and physiological effects of this hormone has been recognized in a wide variety of species ranging from unicellular organisms and plants to higher vertebrates (Cardinali, 1981; Arendt, 1988; Reiter, 1991; Poeggeler, 1993). Analytical methods have been developed and are now widely used for the quantification of melatonin in biological samples. Radioimmunoassays and, more recently, also enzyme-linked immunoassays, became the most widely used techniques for sensitive measurements of melatonin in various body fluids and tissues (Rollag, 1981; Arendt, 1986; Yie et al., 1993). Combined gas-chromatographic/mass spectrometric (GCMS) analysis provides highest specificity and can achieve equal and sometimes better levels of sensitivity (Lewy and Markey, 1978; Skene et al., 1983), HPLC-methods are now increasingly used for the validation of RIA-measurements and for measurements of melatonin in tissues and body fluids (Raynaud and Pevet, 1991; Vieira et al., 1992; Peniston-Bird et al., 1993).

Quantification of melatonin in body tissues and cells by any of these methods requires the previous extraction of the melatonin contained in the sample. It is recognized by most researchers that during the process of sample pretreatment and the extraction procedure, part of the melatonin contained in the original sample is inevitably lost, either through inefficient extraction, irreversible binding or through enzymatic or non-enzymatic degradation. Since melatonin is readily oxidized and acts a potent scavenger of hydroxyl radicals (Reiter et al., 1993), these non-enzymatic losses are of particular significance. The usual procedure which is applied in order to correct the measured values for such losses is the simultaneous determination of the recovery of a certain amount of externally added authentic melatonin standards and the use of these "recovery rates" for mathematical treatment of the measured values. The validity of this strategy is clearly dependent on the validity of its basic assumption. It is tacitly assumed that the loss of melatonin which is externally added is

*Recent Advances in Tryptophan Research,* edited by Graziella Allegri Filippini et al.
Plenum Press, New York, 1996

identic or at least similar to the loss of melatonin which is endogenously contained in the sample. We have tested the validity of this basic assumption by separate measurements of the recovery of exogenously added and endogenously contained melatonin in different tissues under different conditions of sample preparation and melatonin extraction.

## 2. LOSS OF MELATONIN DURING TISSUE PREPARATION AND DURING FREEZING/THAWING OF TISSUE SAMPLES

The amount of endogenously contained melatonin which is lost when an animal is killed and a certain tissue sample is removed is most difficult to assess. In the gut, for instance the measurable amount of melatonin declines substantially; so that sometimes less than 10 % of the original content is left; when the gut was mechanically irritated or vigorously washed. Additional melatonin is lost if pieces of gut tissue are frozen (even in liquid nitrogen) and thawed prior to homogenization. In contrast, melatonin content in the pineal is not much affected by mechanical irritation, washing in buffer of by freezing and thawing. These are clear indications that a certain amount of melatonin is already lost before the tissue sample is analysed for its melatonin content, and, that this loss may be more or less pronounced in different tissues.

## 3. LOSS OF MELATONIN DURING TISSUE HOMOGENIZATION

The earliest possible step for which recovery rates can be estimated by the addition of an external standard (melatonin or a not naturally occurring melatonin derivative which can be separately detected) is the homogenization procedure. For this purpose the standard is added prior to homogenization to the fresh or frozen sample and the homogenization medium (buffer, PCA or organic solvents) and it is tacitly assumed that the losses of endogenously contained melatonin and exogenously added standard are similar. This is obviously not the case.

We were able to demonstrate the preferential loss of endogenously contained melatonin during tissue homogenization using $^{3}$H-melatonin-preloaded pieces of the gut. The upper duodenum of rats was removed, cut longitudinally and, after washing, cut into pieces of about 1 cm. These pieces were incubated either in buffer or in $^{3}$H-melatonin-containing buffer for 5 min at room temperature. After washing, the samples were homogenized with or without the melatonin standard in 0.4 N perchloric acid containing 0.1 % EDTA and 0.05 % sodium bisulfate. Melatonin content was measured by HPLC and electrochemical detection (Huether et al., 1992).

As shown in Table 1, the recovery of the melatonin contained in the sample, if calculated on the basis of the recovery of the externally added standard, appears to be about 43 %. The recovery of endogenously contained melatonin, as derived from the loss of preloaded $^{3}$H-melatonin in the sample is only 16-17 %. Consequently, actual melatonin concentrations in a given tissue sample (as exemplified here for the gut) are about three fold higher than the calculated values based on the recovery of an externally added standard.

## 4. LOSS OF MELATONIN DURING THE EXTRACTION PROCEDURE

With the exception of direct RIA's in saliva or serum/plasma samples and HPLC-analysis in melatonin-rich tissues (pineal, gut), the measurement of melatonin requires the

**Table 1.** Differential loss of endogenously contained and externally added melatonin during tissue homogenization

| Measured values | | Preloaded, no external standard | Preloaded external standard added (20 ng/ml PCA) |
|---|---|---|---|
| Melatonin content (ng/ml homogenate) | | 2.1 ± 0.3 | 9.4 ± 0.9 |
| Radioactivity in homogenate (dpm/100 μl) | | 1233 ± 198 | 1182 ± 213 |
| Radioactivity in the melatonin containing melatonin fraction (injection volume 100 μl) | | 211 ± 54 | 186 ± 31 |
| Recovery of melatonin content based on the recovery | expected: | 22.1 | |
| of the external standard | found: | 9.4 | Recovery = 42.7% |
| Recovery of melatonin derived radioactivity | expected: | 1233 (1182) | |
| endogenously contained in the sample | found: | 211 (186) | Recovery = 17.1 (15.7) % |

The upper duodenum of rats was removed, cut longitudinally and, after washing, cut into pieces of about 1 cm. These pieces were incubated either in buffer or in $^{3}$H-melatonin-containing buffer for 5 min at room temperature. After washing, the samples were homogenized with or without the melatonin standard in 0.4 N perchloric acid containing 0.1 % EDTA and 0.05 % sodium bisulfate (1 : 5 w/v). Melatonin content was measured by HPLC and electrochemical detection. $^{3}$H-melatonin by LSC of the melatonin-containing HPLC-fraction. The addition of $^{3}$H-Melatonin did not cause measurable increases of melatonin content in the sample.

previous extraction and concentration of this hormone by organic solvents. Most often chloroform is used for this purpose.

In order to separately assess the actual efficiency of this extraction procedure for endogenously contained and externally added melatonin, we prepared $^{3}$H-melatonin loaded frozen tissue powders by grinding the frozen tissue in liquid nitrogen. After addition of a certain volume of buffer with or without external melatonin standards to these powders and vigorous shaking, reproducible and identic homogenates were obtained and used for extraction.

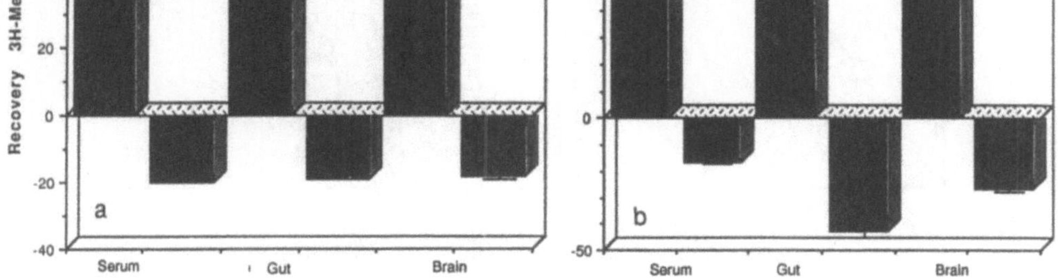


**Figure 1.** Recovery of $^{3}$H-Melatonin during the extraction of serum-, brain- and gut- homogenates using different volumes of chloroform(1:1 v/v in case of Fig. 1a, 1:10 v/v in case of Fig. 1b). Preloaded $^{3}$H-Melatonin tissues were homogenized in phosphate buffer (1:5 w/v), extracted with chloroform, and the radioactivity retained in the chloroform phase (left columns) and in the water phase (right columns) was measured by LSC.

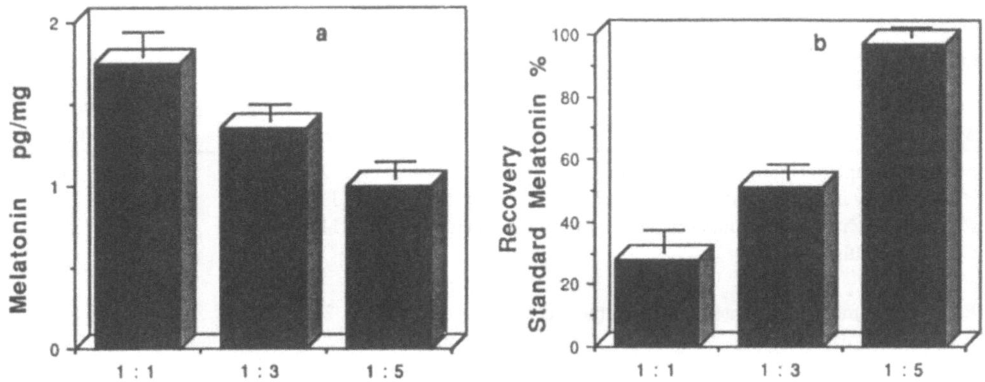


**Figure 2.** Influence of the chloroform volume used for the extraction of tissue samples of the gut on the melatonin content measured by RIA after extraction, freeze-drying and resuspension of the samples (Fig. 2a) and the recovery of an exogenously added melatonin standard by RIA after extraction, freeze-drying and resuspension of the samples (Fig. 2b).

As shown in Fig. 1, the recovery of $^3$H-melatonin from brain and gut decreases when higher amounts of chloroform are used. This effect is not seen in serum and it is less pronounced in brain compared to gut homogenates. As demonstrated in the case of the gut in the absence and in the presence of externally added melatonin, the measurable amounts of endogenously contained melatonin decrease if increasing amounts of chloroform are used for extraction. In contrast, the recovery of externally added melatonin increases if the samples are extracted with increasing amounts of chloroform (Fig. 2). This differential extraction behavior of endogenous and exogenous melatonin impressively shows that recovery-calculations which are based on the loss of externally added melatonin standards are totally misleading. Of all extraction procedures tested, extraction of tissue homogenates with acetone resulted in the highest recovery of the endogenously contained melatonin at least in the gut (Fig. 3).

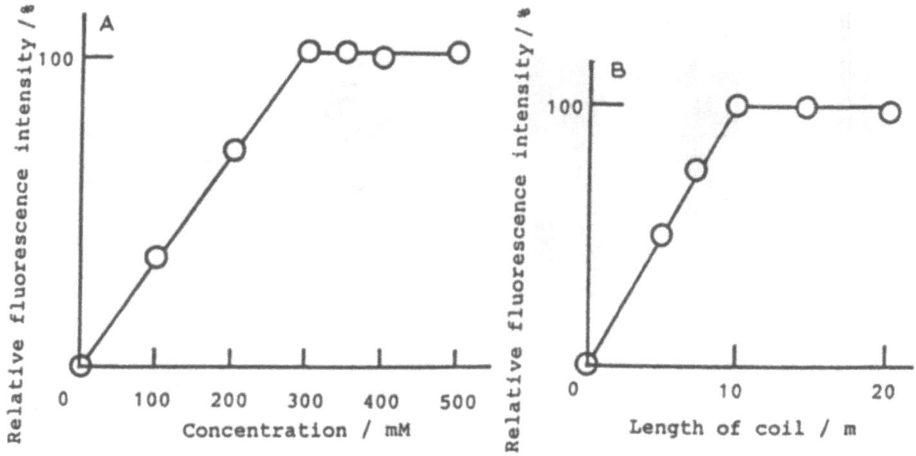


**Figure 3.** Comparison between chloroform (1:1) and acetone (1:4) extraction of tissue samples of the gut on the melatonin content measured by RIA after extraction, freeze-drying and resuspension of the samples (Fig. 3a) and the recovery of an exogenously added melatonin standard by RIA after extraction, freeze-drying and resuspension of the samples (Fig. 3b).

## 5. CONCLUSIONS

In conclusion, the melatonin levels measured by RIA, ELISA, HPLC or GCMS in a certain probe are always lower than the actual content of melatonin in the original sample and may even fall below the detection limits of the assay, e.g., in the gut tissue, (see Brammer, 1994). The measured values can not be corrected based on the recovery of an added melatonin (or related) standard unless it is definitely proven that the relative loss of this standard is identic to the loss of melatonin which is contained in the tissue.

As exemplified in the gut (1) part of the endogenously contained melatonin gets already lost before an external standard can be added, (2) endogenously contained melatonin is much more readily destroyed or lost during the process of tissue desintegration than the externally added standard, and (3) endogenously contained melatonin is much less efficiently extracted by an organic solvent like chloroform ore more readily destroyed during the extraction procedure than the externally added standard. The reason for this different behavior of endogenously contained and externally added melatonin seems to be a greater probability of the endogenously contained melatonin to become subject of reactive oxygen-mediated damage. Melatonin, being a potent scavenger of very short lived hydroxyl radicals is more likely to be destroyed the closer it is located to the sites were such radicals are produced, i.e. in the cells and tissues when they are mechanically or chemically irritated and/or decompartmented during the process of tissue removal, homogenization and extraction. Because of this preferential loss of endogenously contained melatonin, the hitherto published data on the melatonin content in certain tissues, cells or fluids based on the recovery of externally added standards may be much lower than the actual levels contained in the original sample.

## ACKNOWLEDGMENT

The research was supported by a grant of the Deutsche Forschungsgemeinschaft (Hu 351/4-2).

## REFERENCES

Arendt, J., 1986, Assay of melatonin and its metabolites: results in normal and unusual environments. *J. Neural. Transm.* 21:11.

Arendt, J., 1988, Melatonin. *Clin. Endocrinol.* 29:205.

Brammer, G.L., 1994, Duodenum is not a consistent source of melatonin in rats. *Life Sci.* 55:775.

Cardinali, D.P., 1981, Melatonin - a mammalian pineal hormone. *Endocrine Rev.* 2:327-354.

Huether, G., Poeggeler, B., Reimer, A., and George, A., 1992, Effect of tryptophan administration on circulating melatonin levels in chicks and rats: Evidence for stimulation of melatonin synthesis and release in the gastrointestinal tract. *Life Sci.* 51:945.

Lewy, A.J., Markey, S.P., 1978, Analysis of melatonin in human plasma by gas chromatography negative ionization mass spectrometry. *Science*, 201:741.

Peniston-Bird, J.F., Di, W.L., Street, C.A., Kadva, A., Stalteri, M.A., and Silman, R.E., 1993, HPLC assay of melatonin in plasma with fluorescence detection. *Clin. Chem.* 39:2242.

Poeggeler, B., 1993, Melatonin and the light-dark Zeitgeber in vertebrates, invertebrabes and unicellular organisms. *Experientia* 49: 611.

Raynaud, F., Pevet, P., 1991, Determination of 5-methoxyindoles in pineal gland and plasma samples by high-performance liquid chromatography with electrochemical detection. *J. Chromatogr.* 564:103.

Reiter, R.J., 1991, Pineal Melatonin: cell biology of its synthesis and of its physiological interactions. *Endocrine Rev.* 12:151.

Reither, J.R., Poeggeler, B., Tan, D.-X, Chen, L.D., and Manchester, L.C., 1993, Antioxidant capacity of melatonin: a novel action not requiring a receptor. *Neuroendocrinol. Lett.* 15:103.

Rollag, M.D., 1981, Methods for measuring pineal hormones. *in*: The pineal gland, anatomy and biochemistry. R.J. Reiter, ed. Vol 1. Boca Raton, FL: CRC Press, 273-302.

Skene, D.J., Leone, R.M., Young, I.M., Silman, R.E., 1983, The assessment of a plasma melatonin assay using gas chromatography mass spectrometry. *Biomed Mass Sepctrometry*, 10:655.

Vieira, R., Miguez, J., Lema, M., Aldegunde, M., 1992, Pineal and plasma melatonin as determined by high-performance liquid chromatography with electrochemical detection. *Anal. Biochem.* 205:300.

Yie, S.M., Johansson, E., and Brown G.M., 1993, Competitive solid-phase enzyme immunoassay for melatonin in human and rat serum and rat pienal gland. *Clin Chem.* 33:2322.

# DETERMINATION OF QUINOLINIC ACID BY LIQUID CHROMATOGRAPHY WITH FLUORIMETRIC DETECTION

K. Mawatari, K. Oshida, F. Iinuma, and M. Watanabe

Department of Analytical Chemistry
Faculty of Pharmaceutical Sciences
Sagamiko, Kanagawa 199-01, Japan

## 1. INTRODUCTION

We found that some metabolites of tryptophan and medicines induce fluorescence by photoirradiation in the presence of idrogen peroxide and established systems based on liquid chromatography with detection by the fluorescence reactions. Determination of kynurenic acid (Mawatari et al., 1988; 1990a) , kynurenine (Mawatari et al., 1989), nicotinic acid (Mawatari et al., 1991), nicotinamide (Mawatari et al., 1991), nalidixid acid (Mawatari et al., 1992) and isonicitinic acid hydrazide (Mawatari et al., 1990b) have been already reported. These fluorescence reactions by photoirradiation are simple and sensitive, and available for microdetermination of these compounds. In this work, a system of determination of quinolinic acid (QIA), a metabolite of tryptophan, by liquid chromatography and the photochemical fluorescence reaction was established (Mawatari et al., 1995).

QIA is known as a precursor of nicotinamide. It has been also reported to have neurotoxic effect and to be an agonist of the N-methyl-D-aspartate receptor. The determination of QIA in biological samples has been carried out by gas chromatography, gas chromatography-mass spectrometry, liquid chromatography, thin-layer chromatography and radioimmunoassay. However, these methods are time-consuming and/or lack sensitivity. We determined QIA by the photochemical fluorescence reaction after separation with a Unisil QC18 column and the mobile phase containing hydrogen peroxide (pH 3.8) and applied the determination of QIA in urine.

## 2. MATERIALS AND METHODS

### 2.1 Manual Methods

To 0.5ml of aqueous solution of QIA(30.0nmol/ml), was added 2.5ml of 70mM potassium dihydrogenphosphate containing 200mM hydrogen peroxide. The mixture was irradiated with a black light(FL-15BL, characteristic wave length 300-400nm, 20W, NEC, Tokyo) for 20min.

*Recent Advances in Tryptophan Research*, edited by Graziella Allegri Filippini et al.
Plenum Press, New York, 1996

### 2.2. Chromatographic System

The system comprises a high pressure pump(Model LC-6AD, Shimadzu, Kyoto), a sample injector (Model 7125, Rheodyne, Berkeley, CA) fitted with 20 μl loop, a cartrige guard column E (10mmx4.6mm i.d.) and an analytical column(250mmx4.6mm i.d.) with Unisil Q C18 (particle size 5 μm, GL Science, Tokyo). UV irradiation was carried out in PTFE tube(14m x 0.25mm i.d. x 1.5mm o.d.) which was wounded arund the black light. After the eluate passed through the reaction coil, it was cooled by use of a PTFE tube(1.0m long) and inserted ice bath. The mobile phase, 35mM potassium dihydrogenphosphate(adjusted to pH 3.8 with 0.2M citric acid) containing 350mM hydrogen peroxide and 0.05mM tetramethylammonium hydroxide(TMA), was delivered at a flow rate of 0.6ml/min. The fluorescence(ex. 326nm, em. 380nm) was measured with a Model F1000 fluorescence spectrophotometer (Hitachi, Tokyo). The retention time of QIA was about 9 min.

The complete separation of QIA became possible by use of the Unisil QC18 column. The specificity of the fluorescence reaction depends upon the concentration of hydrogen peroxide. In this condition, QIA produces strong fluorescence and peaks of nicotinic acid or nicotinamide are very small. When the concentration of hydrogen peroxide is lower, for example 75nM, nicotinic acid or nicotinamide show strong peaks and a peak of quinolinic acid is small (Mawatari et al., 1991).

### 2.3 Pretreatment of Human Urine

A 24h urine sample was diluted 10 times with 35mM potassium dihydrogenphosphate adjusted to pH 3.8. The sample was filtered using a Chromatodisc 25A(Kurabou, Osaka). An aliquot (10μl) of the filtered sample was injected into the present system.

## 3. RESULTS AND DISCUSSION

### 3.1. Fluorescence Spectra

Fig. 1 shows the excitation and emission spectra of the fluorescence produced from QIA by the mannual method. The excitation and emission maxima were at 326 and 386nm,

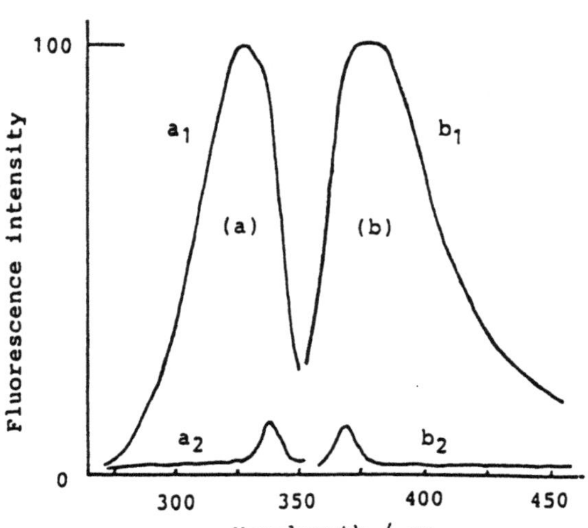


**Figure 1.** Fluorescence excitation (a) and emission (b) spectra of quinolinic acid obtained by a manual method. $a_1$ and $b_1$: QIA solution; $a_2$ and $b_2$: blank for $a_1$ and $b_1$.

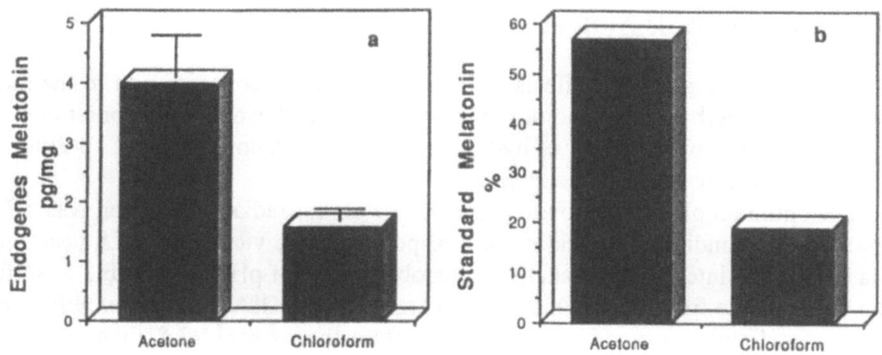


**Figure 2.** (A) effect of hydrogen peroxide concentration on the fluorescence intensity. (B) Effect of coil length on the fluorescence intensity. The photoirradiation reaction was carried out as described in the Experimental section. The amount of QIA was 35 pmol.

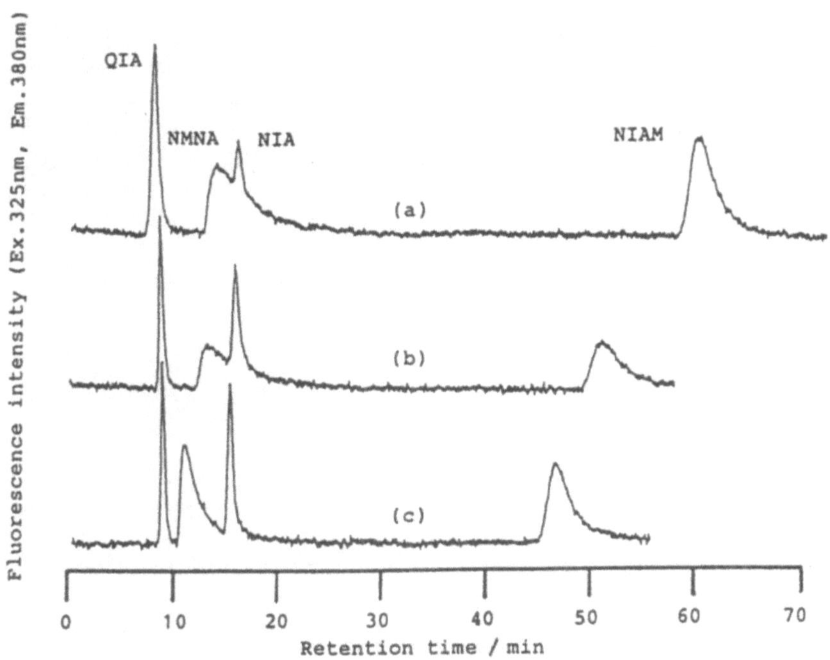


**Figure 3.** Effect of pH and tetramethylammonium (TMA) added in the mobile phase on the chromatogram. (a) pH 4.5, (b) pH 3.8 and (c) pH 3.8 and 0.05 mM TMA added.

respectively. Although the UV irradiation spectrum resembled a bipyridine formed by nicotinic acid at pH 6-8, the fluorophore is yet unknown. The black light was used as a light source, since QIA was not reactive at about 254nm and irradiation at 254nm caused foam to appear in the reaction coil.

## 3.2. Chromatographic System

Fig. 2A and B show the effects of hydrogen peroxide concentration in the mobile phase and coil length on the photoirradiation reaction. The optimum conditions were 300-500mM for the concentration of hydrogen peroxide on a coil length of 10-20m long; 350mM and 14m were adopted in experiment.

The optimum pH of the mobile phase for the photoirradiation reaction was pH 4.5. However, in this condition, a histidine peak appeared in the vicinity of QIA peak on the chromatogram and interfered the analysis. A mobile phase of pH 3.8 was able to shift the histidine peak to the front of the QIA peak and to make the QIA peak sharp, although the sensitivity was slightly lowered (about 10%). Therefore, the lower pH (3.8) than the optimum pH was adopted.

In this system, some nicotinic acid derivatives could also be detected although the sensitivity was far lower than that of QIA. Fig. 3 shows the peaks of N-methylnicotinamide (NMNA), nicotinic acid (NIA) and nicotinamide (NIAM). Kynurenine and tryptophan were

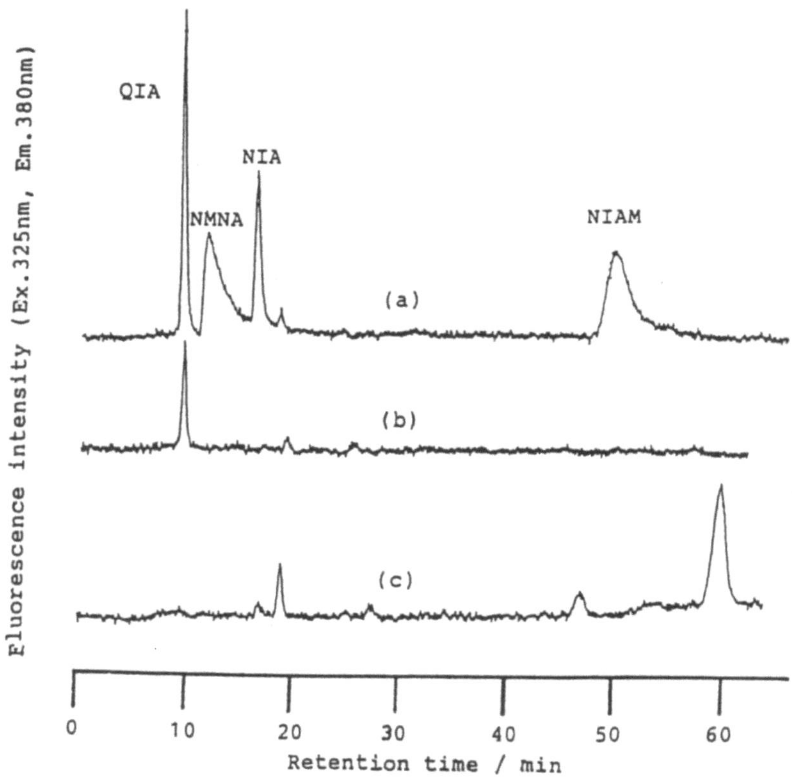


**Figure 4.** Chromatograms obtained with urine samples. Irradiation with UV light: (a) an urine sample spiked with QIA, NMNA, NIA and NIAM; (b) the urine sample. (c) the same sample as in (b) but without irradiation. Conditions as in the Experimental section. The amounts of QIA, NMNA, NIA and NIAM added were 21.0, 2410, 474 and 290 nmol/10 ml of diluted urine, respectively.

also detectable and were eluted behind QIA peak. Therefore, before the repetition of the analysis of QIA in the urine sample, ethanol (1000 μl) was injected into the chromatography before each assay and several compounds were removed from the column. The calibration graph of QIA was linear over the range 0.36-68.8 nmol/ml upon injecting 10 μl of the standard solution. The detection limits (S/N=3) was 1.1 pmol. The R.S.D. was 1.4% at 23.3 pmol/10 μl (n=6) and 1.0% at 34.1 pmol/10 μl (n=8).

## 3.3 Treatment of Urine

Fig. 4 shows chromatograms of urinary QIA and a urine sample. The QIA peak was observed in chromatogram a and b at a corresponding retention time, whereas the chromatogram c of the unirradiated sample does not show the peak at that position. The recovery of QIA added to urine (35 nmol/10ml of urine diluted) was 99.2±1.7% (S.D., n=8).

In urine of a normal person, the range and mean ±S.D. were 10.8-65.0 and 32.4±15.1 mol/day (n=12), respectively.

This system is very simple and easily assembled. It does not require a reagent delivery pump and simplified the procedure of pretreatment. This method is sensitive and sufficiently specific to estimate QIA in human urine and should be useful in nutritional and biological studies.

## REFERENCES

Mawatari K., Iinuma F. and Watanabe M., 1988, Fluorometric determination of kynurenic acid in human serum by high performance liquid chromatography coupled with postcolumn photochemical reaction with hydrogen peroxide, *Anal.Sci.*, 4:195.

Mawatari K., Iinuma F. and Watanabe M., 1989, Fluorometric determination of kynurenine in human serum by high performance liquid chromatography coupled with postcolumn photochemical reaction with hydrogen peroxide, *J.Chromatog.*, 488:349.

Mawatari K., Iinuma F. and Watanabe M., 1990a, Fluorometric determination of urinary kynurenic acid by flow injection analysis equipped with a "bypass line", *Anal.Biochem.*, 190:88.

Mawatari K., Iinuma F. and Watanabe M., 1990b, Determination of isoniazid, acetylisoniazid and nicotinic acid in human urine by high performance liquid chromatography coupled with postcolumn photochemical reaction and fluorescence detection, *Anal.Sci.*, 6:515.

Mawatari K., Iinuma F. and Watanabe M., 1991, Determination of nicotinic acid and nicotinamide in human serum by high performance liquid chromatography ultraviolet-irradiation and fluorescence detection, *Anal.Sci.*, 7:733.

Mawatari K., Iinuma F. and Watanabe M., 1992, Fluorometric determination of nalidixid acid in human urine using high performance liquid chromatography with postcolumn photochemical reaction, *Jpn.J.Clin.Chem.*, 21:13

Mawatari K., Oshida K., Iinuma F. and Watanabe M., 1995, Determination of quinolinic acid in human urine by liquid chromatography with fluorimetric detection, *Anal.Chim.Acta*, 302:179.

# 113

# TRYPTOPHAN CONTENT OF FOODS AND FEEDS

## Conditions for a Reliable Quantitation

J. Landry and S. Delhaye

Laboratoire de Chimie Biologique
INRA, INA-PG
F-78850 Thiverval Grignon, France

## ABSTRACT

A procedure for the quantitative determination of tryptophan from complex material was developed. It involved the hydrolysis of samples in the presence of baryta. Data were presented, revealing its quantitativeness and pointing out its efficiency when it was compared with diverse methods developed by other laboratories as a part of collaborative studies. The influence of the way of treating and heating of (sample + baryta) mixture upon tryptophan recovery was also investigated. The placement of (sample + baryta) mixture in an autoclave, when water was boiling, or the careful deaeration of the mixture when it was heated in an oven was essential for quantitative results. These specific conditions can be explained by differences of the protein solvatation, related to the way by which the (sample + baryta) mixture was heated.

## 1. INTRODUCTION

The current methodology adopted for quantitating amino acids present in complex material such as foods and feeds involves the hydrolysis of sample, the isolation of released amino acids by chromatography and their detection after derivatization. The majority of amino acids is quantitated from hydrolysates obtained by heating samples at 110°C in the presence of 6M HCl for 24hr. Tryptophan, which is extensively destroyed under these conditions, must be released by alkaline hydrolysis. However, tryptophan undergoes partial destruction up to 10 - 30% when it is heated in the presence of alkali, oxygen and some compounds co-existing with proteins in complex materials. Therefore, the perequisite for circumventing this shortcomming involves an exhaustive elimination of oxygen from the medium subjected to hydrolysis. A study of procedures reported the last fifteen years shown that the conditions used for oxygen removal, and for detection vary considerably. Some were

*Recent Advances in Tryptophan Research,* edited by Graziella Allegri Filippini et al.
Plenum Press, New York, 1996

selected to maximize the tryptophan recovery other were arbitrarly choosen. Oxygen was excluded by addition of scavengers (starch, maltodextrin, lactose, antioxydant), or pumping-off or ultrasonic degassing or purging with nitrogen or argon. Hydrolysis was performed in the presence of 4M NaOH or LiOH, or1.35M $Ba(OH)_2$ at 110°C or 120°C or 140°C. Tryptophan was isolated by ion exchange or gel filtration chromatography or by reverse-phase high performance liquid chromatography and detected by spectrophotometry or fluorimetry after or without derivatization.

The validity of proposed procedures was assessed by evaluating the recovery of tryptophan present in the sample or added to it. Thus Lucas and Sotelo (1980), and Sato et al. (1984) have reported conditions which led to the highest yields of tryptophan. However caution must be exercised regarding the conclusions drawn by these authors since they have used ion exchange chromatography and have not considered the possible coelution of tryptophan with lysinoalanine, generated during alkaline hydrolysis. The claim of Lucas and Sotelo (1980) that LiOH was a better hydrolysis agent has not been confirmed by Nielsen and Hurrell (1985), their finding that hydrolysis time required for maximum tryptophan value was inversely to the protein content of sample has been invalidated by Landry and Delhaye (1993a). Similarly, the high recovery of tryptophan added to 9 feedstuffs (98.9 ± 1.2%), as reported by Rogers and Pesti (1990), agrees with the percentages observed by Nielsen and Hurrell (1985) for Trp heated with starch (98.1%) but not for Trp heated with a chocolate drink (86.0%). This suggests that under the conditions used by Rogers and Pesti (1990), free tryptophan added to sample did not migrate at the level of proteins present in the particles of sample and was heated in an air free medium obtained from repeated evacuation and purging with nitrogen.

Another technique routinely used to correct for the low yield of Trp is to add an internal standard, close to tryptophan in its ability to degradation. 5-methyltryptophan, compared with α-methyltryptophan, was found by Nielsen and Hurrell to be the more adequate to satisfy to this criteria. The same held for the data reported by Hagen and Augustin (1989). But, the inverse was concluded by Bech - Andersen (1991).

Therefore, the quantitation of tryptophan present in complex material remains a serious problem for the analyst. The data reported in this paper show that this problem can be solved in a simple way using a procedure developed in our laboratory.

## 2. OUTLINE OF THE ASSAY PROCEDURE DEVELOPED IN OUR LABORATORY

Fifteen years ago, S. Delhaye has developed a procedure of tryptophan assay close to that published by Slump and Schreuder (1969). The operating conditions involved:

- vortexing sample (50mg protein) with 4.2g powdered $Ba(OH)_2$, $8H_2O$ and 8mL water in 30mL polymethylpentene tubes,
- transferring the tubes in a bench-top autoclave, as sterilizer or pressure cooker (Delhaye and Landry,1993) when water was boiling and purging autoclave by letting the vapor to escape for 5min from the closing of lid prior to the rise in temperature at 125°C,
- hydrolysis for 16h,
- neutralisation of hydrolysate with HCl 6mol/L
- separation of tryptophan by gel filtration chromatography (Sephadex G10) and colorimetric assay using ninhydrin (Delhaye and Landry, 1992).

Other conditions were further tested then adopted; namely: hydrolysis for 10h, neutralisation of hydrolysate aliquot (Landry and Delhaye, 1992), separation of tryptophan by high performance liquid chromatography and detection by ultraviolet spectrophotometry (Delhaye and Landry, 1986) and spectrofluorimetry (Landry et al, 1988).

# 3. EVIDENCE OF THE VALIDITY OF OUR ASSAY PROCEDURE

## 3.1. Quantitativeness

The quantitativeness of our procedure was tested by evaluating tryptophan recovered from samples of purified proteins whose the number of tryptophan residues in polypeptide chain was known or from the known amount of lysozyme added to samples of feedstuffs.

The recovery was found to range:

- from 98% (ovalbumin) to 101.2% (cytochrome C) with a mean of 99.6 ± 1.3% for samples of 4 purified proteins (Delhaye and Landry, 1986, 1992)
- from 98.8% (pea) to 100.9% (barley) with a mean of 99.7 ± 0.7% for lysozyme added to 11 samples of diverse feeds (Delhaye and Landry, 1986; Landry et al., 1988).

On the other hand a recovery averaging 99% was achieved for tryptophan added as free or incorporated into tripeptide Gly-Trp-Gly to wheat flour (Delhaye and Landry, 1992).

Therefore conditions used for hydrolysis samples did not promote any degradation of tryptophan, whether it was free or incorporated into peptide or polypeptide and whether it was in the absence or in presence of non protein compounds.

## 3.2. Effectiveness as Assessed by Comparison with Other Procedures

A further insight into the effectiveness of our procedure was provided by comparing the results with those got using diverse methods developed in our laboratory or in others as a part of collaborative studies. All the comparisons were made by evaluating a relative recovery of tryptophan on the basis of 100% recovery for the same sample analysed using our procedure.

The tryptophan recovery averaged:

- 100.05 ± 1.19% for 18 feedstuffs hydrolysed in the presence of baryta and thiodiglycol in glass test tubes evacuated then filled up with nitrogen (3X), and heated in an oven at 110°C for 16h (Landry et al., 1988);
- 88.8 ± 8.6% and 102.5 ± 3.7% using an external and an internal (5-methyltryptophan) standard, respectively for 9 samples autoclaved in the presence of NaOH 4mol/L and 5-methyltryptophan at 125°C for 16h (Landry and Delhaye, 1994);
- 100.6 ± 1.9% for 9 samples hydrolysed in the presence of LiOH 4 mol/L, maltodextrin and 5-methyltryptophan in an oven at110°C for 20h (Landry et al., 1992).

Therefore, the above mentioned data agreed with those reported by Nielsen and Hurrell (1985) as to the efficiency of 5-methyltryptophan to correct for degradation losses of tryptophan in the presence of alkali. But such a procedure applied to some samples (pea for instance) led to an overcorrection, with recoveries reaching up to 115%.

The reliability of experimental conditions used in our procedure on the recovery of tryptophan was elicited also by comparing:

- the tryptophan content deduced from the regression equations relating to the levels of tryptophan and nitrogen in the dry matter of seeds. From the relative tryptophan content of diverse seeds, as calculated from a given nitrogen content and the equations, determined by three laboratories including ours, mean recoveries were found to amount 90.1 ± 3,8% or 91.1 ± 4.1%, depending on whether $Ba(OH)_2$, used under conditions slightly different from ours, or LiOH were employed as hydrolysis agent (Landry and Delhaye, 1993b; Delhaye and Landry, 1995). Similar percentages were found by Slump et al. (1991) in their comparative study involving $Ba(OH)_2$ and LiOH as hydrolysis agents.
- the data obtained as a part of 5 ringtests on tryptophan determination to which our laboratory has participated. Table 1 summarizes the mean recoveries from hydrolyses using LiOH and $Ba(OH)_2$. When $Ba(OH)_2$ was used, two subsets of percentages were observed. One averaged 86%, identical to that got using LiOH (87%). The second was made up of percentages equal or close to 100%: the values stated for the first three ringtests corresponded to the data obtained by our laboratory, those of the 3rd european ringtest to the mean of values obtained by three laboratories, and those of the 4th european ringtest to the mean of values got by all the laboratories whose the majority used our procedure with autoclaving.

## 3.3. Quantitative Study of Some Treatments Performed on (Sample + Baryta) Mixture Prior to Hydrolysis

To confirm the suitability of conditions used in our procedure, the effects of parameters concerning the treatment of the (sample + baryta) mixture prior to hydrolysis were

**Table 1.** Recovery of Tryptophan: comparison of values estimated from diverse ringtests

| | | Alkali | | |
|---|---|---|---|---|
| Ringtest | | LiOH | $Ba(OH)_2$ | |
| 4th BCR | Trp%[a] | 91.0 | | 100[d] |
| (1995) | S E[b] | 6.7 | | 6.5 |
| | n[c] | 36 | | 32 |
| 3rd BCR | Trp% | 86.3 | 84.5 | 98[e, f] |
| (1994) | S E | 7.2 | 5.8 | 2.7 |
| | n | 32 | 12 | 9 |
| 2nd BCR | Trp% | 86.5 | 83.4 | 100[e] |
| (1993) | S E | 6.5 | 5.0 | 0 |
| | n | 32 | 15 | 5 |
| 1st BCR | Trp% | 84.6 | 88.7 | 100[e] |
| (1992) | S E | 4.5 | 5.8 | 0 |
| | n | 44 | 10 | 5 |
| French | Trp% | | 87.1 | 100[e] |
| (1989) | S E | | 7.2 | 0 |
| | n | | 24 | 3 |

[a]recovered tryptophan.
[b]standard deviation estimate.
[c]number of determinations (number of labs x number of samples).
[d]100% recovery corresponds to the mean of values got by 12 laboratories.
[e]Footnote e here.
[f]Footnote f here.

**Table 2.** Influence of the way of treating and heating the (sample + baryta) mixture on the recovery of tryptophan[a]

Heating in autoclave

1. the solid (sample + baryta) mixture was evacuated, vacuum was broken with:
   1.1. nitrogen and boiled water was added
   $(Trp)^{b}_{recov} = 102.4 \pm 1.3$ %  $n^{c} = 12$
   1.2. oxygen and boiled water was added
   $(Trp)_{recov} = 100 \pm 2$ %  n = 3
2. the (sample + baryta + water) mixture was:
   2.1. immersed in ultrasonic bath
   $(Trp)_{recov} = 91.4 \pm 2.5$ %  n = 10
   2.2. placed in autoclave when water was at room temperarure
   $(Trp)_{recov} = 92.3 \pm 2.7$ %  n = 21

Heating in block or oven

3. the mixture was brought to a boiling and the glass tube was closed
   $(Trp)_{recov} = 88.0 \pm 4.9$ %  n = 9
4. the mixture was placed in polypropylene tube used as a liner for glass tube containing boiling water. the tube was closed, then cooled to room temperature for 1h and transfer in oven (105°C, 22h)
   $(Trp)_{recov} = 93.5 \pm 2.5$ %  n = 10
5. the [sample + baryta + thiodiglycol (5%) + water] mixture was heated on the boiling bath for 15 min and the glass tube transferred in oven (105°C, 20h)
   $(Trp)_{recov} = 96.5 \pm 1.6$ %  n = 20

[a]100% recovery corresponds to the tryptophan contents determined using the classical procedure, involving the placement of (sample+baryta) mixtures in an autoclave, when water was boiling.
[b]Tryptophan recovered.
[c]number of samples X number of determinations.

investigated. The results of this study, reported in Table 2, led to the following observations to be made:

- removing air from mixture to be hydrolysed using evacuation and purging with nitrogen did not improve tryptophan recovery;
- increasing oxygen concentration in hydrolysis medium did not promote a decrease of recovery, contrary to anticipation;
- treating mixture by ultrasonics, or its transferring when water was at room temperature, decreased recovery, suggesting a marked influence of the way of heating upon recovery;
- hydrolysing in closed tubes heated in a block or oven decreased recovery irrespective of the studied conditions. However, it is worth recalling that an air free obtained by repeated evacuation and purging with nitrogen applied to the (sample + baryta + thiodiglycol + water) mixture led to quantitative results (Landry et al., 1988)

## 4. DISCUSSION

From the foregoing, the highest yields of tryptophan were achieved when the (sample + baryta) mixtures:

- having no undergone any treatment to exclude air, were placed in autoclave when water is boiling, then heated;
- or were treated for an exhaustive removal of air, then heated in oven or block.

Such salient differences in the operating conditions to reach the same result can be explained from the following assumptions. When autoclave is used, the rapide heating of (sample + baryta) mixture by water vapour would result in salting out of proteins, isolating them from air or oxygen (see Table 2) present in hydrolysis medium. On the contrary, the slow heating up of mixture placed in oven would promote the solvatation of proteins by baryta, increasing the possibilities of interferences between tryptophan residues and air present in the mixture. In this case the exhaustive removal of air from the (sample + baryta) mixture is a perequisite for high tryptophan yields. On the other hand sodium and lithium hydroxides, with respect to barium hydroxide, have a higher solvating power for protein and non-protein compounds at room temperature. Their use as hydrolysis agent must result in losses in tryptophan, irrespective of the conditions used for hydrolysis.

## 5. CONCLUSION

The data presented here afford a further insight into the quantitative determination of tryptophan in complex matrixes. Quantitativeness is obtained using barium hydroxide as hydrolysis agent under definite conditions involving the placement of the mixture to be hydrolysed in autoclave when water is boiling, or the careful degassing of mixture when it is heated in an oven. If these requirements are not fulfilled, tryptophan degradation takes place as does with the use of sodium or lithium hydroxide.

## 6. REFERENCES

Bech-Andersen, S., 1991, Determination of tryptophan with HPLC after alkaline hydrolysis in autoclave using $\alpha$-methyltryptophan as internal standard, *Acta Agric. Scand.*, 41:305

Delhaye, S. and Landry, J.,1985, High performance liquid chromatography and ultraviolet spectrophotometry for quantitation of tryptophan from baryta hydrolysates, *Anal. Biochem.*, 159: 175

Delhaye, S. and Landry, J., 1992, Determination of tryptophan in pure proteins and plant material by three methods, *Analyst*, 117: 1875

Delhaye, S. and Landry, J., 1993, Quantitative determination of tryptophan in foods and feedstuffs. Practical considerations on autoclaving samples for hydrolysis, *J. Agric. Food Chem.*, 41: 1633

Delhaye, S. and Landry, J., 1995, The tryptophan content in proteins of cereal grains and legume seeds as a function of nitrogen content: a reappraisal of tryptophan score, *in* "Recent Advances in Tryptophan Research", Edited by G. Allegri Filippini and C.V.L. Costa

Hagen, R. S. and Augustin, J., 1989, Determination of tryptophan in foods by isocratic reversed-phase high performance liquid chromatography, *J. Micronutr. Anal.*, 5: 303

Landry, J. and Delhaye, S., 1992, Simplified procedure for the determination of tryptophan in foods and feedstuffs from barytic hydrolysis, *J. Agric. food Chem.*, 40: 776

Landry, J. and Delhaye, S., 1993a, HPLC determination of tryptophan in foodstuffs using barium hydroxide at 140°C, *Amino Acids*, 5: 317

Landry, J. and Delhaye, S., 1993b, The tryptophan contents of wheat, maize and barley grains as a function of nitrogen content, *J. Cereal Sci.*, 18: 259

Landry, J. and Delhaye, S., 1994a, Determination of tryptophan in feedstuffs: Comparison of sodium and barium hydroxides as hydrolysis agents, *Food Chem.*, 49: 95

Landry, J. and Delhaye, S., 1994b, Tryptophan determination in feedstuffs: a critical examination of data from two collaborative studies through the evaluation of tryptophan recovery, *J. Agric. Food Chem.*, 42: 1717

Landry, J., Delhaye, S. and Jones, D.G., 1992, Determination of tryptophan in feedstuffs. Comparison of two methods of hydrokysis prior to HPLC analysis, *J. Sci. Food Agric.*, 58: 439

Landry, J., Delhaye, S. and Viroben, G., 1988, Tryptophan content in feedstuffs as determined from three procedures using chromatography of barytic hydrolysates, *J. Agric. Food Chem.*, 36: 51

Lucas, B. and Sotelo, A.,1980, Effect of different alkalies, temperature and hydrolysis times on tryptophan determination of pure proteins and foods, *Anal. Biochem.*, 109: 192

Nielsen, H.K. and Hurrell, R.F., 1985, Tryptophan determination of food proteins by HPLC after alkaline hydrolysis, *J. Sci. Food Chem.*, 36: 893

Rogers, S.R. and Pesti, G.M., 1990, Determination of tryptophan from feedstuffs using reverse phase high performance liquid chromatography, *J. Micronutr. Anal.*, 7: 27

Sato, H., Seino, T., Kobayashi, T., Murai, A., and Yugari, Y., 1984, Determination of the tryptophan content of feed and feedstuffs by ion exchange liquid chromatography, *Agric. Biol. Chem.*, 48: 2961

Slump, P. and Schreuder, H.A.W., 1969, Determination of tryptophan in foods, *Anal. Biochem.*, 27: 182

Slump, P., Flissebaalje, T.D. and Haaksman, I.K., 1991, Tryptophan in food proteins: a comparison of two hydrolytic procedures, *J. Sci. Food Agric.*, 55: 493

# FORMATION AND DETERMINATION OF ALDEHYDE-ADDITION PRODUCTS OF TRYPTOPHAN

K. Meyer, A. Möller, C. Karg, and H. Steinhart

Institute of Biochemistry and Food Chemistry
University of Hamburg, Grindelallee 117
20146 Hamburg, Germany

## INTRODUCTION

The addition of aldehydes is in addition to the oxidation the main reaction of the essential amino acid tryptophan (TRP). This addition leads to the formation of the so called 1,2,3,4-tetrahydro-β-carboline carboxylic acids (THCC's) via the Pictet-Spengler-Reaction, as shown in figure 1.

The THCC's show many physiological effects. Ogawa et al. (1993) described an accumulation of THCC's in the kidney, liver and blood of rats. They also proved that THCC's are able to pass the blood-brain-barrier. In the brain the THCC's might interact with the serotonergic neurotransmitter system affecting the behaviour of people.

THCC's have been discussed as possible causative substances of the Eosinophilia Myalgia Syndrome which has been shown to have an epidemic association with the ingestion of TRP (Adachi et al., 1991b).

Moreover, the THCC's are precursors for the β-carbolines, which are inhibitors of the benzodiazepine receptors, formed by decarboxylation and aromatization (Lippke et al., 1983). Furthermore, the THCC's are also precursors for inhibitors of the monoaminoxidase A, which also affects the behaviour of people, because the monoaminoxidase participates in the metabolism of neurotransmitters.

The aldehydes necessary for the formation of THCC's in food come mainly from fermentation or from smoking. Therefore determinations of the THCC's-content in food have focused on alcoholic beverages (Adachi et al., 1991; Bosin et al., 1986; Herraiz et al., 1993) and on smoked meat (Papavergou and Clifford, 1992).

The purpose of our study was to examine the parameters of the formation of the THCC's in a model system with a focus on the physiological conditions, which are given in fermentation, and the conditions during the processing of food. With these results it should be possible to predict whether a food may contain THCC's or not. The second aim was to prove the detectability of the THCC's in cheese, which is also a fermented food, and to highlight their amounts.

*Recent Advances in Tryptophan Research,* edited by Graziella Allegri Filippini et al.
Plenum Press, New York, 1996

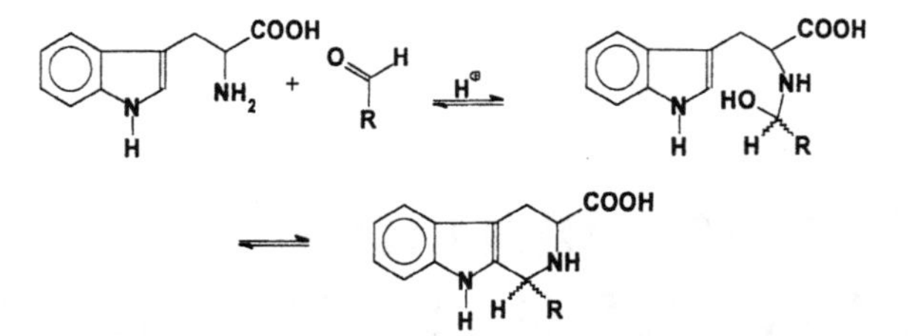


R = $CH_3$ 1-Methyltetrahydro-β-carboline-3-carboxylic acid (MTHCC)

**Figure 1.** Formation of tetrahydro-β-carboline carboxylic acids.

# MATERIAL AND METHODS

## HPLC-System

A L6200A Intelligent pump (Fa. Merck/Hitachi) was used, equipped with a F-1080 fluorescence detector (Ex: 270nm, Em: 330nm, Fa. Merck/Hitachi). The column used (125x4mm id.) was filled with Nucleosil 120-3C18. The flow was 0.8ml/min.

Eluent A was 0.1% trifluoracetic acid in water, eluent B was acetonitrile. The gradient used was:

| | | | | | | | | | |
|---|---|---|---|---|---|---|---|---|---|
| t [min] | 0 | 2 | 6 | 10 | 14 | 17 | 19 | 24 | 26 |
| Eluent A: | 90 | 90 | 85 | 85 | 80 | 75 | 50 | 50 | 90 |
| Eluent B: | 10 | 10 | 15 | 15 | 20 | 25 | 50 | 50 | 10 |

## Model System

Formation of MTHCC: Model samples were designed to contain TRP in concentrations similar to those existing in cheese (0.015 mmol/l).

## Study of Molar Ratio TRP/Acetaldehyde

Samples were prepared by adding various amounts of an aqueous solution of TRP (0.015 mmol/l) in 0.1 M citrate buffer at pH=3.5 and a temperature of 40°C. The molar ratio for TRP to acetaldehyde was adjusted to 1:10, 1:100, 1:500 and 1:1000.

## Study of pH

Four buffer systems were used to establish pH values of 3.0, 3.5, 4.0 (citrate buffer 0.1 M), 7.0 (phosphate buffer, 0.7 M), 8.0 (borate buffer, 0.1 M) and 10.0 (carbonate buffer, 0.1 M). Each sample contained 0.015 mmol TRP/l and 1.5 mmol acetaldehyde/l (molar ratio of 1:100). The reaction was performed at 40°C.

## Study of Temperature

Four different temperatures were investigated. 4°C simulated the temperature of a refrigerator, 22 °C simulated room temperature, and 40°C and 60°C simulated the temperature of fermentation processes. The incubated samples contained 0.015 mmol TRP/l and 1.5 mmol acetaldehyde/l at pH 3.5 (citrate buffer 0.1 M).

## Study of the Reactivity of Different Aldehydes

Samples containing 0.015 mmol TRP/l, 1.5 mmol formaldehyde, acetaldehyde, glyceraldehyde or hexanal in 0.1 M citrate buffer (pH 3.5) were monitored for decrease of TRP at 40°C.

## Determination of Tetrahydro-β-Carboline Carboxylic Acids in Cheese

The cheese samples were cut and homogenised. 30 g of the samples were weighed, freezed at - 40°C and lyophilised. The freeze dried samples were dissolved in 200 ml water and homogenised once more. After centrifugation (20 min, 0°C, 5000 t/min), the solutions were first adjusted to pH=12 with 10 M KOH and then to a pH-value of 3 with perchloric acid to precipitate the proteins. After further centrifugation (20 min, 0°C, 5000 t/min) and filtration the supernatant was loaded into a XAD-2 column (25x1, 5 cm, 5 ml/min, Fa. Sigma Chemicals) and washed with 200 ml of water. With 100 ml methanol the THCC's were eluted in a graduated flask containing a spatula-tipfull of semicarbazide hydrochloride. 25 ml of the eluates were evaporated and acidified with 1 ml of 0.5 M HCl. The solutions were loaded into a SCX-solid-phase-column (Chromabond SA, capacity: 3 ml, Fa Macherey & Nagel), previously conditioned with 2x3 ml methanol and 2x3 ml of 0,1 M HCl. After addition of the sample the column was washed with 2x2 ml 0,1 M HCl, 3 ml methanol and 2x2 ml 0,1 M HCl containing 30 g/l NaCl. The THCC's were eluted with 3 ml of a mixture of 0,5 M NaCl-solution and methanol (1:1, v:v), filled to 5 ml with water and used for HPLC-analysis.

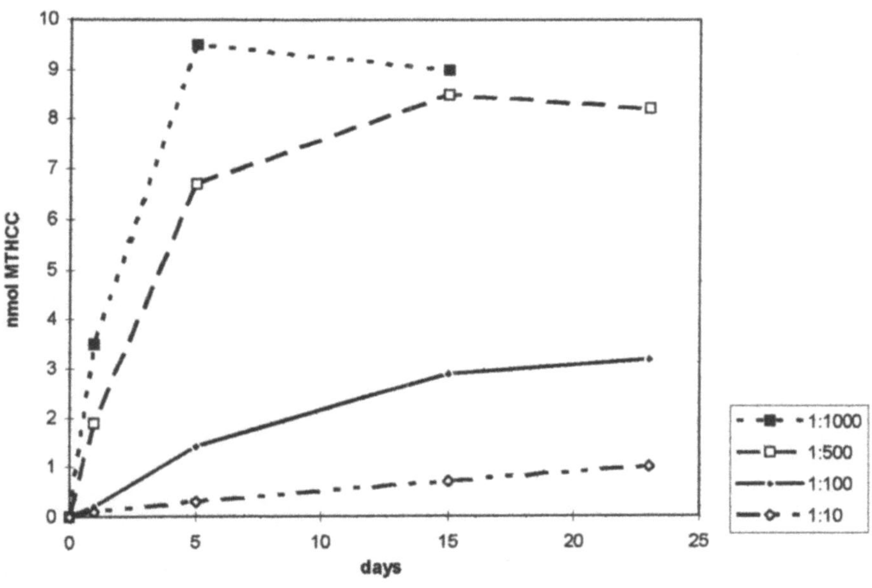


**Figure 2.** Influence of the molar ratio of tryptophan to aldehyde on MTHCC formation.

# RESULTS AND DISCUSSION

## Model System

The amount of MTHCC formed increased with an increasing amount of aldehyde (figure 2).

A conversion of 70 % of TRP and acetaldehyde to MTHCC was reached at the very high molar ratio of 1 to 1000.

In the physiological range of pH the formation of MTHCC increased due to the protonation of aldehyde (figure 3). Further investigations show that with a pH value below 1 the electrophilic attack follows mainly on indol-N (Steinhart et al., this issue).

The formation of MTHCC increased with increasing temperature (figure 4). Temperatures of 40 °C or 60 °C could occur easily during the processing of food or during a storage at incorrect parameters. During the storage of food, a formation of MTHCC will not take place if the food is cooled or stored at room temperature.

Formaldehyde, which is a product of metabolism, is the most reactive aldehyde examined and caused the fastest and greatest lost of TRP (figure 5). Hexanal, which appears as a product of the autoxidation of fatty acids, is more reactive though it is more hydrophobic than acetaldehyde. Acetaldehyde, which is the main product of alcoholic fermentation and a product of metabolism, caused losses of TRP of almost 50 %. The intermediate of the carbohydrate metabolism, glycerinaldehyde, is the most hydrophilic but the least reactive aldehyde studied.

## Determination of Tetrahydro-β-Carboline Carboxylic Acids in Cheese

Despite to the description in literature (Papavergou and Clifford, 1992), in cheese matrices two basic clean up steps, adsorbtion on XAD-2 resin and cation exchange on Chromabond SA, are necessary to archieve a good resolution of MTHCC and THCC

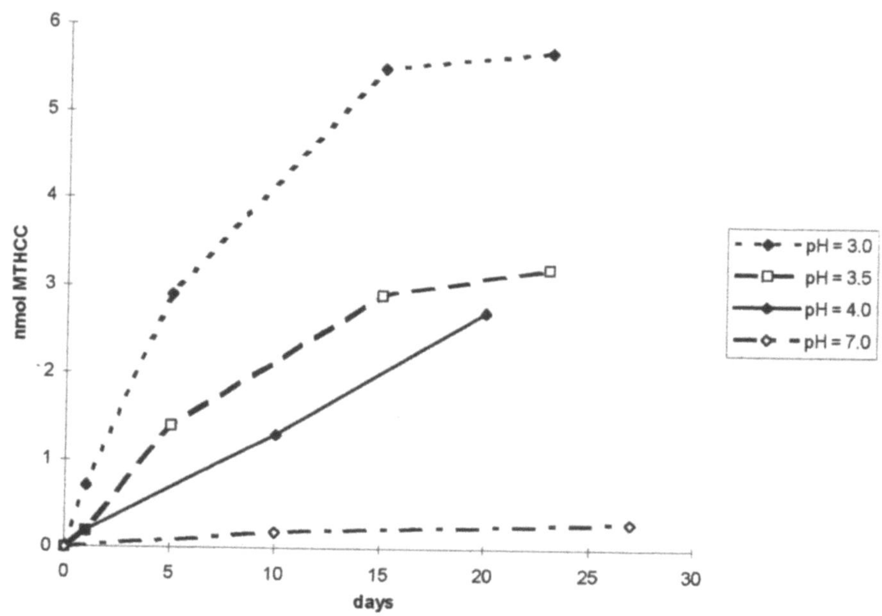


**Figure 3.** The influence of the pH value on the formation of MTHCC.

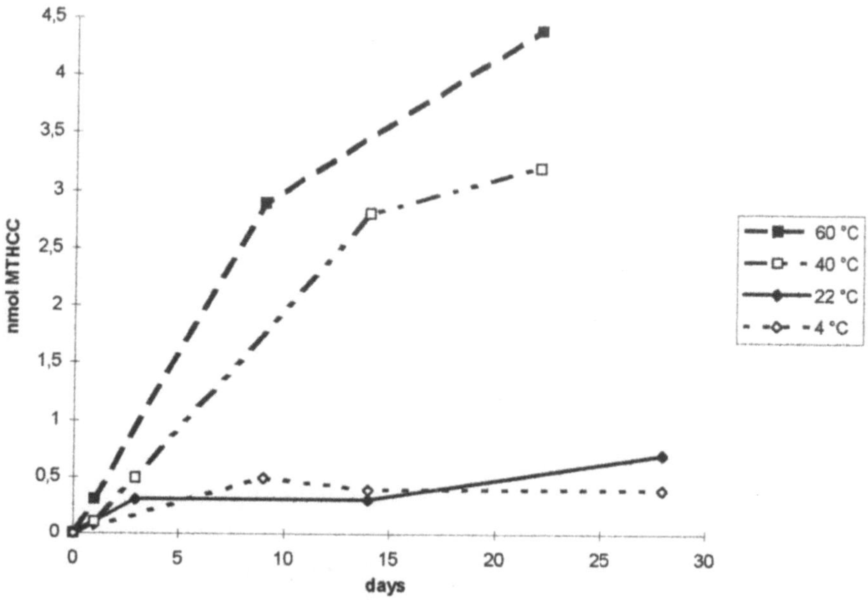


**Figure 4.** Influence of the temperature on the formation of MTHCC.

(tetrahydro-β-carboline-3-carboxylic acid) from disturbing peaks (figure 6). Unfortunately recoveries of added THCC standard solutions displayed great variations and the amounts of THCC could not, therefore, be determined.

The result of our study is given in table 1. The amounts of MTHCC in gouda were lower than 1 μg/kg. The amounts in the other cheese studied were between 20 and 200 μg/kg.

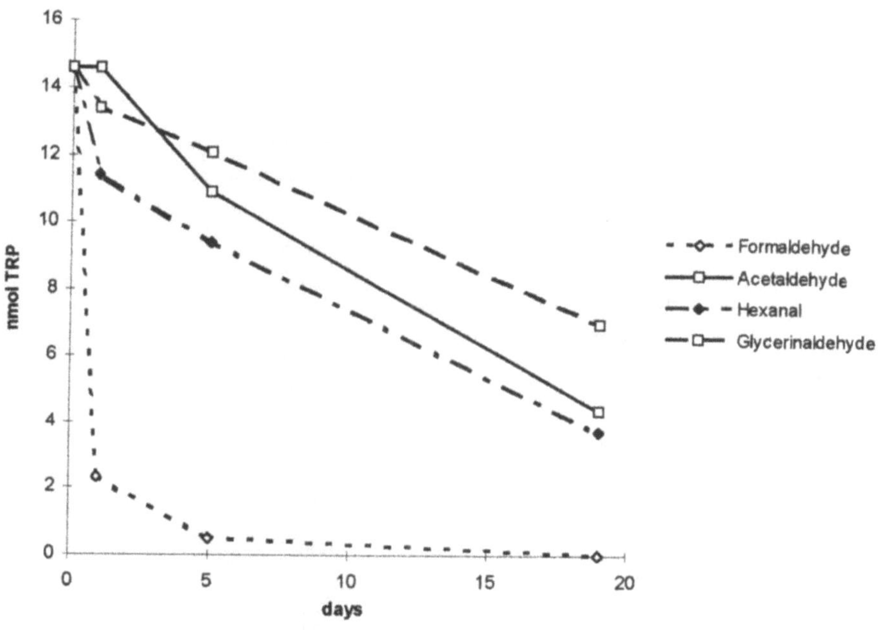


**Figure 5.** Reactivity of the different aldehydes studied.

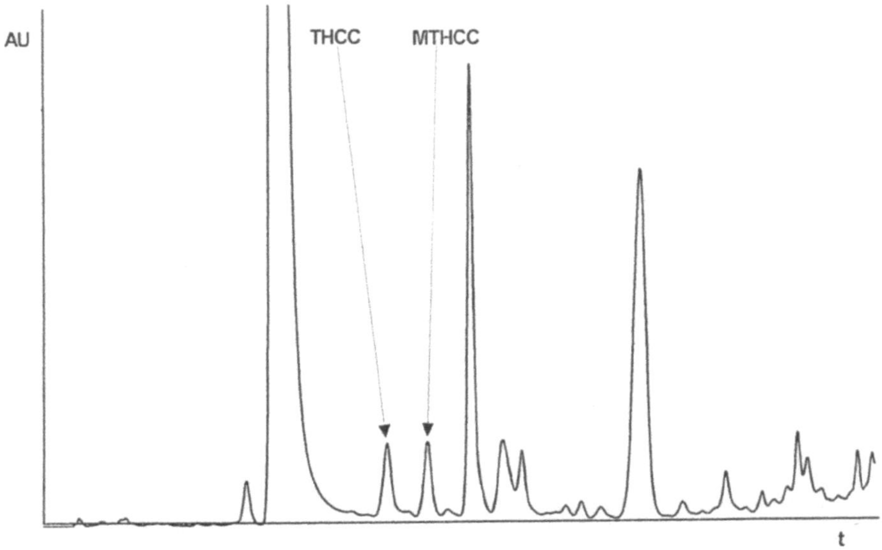


**Figure 6.** Chromatogram after modified clean up.

The difference between the emmentalers I and II is conspicuous. These results show that samples of the same kind of cheese but from different sources can exhibit great variations in the MTHCC content. The MTHCC amount of yoghurt corresponds to the amount in cheese.

The comparison of our result with the results of other workers demonstrates that the content of MTHCC in cheese is very low in comparison to that in soy sauce or salami, wine or ale.

**Table 1.** Content of MTHCC in different food

| | MTHCC [ μ/kg ] |
|---|---|
| Own results: | |
| Gouda | < 1 |
| Emmentaler I | 102.2 |
| Emmentaler II | 21.7 |
| Bergkäse (3 months matured) | 171.4 |
| Bergkäse (6 month matured) | 103.0 |
| Schmelzkäse | 44.3 |
| Yogurt | 31.0 |
| Papavergou | |
| Smoked Cheese | 237 |
| Salami | 1,231–22,632 |
| Smoked Meat | 16–680 |
| Adachi et al. | |
| Ale | 1000 |
| Wine | 3900 |
| Soy Sauce | 502,000 |
| Yogurt | 24 |

# CONCLUSION

In a model system we are able to show that the formation of THCC's is prefered at low pH values, elevated temperature and a high content of aldehydes. In the investigated types of cheese MTHCC was found in amounts up to 200 µg/kg, which can be evaluated as physiologically negligible.

# REFERENCES

Adachi, J., Mizoi, Y., Naito, T., Ogawa, Y., Uetani, Y., Ninomiya, I., 1991a, Identification of tetrahydro-β-carboline-3-carboxylic acid in foodstuffs, human urine and human milk, *J. Nutr.* 121:646.

Adachi, J., Yamamoto, K., Ogawa, Y., Ueno, Y., Mizoi, Y., Tatsuno, Y.,1991b, Endogenous formation of 1-methyl-1,2,3,4-tetrahydro-β-carboline-3-carboxylic acid in man as a possible causative substance of eosinophilia-myalgia syndrome associated with ingestion of L-tryptophan, *Arch. Toxicol.* 65:505.

Bosin, T.R., Krogh, S., Mais, D.,1986, Identification and quantification of 1,2,3,4-tetrahydro-β-carboline-3-carboxylic acid and 1-methyl-1,2,3,4-tetrahydro-β-carboline-3-carboxylic acid in beer and wine, *J. Agric. Food Chem.* 34:843.

Herraiz, T., Huang, Z., Ough, C.S.,1993, 1,2,3,4-Tetrahydro-β-carboline-3-carboxylic acid and 1-methyl-1,2,3,4-tetrahydro-β-carboline-3-carboxylic acid in wines, *J. Agric. Food Chem.* 41:455.

Lippke, K.P., Schunack, W.G., Wenning, W., Müller, W.E., 1983, β-Carbolines as benzodiazepine receptor ligands. 1. Synthesis and benzodiazepine receptor interaction of esters of β-carboline-3-carboxylic acid, *J. Med.Chem.* 26:499.

Ogawa, Y., Adachi, J., Tatsuno, Y.,1993, Accumulation of 1-methyl-tetrahydro-β-carboline-3-carboxylic acid in blood and organs of rats. A possible causative substance of eosinophilia-myalgia-syndrom associated with ingestion of L-tryptophan, *Arch. Toxicol.* 67:290.

Papavergou, E., Clifford, M.N.,1992, Tetrahydro-β-carboline-carboxylic acids in smoked food / chemical reactions in foods II, proceedings of the symposium on chemical reactions in foods, *Food Addit. Contam.* 9:83.

Steinhart, H., Meyer, K., Simat, T., Synthesis and analysis of contaminants in EMS-related tryptophan, this issue

# TRP OXIDATION BY COPPER-ASCORBATE UNDER PHYSIOLOGICAL CONDITIONS

V. Moreaux,[1] I. Birlouez-Aragon,[1] F. Tessier,[1] H. Mondon,[2] P. Junes,[2] and J. M. Lecerf[3]

[1] Institut National Agronomique, Laboratoire de Chimie Analytique
16 rue Claude-Bernard, 75005 Paris, France
[2] Ophtalmology Department, Hôpital Cochin
Paris, France
[3] Nutrition Department, Institut Pasteur
Lille, France

Tryptophan (Trp) is an essential amino acid which, beside its role in protein synthesis, is involved in a great diversity of biological functions (precursor of serotonine, of UV filters (Wood and Truscott, 1993)). In the field of the radical theory of aging, Trp is considered as a possible target of oxygen radicals. If occuring, in vivo Trp oxidation would not only affect these different functions but also produce highly reactive and toxic species. 3-OH-Kyn (hydroxy-kynurenine) and 3-OH-AA (hydroxy-anthranilic acid) have been reported to have a cataractogenic activity (Truscott and Martin, 1989; Tomoda et al., 1990; Truscott et al., 1991; Benavente and Truscott, 1991) in vitro by inducing protein cross-linking, and kynurenine derivatives are suspected to promote bladder carcinogenesis in mice and induce metal-mediated damages on DNA (Hiraku et al., 1995).

However, there is no clear evidence of Trp oxidation in biological tissues during aging or radical-associated diseases. Most data concerning Trp oxidation are reported from in vitro studies. Free Trp has been shown to oxidize after radiolysis or incubation with hydroxyl radical-generating systems such as ascorbate (ASC) (Uchida and Kawakishi, 1988; Maskos et al., 1992; Steinhart et al., 1993), hydrogen peroxide ($H_2O_2$) (Kell and Steinhart, 1990) or poly-unsaturated fatty acid (PUFA) (Krogull and Fennema, 1987) in presence of metal ions. In these conditions, N-formyl Kyn, Kyn, amino-benzoic acid and hydroxylated Trp are the main products formed from Trp. However the conditions used in vitro to oxidize Trp are usually extreme and far from the physiological conditions. It is therefore necessary to verify if Trp oxidation is still possible under physiological concentrations of oxidants.

## I. WHICH PHYSIOLOGICAL CONDITIONS FOR TRP OXIDATION IN VITRO?

The main chemical reactions suceptible to produce oxygen radicals in vivo involve metal ions and ASC, $H_2O_2$ or PUFA, as prooxidants. However, metal ions are generally

*Recent Advances in Tryptophan Research,* edited by Graziella Allegri Filippini et al.
Plenum Press, New York, 1996 

chelated by specific proteins in vivo and only traces of metals should be available for uncontrolled oxidation. Therefore, in vitro experiments using copper or iron concentrations higher than the µmolar range should strongly overestimate the Trp oxidation rates. On the other hand, ASC concentration is maximal in the lens where it reaches 2 mmol/L and varies between 50 and 150 µmol/L in plasma. The concentrations used in most in vitro studies are much higher, around 5 mmol/L, and metals are added up to the mmolar range. Peptidic Trp has also been reported to be preferentially attacked by hydroxyl radicals (Uchida et al., 1985) resulting from ascorbate oxidation, but the vit C to albumin ratio in this experiment was 2000 times higher than in plasma.

We therefore propose in this study to evaluate the Trp susceptibility to vit C-copper mediated oxidation under physiological conditions. Trp and ascorbate concentrations in phosphate buffer (10 mmol/L, pH 7.4) were in the 10 to 5000 µmol/L and 113 to 1130 µmol/L concentration range respectively, values corresponding to the lower and upper levels in vivo. Copper sulfate was added at the lowest concentration compatible with a good reproducibility of the experiment, ie 1.5 µmol/L. The samples were incubated at 37°C up to 13 days.

The reactions involved in such a system are expected to be the following:

$$\text{Vit C} + Cu^{2+} \rightarrow \text{DHA} + Cu^{+} + H_2O_2 \qquad \text{(eq.1)}$$

$$H_2O_2 + Cu^{+} \rightarrow {}^{-}OH + {}^{\circ}OH + Cu^{2+} \qquad \text{(eq.2)}$$

$$\text{Trp} + {}^{\circ}OH \rightarrow \text{Trp Ox} \qquad \text{(eq.3)}$$

Trp concentration was measured by HPLC-UV wiht reverse phase column with ammonium acetate 11.6 mmol/L and 20 % methanol as eluant. A decrease in Trp concentration was observed in all samples, whatever the concentration of reactants, but to a variable extent and at different times depending on Trp and vit C concentrations. In all cases, the rate of Trp oxidation was maximal during the first hours of incubation, time corresponding to ascorbate oxidation and hydrogene peroxide production. Trp oxidation rate was positively related to vit C concentration, ie to $H_2O_2$ production (eq 1), and inversely associated to the initial Trp level.

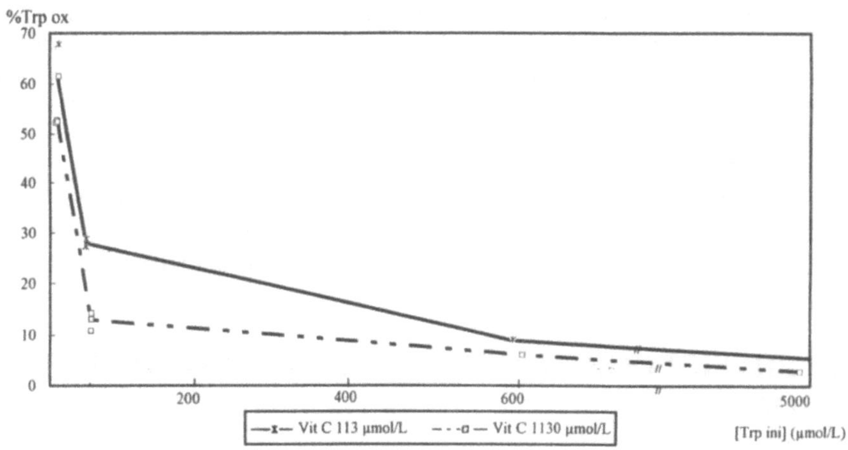


**Figure 1.** Trp oxidation rate as a function of initial Trp concentration after 24 h of incubation at 37 °C.

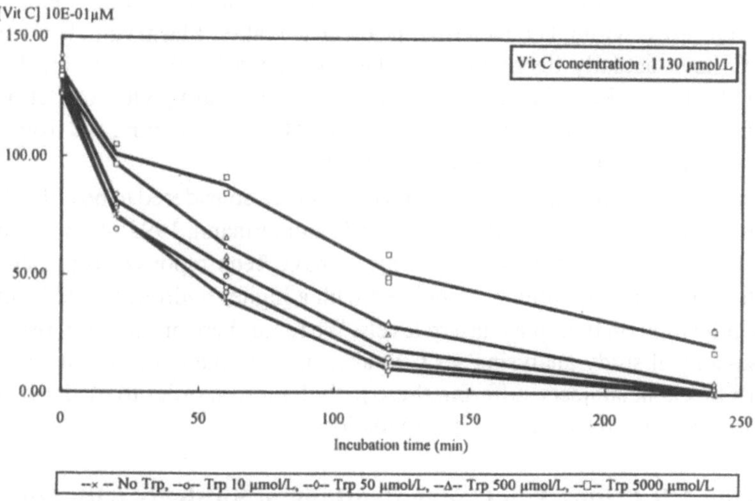


**Figure 2.** Vit C oxidation as a function of Trp concentration.

We parallely observed that Trp inhibits copper-mediated vit C oxidation in a dose-dependent way (Fig 2). This observation could be explained by the ability of Trp to chelate copper, which increases with Trp concentration. The slower hydrogen peroxide formation from eq 1 should therefore be associated to a lower hydroxyl radical production via the copper-catalysed Fenton reaction (eq 2). As a consequence, Trp oxidation (eq 3) should parallely be inhibited, and this is in agreement with our observations.

**↑[Trp]⇒↑copper chelation ⇒↓VitC oxidation, ↓ Fenton reaction⇒↓[·OH] ⇒↓[oxTrp]**

We conclude from these results that in vivo oxidation of Trp by the copper-ascorbate system should be possible in tissues where at least traces of metals are active and where the Vit C to Trp ratio is high. For similar ratios, Trp susceptibility to oxidation will be inversely related to its concentration in the tissue. However, metals are actively chelated in vivo, not only by specific proteins but also by many other amino acids in proteins (His, Ala...). Furthermore hydroperoxide-degrading enzymes should impede the Fenton reaction to take place. These restrictive conditions suggest that Trp oxidation via oxidation of Vit C should be marginal in vivo. Nevertheless, during aging (Nath et al., 1969) and diabetes (Noto et al., 1983), total copper concentration has been observed to be higher in plasma and in lenses, so that the chelation capacity of proteins might be overpassed.

## II. VIT C OXIDATION SEEMS NOT SUFFICIENT TO INDUCE TRP OXIDATION IN VIVO

The previous results indicate that when vit C is actively oxidized in vivo and if any traces of metal ions are active, Trp oxidation could be expected. Vit C is the main antioxidant in plasma (Frei et al., 1988) and in lens (Garland, 1991). When exposed to an oxidative stress, plasma and lens vit C is rapidly oxidized into dehydroascorbic acid (DHA). If the rate of vit C oxidation is not compensated by an active non-enzymatic (GSH) and enzymatic reduction of DHA (by DHA reductase or radical ascorbyl reductase), a decrease in ASC concentration should be observed. However, tissue vit C concentration cannot be interpreted as a marker of an oxidative stress without considering the dietary intakes. Plasma and lens (Taylor et al., 1991) vit C have actually been shown to be influenced by the vit C level in the diet. Evaluation of vit C intake is therefore a prerequisite for estimating vit C oxidation in tissues. A major difficulty arises from the unprecise dietary vit C quantification, and from the variable vit C absorption, which mainly depends on vit C status.

Cardiovascular diseases, diabetes and cataract are considered to be radical-associated pathologies. Diabetic (Som et al., 1981; Lysy and Zimmermann, 1992; Asayama et al., 1993) and cardiac (Gey et al., 1987a; 1987b) subjects have been reported to present decreased plasma antioxidant concentrations, associated with a higher hydroperoxide level. However there is no indication on their plasma Trp levels. We report here preliminary results obtained from a case-control study analysing vit C intakes and plasma concentrations and free and peptidic Trp levels in subjects suffering these pathologies, in order to evidence an eventual Trp oxidation associated to plasma vit C oxidation.

## II.1. Plasma Vit C and Trp Concentrations in Subjects with Diabetes or Cardiovascular Disease and Age-Matched Controls

11 volunteer patients with diabetes were selected in the Hôtel Dieu Diabetology Department (Paris) and compared to 10 age-matched healthy controls consulting for routine analysis (De Mouy laboratory, Paris). 10 subjects suffering from cardiovascular disease and hospitalized for control exams in the Cardiology Department of Calmette hospital (Lille) were compared to 19 age-matched healthy subjects consulting for routine analysis in Pasteur Institut (Lille).

5 mL blood samples were taken on heparin and immediately centrifuged at 4°C. Plasma were rapidly stabilized with metaphosphoric acid (3.6%) and ASC concentration was measured within 10 days of conservation at -80°C by an HPLC-fluorescence method (Tessier and Birlouez-Aragon, submitted). Total free Trp was quantified by HPLC-UV and total plasma Trp was evaluated fluorimetrically (290 nm excitation, 340 nm emission). Vit C intakes were evaluated by a questionnaire on the frequency of fruit and vegetables intakes during the 15 previous days.

The questionnaire was validated on 61 healthy subjects aged 15 to 73 years from Lille and Paris. A significant correlation was found between vit C intakes and plasma concentrations ($r=0.40$; $p<0.01$). However, the slope of the regression line was weak (for 50 mg supplementary vitC, mean plasma levels increased by 6 μmol/L) suggesting that intestinal absorption rates partially compensate for the different intakes.

There was no significant difference in the vit C intake between cases and controls even if some subjects with cardiovascular diseases were observed to have a lower intake than controls (lower quartile and median intakes twice lower). Plasma vit C concentrations were non significantly lower in diabetic subjects compared to their controls, but cardiac persons had twice lower plasma vit C concentrations ($p<0.005$) than control subjects (Table 1). Considering the low relationship between intakes and plasma vit C levels, the low concentrations in cardiac subjects should only partially result from the lower intakes, and could be interpreted as a consequence of a higher oxidative stress.

Plasma free Trp and total Trp fluorescence were similar in all groups. Diabetic subjects had even a significantly higher plasma Trp fluorescence (Table 1).

**Table 1.** Plasma free and peptidic Trp and Vit C concentraions in diabetic and cardiac patients compared to healthy subjects

| | Controls (Paris) | Diabetics (Paris) | Controls (Lille) | Cardiacs (Lille) |
|---|---|---|---|---|
| n | 10 | 11 | 19 | 11 |
| Median age (years) | 51 | 53 | 64 | 66 |
| (min-max) | (46–73) | (32–70) | (45–69) | (38–72) |
| Plasma Vit C (μmol/L) | 43.8 ± 10.45 | 34.20 ± 13.92 | 43.46 ± 16.42 | 21.36 ± 9.66* |
| Dietary Vit C (mg) | 76.5 | 115 | 126 | 68 |
| Quartile (lower,upper) | (53.3–95.6) | (86.3–133.7) | (83.8–138) | (45–120) |
| Free Trp (μmol/L) | 43.00 ± 12.56 | 36.85 ± 8.23 | 35.17 ± 9.02 | 34.80 ± 6.88 |
| Total Trp fluorescence (UR/mgP) | 12.76 ± 1.28 | 15.44 ± 1.49 | 13.79 ± 2.08 | 12.56 ± 2.30 |

*p<0.005, Mann-Whitney Test.

No conclusion can be deduced from the results in diabetic subjects as the oxidative stress, evaluated in terms of vit C levels was not visible. However our results are in agreement with those reported by Cotlier and al. (1981) who did not observed any difference in plasma Trp in diabetic and control subjects. Conversely, the twice lower plasma vit C concentration observed in patients with cardiovascular disease compared to healthy controls could result from a higher ASC oxidation rate, and should therefore be associated to a greater hydroperoxide production. Free Trp oxidation could therefore be expected from the in vitro experiments. But this was not the case and various reasons can be put forward to explain the inefficiency of vit C oxidation to induce Trp oxidation in vivo: 1) the hydroperoxide concentration might be too low, if considering the reducing activities of glutathione and catalase ; 2) metals should not be active enough to allow the Fenton reaction to produce hydroxyl radicals and 3) Trp turn-over in plasma could be too rapid to allow a significant oxidation.

## II.2. Vit C and Trp Concentrations in Cataractous Lenses

In long-lived proteins, such as in lens, any irreversible alteration of peptidic Trp will accumulate during aging. Trp oxidation in lens could be of particular interest as OHAA formation could explain the fluorescent cross-linking observed in cataractous lenses. Pirie (1969) reported a decrease of Trp fluorescence in cataractous lenses. However, Dilley and Pirie (1974) found no significant difference in peptidic Trp content measured by amino acid analysis. Truscott and Augusteyn (1977) confirmed these last results. However, Allegri et al. (1982) reported a decrease of free Trp in cataractous lenses, as compared to clear lenses.

Cataractous nuclei (n=18) were provided by the Ophthalmologic Department from Cochin hospital (Paris) after classification of the lenses into 3 groups according to the opacity grade (1 to 3) evaluated with the slit-lamp. Only one lens had a total white cataract and was classified separately. Nuclei were placed in liquid nitrogen after cataract removal, homogeneized in bicarbonate buffer (pH 8.0) and separated after centrifugation into water-soluble (WS) and -insoluble proteins(WIS). We observed a non significant decrease in the water soluble protein content with the degree of cataract. The white total cataractous lens had however a significantly lower concentration of soluble protein. The decrease in water soluble fraction is due to a protein aggregation during the cataract process. Vit C concentration was quantified by the same technic as for plasma. It was lower than that reported by Varma (1987) for clear lenses and significantly decreased with the degree of cataract (p=0.0141, ANOVA test) (Table 2). We conclude that Vit C is actively oxidated during the cataract process.

**Table 2.** Vit C, free and peptidic Trp in water-soluble (WS) and -insoluble (WIS) fractions of cataractous lenses.

| | Opacity grade | | | Total white cataract |
|---|---|---|---|---|
| | I* | II | III | |
| n | 5 | 9 | 3 | 1 |
| Median age (years) | 81 | 70 | 98 | 69 |
| (min-max) | (68-89) | (66-90) | (69-98) | |
| Vit C (µmoles/100g lens)** | 75.57 ± 28.40 | 62.21 ± 9.66 | 32.55 ± 9.32 | 38.0 |
| Trp (µmoles/100 g lens) | 11.52 ± 2.15 | 12.20 ± 4.07 | 12.64 ± 7.69 | 1.74 |
| PS (mg/L)*** | 14.49 ± 3.08 | 13.74 ± 2.78 | 11.32 ± 3.25 | 4.54 |
| Trp fluorescence (RU/mg PS) | 3.04 ± 0.47 | 3.05 ± 0.59 | 3.45 ± 0.26 | 2.78 |
| Trp fluorescence (RU/mgPIS) | 5.27 ± 1.06 | 4.50 ± 1.34 | 3.09 ± 1.16 | 7.34 |

*Lenses were classified according to the density measured on a slit lamp.
**significant decrease p=0.0141.
***p=0.08 (ANOVA test).

Despite active Vit C oxidation in the cataractous lenses, neither free Trp (measured by HPLC-UV) nor total Trp fluorescence in the soluble fraction were found lower with the grade of cataract. Free Trp was however very low in the total white cataractous lens but vit C concentration was similar to that of group 3 lenses. A decrease in total Trp fluorescence (p=0.088, anova, table 2) was evidenced in the water-insoluble fraction, confirming previous observations (Pirie, 1969; Liang and Chylack, 1987) . Nevertheless, direct HPLC-UV analysis after alkaline hydrolysis (Landry and Delhaye, 1992) failed in showing any change in WIS peptidic Trp content: the Trp to Tyr ratio was 0.3636±0.0340, n=3, for the lenses of opacity grade 1 and 0.3666±0.0147, n=3, for the lenses of opacity grade 3. The decrease in Trp fluorescence could be explained by an energy transfer between Trp and 430 nm-emitting fluorophores in the membrane-protein aggregates complex (Scalbert-Menanteau, 1995), which could be more efficient when aggregates accumulate.

The free Trp levels were found to be 18 times lower in our lenses than those observed by Allegri et al. (1982) in clear lenses but also 5 to 15 times lower than those reported for cataractous lenses. As we could not provide us with clear lenses, it cannot be concluded if the difference in Trp concentration between the two studies is due to analytical discrepancies or to a real effect of the cataract process.

## III. CONCLUSION

We show in this study that vit C or $H_2O_2$-mediated oxidation of Trp is strongly inhibited by its ability to chelate copper. Therefore, the prooxidant activity of vit C is lower and the Fenton reaction is inhibited in presence of Trp. In vivo, other amino-acids chelate copper, reinforcing the protection of proteins and particularly of Trp against metal-mediated oxidation. This study failed actually in evidencing any Trp oxidation neither in plasma of subjects with cardiovascular disease nor in cataractous lens, despite an active Vit C oxidation , suggesting an oxidative stress in these tissues. Furthermore, most studies which had shown Trp oxidation in vivo were based on fluorescence measurements. We show here that fluorescence does not allow a quantitative evaluation of peptidic Trp, and further studies are needed to directly evaluate free and peptidic Trp to prove any oxidation in vivo.

## ACKNOWLEDGMENT

We are grateful to Prof. Slama and Dr. De Mouy for having provided us with blood samples of healthy and diabetic subjects. The dietary Vit C were evaluated by the dieteticiens S. Roussot and S. Voisin.

## REFERENCES

Allegri G., Angi MR, Costa C. and Vanzan S., 1982, Tryptophan lens levels in senile cataract, *Boll. Soc.Ital.Biol.Sper*. 28:340.

Asayama K., Uchida N., Nakane T., Hayashibe H., Bobashi K., Amemiya S., KAto K and Nakazaa S., 1993, Antioxidants in the serum of children with insulin-dependent diabetes mellitus, *Free Radical Biol.Med*, 15:597.

Benavente M.G. and Truscott RJW., 1991, Modification of proteins by 3-hydroxyanthranilic Acid: the role of lysine residues, *Arch.Biochem.Biophys*., 290:451

Cotlier E, Sharma Y.R., Zuckerman J., Pucklin, B., 1981, Teasley and J.Irvine Plasma Tryptophan in Humans with Diabetic and Senile cataracts *Exp.Eye Res*. 33:247.

Dilley K.J. and Pirie A., 1974, Changes to the proteins of the human lens nucleus in cataract, *Exp Eye Res*, 19:59.

Frei B., Stocker R. and Ames B N, 1988, Antioxidant defenses and lipid peroxidation in human blood plasma. *Proc. Natl.Acad Sci USA*, 85:9748.

Garland DL, 1991, Ascorbic acid and the eye , *Am. J.Clin.Nutr*., 54:1198s.

Gey K.F., Brubacher G.B., and Stahelin H.B., 1987a, Plasma levels of antioxidant vitamins in relation to ischemic heart disease and cancer, *Am.J.Clin.Nutr*., 45: 1368.

Gey K.F., Stahelin H.B. and Puska P., 1987b, Relatioship of plasma level of Vit C to mortality from ischemic heart disease, *Ann.NY Acad.Sci*. , 498:110.

Hiraku Y., Inoue S., Oikawa S., Yamamoto K., Tada S., Nishino K. and Kawanishi S., 1995, Metal-mediated oxidative damage to cellular and isolated DNA by certain tryptophan metabolites, *Carcinogenesis*, 16:349.

Kell G. and Steinhart H., 1990, Oxidation of Tryptophan by $H_2O_2$ in model systems, *J.Food Sci*. 55:1120

Krogull M.K. and Fennema O., 1987, Oxidation of tryptophan in the presence of oxidizing methyl linoleate , *J.Agric.Food Chem*., 35:66.

Landry J. and Delhaye S., 1992, Simplified procedure for the determination of tryptophan in foods and feedstuffs from barytic hydrolysis, *Am. Chem.Soc*, 40:776.

Liang J.N. and Chylack L.T., 1987, Spectroscopic study on the effects of nonenzymatic glycation in human a crystallin, *Invest.Ophtalm.Vis Sci*. 28:790.

Lysy J., and Zimmermann J., 1992, Ascorbic acid status in diabetes mellitus, *Nutr.Res*., 12:713.

Maskos Z., Rush J.D. and Koppenol W.H., 1992, The hydroxylation of Tryptophan, *Arch.Biochem.Biophys*., 296:514.

Nath R., Srivastava SK. and Singh K., 1969, Accumulation of copper and inhibition of lactate dehydrogenase activity in human senile cataractous lens, *Indian J. Exp.Biol*., 7: 25.

Noto R, Alicata R., Sfogliano L., Neri S. and Bifarella M., 1983, A study of cupremia in a group of elderly diabetics, *Acta Diabet lat*. 20:81.

Pirie A, 1969, Insoluble protiens of the lens, *Exp.Eye Res*., 8: 233.

Scalbert-Menanteau P., 1995, Role des oses dans les modificatiions post-traductionnelles des proteines membranaires au cours du vieillissement et de l'opacification du cristallin humain, Thesis of Institut National Agronomique Paris.

Som S., Basu S., Murherjee D., Deb S., Choudhury P, Murkherjee S., Chatterjee S.N., Chatterjee IB., 1981, Ascorbic Acid metabolism in diabetes mellitus, *Metabolism* , 30: 572.

Steinhart N, Vollmar M. and Sailer C.,1993, Pro-and Antioxidative effect of ascorbic acid on L-tryptophan in the system fe3+/ascorbic acid/O2, *J.Agric.Food Chem*., 41:2275.

Taylor A., Jacques PF, Nadler D., Morrow F, Sulky SI and Shepard D., 1991, Relationship in humans between ascorbic acid consumption and levels of total and reduced ascorbic acid in lens, aqueous humor and plasma, *Curr. Eye Res*., 10 : 751.

Tessier F and Birlouez-Aragon I, Validation of a micromethod for determining oxidized and reduced Vitamin C in Plasma by HPLC-Fluorescence, *Int.J.Vit.Nutr. Res*.,submitted for publication

Tomoda A., Yoneyama Y., Yamaguchi T., Shirao E. and Kawakishi K., 1990, Mechanism of coloration of Human Lenses Induced by Nera-Ultraviolet-photo-oxidized 3-hydroxykynurenine, *Ophtalm Res.*, 22:152

Truscott RJW. and Augusteyn R C., 1977, Oxidative changes in human lens proteins during senile nuclear cataract formation, *Biochim.Biophys.Acta* 492:43.

Truscott RJW. and Martin A., 1989, The reaction of Proteins with 3-hydroxyanthranilic acid as a possible model for senile nuclear cataract in man, *Exp.Eye Res*, 49:927.

Truscott R.J.W., Pyne S.G. and Manthey M., 1991, Reactive metaoblite Hypothesis for Human senile Cataract, *Lens Eye Tox.*, 8:251.

Uchida K., Enomoto N., Itakura K. and Kawakishi S., 1985, The hydroxyl radical generated by an Iron(II)/EDTA/ascorbate system preferentially attacks trpyptophan residues of the protein, *Agric.Biol.Chem*, 53: 3285.

Uchida K. and Kawakishi S., 1988, Selective Oxidation of Tryptophan and Histidine residues in protein through the copper-catalyzed autoxidation of L-ascorbic acid, *Agric. Biol. Chem.*, 52:15299.

Varma S.D., 1987, Acid ascorbic and the Eye with special reference to the lens, *Anal.NY Acad. Sci.* 498:280.

Wood A.M. and Truscott R.J., 1993, UV Filters in Human lenses : Tryptophan catabolism, *Exp.Eye Res.*,56:317.

116

# TIME-RESOLVED PHOTOACOUSTIC CALORIMETRY OF PEPTIDES AND PROTEINS CONTAINING A SINGLE TRYPTOPHAN

L. Brancaleon

Dipartimento di Fisica
Universita' di Parma
Viale delle Scienze, 43100, Parma (Italia)

## 1. INTRODUCTION

The photophysiscs of tryptohan has been widely studied in the past (Bent and Hayon, 1975; Valeur and Weber, 1977; Volkert et al., 1977; Szabo and Rayner, 1980; Petrich et al., 1983; Chang et al., 1983; Willis et al., 1991; McMahon et al., 1992; Gai et al., 1994) in order to use this molecule as an intrinsic probe to investigate the structural properties of several proteins (Orstan and Ross, 1987; Abinson et al., 1989; Gordon et al., 1992; Bajzer and Prendergast, 1993). For this purposes the radiative parameters (fluorescence quantum yield, fluorescence lifetimes, quenching constants, energy transfer rates past (Bent and Hayon, 1975; Szabo and Rayner, 1980; Eftink and Ghiron, 1981; Chang et al., 1983; Petrich et al., 1983; Creed, 1984; Lee and Robinson, 1984; McMahon et al., 1992; Gai et al., 1994) have been detailed determined in order to reliably explain the changes in the microenvironment surrounding the indole ring; these changes usually reflect conformational modification of the protein. Besides the fluorescence other non radiative relaxation mechanisms occur starting from the first excited singlet state. The basic idea of this work is to study the photophysics of Trp (and peptides) from a reverse angle, namely investigating the non radiative relaxations that proceed from the first excited state; since the radiative quantum yield of Trp in aqueous solution is only about 0.15, the energy absorbed is mainly released through radiationless pathways. The time-resolved photoacoustic calorimetry (LIOAC) is the most suitable technique to detect radiationless processes occurring in liquid solutions (Rudzky et al., 1985; Small et al., 1989; Peters and Snyder, 1991; Braslavsky and Heibel, 1992). The LIOAC can detect radiationless decays of the excited states and volumetric contribution accompanying these decays allowing the temporal resolution of the non-radiative transients in a time window basically dependent on the experimental set up (10 nsec-5msec for our apparatus). The main goal of the project is the use of LIOAC to detect photoinduced structural changes in proteins and peptides. The detection of photoinduced

*Recent Advances in Tryptophan Research*, edited by Graziella Allegri Filippini et al.
Plenum Press, New York, 1996

conformational changes is one of the opportunities available using LIOAC (Peters and Snyder, 1991; Braslavsky and Heibel, 1992; Herman and Goodman, 1989; Yurela et al., 1994; Van Brederode et al., 1995). The work presented reports the results obtained with Trp and single tryptophan containing dipeptides. N-Trp and C-Trp terminal peptides have been considered for this work as according to previous works (Chang et al., 1983; Petrich et al., 1983; Chen et al., 1991) these peptides show different relaxation pathways. The study has been carried on using two different excitation wavelengths (266 nm and 304 nm) because the photophysics of the indole containing molecules is also affected by the excitation wavelength (Valeur and Weber, 1977; Szabo and Rayner, 1980; Chang et al., 1983; Petrich et al., 1983; Gai et al., 1994). Tryptophan and N-Trp terminal peptides show three radiationless transient (a fast lifetime < 1 ns; a second lifetime about 40 ns and a third one about 800 ns) with large positive volumetric change, while the radiationless relaxation of the singlet state of C-Trp terminal peptide occurs with two lifetimes and negligible volumetric contribution.

## 2. PHOTOACOUSTIC CALORIMETRY

The time-resolved photoacoustic calorimetry technique has been developed during the last fifteen years thanks to the development of pulsed lasers and piezoelectric transducers (Rudzky and Goodman, 1985; Herman and Goodman, 1989; Small et al., 1989; Peters and Snyder, 1991; Braslavsky and Heibel, 1992; Yurela et al., 1994; Van Brederode et al., 1995).

The experimental set up is shown in **Figure 1.** The piezoelectric transducer is the basic device of this pulsed technique because it provides the transduction of a pressure wave generated in the sample by the radiationless relaxation of the excited states, to an electrical signal that can be digitized; moreover the piezoelectric crystals allow the temporal resolution

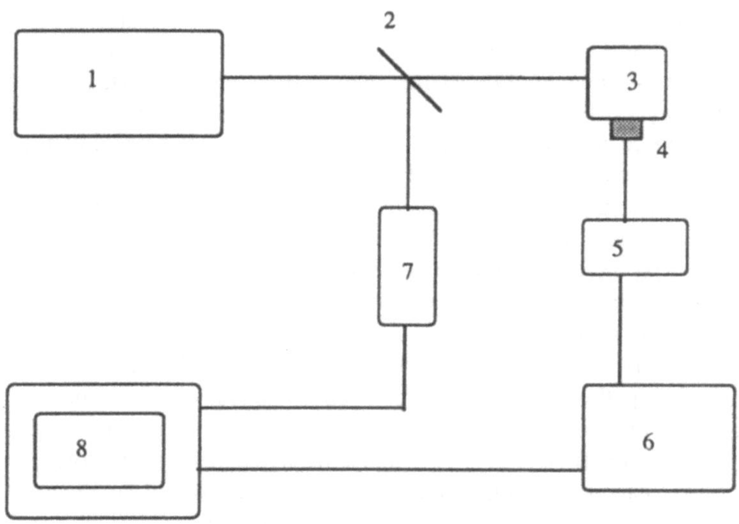


**Figure 1.** Time-resolved photoacoustic calorimetry experimental set-up: 1) Q-switched laser Nd:YAG (Quantel), 5-7 ns pulse width; 2) Beam splitter; 3) Temperature controlled photoacoustic cell composed by a fluorescence cuvette (1cm x 1cm). The temperature has been controlled using a Heto water bath; 4-5) Piezoelectric transducer Panametrics V103 and preamplifier; 6) Digitizer LeCroy 9450A; 7) Energy detector and energy meter RJ-765 and RJ-7620 Laser Precision; 8) Computer IBM/PS2, 80286 processor.

of the non radiative events as we shortly see. The PZT piezoelectric transducer like the one used in our laboratory is a ringing crystal with a leading frequency of about 1MHz. This kind of transducers ideally provide a temporal resolution window between 1 ns and 1 ms; because of the electronic devices and the acoustic transit time (Braslavsky and Heibel, 1992) the actual window of our instrumentation reduces to 10 ns through 10 μs. As mentioned above the relaxation of the excited electronic states generates a volume change in the solvent surrounding the absorbing molecule; mainly we have two different origins for these volume changes.

*Thermal Volume Change:* **Figure 2** shows a typical Jablonsky diagram representing electronic transitions occurring after the absorption of a photon. After the absorption event (occurring faster than one femtosecond) the electron is in the vibronic excited state $V_n$ and it undergoes a fast (t< 1 ps) relaxation through the vibronic levels to the state $S_1$; from this state several events can occur: a) the electron with a certain efficiency can undergo a radiative transition to $S_0$ emitting a photon (fluorescence); b) it can relax through the vibronic levels to $S_0$ (internal conversion); c) it can undergo to an intersystem crossing to the triplet state ($T_1$) and this event is again occurring as a relaxation through the vibrational levels. Other relaxations can occur starting from the excited states such as photoinduced chemical reaction leading to different molecular species, production of radicals and other photochemical events. Each radiationless process occurs with the emission of heat to the surrounding solvent. As these transitions occur under the adiabatic conditions the release of heat from the molecule to the solvent induces an adiabatic temperature change in the solvent surrounding the molecule; the heat released produces a volumetric expansion of the solvent. The volume change eventually generates a pressure wave in the solvent. Usually two thermal contributions are distinguished according to the lifetimes of the relaxation: a prompt heat (heat released in a time scale much faster then the transducer rise time) and delayed heat which is heat released by transients on a longer time scale (Braslavsky and Heibel, 1992).

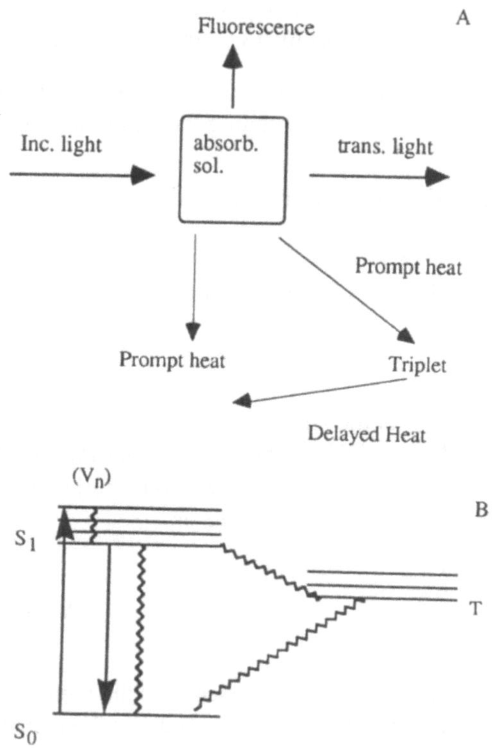


**Figure 2.** Schematic description of the relaxation pathways of the first excited singlet state. A) the absorbed light can be released as a promt heat through the internal conversion and intersystem crossing to the triplet state. The delayed heat occurs starting from long lived states such as triplet states. B) The same model described by a Jablonsky diagram. The solids arrows show the absorption and fluorescence event. The broken lines show the non radiative relaxations. The internal conversions occur through the vibronic states of $S_1$ and through the transition $S_1 \rightarrow S_0$. The intersystem crossing occurs in the transition $S_1 \rightarrow T$ . The radiationless decay occurring from the triplet state is the resposible for the delayed heat.

*Conformational Volume Change:* The relaxation of the excited states can also involve, at different stages, the reorganization of the solvent surrounding the molecule. The rearrangement of the solvent very often implies a change of the volume of the solvation shell or the displacement of few molecules of solvent; this volume reorganization once again generates a pressure wave travelling in the solvent and can be monitored by the piezoelectric transducer. Several mechanism can induce the solvent reorganization:

a. change of the dipole moment. The transition from one excited state can generally be related to a change in the dipole moment of the molecule (Yurela et al., 1994; Van Brederode et al., 1995). A change of dipole moment of the solute in polar solvents is expected to change the interaction solute-solvent changing consequently also the solvation shell surrounding the solute. Sometimes it is also possible that the interaction solute-solvent is stabilized by O-H bonds and the absorption of light could lead to break the hydrogen bonds with a consequent "displacement" of the solvent molecules. This mechanism could for instance occur when photisomerizations are involved (Yurela et al., 1994).

b. Van der Waals volume change: the electronic transitions (absorption or relaxations) can also induce direct or indirect change of the Van der Waals volume of the molecule. A Van der Waals volume change implies a change of the volume occupied by the molecule which means that the bond lengths and the bond angles have been modified by the electronic transition. Though this mechanism could sometimes occur in single chromophores, is actually more common when large macromolecules such as proteins are involved (Peters and Snyder, 1991); when a protein is involved the volume change could be related to a "real" conformational change induced directly or indirectly by the absorption of a photon.

c. photoionization: if one of the relaxation mechanism is the emission of a proton or an electron this produces volumetric changes in the surrounding solvent because the emitted particle must be solvated and the solvation process is accompanied of course by a volumetric change.

These are just some of the examples about the mechanism related to volume changes of the solvent surrounding the chromophore.

## 2.1 Photoacoustic Signal

As above mentioned the photoacoustic signal is represented by the sum of two contribution (Peters and Snyder, 1991; Braslavsky and Heibel, 1992):

$$S = k(\Delta V_{th} + \Delta V_{con}) \tag{1}$$

where S is the signal amplitude, k is the coupling constant dependent on the experimental conditions, $DV_{th}$ is the thermal contribution and $DV_{con}$ is the conformational contribution.

The conformational contribution is a pure volumetric signal while the thermal contribution can be written as:

$$\Delta V_{th} = Q \frac{\beta}{C_p \rho} \Bigg|^{T} \tag{2}$$

where Q is the heat released, β is the thermal expansion coefficient of the solvent, Cp is the specific heat of the solvent and ρ is the density of the solvent.

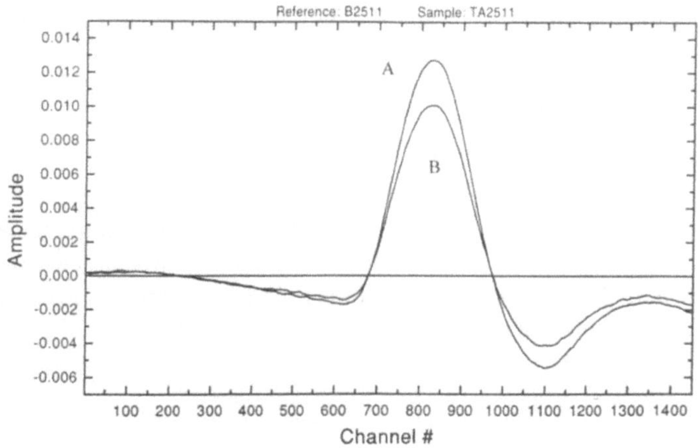


**Figure 3.** Example of photoacoustic signal at 25°C. The waveform A is the signal of the sample (Trp-Ala) and B is the waveform of the reference compound (Bromocresol purple). Amplitude (mV) versus channel number (channel with = 2.5 ns).

A photoacoustic signal is represented in **Figure 3**. The time profile shown in figure 3 is actually the convolution of the transducer response (or instrumental response) with the time profile of the heat released which supposed to be exponential, like the corresponding radiative relaxation process. The protocol used to acquire and analyze the photoacoustic signal is rather long and its detailed description can be found in previous works (Rudzky, 1985; Peters and Snyder, 1991; Braslavsky and Heibel, 1992), here I will very shortly describe the main steps. The acquisition of the signal of a calorimetric compound is needed; a calorimetric compound is basically a compound that releases all the energy absorbed as prompt heat (the emission of heat in these compounds occurs in few picoseconds and such event can be considered as δ function compared to the rise time of the transducer which is of about 200 ns). The response digitized is considered the pure instrumental response to a prompt and unitary radiationless relaxation event. Several calorimetric compounds can be used depending on the excitation wavelength and on the solvent used (Braslavsky and Heibel, 1992). The signal from the sample is then acquired. Reference solution and sample solution must undergo two strict conditions: first of all the optical densities of the two solutions must match exactly in order to avoid distortions due to differences in the generation of the signal; secondly since the signal amplitude depends on β which is temperature dependent, the temperature of the two solutions must match within 0.2 °C. The analysis of the sample waveform basically consists in a deconvolution which is described in other works (Rudzky, 1985; Small et al., 1989). The deconvolution of the sample waveform with respect to the reference signal provides the parameters of the radiationless decay α and τ.

Where τ represents the radiationless lifetime of the process detected and α is the parameter as expressed in equation 4. As mentioned before the time-resolved photoacoustic calorimetry provides the direct detection of photoinduced volumetric changes; this opportunity is particularly accessible in water solutions. Using equation 2, the equation 1 can be written in the following way:

$$S_{sam} = k(Q\frac{\beta}{C_p\rho} + \Delta V_{con}) \qquad (3)$$

In water solutions β goes to 0 below 4°C depending on the concentration of the buffer. It follows that at the temperature where the thermal expansion coefficient is zero the remaining signal is a pure conformational change. It is therefore possible to estimate the photoinduced volumetric changes. In fact the photocaoustic signal of the reference compound is supposed to be given only by the thermal contribution and can be written as

$$S_{ref} = kQ \left. \frac{\beta}{C_p\rho} \right|^{T}$$

at each temperature, where k is the coupling constant of the system. At the same temperature the signal of the sample is given by equation 3. From the equation above it follows that at the temperature where β vanishes the signal of the reference compound disappears. In terms of photoacoustic parameters we can write that at each temperature the value of $\alpha_i$ is the sum of a thermal and a volumetric contribution:

$$\alpha_i^T = \alpha_i^{th} + \alpha_i^{vol} \tag{4}$$

at the temperature $T^0$ where β vanishes (below 3.9°C in water solutions):

$$\alpha_i^{T^0} = \alpha_i^{vol} \tag{5}$$

Assuming the volumetric contribution independent from the temperature (Rudzky, 1985; Peters and Snyder, 1991), by acquiring the waveforms at the temperatures T and $T^0$ it is possible to obtain the pure thermal and the pure volumetric contributions according to equations 4 and 5. In fact using a simple energy balance we can write:

$$E_{h\nu}\alpha^T = E_{h\nu}\alpha_{th} + \Delta V_{con} \left. \frac{C_p\rho}{\beta} \right|^{T} \Rightarrow \Delta V con = (E_{h\nu}\alpha^T - E_{h\nu}\alpha_{th}) \left. \frac{\beta}{C_p\rho} \right|^{T}$$

To carry on this kind of approach it is necessary that the temperature T be close enough to $T^0$ to consider the change of the acoustic transient time (which is temperature dependent) negligible compared to the rise time of the transducer. For this purpose the temperature T is usually less than five degrees above $T^0$.

## 3. RESULTS

The photoacoustic data have been completed by comparing them to nanosecond flash photolysis. This technique provides the optical detection of the photoinduced transients (Bensasson et al., 1990) and has been utilized as a tool to obtain the quantum yields of the processes. A typical photoacoustic signal obtained by our peptides is shown in figure 3 and the summary of the photoacoustic parameters is represented in **Table 1.**

From the values reported in the table 1 some features are evident: Tryptophan, TryptophylGlycine and TryptophylAlanine show three components (the prompt heat contribution, and two lifetimes of about 40ns and 500ns) both for excitation at 266 nm and 308 nm, while GlycilTryptophan is charcterized by only two components (the prompt heat and a lifetime of 500 ns). The difference between the peptides is not surprising, as in previous works (Chang et al., 1983; Petrich et al., 1983; Braslavsky and Heibel, 1992) has been shown that Trp-Gly and Gly-Trp undergo completely different photophysics. Trp, Trp-Gly and Trp-Ala show other common features like similar values of a and the presence of a transient of some tens of nanoseconds. Another very important feature is represented by the differences in the volumetric contributions: Trp, Trp-Gly and Trp-Ala are characterized by a large

**Table 1.** Summary of the photoacoustic data. $\alpha_i$ is the fraction of energy released as heat. $\alpha_{th}$ is the difference between $\alpha_7$ and $\alpha_4$

| | 266 nm | | | | 304 nm | | | |
|---|---|---|---|---|---|---|---|---|
| | $a_1$ | $t_1$ | $a_2$ | $t_2$ | $a_1$ | $t_1$ | $a_2$ | $t_2$ |
| Gly-Trp | | | | | | | | |
| 25°C | 0.91 | < 10 ns | 0.05 | 1.03 ms | 0.96 | < 10 ns | 0.23 | 850 ns |
| Trp | | | | | | | | |
| 7°C | 0.33 | " | 1.39 | 39 ns | 0.52 | " | 1.47 | 54 ns |
| 4°C | 0.03 | " | 1.11 | 57 ns | 0.22 | " | 0.6 | 70 ns |
| $a_{th}$ | 0.3 | | 0.28 | | 0.3 | | 0.87 | |
| Trp-Gly | | | | | | | | |
| 7°C | 0.18 | " | 1.18 | 48 ns | | | | |
| 4°C | –0.1 | " | 0.78 | 75 ns | | | | |
| $a_{th}$ | 0.28 | | 0.4 | | | | | |

volume change (**see figure 4**) where the volumetric contribution is large and positive for the intermediate transient and large and negative for the longer lived transient. The volumetric contribution accompanying the prompt heat is smaller and usually positive. The volumetric contribution in Gly-Trp is, on the other hand, virtually negligible. The photoacoustic parameters cannot be rationailized on their own because the time-resolved photoacoustic calorimetry does not provide informations about the nature of the transients.

In order to provide additional information about the transients detected by our photoacoustic experiments, a nanosecond flash photolysis experiment has been developed on the same compounds in the same conditions adopted for the photoacoustic measurements. The flash photolysis measurements were carried on at 266 nm and 308 nm. Through the characteristic absorption bands, the photoinduced transients have been detected and their lifetimes have been obtained. Moreover the quantum yield of the formation of these transients have been determined. The quantum yields are summarized in **table 2**.

The flash photolyisis data allowed to draw a model, for the peptides, which is schematically represented in **Figure 5**. According to this model Trp, Trp-Gly and Trp-Ala

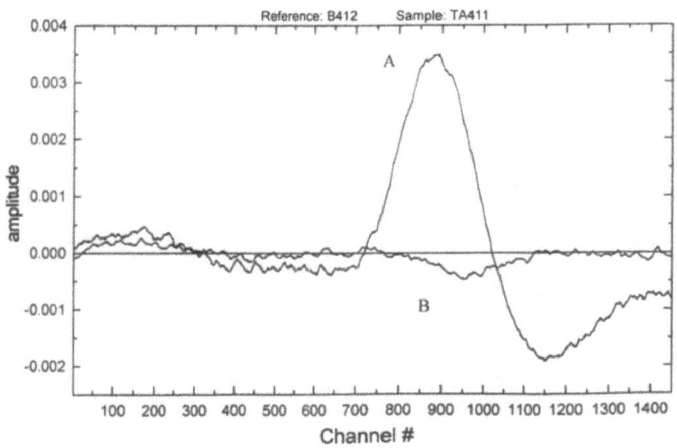


**Figure 4.** Example of volumetric detection. B is the signal of the reference compound at 2.6 °C where the thermal expansion coefficient of the solution is 0. A is the signal of the sample (Trp-Gly) at the same temperature (2.6°C). Amplitude (mV) versus channel number (channel with = 2.5 ns).

**Table 2.** Summary of the flash photolysis data on Trp and peptides in aerated solution. Φ is the quantum yield of the transient product

| | 266 nm | | 308 | |
|---|---|---|---|---|
| | Φ | τ | Φ | τ |
| Trp | | | | |
| triplet | 0.28 | 640 ns | 0.17 | 700 ns |
| electron | 0.26 | 158 ns | 0.08 | 132 ns |
| Trp-Gly | | | | |
| triplet | 0.19 | 671 ns | 0.11 | 750 ns |
| electron | 0.2 | 160 ns | 0.05 | 141 ns |
| Gly-Trp | | | | |
| triplet | 0.15 | 538 ns | 0.08 | 800 ns |
| electron | 0.15 | 172 ns | 0.12 | 118 ns |

undergo three different subnanosecond relaxation pathways from the first excited singlet state: 1) photoionization with the consequent production of cation and free radicals; 2) intersystem crossing to the triplet state; 3) intersystem crossing to another transient whose nature is still uncertain and that we will call x state. The photoejected electron in air saturated solution decays with an apparent lifetime of 180 ns, in nitrogen saturated solution this

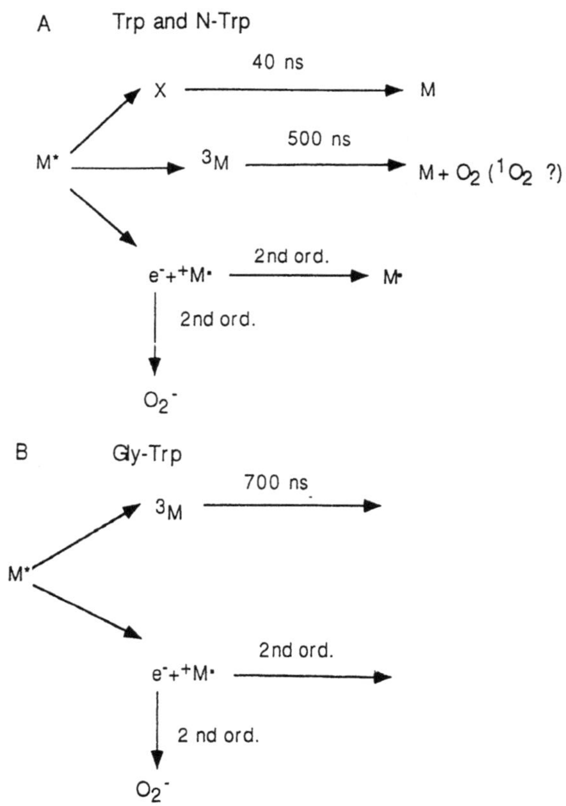


**Figure 5.** A) Model proposed for Trp and N-Trp terminal peptides. B) Model proposed for Gly-Trp. M* is the vibronic excited state of the first excited singlet of the chromophore.

apparent lifetime is much longer (≅1 ms) and in oxygen saturated solution is almost completely quenched. This evidence leads us to assume that the electron reacts with molecular oxygen to give superoxide anion. In air saturated solutions the triplet state decays with a lifetime of about 500 ns; as the lifetime is longer (> 1 μs) in nitrogen saturated solution and is quenched in oxygen saturated solution, it is therefore evident that the triplet lifetime in air saturated solutions is determined to the diffusion of oxygen. The quenching by the molecular oxygen could lead both to singlet oxygen or to collisional quenching without production of singlet oxygen. The x state decays with a lifetime of about 40 ns and is not affected by oxygen or nitrogen saturation. The photophysics of Gly-Trp is similar but presents a fundamental difference: in this peptide there is no evidence of formation of the x state. The presence of this x state in Trp and Trp-Gly was already known by previous works (Bent and Hayon, 1975) but its characteristic is still unknown. Comparing the flash photolysis data with the photoacoustic ones we have been able to obtain some quantitative estimate of the photophysical parameters related to the transient of Trp and Trp containing peptides. Applying the model represented in figure 5 to the photoacoustic data we are able to obtain equations that are reported below. The fraction of energy released in the prompt event ($\alpha_1^{th}$) can be considered as the sum of the fast events: photoionization, internal conversion among the vibronic levels of $S_1$, intersystem crossing to the triplet and to the x state. In Gly-Trp where the x level is not present the intersystem crossing term is not considered while is introduced the internal conversion $S_1 \rightarrow S_0$; the internal conversion has not been considered in Trp and N-Trp peptides as in many previous works (Chang et al., 1983; Petrich et al., 1983; Chen et al., 1991) the internal conversion to the ground state is reported to be negligible. The fraction of heat released in about 50 ns ($\alpha_2^{th}$), in Trp and N-Trp peptides, has been attributed to the decay of the x level. Starting from these assumptions the following equations can be written:

Gly-Trp:

$$\alpha_1^{th} = \Phi_d \frac{\Delta E}{E} + (1 - \Phi_d) \frac{E - E_0}{E} + (1 - \Phi_d)(1 - \Phi_f - \Phi_T) \frac{E_0}{E} + (1 - \Phi_d) \Phi_T \frac{E_0 - E_T}{E} \quad (6)$$

Trp, Trp-Gly and Trp-Ala:

$$\alpha_1^{th} = \Phi_d \frac{\Delta E}{E} + (1 - \Phi_d) \frac{E - E_0}{E} + (1 - \Phi_d) \Phi_x \frac{E_0 - E_x}{E} + (1 - \Phi_d) \Phi_T \frac{E_0 - E_T}{E}$$

$$\alpha_2^{th} = \Phi_x \frac{E_x}{E} \quad (7)$$

where $\Phi_d$ is the quantum yield of photoionization, $\Delta E$ is the energy released in the photoionization, E is the excitation energy, $E_0$ is the energy of the 0-0 transition, $\Phi_f$ is the fluorescence quantum yield, $F_T$ is the triplet quantum yield, $E_T$ is the triplet energy, $\Phi_x$ is the quantum yield of the unknown intermediate and $E_x$ is its energy level. From the equation 6, using the quantum yield as obtained from the flash photolysis, we obtain the value of $\Delta E$, e.g. of the energy released in the photoionization: $\Delta E$ is about 100 kcal/mol for excitation at 266 nm and about 101 kcal/mol for excitation at 304 nm. Since the corresponding excitation energies are 108 and 94 kcal/mol the value obtained for $\Delta E$ leads to the conclusion that in the photoionization process the energy absorbed energy is completely released. Taking the values of $\Delta E$ obtained for Gly-Trp into the equations for the Trp and the other peptides we obtained $\Phi_x$ and $E_x$. For the compound studied the value of $E_x$ is very close to the energy of the first excited singlet state. Considering the volumetric contributions, e.g. the values of

$\alpha_i^{T_0}$, we can, for analogy, write the equations for the volumes (since the volumetric contribution is negligible in Gly-Trp the equations are applicable to Trp and N-Trp peptides):

$$\Delta V_1 = \Phi_d \Delta V_d + (1 - \Phi_d)\, \Phi_x \Delta V_x + (1 - \Phi_d)\, \Phi_T \Delta V_T$$

$$\Delta V_2 = -\Phi_x \Delta V_x$$

$$\Delta V_3 = \Phi_T \Delta V_T \qquad (8)$$

where the quantum yields are the same used in the thermal equations and ΔV are the volume changes related to the decays. Comparing the results obtiained for the peptides and tryptophan there are many significative diffrences: Gly-Trp as already mentioned is the only peptide that does not show any significative volume change; in Trp and N-Trp peptides the same radiationless mechanism is accompanied by different volume changes. The volume change due to the photionization is positive and of about 14 ml/mol in Trp, while is 6 ml/mol in the peptides; the volume associated to the formation of the x state is negative and of about -20 ml/mol in Trp and -10 ml/mol in the peptides; the formation of the triplet state seems to imply a huge positive volumetric change of about 20 ml/mol in N-Trp terminal peptides while is reduced in Trp. All these volume changes are presumably due to rearrangement of the solvents around the different intermediate states. There are no conclusive arguments yet for the interpretation of the data obtained, but the photoacoustic calorimetry, as expected, is giving a new insight to a very complicated problem like the photophysics of Trp containing molecules.

## ACKNOWLEDGMENTS

The photoacoustic experiment has been performed also by Dr. Cristiano Viappiani and Davide Diemmi; the photoacoustic measurements have been carried on at the Laboratory for Non-Linear Spectroscopy of Florence thanks to the grant of the European Community: GE1-CT92-0046.

I wish to thank Dr. Stefano Cavalieri and Dr. Paolo Foggi of the LENS for the help provided.

The flash photolysis measurements have been performed at the Wellman Laboratories of Photomedicine in Boston,MA with the extremely useful collaboration of Dr. Robert W. Redmond and Dr. Paul Wood.

## REFERENCES

Albinson, B., Kubista, M., Norden, B, Thulstrup, E.W., 1989, Near ultraviolet electronic transitions of the Tryptophan chromophore: linear dichroism, fluorescence anysotropy and magnetic circular dichroism spectra of some indole derivatives, *J. Phys. Chem.*, 93: 6646.

Bajzer, Z., Prendergast, F.G., 1993, A model for multiexponential Tryptophan fluorescence intensity decay in proteins, *Biophys. J.*, 65: 2313.

Bensasson, R.V, Land, E.J., Truscott, T.G., 1990, Excited state and free radicals in biology and medicine, Oxford University Press ed.

Bent, D.V. and Hayon, E., 1975, Excited state chemistry of ariomatic amino acids and related peptides, *J. Am. Chem. Soc.*, 97: 2612.

Braslavsky, S.E. and Heibel, G.E., 1992, Time-resolved photothermal and photoacoustic methods applied to photoinduced processes in solution,*Chem. Rev.*, 92: 1381.

Chang, M.C., Petrich, J.W., McDonald, D.B., Fleming, G.R., 1983, Nonexponential fluorescence decay of Tryptophan, Tryptophylglycine and Glycyltryptophan, *J. Am. Chem. Soc.*, 105: 3819.

Chen, R.F, Knutson, J.R., Ziffer, H., Porter, D., 1991, Fluorescence of Tryptophan dipeitides: correlations with the rotamer model, *Biochemistry*, 30: 5184.

Creed, D., 1984, The photophysics and photochemistry of the near-UV absorbing aminoacids- I. Tryptophan and its simple drivatives, *Photochem. Photobiol.*, 39: 537.

Eftink, M.R., Ghiron, C.A., 1981, Fluorescence quenching studies with proteins *Anal. Biochem.*, 114: 189.

Gai, F, Rich, R.L., Petrich, J.W., 1994, Monophotonic ionization of 7-Azaindole, Indole and their derivatives and the role of overlapping excited states, *J. Am. Chem. Soc.*, 116: 735.

Gordon, H.L., Jarrell, H.C., Szabo, A.G., Willis, K.J., Somoraj, R.L., 1992, Molecular dynamics simulations of the conformational dynamics of Tryptophan, *J. Phys. Chem.*, 96: 1915.

Herman, M.S. and Goodman, J.L., 1989, Determination of the enthalpy and reaction volume changes of organic photoreactions using photoacoustic calorimetry, *J. Am. Chem. Soc.*, 111: 9105.

Lee, J. and Robinson, W.G., 1984, Indole: a model system for one photon threshold photoionization in polar media, *J. Chem. Phys.*, 81: 1203.

McMahon, L.P., Colucci, W.J., McLughlin, M.L., Barkley, M.D., 1992, Deuterium isotope effects in constrained Tryptophan derivatives: implication for Tryptophan photophysics, *J. Am. Chem. Soc.*, 114: 8442.

Orstan, A., Ross, A.J.B., 1987, Investigation of the b-Cyclodextrin-Indole inclusion complex by absorption and fluorescence spectroscopy, *J. Phys. Chem.*, 91: 2739.

Peters, K.S., Watson, T, Marr,K., 1991, Time-resolved photoacoustic calorimetry: a study of myoglobin and rhodopsin, *Ann. Rev. Biophys. Biphys. Chem.*, 20: 343.

Petrich, J.W., Chang, M.C., McDonald, D.B., Fleming, G.R., 1983, On the origin of nonexponential fluorescence decay in Tryptophan and its derivatives, *J. Am. Chem. Soc.*, 105: 3824.

Rudzky, J.E., 1985, The photochemical dynamics and energetics of rhodopsin and myoglobin, Ph.D. Thesis, Harvard University, Boston,MA

Rudzky, J.E., Goodman, J.L., Peters, K.S., 1985, Simultaneous determination of photoreaction dynamics and energetics using pulsed time-resolved photoacoustic calorimetry, *J. Am. Chem. Soc.*, 107: 7849.

Small, J.E., Hutchings, J.J., Small, E.W., 1989, Determination of the fluorescence quantum yields using pulsed-laser photoacoustic calorimetry, *SPIE Fluorescence detection III*, 1054: 26.

Szabo, A.G. and Rayner, D.M., 1980, Fluorescence decay of Tryptophan conformers in acqueous solutions, *J. Am. Chem. Soc.*, 102: 554.

Valeur, B. and Weber, G., 1977, Resolution of the fluorescence excitation spectrum of Indole into the 1LA and 1LB excitation bands, *Photochem. Photobiol.*, 25: 441.

Van Brederode, M.E., Gensch, T., Hoff, W.D., Hellingwerf, K.J., Braslavsky, S.E., 1995, Photoinduced volume change and energy storage associated with early transformations of the photoactive yellow protein from Ectothiorhodospira halophila, *Biophys. J.*, 68: 1.

Volkert, W.A., Kuntz R.R., Ghiron, C.A., Evans, R.F., 1977, Flash photolysis of Tryptophan and N-Acetyl-Tryptophanamide; the effect of bromide on transient yields, *Photochem. Photobiol*, 26: 3.

Willis, K.J., Szabo, A.G., Krajcarski, D.T., 1991, Excited state reaction and the origin of the biexponential fluorescence decay of Tryptophan zwitterion, *Chem. Phys. Lett.*, 182: 614.

Yurela, I., Churio, M.S., Gensch, T., Braslavsky, S.E., Holzwarth, A.R., 1994, Optoacoustic and singlet oxygen near-IR emission study of the isolated D1-D2-Cyt B559 reaction center complex of photosystem II. Protein movement associated with charge separation, *J. Phys. Chem.*, 98: 12789.

# 117

# TIME-RESOLVED FLUORESCENCE

## An Approach in Protein Analysis

A. Villari,[1] N. Micali,[2] M. Fresta,[3] S. Trusso,[2] and G. Puglisi[3]

[1] Dipartimento Farmaco-Chimico, Facoltà di Farmacia, Università di Messina
Viale Annunziata, 98010 Messina, Italy
[2] Istituto di Tecniche Spettroscopiche CNR Messina
Salita Sperone 31, Villaggio S. Agata, 98126 Messina, Italy.
[3] Istituto di Chimica Farmaceutica e Tossicologica, Università di Catania
Viale Andrea Doria n. 6, 95125 Catania, Italy

## ABSTRACT

This paper deal with the spectroscopic analysis of proteins, i.e., human serum albumin (HSA) and bovine serum albumin (BSA). These two proteins present very similar UV absorption and fluorescence spectra. The UV-spectra of both proteins result from the sum of the absorption spectra of the aromatic amino acid, mainly tryptophan and tyrosine, present in the biological molecules. On the contrary, the fluorescence HSA and BSA spectra are not the sum of the emission fluorescence spectra of tyrosine and tryptophan, but they are mainly constituted by the tryptophan fluorescence with a poor contribution arising from tyrosine. This fact may be due to inter- and intramolecular quenching phenomena. Significant difference were recorder in the life time fluorescence decay of HSA and BSA, which showed $\tau$ values of 2.3 and 4.5 nanoseconds, respectively. The time-resolved fluorescence spectroscopy was performed with an experimental apparatus assembled in our laboratories, which is able of measuring the fluorescence decay in the range $2\times10^{-10}$-$2\times10^{-8}$ sec.

## 1. INTRODUCTION

Spectroscopic analysis of proteins consist mainly in the analysis of three amino acids, i.e. tryptophan, tyrosine and phenylalanine, normally present in the primary structure of proteins (Hochstrasser and Johnson, 1988; Mattioli et al., 1991). In particular, quantitative determination of tryptophan in peptides and proteins involves methodological and analytical problems that have been the object of many investigation because this residue plays a crucial role in a number of peptides and proteins of biological interest (Teutlein et al., 1992; Breton et al., 1986; Creighton et al., 1988; Parson et al., 1990; Woodbury et al., 1985). The

*Recent Advances in Tryptophan Research*, edited by Graziella Allegri Filippini et al.
Plenum Press, New York, 1996

chromatographic procedures represent one of the most used method in the tryptophan analysis, which is often carried out after an hydrolysis process of the peptides. The high lability of tryptophan residues under the normal reaction conditions (Chang et al., 1983) requires special treatment (Jones et al., 1981; Simpson et al., 1986), which do not always ensure suitable results. In addition, the hydrolysis procedures, routinely employed, have frequently led to a non-quantitative recovery of the various amino acids (Jones et al., 1981), triggering an imprecision in the determination of the peptide content.

Another class of analytical methods consists of the spectrophotometric measurements of tryptophan after direct colour reaction (Horton and Koshland, 1965; Spies, 1967; Hassan, 1975; Molnàr-Perl and Pìnter-Szakàcs, 1989). These procedures, particularly those carried out onto intact peptides (Molnàr-Perl and Pìnter-Szakàcs, 1989), although more rapid than the methods based on hydrolysis and chromatography, are still tedious and time-consuming, requiring carefully controlled reaction conditions, long reaction time, and sometimes thermostabilization.

The direct determinations by UV-Vis spectrophotometric analysis in the 250-300 nm region consist in the measuring of a band arising from the overlapping of the bands of two different compounds, that is, tryptophan and tyrosine (Goodwin and Morton, 1946; Bencze and Schmid, 1957). All these methods need a considerably high amount of protein and are affected by other UV absorbing components, including other amino acids like phenylalanine. Consequently, mathematical methods are needed to resolve complex band contours into their individual components. Thus, second-derivative spectroscopy has been proposed for the determination of aromatic amino acids in proteins (Ballestrieri et al., 1978; 1980). It was also reported the determination of the tryptophan/tyrosine ratio with the possibility to obtain the absolute values by spectrophotometric titration (Servillo et al., 1982).

Although these techniques are quite suitable in the discrimination between the various aromatic amino acids (Bertini et al., 1993), are not sufficient for recognising similar protein or high molecular weight peptides. In fact, tryptophan, tyrosine and phenylalanine are almost ubiquitariously contained. In any case, the sensitivity is the main problem of the UV spectrophotometric analysis, which could be overcome by spectrofluorimetric determinations. The latter, even if much more sensitive than UV spectroscopy, is not able to discriminate among similar protein. Therefore, the characterisation of whole protein is very difficult and represent an arduous challenge. An attempt in the differentiation among similar proteins involves fine analytical methods. In fact, complications arise from the secondary, ternary and quaternary structure of proteins, which provided interaction among the various chromophores, such as energy transfer process. An example is provided from albumin of different origins.

We have investigated and characterised with a time resolved fluorescence analytical method two albumins of different origins: human albumin and bovine albumin. These albumin present similar UV-Vis and fluorescence spectra. The time-resolved fluorescence analysis provided different decay time as a function of the type of investigated albumin. Therefore, this fluorescence approach allowed not only the accurate and sensitive determination of the compounds, but also the differentiation between the two albumins, employing no mathematical (derivative method) procedure. A sensitive and accurate quali- quantitative analytical procedure, worth of further investigation, was employed.

## 2. THEORETICAL BACKGROUND

The fluorescence spectroscopy plays an important rule in the analysis of biological (Godik et al., 1993) and pharmaceutical compounds (Villari et al., 1992; 1994). The fluorescence phenomenon is based on two different processes: the absorption of a beam light

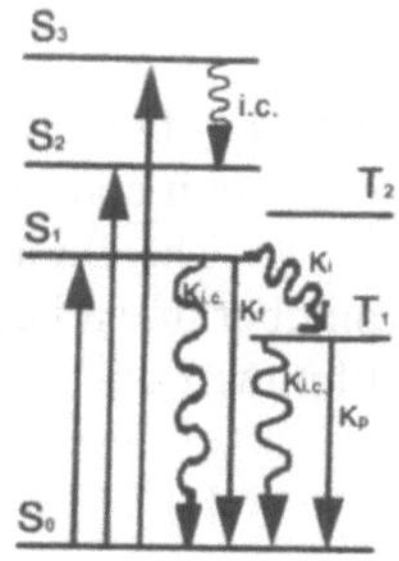


**Figure 1.** Schematic representation of light absorption and fluorescence emission. $\mathbf{S_0}$ = fundamental state; $\mathbf{S_1}$, $\mathbf{S_2}$ and $\mathbf{S_3}$ = excited states (singlet); $\mathbf{T_1}$ and $\mathbf{T_2}$ = excited states (triplet); **i.c.** = internal conversion; $\mathbf{K_f}$ = rate constant of the fluorescence process; $\mathbf{K_{i.c.}}$ = rate constant of the non-radiative decay; $\mathbf{K_i}$ = rate of inter-systemic exchange; $\mathbf{K_p}$ = rate constant of the phosphorescence process.

at a certain wavelength and the emission of light at wavelength higher than absorption (see Figure 1).

The absorption process is based on the energy transfer between the incident light and particular atomic groups present in the analysed molecule (chromophores). The energy is exchanged in finite amount, i.e., quantum energy. During this process, molecules can modify the lowest vibrational level of the electronic fundamental state ($S_0$), in which the couples of electrons present opposite spin, reaching in $10^{-15}$ sec an excitation state ($S_1$) or ($S_2$). These excitation state are reached with electron movement from bond-molecular orbitals to higher energy orbitals in which the original electron parity and spin direction are maintained.

The absorption spectra should be constituted by a number of well distinct band, one for each vibrational level. Practically, organic compounds present absorption spectra with very broaden bands owing to the presence of the rotational levels accompanying each vibrational level. For these reasons, the specificity of the UV-Vis spectra among different organic compounds with identical chromophore groups is very low.

After the energy absorption, the emission process begins. In this case the molecule loose rapidly energy to reach again the fundamental state. Part of the vibro-rotational energy of the molecule is lost by means of molecular collisions and/or internal conversion, obtaining a lower level of molecular energy ($S_1$). This process takes place in about $10^{-12}$ sec, without energy emission. The following step, in which the fundamental state is reached (from $S_1$ to $S_0$), takes place in about $10^{-9}$-$10^{-6}$ sec with energy emission, the so called luminescence.

A strong correlation exists among molecular structures, absorption spectra and fluorescence emission spectra. Namely, similar molecular structures with similar absorption spectra provide similar fluorescence emission spectra owing to particular electronic arrangements. Thus, the possibility of measuring the lifetime of the fluorescence excited state may offer a tool to discriminate among molecules with a similar structure, while maintaining high sensitivity in molecule determination (fluorescence technique).

The fluorescence lifetime $\tau$ of $N_0$ excited states is defined as the mean lifetime of the excited state, that is, the probability of finding $N_0/e$ excited states still in the excited state after time $t$. This condition is verified when only a fluorescence decay mechanism exists. In such a system each excited fluorophore is characterised by a probability $h$ of surviving in the excited state at a time $t$ independent from $t$ itself. In this case the number of excited states varies as a function of the time, in agreement with the following equation:

$$N=N_0 \exp(-ht) \qquad \text{(eq.1)}$$

where $N$ is the number of excited states at time $t$, while $N_0$ is the number of excited states at time 0. The endurance of the excitation is different from a fluorophore to another, but the mean endurance of the excited state $\tau$ will be determined, being $\tau \alpha\ 1/h$. Thus, the mean excited lifetime equals:

$$\tau = 1/[k_2+K_{n2}+(K_q[Q])] \quad \text{(eq.2)}$$

where $k_2$ the radiative decay constant, $k_{n2}$ the non-radiative decay constant, $k_q$ bimolecular extinction constant and $[Q]$ molar concentration of extinguishing agents.

The fluorescence intensity obeys the same law, because it is proportional to the number of excited states. The general equation relating the fluorescence intensity $I$ and the lifetime $\tau$ is:

$$I = I_0\, e^{-t/\tau} \quad \text{(eq.3)}$$

where $I$ is the fluorescence intensity at time $t$, $I_0$ is the maximum fluorescence intensity during excitation, $t$ is the time after removing source of excitation, and $\tau$ is the average lifetime of the excited state.

Precise measurements of the fluorescence lifetime is important, since it can be used to calculate the natural lifetime $\tau_0$ or the absolute quantum efficiency $\phi_0$, if one or the other is known:

$$\tau = \phi_0\, \tau_0 \quad \text{(eq.4)}$$

The availability of the picosecond-flash fluorometers allows the direct measurement of $\tau$. These apparata are more precise than any other previously made.

## 3. EXPERIMENTAL APPARATUS

The compounds that we have analysed presented fluorescence lifetime even shorter than $10^{-8}$ sec. In order to make possible a precise and accurate determination of the various fluorescence lifetime, an experimental apparatus capable of measuring the fluorescence decay in the range $2\times10^{-10}$-$2\times10^{-8}$ sec (200 ps - 20 ns) was assembled in our laboratories. A schematic representation is reported in Figure 2.

The experimental apparatus is made up of a pulsed laser source and a coincidence electronic system TAC (Time Amplitude Converter) as detector. The excitation system is constituted of a Argon ion mode-locked laser operating at the wavelength of 514.5 nm with an impulse width of 150 ps at a repetition rate of 82 Mhz. The impulse train excites in a synchronous way a dye laser (Rhodamine 6 G), which has an output wavelength of ~600 nm (changing suitably the dye in the range 540-740 nm) and an impulse width of 2 ps. Finally,

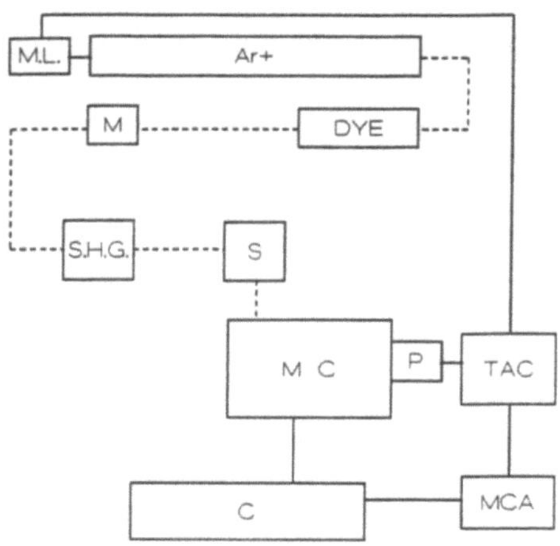


**Figure 2.** The experimental apparatus for the measurement of the fluorescence lifetimes was assembled in our laboratories. Abbreviation of the various parts: **MC** = Mode-locked system; **Ar$^+$** = Argon ion laser; **DYE** = Dye laser; **M** = Mirror; **SGH** = Second armonic generator crystal; **S** = sample; **MC** = monochromator; **P** = fast photomultiplier; **TAC** = Time amplitude converter; **MCA** = Multichannel analyzer; **C** = computer.

a KDP crystal by means of a second harmonic generator provided the UV-excitation wavelength at ~300 nm (270-370 nm).

The detection system is made up of a monochromator with a fast photomultiplier set at the wavelength of fluorescence emission. The impulse generated by the photomultiplier works as start for the clock of the TAC, while the electronic synchronization of the mode-locking works as stop. In this way the time between the excitation and the detection of the first fluorescence photon is measured. Then, by using a multichannel, it is possible to obtain the distribution of the number of photons as a function of time. The $\tau$ values were obtained by deconvolving the instrumental response, obtained by measuring the apparent decay of Rayleigh scattering, from the measured decay functions.

## 4. RESULTS AND DISCUSSION

The fluorescence decay time $\tau$ is a significant parameter of fluorescence, but the direct measurement of this quantity was difficult up to now since, for organic molecules, $\tau$ is in the order of 1 to 50 ns. Most $\tau$ values of organic compounds have been obtained in non aqueous solvents, such as cyclohexane, which favour high quantum yields and are suitable for studying highly structured spectra. Few decay times have been measured in water, especially those whose emission occurs in the ultraviolet region. There has been great interest in fluorescence and other manifestation of excited electronic states in biological systems (Godik et al., 1993), not only because these state occurs naturally (as in the case of photosynthesis and firefly luminescence) but also because luminescence techniques are valuable for studying the structure of biopolymers and their interaction with small molecules (Mantsch et al., 1988). Therefore, it is important to measure all basic fluorescence properties, such as $\tau$, in aqueous solutions with equipment sensitive to UV-radiation, since many biological materials, such as proteins, emit in this region. We now report the spectroscopic characterization and the $\tau$ values of two albumins of different origin: human serum albumin (HSA) and bovine serum albumin (BSA).

The UV-absorption spectra of HSA and BSA seem to be the sum of the absorption spectra of the various aromatic amino acids, mainly tyrosine and tryptophan (Figure 3). Unlike the UV-absorption spectra, the fluorescence spectra of these two proteins are not the simple sum of the contributions of the aromatic amino acids in neutral water solutions (Figure 4). This phenomenon could be ascribed to the fact that tyrosine, tryptophan and phenylalanine show characteristic fluorescence in water solution. In the protein molecules the environment of the aromatic residues is modified in several ways: the surrounding medium is less polarizable than water, rotational freedom is restricted and adjacent groups of the polypeptide framework may take part in specific quenching processes. Furthermore, the average distance among the aromatic residues is about 10 Å, a distance at which energy transfer is know to take place in other systems.

The examination of the fluorescence spectra (Figure 4) of HSA and BSA, containing both tyrosine and tryptophan, has shown no strong indication of tyrosine fluorescence. The use of a deconvolution method (Wuite, 1959) in the analysis of the spectrofluorimetric data showed the presence of two components alone in the albumin spectra. One component arose from the fluorescence emission of tryptophan. The second component may be shown to have a maximum of emission at 303 ± 2 nm and a quantum yield of 0.018-0.022. The second component is excited with a lesser extent than the main tryptophan fluorescence by wavelength shorter or longer than 280 nm; thus, the emission spectrum could be identified as due to tyrosine. The scarce contribution of tyrosine in the albumin fluorescence may be probably due to quenching effects by processes arising from secondary bonding involving the phenol

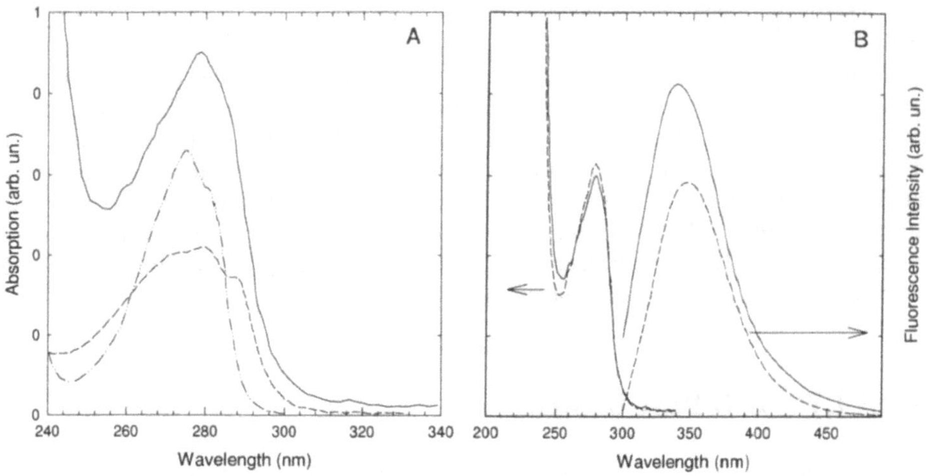


**Figure 3.** UV and fluorescence spectroscopic characterization of HSA and BSA. In part A it is reported the UV spectrum of HSA (——), which results from the sum of the absorption spectra of tyrosine (- - - -) and tryptophan (–•–•–); the same result was obtained for BSA (data not reported). In part B the UV absorption spectra and the emission fluorescence spectra of HSA (——) and BSA (- - - -) are shown. The UV absorption spectra were recorded with a Uvikon 820 spectrophotometer (Kontron Instrument, Zurich). The fluorescence spectra were recorded with a Hitashi F-2000 spectrofluorimeter.

group. Also in our experiments a poor fluorescence contribution of tyrosine was observed, as the excitation wavelength was 295 nm (beyond the absorption $\lambda_{max}$ of tyrosine).

It is worth noting that no appreciable difference was observed in the UV and fluorescence spectra of HSA and BSA (Figure 3b), which presented bands centred at 279 and 342, and 278 and 343nm, respectively. The absence of spectroscopic differences is accompanied by a difference in the structure of HSA and BSA, which are constituted by 609 and 607 amino acids, respectively. Another difference lies in the quali- quantitative compo-

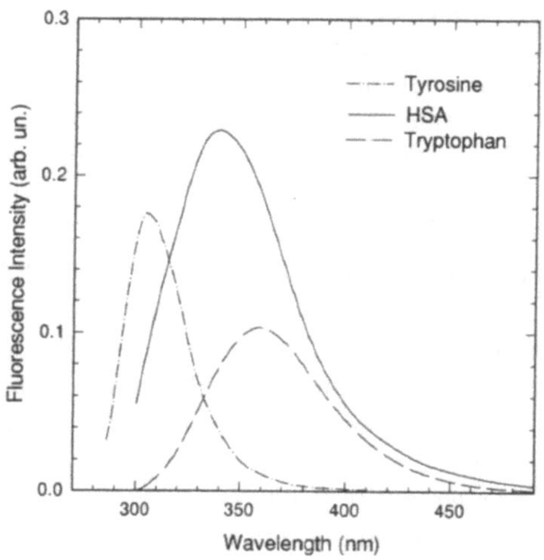


**Figure 4.** Fluorescence spectra of HSA (——), tyrosine (–•–•–) and tryptophan (- - -) in aqueous solution.

**Table 1.** Differences in the amino acids constituting the human serum albumin (HSA) and the bovine serum albumin (BSA)

| Sample | Number of amino acids | Percentage of tryptophan | Percentage of tyrosine |
|---|---|---|---|
| HSA | 609 | 0.33 | 3.12 |
| BSA | 607 | 0.50 | 3.46 |

sition of the various amino acids, and particularly in those of the aromatic ones (tryptophan and tyrosine), present in the two serum albumins, as summarized in table 1.

Common spectroscopic investigations were not able to find out a significant difference between HSA and BSA, while with our experimental set-up it was possible to evidence a remarkable difference in the fluorescence lifetime of HSA and BSA (Figure 5). The experimental values of the fluorescence lifetime of HSA and BSA were 2.3 ns and 4.5 ns, respectively. These values are very different from those previously reported (Chen et al., 1967) for HSA ($\tau$ = 4.5 ns) and for BSA ($\tau$ = 4.6 ns). This difference is probably due to the experimental apparatus used in the fluorescence lifetime measurement, that is, our apparatus is much more sensitive, accurate and precise than those commonly available. In fact, up to now, no significant difference was reported among the various albumins of different origins. Considering that many fluorescence lifetimes of organic compounds and proteins fall in the lower detection limit of the common instruments, a levelling effect is observed among similar substances, whose $\tau$ values difference cannot be discriminated. Evidently, the differences in the structure and composition of HSA and BSA lead to different fluorescence lifetimes, likely generated not only by the quali- quantitative differences but also by the different spatial

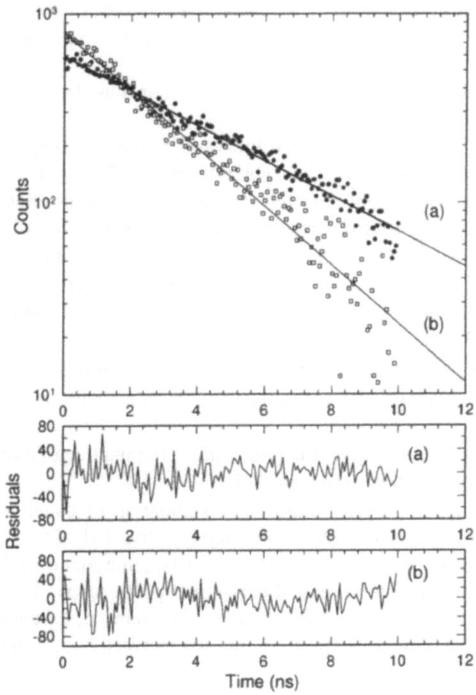


**Figure 5.** Fluorescence decay of HSA (a) and BSA (b). The continuous line is the best fit Eq. 3.

disposition of the aromatic amino acids eliciting different phenomena of energy transfer and fluorescence quenching.

Then, the determination of $\tau$ values with a suitable experimental apparatus represent a sensitive, precise and accurate quali- quantitative analytical procedure for the characterization of organic and biological compounds. With regards to biological application, besides this paper, the time-resolved fluorescence spectroscopy was employed in the characterization of protein moiety involved in the photosynthetic reactions (Godik et al., 1993). This kind of studies are worth of further investigations.

## 5. REFERENCES


Ballestrieri C., Colonna G., Giovane A., Irace G., and Servillo L., 1978, Amino acid composition and physico-chemical properties of bluefin tuna (Thunnus thynnus) myoglobin, *Eur. J. Biochem.* 90:433.

Ballestrieri C., Colonna G., Giovane A., Irace G., and Servillo L., 1980, Second-derivative spectroscopy of proteins: studies on tyrosyl residues, *Anal. Biochem.* 106:49.

Bencze W.L., and Schmid K., 1957, Deth of tyrosine and tryptophan in proteins, *Anal. Chem.* 29:1193.

Bertini J., Mannucci C., Noferini R., Perico A., and Rovero P., 1993, Rapid simultaneous determination of tryptophan and tyrosine in synthetic peptides by derivative spectroscopy, *J. Pharm. Sci.* 82:179.

Breton J., Martin J.L., Migus A., Antonetti A., and Orsag A., 1986, Femtosecond spectroscopy of excitation energy transfer and initial charge separation in the reaction centre of the photosynthetic bacterium *Rhodopseudomonas viridis*, *Proc. Natl. Acad. Sci. USA* 83:5121.

Chang J.Y., Knecht R., and Braun D.G., 1983, Isolation and sequence determination of polypeptides at the picomole level, *Methods Enzymol.* 91:41.

Chen R.F., Vurex G.G., and Alexander N., 1967, Fluorescence decay times: proteins, coenzymes, and other compounds in water, *Science* 156:949.

Creighton S., Hwang J.K., Warshell A., Parson W.W., and Norris J., 1988, Simulating the dynamics of the primary charge separation process in bacterial photosynthesis, *Biochemistry* 27:774.

Godik V.I., Blankenship R.E., Causgrove T.P., and Woodbury N., 1993, Time-resolved tryptophan fluorescence in photosynthetic reaction centres from *Rhodobacter sphaeroides*, *FEBS Letters* 321:229.

Goodwin T.W., and Morton R.A., 1946, Deth of carotene and vitamin A in butter and margarine, *Biochem. J.* 40:628.

Hassan S.S.M., 1975, New spectrophotometric method for simultaneous determination of tryptophan and tyrosine, *Anal. Chem.* 47:1429.

Hochstrasser R.M. and Johnson C.K., 1988, Biological Processes Studied by Applied Physics, *in*: Ultrashort Laser Pulses and Application, W. Kaiser, Ed., vol. 60, Springer-Verlag, Berlin, p. 357.

Horton H.R., and Koshland D.E., 1965, Highly reactive coloured reagent with selectivity for the tryptophan residue in proteins -2-hydroxy-5-nitrobenzyl bromide, *J. Am. Chem. Soc.* 87:1126.

Jones B.N., Paabo S., and Stein S., 1981, Amino acid analysis and enzymatic sequence determination of peptides by an improved o-phtaldialdehyde percolum labelling procedure, *Liq. Chromatogr.* 4:565.

Mantsch H.H., Moffatt D.J., and Casal H.L., 1988, Fourier transform methods for spectral resolution enhancement, *J. Mol. Struct.* 173:285.

Mattioli T.A, Gray K.A., Lutz M., Oesterhelt D., and Bruno R., 1991, Resonance raman characterization of *Rhodobacter sphaeroides* reaction centres bearing site-directed mutations at tyrosine M210, *Biochemistry* 30:1715.

Molnàr-Perl I., and Pintèr-Szakàcs M., 1989, Spectrophotometric determination of tryptophan in intact proteins by the acid ninhydrin method, *Anal. Biochem.* 177:16.

Parson W.W., Chu Z., and Warshell A., 1990, Microscopic simulation of quantum dynamics and nuclear funnelling in bacterial reaction centres, *Biochim. Biophys. Acta* 1017:251.

Servillo L., Colonna G., Ballestrieri C., Ragone R., and Irace G., 1982, Simultaneous determination of tyrosine and tryptophan residues in proteins by second-derivative spectroscopy, *Anal. Biochem.* 126:251.

Simpson R.J., Moritz R.L., Nice E.C., Grego B., Yoshizaki F., Sugimura Y., Freeman H.C., and Murata M., 1986, Complete amino acid sequence of plastocyanin from a green alga, *Enteromorpha prolifera*, *Eur. J. Biochem.* 157:497.

Spies J.R., 1967, Allergeus (XIX) no. of antigens and the homogeneity of the isolated antigens of fraction CB-1A from castor beans, *Anal. Chem.* 39:1412.

Teutlein H., Shulten K., Deisenhofer J., Michel H., Brunger A., and Karplus M., 1992, *in*: The Photosynthetic Bacterial Reaction Centre, Structure and Dynamics, NATO ASI Series A 149, J. Breton and A. Vermeglio, Eds., Plenum Press, New York, p. 139.

Villari A., Micali N., Fresta M., and Puglisi G., 1992, Simultaneous spectrophotometric determination in solid phase of aspirin and its impurity salicylic acid in pharmaceutical formulations, *J. Pharm. Sci.* 81:895.

Villari A., Micali N., Fresta M., and Puglisi G., 1994, Spectrofluorimetry at zero angle: determination of salicylic acid in acetylsalicylic acid pharmaceutical formulations, *The Analyst* 119:1561.

Woodbury N.W., Becker M., Middendorf P., and Parson W.W., 1985, Nanosecond fluorescence from isolated photosynthetic reaction centres of *Rhodopseudomonas sphaeroides*, *Biochemistry* 24:7516.

Wuite A., 1959, Effect of pH on fluorescence of tyrosine, tryptophan and related compounds, *Biochem. J. 71:217.*

# FLUOROMETRIC DETERMINATION OF L-KYNURENINE WITH GLYCOLALDEHYDE BY HIGH PERFORMANCE LIQUID CHROMATOGRAPHY

H. Iizuka, H. Sugano, and T. Yajima

School of Pharmaceutical Sciences
Toho University
2-2-1Miyama, Funabashi, Chiba 274, Japan

L-Kynurenine (KYN) is one of the metabolites of L-tryptophan(Trp) on the kynurenine pathway. The first step of the kynurenine pathway is an oxidative cleavage of the indole ring of Trp, which is catalyzed by tryptophan 2, 3-dioxygenase (EC 1.13.11.11, TDO) to form N-formyl-L-kynurenine. In mammalian system, TDO is found only in liver and N-formyl-L-kynurenine is hydrolyzed instantly to KYN by an excess of formamidase present in liver. Therefore, in many cases assay of the TDO activity is based on the determination of KYN formed by TDO. We have developed a new fluorometric method for the determination of KYN by a single-pump high performance liquid chromatography (HPLC) using glycolaldehyde(GLA) as fluorogenic reagent. The method was successfully applied to the assay of the TDO activity.

## EXPERIMENTAL

### Materials

Glycolaldehyde dimer was purchased from Fluka (Buchs, Switzerland); L-tryptophan was from Peptide Institute Inc.(Osaka, Japan). L-Kynurenine, methemoglobin(bovine), L-methyl-dopa(MDP), other tryptophan analogs were from Sigma Chemical Co.(St. Louis, MO, USA). All other reagents and solvents were of analytical reagent grade.

### Apparatus

The HPLC system was as follows: pump: Shimadzu LC-9A; injector: Rheodyne Model 8125 (5 μl loop); detector: Shimadzu RF-535 (Ex. 240 nm, Em. 405nm); column: Nucleosil 5C18 (2.1 mm I.D.× 15 cm). A jointless stainless steel tube (0.25 mm I.D.) was used as a reaction coil (30 m) and a cooling coil (3m) which were kept at 100°C and at room temperature, respectively. An eluent consisted of 4% acetonitrile buffer solution (pH6) containing 6 mM GLA. The flow rate was maintained at 0.1 ml/min.

## Crude Extracts

Male Wister rats (about 200g, 8 weeks old) were stunned, bled from the neck, and the livers were removed after perfusion with saline in situ. The livers were immediately homogenized or frozen. Crude extracts were prepared as follows: Minced liver tissue(3 g tissue per 10 ml) was homogenized with a Teflon homogenizer kept in a cold bath, and the homogenate was centrifuged for 15 min at 10,000 r.p.m.(4°C). To the supernatant was added 63 g/l (45% saturation) of ammonium sulfate and the solution was centrifuged under the same condition mentioned above.

The precipitate was dissolved in 0.01M sodium phosphate (pH 7) containing 1 mM Trp and dialyzed for about 18 hours against the same buffer solution (Schimke et al., 1965). All samples were stored at -35°C until use.

## Standard Assay of TDO Activity

The conditions for the assay are the modification of the methods of Metzler et al. (1982) and Seglen and Jervell (1969). Briefly, 3.8 ml of 0.02M sodium phosphate(pH7) containing 2.5mM Trp and 0.136 mg/ml methemoglobin were preincubated reaction for 10 min at 37°C. Two hundred ml of the crude extracts was added to a preincubated mixture and the mixture was incubated for additional 60 min at 37°C with shaking. The enzyme reaction was terminated by adding 1 ml of 25 % trichloroacetic acid. After removal of the protein precipitate by centrifugation for 10 min at 3000 r.p.m., 4 ml of the supernatant was neutralized with 1N NaOH and water was added to the solution to make a total volume of 5.5 ml. Five hundred μl of 0.5 μM MDP as an internal standard was added to 1 ml of the neutralized solution and the mixture was made up to 5 ml with water. A 5 μl portion of the resulting solution was injected into HPLC for analysis.

# RESULTS AND DISCUSSION

## Eluent

Six mM GLA in the eluent gave the highest peak of KYN. A base-line separation of Trp metabolites (Fig. 1) was obtained with 5% acetonitrile, and the peak of MDP was separated from a shock of injection using 4% acetonitrile. As the fluorescent reaction between GLA and KYN was found to be optimum at pH6 by a manual method, various buffers of pH6 were examined by an HPLC method. Buffers examined are listed in Table 1. Among the buffers examined, best result was obtained either with 0.1M citric acid-0.2M $Na_2HPO_4$ buffer or with 0.15M citric acid-0.3M $Na_2HPO_4$ buffer. In this experiment, further studies were carried out with 0.1M citric acid-0.2M $Na_2HPO_4$ buffer.

## Working Curve

A good linear working curve was obtained between 0.5 pmol/5μl and 500 pmol/5 μl. The correlation coefficient was caluculated to be 0.999. A detection limit was 0.2 pmol/5 μl(S/N=3), and the relative standard deviation at 25 pmol/5μl of KYN was 3.2%(n=6).

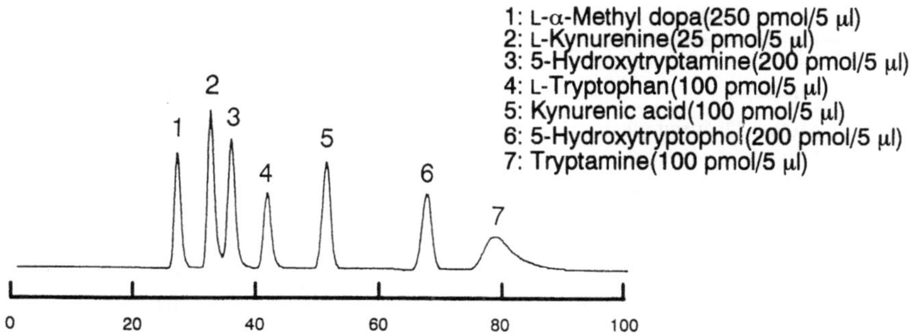


**Figure 1.** Typical chromatogram of kynurenine, tryptophan and their major metabolites.

## Typical Chromatogram

Fig.1 shows a typical chromatogram of the mixture of KYN, Trp and its major metabolites. Peaks of 5-hydroxy-L-tryptophan(5HTP) and anthranilic acid were not separated from that of KYN. Xanthurenic acid produced two peaks (retention time: 44 min and 49 min), and the retention time of the major peak (49 min) was almost the same as that of kynurenic acid.

## TDO Activity

It was shown that rat liver TDO activity could be assayed by determining KYN by the proposed method. KYN formation was found to be linear up to 90 min of incubation after 10 min lag period. The Michaelis constant was calculated to be 278 μM for Trp, which was in good agreement with the value reported in the literature (Cho-Chung and Pitot, 1967; Feigelson and Brady, 1974). There was no formation of KYN when the crude extract was incubated with D-tryptophan or with Trp in the presence of trichloroacetic acid.

## Effects of Tryptophan Analogs on TDO Activity

Effects of various tryptophan analogs on TDO activity are summarized in Table 2. Among these compounds, 5HTP was the most effective inhibitor. With regard to 5-fluoro-DL-tryptophan, it could also serve as substrate and its metabolite was detected in the reaction mixture. However, the metabolite could not be separated well from KYN under the conditions employed. Niether 5HTP nor 5-methoxy-DL-tryptophan could be a substrate in spite of the presence of a ring substitution at 5 position. It is inferred that 5HTP is inactive as

**Table 1.** Buffers examined

| Buffer (pH6) | RPH* |
|---|---|
| 0.1 M $Na_2HPO_4$–$KH_2PO_4$ | 100 |
| 0.15 M $Na_2HPO_4$–$KH_2PO_4$ | 93 |
| 1/15 M $Na_2HPO_4$–$KH_2PO_4$ | 141 |
| 0.05 M Citric acid–0.1 M $Na_2HPO_4$ | 330 |
| 0.1 M Citric acid–0.2 M $Na_2HPO_4$ | 413 |
| 0.15 M Citric acid–0.3 M $Na_2HPO_4$ | 415 |
| 0.1 M Disodium citrate–0.1 M NaOH | 259 |

*Relative peak height.

**Table 2.** Effects of tryptophan analogs on TDO activity

| Compounds | Concn. (mM) | Inhibition (%) | Metabolite retention time |
|---|---|---|---|
| 1-Methyl-DL-tryptophan | 0.4 | 6.1 | ND |
| 4-Methyl-DL-tryptophan | 0.25 | 7.8 | ND |
| 5-Methyl-DL-tryptophan | 0.4 | 2.5 | 50 min |
| 6-Methyl-DL-tryptophan | 0.4 | 16.6 | 50 min |
| 7-Methyl-DL-tryptophan | 0.4 | 11.6 | ND |
| 5-Hydroxy-L-tryptophan | 0.02 | 32.1 | ND |
| 5-Methoxy-DL-tryptophan | 0.25 | 14.0 | ND |
| 5-Fluoro-DL-tryptophan | 0.25 | 12.3 | 30 min |
| N-Acetyl-L-tryptophan | 0.25 | 10.2 | ND |
| L-Abrine | 1.0 | 8.2 | ND |
| N-Formyl-DL-tryptophan | 0.4 | 10.0 | ND |

L-Abrine: N-Methyl-L-tryptophan.
ND: Not detected.
Kynurenine retention time: 30 min.

substrate of TDO due to a specific interaction between its hydroxy group and the surrounding amino acid side chain of the enzyme (Uchida et al., 1992).

## Ammonium Sulfate Fractionation

In order to examine the effect of 5-methyl-DL-tryptophan (5MTP) and 6-methyl-DL-tryptophan (6MTP) on TDO further, a partial purification of TDO was attempted. A 35-45% ammonium sulfate fraction was chosen as enzyme source because it showed sufficient specific activity and stability for the purpose. It was found that the enzyme activity of the ammonium sulfate fraction was increased by adding ascorbic acid to the incubation mixture. When the ammonium sulfate fraction was used, freshly dissolved 7.5 mM ascorbic acid in 0.02M phosphate buffer (pH7.0) and neutralized with 1.4M NaOH was added (Knox et al., 1970) so that TDO was activated by reduction of heme (Frieden et al., 1961). On the other hand, the addition of ascorbic acid was not necessary for the crude extracts. It seems that the heme of TDO in the ammonium sulfate fraction was oxidized during a purification process.

## 5MTP and 6MTP

Table 3 shows the effects of concentration of 5MTP and 6MTP on TDO with the ammonium sulfate fraction. It was found that the higher the concentration of 5MTP and 6MTP, the higher their metabolites peaks. There was no formation of their metabolite peaks when TDO was incubated in the presence of trichloroacetic acid. Therefore, it is believed that the obtained peaks correspond to those of analogs of KYN formed by the enzyme reaction with the Trp analogs as substrates.

**Table 3.** 5MTP and 6MTP as inhibitors

| Concn. | Inhibition (%) | | Metabolite peak height | |
|---|---|---|---|---|
| (mM) | 5MTP | 6MTP | 5MTP | 6MTP |
| 0.2 | 3.56 | 10.4 | 17 | 21 |
| 0.4 | 4.44 | 14.2 | 45 | 59 |
| 0.6 | 7.74 | 18.9 | 55 | 85 |

# CONCLUSION

A highly sensitive fluorometric method for the determination of KYN was established. Moreover, the proposed method was also shown to be the highly sensitive metod for the assay of TDO. The metod was sensitive enough to detect the enzymatic reaction products formed from the compounds known as nhibitors of TDO. It is also feasible to apply the proposed method for the assay of indoleamine 2,3-dioxygenase (IDO).

# REFERENCES

Cho-Chung Y.S. and Pitot H.C., 1967, Feedback control of rat liver tryptophan pyrrolase. I. End products inhibition of tryptophan pyrrolase activity, *J. Biol. Chem.*, 242: 192.

Feigelson P. and Brady F.O., 1974, " Molecular Mechanisms of Oxygen Activation," ed.by O.Hayaishi. Academic Press, New York

Frieden E., Westmark G.W. and Schor J.M., 1961, Inhibition of tryptophan pyrrolase by serotonin, epinephrine, and tryptophan analogs, *Arch. Biochem. Biophys.*, 92: 176.2)

Knox W.E., Yip A. and Reshef L., 1970, L-Tryptophan 2,3-dioxygenase (tryptophan pyrrolase) (rat liver), in: " Methods in Enzymology," Vol. 17A, ed. by H. Tabor, C.W. Tabor, Academic Press, New York, pp. 415-428.

Metzler H., Gebhardt R., Oberrauch W. and Mecke D., 1982, A convenient and highly sensitive spectrophotometric assay for tryptophan 2,3-dioxygenase, *Anal. Biochem.*, 121:10.

Schimke R.T., Sweeney E.W,. and Berlin C.M., 1965, The roles of synthesis and degradation in the control of rat liver tryptophan pyrrolase, *J. Biol. Chem.*, 240: 322.

Seglen P.O. and Jervell K.F., 1969, Tryptophan oxygenase activation and ascorbate oxidation in whole homogenates from rat liver, *Biochim. Biophys. Acta.*, 171: 47.

Uchida K., Usami M., Bandow H. and Harada I., 1992, Characteristics of substrates and inhibitors in binding to rat liver L-tryptophan 2,3-dioxygenase: a Fourier transform infrared and kinetic study, *Biochim. Biophys. Acta.*, 1121: 153.

# MASS SPECTROMETRIC STUDY OF 5,6- AND 5,7- DIHYDROXYTRYPTAMINES AND THEIR BIOLOGICAL PRECURSORS

A. Bertazzo,[1] S. Catinella,[2] and P. Traldi[2]

[1] Department of Pharmaceutical Sciences
University of Padova
Via Marzolo 5, I-35131, Padova, Italy
[2] CNR
Area della Ricerca, Padova, Italy

## INTRODUCTION

It has been proposed that a defect in the metabolism of 5-hydroxytryptamine or serotonin (5-HT) leads to the formation of more highly hydroxylated indoleamines (Udenfriend et al., 1956; McIsaac and Page, 1959; Eriksen et al., 1960) such as 5,6- and 5,7-dihydroxytryptamine (5,6-DHT and 5,7-DHT), powerful serotoninergic neurotoxins (Baumgarten et al., 1971; 1972a; 1972b; Baumgarten and Lachenmeyer, 1972; Bjorklund et al., 1973a; 1973b; Saner et al., 1974). The discovery of this oxidative pathway from 5-HT may provide a chemical basis for understanding certain neurological and psychotic disorders.

It is generally assumed that neuromelanin originates from catecholamines. However, the implication of tryptamines in neuromelanogenesis cannot be excluded *a priori*. For this reason, we undertook this preliminary study on the mass spectrometric behaviour of tryptamines, for information on the intrinsic reactivity of these substrates to explain neuromelanin formation.

The present study reports the mass spectrometric behaviour of tryptamine (**1**), 5-HT (**2**), 5,6- and 5,7-DHT (**3** and **4** respectively) as obtained by various ionization methods, i.e., electron ionization (EI), fast atom bombardment (FAB), and mass analysed ion kinetic energy (MIKE) spectroscopy. Under EI and FAB conditions, tryptamine and 5-HT generate abundant molecular ions whose related MIKE spectra show numerous fragments clearly related to the structure of the neutral molecule.

Regarding the two isomeric 5,6- and 5,7-dihydroxyindoles, little information could be achieved by applying the above techniques. In fact, under FAB conditions, overlapping spectra were obtained, while their EI spectra mainly differed in terms of pyrolysis products. Thus, in order to distinguish these ionic isomers better, tandem mass spectrometry experiments were performed by ion-trap mass spectrometry (ITMS) together with energy-resolved mass spectrometric measurements (ERMS). With this approach, clear differences were revealed between 5,6- and 5,7-dihydroxyindoles.

*Recent Advances in Tryptophan Research*, edited by Graziella Allegri Filippini et al.
Plenum Press, New York, 1996

# MATERIALS AND METHODS

Compounds **1-4** were purchased from Sigma Chimica (Sigma-Aldrich, Milano, Italy). EI and FAB measurements were performed on a VG ZAB 2F instrument (Morgan et al., 1979). The electron impact spectra were obtained with an electron energy of 70 eV (220 mA) and with a source temperature up to 300 °C; fast atom bombardment (Barber et al., 1981) spectra were obtained bombarding glycerol solutions of **1-4** with 8 keV Xenon atoms. Metastable ion studies were performed by mass analysed ion kinetic energy (MIKE) spectroscopy (Cooks et al., 1973).

Energy-resolved mass spectra were obtained using an ion trap mass spectrometer (ITMS) (Ardanaz et al., 1991) (Finnigan MAT, Ca, USA) working in presence of helium as buffer gas until to reach a final pressure of $1.0 \times 10^{-4}$ torr and with a trap temperature of 120°C.

The selected ions were isolated by a two-step procedure (Todd, 1991) by application of consecutive positive and negative DC voltages.

Collisional experiments were performed by application of a supplementary A.C. voltages ("tickle voltage") on the two end-caps electrodes with a voltage ranging from 500 to 750 mV.

# RESULTS

## EI Mass Spectrometry

The mass spectra of compounds **1-4** are reported in Table 1. The MIKE spectra of $M^{+\cdot}$ species (when present) are reported in Figure 1.

Tryptamine (**1**) undergoes the fragmentation pattern shown in Scheme 1. The most abundant peak is at m/z 131 and originates from $CH_2$-$CH_2$ bond cleavage with H rearrangement, leading to the molecular ion of 3-methylindole. The fragment due to simple cleavage is also present at m/z 130. A further primary fragment, favoured for molecular ion with low internal energy (as detected by MIKE spectra) is that at m/z 133, originating from cleavage of the indole ring with loss of HCN. Further less abundant ions are due to the loss of the whole side-chain giving rise to the ions at m/z 116.

In the case of serotonin (**2**), analogous fragmentation channels are activated. The ions due to $CH_3N$ and $CH_4N$ losses, leading to ions at m/z 147 and 146, are still present and are

**Table 1.** EI mass spectra of compounds **1-4***

| | |
|---|---|
| **1.** | 160 (20), 132 (7.5), 131 (80.2), 130 (100), 129 (6.3), 128 (4), 117 (2.2), 115 2.2), 103 (13.6), 102 (9), 101 (2.1), 89 (2.2), 78 (2.6), 76 (22.1), 75 (4.5), 73 (2.3), 63 (3.8), 52 (2.5), 51 (5.9) |
| **2.** | 176 (28), 174 (2.4), 160 (2), 158 (2.7), 148 (10.2), 147 (78.4), 146 (100), 145 (7), 133 (2.1), 131 (12.2), 129 (3.0), 119 (4.9), 117 (8.9), 116 (2.1), 105 (4.6), 104 (2), 102 (2.5), 91 (10.6), 90 (4.3), 89 (6.8), 81 (6.2), 80 (6.2), 79 (2.7), 78 (2.5), 77 (8.0), 76 (2.1), 75 (2.4), 65 (9.4), 64 (3.6), 63 (7.3), 62 (2.8), 57 (2.8), 55 (3.0), 53 (2.9), 52 (4.7), 51 (7.4), 50 (4.5) |
| **3.** | 192 (25), 191 (3.0), 186 (7.0), 178 (5.7), 177 (4.5), 176 (4.8), 171 (3.7), 170 (4.2), 164 (5), 163 (66.4), 162 (100), 161 (7.1), 150 (8.8), 149 (70), 148 (11.4), 146 (6.9), 133 (6.8), 132 (8.0), 131 (5.2), 130 (26.3), 129 (4.1), 120 (9.3), 119 (4.3), 118 (7.0), 116 (12), 114 (90), 113 (60), 111 (95), 105 (16.7), 103 (33.7), 99 (6.7), 97 (32.1), 96 (7.1), 93 (16.2), 92 (6.2), 91 (9.1), 89 (12.2), 86 (16), 84 (66.4), 83 (92), 81 (10.1), 78 (15.1), 76 (87), 75 (31.2) |
| **4.** | 131 (2.2), 130 (6.8), 114 (100), 112 (55.7), 105 (6.1), 104 (4.9), 99 (3.0), 98 (6.4), 96 (3.7), 95 (4.5), 94 (12.5), 91 (4.7), 85 (12.3), 83 (53.6), 81 (5.6), 78 (6.0), 77 (19.7), 76 (100), 73 (6.6), 71 (10.1), 70 (17), 69 (34.8), 68 (11.4), 67 (8.7), 66 (96), 64 (33), 63 (98), 62 (50.1), 60 (91), 59 (23.7), 57 (16.0), 56 (29.6), 55 (19.9), 54 (37.8), 53 (15.1), 52 (26.3), 51 (10.5) |

*Values are m/z (relative intensity, %); relative intensities>2% are listed.

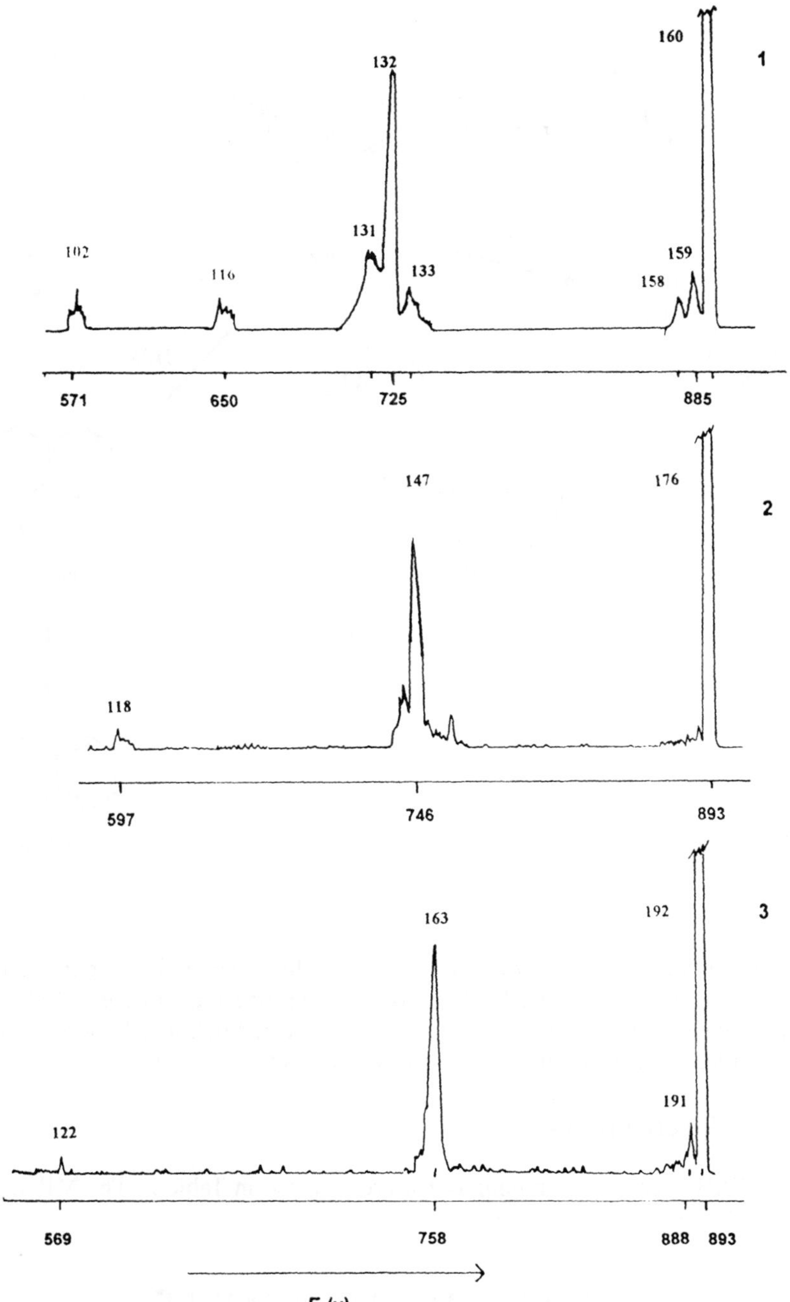


**Figure 1.** MIKE spectra of $M^{+\cdot}$ of compounds **1-3**.

the most abundant. The HCN loss observed for compound **1** is in this case replaced by the loss of $CN^{\cdot}$ radical, leading to the ion at m/z 150.

A clear difference exists in the EI mass spectra of the two isomeric compounds **3** and **4**. While for **3** the molecular ion at m/z 192 is clearly detectable, together with the primary fragments due to $CH_3N$ and $CH_4N$ losses at m/z 163 and 162, in the case of compound **4** it was never possible to obtain a significant EI spectrum. This may reasonably be ascribed to

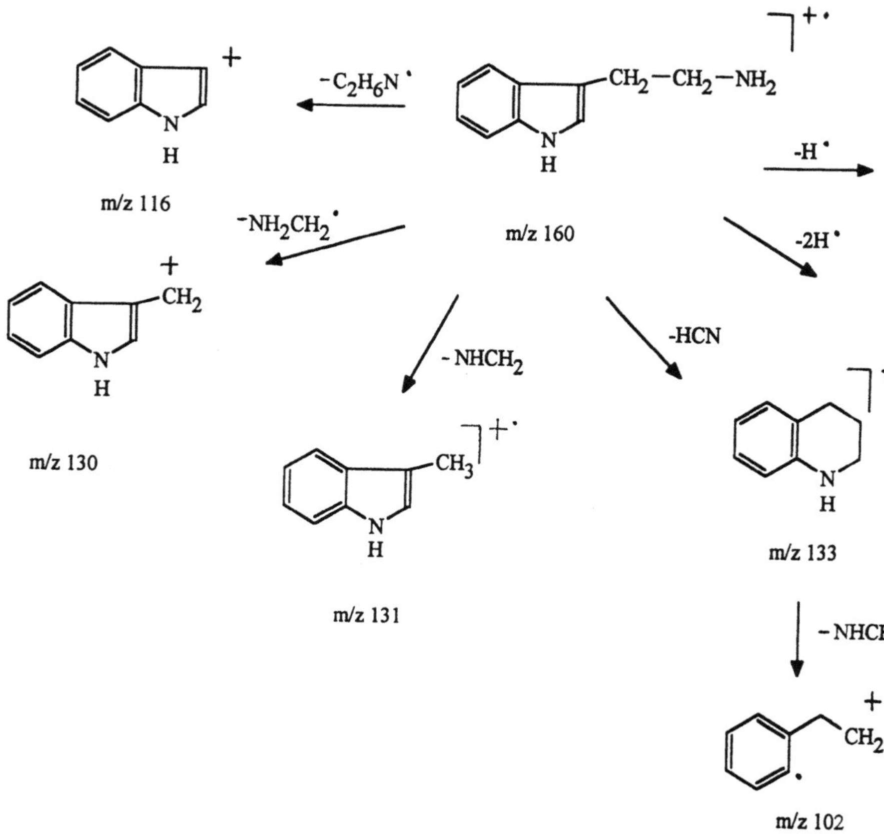


Scheme 1

the different volatility of the two compounds. While compound **3** vaporizes at a probe temperature of 150°C, compound **4** only vaporizes at a probe temperature of 300 °C, together with thermal decomposition. Hence, the spectrum reported in Table 1 for compound **4** can only be considered the spectrum of its pyrolysis products.

## FAB Mass Spectrometry

The FAB spectra of compounds **1-4** are reported in Table 2. The MIKE spectra of $[M+H]^+$ species are shown in Figure 2.

**Table 2.** FAB mass spectra of compounds **1-4***

| | |
|---|---|
| **1.** | 162 (20), 161 (100), 160 (15), 145 (18), 144 (98), 143 (20), 132 (18), 131 (30), 130 (33), 118 (7), 117 (15), 116 (5), 115 (10), 77 (8), 76 (4), 75 (12) |
| **2.** | 179 (5), 178 (30), 177 (100), 176 (25), 175 (10), 161 (17), 160 (75), 159 (14), 148 (11), 147 (10), 146 (25), 145 (17), 144 (2), 143 (6), 135 (2), 134 (4), 133 (7), 132 (2), 131 (4), 130 (5) |
| **3.** | 194 (4), 193 (16), 192 (5), 177 (3), 176 (15), 175 (2), 163 (5), 162 (5), 149 (5), 115 (12), 114 (100), 113 (8) |
| **4.** | 194 (24), 193 (80), 192 (28), 191 (12), 177 (15), 176 (60), 175 (20), 150 (15), 149 (100), 148 (3), 147 (14), 132 (20), 131 (30), 130 (12), 129 (21) |

*Values are m/z (relative intensity, %); relative intensities >2% are listed.

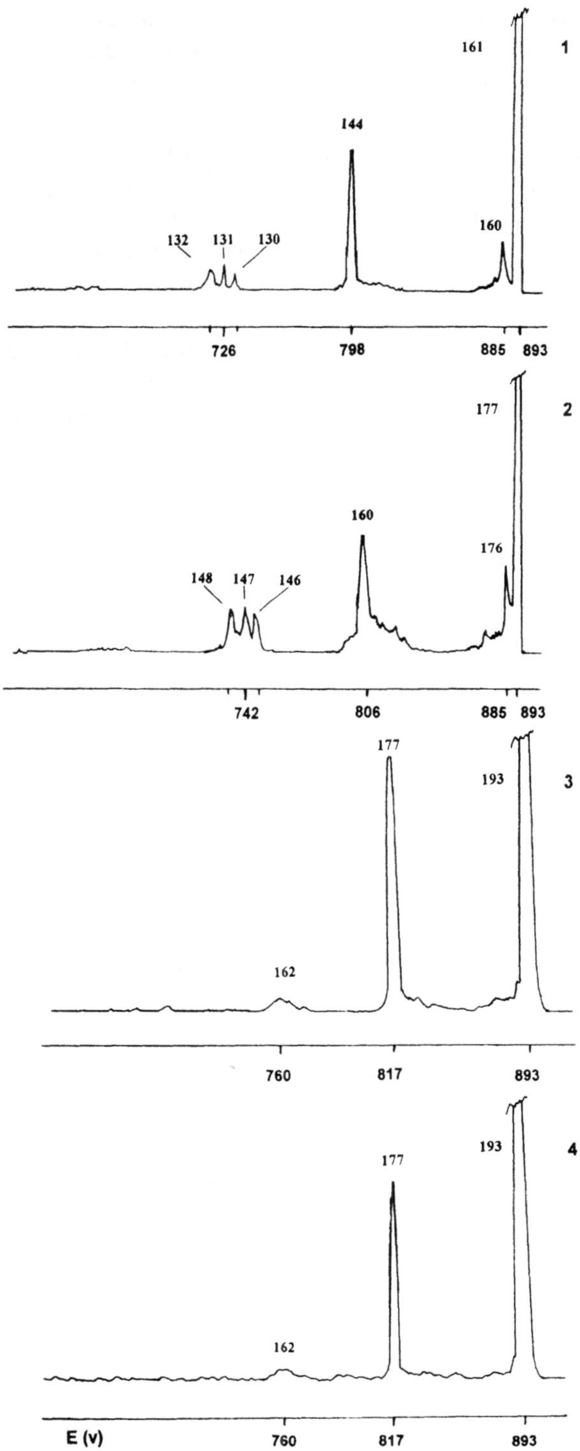


**Figure 2.** MIKE spectra of $[M+H]^+$ of compounds **1-4**.

In FAB conditions, compounds **1-4** behave quite differently from what is observed in EI conditions. While the fragmentation pathways related to cleavage of the side-chain are still present, leading to fragment ions with the same m/z values as those observed in EI conditions (m/z 130 and 131), new decomposition pathways become operative, consisting of $NH_2$ and $NH_3$ losses. The $[M\text{-}NH_3]^{+\cdot}$ species is responsible for a series of new decomposition processes.

In these operative conditions, both isomers **3** and **4** give rise to abundant protonated molecules. However, both FAB mass spectra and MIKE of $[M+H]^+$ species of **3** and **4** are practically identical, not allowing structural characterization of the two isomers. In order to obtain such characterization, new experiments by ion trap mass spectrometry were undertaken, and $M^{+\cdot}$ ions for **3** and **4** were easily produced. However, no differences were found between the spectra of the two isomers.

Their characterization was achieved by energy-resolved mass spectrometry e.g., by selecting molecular species and decomposing them by collisions of increasing strength. The plots of the abundances of the product ions vs the collision energy obtained for **3** and **4** (Figure 3) show that, in the case of compound **4**, the formation of the ion at m/z 122 is

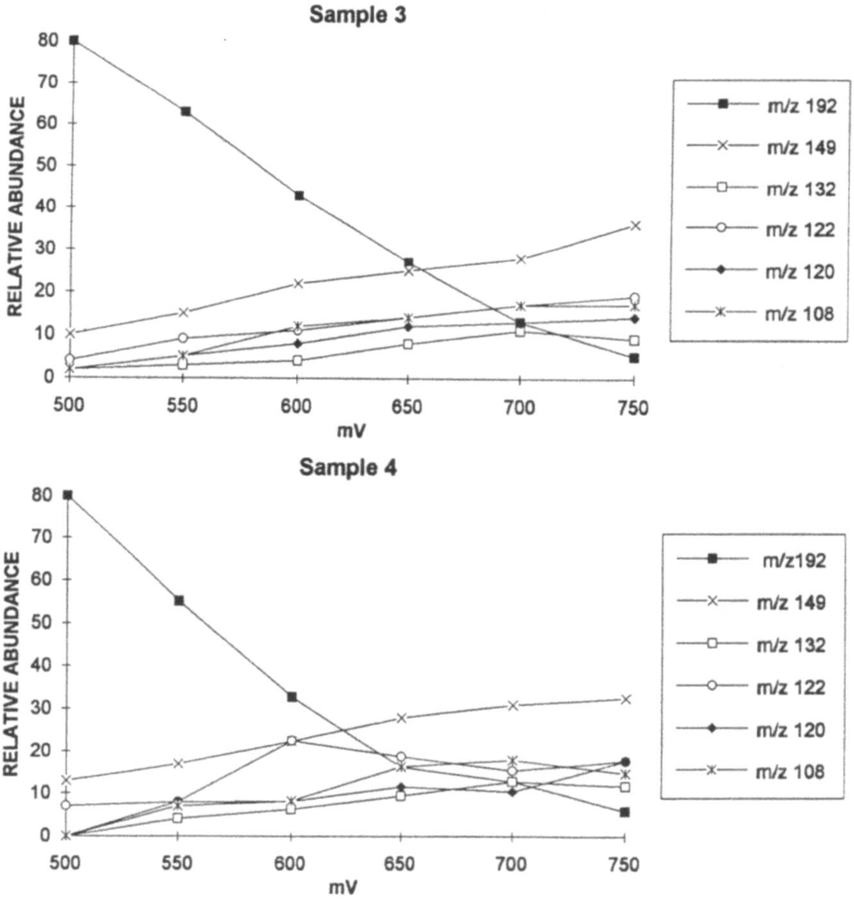


**Figure 3.** Energy-resolved mass spectra obtained on molecular cation ($M^{+\cdot}$) for dihydroxyindole isomers **3** and **4**, at various collision energy amplitudes.

energetically more favoured than for **3,** thus allowing a clear distinction between the two isomers.

In conclusion, most of the fragmentation processes of the title compounds are related to the side-chain. Characterization of the two isomeric compounds **3** and **4** was easily attained by collision experiments in an ion-trap mass spectrometer.

## REFERENCES


Ardanaz C.E., Kavka J., Guidugli F., Traldi P., and Vettori U., 1991, The ion-trap mass spectrometer in ion structure studies. The case of $[M-H]^+$ ions from chalcone, *Rapid Commun.Mass Spectrom.*, 5:5.

Barber M., Bordoli R.S., Sedgwick R.D., and Tyler A.N., 1981, Fast atom bombardment of solids (F.A.B.): a new ion source for mass spectrometry, *J.Chem.Soc.Chem.Commun.*, 7:325.

Baumgarten H.G., Bjorklund A., Lachenmeyer L., Nobin A., and Stenevi U., 1971, Long-lasting selective depletion of brain serotonin by 5,6-dihydroxytryptamine, *Acta Physiol.Scand.Suppl.*, 373:1..

Baumgarten H.G., Evetts K.D., Holman R.B., Iversen L.L., and Wilson G., 1972a, Effects of 5,6-dihydroxytryptamine on monoaminergic neurones in the central nervous system of the rat, *J.Neurochem.*, 19:1587.

Baumgarten H.G., Goether M., Holstein A.F., and Schlossberger H.G., 1972b, Chemical sympathectomy induced by 5,6-dihydroxytryptamine, *Z.Zellforsch.*, 128:115.

Baumgarten H.G., and Lachenmeyer L., 1972, 5,7-dihydroxytryptamine: improvement in chemical lesioning of indoleamine neurons in mammalian brain, *Z.Zellforsch.*, 135:399.

Bjorklund A., Nobin A., and Stenevi U., 1973a, Effects of 5,6-dihydroxytryptamine on nerve terminal serotonin and serotonin uptake in the rat brain, *Brain Res.*, 53:117.

Bjorklund A., Nobin A., and Stenevi U., 1973b, The use of neurotoxic dihydroxytryptamines as tools for morphological studies and localized lesioning of central indoleamine neurons, *Z.Zellforsch.*, 145:479.

Cooks R.G., Beynon J.H., Caprioli R.M., and Lester G.R, 1973,. "Metastable Ions", Elsevier, Amsterdam.

Eriksen N., Martin G.M., and Benditt E.P., 1960, Oxidation of the indole nucleus of 5-hydroxytryptamine and the formation of pigments: isolation and partial characterization of a dimer of 5-hydroxytryptamine, *J.Biol.Chem.*, 235:1662.

McIsaac W.M., and Page I., 1959, The metabolism of serotonin (5-hydroxytryptamine), *J.Biol.Chem.*, 234:858.

Morgan R.P., Beynon J.H., Bateman R.H., and Green N.B., 1979, The MM-ZAB-2F double-focussing mass spectrometer and Mike spectrometer, *Int.J.Mass Spectrom.Ion Phys.*, 28:171.

Saner A., Pieri L., Moran J., DaPrada M., and Pletscher A., 1974, Decrease of dopamine and 5-hydroxytryptamine after intracerebral application of 5,6-dihydroxytryptamine, *Brain Res.*, 76:109.

Todd J.F.J., 1991, Ion trap mass spectrometer - past, present and future, *Mass Spectrom. Rev.*, 10:3.

Udenfriend S., Titus E., Weissbach H., and Peterson R.E., 1956, Biogenesis and metabolism of 5-hydroxyindole compounds, *J.Biol.Chem.*, 219:335.

# INDEX

MIX
Papier aus verantwortungsvollen Quellen
Paper from responsible sources
FSC® C105338

If you have any concerns about our products,
you can contact us on
**ProductSafety@springernature.com**

In case Publisher is established outside the EU,
the EU authorized representative is:
**Springer Nature Customer Service Center GmbH**
**Europaplatz 3, 69115 Heidelberg, Germany**

Printed by Libri Plureos GmbH
in Hamburg, Germany